IFIP Advances in Information and Communication Technology 394

IFIP – The International Federation for Information Processing

IFIP was founded in 1960 under the auspices of UNESCO, following the First World Computer Congress held in Paris the previous year. An umbrella organization for societies working in information processing, IFIP's aim is two-fold: to support information processing within its member countries and to encourage technology transfer to developing nations. As its mission statement clearly states,

> *IFIP's mission is to be the leading, truly international, apolitical organization which encourages and assists in the development, exploitation and application of information technology for the benefit of all people.*

IFIP is a non-profitmaking organization, run almost solely by 2500 volunteers. It operates through a number of technical committees, which organize events and publications. IFIP's events range from an international congress to local seminars, but the most important are:

- The IFIP World Computer Congress, held every second year;
- Open conferences;
- Working conferences.

The flagship event is the IFIP World Computer Congress, at which both invited and contributed papers are presented. Contributed papers are rigorously refereed and the rejection rate is high.

As with the Congress, participation in the open conferences is open to all and papers may be invited or submitted. Again, submitted papers are stringently refereed.

The working conferences are structured differently. They are usually run by a working group and attendance is small and by invitation only. Their purpose is to create an atmosphere conducive to innovation and development. Refereeing is also rigorous and papers are subjected to extensive group discussion.

Publications arising from IFIP events vary. The papers presented at the IFIP World Computer Congress and at open conferences are published as conference proceedings, while the results of the working conferences are often published as collections of selected and edited papers.

Any national society whose primary activity is about information processing may apply to become a full member of IFIP, although full membership is restricted to one society per country. Full members are entitled to vote at the annual General Assembly, National societies preferring a less committed involvement may apply for associate or corresponding membership. Associate members enjoy the same benefits as full members, but without voting rights. Corresponding members are not represented in IFIP bodies. Affiliated membership is open to non-national societies, and individual and honorary membership schemes are also offered.

Luis M. Camarinha-Matos Slavisa Tomic
Paula Graça (Eds.)

Technological Innovation for the Internet of Things

4th IFIP WG 5.5/SOCOLNET Doctoral Conference on
Computing, Electrical and Industrial Systems, DoCEIS 2013
Costa de Caparica, Portugal, April 15-17, 2013
Proceedings

Springer

Volume Editors

Luis M. Camarinha-Matos
Slavisa Tomic
Paula Graça
Universidade Nova de Lisboa, FCT-DEE
2829-516, Monte Caparica, Portugal
E-mail: cam@uninova.pt
s.tomic@campus.fct.unl.pt
mpaulagraca@gmail.com

ISSN 1868-4238 e-ISSN 1868-422X
ISBN 978-3-642-37290-2 e-ISBN 978-3-642-37291-9
DOI 10.1007/978-3-642-37291-9
Springer Heidelberg Dordrecht London New York

Library of Congress Control Number: 2013934266

CR Subject Classification (1998): J.2, C.2, I.2.9, I.2.4, I.2.6, C.3, C.4, G.1.6

Typesetting: Camera-ready by author, data conversion by Scientific Publishing Services, Chennai, India

Printed on acid-free paper

Springer is part of Springer Science+Business Media (www.springer.com)

Preface

This proceedings book, which is based on research results produced in engineering doctoral programs, brings the importance of the Internet of Things (IoT) under the spotlight. The Internet and mobile connectivity together with the vision of the IoT promise exciting challenges for researchers and industry and benefits for society. The IoT is an integrated part of the Future Internet and could be defined as a dynamic global network infrastructure with self-configuring capabilities based on standard and interoperable communication protocols, enabling the interlinking of the physical world with cyberspace.

A key challenge for successful usage of IoT in the future is to identify the various areas which will be influenced by or contribute to the future development of the IoT — energy-efficient communication, scalable architectures of networks, bio-inspired algorithms, energy-aware systems, energy harvesting, smart and micro devices, biomedical devices, service-triggered security, self-organizing systems, distributed decision making, collaborative networks and ecosystems, connected machines and robotics, remote operation and supervision, collective adaptive systems, nano-electronics, etc.

Although typical PhD students are not experienced researchers, but rather in the process of learning how to do research, observation of worldwide publications shows that a high number of technologically innovative ideas are produced in the early careers of researchers. The DoCEIS series of Doctoral Conferences on Computing, Electrical and Industrial Systems aim at creating a space for sharing and discussing ideas and results from doctoral research in these inter-related areas of engineering. Innovative ideas and hypotheses can be better enhanced when presented and discussed in an encouraging and open environment. DoCEIS aims to provide such an environment, releasing PhD students from the pressure of presenting their propositions in more formal contexts.

The fourth edition of DoCEIS, which was sponsored by SOCOLNET, IFIP, and the IEEE Industrial Electronics Society, attracted a considerable number of paper submissions from a large number of PhD students (and their supervisors) from 24 countries. This book comprises the works selected by the International Program Committee for inclusion in the main program and covers a wide spectrum of topics, ranging from collaborative enterprise networks to microelectronics. As such, research results and on-going work are presented, illustrated, and discussed in areas such as:

- Collaborative enterprise networks
- Service orientation
- Intelligent computational systems
- Perceptional systems
- Robotics and manufacturing
- Embedded systems and Petri nets

- Control and decision
- Energy systems and smart grid
- Electronics and telecommunications

As a gluing element, all authors were asked to explicitly indicate the (potential) contribution of their work to the IoT.

We expect that this book will provide readers with an inspiring set of promising ideas and new challenges, presented in a multidisciplinary context, and that by their diversity these results can trigger and motivate richer research and development.

We would like to thank all the authors for their contributions. We also appreciate the efforts and dedication of the DoCEIS Program Committee members who both helped with the selection of articles and contributed with valuable comments to improve their quality.

April 2013

Luis M. Camarinha-Matos
Slavisa Tomic
Paula Graça

Organization

4th IFIP / SOCOLNET Doctoral Conference on COMPUTING, ELECTRICAL AND INDUSTRIAL SYSTEMS
Costa de Caparica, Portugal, April 15–17, 2013

Conference and Program Chair

Luis M. Camarinha-Matos, Portugal

Organizing Committee Co-chairs

Luis Gomes, Portugal
João Goes, Portugal
João Martins, Portugal

International Program Committee

Marian Adamski, Poland
Amir Assadi, USA
José Barata, Portugal
Olga Battaia, France
Marko Beko, Portugal
Luis Bernardo, Portugal
Vedran Bilas, Croatia
Theodor Borangiu, Romania
Xavier Boucher, France
Erik Bruun, Denmark
Giuseppe Buja, Italy
Luis M. Camarinha-Matos, Portugal
António Cardoso, Portugal
João Catalão, Portugal
Wojciech Cellary, Poland
Naoufel Cheikhrouhou, Switzerland
Alok Choudhary, UK
Fernando J. Coito, Portugal
Luis Correia, Portugal
Ed Curry, Ireland
Ridha Derrouiche, France
Jorge Dias, Portugal
Pedro Encarnação, Portugal
Ip-Shing Fan, UK
Florin G Filip, Romania
Maria Helena Fino, Portugal
José M. Fonseca, Portugal
Leopoldo Garcia Franquelo, Spain
Liljana Gavrilovska, Macedonia
Paulo Gil, Portugal
João Goes, Portugal
Luis Gomes, Portugal
Antoni Grau, Spain
Ricardo Jardim-Gonçalves, Portugal

Bernhard Katzy, Germany
Hans-Jörg Kreowski, Germany
Johann Walter Kolar, Switzerland
Paulo Leitão, Portugal
Ratko Magjarevic, Croatia
Veljko Malbasa, Serbia
João Martins, Portugal
Maria do Carmo Medeiros, Portugal
Paulo Miyagi, Brazil
Jörg Müller, Germany
Milica Naumovic, Serbia
Rudy Negenborn, The Netherlands
Horacio Neto, Portugal
Rui Neves-Silva, Portugal
Henrique O'Neill, Portugal
Luis Oliveira, Portugal
Manuel D. Ortigueira, Portugal
Angel Ortiz, Spain
Gordana Ostojic, Serbia
Peter Palensky, Austria
Luis Palma, Portugal
Nuno Paulino, Portugal
Carlos Eduardo Pereira, Brazil
Willy Picard, Poland
Paulo Pinto, Portugal
José M. Quintana, Spain
Ricardo Rabelo, Brazil
Sven-Volker Rehm, Germany
Rita Ribeiro, Portugal
Rolf Drechsler, Germany
Juan Rodriguez Andina, Spain
Enrique Romero, Spain
Jose de la Rosa, Spain
Gheorghe Scutaru, Romania
Fernando Silva, Portugal
Adolfo Steiger-Garção, Portugal
Sasu Tarkoma, Finland
J. Tenreiro Machado, Portugal
Klaus-Dieter Thoben, Germany
Stanimir Valtchev, Portugal
Manuela Vieira, Portugal
Vasos Vassiliou, Cyprus
Dmitri Vinnikov, Estonia
Ljiljana Zivanov, Serbia

Organizing Committee (PhD Students)

Fàbio Silva
João Borges
Joaquim Moreira Lima
Luis Pires
Manuel Carvalho
Maria Paula Graça
Massimiliano Zanin
Miguel Campilho Gomes
Nuno Amaro
Rui Medeiros
Sebastian Scholze
Slavisa Tomic
Sudeep Ghimire
Tahereh Nodehi
Vasco Gomes

Technical Sponsors

Society of Collaborative Networks

IFIP WG 5.5 COVE
Co-Operation infrastructure for Virtual Enterprises
and electronic business

IEEE-Industrial Electronics Society

Organizational Sponsors

UNINOVA

Organized by:

PhD Program on Electrical and Computer Engineering FCT-UNL.

Table of Contents

Part V: Computational Systems

Part VI: Computational Systems Applications

Part VII: Perceptional Systems

Part VIII: Robotics and Manufacturing

Part XIII: Energy Generation

Part XIV: Energy Distribution

Part XV: Energy Transformation

Part XVI: Optimization Techniques in Energy

Part XVII: Telecommunications

Part XVIII: Electronics: Devices Design

Part XIX: Electronics: Amplifiers

Part XX: Electronics: RF Applications

Part XXI: Electronics: Applications

Part I
Introduction

Contributing to the Internet of Things

Luis M. Camarinha-Matos, João Goes, Luís Gomes, and João Martins

Departamento de Engenharia Electrotécnica, Faculdade de Ciências e Tecnologia, Universidade Nova de Lisboa, 2829-516 Caparica, Portugal
cam@uninova.pt

Abstract. The fast development of networked smart devices equipped with sensors and radio-frequency identification, connected to the Internet, is enabling the emergence of many new applications and the redesign of traditional systems towards more effective operation. Raising awareness among engineering PhD students for the potential of this new wave in their research work is a crucial element in their education. With this aim, the doctoral conference DoCEIS'13 focused on technological innovation for the Internet of Things, challenging the contributors to analyze in which ways their technical and scientific work could contribute to or benefit from this paradigm. The results of this initiative, which was reasonably successful, are briefly analyzed.

Keywords: Internet of Things, Cyber Physical Systems.

1 Introduction

The fast development of the Internet of Things is enabling the emergence of many new applications and the redesign of traditional systems towards more effective operation. In fact, more and more objects are becoming embedded with sensors and gaining the ability to communicate. New smart devices, but also traditional machinery are "joining the Internet", facilitating the development of more integrated services and optimization of existing systems.

Some of the elements contributing to the wide potential of this area include: remote access / control, more effective monitoring and supervision thus allowing better performance, real-time access to data which supports timely decision making, wider systems integration, complemented with access to cloud-based resources, mobility without losing access to systems, access to large amounts of sensorial data, etc.

Nowadays, a substantial amount of technological innovation is the result of the research works of engineering PhD students. The very nature of the Internet of Things, combining the physical and the cyber worlds, requires the combination of a set of competencies typically covered by the area of Electrical and Computer Engineering. It is however necessary to call the attention of students in this area, which typically tend to focus on a specific research topic, for the potential of the "interconnected things" and the role they can play in this process.

The DoCEIS'13 conference was thus organized with this mission and some of the results are summarized in this paper.

L.M. Camarinha-Matos, S. Tomic, and P. Graça (Eds.): DoCEIS 2013, IFIP AICT 394, pp. 3–12, 2013.

2 Current Trends

The term "Internet of Things" was first proposed by Kevin Ashton in 1999 to describe a system where the Internet is connected to the physical world via ubiquitous sensors. The concept emerged in the context of the developments at the MIT Auto-ID Center on identification technologies.

According to some European views [1], **Internet of Things** (IoT) can be defined as “a dynamic global network infrastructure with self configuring capabilities based on standard and interoperable communication protocols where physical and virtual “things” have identities, physical attributes, and virtual personalities and use intelligent interfaces, and are seamlessly integrated into the information network”.

Other definitions put more emphasis on technology: networked smart devices equipped with sensors and radio-frequency identification, connected to the Internet, all sharing information with each other without human intervention [2].

In this context, a “**thing**” could be defined as a real/physical or digital/virtual entity that exists and moves in space and time and is capable of being identified. Things are commonly identified either by assigned identification numbers, names and/or location addresses.

A related concept is represented by the term **Cyber-Physical Systems** (CPS). According to the US National Science Foundation (NSF), CPS are engineered systems that are built from and depend upon the synergy of computational and physical components [3].

Although some literature seems to use the two terms as synonyms, depending on the geographical origin of the authors, IoT can more properly be seen as a subset of CPS (Fig. 1). In fact, the notion of CPS includes not only things connected to the Internet, but also other physical systems embedding computational power. When these systems are geographically distributed and interconnected, they will likely resort to Internet.

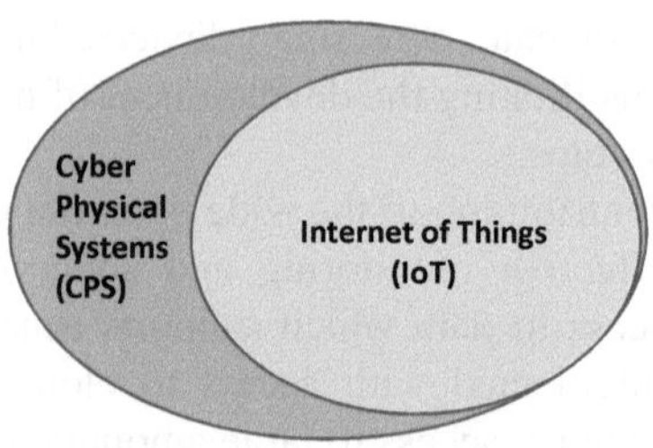

Fig. 1. Relationship between CPS and IoT

A number of other related terms have emerged in the last decade to represent partial perspectives or focused application contexts. Some examples:

- ***Industrial Internet***. While focusing on industrial applications, it aims a global network connecting people, machines and data with the purpose of facilitating management, operation and maintenance of industrial facilities, and their improving performance. According to General Electric [4], "the full potential of Internet-based digital technology has yet to be fully realized across the global industry system. Intelligent devices, intelligent systems, and intelligent *decisioning* represent the primary ways in which the physical world of machines, facilities, fleets and networks can more deeply merge with the connectivity, big data and analytics of the digital world". In this sense, the Industrial Internet can be seen as a subset of the IoT.

- ***Internet of Events***. A perspective of the IoT that puts the emphasis on time dependency and discrete events handling [5]. As such, events modeling and management, time critical reactivity, and process modeling and supervision are the relevant issues here.
- ***Sensing Enterprise***. A concept introduced in the FInES (Future Internet Enterprise Systems) cluster of projects [6] to refer to an enterprise anticipating future decisions by using multi-dimensional information captured through physical and virtual objects and providing added value information to enhance its global context awareness. In other words, it particularly focuses on enriching enterprises' context awareness through intelligent, interconnected and interoperable smart components and devices to power enterprise systems, making them responsive to events in real time and aiming at reaching seamless transformation of (raw) data to (tailored) information and (experienced) knowledge.
- ***Ambient Intelligence***. A concept that represents electronic-enhanced environments which are sensitive and responsive to the presence of people [7]. Therefore it builds upon the notions of pervasive computing, embedded systems, context awareness, and human-centric computer interaction. One of the relevant application areas is the so-called ambient assisted living, which uses technology to assist elderly. IoT is naturally a way to materialize the vision of a technology that becomes invisibly embedded in our natural surroundings, present whenever we need it.

Due to its genesis, closely associated to communities involved in objects identification and logistics applications, IoT has been quite biased by RFID (Radio Frequency Identification) developments. Although RFID is an important enabling technology, it might also limit the vision, making it difficult to harness the full potential of the IoT. In fact, the "things" that can be connected to Internet are not only simple (passive) objects, but rather progressively more intelligent and autonomous "entities" with rich sensorial and acting capabilities. In terms of communications, a mix of wireless and wired forms co-exists. On the other hand, not all RFID applications represent cases of IoT. Under this more general perspective of "things", earlier examples of connection of devices to Internet - toaster, robots, like the Robotgarden, refrigerators, etc. - are important initial cases.

Recent advances in other enabling technologies, e.g. NFC (Near Field Communications), low power devices, embedded intelligence, sensor networks, and novel architectures for the support infrastructures, combined with easier access to resources through cloud computing, are creating the conditions for truly pervasive computing and ambient intelligence, and thus the emergence of a wave of novel applications with wide impact in all sectors of life.

A brief overview of the main milestones in the development of this area is represented in Fig.2.

The area of Internet of Things has been growing in the last years, as represented by the large number of related projects (see [1]) and the Google trends graph (Fig.3).

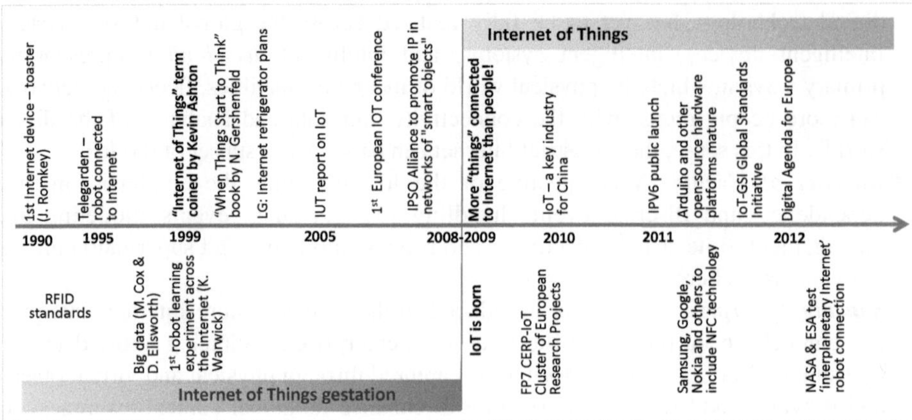

Fig. 2. Some milestones in the development of the IoT

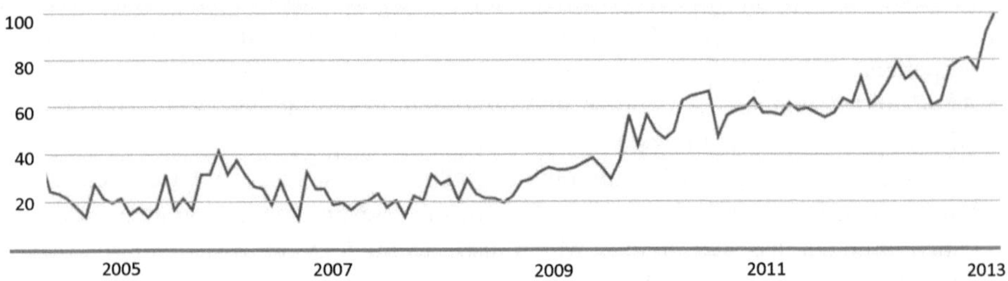

Fig. 3. Google trends graph for Internet of Things

In terms of application cases, many implementations and futuristic scenarios have been designed for a large spectrum of domains. Some relevant examples are summarized in Table 1.

Table 1. Applications examples

Application domain	Example features
Aerospace and aviation	▪ Fighting counterfeiting - introducing electronic pedigrees for certain categories of aircraft parts: decentralized database, RFID tags, etc. ▪ Wireless monitoring of the aircraft – sensors network detecting various conditions such as pressure, vibrations, temperature etc.
Airports and other hubs	▪ Safety – RFID tags associated to luggage; building security monitoring. ▪ Indoor location systems. ▪ Interaction with customers gadgets: NFC for check-in, payments, etc.
Automotive	▪ Real-time locating systems and connecting with other IoT sub networks, improving vehicle tracking and management. ▪ Dedicated Short Range Communication – Vehicle-to-vehicle and vehicle-to-infrastructure communications (Intelligent Transportation Systems). ▪ The vehicle itself as a 'thing', enabling it to make automatic emergency calls or breakdown calls when appropriate, collecting as much data as possible from surrounding 'things'. ▪ Interactions between the vehicle and user's gadgets.

Table 1. *(continued)*

Driving insurance	▪ Pay-as-you-drive: Electronic recorders in the car, which are able to record *acceleration*, *speed*, and other parameters, and communicate this information to the insurer -> cheaper rate or premium. ▪ Similar approach applied to buildings, machinery, etc.
Environment monitoring	▪ Environment surveillance: earth quakes, tsunami, forest fires, floods, pollution (water and air).
Food traceability and agribusiness	▪ Food traceability. ▪ Traceability of agricultural animals and their movements. ▪ Real time detection of animals during outbreaks of contagious disease. ▪ Counting animals in a farm (for subsidies).
Independent Living support	▪ Detecting the activities of daily living, particularly in the case of elderly or people with special needs, using wearable and ambient sensors, monitoring social interactions using wearable and ambient sensors, monitoring chronic diseases using wearable vital signs sensors, and in body sensors. ▪ "Things" can learn regular routines and raise alerts or send out notifications in anomaly situations.
Intelligent / smart cities	▪ Environment monitoring, public safety, urban analytics, emergencies handling, participatory governance, etc. ▪ Integrated multi-modal transportation management (mobility networks), Energy and water management, Waste collection management, etc. ▪ City level social networks (& interaction with gadgets).
Intelligent buildings	▪ Home automation, operation of home appliances. ▪ Sensors for temperature, humidity provide the necessary data to automatically adjust the comfort level and to optimize the use of energy for heating or cooling (ubiquitous sensor networks). ▪ Smart metering for measuring energy consumption and transmitting this information to the energy provider electronically. ▪ Monitoring and reacting to human activity, such that exceptional situations could be detected and people can be assisted in everyday activities e.g. supporting the elderly. ▪ Building security infrastructure.
Intelligent transportation infrastructures	▪ Tolling and vehicle monitoring. ▪ Roads and highways with warning messages and diversions according to traffic and climate conditions and unexpected events like accidents or jams. ▪ Smart parking, traffic congestion management, etc.
Manufacturing & Product Life Cycle management	▪ Embedded smart devices or the use of unique identifiers and data carriers that can interact with an intelligent network infrastructure and information systems: optimization of processes, refine schedules, and improve logistics. ▪ Product's usage history along its life cycle (from cradle to grave). ▪ Remote maintenance and updating of manufacturing equipment.
Medical technology, healthcare	▪ Measurement and monitoring methods of vital functions (temperature, blood pressure, heart rate, cholesterol levels, blood glucose etc). ▪ Implantable wireless identifiable devices could be used to store health records that could save a patient's life in emergency situations. ▪ Edible, biodegradable chips could be introduced into the body and used for guided action.
Operation in dangerous environment	▪ Tele-operation and tele-presence, Tele-supervisory control, etc. ▪ Detection of gas levels and leakages in industrial environments. ▪ Under water exploration, fire fighting, etc.
Retail, Logistics, Supply Chain Management	▪ RFID-equipped items and smart shelves: automatically checking of goods receipt, real time monitoring of stocks, tracking out-of-stocks or the detection of shoplifting. ▪ Improving logistic processes with RFID data. ▪ Guidance in the shop according to a preselected shopping list, fast payment solutions like automatically check-out using biometrics.

Table 1. *(continued)*

Safety and security monitoring	▪ Building monitoring: water leaks, gases, vibrations, fire, un-authorized entry, vandalism. ▪ Personnel: mugging alarm, equipment surveillance, payment systems, identity security.
Smart grid	▪ Advanced metering, supporting more effective energy management. ▪ Interaction with smart appliances. ▪ Intelligent monitoring. ▪ Demand response and value added services

As it becomes clear in the above application examples, the Internet of Things is a multi-disciplinary subject that requires the integration of contributions from multiple technologies. Table 2 lists the most relevant ones and summarizes the key issues to be addressed in each one.

Table 2. Involved technologies

Technology	Illustrative key issues
Communication technology	Energy efficient bi-directional communications, Multi-frequency, wireless sensor networks, message-queue based communications targeting cloud environment, etc.
Network technology	Protocol gateways, Scalable architectures, Secure & reliable wireless communication protocols, service based network, etc.
Network discovery	Automated discovery mechanisms and mapping capabilities (new "things" continuously appear / disappear, some "things" evolve)
Software and algorithms	Micro operating systems, Service oriented computing, Applications in cloud environments, Self-adaptive software (autonomic systems), Bio-inspired algorithms, Energy-aware systems, Context aware software, Events management, Balancing local vs. cloud intelligence and decision making, Objects' representation (services, agents), etc.
Hardware devices	Nano-electronic smart devices, Energy harvesting, Polymers electronics, Embedded intelligence, Low cost tags, Smart devices, Multi-standard protocols, Heterogeneous architectures, Low cost devices, etc.
Data and signal processing technology	Semantic interoperability, Data sharing, Data aggregation, Stream processing, Big data processing, etc. Making sense of the massive amounts of data that can be generated by intelligent devices (big data) is one of the key components of the IoT.
Discovery and search engine technologies	Device discovery, Distributed repositories, Positioning and localization, Terrestrial mapping, Location awareness, etc.
Power and energy storage technologies	Batteries / micro-batteries, Energy harvesting, Energy consumption mapping, Energy-based priority scheduling, dynamic tariffs, etc.
Security and privacy technologies	Privacy for heterogeneous devices, Decentralized authentication and trust, Energy efficient encryption, Anonymity mechanisms, Data ownership, etc.
Standardisation	Standards for cross interoperability, Standards for intelligent devices, Languages for things interaction, Standard infrastructure architectures, etc.
Relationship network management technologies	Identity, relationship and reputation management, Organizational structures, Collaborative models and functions, Collective intelligence, Traceability of distributed decision making, etc.

A comprehensive research and development agenda for IoT can be found in [1]. In addition to the purely technological issues, there are other relevant aspects to be properly addressed in order to let the full potential of IoT be achieved. These include: Legal and regulatory aspects, Socio ethical aspects, and Economical aspects.

Although considerable research and practical developments were made during the last two decades, it is likely that substantial efforts are still needed in this area, as illustrated by the Gartner's hype cycle (Fig. 4). According to this forecast, it will still take more than 10 years to reach the "plateau of productivity" in this area.

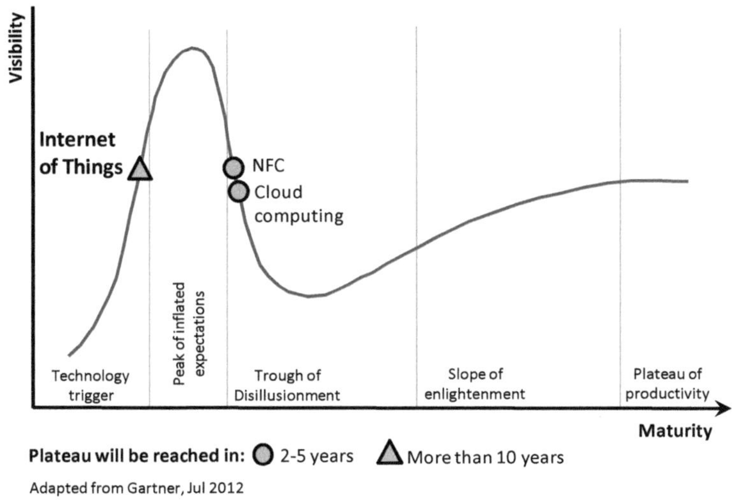

Fig. 4. Hype cycle for Internet of Things (adapted from Gartner's [8])

3 Example Contributions

IoT is a particularly relevant topic for Electrical and Computer Engineering (ECE) researchers and professionals. In the last decades the scope of ECE has expanded so widely that it risks some "fragmentation". Most professionals (and students) focus on a specialization sub-field, rarely mastering a comprehensive view of the whole field. Since IoT requires a "strong dialogue" among the various sub-fields of electrical and computer engineering (and other areas), it represents a potential gluing element to bring them together.

Under this assumption, a challenge was presented to DoCEIS'13 conference participants [9] (doctoral students from various countries and different sub-fields) as summarized in the following questions:

- *In which aspects your research can contribute to the development of the Internet of Things?*

or

- *In which aspects your area of work could be affected / influenced in the future by the development of the Internet of Things?*

As a result, all contributors made an effort to analyze the relationship between their specific research work and the IoT. Among the accepted papers there is an almost balanced distribution between those that contribute to the development of support technologies for IoT and those that can benefit from IoT adoption [9], as summarized in Fig. 5. The size of the rectangles is proportional to the number of contributions in each specific sub-field.

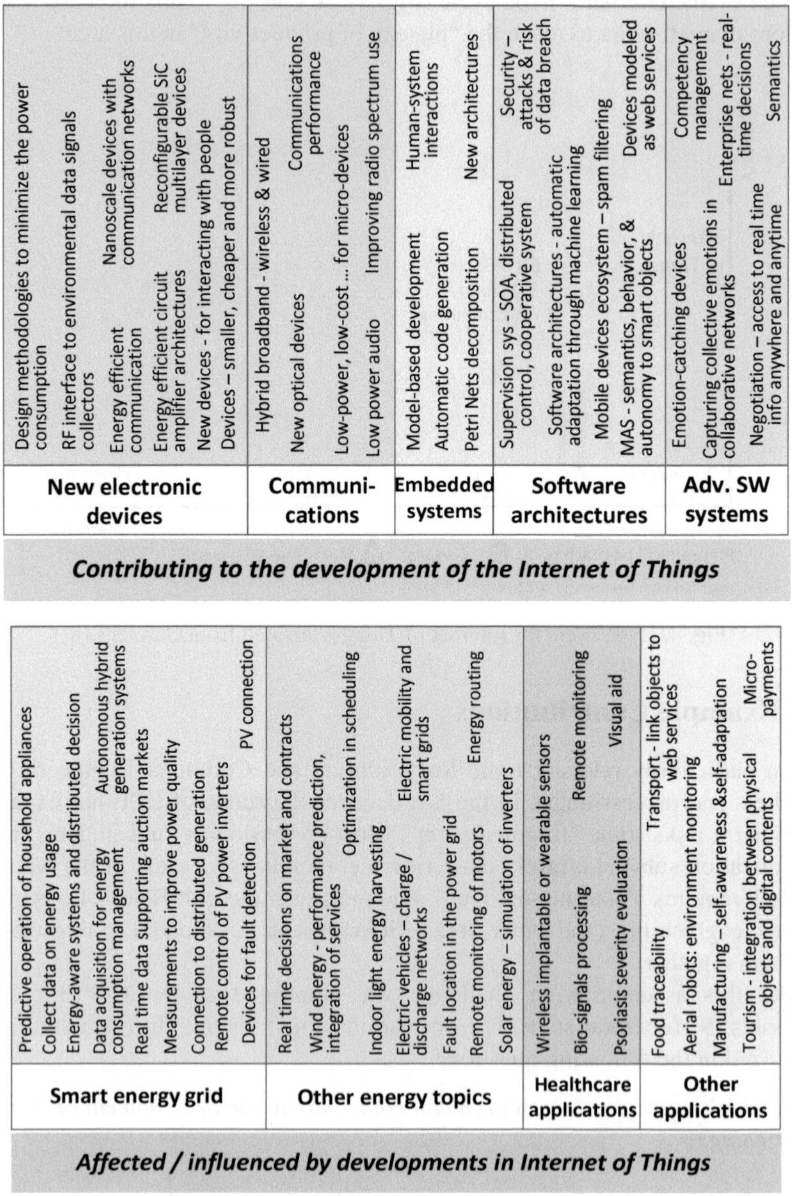

Fig. 5. DoCEIS'13 contributions to IoT

4 Open Challenges

From the analysis of the state of the art and also considering the sample of contributions to DoCEIS'13, it becomes evident that more attention needs to be devoted to the system level. Such system perspective is needed in order to better understand and manage complex IoT environments.

More and more we live surrounded by large numbers of devices with growing (embedded) intelligence / sensing and computational power, connected to Internet, and forming complex inter-dependent systems. These devices can typically be "represented" in the cyber-space by service entities (or agents). The combination of physical devices with their "cyber representatives" constitutes a cyber-physical system. In this environment, devices / representatives can "appear and disappear" from the cyber-space.

Devices / sub-systems typically have an "owner"; therefore, in addition to the growing autonomy of such entities (which comes from the growing intelligence / cognitive capabilities), they also have to "obey" to their owners, which introduces a new dimension to the problem of designing such systems. When systems involve a large number of entities (hundreds? thousands? millions?), flat organizational structures are not appropriate. Therefore some "structural thinking" is necessary, leading to the organization of such entities in "communities" or "societies" ("ecosystems") of cyber-physical artifacts (Fig. 6). Important issues in these "communities" are the definition of "borders" / membership, roles, and evolution.

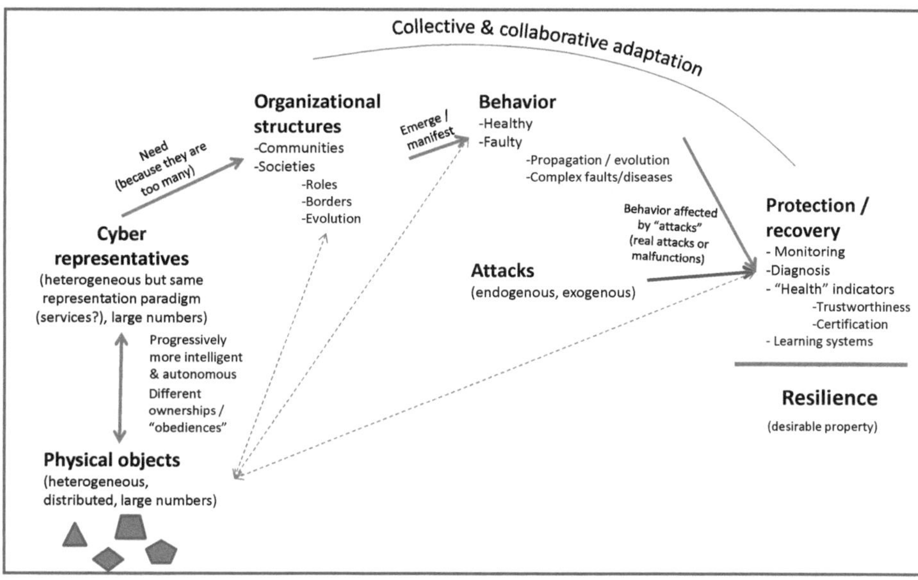

Fig. 6. Toward sustainable IoT-based systems

In organized communities – and depending on their design / purpose – different behaviors can emerge. Some behaviors are consistent with the system's purpose (healthy behaviors). But we can also have faulty / deviating behaviors. These are particularly critical as complexity increases and we become more dependent on such systems. Understanding (and detecting) faulty behaviors is thus critical. How do these behaviors propagate / extend over "different regions" of the communities? How do they evolve? How can systems adapt to faulty situations? Collaborative networks and collective adaptive systems principles are important here.

Faulty behaviors can have an endogenous source (component's malfunctioning, interoperability "frictions", non-collaborative behavior, etc.) or result from exogenous attacks (e.g. terrorist cyber-attacks). It is therefore necessary to develop adequate protection / recovery approaches and mechanisms. This involves distributed monitoring, (collective) diagnosis of detected faulty behaviors, and launching recovery / self-healing processes. Considering the complex nature of such systems, it is important to first elaborate some "health indicators", including system's trustworthiness indicators, system's certification, etc. How much can we trust in systems that we do not fully understand and over which we do not have full control (as they are complex, evolving, components belonging to different owners, etc.)? Learning mechanisms should be an intrinsic functionality here.

5 Concluding Remarks

Developments in the Internet of Things are having a strong impact in all sectors of society and their importance is likely to grow in the coming years.

The needed technological components and approaches to connect the physical and the cyber worlds open important expansion opportunities for the area of Electrical and Computer Engineering. This is clearly reflected in the contributions to the DoCEIS'13 (4th Doctoral Conference on Computing, Electrical and Industrial Systems).

References

1. Sundmaeker, H., Guillemin, P., Friess, P., Woelfflé, S. (eds.): Vision and Challenges for Realising the Internet of Things. CERP-IoT, European Commission (2010)
2. Pretz, K.: The Next Evolution of the Internet. The Institute, IEEE (January 7, 2013), http://theinstitute.ieee.org/technology-focus/technology-topic/the-next-evolution-of-the-internet
3. NFS. Cyber-Physical Systems (CPS) (2012), http://www.nsf.gov/funding/pgm_summ.jsp?pims_id=503286 (accessed on February 4, 2013)
4. Evans, P.C., Annunziata, M.: Industrial Internet: Pushing the boundaries of minds and machines. GE report (March 2012), http://www.ge.com/docs/chapters/Industrial_Internet.pdf
5. Ortner, E., Schneider, T.: Temporal and Modal Logic Based Event Languages for the Development of Reactive Application Systems. In: Proc. 1st International Workshop on Complex Event Processing for the Future Internet, Vienna, Austria, September 28-30 (2008)
6. FInES, A European Innovation Partnership for Catalysing the Competitiveness of European Enterprises. Position Paper on Orientations for FP8, European Commission (2011), http://cordis.europa.eu/fp7/ict/enet/documents/fines-position-paper-fp8-orientations-final.pdf
7. Cooky, D., Augustoz, J., Jakkula, V.: Ambient Intelligence: Technologies, Applications, and Opportunities. Pervasive and Mobile Computing 5(4), 277–298 (2009)
8. LeHong, H., Fenn, J.: Key Trends to Watch in Gartner 2012 Emerging Technologies Hype Cycle (2012), http://www.forbes.com/sites/gartnergroup/2012/09/18/key-trends-to-watch-in-gartner-2012-emerging-technologies-hype-cycle-2/
9. Camarinha-Matos, L.M., Tomic, S., Graça, P. (eds.): DoCEIS 2013. IFIP AICT, vol. 394. Springer, Heidelberg (2013)

Part II
Collaborative Enterprise Networks

Negotiation Support and Risk Reduction in Collaborative Networks

Ana Inês Oliveira[1] and Luis M. Camarinha-Matos[1,2]

[1] CTS, Uninova, Departamento de Engenharia Electrotécnica, Faculdade de Ciências e Tecnologia, FCT, Universidade Nova de Lisboa, 2829-516 Caparica, Portugal
[2] Faculdade de Ciências e Tecnologia, Universidade Nova de Lisboa, Campus da Caparica, Quinta da Torre, 2829-516 Monte Caparica, Portugal
{aio,cam}@uninova.pt

Abstract. In face of the current economic turbulence, companies face new challenges. In order to respond to new business opportunities, it is crucial that companies attain strategic alliances so that they can obtain or maintain market competiveness. The formation of alliances and partnerships for collaborative problem solving is of extreme importance, being therefore essential to understand their structures and requirements. To overcome a number of difficulties that may appear in the formation of such alliances, it is necessary to properly model the elements that constitute the alliance agreements through a suitable negotiation support environment that besides the basic functionalities of data storage and alerts can also conduct the entire negotiation process making it traceable. In this context, this paper presents the main requirements of an electronic negotiation support environment in a collaborative network, identifies the main risk sources and drivers in collaborations, and analyses how a negotiation support system can help in reducing the potential risk in collaboration.

Keywords: collaborative networks, negotiation support environment, risks, agreement.

1 Introduction

In today's unstable and highly competitive business environments companies and organizations have to shift their modus operandi if they want to obtain or maintain market competiveness. It is therefore essential to move to a new business paradigm where the creation of strategic alliances is vital[1, 2].Past "traditional" enterprises are increasingly replaced by new collaboration structures of enterprises, forming temporary networks of independent companies that share skills, costs and access to each other's market[3].In this context, the possibility of rapidly forming a consortium to respond to a business or collaboration opportunity gives companies an expression of agility and survival mechanisms in face of this market turbulence. Moreover, the topic of collaborative networks (CNs) appears as significantly promising because if enterprises or organizations share a common interoperable infrastructure, common operating principles, common cooperation agreements, and a base of trust among

L.M. Camarinha-Matos, S. Tomic, and P. Graça (Eds.): DoCEIS 2013, IFIP AICT 394, pp. 15–24, 2013.

them, then their ability to rapidly form virtual organizations(VOs) is increased [4]. Nevertheless, this kind of business formation is not without risks, so besides the important and classical task of selecting the adequate partners with the most suitable competencies to form a consortium able to respond to the requirements of the business or collaboration opportunity (BO/CO), it is also of extreme importance to consider the potential risks in collaboration. If there is a robust and reliable negotiation environment that supports the potential VO partners in achieving agreements during the VO creation process considering the involved collaboration risks, then that may lead to a reduction of risks and amount of time spent in this process. These VO agreements will be the basis for the governing principles of the VO during its operation phase[5].Considering this background and a virtual organization breeding environment (VBE) [6] context, that supports and fosters the creation of dynamic VOs, the main research question that emerges is:

How can an electronic negotiation support environment increase the agility in the creation process of successful dynamic virtual organizations?

One important motivation for this work is that by contextualizing the VO creation process in the VO breeding environment (VBE), making use of all its infrastructures and functionalities, the process may become more agile in terms of time. Furthermore, identifying and characterizing the involved potential collaboration risks can also make the process more agile in terms of adaptation to unexpected events and thus reducing the risks in collaboration. Moreover, managing risks in VOs is a rather underexplored and unstructured scientific area [3].

2 Technological Innovation for the Internet of Things

The Internet of Things (IoT) is a novel paradigm that is rapidly gaining ground in the scenario of modern wireless telecommunications[7]. This paradigm is based on the concept of pervasive presence of a variety of things or objects (such as RFID tags, sensors, actuators, mobile phones, etc.) which are able to interact with each other and even cooperate to reach common goals[8]. As a result, the IoT causes a natural impact on the behavior of its users, either in everyday-life or business aspects.

In the context of a negotiation process towards the creation of a VO, it is of extreme importance that the potential partners can have quick access to relevant information that is acquired through devices or other objects of their business environment so that it can contribute to boost the VO creation process. For that reason, if the potential partners of the VO have access to information, independently of the geographical place where they are (through different devices, e.g. smart phones), and on top of that have access to the required information of their infrastructure and related devices in real time, then the negotiation of the VO creation process it could become more agile in terms of time, and therefore the IoT appears as a suitable paradigm.

On the other hand, in collaborative networks, while the IoT can help to overcome some technical problems, in terms of finding accurate information and providing anywhere and anytime access, there are some business aspects that are still open issues. Therefore, a negotiation support environment that deals with this information can also contribute to elaborate on new business models related to the IoT [9].

3 Related Literature

Being the main topic of this work a negotiation support environment to assist organizations to increase the agility of a VO creation process, it is then essential to mention a brief summary of related literature that has directly influenced the current work. More detailed review of related literature can be found in[5, 10, 11].

VOs and Their Related Environments. Collaborative networks is a scientific discipline that covers the study of networks consisting of a variety of entities that are largely autonomous, geographically distributed and heterogeneous, and that collaborate to better achieve common or compatible goals, and whose interactions are supported by computer network [4, 12]. The Virtual Organization paradigm constitutes one of the first manifestations of the collaborative networks. Being the concept developed and applied in several domains and areas, many contributions for the characterization and modeling of the paradigm can be found in the literature, as exemplified by[1, 2, 4, 13]. The main idea behind this concept is basically of a temporary consortium of enterprises and/or organizations, geographically dispersed, that strategically join their competencies to rapidly respond to a business or collaboration opportunity. Previous works have established some reasons for companies or organizations to establish cooperative/collaborative networks, such as [14]: flexibility, capacity, speed, skills and competences, etc. The main outcome here is always the risk-sharing, but also, the sharing of gains either in terms of profit or in terms of market information. Nevertheless, in face of a new business opportunity, if companies can take advantage of a common interoperable infrastructure, common operating rules, common cooperation agreement, and a base trust level among the organizations, then they become more ready and prepared to collaborate. Therefore, an approach is to consider that dynamic VOs are mostly created in the context of a VO Breeding Environment (VBE)[4, 6].

Risks in Collaborations. Risk is an ambiguous concept, thus there are many definitions of risk depending on the specific application and on the situational context[15]. One of the most common meanings for risk is a threat or danger and often implies the probability of a negative outcome. Harland, Brenchley & Walker [16]have done an exhaustive review of definitions and classifications of types of risk and defined risk (R) as the product of the probability (P) of a loss (loss) and the significance or impact (I) of the loss, related to an event n (n): Rn = P(loss)n x I(loss)n. Other authors claim that risk, *per se*, has neither a positive nor a negative value and is perhaps more related to uncertainty, where eventualities can be either beneficial or adverse[3].Considering the VO life-cycle, there are of course some risks that traditional risk management does not deal with, in particular when considering the sharing of skills, costs and access to each other's markets. Also, risks may change from project to project or opportunity to opportunity. Therefore, new challenges in how to manage VO risks are enforced.

Negotiation and Contracting. Negotiation is an iterative communication and decision-making process between two or more parties who seek a consensus decision and cannot apply unilateral actions to achieve their objectives [17, 18].As a negotiation processes involve a transversal, multi- and inter-disciplinary approach, it is necessary to have a holistic view of the problem, making use of multiple methodologies and paying attention to the practical details [19]. A negotiation process can rely on several mechanisms such as: auctions, game theory, intelligent agent mechanisms, etc. [20]. Nevertheless, such process if often conducted by human actors that in the last instance are the ones responsible for decision-making.Although some works try to insert some automation into the negotiation process [21, 22], this continues to be a rather difficult issue. Progress in this area during the last years has highlighted a number of important topics that need to be considered when developing processes and methodologies for negotiation, including Contract Models, Ontology, Contract Framework, Electronic Institutions, Digital Signature, etc [5].

Considering the above rich conceptual inheritance but also practical requirements attained through the interaction with existing enterprises networks, the next section describes the main requirements for an electronic negotiation support environment that considers the potential collaboration risks, and allows members of a collaborative network to conduct their negotiation towards the internal consortium agreement. As a result, the VO creation process can become more agile in terms of adaptation to unexpected events and reduce possible risks in collaboration.

4 Research Contribution and Innovation

Due to the continuous changes in socio-economic markets, companies and organizations have to constantly adapt themselves to survive. One option is to strategically join their competencies to rapidly respond to a business or collaboration opportunity through a goal-oriented network that is an association of individuals and/or organizations. Nevertheless, besides taking advantage of clustering themselves, companies and organizations have to find better solutions to meet the customer needs. One trend in manufacturing is to move towards highly customized products and ultimately one-of-a-kind products [23]. In this context, this trend is reflected in the term mass-customization that refers to a customer co-design process of products and services which meet the needs/choices of each individual customer with regards to the variety of different product features. Moreover, in all cases, even in the case of new product orders or co-creation teams for innovative services, the goal-oriented collaborative network can be labeled virtual organization (VO). It should be noted that different VOs, with very different time durations, can be needed during the life cycle of a complex product, such as solar energy parks, intelligent buildings, etc.

This consortium formation mainly consists in planning and scheduling the work order and selecting the appropriated partners to join the VO. But, and according to innumerous authors in literature (as mentioned in the related literature section), it is desirable that a virtual enterprise or organization is able to deal with the associated risks, vulnerability, robustness, flexibility, resilience and business continuity. Being *agility*, the quality or state of the organization of being able to have a quick resourceful and adaptable response, then the time and amount of resources consumed during the VO creation process give a good indication of the level of agility of a collaborative network. To achieve that agility, it is essential that some requirements are provided to its members, namely in terms of common infrastructures, governance models and rules etc. For that, the existence of a VO breeding environment context (VBE), enables a partial fulfillment of those requirements [6].Furthermore, due to the heterogeneous contexts of the VO breeding environments that usually companies or organizations belong to, it is possible to create value among VBE members if there exists an electronic negotiation support environment that contributes to boost the participation in consortia creation. Thus, the hypothesis adopted for this work is:

The process of creating dynamic virtual organizations can become more agile if an appropriate electronic negotiation wizard environment is established with the necessary soft modeling characteristics to structure and conduct the entire negotiation process, making it traceable, reducing the collaboration risks, and managing the participants' expectations. Moreover, the negotiation environment should be customizable according to different collaboration levels, either in terms of commitment or in terms of duration.

The key reason why it is important to have a consistent negotiation support environment for the formation of VOs is essentially to improve the entire process of establishing the VO agreement that will induce the governing rules and principles of the consortium during the operation phase and as mechanism to reduce the potential collaboration risk. Therefore, it is of extreme importance to make a comprehensive analysis of the important characteristics that such support environment shall involve.

Risk sources and drivers in collaboration. Having into account the environmental characteristics of collaboration, one topic that has considerable importance and can influence the negotiation process is the related associated risks and their prediction. For that, the current work relies on the characterization of risks and failures in collaboration, so that the negotiation model can also support risk analysis of potential risk of members' behavior during VO operation. Fundamentally it is important to consider the sources and drivers of such risks. Elementary questions arise such as: what can happen and what can be the case?; how likely is it that it will happen?; and if it does happen, what are the consequences?

Table 1. Risks in collaborative network environments

Risks in Collaborative Network Environments			
Drivers	***Types***		***Sources***
Internally-driven risks	Operational	Human error or staff failing to perform	Organizational risk
	Commercial	Risk of business relationships failing	Organizational/Network risks
	Technical	Risk of physical assets failing or being damaged	Organizational risk
Externally-driven risks	Financial	Uncertainty of financial controls failing or succeeding	Organizational/Network/environ mental risk
Decision driven risks	Strategic	Related to the uncertainty of plans failing or succeeding	Network risk

According to some authors in risk management, and collaborative environment characteristics, it is assumed that risks in collaborative environment can be categorized according to internally, externally and/or decision driven risks. Also, risks may have different sources, namely organizational, network, and/or environmental. Therefore, Table 1 includes a mapping of risks drivers and sources in collaborative network environments.

In view of the many different VO topologies and various domains where they can be, different risks can exist. So, having into account the above risk drivers and sources in CNs, whenever there is a formation of a new VO, those risks have to be considered. Fig. 1 illustrates the case of a consortium formation with VBE members, local support entities and suppliers and customer (VO co-creation space) where there are some risks that have to be considered, as well as its drivers.

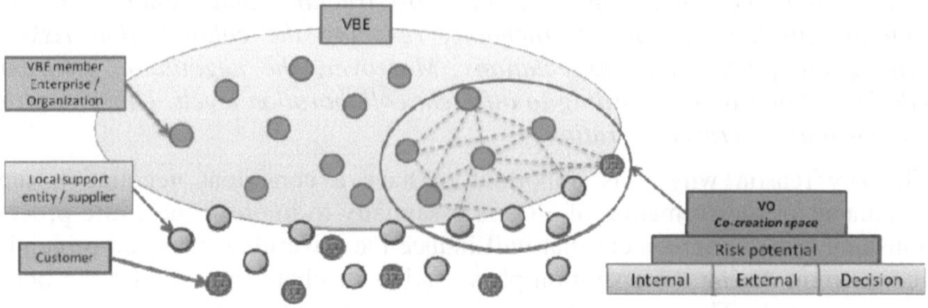

Fig. 1. Co-creation space formation

As a result, if there is a support mechanism for the consortia formation that allows the negotiation process to be conducted with mechanisms that allow risk identification and characterization, then the VO creation process could in fact become more agile in terms of adaptation, and failures in the operation phase also reduced.

Main requirements of a negotiation environment. To support some of the developed concepts, and through the interactions with various end-users networks various critical negotiation activities were identified:

- The contract/agreement should follow a basic set of standard templates: It is important to depart from common templates, selected for each kind of BO/CO, and extend the selected template to cope with the detailed agreement specifications using "add-on" clauses;
- Reaching agreements concerning coordination aspects;
- Reaching agreements on the detailed activities and scheduling;
- Information exchange agreement: i.e. how should information be exchanged among partners, and also which kind of information should be exchanged;
- Support for privacy of proposals, where only the involved partners have access to the information being negotiated;
- Considering the potential risks collaboration, reaching agreement concerning the sharing of risks among the involved partners; and
- Provide a mechanism for tracing the history of the negotiation.

Having into account this list, it is evident that these types of agreements require fundamentally decision making by human actors rather than fully-automated decision-making. Therefore, in this case, what is addressed is not a complex e-contracting process where the system is capable of automatically generate, interpret, execute, and manage a contract or agreement, but to a certain extent, a system that is capable of storing and receiving inputs into an electronic source for later interpretation and user guidance through the process. Therefore, considering the VBE context [6],the main actors involved in this process are described in Table 2.

Table 2. Actors in the negotiation support environment

Actors	*Description*
Potential Customer	Individual or organization that makes a new product order or innovative service request.
VO Planner	VBE Member that in face of a new collaboration opportunity identifies the necessary competences and capacities, selects an appropriate set of partners (VBE Members and/or outsiders), and structures the new VO.
Potential VO Internal VBE Partner	VBE Member that is a possible partner of the VO that is being created.
Potential VO Local Partner	External interested stakeholder (e.g. Local suppliers and Local Support Entities), that is a possible partner of the VO that is being created. Nevertheless, depending on the nature of the Potential VO Local Partner, there might be different levels of involvement in the collaboration process.

So that the negotiation environment can be properly implemented, it is necessary that it has access to the VBE information system mainly to the VBE Members' profile and competences, as well as, access to collaboration history so that risk forecasting and partners' expectations assessment can be performed. Thus, the basic modules of the negotiation support environment are:

Table 3. Basic modules of the negotiation support environment

Modules	*Description*
Negotiation support modules	**Negotiation Template Management (NegTM).** Collection of contract templates and negotiation topic templates to support the VO creation. In the contract construction process it should be possible to build or edit the contract skeleton or template and add them to the collection.
	Negotiation Contract Editor (NegCE). The main point of interaction with the user, allowing the initiation, conduction and monitoring of the entire negotiation process in the VO creation. The main users are the VO Planner and the Potential Partners (that have different authorization levels depending if they belong to the VBE or are potential local partners). More details in [5].
	Negotiation Virtual Negotiation Rooms (NegVNR). It is the virtual space where the potential partners of the VO are invited to join in order to negotiate and/or discuss the necessary topics/clauses that need an agreement. More details in [5].
	Negotiation Support for Agreement Establishment (NegSAE). Like an e-Notary that allows clients to exchange information with warranty of authenticity and validity as well as providing a safe repository for saving and requesting documentation (through digital signatures and encryption techniques).
	Negotiation Assessment and Potential Risk Forecasting (NegAPRF). Through interaction with the VBE management system it will be made an assessment of potential partners' expectations towards collaboration and the forecasting of potential risks of the collaboration.
Interaction with other systems	The main interaction is basically with the VBE information management system to have access to the VBE members profile and competencies as well as access to collaboration history.

Having into consideration the main requirements of the negotiation support environment as well as the identified mechanisms, Fig.2illustrates the strategic dependency model for the negotiation of a goal-oriented collaborative network (VO) where the dependences between actors and system modules are represented.

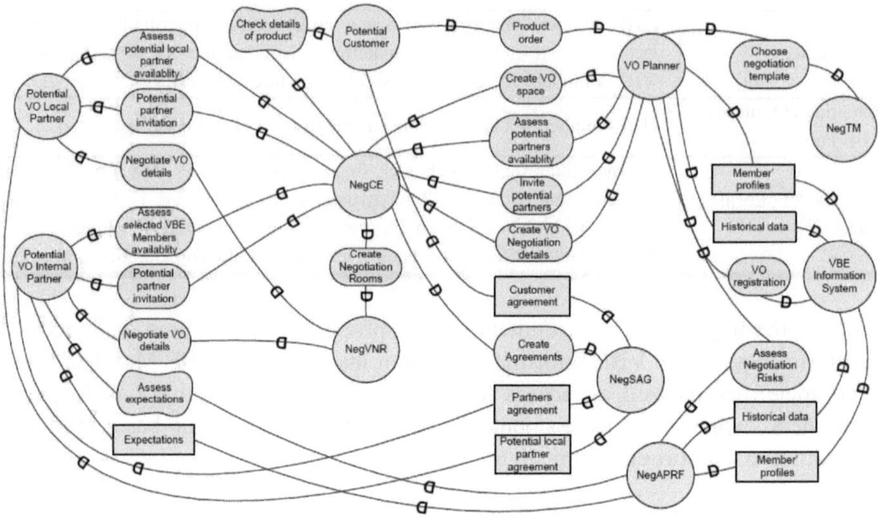

Fig. 2. Strategic dependency model for the negotiation support environment

Discussion of Results. As a first result, to support some of the previous concepts and basic functionalities, an Agreement Negotiation Wizard (WizAN) was designed and developed [5, 11]. The prototype aimed to assist the human users in their decision making process of consortia creation, structuring the negotiation process and making it traceable. The basic functionalities that were implemented were synchronism, negotiation editor, CSCW functionalities, privacy, and some other basic functionalities. Through interaction with some real VBE networks, the prototype was positively validated in a real scenario with a Swiss and a Chinese VBE supporting negotiations between partners from the two geographical areas [5].As a result, it is possible to draw some positive conclusions, namely in terms of preventing misunderstandings (particularly cultural) due to focused negotiation and the possibility to attach (electronic) documents. Moreover, a degree of *authenticity* is also guaranteed due to the existence of an eNotary service. Also, the system ensures the privacy of the information exchanged during negotiations, guaranteeing that partners have access only to authorized information. Therefore, by using such system, it is possible to reduce the negotiation time of the VO creation process, which increases the agility indicator.

Through the on-going PhD research and participation in European research projects, the topic of participants' expectation assessment and collaboration risk forecasting during a VO creation phase, gains more significance also for the agility indicator because it is possible to reduce the risks in collaboration and increase adaptation to unexpected events. So a first identification of risk drivers and sources is prepared and summarized in this paper. The test bed to assess the agility indicator in VO creation phase is provided by the GloNet project in a solar park context.

5 Conclusions and Further Work

As it has been discussed for several years, the virtual organizations paradigm is of extreme importance for companies to survive in a competitive market. It is therefore vital to provide mechanisms to allow more dynamism in the process of consortia creation for VOs. If during the creation process the VO is aware of the potential risks that such collaboration might have during the operation phase, it will indeed be much simpler to overcome or even to avoid some execution/operation problems. Moreover, if the VO has its governing rules and principles as well as knowledge of the potential collaboration risks well defined in its collaboration agreement, then the chances of success will certainly increase. If there is an electronic negotiation support environment to help organizations to structure and conduct the entire negotiation process, then the VO creation process can become more agile.

This paper summarizes an on-going research work that tries to accomplish the above requirements. As some of the concepts and preliminary results have already been positively validated, it is now foreseen to achieve a more advanced environment that comprehends most of the described areas with the needed adaptations to support the aimed negotiation support for dynamic VOs with "smart" characteristics, such as: collaboration risks forecast, assessment of participants' expectations, traceability, etc.

A first identification of drivers and sources of those risks has already been achieved and presented in this paper. The emphasis of the current work is now on a taxonomy for potential risks in collaboration and how they are directly related to the potential participants expectations and competences. The validation process is then intended to consist in peer validation and supported by the EU project GloNet.

Acknowledgments. This work has been supported by the *Collaborative Networks and Distributed Industrial Systems* Research Group of Uninova and partly by the GloNet project (FP7 programme) funded by the European Commission.

References

1. Bititci, U., Turner, T., Mackay, D., Kearney, D., Parung, J., Walters, D.: Managing synergy in collaborative enterprises. Production Planning & Control 18(6), 454–465 (2007)
2. Camarinha-Matos, L.M., Afsarmanesh, H.: Related Work on Reference Modeling for Collaborative Networks. In: Camarinha-Matos, L.M., Afsarmanesh, H. (eds.) Collaborative Networks: Reference Modeling, pp. 15–28. Springer (2008)
3. Husdal, J.: A Conceptual Framework for Risk and Vulnerability in Virtual Enterprise Networks. In: Managing Risk in Virtual Enterprise Networks: Implementing Supply Chain Principles, p. 1 (2010)
4. Camarinha-Matos, L.M., Afsarmanesh, H., Ollus, M.: ECOLEAD and CNO base concepts. In: Camarinha-Matos, L.M., Afsarmanesh, H., Ollus, M. (eds.) Methods and Tools for Collaborative Networked Organizations, pp. 3–32. Springer (2008)
5. Oliveira, A.I., Camarinha-Matos, L.M., Pouly, M.: Agreement negotiation support in virtual organisation creation–an illustrative case. Production Planning & Control 21(2), 160–180 (2010)

6. Afsarmanesh, H., Camarinha-Matos, L.M., Ermilova, E.: VBE Reference Framework. In: Methods and Tools for Collaborative Networked Organizations, pp. 35–68. Springer (2008)
7. Atzori, L., Iera, A., Morabito, G.: The internet of things: A survey. Computer Networks 54(15), 2787–2805 (2010)
8. Giusto, D., Iera, A., Morabito, G., Atzori, L.: The Internet of Things: 20th Tyrrhenian Workshop on Digital Communications. Springer (2010)
9. Bucherer, E., Uckelmann, D.: Business Models for the Internet of Things. In: Architecting the Internet of Things, pp. 253–277. Springer (2011)
10. Camarinha, L.M., Oliveira, A.I., Sesana, M., Galeano, N., Demsar, D., Baldo, F., Jarimo, T.: A framework for computer-assisted creation of dynamic virtual organisations. International Journal of Production Research 47(17), 4661–4690 (2009)
11. Oliveira, A.I., Camarinha-Matos, L.M.: Agreement Negotiation Wizard. In: Camarinha-Matos, L.M., Afsarmanesh, H., Ollus, M. (eds.) Methods and Tools for Collaborative Networked Organizations, pp. 191–218. Springer (2008)
12. Picard, W., Rabelo, R.J.: Engagement in collaborative networks. Production Planning & Control: The Management of Operations 21(2), 101–102 (2010)
13. Parung, J., Bititci, U.S.: A metric for collaborative networks. Business Process Management Journal 14(5), 654–674 (2008)
14. Child, J., Faulkner, D., Tallman, S.: Strategies of cooperation: Managing alliances, networks, and joint ventures. Oxford University Press, USA (2005)
15. Jüttner, U., Peck, H., Christopher, M.: Supply chain risk management: outlining an agenda for future research. International Journal of Logistics: Research and Applications 6(4), 197–210 (2003)
16. Harland, C., Brenchley, R., Walker, H.: Risk in supply networks. Journal of Purchasing and Supply Management 9(2), 51–62 (2003)
17. Turel, O., Yuan, Y.: User acceptance of Web-based negotiation support systems: The role of perceived intention of the negotiating partner to negotiate online. Group Decision and Negotiation 16(5), 451–468 (2007)
18. Strobel, M., Weinhardt, C.: The Montreal Taxonomy for Electronic Negotiations. Group Decision and Negotiation 12(2), 143–164 (2003)
19. Gimpel, H., Jennings, N.R., Kersten, G.E., Ockenfels, A., Weinhardt, C.: Negotiation, Auctions, and Market Engineering. LNBIP, vol. 2. Springer, Heidelberg (2006)
20. Rocha, A.P., Oliveira, E.: An Electronic Market Architecture for the Formation of Virtual Enterprises. In: Infrastructures for Virtual Enterprises - Networking Industrial Enterprises. Kluwer Academic Publishers, Boston (1999)
21. Jennings, N.R., Norman, T.J., Faratin, P., O'Brien, P., Odgers, B.: Autonomous Agents for Business Process Management. Journal of Applied Artificial Intelligence 14, 145–189 (2000)
22. Bartolini, C., Preist, C., Jennings, N.R.: A software framework for automated negotiation. In: Choren, R., Garcia, A., Lucena, C., Romanovsky, A. (eds.) SELMAS 2004. LNCS, vol. 3390, pp. 213–235. Springer, Heidelberg (2005)
23. Camarinha-Matos, L.M., Macedo, P., Ferrada, F., Oliveira, A.I.: Collaborative Business Scenarios in a Service-Enhanced Products Ecosystem. In: Camarinha-Matos, L.M., Xu, L., Afsarmanesh, H. (eds.) PRO-VE 2012. IFIP AICT, vol. 380, pp. 13–25. Springer, Heidelberg (2012)

Inter-Enterprise Architecture and Internet of the Future

Alix Vargas[1], Andrés Boza[1], Llanos Cuenca[1], and Ioan Sacala[2]

[1] Centro de Investigación en Gestión e Ingeniería de Producción (CIGIP), Universitat Politècnica de València, Camino de Vera s/n Ed 8G -1° y 4° planta Acc D (Ciudad Politécnica de la Innovación) Valencia Spain
alvarlo@posgrado.upv.es, {aboza,llcuenca}@cigip.upv.es
[2] University Politehnica Bucharest, Faculty of Automatics and Computer Science, 313, Splaiul Independetei, 60042 Bucharest Romania
sacalaioan@yahoo.com

Abstract. This paper proposes the concept of Inter-Enterprise Architecture (IEA), which seeks the application of tools and methodologies developed in the Enterprise Architecture (EA) field for the individual firm, adapting to an environment of collaboration between several companies that make networks or supply chains, in order to facilitate the integration and interoperability of their collaborative processes in line with its IS/IT (Information Systems/ Information Technology) to harmonize the joint processes, reduce risk and duplication, increase service and customer responsiveness, reduce technology costs and align the joint business to IS/IT.

Keywords: strategic alignment, enterprise collaboration, enterprise architecture.

1 Introduction

The current business environment is dominated by globalization, and the biggest consequence of this effect is the increase of competition, not only between companies but also between supply chains and networks. As a result, supply chains and networks are now looking to enforce collaborative agreements, which would produce more efficient workflow, flexibility, effectiveness, agility and coordination between chain links. In order to achieve proper synchronization and organization, every chain link must be provided with sufficient information and communication technology to further facilitate their operational integration. In achieving this level of cooperation, there has been an increase in the number of publications that combines topics of strategic alignment and enterprise collaboration. To fuse these principles, current research proposes the use of enterprise engineering through utilizing enterprise architecture.

In order to explain the origin of that formulation, state of the art research has been conducted in three areas: strategic alignment, enterprise architecture, and enterprise collaboration; this has helped to identify trends in current literature, studying the models that have been presented and their relationships with one another, structuring main concepts and associated issues, analysing main ideas and common points, and

L.M. Camarinha-Matos, S. Tomic, and P. Graça (Eds.): DoCEIS 2013, IFIP AICT 394, pp. 25–32, 2013.

identifying a large gap in the literature, due to the fact that there is currently not any documentation tying these three fields of research together. As a result of this gap, the concept of Inter-Enterprise Architecture (IEA) is proposed, using the tools and methodologies of enterprise architecture, which have been developed for the individual enterprise, but adapting them in a collaborative environment between several enterprises making up supply chains and networks. This will facilitate the collaboration and integration process between enterprises and their information and technology systems, supporting joint processes, reducing risks and redundancies, increasing customer service and responsiveness, reducing technology costs and allowing for alignment on multiple levels: joint business processes, and IS/IT. This level of integration has made the Internet vitally important to all business environments. Therefore, the link between the topics of IEA and Internet of the future is evident.

2 Future Internet and Internet of Things

Currently, it is well known that the Internet is used by everyone in many different ways. Over time, the Internet has become a necessity in business environments, but many SMEs cannot use it as a management and operational tool. In order for the Internet to become useful as a real universal business system used by collaborative networks, it is necessary than society research and industry work jointly and coordinately.

According to the European Commission, Internet of Things (IoT) is an integrated part of the future Internet and could be defined as a dynamic global network infrastructure with self configuring capabilities based on standard and interoperable communication protocols where physical and virtual "things" have identities, physical attributes, virtual personalities and use intelligent interfaces, and are seamlessly integrated into the information network[1].

The current research proposes that the concept of an IEA describes how an instrument assures strategic alignment between business and IT in collaborative networks. Thus, it is necessary to also understand the interactions between virtual and physical worlds. Enterprise Architecture and Modelling based on the Internet of Things makes sense because of the proliferation of cheap sensors heralded by the Internet of Things. With this kind of technology being utilized, collaborative enterprise efforts will achieve a better understanding of the joint process of allowing business decisions to be made in real time, even without direct human intervention.

3 Conceptualization

So far, in the context of the current research, three papers have been published that consolidate the state of the art research in three main topics: strategic alignment, enterprise collaboration and enterprise architecture [2, 3, 4]. In this section, we are going to brief the reader in a general way about the principal conclusions deduced in these papers, after a deep analysis of the literature. We invite the reader of this paper to check these papers in order to enhance their knowledge of these topics.

3.1 Conceptualization of Strategic Alignment

According to the most important elements provided for several authors, in [4], we suggested the following definition for strategic alignment: *"The strategic alignment of business and IS/IT is a dynamic and continuous process that enables integration, adjustment, consistency, understanding, synchronization and support between business strategies and strategies of IS / IT, in order to contribute and maintain the correct performance of the organization, creating a competitive advantage that is sustained over time".*

The principal models in strategic alignment have been proposed for [5] and [6], Strategic Alignment Model (SAM) and Alignment Maturity Model (AMM), respectively. The field of strategic alignment is reinforced by SAM but their bases are theoretical and not practical. AMM, presents a practical component that SAM does not have, this model is based on the concepts that are treated conceptually in the SAM, but incorporating a practical aspect to the field of strategic alignment, proposing a model to measure the degree of maturity of the business and IS/ IT alignment, enabling the company that applied to identify, how it is, where and how to improve [2].

The principal elements identified in this field are: business strategy, organizational and processes infrastructure, IS infrastructure, IS / IT strategy, IT conceptualisation, maturity model alignment, applications and services portfolio, alignment heuristics, strategic dependencies model and portfolio applications [5, 6, 7].

3.2 Conceptualization of Collaboration

In [3],business collaboration is defined as: "*A joint process between members of the SC, where the decisions are made jointly, based on the information shared and exchanged on a bilateral form, achieving coordinate and synchronize joint activities to meet customer requirements and achieve process efficiency sets to generate a mutually beneficial*".

According to [8], the collaboration process consists of six activities. However, this generic process has not taken a certain crucial aspect into account: the definition of how to share benefits equitably to ensure the stability of the collaboration [9]. The solution to this problem is provided by [10], which proposed the definition of a system of compensatory payments, which may agree with the definition phase of the negotiation and exception handling and can be implemented when evaluating the results. Another previously unconsidered aspect in this generic process is the need for feedback between the parties once it has completed the process of collaboration in the stipulated horizon, which also must review the plan and modify if necessary. Thus, we proposed in [4], the following enterprise collaboration process: 1) Definition and collaboration agreement, 2) Planning in the local domain, 3) Plan of exchange, 4) Negotiation, exception handling and compensation system, 5) Execution, 6) Measurement of results and implementation of compensation plan, and 7) Feedback and review of the plan.

3.3 Conceptualization of Enterprise Architecture

In [3], Enterprise Architecture is defined as: *"A discipline that provides a set of principles, methods, models and tools used for analysis, design and redesign of a company, thus allowing to represent and document the elements that form the company (such as organizational structure, business processes, systems information and technology infrastructure) and the relations, organization and joints between these elements, allowing the company to be represented in a holistic and integrated perspective, in order to achieve the business objectives and facilitate decision-making processes"*

In recent years, several researchers have proposed enterprise architectures, among which stand out: CIMOSA [11], GIM-GRAI [12], PERA and GERAM [13], IE-GIP [14, 7], TOGAF-ADM [15], ARDIN [16] and ARIS [17].

The common elements that handle these enterprise architectures are: methodology, framework and language modeling [3].The definition of a methodology facilitates the implementation of the architecture [18]; the framework allows a graphic and simple structure of the elements that make up the enterprise [19] and how these elements are related [18]; furthermore, modeling language allows for modeling, organization and understanding of the relationships between elements of the enterprise[3].

4 Research Contribution

It has been crucial to prepare state of the art research in three areas: strategic alignment, enterprise architecture, and enterprise collaboration; this has helped identify trends in current literature, studying the models that have been presented and their relationships with one another, structuring main concepts and associated issues, analysing main ideas and common points, and identifying a large gap in the literature, due to the fact that there is not currently any documentation tying these three fields of research together.

As a result of this gap, the concept of Inter-Enterprise Architecture (IEA) is proposed, looking for applications of the tools and methodologies of enterprise architecture, which have been developed for the individual enterprise, but adapting them in a collaborative environment between several enterprises that make up supply chains and networks. This will facilitate integrative collaboration processes among enterprises with their information systems and technology systems, supporting joint processes, reducing risks and redundancies, increasing customer service and responsiveness, reducing technology costs and allowing for alignment on multiple levels: joint business processes, Information and Communication Technology (ICT). In Fig. 1, currently complete work is summarized, and in the spotlight, the field of intended study is represented.

The implementation of an IEA is part of establishing a set of architectural guidelines that ensure comprehensive development between models and inter-enterprise needs, with joint business processes and ICT. This set of ICT strategic guidelines must be based on joint strategic planning and corporate recognition of business strategies and activities that support such planning. These guidelines will also be responsible for spawning necessary information for the joint operation of organizations and technologies required to support joint operations and processes for implementing new technologies in response to changing needs and jointly involved organizations.

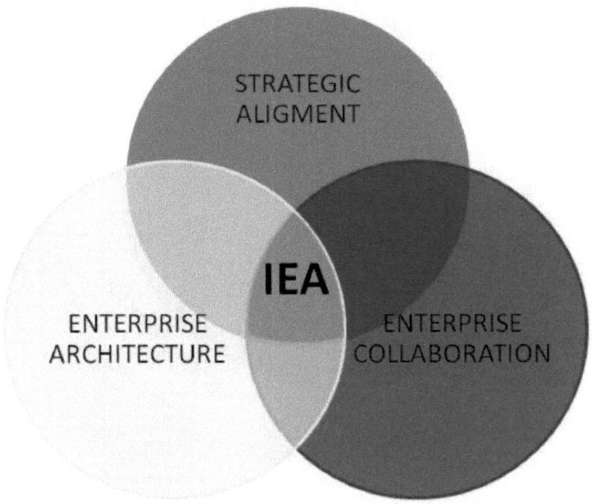

Fig. 1. Origin of the IEA concept

The importance of the IEA resides ultimately in its utility for the joint organization, which will remain there as long as it will be updated every time there are changes in joint strategic planning, changes in vital business processes, or changes in ICT that support vital processes.

Fig. 2 describes the main conceptual elements of an IEA. The core of the IEA is the collaborative process (CP), which is triggered by the necessity of collaboration between network partners. As a result, the collaborative process begins following a minimum of seven activities identified as vital. If the joint organization really wants to take advantage of that collaborative process, then it is necessary to involve the joint organization in an alignment process, taking into account the elements of the strategic alignment model (SAM).Thus, it is also important to adapt these to a collaborative organization, where joint business and IT dimensions have to have a clear functional integration, and the external and internal elements of business and IT have to have a strategic fit. To achieve this alignment, it is necessary to enforce the utilization of enterprise architecture and its main elements (EEA): methodology, framework and language modelling.

On the other hand, it is evident that technology plays a crucial role in this conceptual model. It has been discussed in previous paragraphs through the utilization of technology as Internet of Things or Web of Services. Therefore it is possible to enable collaborative networks to make decisions in real time, allowing agility, accuracy and effectiveness in the joint processes.

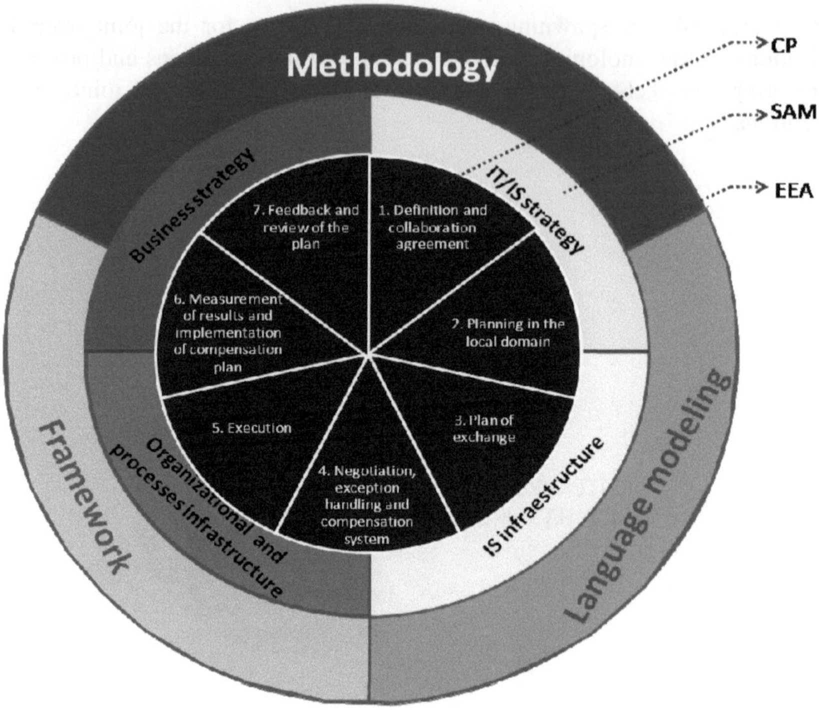

Fig. 2. Conceptual model of IEA

According to the European Commission, in 2025, the Internet will become entirely universal in all business environments, and new business models will be designed to support competition and enterprise collaboration. This trend is known as "internet of the future based on enterprise systems". The Future Internet networked enterprises [FInES] involves the following basic pillars: Internet of Things (IoT), Internet of Services (IoS), Internet of People (IoP), Internet of Knowledge (IoK) [20].

5 Critical View

The fields of enterprise collaboration, strategic alignment and enterprise architecture have made significant progress in the last years in terms of research and applications in the industry. More recently, each one of these fields is seeking to collaborate in a complementary way. Current research proposes the fusion of these three fields in order to extract the main elements that give origin to the concept of IEA. Due to the existence of different perspectives in the literature, this research is looking to contribute to existing research with a more theoretical foundation.

6 Conclusions and Further Work

In order to fuse and implement the principles of collaboration and strategic alignment, we propose the use of enterprise engineering through the use of enterprise architectures.

Enterprise architectures are tools that provide business concepts, models and tools that enable the integration of the elements that comprise it. The supply chains or networks are a set of related companies where the principles of enterprise architectures can be extended to this type of collaborative organizations.

In relation to research that combines the use of strategic alignment, business collaboration and enterprise architectures, there are no references found that take into simultaneous account these three fields of knowledge, but there are only articles that deal with different ways to combine two of these three disciplines.

The line of research that arises from this analysis is to make a proposal for inter-enterprise architecture in the context of collaborative networks based in Internet of the future, looking to achieve a real strategic alignment between joint business and joint information technology. This line approaches and combines the three disciplines analysed: enterprise architecture, strategic alignment and business collaboration.

We want to continue working in this line of research seeking a proposal in future papers of a Reference Architecture for Collaborative Networks based in Future Internet.

References

1. Information Society and Media European Commission: FInES Cluster Position Paper. Future Internet Enterprise Systems (FInES) Cluster (September 2009)
2. Vargas, A., Boza, A., Cuenca, L.: Lograr la alineación estratégica de negocio y las tecnologías de la información a través de Arquitecturas Empresariales: Revisión de la Literatura. In: Cartagena, XV Congreso de Ingeniería de Organización, pp. 1061–1070 (2011a)
3. Vargas, A., Boza, A., Cuenca, L.: Towards Interoperability through Inter-enterprise Collaboration Architectures. In: Meersman, R., Dillon, T., Herrero, P. (eds.) OTM 2011 Workshops. LNCS, vol. 7046, pp. 102–111. Springer, Heidelberg (2011)
4. Vargas, A., Boza, A., Cuenca, L., Ortiz, A.: The importance of strategic alignment in enterprise collaboration. In: 6th International Conference on Industrial Engineering and Industrial Management, Vigo (2012)
5. Henderson, J., Venkatraman, N.: Strategic alignment: Leveraging information technology for transforming organizations. IBM Systems Journal 32(1), 472–484 (1993)
6. Luftman, J.: Assessing Business-IT alignment maturity. Communications of the Association for Information Systems 4 (2000)
7. Cuenca, L., Boza, A., Ortiz, A.: An enterprise engineering approach for the alignment of business and information technology strategy. International Journal of Computer Integrated Manufacturing 24(11) (2011)
8. Kilger, C., Reuter, B., Stadtler, H.: Collaborative Planning. In: Stadtler, H., Kilger, C. (eds.) Supply Chain Management and Advanced Planning—Concepts, Models Software and Case Studies, pp. 263–284. Springer, Heidelberg (2008)
9. Audy, J., Lehoux, N., D'Amours, S.: A framework for an efficient implementation of logistics collaborations. International Transactions in Operational Research, 1–25 (2010)
10. Stadtler, H.: A framework for collaborative planning and state-of-the-art. OR Spectrum 31, 5–30 (2010)
11. CIMOSA Association: CIMOSA Primer on key concepts, purpose and business value, `http://cimosa.cnt.pl/`

12. Chen, D., Vallespir, B., Doumeingts, G.: GRAI integrated methodology and its mapping onto generic enterprise reference architecture and methodology. Computers in Industry 33, 387–394 (1997)
13. Williams, T., Li, H.: PERA and GERAM enterprise reference architectures in enterprise integration. Information Infrastructure Systems for Manufacturing, 1–27 (1998)
14. Ortiz, A.: Propuesta para el Desarrollo de Programas de Integración Empresarial en Empresas Industriales. Aplicación a una Empresa del Sector Cerámico. Ph.D Dissertation. Universidad Politécnica de Valencia (1998)
15. The Open Group, `https://www.opengroup.org/index.htm`
16. Chalmeta, R., Grangel, R.: ARDIN extension for virtual enterprise integration. The Journal of Systems and Software 67 (2003)
17. Scheer, A., Schneider, K.: ARIS – Architecture of Integrated Information. In: Handbook on Architectures of Information Systems. International Handbooks on Information Systems, s.l., vol. 3, pp. 605–623 (2006)
18. Bernard, S.: An introduction to enterprise architecture. Author House, Bloomington (2005)
19. Cuenca, L., Ortiz, A., Boza, A.: Arquitectura de Empresa. Visión General. In: IX Congreso de Ingeniería de Organización, Gijón, pp. 1–8 (2005)
20. Burlacu, G., Stanescu, A., Sacala, I., Cojocaru, L.: Development of a Modeling Framework for Future Internet Enterprise Systems. In: IEEE 16th International Conference on System Theory, Control and Computing, Sinaia, October 12-14 (2012)

Collective Emotions Supervision in the Product-Servicing Networks

Filipa Ferrada[1] and Luis M. Camarinha-Matos[1,2]

[1] CTS, Uninova, Departamento de Engenharia Electrotécnica, Faculdade de Ciências e Tecnologia, FCT, Universidade Nova de Lisboa, 2829-518 Caparica, Portugal
[2] Departamento de Engenharia Electrotécnica, Faculdade de Ciências e Tecnologia, FCT, Universidade Nova de Lisboa, 2829-518 Caparica, Portugal
{faf,cam}@uninova.pt

Abstract. The GloNet network is an agile virtual enterprise environment for networks of SMEs involved in highly customized and service-enhanced products through end-to-end collaboration with customers and local suppliers (co-creation) in the solar park construction sector. A challenge that arises in this network is the effective management of the interactions among participating organizations and their customers. Some participants' problems, that are not solvable with the working hard procedures of the project execution, might arise leading to some potential tension. Most of the times this tension is a result of soft "emotional" issues experienced in the collaboration "environment". Considering these challenges and having in mind enhancing the quality of the interactions between the involved participating organizations, this paper describes a scenario for supervising collective emotions in the GloNet network context.

Keywords: Emotions, Collective Emotions, Collaborative Networks, Product-servicing Networks, Collaboration Health.

1 Introduction

In the current global and dynamic business environments we have been assisting to a growing trend towards developing highly customized products and services. This assumes the *mass customization* paradigm of providing individually designed products and services to customers by means of process flexibility and quick responsiveness [1]. Furthermore, besides the physical product, a number of business services add value to the product.

The GloNet network aims to support an agile virtual enterprise environment for networks of SMEs interested in pursuing mass customization businesses through implementing the *glocal* [2] enterprise notion. This involves highly customized and service-enhanced products through end-to-end collaboration with customers and local suppliers (co-creation).

A challenge in this network is the effective management of the interactions among participating organizations themselves and the customers, namely in what concerns the management of "soft" issues that are not solvable with the hard processes of the

L.M. Camarinha-Matos, S. Tomic, and P. Graça (Eds.): DoCEIS 2013, IFIP AICT 394, pp. 33–42, 2013.

project execution. These issues can include inter- and intra-organizational abilities, problems in keeping team cohesion, leadership, customers and the network as a whole[3]. It might also include external factors such as the participating organization itself, the people that stand for the organization, etc. In addition, one cannot forget that organizations are composed of people, and people have intrinsically associated emotions which are present in all interactions, so most of the times these conflicts are a result of emotions speaking for themselves. Considering these challenges, it is important to understand these psychological aspects and effectively manage the participating organizations by avoiding emotional conflicts[4].

In this context the main research question is: *What could be a suitable set of models, methods and tools to promote emotional health in collaborative networks, namely allowing the diagnosis of the networks' emotional state and assisting in conflicts resolution?*

In order to give an answer to this research question, this paper presents a business scenario for supervising collective emotions in the context of the GloNet project. The reminder of this paper is organized as follows: Section 2 identifies the relationship of this work to Internet of Things; Section 3 gives a brief description of the related literature namely the collective emotions in the collaborative networks context, the GloNet network and the characteristics inherent to business scenarios used in this application case; Section 4 presents the business scenario on collective emotions supervision services; and finally Section 5 concludes.

2 Relationship to Internet of Things

More than ever, "things" are more actively present in our daily routines either at our homes or at business. According to Sandmaeker et al. [5]*Internet of Things* (IoT) "is a integrated part of the Future Internet and could be defined as a dynamic global network infrastructure with self-configuring capabilities based on standard and interoperable communication protocols where physical and virtual "things" have identities, physical attributes, and virtual personalities and use intelligent interfaces, and are seamlessly integrated into the information network".

In product-servicing collaborative networks IoT might be applied through embedding smart devices able to interact and communicate among them and with the environment by exchanging data and information and reacting autonomously to events (e.g. the data exchanged by the embedded sensors in the solar plant help in the optimization of processes, refinement of schedules, improvement of logistics, etc.).In such networks, the communication, which may be remote, between participants (people on behalf of a company) and the system might be through multiple smart devices such as mobile phones, tablets, laptops, etc. These will give rise to a large amount of data from processes, products, communication interactions and business services systems. Here emotional information might be identified through communication flows and peer interactions intensity which will be then translated into emotional evidences to the emotions-oriented supervision system. In addition, the work presented in this paper might contribute to the future developments of IoT in what concerns network context awareness and distributed intelligence problem solving.

3 Related Literature

3.1 Collective Emotions in the Collaborative Networks Context

Collective emotions have been defined in a relatively general way as emotions that are shared by large numbers of individuals in a certain society [6] while group-based emotions are defined as emotions that are felt by individuals as a result of their membership in a certain group or society [7]. According to Bar-Tal, et al. [8], both concepts suggest that individuals may experience emotions, not necessarily in response to their personal life events, but also in reaction to collective or societal experiences in which only a part of the group members have taken part. But while the former concept suggests that group members may share the same emotions for a number of different reasons, the latter refers only to emotions that individuals experience as a result of identifying themselves with their fellow group members [8].

Emotions are undeniably present in Collaborative Networks, because they are composed of people (representing the CN member organization) interacting, collaborating and sharing common goals. Each participant brings their own emotional tendencies to the network, and the composition and interaction of those emotions produce an emotional energy that goes beyond the tendencies of individual members. On the other hand, individual members may also experience emotions in reaction to the collective emotion shared by the collaborative network.

In this context, it is interesting to inquire to what extent collective emotions influence the normal operation of CNs.

3.2 Product-Servicing Network – GloNet Network

GloNet focuses on collaborative environments for networks of SMEs involved in highly customized and service-enhanced products. The project aims at supporting the notion of *glocal* enterprise with value creation from global networked operations and involving product-service linkage, and management of distributed manufacturing units, in collaboration with customers and local suppliers (co-creation)[2, 9].

Further to service-based enhancement, there is a growing trend in manufacturing to move towards highly customized products, ultimately one-of-a-kind, which is reflected in the term *mass customization*[1]. In fact, and according to Piller [10], mass customization refers to a customer co-design process of products and services which meet the needs/choices of each individual customer with regard to the variety of different product features. Important challenges in such manufacturing contexts can be either obtained from the requirements of complex technical infrastructures, solar parks, intelligent buildings, etc.

The guiding use case in GloNet is focused on the production and lifecycle support of solar energy parks. The norm of operation in this industry is that of one-of-a-kind production. The results (products and services) are typically delivered through complementary competences shared between different project participants (organizations can range from mechanical and electrical companies to software product development enterprises in the area).A key challenge is the design and delivery of multi-stakeholder complex services along the product life cycle (typically 20 years).

3.3 Business Scenarios

The development of business scenarios is an important technique that helps in better characterizing requirements, identifying and understanding business needs, and thus provides important inputs for the next phase of the emotions-oriented supervision system development: the design of the system architecture [11].

Although the concept is not precisely defined in the literature, the adopted notion here is that a **business scenario** represents a **significant business need or problem** in the target domain. In other words, it provides a reasonably extensive description of a business problem, which enables individual requirements to be viewed in relation to one another in the context of the overall use case / target domain [12].

Taking into consideration that GloNet aims at designing and developing support environments for SME networks, a number of networked structures based on the foundations of collaborative networks [13] are identified: (i) *Long-term strategic alliances*; (ii) *Customer related communities*; and (iii) *Goal-oriented networks.* For the business scenario developed in this paper the focus will be on the Long-term strategic alliances, which typically involve product/project designers, manufacturers, service providers, and some support entities, configuring a kind of virtual organizations breeding environment (VBE). A VBE represents an association of organizations and a number of related supporting institutions, adhering to a base long term (formal or informal) cooperation agreement, and adopting common operating principles and infrastructures, with the main goal of increasing their preparedness towards rapid configuration of goal-oriented networks (Virtual Organizations/Virtual Enterprises - VO/VE).

In this way, a business scenario to supervise collective emotions in the context of the GloNet network was considered and developed. The following main elements are considered: (i) Description and purpose; (ii) Goals, outcomes and main features, (iii) Environment and actors; (iv) Details on actors, roles and responsibilities; and (v) Business processes. In addition to tables and textual descriptions, the following formalisms are adopted to help characterizing the business scenario: i* (i-star) to describe actors, individual and common goals, tasks and their inter-relationships[14]; and BPMN – to represent business processes[15].

4 Business Scenario: Collective Emotions Supervision Services

4.1 Description and Purpose

The purpose of this scenario is to explore the feasibility of supervising collective emotions in the context of product-servicing networks. The main idea underneath this scenario is to illustrate how “emotional issues” of collaboration might be handled and supported by ICT.

Supervising the emotional interactions within the network will provide multi-modal emotional input for achieving awareness of the participants as well as the collective emotional state. The supervision system should offer mechanisms to perceive what is not emotionally working and, with or without the administrator intervention, make the necessary adjustments to promote the level and “health” of collaboration.

4.2 Goals, Outcomes and Main Features

Table 1. Goals, outcomes and main features of the collective emotions supervision services

Goals	Description	Outcomes	Main Features
Monitor network collective emotions	Monitoring all the participants' interactions, seeking for emotional anomalies. These malfunctions might occur when participants are socializing, negotiating a new business opportunity or when co-working in a project.	A service that provides mechanisms to identify and monitor the emotional state of the network. This service will provide emotional information according to the overall state of the network and will support an alert mechanism whenever something unusual is detected.	- Mechanisms to collect and configure member emotional evidences. - Mechanisms to identify and characterize collective emotions accordingly. - Mechanisms to scrutinize all interactions looking for emotional flaws. - Alerts are fired when changes in the emotional state are noticed.
Collective emotion fault analysis	After detecting that something is not properly working, a first analysis of the occurrence, identifying the emotional fault, should be done.	A service that provides mechanisms to conduct a comprehensive diagnosis in order to identify the emotional fault.	- Mechanisms to identify and characterize the detected emotional anomaly. - Delivery of the identified emotional fault.
Suggest healing actions	Once identified the emotional problem, a healing process is suggested with the aim to restore the emotional state of the network to normalcy.	A service that provides healing actions to the diagnosed emotional fault.	- Mechanisms to evaluate and analyze the "situation" and assess its potential impact on the network. - Creation of an action plan to heal the identified problem. - Healing action logs.
Suggest preventive actions	A continuous assessment of the possible future emotional conflicts among participants is assured by suggesting preventive actions.	A service that predicts potential conflicts or tension inside the network. It aims to suggest preventive actions for the sake of the network emotional equilibrium.	- Mechanisms to evaluate and analyze the possible emotional conflicts and assess its potential impact on the network. - Creation of an action plan to prevent the potential problem. - Preventive action logs.

4.3 Environment, Actors and Their Roles

This scenario envisages the long term strategic alliances (represented as VBE). The involved actors may be individuals and/or organizations that can play the following roles:

VBE Administrator. The VBE Administrator is responsible to supervise the collaborative health of the network by taking the necessary decisions in order to ensure the collective emotional equilibrium.

VBE Member. VBE Members are responsible to make emotions evidences available to the VBE Administrator allowing in this a proper emotions supervision.

VBE Emotions Critical Situation Manager. The VBE Emotions Critical Situation Manager is responsible to diagnose and propose healing solutions whenever the VBE Administrator is informed by the monitoring system that an emotional anomaly occurred. It is also responsible to suggest preventive measures whenever the VBE Administrator perceives a potential thread in the emotional state of the network.

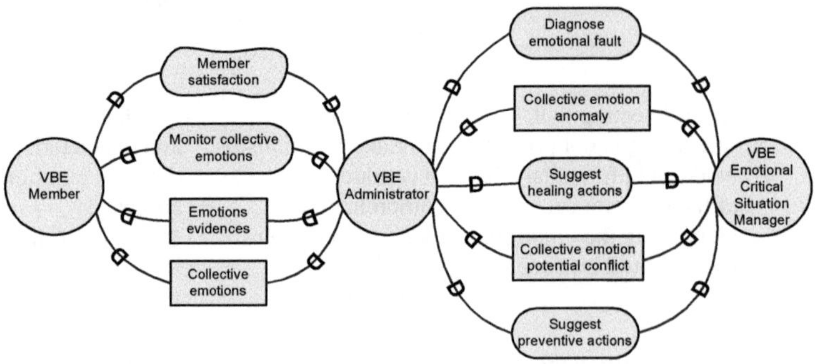

Fig. 1. i* Strategic Dependency Model for Collective Emotions Supervision

4.4 Details on Actors, Roles and Responsibilities

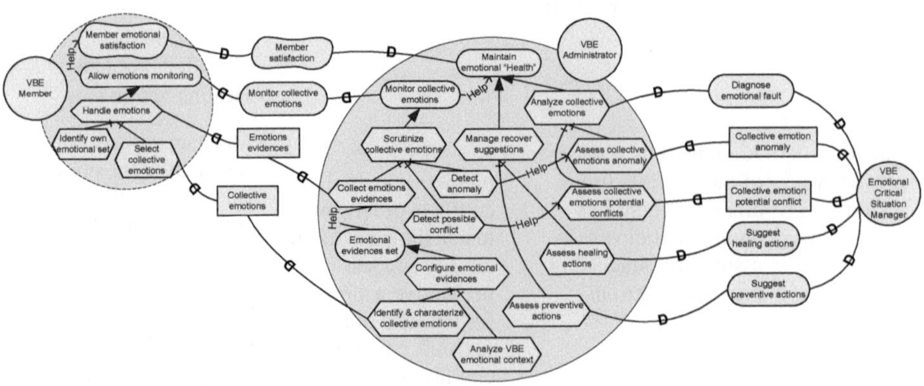

Fig. 2. i* Strategic Rationale Model for Collective Emotions Supervision (partial view 1)

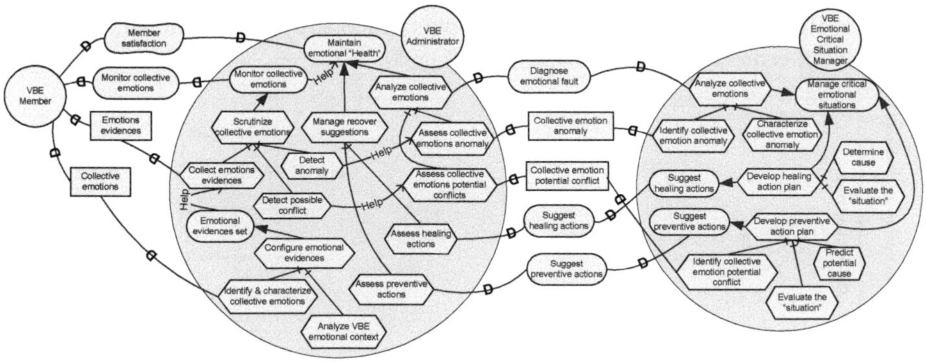

Fig. 3. i* Strategic Rationale Model for Collective Emotions Supervision (partial view 2)

4.5 Business Processes

The collective emotions supervision services scenario envisages the following main business processes:

1) Collective emotions characterization and evidences configuration process
2) Collective emotions monitoring process
3) Collective emotion anomaly diagnosis process
4) Healing actions process
5) Preventive actions process.

Collective emotions characterization and evidences configuration process. This process is conducted by the VBE Administrator and interacts with the VBE Members in order to identify and characterize the collective emotions associated to the VBE. In a second phase of the process the VBE Administrator identifies and configures what are the emotional evidences necessary to keep under monitoring in order to maintain the emotional health of the VBE (Fig. 4).

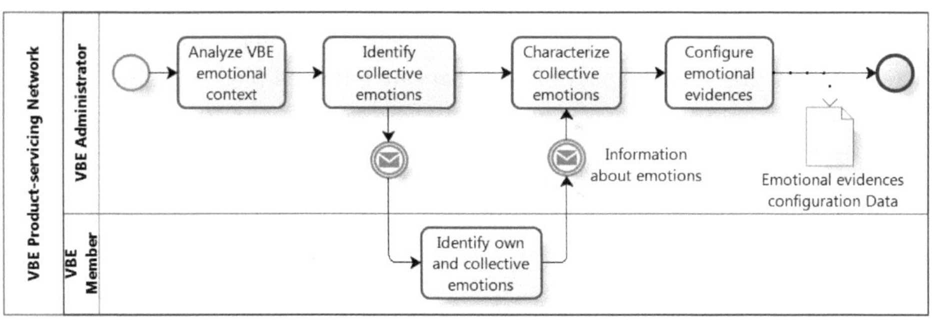

Fig. 4. BPMN diagram of collective emotions characterization and evidences config process

Collective Emotions Monitoring Process. This business process, conducted by the VBE Administrator, comprises the continuous monitoring of emotion evidences within a VBE and, when an anomaly that puts in jeopardy the emotional "health" of the network is detected alerts are triggered (Fig. 5).

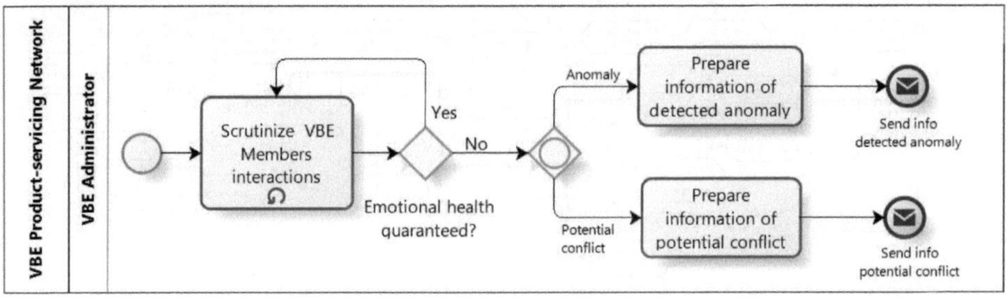

Fig. 5. BPMN diagram of collective emotions monitoring process

Collective Emotion Anomaly Diagnosis Process. This process, which is activated by the monitoring process whenever an anomaly is detected, is carried out by the VBE Emotional Critical Situation Manager and aims to identify and characterize and finally deliver the emotion anomaly (Fig. 6).

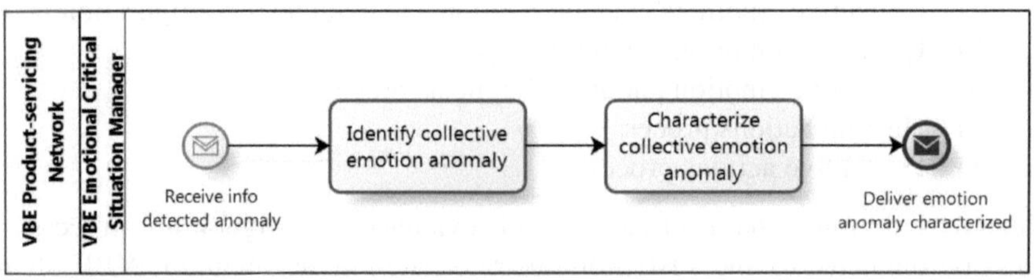

Fig. 6. BPMN diagram of the collective emotion anomaly diagnosis process

Healing Actions Process. This process is activated, with a message containing the characterized emotion anomaly, to the VBE Emotional Critical Situation Manager which is in charge of developing a healing action plan and delivers it to the VBE Administrator. The latter, evaluates the suggested action plan and implements the actions accordingly. Finally the VBE Members are informed (Fig. 7).

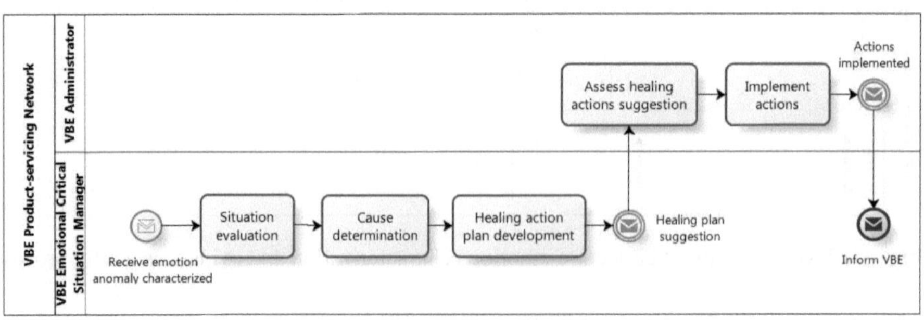

Fig. 7. BPMN diagram of the healing actions process

Preventive Actions Process. This process, which is activated by the monitoring process whenever a potential conflict is detected, is carried out by the VBE Emotional Critical Situation Manager which firstly identifies the collective emotion potential conflict, evaluates the situation and judges if the detected potential conflict is a thread to the emotional health of the network. If the decision is positive, then a preventive plan is developed and delivered to the VBE Administrator to be implemented. If the decision is negative, i.e., if the identified potential conflict does not put in jeopardize the emotional well being of the network then the process is terminated (Fig. 8).

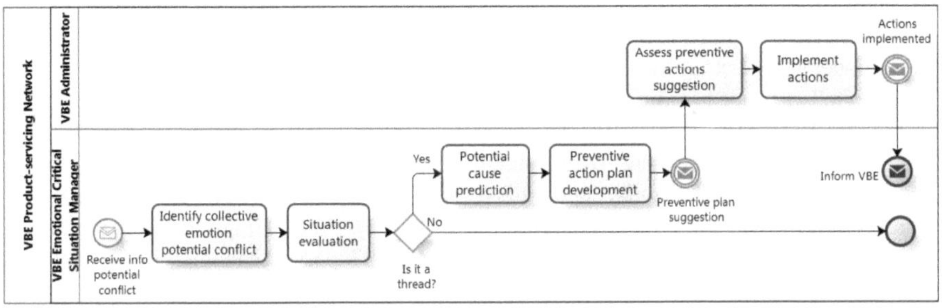

Fig. 8. BPMN diagram of the preventive actions process

5 Conclusions and Future Work

Product-servicing networks of SMEs aiming at develop customizable serviced-enhanced products through the involvement of customers in the creation of product and services (co-creation) are being developed in the GloNet project. Such collaborative networks incorporate global networks of manufacturers and local supply chains in order to attack global business opportunities.

Managing these networks is per se a great challenge, especially because several types of conflicts might arise among partners which might lead to a tension environment. This paper addresses collective emotions as potential drivers to generate conflicts in the GloNet network and presents a business scenario of a supervision system consisting on services to monitor, identify and heal the emotional state of the product-servicing collaborative network.

This work intends to understand what should be the requirements to design and develop an emotions-oriented supervision system in collaborative networks. As such it represents only the design phase of the research project. Future work is planned on building a conceptual model of collective emotions in collaborative networks as well as building a conceptual framework aimed at developing a working definition of the involved concepts and their relationships, developing the purposed processes and validating the system.

Acknowledgments. This work has been supported by the Collaborative Networks and Distributed Industrial Systems (CoDIS) Research Group of Uninova and partly by the GloNet project funded by the European Commission.

References

1. Pollard, D., Chuo, S., Lee, B.: Strategies for Mass Customization. Journal of Business & Economics Research 6(7), 10 (2008)
2. Camarinha-Matos, L.M., Afsarmanesh, H., Koelmel, B.: Collaborative Networks in Support of Service-Enhanced Products. In: Camarinha-Matos, L.M., Pereira-Klen, A., Afsarmanesh, H. (eds.) PRO-VE 2011. IFIP AICT, vol. 362, pp. 95–104. Springer, Heidelberg (2011)
3. Kerr, B.A.: Connecting Emotional Intelligence to Success in the Workplace (2009)
4. Glinow, M.A.V., Shapiro, D.L., Brett, J.M.: Can we talk, and should we? Managing emotional conflict in multicultural teams. Academy of Management Review 29(4), 578–592 (2004)
5. Sundmaeker, H., et al. (eds.): Vision and Challenges for Realising the Internet of Things, CERP-IoT. European Commission - Information Society and Media, Luxembourg (2010)
6. Stephan, W.G., Stephan, C.W.: An integrated threat theory of prejudice. In: Oskamp, S. (ed.) Reducing Prejudice and Discrimination, pp. 225–246. Erlbaum, Hillsdale (2000)
7. Smith, E.R.: Social identity and social emotions: Toward new conceptualization of prejudice. In: Mackie, D.M., Hamilton, D.L. (eds.) Affect, Cognition and Stereotyping: Interactive Processes in Group Perception, pp. 297–315. Academic Press, San Diego (1993)
8. Bar-Tal, D., Halperin, E., de Rivera, J.: Collective Emotions in Conflict Situations: Societal Implications. Journal of Social Issues 63(2), 441–460 (2007)
9. Camarinha-Matos, L.M., Macedo, P., Ferrada, F., Oliveira, A.I.: Collaborative Business Scenarios in a Service-Enhanced Products Ecosystem. In: Camarinha-Matos, L.M., Xu, L., Afsarmanesh, H. (eds.) PRO-VE 2012. IFIP AICT, vol. 380, pp. 13–25. Springer, Heidelberg (2012)
10. Piller, F.T.: Mass Customization: Reflections on the State of the Concept. The International Journal of Flexible Manufacturing Systems 16(4), 313–334 (2004)
11. Ferrada, F., Camarinha-Matos, L.M.: Emotions in Collaborative Networks: A Monitoring System. In: Camarinha-Matos, L.M., Shahamatnia, E., Nunes, G. (eds.) DoCEIS 2012. IFIP AICT, vol. 372, pp. 9–20. Springer, Heidelberg (2012)
12. OpenGroup. Business Scenarios (2011), `http://pubs.opengroup.org/architecture/togaf7-doc/arch/p4/bus_scen/bus_scen.htm` (cited April 5, 2012)
13. Camarinha-Matos, L.M., Afsarmanesh, H. (eds.): Collaborative Networks: Reference Modeling. Springer, New York (2008)
14. Yu, E.S.K.: Modelling Strategic Reltionships for Process Reengineering, Department of Computer Science, University of Toronto, Toronto (1995)
15. OMG, Business process modeling notation. Version 1.1 (2008)

Part III
Service Orientation

Web Service Composition with Uncertain Non-functional Parameters

Łukasz Falas and Paweł Stelmach

Wrocław University of Technology, Institute of Computer Science, Wybrzeże Wyspiańskiego 27, 50-370 Wrocław, Poland
{Lukasz.Falas,Pawel.Stelmach}@pwr.wroc.pl

Abstract. The problem of service composition is of key importance in the Internet of Things paradigm. A composite Web Service can clearly determine how real life objects described as Web Services could interact with each other. This article proposes a non-functional parameters aggregation algorithm, used in composite service plan optimisation stage, taking into account the uncertainty of the non-functional parameters. Also, an optimisation algorithm, using the proposed aggregation method, is presented to show the complete solution for the problem. In this proposition, the algorithm focuses on composite service plan optimisation with uncertain parameters with the given uncertainty description on example of uncertainty parameter. However, the proposed algorithm can be generalized, so it can be used also with other uncertain parameters.

Keywords: QoS-oriented service composition, uncertain non-functional parameters, Service Oriented Architecture.

1 Introduction

In the Internet of Things paradigm real life objects can be described with Web Services. Web Services are often considered building blocks for composite services performing complex tasks. The same approach could be used in the Internet of Things paradigm to show how Internet users can interact with a composition of objects, which is more than simply a sum of its parts. This composability aspect of Web Services enables us to fulfil user requirements, which cannot be met by any single available Web Service. The service composition task in general aims to fulfil both functional and non-functional user requirements (the latter are known also as Quality of Service parameters). This paper focuses on the QoS-oriented composition and defines the problem as composite service plan optimisation with uncertain non-functional parameters for each Web Service.

The aspect of uncertainty of non-functional parameters is often omitted in the general service composition problem, however it is crucial to ensuring that the Web Service model is properly describing the real behaviour of the Web Service. In real life scenarios Web Services exist in dynamically changing environments and their parameters – like execution time or cost – vary, depending on other parameters values. Therefore Web Services, as real life objects, cannot be described with pre-determined non-functional parameters values.

L.M. Camarinha-Matos, S. Tomic, and P. Graça (Eds.): DoCEIS 2013, IFIP AICT 394, pp. 45–52, 2013.

The subject of Web Services composition is often raised in literature. In various papers approaches like AI Planning [1][2], workflow based service composition[3] or template based service composition[4][5] were suggested. Those approaches mainly focus on satisfying the functional requirements for the service. With time more and more authors point that the semantics of composition, user requirements and of Web Services are not regarded sufficiently. Many take on the topic of QoS-Aware service composition focusing on fulfilment of non-functional requirements [6][7].

However in the QoS-Aware service composition approach the problem of non-functional parameters uncertainty was often omitted. Papers where non-functional parameters uncertainty is only a side-topic to the general service QoS optimisation problem often use a simple probabilistic approach based on mean calculation for uncertain parameters, which is later used as a fixed value of given non-functional parameters[8]. Papers, which focus on the problem of non-functional parameters uncertainty, also utilize other probabilistic measures (i.e. skewness or curtosis) [9][10] to achieve reliable results. Some of them even propose a comprehensive solution for the problem of non-functional parameters uncertainty [11].

2 Relationship to Internet of Things

Service Oriented Architecture (SOA) is an approach, in which some functionality of a web application can be provided as Web Services. Nowadays Web Services are still mostly useful to programmers, offering general operations that each can provide actions on multiple objects. In the future however, it is more likely that a single Web Service could be responsible for a single real-life object, like a light in a house or a TV. The latter is already possible but not popular as programmers still think about more general methods that help them handle multiple problems with a single snippet of code. However, this creates a demand for more monolithic environments, which stands in opposition to the SOA paradigm. It is true that operations (like "show device state") will be available via Web Services and accessible remotely from different locations, thus making the system distributed, but single objects like TV, coffee express, fridge or washing machine will be only data in a database in a single centralized management system. In contrast, each device could be available via a Web Service and managed independently of another by various management systems. In fact, as in the Internet of Things paradigm, we could manage them in a direct manner, looking at Web Services as wrappings for real-life objects.

In this context we cannot use a typical approach when a Web Service could be executed hundreds of times at the same time until the server can't process any more requests. Here, the "execution time" of the object-based Web Service is not equal to duration of single use of that object but should rather be described in general (for instance with uncertainty), considering the demand on that object as usually two clients cannot use one Web Service at the same time.

3 Formal Problem Definition

3.1 Service Composition

In general, service composition process consists of two steps. First, the required functionalities and their interactions – i.e. control and data flow – are identified. Second, for a set of the functionalities appropriate candidate services are discovered from the repository and then selected in an optimisation task resulting in a composite service execution plan describing a required composite service.

In this paper we will not describe a detailed approach on how candidate services are discovered and assume that a composite service execution plan with service candidate for each step of the plan is available. In further sections we will describe an algorithm for composite service plan optimisation, in which from sets of candidates for each step of the plan only one Web Service is selected in such a manner that the quality of the whole composite service, determined by QoS parameters, meets the non-functional requirements.

A composite service plan, denoted by **S**, is described with a graph, in which nodes denote steps of the plan, which are in fact users functional requirements. We denote them by:

$\boldsymbol{\varphi_i}$ – i-th functionality (functional requirement, node in a graph)

Candidate services are services that meet the functional requirements but differ in non-functional parameters. For each functionality$\boldsymbol{\varphi_i}$ there are$\boldsymbol{m_i}$candidate services denoted by:

$\boldsymbol{s_{ij}}$ – j-th candidate service for functionality φ_i

In the process of composite service plan optimisation, for each functionality $\boldsymbol{\varphi_i}$, only one Web Service will be selected from a corresponding set of candidate services.

The selection is in general based on non-functional parameters of each candidate, (referred to as Quality-of-Service – QoS attributes) like service execution time, service execution cost, service availability, service execution success rate, service reputation or service execution frequency. The quality of the composite service is determined by the quality of selected services, meaning that each services' QoS property contributes to the QoS calculated for the whole composite service. Depending on the composite service structure (determined by the graph that represents the service plan) the non-functional parameters of the composite service will be calculated differently, with respect to the functional parameter type. For example *cost* of a composite service could be a simple sum of a cost of each selected service, whereas *service execution time* is a sum of execution times whenever services are executed one after another (in a series) and a maximum when services are executed in parallel.

Non-functional parameters of a composite service are denoted by $A_k(S)$ to highlight that they we calculated in the process of aggregation of QoS parameters of services selected in the plan S, where k is an index determining for which non-functional parameter of the QoS parameters vector (k = 1: execution cost, k = 2: execution time, k = 3: availability, k = 4: success rate, k = 5: reputation, k = 6: execution frequency), the aggregation is calculated.

Additionally a constraint (C_k)can be defined for each non-functional parameter:

$$\begin{cases} A_k(S) \leq C_k & \text{if } k = 1,2 \text{ (time, cost)} \\ A_k(S) \geq C_k & \text{if } k = 3,4,5,6 \text{ (availability, etc.)} \end{cases}$$

Finally, the quality criterion is defined as follows:

$$Q(S) = \min\left[\left\{\sum_{k=1}^{2} w_k \max\left(\frac{A_k(S) - C_k}{C_k}, 0\right)\right\} + \left\{\sum_{k=3}^{6} w_k \max\left(\frac{C_k - A_k(S)}{C_k}, 0\right)\right\}\right]$$

where$w_k$is a weight of the *k*-th QoS constraint defined by the user (k = 1,2, …, 6)

3.2 Uncertainty

As a base for our uncertainty model we use uncertain variable definition introduced by Bubnicki in [12]. In this approach one cannot define a certain value for a specific parameter but instead an expert defines a confidence index for a set of values, expressing his confidence that those values could be obtained. In general, the expert can define a function ***h*** that assigns parameters values to appropriate confidence index values.

Typically, one would ask the expert if the value of a parameter belongs to some set ***D*** and then with function ***h***, a certainty of that statement could be determined. However, in presented case we want to ask the opposite question. Assuming a certain degree of confidence (value for confidence index) we would like to determine what are the most certain values for the parameter.

To be exact, in service composition, we would like to know what is the uncertain availability of a given candidate service $\boldsymbol{s_{ij}}$, assuming that its confidence index is greater than $c(s_{ij}) \stackrel{\text{def}}{=} c_{ij}$.Taking into consideration a certain demand ***d*** for the service $\boldsymbol{s_{ij}}$and confidence index $\mathbf{c_{ij}}$(stating that the demand was in the same timeframe), the approximated value of the *k*-th non-functional parameter of service $\boldsymbol{s_{ij}}$is denoted as:

$$\bar{q}_{ij}^{k}(c_{ij}, d) = [q_{ij,s}^{k}, q_{ij,e}^{k}]$$

In the following sections we focus on the uncertainty coming from uncertain demand for a service and we give an example of uncertainty based on availability.

4 Research Contribution and Innovation

4.1 Service Plan Optimisation Algorithm

The service plan optimisation algorithm presented in this paper is based on a combined approach using simulated annealing and tabu search. The general frame of the algorithm is as follows: first greedily find an initial plan, then if it doesn't meet the requirements search for a neighbour plan. Innovations presented in our approach are situated especially in determining values for uncertain non-functional parameters, in determining locally the best candidate service for a given functionality with a use of a fitness function (which is used to generate both initial and neighbour plans) and in so called QoS aggregation (used to verify the quality of the generated plan: both initial and neighbour). Limitations of this paper did not allow presenting all algorithms of the approach, focusing only on those algorithm improvements relevant to the uncertainty aspect of service composition.

4.2 Determining Uncertainty in Non-Functional Service Parameters

With varying demand for a specific service the function $\boldsymbol{h}$, defining what certainty can be attributed to which parameters values, will also be different (this is depicted in Fig.1), thus we use denotation $\boldsymbol{h_d}$ to indicate that the function $\boldsymbol{h}$ is determined for a specific demand d.

The confidence index $\mathbf{c}_{ij}$ is determined based on the certainty that requests for the service $\boldsymbol{s_{ij}}$ were made in the same timeframe, thus it is greatly influenced by the timeframe in which the demand for a service was calculated.

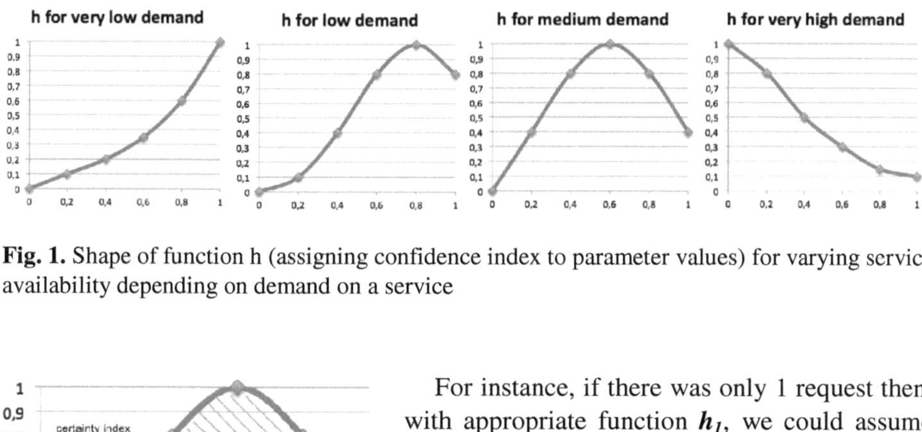

Fig. 1. Shape of function h (assigning confidence index to parameter values) for varying service availability depending on demand on a service

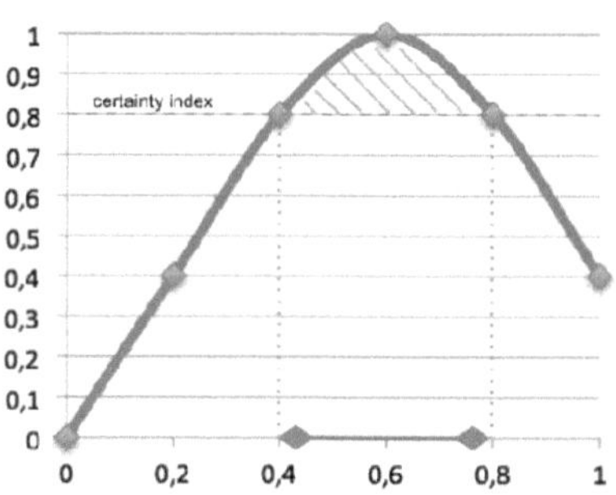

Fig. 2. Method for determining the uncertain value (of availability) assuming a certainty index

For instance, if there was only 1 request then, with appropriate function $\boldsymbol{h_1}$, we could assume that current confidence index would be equal to 1 (only 1 service request was made in that time so we are sure that the timeframe is “the same”) and we would know the exact availability. Were there $\boldsymbol{d}$ requests, we would use function $\boldsymbol{h_n}$ and assuming the level of certainty $\boldsymbol{c_{ij}}$ (that those requests were made in the same timeframe) we could calculate the uncertain value of availability parameter (also see Fig.2):

$$\bar{q}^k_{ij}(c_{ij}, d) = q^k_{ij} * \arg h_d(c_{ij}) = [q^k_{ij,s}, q^k_{ij,e}]$$

The above is true assuming that function $\boldsymbol{h_d}$ is dependent only on popularity $\boldsymbol{d}$ and not varying for different services.

$q_{ij}^{k}$is a typical value of *k-th* non-functional parameter for a service s_{ij} if no uncertainty is present.

4.3 Atomic Service Fitness Function

The goal of a fitness function is to establish a local fitness score according to weighted QoS requirements and thus to define a single order on candidate services (for a given functionality) to jointly represent their various non-functional parameters. In initial execution plan generation algorithm this function is used directly for each functionality to determine greedily which candidate service is locally the best.

Input: q_{ij}^{k} – value of the *k*-th QoS parameter of service s_{ij}; $\{q_{il}^{k} | 1 \le l \le m_i\}$ – values of the k-th QoS parameter for candidate services for functional requirement φ_i (there are m_i of those services), set of candidate services SC

Output: Aggregated QoS (fitness) estimate for the s_{ij} service

for each service s_{ij} **in** SC **do**

  for k=1 **to** 6 **do**

$$q_{ij}^{k} = \begin{cases} -\dfrac{q_{ij}^{k} - \text{mean of}\{q_{il}^{k} | 1 \le l \le m_i\}}{\text{std deviation of } \{q_{il}^{k} | 1 \le l \le m_i\}} & \text{for } k = 1,2 \text{ (time, cost)} \\ \left[\dfrac{q_{ijs}^{k} - \text{mean of}\{q_{ils}^{k} | 1 \le l \le m_i\}}{\text{std deviation of } \{q_{ils}^{k} | 1 \le l \le m_i\}}, \dfrac{q_{ije}^{k} - \text{mean of}\{q_{ile}^{k} | 1 \le l \le m_i\}}{\text{std deviation of } \{q_{ile}^{k} | 1 \le l \le m_i\}}\right] & \text{for } k = 3 \text{ (availability)} \\ \dfrac{q_{ij}^{k} - \text{mean of}\{q_{il}^{k} | 1 \le l \le m_i\}}{\text{std deviation of } \{q_{il}^{k} | 1 \le l \le m_i\}} & \text{for } k = 4,5,6 \text{ (other)} \end{cases}$$

  end

end

return $[w_3 q_{ijs}^{k} + \sum_{k \text{ in } \{1,2,4,5,6\}} w_k q_{ij}^{k}, w_3 q_{ije}^{k} \sum_{k \text{ in } \{1,2,4,5,6\}} w_k q_{ij}^{k}]$

In the above algorithm availability is a non-functional parameter with uncertain values, thus its normalization and summation with other parameters is not a typical sum but instead the fitness score for the atomic service is also an uncertain number.

Then, depending on user approach (optimistic, pessimistic, average) appropriate order on candidate services can be imposed (based on maximum, minimum or average value).

4.4 Service Plan Aggregation Algorithm

This aggregation algorithm recursively establishes a vector of non-functional parameters for any composite service provided with a service execution plan $\boldsymbol{S}$ with QoS parameters for each of atomic services in that plan.

The algorithm replaces sequence and parallel service structures with single services which QoS parameters are aggregated so that the new aggregated value is calculated according to formulas in Table 1 (where $q^k(\varphi_i)$is a value of *k-th* non-functional parameter of a service currently selected for functionality φ_i). The general concept for such a Table is known in literature for composite service plan optimisation. Here, formulas for availability aggregation take into account the uncertain nature of this non-functionality.

Table 1. QoS aggregation formulas for various non-functional parameter types

Param.	Cost	Execution Time	Availability (uncertain)	Success rate	Reputation	Execution Frequency
Sequential	$\sum_{i=1}^{n} q^1(\varphi_i)$	$\sum_{i=1}^{n} q^2(\varphi_i)$	$[\prod_{i=1}^{i} q_s^3(\varphi_i), \prod_{i=1}^{i} q_e^3(\varphi_i$	$\prod_{i=1}^{i} q^4(\varphi_i)$	$\frac{1}{n}\sum_{i=1}^{n} q^5(\varphi_i)$	$\frac{1}{n}\sum_{i=1}^{n} q^6(\varphi_i)$
Parallel	$\max q^1(\varphi_i)$	$\max q^2(\varphi_i)$	$[\min q_s^3(\varphi_i), \min q_e^3(\varphi_i$	$\min q^4(\varphi_i)$	$\frac{1}{n}\sum_{i=1}^{n} q^5(\varphi_i)$	$\frac{1}{n}\sum_{i=1}^{n} q^6(\varphi_i)$

5 Discussion of Results

The algorithms presented in previous sections allow for composite service optimisation, considering that some parameters are described with uncertain values. Regarding limitations of this paper only availability parameter was described in more detail, as its dependency on users demand on a service is most clear. Other non-functional parameters like execution time or cost could also be presented as uncertain parameters and the main approach to optimisation wouldn't have to be changed, however more in depth discussion would be needed. For instance uncertain execution time leads to significantly different ***h*** function shapes compared to availability parameter, depending on demand on that service. With each request for a certain service in a specific timeframe the uncertainty of expected experienced execution time (considered as waiting time and actual execution time) increases significantly, ranging from ***t*** to ***d*t***, where ***d*** is the demand and ***t*** is the typical execution time.

6 Conclusions and Further Work

In this paper a service composition problem in the Internet of Things paradigm was discussed. We presented an algorithm for composite service plan optimisation with uncertain non-functional parameters. In the presented works some parts of the algorithm were presented in more detail, namely a method for calculating uncertain availability parameter value, estimation of fitness function for service candidates with uncertain parameter and a method for aggregation of non-functional parameters of services considering series and parallel composition structures.

Algorithms presented in this paper will be included in the PlaTel platform for composition and management of telecommunication services [13] and are a step in an on-going research on composite service composition and its application to various domains like telecommunication sports or transport. Further testing of the algorithms will be performed in the PlaTel framework, using an advanced testing framework for Web Services currently in development.

Further work will focus on uncertainty description of other non-functional parameters besides availability. Finally, more research will be done in the matter of demand for a service at a specific timeframe, how that timeframe is established and how it can influence uncertainty of parameters in other timeframes.

Acknowledgments. The research presented in this paper has been co-financed by the European Union as part of the European Social Fund.

References

1. Xin, L., Xinhuai, T., Zhaoteng, S., Xiaozhou, Y., Delai, C.: AFlow: An Automated Web Services composition system based on the AI planning and workflow. In: Proceedings of the 2010 International Conference on Progress in Informatics and Computing, pp. 1067–1071 (2010)
2. Wu, Z., Ranabahu, A., Gomadam, K., Sheth, A.P., Miller, J.A.: Automatic Composition of Semantic Web Services using Process and Data Mediation. In: Proceedings of the 9th Intl. Conf. on Enterprise Information Systems (2007)
3. Pathak, J., Lutz, S.B.R., Honavar, V.: MoSCoE: A Framework for Modeling Web Service Composition and Execution. In: 22nd International Conference on Data Engineering Workshops. IEEE Computer Society (2006)
4. Sirin, E., Parsia, B., Hendler, J.: Template-based Composition of Semantic Web Services. In: AAAI Fall Symposium on Agents and the Semantic Web, Virginia, USA (2005)
5. Jie, G., Bo, C., Junliang, C., Lei, Z.: A Template-Based Orchestration Framework for Hybrid Services. In: 2008 Fourth Advanced International Conference on Telecommunications, Athens, Greece, pp. 315–320 (2008)
6. Ko, J.M., Kim, C.O., Kwon, I.-H.: Quality-of-service oriented web service composition algorithm and planning architecture. Journal of Systems and Software 81(11), 2079–2090 (2008)
7. Yu, T., Zhang, Y., Lin, K.-J.: Efficient algorithms for Web services selection with end-to-end QoS constraints. ACM Transactions on the Web 1(1), 6-es (2007)
8. Zeng, L., Benatallah, B., Ngu, A.H.H., Dumas, M., Kalagnanam, J., Chang, H.: QoS-Aware Middleware for Web Services Composition. IEEE Transactions on Software Engineering 30(5), 311–327 (2004)
9. Wan, C., Wang, H.: Uncertainty-aware QoS Description and Selection Model for Web Services. In: Proceedings of the 2007 International Conference on Services Computing, pp. 154–161 (2007)
10. Wiesemann, W., Hochreiter, R., Kuhn, D.: A Stochastic Programming Approach for QoS-Aware Service Composition. In: Proceedings of the 8th IEEE International Symposium on Cluster Computing and the Grid, Lyon, France (2008)
11. Hwang, S.Y., Wang, H., Tang, J., Srivastava, J.: A probabilistic approach to modeling and estimating the QoS of web-services-based workflows. Information Sciences 177, 5484–5503 (2007)
12. Bubnicki, Z.: Uncertain variables in the computer aided analysis of uncertain systems. In: Pichler, F., Moreno-Díaz, R., Kopacek, P. (eds.) EUROCAST 1999. LNCS, vol. 1798, pp. 528–542. Springer, Heidelberg (2000)
13. Świątek, P., Stelmach, P., Juszczyszyn, K., Prusiewicz, A.: Service Composition in Knowledge-based SOA Systems. New Generation Computing 30(2), 165–188 (2012)

Service Composition Scenarios in the Internet of Things Paradigm

Paweł Stelmach

Wrocław University of Technology, Institute of Computer Science, Wybrzeże Wyspiańskiego 27, 50-370 Wrocław, Poland
Pawel.Stelmach@pwr.wroc.pl

Abstract. In this work the topic of service composition scenarios is introduced and discussed with the use of leading example of service composition in the transport domain. The example is used to visualize the connection to the Internet of Things paradigm. The service composition itself is defined with a holistic approach, describing all steps of the proposed composition process like: requirements definition, requirements decomposition or aggregation using domain ontology, composite service structure construction, service discovery, structure and service plan optimization. Lastly, implementation of service composition with service composition scenarios is discussed.

Keywords: service composition, Service Oriented Architecture, service composition scenarios.

1 Introduction

The demand for quick delivery of new functionalities to applications is increasingly more significant in today's world. Sometimes it can determine the future of a big company that cannot quickly adapt to new market requirements. To give an example, one can think of banks that needed to add Internet capabilities for their clients and their monolithic, enormous applications put them in a disadvantage, compared to new Internet banking companies.

Web services have received much interest due to their potential in facilitating seamless business-to-business or enterprise application integration, allowing for building smaller component-based applications that could be easily replaced or extended with new functionality.

When applications are distributed a new problem arises – how to connect their capabilities. One could hardwire services but this, assuming existence of hundreds of services in a company, would inevitably lead to problems with clarity of so designed systems. Functionalities obtained via service compositions allow for loose coupling of services, connecting them via their interfaces and allowing for easy interchangeability. Much work on the use of web services has been done in the context of Service Oriented Architecture paradigm [1].

Still, the trend of outsourcing more and more tasks to outside companies and personalization of systems, created a need for more automation and thus, focusing

L.M. Camarinha-Matos, S. Tomic, and P. Graça (Eds.): DoCEIS 2013, IFIP AICT 394, pp. 53–60, 2013.

interest on automated service composition. In the last decade many researchers have presented their approaches to service composition. Basic approaches presented the use of AI Planning techniques ([2]) or, on top of it, introduce other ideas like two-staged composition with abstract processes ([4]) and more recently [5]. Through the last years various other approaches were introduced, among them workflow-based ([6]) and template-based approaches ([7], [8]). More recent take on service composition is visible in emergence of semantic match making in service composition ([9]) and fulfillment the Quality of Service (QoS) requirements ([10], [11], [12], [13]).

This paper focuses on another perspective. Rather than analyzing different methods for building abstract composite services, it is investigated what really are the requirements for the service composition and how a composite service can be obtained. To do this we must begin with the analysis of the leading example of service composition, namely, composition in the transport domain and present approaches of some other authors to this subject.

The remainder of this paper is structured as follows. In Section 2 relation to the Internet of Things is discussed. Section 3 defines service composition problem definition and motivation for research. Section 4 presents research contribution and innovation. Section 5 discusses the applicability of proposed innovations. Finally, in section 6 concluding remarks and future research plans are presented.

2 Relationship to Internet of Things

Web Services are crucial to the Internet of Things paradigm. They allow for precise definition of capabilities of interfaced objects and interaction with them. Appropriate web description of real world objects will allow Internet users to define their requirements of those objects and their capabilities. Those requirements could be increasingly complex, leading to defining requirements of composite interactions between specific objects or, generalizing, requesting strictly capabilities and not specific objects. Such process of designing requirements and finding optimal selection of objects available via web services could be defined as service composition problem.

In transport domain applicability of web services and service composition to the Internet of Things paradigm can be twofold. Typically, services could be divided into travel search and reservation services. However, in the context of the Internet of Things paradigm Web Services and service composition can look quite different than a typical operations-based service composition. If, in the travel domain, we define a task of travel composition then one could use services that interface real planes or flights to be taken and a sequence of flights from consecutive cities could be looked at as a composite service that consists of a series of Web Services. In the context of Internet of Things each flight or plane could be responsible for its own information management, registration etc., making it more natural to have separate services for each "thing" in the world. However, regardless of the choice of approach, both models recognize the existence of real world objects and define Web Services that mediate their use – whether it is a single service for one object or managing a set of objects of one owner.

3 Problem Definition

3.1 Service Composition

The service composition problem can be presented as transformation of user requirements into a composite service execution plan that fulfills them (Fig.1). Typically user will present both his functional and non-functional requirements, sometimes referred to as Service Level Agreement (SLA). To find a service that fulfills those requirements a three-stage approach is proposed:

- **structure:** SLA with all required functionalities is defined and given a structure of a graph, where each node contains a functional requirement and edges that connect graph nodes define the required control flow,
- **scenario:** the set of edges of the graph is extended so that the graph is consistent (complete information about order of execution of all functionalities is known) and service repository is searched in order to find services capable of providing requested functionalities; the main goal of this stage is to find services that fulfill each of the requirements but may differ in non-functional parameters (this part is usually performed using semantic filters)
- **optimal execution plan:** for each vertex representing a required functionality only one service is selected (from candidates gathered in the previous step), in such a manner that all the services in the structure fulfill the non-functional requirements (this is determined according to the aggregated QoS parameter value of the whole composite service)

More details on the three-stage service composition and algorithms for this approach can be found in [14].

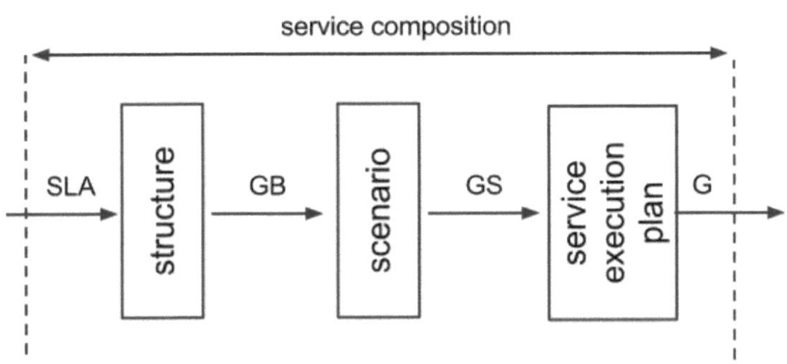

Fig. 1. Stages of service composition approach

3.2 Motivation

What is a service? In aerial transport domain, is it a single flight from A to B or a flight-searching component that is connected to a database of "A to B"-like flights. A programmer would more likely choose the latter, however, in the Internet of Things paradigm it is not so obvious. The distinction is deeply dependent on when and how would we want to use such a service and, getting closer to the topic of service composition, what would a composite service look like and how or when would we want to compose it. Those questions are very often omitted in literature and authors propose a real life motivating example but quickly switch to abstract ideas mixing all types of composition approaches and neglecting the notion of describing in which situations composition algorithms should be used.

For example in [15] (see Fig. 2) a composite service is comprised of a TravelPlanner service and services like AirlineBooking, HotelReservation and CarRental. However, this shows how often two different roles are mixed together: of a designer and a final user. The designer does not need the automated composition to create such a service and the user does not want to know that a TravelPlanner service exists and will not demand it. Thus, no one needs such composition. This happens because TravelPlanner service and HotelReservation service are not of the same kind and it should not be the goal of service composition to achieve such a service in one end-to-end process at a single time.

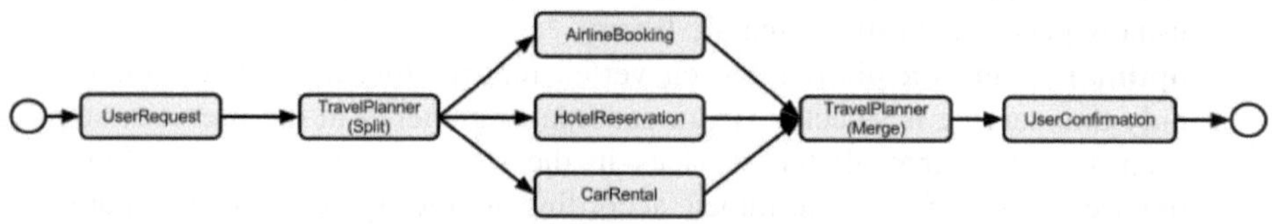

Fig. 2. Travel Planner composite service

4 Research Contribution and Innovation

4.1 Client and Provider Perspective Distinction

From the client perspective the actual focus is put on the *result* of the requested service and not the service itself. When we imagine our clients' actions, they do not originate in a Google-like web service search engine. In fact it is more probable that the client already is on a travel-planning site and so the service composition request is not at all random and in should not be treated as generic with a generic composition approach. In this context we should rather empower the provider with the right tools to build appropriate, simple and domain-specific web forms, rather than to use some kind of generic tools e.g., graph-based composition user interface to generate requests similar to one presented in Fig.2.

When describing the providers' perspective, it is necessary to understand that the composite service provider does not have to be the provider of web services used in the composite service. In fact, he could be seen as an intermediary between a client and the actual web service provider. His goal is to create such a service that the client would want to use and to that purpose he would typically use manual composition approach. He would not want to automatically compose a service as in Fig. 2 and definitely would not want to burden the user with such a task or risk that the composition algorithm will omit crucial parts of the composite service like TravelPlanner service. However, he does not want to create and maintain dozens of possible composite services that differ in the selection of airline, bus, train, car or hotel reservation depending on what are the users' current needs. What he would like is to *manually* define some kind of template that would want to allow a controlled composition to be dynamically performed for the user, while determining the limits and behavior of such a controlled composition mechanism.

4.2 Service Composition Scenario Definition

For this purpose we present a scenario driven dynamic and data driven service composition approach. Scenarios, that define the scope of the composition, are defined manually by the composite service provider along with the basic template describing what parts of the composite service should remain constant.

Fig. 3 presents such a scenario where the user actions and travel planning services are statically bound at design time and other parts of the composite service are represented as dynamic composition requests that will result in the composition and execution of a composite service so that the service composition is dependent on the data received from the previous step.

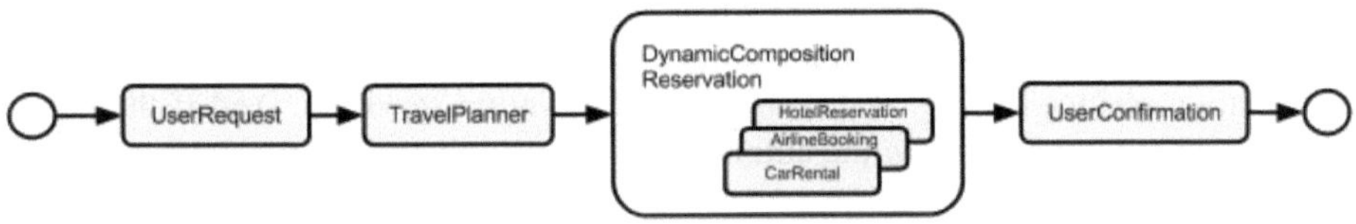

Fig. 3. An example with a composition scenario

The stages of dynamic composition can be further controlled in the service composition scenario with:

- limiting the set of candidate services,
- choice of a single-step approach, selecting Web Services directly, as in AI Planning techniques or a choice of a two-step approach, first selecting abstract services and then performing dynamic service discovery and binding,
- selecting a goal-based technique over a keyword-based technique,
- if possible, preferring parallel structures over series or series when some services need data from the previous services and should be composed in such a way.

With such approach the composite service provider can deliver a *single* composite service that will be partially recomposed at runtime, assuming that he can deliver appropriate data with the UserRequest or TravelPlanner service.

4.3 Advanced Composition Scenarios

In previous sections, in order to avoid confusion, a simple example taken from literature was used. However, even that example is in fact not what the composite service provider would want. The TravelPlanner service would have to have access the data from all providers i.e., airline, hotel and car provider or more if the user would want to use train instead of a plane. Also, the data would be accessed via Web Services and this would stand in contradiction with the SOA paradigm that services should not execute other services directly. In fact, the travel planner should be a composition of search services, dynamically composed based on users requirements (see Fig. 4). Only then a merging of results may happen and a suggested travel plan that is valid in each option should be delivered to user interface.

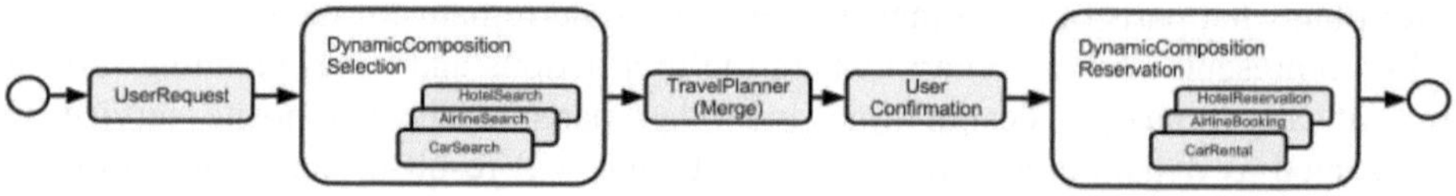

Fig. 4. Example of an advanced composition scenario

The example in Fig. 4 also shows that the difference between selection and reservation could be easily neglected when designing a domain example, however, it presents itself with important implications such as wrong placement of the user confirmation service (in Fig. 2 it is *after* the reservation services). And if the human operator can get it wrong, how the automated composition could know it.

5 Discussion of Results

In the motivation in section 3 it has been briefly mentioned that it is possible to compose services using them as immediate interfaces for singular real world objects (namely one service corresponds to one object or operation like flight from A to B). However, most examples presented a typical approach with generic services. This does not mean, of course, that the general-purpose services don't implement the Internet of Things paradigm. They still do but in such a scenario they are responsible for sometimes hundreds of such real world objects.

Still, it would be valid to present a service composition of such "atomic" services. One could use composition scenarios presented in previous sections and in a dynamic composition component use AI Planning techniques for composition of a series of "A to B" travel services. However, to bring back the focus of this paper, we should consider why such a composition should be delivered. As an effect of such considerations one has to remember that for the user the reason for using service composition is in fact its results and, indirectly, its execution. Here, the execution of the travel service itself should be delayed in time and prior to that the reservation should be performed. Such composition could be useful with composing or recomposing long running processes, when the user is in fact "inside" the process as it executes but this topic is beyond the scope of this paper.

6 Conclusions and Further Work

In this paper a topic of service composition in the Internet of Things paradigm was discussed. A motivating example for the need of different approach to service composition was introduced and, in that context, service composition scenarios based approach was presented and discussed.

Further work will focus on detailed definition and systematization of various composition mechanisms to structure, scenario and optimal plan stages and their adaptation to the composition scenarios format.

Research presented in this paper will be included in the PlaTel platform for composition and management of telecommunication services [16] as part of the composite service execution engine dynamic behavior component. This research is a part of ongoing research on service composition and its application to various domains like telecommunication, sports or transport.

Acknowledgments. The research presented in this paper has been co-financed by the European Union as part of the European Social Fund.

References

1. SOA Reference Model Technical Committee. A Reference Model for Service Oriented Architecture, OASIS (2006)
2. Ponnekanti, S.R., Fox, A.: SWORD: A Developer Toolkit for Web Service Composition. In: Proceedings of the 11th World Wide Web Conference, Honolulu, HI, USA (2002)
3. Aggarwal, R., Verma, K., Miller, J., Milnor, W.: Constraint Driven Web Service Composition in METEOR-S. In: Proceedings of the 2004 IEEE International Conference on Services Computing, pp. 23–30 (2004)
4. Xin, L., Xinhuai, T., Zhaoteng, S., Xiaozhou, Y., Delai, C.: AFlow: An Automated Web Services composition system based on the AI planning and workflow. In: Proceedings 2010 International Conference on Progress in Informatics and Computing, pp. 1067–1071 (2010)
5. Wu, Z., Ranabahu, A., Gomadam, K., Sheth, A.P., Miller, J.A.: Automatic Composition of Semantic Web Services using Process and Data Mediation. In: Proceedings of the 9th Intl. Conf. on Enterprise Information Systems (2007)
6. Pathak, J., Lutz, S.B.R., Honavar, V.: MoSCoE: A Framework for Modeling Web Service Composition and Execution. In: 22nd International Conference on Data Engineering Workshops. IEEE Computer Society (2006)
7. Sirin, E., Parsia, B., Hendler, J.: Template-based Composition of Semantic Web Services. In: AAAI Fall Symposium on Agents and the Semantic Web, Virginia, USA (2005)
8. Jie, G., Bo, C., Junliang, C., Lei, Z.: A Template-Based Orchestration Framework for Hybrid Services. In: 2008 Fourth Advanced International Conference on Telecommunications, Athens, Greece, pp. 315–320 (2008)

9. Klusch, M., Fries, B., Sycara, K.: OWLS-MX: A hybrid Semantic Web service matchmaker for OWL-S services. Web Semantics: Science, Services and Agents on the World Wide Web 7, 121–133 (2009)
10. Anderson, S., Grau, A., Hughes, C.: Specification and satisfaction of SLAs in service oriented architectures. In: 5th Annual DIRC Research Conference, pp. 141–150 (2005)
11. Ko, J.M., Kim, C.O., Kwon, I.-H.: Quality-of-service oriented web service composition algorithm and planning architecture. Journal of Systems and Software 81(11), 2079–2090 (2008)
12. Yu, T., Zhang, Y., Lin, K.-J.: Efficient algorithms for Web services selection with end-to-end QoS constraints. ACM Transactions on the Web 1(1), 6-es (2007)
13. Blanco, E., Cardinale, Y., Vidal, M.-E., Graterol, J.: Techniques to Produce Optimal Web Service Compositions. In: IEEE Congress on Services, pp. 553–558 (2008)
14. Stelmach, P., Grzech, A., Juszczyszyn, K.: A Model for Automated Service Composition System in SOA Environment. In: Camarinha-Matos, L.M. (ed.) DoCEIS 2011. IFIP AICT, vol. 349, pp. 75–82. Springer, Heidelberg (2011)
15. Jeong, B., Cho, H., Kulvatunyou, B., Jones, A.: A multi-criteria web services composition problem. In: IEEE International Conference on Information Reuse and Integration, IRI 2007. IEEE (2007)
16. Świątek, P., Stelmach, P., Juszczyszyn, K., Prusiewicz, A.: Service Composition in Knowledge-based SOA Systems. New Generation Computing 30(2), 165–188 (2012)

Automatic Adaptation of SOA Systems Supported by Machine Learning

Kornel Skałkowski and Krzysztof Zieliński

AGH University of Science and Technology, Faculty of Computer Science, Electronics and Telecommunication, Department of Computer Science, al. Mickiewicza 30, 30-059 Kraków, Poland
{skalkow,kz}@agh.edu.pl

Abstract. Recent advances in the development of information systems have led to increased complexity and cost in terms of the required maintenance and management. On the other hand, systems built in accordance with modern architectural paradigms, such as Service Oriented Architecture (SOA), posses features enabling extensive adaptation, not present in traditional systems. Automatic adaptation mechanisms can be used to facilitate system management. The goal of this work is to show that automatic adaptation can be effectively implemented in SOA systems using machine learning algorithms. The presented concept relies on a combination of clustering and reinforcement learning algorithms. The paper discusses assumptions which are necessary to apply machine learning algorithms to automatic adaptation of SOA systems, and presents a machine learning-based management framework prototype. Possible benefits and disadvantages of the presented approach are discussed and the approach itself is validated with a representative case study.

Keywords: SOA, adaptive manager, machine learning.

1 Introduction

The growing complexity of modern IT systems hinders effective administration, resulting in increased maintenance costs. Geographical distribution of services, dynamic workflow enactment and on-demand service selection improve the systems' scalability and flexibility but do not foster their overall manageability. It should be noted, however, that contemporary architectural paradigms such as Service Oriented Architecture (SOA) [1] or Internet of Things (IoT) [2], provide sophisticated adaptation features [3,4]. Flexibility in terms of service/sensor coupling, instant binding or semantic message routing can be used to modify information flow between system components during runtime, affecting processing speed. Such extensive adaptation opportunities are characteristic of modern design approaches and can be leveraged to solve problems associated with system management and administration through automatic or semi-automatic adaptation.

Issues involved in the adaptation process of enterprise systems are addressed by the well-known MAPE adaptation pattern [5], which introduces four elements

L.M. Camarinha-Matos, S. Tomic, and P. Graça (Eds.): DoCEIS 2013, IFIP AICT 394, pp. 61–68, 2013.

(Monitoring, Analysis, Processing, Execution) necessary in every adaptation framework. This paper presents a new approach to implementation of the Analysis and Processing elements of the MAPE pattern based on a combination of two types of machine learning methods. A clustering algorithm is used to provide automatic recognition of similar system states and grouping them into subsets (called clusters), based on information provided by the Monitoring element interface (e.g. regarding a system load or observed bottlenecks). The goal of further processing is then to find a mapping between the clusters and adaptation actions provided by the Execution element interface (e.g. a service replication, routing changes or resources allocation). These actions should be assigned to clusters in such a way that execution of actions attributed to a cluster to which a current system state has been assigned increases the overall system QoS (Quality of Service). In order to find a mapping which satisfies this condition, a reinforcement learning algorithm has been devised. The paper explains how such a combination of machine learning methods can effectively and flexibly implement the MAPE pattern in service-based systems and discusses assumptions which have to be met by an adaptable system in order to be applicable to our solution. The proposed approach to MAPE is evaluated on the basis of a proof-of-concept implementation.

The paper is organized as follows: in Section 2 the relationship between the proposed approach and IoT architectures is discussed. Section 3 briefly presents related approaches to MAPE implementation. In Section 4 a machine learning-based approach to implementation of the MAPE pattern is elucidated. Section 5 shows how the concept has been implemented in a prototype framework and which algorithms have been chosen. Section 6 discusses evaluation results while Section 7 concludes the paper and discusses future work.

2 Relationship to Internet of Things

Adaptation issues are widely present in various aspects of IoT systems. The vast quantities of objects involved in such systems, huge amounts of information produced, chaotic working environments and the need for autonomous control make efficient and flexible adaptation a crucial part of many IoT solutions. Implementation of the MAPE pattern in IoT architectures requires dedicated monitoring and execution layers which can cope with such issues. Since the approach presented in this paper does not impose any specific monitoring and management framework, it can be applied to IoT infrastructures as well as to other manageable systems.

3 Related Work

Existing approaches to implementation of the MAPE pattern are based on rule/policy engines, decision theory or fuzzy logic. The use of machine learning techniques for SOA system adaptation is only partially covered in existing papers. An approach to context-based adaptation in production systems based on data mining techniques has been proposed in the Self-Learning project [6], which bases in part on learning and adapter modules. Nevertheless, recent publications released by this project do not clearly point to any particular data mining algorithms and do not present any evaluation results. Other existing papers focus on machine learning-based selection of web services [7,8] and reliability assessment in SOA systems [9]. Although these approaches can partially solve the issue of automatic management of service-based systems, they do not constitute a complete implementation of the MAPE pattern.

4 Machine Learning-Based Approach to MAPE Pattern Implementation

Most existing machine learning algorithms operate on sets of n-dimensional real valued vectors $x \in \mathbf{R}^n$. Unsupervised learning methods, i.e. clustering algorithms, operate directly on such sets, whereas in the case of supervised learning methods or reinforcement learning algorithms an additional value, y, is assigned to every vector and interpreted as the "correct answer" to it. The goal of the analysis and processing elements of the MAPE pattern is to find out which action offered by the execution layer should be invoked in a specific system state. To achieve this goal using machine learning methods we have to represent system state as an n-dimensional vector, while the action suitable for a given system state is equivalent to the "correct answer" value. Based on these observations, the system state at point t is represented as $x^{(t)} \in \mathbf{R}^n$, whereas the set of all observed system states at various points in time ($x^{(t1)}, x^{(t2)}, ..., x^{(tm)}$) will be called the system state space **X**. It is important to stress that vectors $x^{(t)}$ should contain all available information about the system which should be taken into account during management, including the working context and current configuration. Representing the state of a system in the form of a vector of real values may seem somewhat constraining, yet even those parameters which are expressed in non-numeric form (e.g. strings or enumerations) can usually be converted to numeric values by applying appropriate mappings. Since the main goal of our approach is to manage a complex system in a way which increases its overall QoS level, the system state vectors are assumed to reflect the QoS experienced by users in some way. Certain system parameters directly reflect QoS (e.g. processing time), whereas in more sophisticated cases the QoS level can be calculated with evaluation function e(x).

The management interface which constitutes the execution element of the MAPE pattern, is assumed to be represented as set of values: $\mathbf{A}=\{a_0, a_1, .., a_k\}$, consisting of the available adaptation actions. In order to avoid contradictions and discrepancies during learning we have to assume the action set **A** meets several conditions. First of all, there are *no duplicate actions* in the set, since most learning algorithms use injective functions to produce the "answer value" a_i. Furthermore, we assume that every action a_i can be *repeated any number of times* and the actions are *stateless* (i.e. no action has a different effect when invoked several times in the same system state). Finally, it is necessary for the actions to be *independent of each other*, meaning that no action should require prior execution of any other action. If some actions have to be invoked in a specific sequence, they should be represented as a single action. These assumptions are not challenging and every well designed management framework usually satisfies them all. In order to make actions comparable to "correct answer" values returned by learning algorithms, the actions should be bijectively mapped to numbers, e.g. simply enumerated. In order to facilitate implementation, both sets (**X** and **A**) are assumed to be fixed for each given adaptable system.

Given the system state and management interface we can precisely define the overall framework goal. Roughly speaking, the framework should perform actions from the space **A** (e.g. service launch or migration) so that the adaptable system provides the best possible QoS level for end users. From a mathematical view point, this problem can be divided into two subproblems. The first subproblem is clustering the system state space **X** into a set of non-empty sets $\{C_1, ..., C_l\}$ which should be characterized by the maximum possible homogeneity of elements within each set (e.g. lowest sum of distances between the elements of C_i) and the maximum possible diversity between sets (e.g. greatest sum of distances between the $\{C_i\}$ sets' centroids). The C_i sets can evolve during system runtime, reflecting changes in the system and its working environment. The second subproblem is mapping the clustered system state space $\{C_i\}$ onto actions: the framework has to find a mapping $\forall_{i=1, ..., l}$ **F**: $C_i \rightarrow (a_j$: $j=0, ..., k_{(Ci)})$ such that the execution of actions returned by the mapping **F** when the system state belongs to the cluster C_i causes $\sum_i e(x^{(ti)})$ to assume its lowest possible value. The function $e(x^{(t)})$ is the overall system QoS evaluation metric calculated using state vectors $x^{(t)}$ whose values are inversely proportional to the condition of the system. Applying **F** yields a sequence of actions with length $k_{(Ci)}$, sorted from the most appropriate to the least appropriate one (for a given system state) – thus we can say that mapping **F** reflects the adaptable system model. The first subproblem may seem unnecessary as one might claim that actions could be assigned directly to system states $x^{(t)}$. In reality, however, this assumption is only satisfied by very small systems, where **X** can be modeled e.g. as a small finite state machine. In most real systems – especially complex enterprise SOA solutions – this assumption is no longer valid. In such cases the space **X** is usually infinite and multidimensional, so that both elements are essential in order to accomplish the framework objectives.

The first task is a well-known clustering problem, the only major issue being that the clustered space **X** is not known a priori, but is instead constructed during runtime by aggregating $x^{(t)}$ vectors. This issue can be solved using online clustering methods which are designed to cluster data streams. In turn, the second task leads us to the area of reinforcement learning algorithms which are used to teach computer systems how to act in different situations in order to achieve a given goal. The learning mechanism in such algorithms is based on rewards, usually represented as a single real number. In our case the situations are represented as state vectors $x^{(ti)}$ at different points in time t_i, whereas the reward constitutes the system evaluation metric e(x). The reinforcement learning algorithm returns a function, $h_\theta(x)$, called the hypothesis, which provides "correct answer" values for different vectors x. In our approach this function is equivalent to mapping **F** – the returned sequence of actions comprises set **A**, which is calculated on the basis of differences between actions from **A** and the $h_\theta(x)$ function results. The final necessary element is normalization of state vectors $x^{(t)}$. Since most machine learning algorithms require input vectors to have all elements normalized to a common range of values, a normalization function has to be applied to all state vectors prior to clustering. The approach is depicted in Fig. 1.

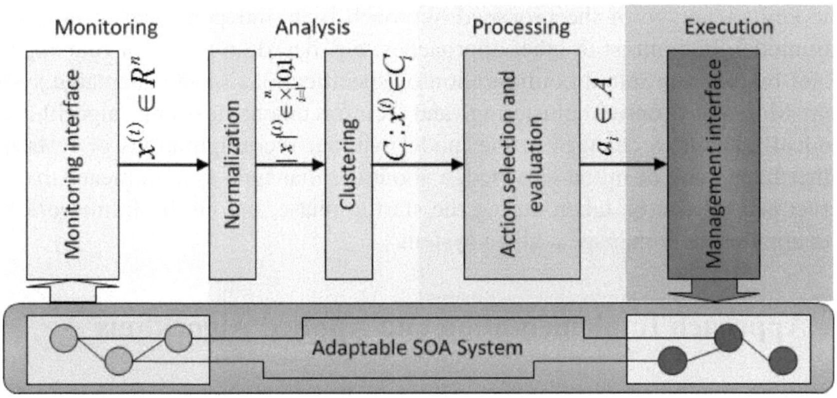

Fig. 1. The machine learning-based approach to the MAPE pattern

The bottom part of Fig. 1 presents an SOA system subjected to adaptation. Above, four elements of the MAPE pattern are shown. The monitoring and management interfaces are assumed to be provided by external frameworks which can be tuned to a specific adaptable system (provided that they meet the assumptions previously stated). The monitoring interface returns vectors $x^{(t)}$ composed of real values representing system states. The management interface exposes **A** – the set of management actions available in the system. The analysis layer consists of two services: the normalizer service, responsible for mapping state vectors to an n-dimensional [0,1] hypercube, and the clustering service which, based on normalized $|x|^{(t)}$ vectors, extracts clusters $\{C_i\}$ representing groups of similar system states. Each cluster represents a pool of system states which significantly diverge from all other states. By assigning vector $|x|^{(t)}$ to an appropriate cluster, the framework can check whether the adaptable system currently belongs to the best possible cluster. This information, along with the current QoS evaluation metric derivative $\partial e(x)/\partial t$ calculated as a differential approximation from several successive observations of e(x), is used by the processing layer to select actions. When $\mathrm{sgn}(\partial e(x)/\partial t) < 0$ no management action is performed because even if the system state is not in the best possible cluster, it is improving and this trend should be maintained. If, however, $\mathrm{sgn}(\partial e(x)/\partial t) \geq 0$ and the system state does not belong to the best cluster, execution of a management action is necessary. In order to do so, a sequence should be returned by the reinforcement learning algorithm. A third case should be distinguished, with $\mathrm{sgn}(\partial e(x)/\partial t) \geq 0$ and the system already assigned to the best cluster. In this case we may not know why the system condition is deteriorating – whether due to brief fluctuations (e.g. caused by a slightly higher load) or more permanent reasons. Thus, a prediction algorithm could be applied to estimate the likely evolution of the system state. Every executed action is evaluated, and, based on the evaluation result, the reinforcement learning algorithm's hypothesis function $h_\theta(x)$ is up- or downregulated in order to improve future decisions. Evaluation bases on observation of system state changes reflected in the QoS evaluation metric derivative $\partial e(x)/\partial t$ over a period of time. If the system state remains poor and shows no signs of improvement ($\mathrm{sgn}(\partial e(x)/\partial t) < 0$), another action from the list returned by the algorithm is executed and evaluated. This process repeats until the system state begins to improve.

The key advantage of the proposed approach is its independence of any specific system model. In contrast to other approaches, e.g. based on policy or rule engines, it does not require any initial configuration or specifications of the adaptable system's model. Moreover, online clustering and reinforcement learning algorithms can dynamically adapt to changes in the model without reconfigurations or restarts. On the other hand, lack of initial knowledge about the managed system means that many incorrect actions can be taken during the startup phase, before the framework learns how to appropriately manage a given system.

5 Approach Implementation and Applied Algorithms

A prototype implementation of the approach described in the previous section is currently being developed. Its most recent version has been implemented as a set of OSGi [10] services providing the above mentioned features. The implementation consists of four services: the normalization service (responsible for state vector normalization), the clustering service (responsible for clustering), the strategy service (responsible for action selection and evaluation) and the evaluation service (providing the e(x) function values).

Since the aim of the implementation is to validate the proposed concept rather than provide sophisticated functionality, the prototype relies on simple machine learning algorithms. Specifically, the clustering service implements a standard k-means algorithm to cluster state vectors collected over a period of time. In the future this algorithm will be swapped for an online clustering algorithm based on the PCA method [11]. As the reinforcement learning algorithm, a simple adaptive gradient descent implementation with a polynomial hypothesis function has been used. The main disadvantage of this algorithm is slow convergence – in the future we intend to apply a more efficient reinforcement learning algorithm.

6 Prototype Evaluation Results – Preliminary Study

The objective of evaluation of the prototype framework was to check whether it properly accomplishes its goals, i.e. invokes appropriate management actions when the overall system QoS level decreases, and to verify if the learning method is appropriate, i.e. whether the hypothesis function properly converges regardless of its initial coefficients. Both goals were evaluated on a load balancing case study in an SOA system. The simulated system consisted of three services. The first service had to invoke either the second or the third service in order to accomplish its functionality. By default, the first service used only the second service – thus the third service remained idle. The response time of the second service was highly dependent on the number of simultaneous invocations. As the number of concurrent requests grew, the service's response time increased noticeably, affecting the overall system QoS. In such cases, the first service was expected to begin using the third service in order to balance load and avoid a decrease in the overall QoS. The simulated services have been implemented using the OSGi technology and deployed in an OSGi monitoring and management framework provided by the AS3 Studio [12] toolkit. The monitoring interface was configured to monitor two parameters of the system: average processing time (**APT**) and invocation rate (**IR**). These two parameters were passed to the framework prototype as a single vector $x^{(t)}$ = [**APT**, **IR**]. The evaluation function was calculated as e(x) = **APT** + **IR**/2. The management interface exposed two actions: "do-nothing", whose invocation did not affect the simulated system in any way, and "balance-load", which activated load balancing in the first service for a period of time. As a result of the second action, the first service would begin dispatching its requests to both the second and the third service. The evaluation was performed on a computer with an Intel Core 2 Duo 2.80 Ghz CPU and 4 gigabytes of RAM. The hypothesis function was a simple linear polynomial of two variables $h_\theta(x^{(t)}) = \theta_0 + \theta_1 x_1 + \theta_2 x_2$, where x_1 = **APT** and x_2 = **IR**. Tab. 1 presents evaluation results from three test runs.

Table 1. Evaluation results of the prototype framework

Processing time speedup	Initial $h_\theta(x)$	Final $h_\theta(x)$	Convergence time	Invalid actions
24%	$0.64+0.69x_1+0.57x_2$	$0.68+0.71x_1+0.59x_2$	0:20 [h]	2
21%	$0.23-0.44x_1+0.01x_2$	$1.06+0.07x_1-0.21x_2$	2:00 [h]	7
19%	$-0.55-0.23x_1-0.03x_2$	$1.12+0.13x_1-0.53x_2$	2:30 [h]	21

Evaluation results confirm that the proposed approach to the MAPE pattern implementation is viable and properly accomplishes the stated goals. Processing time speedup was in the 19% - 25% range, depending on initial coefficients of the hypothesis function. The greatest speedup was observed for near-optimal initial hypothesis coefficients, because in this case the framework almost always executed the "balance-load" action when necessary. In other cases the framework executed a greater number of "do-nothing" actions, before it learned that this action was inappropriate for a high system load state. Convergence time was directly dependent on initial hypothesis coefficients. Better coefficients improved the algorithm's convergence; however in all cases convergence was eventually attained (although with differing final hypothesis coefficients).

7 Conclusions and Further Work

The proof-of-concept evaluation of the approach proposed in the paper shows that machine learning methods can be applied to implementation of the MAPE pattern. Our combination of clustering and reinforcement learning algorithms properly identifies disruptions in system QoS and invokes appropriate management actions. The main advantage of the proposed approach is its independence of any specific system – the framework does not require any *a priori* knowledge about the adaptable system. Flexibility offered by online clustering and reinforcement learning methods means that the approach can be applied to SOA system adaptation as well as to IoT system management. The only evident disadvantage is its potentially long convergence time.

Further development will focus on more advanced algorithms for online data clustering and selection of management actions. The framework effectiveness and scalability will also be evaluated on much more complex case studies and real-world systems, e.g. a telemedicine platform, where maintaining a certain level of QoS is crucial. We also intend to improve the efficiency of our approach by implementing a system state prediction algorithm which could invoke management actions in order to prevent QoS disruptions.

Acknowledgments. The research presented in this paper was partially supported by the European Regional Development Fund programs no. POIG.01.03.01-00-008/08 and UDA-POKL.04.01.01-00-367/08-00.

References

1. Arsanjani, A., Zhang, L.-J., Ellis, M., Allam, A., Channabasavaiah, K.: S3: A Service-Oriented Reference Architecture. IT Professional 9(3), 10–17 (2007)
2. Atzori, L., Iera, A., Morabito, G.: The Internet of Things: A survey. The International Journal of Computer and Telecommunications Networking 54(15), 2787–2805 (2010)
3. Zieliński, K., Szydło, T., Szymacha, R., Kosiński, J., Kosińska, J., Jarząb, M.: Adaptive SOA Solution Stack. IEEE Transactions on Services Computing 5(2), 149–163 (2012)
4. André, F., Daubert, E., Gauvrit, G.: Towards a Generic Context-Aware Framework for Self-Adaptation of Service-Oriented Architectures. In: 2010 Fifth International Conference on Internet and Web Applications and Services, ICIW, pp. 309–314 (2010)
5. Kephart, J.O., Chess, D.M.: The vision of autonomic computing. Computer 36(1), 41–50 (2003)
6. Stokic, D., Scholze, S., Barata, J.: Self-learning embedded services for integration of complex, flexible production systems. In: IECON 2011 - 37th Annual Conference on IEEE Industrial Electronics Society, November 7-10, pp. 415–420 (2011) ISSN: 1553-572X
7. Al-Masri, E., Mahmoud, Q.H.: Discovering the best web service: A neural network-based solution. In: IEEE International Conference on Systems, Man and Cybernetics, SMC 2009, October 11-14, pp. 4250–4255 (2009)
8. Mohanty, R., Ravi, V., Patra, M.R.: Web-services classification using intelligent techniques. Expert Systems with Applications 37(7), 5484–5490 (2010) ISSN 0957-4174
9. Challagulla, V.U.B.: A Machine Learning-Based Reliability Assessment Model for Critical Software Systems. In: 31st Annual International Computer Software and Applications Conference, COMPSAC 2007, July 24-27, vol. 1, pp. 79–86 (2007)
10. The OSGi Alliance: OSGi Service Platform Core Specification, Release 4.1 (2007), `http://www.osgi.org/Specifications`
11. McWilliams, B., Montana, G.: Predictive Subspace Clustering. In: 2011 10th International Conference on Machine Learning and Applications and Workshops (ICMLA), vol. 1, pp. 247–252 (2011)
12. Żmuda, D., Psiuk, M., Zieliński, K.: Dynamic monitoring framework for the SOA execution environment. Procedia CS 1(1), 125–133 (2010)

Part IV
Intelligent Computational Systems

Modelling of Things on the Internet for the Search by the Human Brain

Fernando Luis-Ferreira[1,2] and Ricardo Jardim-Gonçalves[1,2]

[1] Departamento de Engenharia Electrotécnica, Faculdade de Ciências e Tecnologia, FCT, Universidade Nova de Lisboa, 2829-516 Caparica, Portugal
[2] Centre of Technology and Systems, CTS, UNINOVA, 2829-516 Caparica, Portugal
{flf,rg}@uninova.pt

Abstract. The Internet has become the main source of information for business and research activities. Despite the value of libraries supported by computational cataloging, there are far more opportunities to retrieve information on the Internet than in paper books. However, when we seek the Internet we get essentially chunks of text with titles and descriptors resulting from search engine's activity. Albeit some information may contain sensorial or emotional contents, the search results come essentially from algorithmic execution over keywords by relevance. Our brain retrieves information about things in real world by capturing sensorial information and storing it with emotional experience. We can question why things in Internet are not represented in a similar way to human brain. The present research aims to support a new type of search by sensations and emotions in a path to model Things in Internet towards a human-like representation of objects and events, based on lessons learned from the human brain.

Keywords: Knowledge Management, Neuroscience, Sensation and Perception, Internet of Things, Interoperability.

1 Introduction

In the last years, the Internet has evolved towards user's needs in terms of expression of ideas with a boom in the blogosphere and then with the social networks allowing a real-time expression of actions and thoughts. With a computer or a smartphone, a person can express ideas and report events putting it online in murals or tweeting events as they happen. We can talk about a freedom of speech as anyone can say what they want and publish it to be available anywhere in the world. Diverse approaches where developed for users to organize their photos, to share whatever they want with friends colleagues or to all people. We may wonder what else could be provided to users that are more than a support for text and image or sound. The research reported in this paper addresses new approaches for users to interact with the web or with each other using the web.

The main vision of the current work is to learn from the human sensorial interaction with the world and the way we perceive the world mediated with emotions. In general,

L.M. Camarinha-Matos, S. Tomic, and P. Graça (Eds.): DoCEIS 2013, IFIP AICT 394, pp. 71–79, 2013.

the authors want to learn and, as much as possible, replicate how the brain and the nervous system handles information and generates knowledge and wisdom. The research presented in this paper seeks for an enlarged sensorial representation of objects in the Internet and, in some cases, mediated with emotional information. The relevance of this work is extensible from the individual representation of episodes of a person's life, as depicted in this paper (Fig. 3), to the opportunity of an enlarged object representation, either in an Internet of Things (IoT) context or, in a broad sense, as a generalization in the Internet with potential impact for Business, Social Networking or Education.

In the next section, a modification is proposed on how Things are handled in the Internet, on an IoT context, by means of an enrichment of the description of those Things that relies on the inclusion of human perceptions and emotions.

In section 3, the presented Research is contextualized in the framework of the ongoing doctoral studies. In section 4 the Methods and Associated concepts developed in the current research are presented highlighting its application in a knowledge management framework, and finally, section 5 presents Conclusions and Future work.

2 Relationship to Internet of Things

Internet of Things proposes an environment littered with communicating objects, with a pervasive presence around us of a variety of things, or objects, such as Radio-Frequency Identification (RFID), tags, sensors, actuators, mobile phones, etc. These networked objects will play its role in both working and domestic fields, with scenarios in domotics, assisted living, e-health and enhanced learning as possibilities for this new paradigm with a leading role in a near future [1]. Considering the proposed different layers for an IoT architecture, it can be designed with a path starting on a perception layer, being its main function to identify objects and gather information. It is formed mainly by sensors and actuators, monitoring stations (such as cell phone, tablet PC, smart phone, PDA, etc.), nano-nodes, RFID tags and readers/writers [2]. But what can those assets tell us about human beings? Can they contribute to our perception of the world? Are they an extension of our sensorial organs? They can tell us about goods, flow of materials and technological devices, but what can they tell about people? From the development of cyber-physical systems (CPS) results a new range of possibilities of interaction with the physical world around us [3] but in order to advance that interaction a step ahead, we propose that computational resources meet our biological perception of the world by considering our sensorial capabilities. To close this loop, the inclusion of emotional information will better characterize objects or events by allowing an emotion-driven management of those objects. Considering the relevance of sensorial and emotional information as well as our relation with the world, by means of sensations and emotional regulation, we propose a model for representing things, supporting sensations and emotional information as depicted in the next figure. It represents a model that includes an

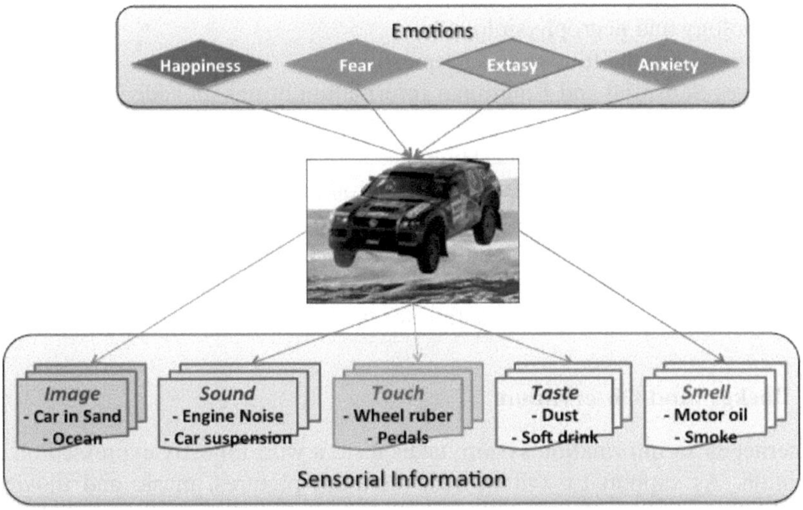

Fig. 1. Modelling things to include Sensations and Emotions

association of sensorial information, as humans perceive Things in real world, along with support for emotional response to those objects or events

The use of tagging technics, either automatic or made by users [4] will allow the description of objects, in a first stage, with either symbolic or textual information. In a later stage, as technology makes available new devices, we envisage that information can be retrieved by portable non-invasive devices. In an ideal scenario, using this model along sensorial and emotion-catching devices in an IoT world, it would be possible to automatically acquire sensorial and emotional information.

3 Followed Research Method

The on-going research is following the traditional research method based on research questions that result from a background observation with an hypothesis that will be tested over a design experiment, leading to results to be analysed and published on its major findings. The research path is focused on improving information systems, especially its aspects over the internet, taking examples from existing neuroscience models, in particular the Two-Stream Hypothesis [5], as a source of guidance for knowledge management for its perspective on a "where" and a "what" streams that also seem to be useful in a IoT vision of our interaction with the connected world.

3.1 Research Question

The research question defines the area of the interest for the authors and the problem they want to address. The selected question;

Q - How to improve information systems based on lessons learned from neuropsychology and neurophysiology?

Is decomposed in the following sub-questions

Sq1 - Does Sensorial and Emotional Information brings an added value to Internet objects for human usage?

Sq2 - Would the two-stream Hypothesis of the brain be suitable for taking advantage of sensorial and emotional information for knowledge management?

Those questions will be addressed with analogous kind of sensorial perception and emotional contextualization in handling Internet Objects. To pursue such approach some scientific challenges will be addressed on how to model things in the Internet with sensorial and emotional information for the search by the human brain.

3.2 Background Observation

The Internet as an information system takes us to a world mostly expressed on textual information. As support to written data there are pictures, music and movies. The human brain uses the five senses to retrieve, store, handle and later, seek for information to think and to establish correlations with newly acquired information.

Many brain models have been constructed based on experiments and observation. From hundreds of years most of the great classical philosophers like Plato, Aristotle, Spinoza, Descartes, Hobbes and Hume, had recognizable theories of emotion [6] tried to develop cognitive models and understand how the mind works. Since those times many things changed. Today with Internet's worldwide diffusion and its vast amount of information, would be desirable to use it as a veritable extension of the human brain [7], but in order to make it possible it is logical to support, along with the growth of information, new ways of representing objects as perceived by the human brain. In our research we consider that an unavoidable method, towards a brain-like representation of objects, is to include sensorial and emotional information.

Considerations about the state of art on sensorial information in the Internet were made in our previous paper [8] where we propose a framework for knowledge management using the two-stream hypothesis. From that framework, based on sensations and emotions, it is necessary to go beyond state of art and develop models that would allow a human-like representation of Objects and Episodes. By Episodes, or i-episodes, we define the possibility of storing interconnected sensorial and emotional information segments, on i-episode, were that information makes sense as a whole to the person who experienced that place at that time. In the same line, we can relate sensorial and emotional experience of a person towards a Thing. In the next section the proposed hypothesis explore research paths to accomplish those goals.

3.3 Hypothesis

If we adopt selected brain models in order to improve knowledge management in particular or in the Internet, then the probability of presenting and finding information that meets our ways of perceiving the world will be higher. Getting more concrete with the following construction; 1) A new conceptualization of knowledge representation can support sensorial and emotional information 2) Support for meta information can foster sensorial and emotional representation methods 3) Semantic mediation can be supported by enlarged information and thus improving knowledge extraction and search results. With this in mind we will extend the sensorial information to the five senses, or at least enable that potential, also associating emotions textually expressed or device captured. Our primary goal is to facilitate better knowledge management by enabling different types of information representation and extraction based in sensorial and emotional tagging.

4 Methods and Associated Concepts

In our first exploration of the question, we took a look at a brain model and explored how it could be valuable to be applied in a technological framework. The basis for that was to observe that we receive images that are formed in the occipital region of the brain [9], then according to the two-stream hypothesis, the dorsal stream goes from the occipital lobe to the temporal lobe and is known as the "what stream" and the Ventral Stream goes from the occipital lobe to the temporal lobe and is known as the "where stream"[5].

Two findings are relevant in this theory; first, the brain tries to identify what is in the visual field, identifying Objects, mostly by comparing with memories. Then the brain tries to give spatial context by establishing a sense of location for the visual information Episodes. So the cognitive process starts with images at the visual cortex, then two streams follow a path thru other regions of the brain where the same process of analysis and comparison to existing records is extended to other recorded sensorial information, thus empowering the what and where context findings. And that is the area where, from our research strategy, technology meets the brain. The proposed approach relies on the establishment of a databank of sensations that could help us identify "what" are the Things in real world as we capture sensations. Then it can result in Episodes, if for a given person, a relation is established between a set of sensations giving us the "where". As in the brain, it is supposed to store emotions associated with those sensations that give relevance to the objects or to events.

Handling Emotions in Internet

Emotions are what make us humans. They give meaning to our existence [10]. They reflect the effect caused on humans by the sensorial experience of the world. An important part of our information gathering behavior has always been to find out what other people think [11], thoughts and our perception of the world result in an emotional response that can be consciously assimilated and can simultaneously produce a physiological response. In affective computing, we can separately examine functions that are not so easily separated in humans [12]. We identified a lack of applications that consider emotion related aspects. This situation is mainly due to the great amount of proposed theoretical models and the complexity of human emotions [13]. So, in order to proceed with our framework design, we need to include emotions along with sensorial information, both represented, in an early stage, with tags.

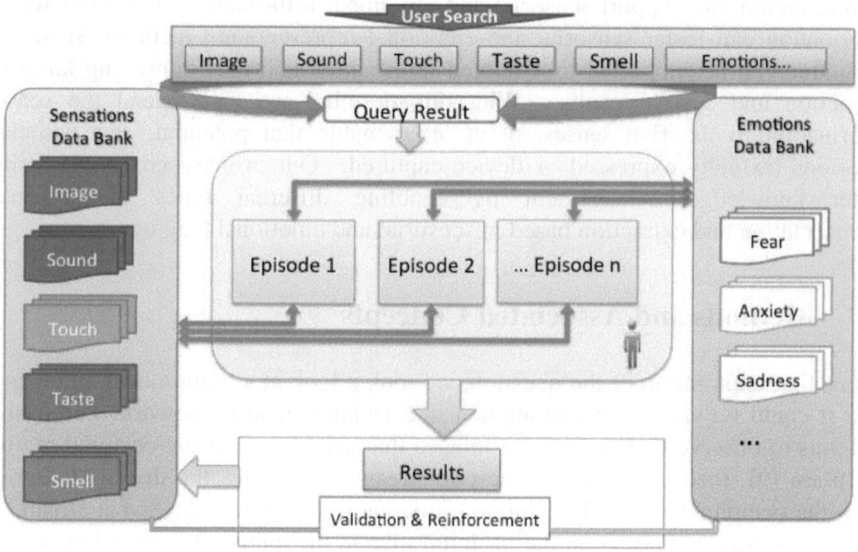

Fig. 2. Proposed Framework for Knowledge Management

This framework proposes an enlarged search using emotions and sensations that will result in episodes, being those unitary or complex and from a person or others. This search will be done in specific databanks with the most diverse sensations and emotions that, at the same time, consider clusters named Episodes. When an episode is validated by one or more users it will be reinforced and progress in relevance.

The next challenge to the ongoing research is to gather and store related pieces of sensorial and emotional information in the form of episodes, a so called i-Episode, that makes sense to a person or to a group of persons. The result of that new type of search may be validated by the user and reinforce a possible classification e.g. beautiful sunset, amazing jump, good scent. The stored information, if publicly available, can be used by others to represent their sensations or reproduce their events.

4.1 Design Experiment

We want to prove that a framework inspired in the two-stream hypothesis, as described above, has a great potential for handling information in innovative and fertile knowledge bases. In order to accomplish that goal we need a new modelling of Things, similar to what our brain does, which is also in the scope of the research reported in this paper. In order to accomplish that goal, we exploit the retrieval and storage of information as it happens based on the two-stream hypothesis. This can be achieved by establishing a connection with 'where' it was and 'what' is represented. Then it will instantiate a knowledge base and harmonise it with existing information, which includes sensorial and emotional information, either by properly instantiate with retrieved knowledge or by semantic annotation. This could be backed by semantically-based Web Map Mediation Services a core of semantic and ontological tools for mapping [14], mediation, annotation and what else found needed for pursuing the most consensual and interoperable solution as possible.

In the next picture the proposed modelling approach supported by both a databank of emotions and other for sensations, with correlations that can be associated as episodes of a person's experience in life. Those data banks can be populated by known sensations and emotions and will permanently be updated with user's contributions that can be used by them or by others. When performing a query it is possible to find our own episodes but also other's episodes if available publicly. The resulting queries, if validated by users can increase the bonds between objects and associated emotions or perceptions.

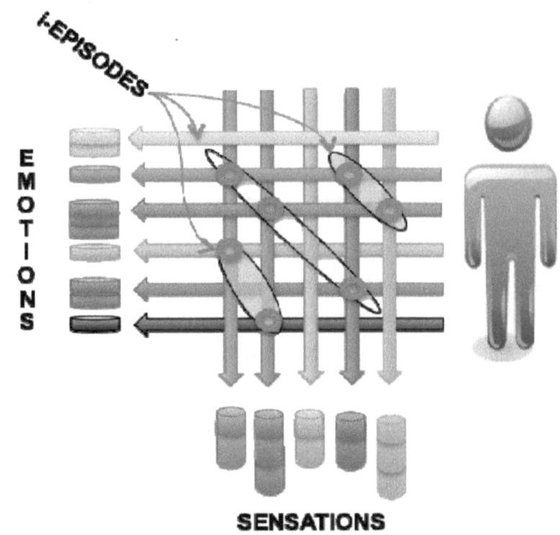

Fig. 3. Emotions and Sensations collected and clustered for a person as i-Episodes

As an example, if surfers associate a beach with good waves, soft sand and smooth sea scent then the occurrence of more validated results will reinforce such associations. In figure 3, three instantiations of the proposed data model result in a person's set of episodes. Those episodes encompass selected emotions and perceptions for the person's i-Episode description. Both emotions and perceptions are stored in specific databases, which use existing content or the user's description, or measurements. As a result, a new form of modelling information allows richer sensorial descriptions and empowers existing knowledge. In terms of functionality, new services can be deployed, making use of sensorial and emotional information, providing more ubiquitously searching and finding of information. Comparing to existing solutions, we propose a framework that supports; 1) a new data model including sensorial an emotional information along with 'traditional' data. 2) a new approach to search on the internet by allowing new specific fields with the proposed data model 3) a modelling approach that supports correlations between sensations and emotions. As a result we want to change the established paradigm of Internet object description, allowing new methods for knowledge management. The first stage consists in presenting a new data model that support sensorial and emotional information starting its filling process with an annotation sequence that is supported by an ontology [15]. On a second stage a framework is being developed to support the new object modelling, its storage, usage and retrieval in a seamless and standardized operating mode. Finally the objective will be to establish a methodology for the association process of sensorial and emotional information in order to promote the existence of i-Episodes. The final result will be a permanently growing knowledge base of Emotions, Sensations and Episodes. As an instantiation of what is described, imagine that a person stores diverse Episodes in life. These episodes can include sport events, music concerts and other events where the person describes sensations and emotions. With this reach variety of information it is possible that other persons attending the same events will able to find them on the Internet even with some disability, e.g. a blind can find the environment of a rock concert by searching the music and the food served at that precise event. People can share memories using sensations and emotions in databank, building their own set of Episodes.

5 Concluding Remarks and Future Work

The present work describes an approach to incorporate sensations and emotions in information available on the Internet but also to establish relations between different objects and its sensorial and emotional relation to persons. This is a challenging question and addresses a person's centric content for the Internet. To accomplish those objectives we want to enable the creation of a person's own episodes that are more than pictures, by the inclusion of their extended sensorial and emotional experience. At the present stage the authors to pursue the establishment of a bank of emotions and sensations that could be accessed by the person or by others when possible. The case applies for people sharing the same type of event, e.g. people attending the same party, the same sports event or even the same natural phenomena. Our objective is gaining consistency as we are enlarging the possibility of representing emotions and sensations, starting with the usage of tags, but allowing other representations of captured information from current, emerging and future technologies. In an ideal situation those episodes would be captured by portable or wearable devices or by communicating with the surrounding IoT connected devices.

The novelty of the present work can be identified by the possibility of retrieve information related with the five senses and user' emotions as part of Object's representation and use it to increase the description of those Things. Then we explore the possibility of interconnecting that information as Episodes as in fact happens with a person's experience. The conceptual goal addresses lessons learned from the Two-Stream Hypothesis trying to find what is searched and where thus it exist or where thus that happened. Future work for the proposed framework will be to support the establishment of relations envisaging the permanence of Events on the Internet. This research may impact in many societal and business aspects. Just to name same, people with sensorial limitation, either blind or deaf, will have more chances to explore the Internet and have a more immersive experience with the expanded emotional and sensorial information. In business, enhanced product description will increase chances to reach a wider range of potential customers and better description of products, from raw material to consumer goods, will empower an advantage over the concurrence.

References

1. Atzori, L., Iera, A., Morabito, G.: The Internet of Things: A survey. Computer Networks 54(15), 2787–2805 (2010)
2. Domingo, M.C.: An overview of the Internet of Things for people with disabilities. Journal of Network and Computer Applications 35(2), 584–596 (2012)

3. (Raj) Rajkumar, R., Lee, I., Sha, L., Stankovic, J.: Cyber-physical systems. In: Proceedings of the 47th Design Automation Conference - DAC 2010, p. 731 (2010)
4. Liu, D., Wang, M., Hua, X.-S., Zhang, H.-J.: Semi-Automatic Tagging of Photo Albums via Exemplar Selection and Tag Inference. IEEE Transactions on Multimedia 13(1), 82–91 (2011)
5. Mishkin, M., Ungerleider, L.G., Kathleen, A.: Object vision and spatial vision: two cortical pathways. Trends in Neuroscience 6, 414–417 (1983)
6. de Sousa, R.: Emotion. The Stanford Encyclopedia of Philosophy (2012), `http://plato.stanford.edu/archives/spr2012/entries/emotion/` (accessed: November 11, 2012)
7. Maurer, H., Tochtermann, K.: On a New Powerful Model for Knowledge Management and its Applications. Journal of Universal Computer Science (2002), `http://www.jucs.org/jucs_8_1/on_a_new_powerful/maurer_h.html` (accessed: November 11, 2012)
8. Ferreira, F., Jardim-Goncalves, R.: Framework for Knowledge Management Based in the Two-Stream Hypothesis. In: Camarinha-Matos, L.M., Shahamatnia, E., Nunes, G. (eds.) DoCEIS 2012. IFIP AICT, vol. 372, pp. 69–76. Springer, Heidelberg (2012)
9. Netter, F.H.: Atlas of Human Anatomy, 2nd edn. Novartis (1990)
10. Hilligsoe, S., Jackobsen, H.: Negotiation: The art of reaching agreement. Academica (2009)
11. Pang, B., Lee, L.: Scientific Commons: Opinion Mining and Sentiment Analysis. Foundations and Trends® in Information Retrieval 2(1-2), 1–135 (2009)
12. López, J.M., Gil, R., García, R., Cearreta, I., Garay, N.: Towards an ontology for describing emotions. In: Lytras, M.D., Damiani, E., Tennyson, R.D. (eds.) WSKS 2008. LNCS (LNAI), vol. 5288, pp. 96–104. Springer, Heidelberg (2008)
13. Pinker, S.: So How Does the Mind Work? Main 20(1), 1–24 (2005)
14. Gahegan, M., Smart, W., Masoud-Ansari, S., Whitehead, B.: A semantic web map mediation service. In: Proceedings of the 1st ACM SIGSPATIAL International Workshop on Spatial Semantics and Ontologies - SSO 2011, pp. 1–8 (2011)
15. Talantikite, H.N., Aissani, D., Boudjlida, N.: Semantic annotations for web services discovery and composition. Computer Standards & Interfaces 31(6), 1108–1117 (2009)

A Neural Network Based Security Tool for Analyzing Software

Adetunji Adebiyi, Johnnes Arreymbi, and Chris Imafidon

School of Architecture, Computing and Engineering, University of East London, London, UK
adetunjib@hotmail.com, {J.Arreymbi,C.O.Imafidon}@uel.ac.uk

Abstract. The need to secure software application in today's hostile computer environment cannot be overlooked. The increase in attacks aimed at software directly in the last decade and the demand for more secure software applications has drawn the attention of the software industry into looking for better ways in which software can be developed more securely. To achieve this, it has been suggested that security needs to be integrated into every phase of software development lifecycle (SDLC). In line with this view, security tools are now used during SDLC to integrate security into software applications. Here, we propose a neural network based security tool for analyzing software design for security flaws. Our findings show that the trained neural network was able to match possible attack patterns to design scenarios presented to it. With the information on the attack pattern identified, developers can make informed decision in mitigating risks in their designs.

Keywords: Threat Modeling, Neural Network, Attack Patterns, Security Tools, Secure software design.

1 Introduction

The dependence of our society today on software systems demand that they are robust, reliable and secured when they are deployed. Therefore, it is very important that the design of software must be that which makes it to function properly in a hostile computer environment, protecting the information and systems on which it runs with minimal risks. However, as software attacks become more sophisticated especially with the increase of attack tools available, today's security solutions are no longer adequate in providing security[5][18] because many of these attacks circumvent the traditional defenses and target the software directly [2]. By retrofitting security into software after development exacerbate the issue as these may lead to significant change in the architectural design and code of the software which often introduces more flaws and increases the cost of production [4].

Reportedly, 50% of security problems in software products today have been found to be design flaws [8]. In this view, many authors argue that it is much better to find and fix flaws during the early phase of software development because it is more costly to fix the problem at a late stage of development and much more costly when the software has been deployed [4][11][18]. To achieve this, a neural network based

L.M. Camarinha-Matos, S. Tomic, and P. Graça (Eds.): DoCEIS 2013, IFIP AICT 394, pp. 80–87, 2013.

tool that would enable software engineers to evaluate their software design for security flaws by matching attack patterns to software design scenarios presented to it is proposed in this paper. Each attack pattern matched to the design helps the software designer to see areas of vulnerability in the design that needs to be secured. Based on this information, software designers can integrate security capabilities that will mitigate the identified risks in their design before the software is coded.

2 Internet of Things (IoT)

One of the core issues in IoT is the connectivity between social, cloud, mobile and everyday objects. While IoT has lots of benefits, it also increases the attack surface of applications running on the connected devices. Many devices that were never intended to be connected to the internet become exposed to software-based attacks while connected to other devices. Furthermore, as devices connected in IoT will be generating a huge amount of data, the risk of data breach also increases. Therefore, the security of software running these devices connected in IoT cannot be overlooked, This paper contributes in this area by proposing a tool based on neural network that can be used to analyze the software design for security flaws.

3 Current Security Tools for Analyzing Software Design

In recent years, various security tools have been developed to enable software developers who are not security experts to scrutinize their software design and identify security flaws in a similar way as a security expert. Microsoft developed two of these security tools [10] [11]. The first was Threat Analysis and Modeling (TAM) tool. This was developed with the aim of enabling non-security expert software developers to use already known data and specific line of business application requirement and architecture to carry out threat modeling in an asset-centric approach. With this tool software developers can focus on protecting the assets within their application by identifying associated threats and counter-measures when it's being designed. The second is SDL Threat Modeling Tool. This is a core element in the design phase of Microsoft Security Development Lifecycle which helps software developers to analyze their software designs prior to its implementation. Also, this tool was not developed for security experts but for software developers to aid the creation and analysis of threat models [10]. In contrast to TAM tool, SDL Threat Modeling tool builds on well-known development activities such as the use of data flow diagram (DFD) for drawing the architecture of the software being designed. Thus, following a software-centric approach, threat modeling with this tool focuses on the software and the analysis of its design [17].

While these tools have lots of useful features that enable software developers to do threat modeling easily, they have a few draw backs. Firstly, the quality of report generated by the tools is still limited by the knowledge of the software developer creating the threat model. Secondly, software developers require the understanding and interpretation of the extensive list of threats identified by the tools. This may become a daunting task especially when the threats are not prioritized as the case is with the use of SDL Threat Modeling Tool. Thirdly the process of threat modeling can increasingly become complex while using the tools due to factors such as number of developers involved in the threat modeling process, the nature of DFD created and potential stakeholders [1][11].

There is now a range of security tools from open source with similar threat modeling approach like that of Microsoft threat modeling tools such as SeaMonster, TRIKE and Coras, which use techniques that software developers are familiar with for the identification and mitigation of threats. There are other threat modeling approaches based on standards such as the Risk Analysis Toolkit (based on ISO 1799) which generates security polices from question and answers [15] and other open security tools like the Common Vulnerability Scoring System (CVSS) that is designed to for rating IT vulnerabilities [3].

4 The Neural Network Tool

Previous researches show various ways through which neural network have been used in the area of security. Neural network based applications has been used successfully in the area of network security as intrusion detection systems, misuse detection systems and firewalls [19] [21] [22]. Also in the field of application security, neural network has been proposed to be used as virus detection system [20]. It would be noticed however, that these neural network based applications can only provide a form security after software deployment.

Our proposed Neural Network tool is based on the abstract and match technique through which software flaws in a software design can be identified when an attack pattern is matched to the design. Using well known approaches such as DFD and sequence diagrams, software developers are able to abstract information about their software designs needed by the Neural Network tool for matching possible attack patterns. When potential attack patterns are matched against the design, the software developers are able to take the necessary steps in mitigating the security flaw identified. Thus software developers are able to integrate security into their software design during the design phase of SDLC when it is easy and cost effective to resolve security problems.

One of the limitations with some of the current approaches is the difficulty of getting software developers to think like attackers during the threat modeling process as this mindset is not native to them [17]. It has been suggested that software developers can instead look at the attack surface of their software design and think of how to build defenses into their application [17]. Our proposed technique achieve this by associating components in the design with attacks that can be performed on them when possible attack patterns are matched to the software design thereby addressing security defenses needed to be put in place.

4.1 The Neural Network Architecture

A three-layered feed-forward back-propagation neural network is used to evaluate scenarios from software designs and identify possible attacks in the design. The back-propagation neural network is a well-known type of neural network commonly used in pattern recognition problems [16]. A back-propagation network has been used in this research because of its simplicity and reasonable speed. The architecture of the back-propagation network is shown in the figure below. This consists of the input layer, the hidden layer and the output layer. Each of the hidden nodes and output nodes apply a tan-sigmoid transfer function (2/(1+exp(-2*n))-1) to the various connection weights. The weights and parameters are computed by calculating the error between the actual and expected output data of the neural network when the training data is presented to it. The error is then used to modify the weights and parameters to enable the neural network to a have better chance of giving a correct output when it is next presented with same input

4.2 Data Collection

Data of attack scenarios from online vulnerability databases such as CVE Details, Security Tracker, Secunia, Security Focus and The Open Source Vulnerability Database were used in this research. From the online vulnerability databases a total of 715 attack scenarios relating to 51 regularly expressed attack patterns by Williams and Gegick [4] were analyzed. This consisted of 260 attack scenarios which were unique in terms of their impact, mode of attack, software component and actors involved in the attack and 455 attack scenarios which are repetition of the same type of exploit in different applications they have been reported in the vulnerability databases. The attacks were analyzed to identify the actors, goals and resources under attack. Once these were identified the attack attributes in Table 1 were used to abstract the data capturing the attack scenario for training the neural network.

4.3 Data Encoding

The training data samples each consist of 12 input units for the neural network. This corresponds to the values of the attributes abstracted from the attack scenarios. The training data was generated from the attack scenarios using the attributes. For instance training data for the attack on webmail (CVE 2003-1192) was generated by looking at the online vulnerability databases to get its details on the attributes we are interested in. This attack corresponds to regularly expressed attack pattern 3. Williams and Gegick [4] describe the attack scenario in this attack pattern as a user submitting an excessively long HTTP GET request to a web server, thereby causing a buffer. This attack pattern is represented as:

```
(User)(HTTPServer)(GetMethod)(GetMethodBufferWrite)(Buffer)
```

In this example, the data generated from the attack scenario using the attribute list is shown in Table 1. Using the corresponding values for the attributes; the data is then encoded as shown in the Table 1. The second stage of the data processing involves converting the value of the attributes in Table II into ASCII comma delimited format before it is used in training the neural network. For the expected output from the neural network, the data used in training network is derived from the attack pattern which has been identified in each of the attack scenarios. Each attack pattern is given a unique ID which the neural network is expected to produce as an output for each of the input data samples. The output data sample consists of output units corresponding to the attack pattern IDs. For instance, the above sample data on Webmail attack which corresponds to regularly expressed attack pattern 3, the neural network is trained to identify the expected attack pattern as 3.

Table 1. Sample of pre-processed training data from attack scenario

S\N	Attribute	Observed data	Value
1	Attacker	No Access	0
2	Source	External	1
3	Target	Buffer	9
4	Attack Vector	Long Get Request	39
5	Attack Type	Availability	5
6	Input Validation	Partial Validation	2
7	Dependencies	Authentication & Input	6
8	Output	None	0
	Authentication	None	0
10	Access Control	URL Access	2
11	HTTP Security	Input Validation	3
12	Error	None	0

4.4 The Neural Network Training

To train the neural network the training data set is divided into two sets. The first set of data is the training data sets (260 samples) that were presented to the neural network during training. The second set (51 Samples) is the data that were used to test the performance of the neural network after it had been trained. At the initial stage of the training, it was discovered that the neural network had too many categories to classify the input data into (i.e. 51 categories) because the neural network was not able to converge. To overcome the problem, the training data was further divided into two sets. The first set contained 143 samples and the second set contained 117 samples. These were then used for training two neural networks. Mat lab Neural Network tool box is used to perform the training. The training performance is measured by Mean Squared Error (MSE) and the training stops when the generalization stops improving or when the 1000th iteration is reached.

4.5 Result and Discussion

It took the system about one minute to complete the training for each the back-propagation neural network. For the first neural network, the training stopped when the MSE of 0.0016138 was reached at the 26th iteration. The training of the second neural network stopped when the MSE of 0.00012841 was reached at the 435th iteration.

To test the performance of the network, the second data sets were used to test the neural network. It was observed that the trained neural network gave an output as close as possible to the anticipated output. The actual and anticipated outputs are compared in the Table 4. The test samples in which the neural network gave a different output from the predicted output when testing the network includes tests for attack patterns 10, 35, 39, 40 and 52. While looking into the reason behind this, it was seen that the data observed for these attack patterns were not much. With more information on these attack patterns for training the neural network, it is predicted that the network will give a better performance. During the study of the results from the neural networks, it was found that the first neural network had 96.51% correct results while the second neural network had 92% accuracy. The accuracy for both neural networks had an average of 94.1%. Given the accuracy of the neural networks, it shows that neural networks can be used to assess the security in software designs

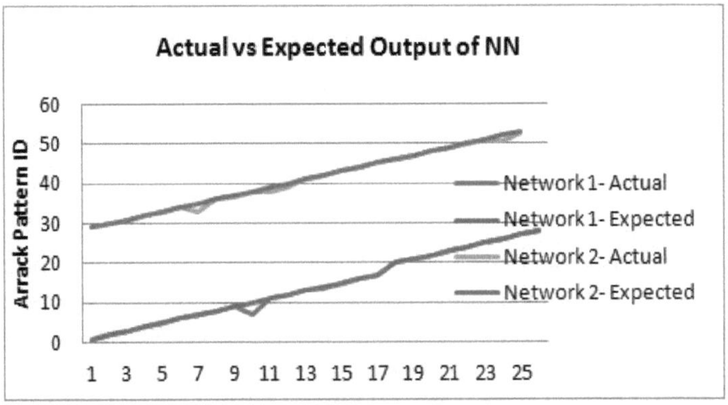

Fig. 1. Actual vs. Expected Output of the Neural Network

5 Conclusion

It cannot be overstated that the cost of fixing security flaws in software applications is very costly after they are deployed. The cost could be 30 times more than the cost of finding and fixing the problem early in the SDLC. Therefore, integrating security into a software design will help tremendously in saving time and money during software development and when the software is deployed. For instance, it is less expensive and less disruptive to discover design-level vulnerabilities during the design, than during implementation or testing, forcing a costly redesign of pieces of the application. Therefore, the use of the proposed neural networks tool for analyzing software design for security flaws will consolidate the efforts of software developers in identifying areas of security weakness in their software design. By fixing the security flaws in design before coding begins will subsequently lead to the development of more secured software applications. Thus, neural networks given the right information for its training will also contribute in equipping software developers to develop software more securely especially in the area of software design.

6 Future Work

The regularly expressed attack pattern used in training the neural network is a generic classification of attack patterns. Therefore, any unknown attack introduced to the neural network will be classified to the closet regularly expressed attack pattern. However, the success of the neural network in analyzing software design for security flaws largely depends on the input data capturing the attributes of the software design introduced to it. As this requires a human endeavor, further work is required in this area to ensure that correct input data is retrieved for evaluation. In addition, the neural network needs to be thoroughly tested before it can gain acceptance as a tool for evaluating software design for security flaws. To further improve the performance of the neural network system as a tool for evaluating software design, we are currently looking into the possibility of the system suggesting solutions that can help to prevent the identified attacks. Current research on solutions to software design security flaws gives a good insight in this area. Suggested solutions such as the use security patterns [6] and introduction of security capabilities into design in the SAT [12]

References

1. Berg, B.: SDL: Threat Modeling tools vs. Threat Analysis tool, http://www.dib0.nl/code/166-sdl-threat-modeling-tool-vs-threat-analysis-tool
2. Burns, S.F.: Threat Modeling: A Process to Ensure Application Security, SANS Institute InfoSec Reading Room, http://www.sans.org/reading_room/whitepapers/securecode/threat-modeling-process-ensure-application-security_1646
3. Common Vulnerability Scoring System (CVSS-SIG), http://www.first.org/cvss
4. Gegick, M., Williams, L.: On the design of more secure software-intensive systems by use of attack patterns. Information and Software Technology 49, 381–397 (2006)
5. Keary, E.: Integration into the SDLC, The OWASP Foundation, https://www.owasp.org/images/f/f6/Integration_into_the_SDLC.ppt
6. Kienzle, D.M., Elder, M.C.: Final Technical Report: Security Patterns for Web Application Development (2002), http://www.scrypt.net/~celer/securitypatterns/final%20report.pdf
7. Kenneth, R., Wyk, V., McGraw, G.: Bridging the Gap Software Development and Information Security. IEEE Security & Privacy 3(5), 75–79 (2005)
8. McGraw, G.: Building Secure Software. A difficult but critical step in protecting your business. Citigal, Inc. (2003), http://www.cigital.com/whitepapers/dl/Building_Secure_Software.pdf
9. McGraw, G.: The Role of Architectural Risk in Software, Inform IT Network, http://www.informit.com/articles/article.aspx?p=446451
10. Microsoft Security Development Lifecycle, SDL Threat Modeling Tool, http://www.microsoft.com/security/sdl/adopt/threatmodeling.aspx
11. Mockel, C., Abdallah, A.E.: Threat Modeling Approaches and Tools for Securing Architectural Designs of E-Banking Application. Journal of Information Assurance and Security 6(5), 346–356 (2010)

12. Mouratidis, H., Giorgini, P.: Security Attack Testing (SAT)- testing the security of information systems at design time. Information Systems 32, 1166–1183 (2007)
13. OWASP Top 10, The Ten Most Critical Web Application Security Risk, http://owasptop10.googlecode.com/files/OWASP%20Top%2010%20-%202010.pdf
14. Pemmaraju, K., Lord, E., McGraw, G.: Software Risk Management. The importance of building quality and reliability into the full development lifecycle. Citigal, Inc. (2000), http://www.cigital.com/whitepapers/dl/wp-qandr.pdf
15. Ricard, R.: ISO 1799 Risk Analysis Toolkit (2011), http://sourceforge.net/projects/ratiso17799
16. Srinivasa, K.D., Sattipalli, A.R.: Hand Written Character Recognition using Back Propagation Network. Journal of Theoretical and Applied Information Technology 5(3), 257–269 (2009)
17. Swigart, S., Campell, S.: Threat Modeling at Microsoft, http://download.microsoft.com/download/6/9/B/69BCB7C6-D158-4073-AD3E-F849E8ACBCE0/SDL_Series_-_4.pdf
18. Spampinato, D.G.: SeaMonster: Providing Tool Support for Security Modeling. In: NISK Conference, http://www.shieldsproject.eu/files/docs/seamonster_nisk2008.pdf
19. Ahmad, I., Swati, S.U., Mohsin, S.: Intrusion detection mechanism by resilient bpck Propagation (RPROP). European Journal of Scientific Research 17(4), 523–530 (2007)
20. Liu, G., Hu, F., Chen, W.: A neural network emsemble based method for detecting computer virus. In: Proceedings of 2010 International Conference on Computer, Mechatronics, Control and Electronic Engineering, vol. 1, pp. 391–393 (2010)
21. Pan, Z., Chen, S., Hu, G., Zhang, D.: Hybrid neural network and c4.5 for misuse detection. In: Proceedings of 2003 International Conference on Machine Learning and Cybernetics, vol. 4, pp. 2463–2467 (2003)
22. Joseph, A., Bong, D.B.L., Mat, D.A.A.: Application of Neural Network in User Authentication for Smart Home Systems. World Academy of Science, Engineering and Technology 53, 1293–1300 (2009)

Semantic Adaptation of Knowledge Representation Systems

Catarina Lucena[1,2], João Sarraipa[1,2], and Ricardo Jardim-Gonçalves[1,2]

[1] Departamento de Engenharia Electrotécnica, Faculdade de Ciências e Tecnologia, FCT, Universidade Nova de Lisboa, 2829-516 Caparica, Portugal
[2] Centre of Technology and Systems, CTS, UNINOVA, 2829-516 Caparica, Portugal
{cml,jfss,rg}@uninova.pt

Abstract. Due to the worldwide diversity of enterprises, a high number of ontologies representing the same segment of reality which are not semantically coincident have appeared. To solve this problem, a possible solution is to use a reference ontology to be the intermediary in the communications between the community enterprises and to outside. Since semantic mappings between enterprise's ontologies are established, this solution allows each of the enterprises to keep internally its own ontology and semantics unchanged. However information systems are not static, thus established mappings become obsolete with time. This paper's presents a PhD research with the objective to identify a suitable approach that combines semantic mappings with user's feedback, providing an automatic learning to ontologies & enabling auto-adaptability and, consequently, dynamism to the information systems.

Keywords: Ontologies, Knowledge Maintenance, Adaptability, Ontology Learning.

1 Introduction

The World Wide Web is a vast and growing source of information and services which need to be shared by people and applications. Ontologies play a major role in supporting the information exchange and sharing by extending syntactic interoperability of the Web to semantic interoperability. However, since information systems are not static, ontologies cannot be thought as an achieved conceptualization of well-delimited and static domain. One of the important aspects in the evolution process is to guarantee the consistence of the ontology when changes occur. That requires a semantic adaptation of its represented knowledge.

Adaptability can be defined as the ability of a system to adapt itself efficiently and fast to changed circumstances. An adaptive system is therefore an open system that is able to fit its behaviour according to changes in its environment or in parts of the system itself [1]. One example is the Internet of Things (IoT) where smart interaction between objects that adapt to the current situation without any human involvement will become the next logical step to people stay connected anytime and anywhere.

L.M. Camarinha-Matos, S. Tomic, and P. Graça (Eds.): DoCEIS 2013, IFIP AICT 394, pp. 88–95, 2013.

2 Relationship to Internet of Things

The Internet of Things (IoT) is a novel paradigm that is rapidly gaining ground in the scenario of modern wireless telecommunications. The basic idea of this concept is the pervasive presence around us of a variety of things or objects - such as Radio-Frequency IDentification (RFID) tags, sensors, actuators, mobile phones, etc. - which, through unique addressing schemes, are able to interact with each other and cooperate with their neighbours to reach common goals [2].

Semantic oriented computing manifests its potential to cope with the challenging problems of heterogeneity and interoperability exposed by the large number of things with different characteristics. The application of semantic technology to the IoT domain will provide systems with the ability to better understand terms and concepts as data is transmitted from one system to another, while preserving the meaning of the content. There are many applications using semantic Web technologies in IoT research [3, 4], however, current work has mostly focused on IoT resources management while not on how to access and utilise information generated in IoT. It is also necessary to consider that a semantic model or ontology is not enough to our data be interoperable. There are/could be *n* ontologies for a domain. In this sense is necessary to exist ontology mappings, reference ontologies and standardisation efforts. This PhD research can contribute to the Semantic Adaptation of Knowledge Representation through identification of a suitable approach that provides an automatic learning to ontologies & enabling auto-adaptability and dynamism to the IoT information systems.

3 Followed Research Method

Research is a systematic process of collecting and analysing information to increase our understanding of the phenomenon under study. It is the function of the researcher to contribute to the understanding of the phenomenon and to communicate that understanding to others [5]. It is argued that the clear definition of a research strategy is a fundamental and necessary requirement for a sound empirical study [6]. The research method adopted by the authors (see Fig. 1) is an instantiation of the 7 classic phases scientific method [7]. The difference is that this has 8 phases, where, for each research question/hypothesis, the last phase aims the technological transfer to industry.

Is possible to see in Fig. 1that the first 5 steps will be repeated cyclically until prove/show the studied theory. By the last, the referred cycle must be carry out the number or times needed in order to mature all the small research question that compose the research overall objective, which in this case is to contribute to the "Semantic Adaptation of Knowledge Representation Systems".

Fig. 1. Adopted Research Method

3.1 Research Questions

A research question is an inquiry that is asked for the purpose of gaining knowledge or useful information on a area of interest to which the authors is intended to participate and contribute for. Research questions are used to determine possibilities and gain valuable insight. Thus, the research questions to be addressed are: 1) How to improve ontology based systems to facilitate its intelligence increase? 2) How to enhance the knowledge acquisition from information system's external users? 3) How to improve the interoperability of software applications and information systems semantics?

Although to contribute to this it is needed to make a first background observation to identify some hypothesis in the resolution of the objective.

3.2 Background Observations

The ability to integrate and apply specialized knowledge of organizational members is fundamental to a firm's ability to create and sustain competitive advantage. This include the ability of organizations to be flexible and respond more quickly to changing market conditions, and the ability to be more innovative as well as improving decision making and productivity [8].

Knowledge Management is the process of capturing the collective expertise and intelligence in an organization, using it to promote innovation through continued organizational learning [9]. In additional, it has become a challenging activity for most competitive business organisations. There is growing recognition in the business community about the importance of knowledge as a critical resource for organisations [10, 11]. Individuals and companies are obliged to focus on maintaining and enhancing their knowledge asset in order to innovate [11,12] and survive in the current competitive markets.

It is now frequently assumed that knowledge is modelled and stored in structures called 'ontologies' which are defined as a *formal and explicit specification of a shared conceptualization* [13] and may be used as a unifying framework to facilitate knowledge sharing and interoperability between independently developed systems . Ontologies are computer implementations of human-like knowledge, for the purpose of describing domains of the world and sharing this knowledge between application programs (and also between people) [14]. Its recognised capacity to represent knowledge, to facilitate reasoning, use and exchange knowledge between systems contribute to increase the computational intelligence [15].

Due to the worldwide diversity of communities, a high number of ontologies representing the same segment of reality which are not semantically coincident have appeared. To solve this problem, a possible solution is to use a reference ontology to be the intermediary in the communications between the community enterprises and to outside. Since semantic mappings between enterprise's ontologies are established, this solution allows each of the enterprises to keep internally its own ontology and semantics unchanged. However information systems are not static, thus established mappings become obsolete with time.

Whether knowledge is stored in ontologies, prepositional knowledge based or simple databases, it must be maintained and kept up to date. There are many reasons for ontology changes: the continual evolution of the modelled domain, the refinement of the ontology conceptualization, the modification of the application by adding functionalities according to new end-user requirements and the reuse of the ontology for others tasks or applications. To take into account all these evolving aspects, ontologies have to be adapted to change requirements [16] in a formal dynamic Knowledge Maintenance (KMa) establishment. KMa is focused on the Knowledge Base improvement to actively be updated, monitored accordingly to the knowledge evolution of its related domain [17].

3.3 Hypothesis

The hypotheses to which the authors will follow to execute experiments are the following: 1) A proper interoperability system based on knowledge representation and reasoning is able to be adapted based on external feedback, facilitating the semantic adaptability on future enterprise systems 2) An ontology based framework integrated with proper operational research methods would facilitate the knowledge acquisition from user's feedback and would increase its ability to KMa 3) The next generation of intelligent systems to assist on interoperability of software applications and information systems needs the support of machine learning and operational research methods.

4 Design Experiment

This chapter presents a preliminary study to better understand the authors' research area, which has been used to properly design the experiments. There are introduced some statements about Artificial Intelligence (AI) and its parallelism to neuroscience addressing how human brain deals with semantic memories and learning. Then, human based learning techniques and Ontology Learning (OL) are introduced since together with some machine learning techniques could be used to facilitate knowledge systems maintenance.

Crucial systems to understand are those involved in memory, but in addition, learning mechanisms are at the heart of how the brain processed information. [18] states that is by modifying the synaptic connection strengths (or weights) between neurons that useful neuronal information processors for most brain functions, including perception, emotion, motivation, and motor function, are built. One example is the study made by Patterson in [19]. This study used basic emotions as a facilitator for learning. Thus, emotions were defined in this case, due to the animal's use, as states elicited by rewards and punishments. A reward is anything for which an animal will work. A punisher is anything an animal will work to escape or avoid. Rewards and punishments can be more formally defined as instrumental reinforcers, i.e. stimuli or events which, if their occurrence, termination, or omission is made contingent upon the making of a response, alter the probability of the future emission of that response.

Following the same idea, artificial intelligent semantic relatedness techniques/ methods can also be considered able to facilitate semantic adaptation ability to its connected system, inspired on brain learning. Since the main idea is to find a system able to learn, but also able to represent and manage complex inputs as concepts like humans do, one possible solution is to use OL. This refers to extracting ontological elements (conceptual knowledge) from input and build ontology from them [20]. Ontology learning can be defined as the set of methods and techniques used for building, semi-automatically or automatically, ontology from scratch, enriching, or adapting an existing ontology using several sources [21]. Compared with manually crafting ontologies, ontology learning is able to not only discover ontological knowledge at a large scale and faster pace, but also mitigate human-introduced biases and inconsistencies [22]. OL uses methods from a diverse spectrum of fields such as machine learning, knowledge acquisition, natural-language processing, information retrieval, artificial intelligence and database management [23]. It is needed to emphasize that the insertion of statistics in ontologies leaded to the formation of the probabilistic ontology concept, which embodies the enabling of ontologies to represent uncertainty knowledge. Such uncertainty is present in knowledge proportionally to its complexity. How much complex the knowledge is how much uncertain it is.

Currently, it is widely accepted that systems that possess knowledge and are capable of decision making and reasoning are regarded as 'intelligent' [24]. There are recognised techniques, such as fuzzy logic, artificial neural networks, machine learning and evolutionary algorithms that contribute to increase a system's 'machine intelligence quotient' [25]. The rationale behind the intelligent label of those techniques is their ability to represent and deal with knowledge [26]. Consequently, in this paper, Artificial Neural Networks (ANN), Fuzzy Logic (FL) are addressed.

An ANN is an information-processing paradigm that is inspired by the way biological nervous system, such as brain, process information. Neural networks can be useful learning from existing data even when humans find it difficult to identify rules. Such as humans, ANN learn from experience and are able to adapt the Knowledge Base when facing new data. Focus on Ontology Learning, one application to consider is [27] where is proposed a method consisting of Projective Adaptive Resonance Theory neural network and Bayesian network probability theorem to automatically construct ontology. One problem related to ANN is if the neural network is implemented as a "black box", then any information "learned" by the network during this training is unavailable. Previous researchers, such as [28, 29] developed design techniques that allow network operation to be decoded after training. This researches made possible the automatic learning and adaptability of ANN with user's feedback related to the information learned.

In order to lead with uncertainty knowledge one solution is the application of Fuzzy Logic in OL. Fuzzy Logic is a multivalued logic able to absorb vague information, usually described in natural language, and convert it into a numerical format for easy computational manipulation, searching for shaping or emulate the human reasoning. In [30] is presented a fuzzy temporal model integrated with an ontology model to allow annotating ontology definitions with time specifications. Another successful application of FL to emulate human behaviour is the application of Fuzzy Logic to measure knowledge sharing, namely the confidence and knowledge complexity level [31].

In this sense, is possible to conclude that the referred learning technologies should be considered to implement a robust and complete intelligent system able to maintain by itself its knowledge. These may conduct authors to the possibility of build an (inspired human) prototype where semantics are provided by domain experts but which could be updated (maintained) by external users in a similar way as humans do when they learn through the others.

KMa is then proposed to be ruled by the analysis of the user's interactions feedback through OL. OL will facilitate the learning from the users usability in order to constantly improve the semantic interoperability between systems. This is accomplished by the insertion of statistics in ontologies leading to the formation of the probabilistic ontology concept, which embodies the enabling of ontologies to represent uncertainty knowledge.

5 Conclusions

The authors' research intended to contribute with a framework that allows the combination of semantic mappings with user's feedback in order to provide an automatic learning capability to ontologies enabling auto-adaptability of the information systems in the advent of dynamics. Such methodologies should be able to: 1) facilitate knowledge acquisition and maintenance increasing the intelligence of ontology based systems 2) work as a semantics mediator between enterprises communications with capacity to adapt to changes dynamics in the enterprises internal models 3) Improve the interoperability of software applications and information systems semantics.

Together all the mentioned methodologies are expected to contribute with ideas/solutions to the system's intelligence increase, which will facilitate semantics adaptability of systems.

In the scope do the research is intended to implement some prototype in order to prove or disprove the hypothesis. This will allow gathering of data and execution of tests according to some pre-established validation methods. Remarks concerning the implementation must be educe, since the research may find evidence that the prototype needs rectifications, the hypothesis failed the test or if it is necessary to reformulate the research questions (see Fig 1). Apart of having a set of main research questions related to the thesis research goal, there will be several sets of "small" research questions. For each of these questions is intended a publications about a specific topic, and sometimes a technology transfer to industry.

References

1. Gronau, N., Andresen, K.: Design and Use Patterns of Adaptability in Enterprise Systems. Reihe Wirtschaftsinformatik, Gito (2005)
2. Giusto, D., Iera, A., Morabito, G., Atzori, L.: The Internet of Things. Springer
3. Russomanno, D.J., Kothari, C., Thomas, O.: Sensor ontologies: from shallow to deep models. In: SSST, pp. 107–112 (2005)
4. W3C, W3c ssn incubator group report (2012), http://www.w3.org/2005/Incubator/ssn/wiki/Incubator_Report
5. Venkataram, P.: An article to clear up some misconceptions about the nature of research (2012), Retrieved from the web: http://cce.iisc.ernet.in/motivationprinciples.pdf
6. Amaratunga, R., Baldry, D., Sarshar, M., Newton, R.: Qualitative and quantitative research in the built environment: application of "mixed" research approach: a conceptual framework to measure fm performance. Work Study (renamed International Journal of Productivity and Performance Management) 20, 17–31 (2002)
7. Camarinha-Matos, L.: Unit 2: Scientific method. In slides of the Scientific Research Methodologies and Techniques course of the PhD Program in Electrical and Computer Engineering of the FCT-UNL (2012), Retrieved from the web: http://www.uninova.pt/~cam/teaching/SRMT/SRMTunit2.pdf
8. Harris, D.: Creating a knowledge centric information technology environment. Harris Training & Consulting Services Inc., Seattle (1996)
9. Carneiro, A.: How does knowledge management influence innovation and competitiveness? Journal of Knowledge Management, 87–98 (2000)
10. Leonard-Barton, D., Leonard, D.: Wellsprings of Knowledge: Building and Sustaining the Sources of Innovation. Harvard Business Press (1998)
11. Matthews, K., Harris, H.: Maintaining knowledge assets. In: Mathew, J., Kennedy, J., Ma, L., Tan, A., Anderson, D. (eds.) Engineering Asset Management, pp. 618–626. Springer, London (2006)
12. Metaxiotis, K.S., Ergazakis, K., Psarras, J.E.: Exploring the world of knowledge management: agreements and disagreements in the academic/practitioner community. J. Knowledge Management 9(2), 6–18 (2005)
13. Gruber, T.R.: A translation approach to portable ontology specifications. Knowl. Acquis. 5, 199–220 (1993)
14. An, Y.J., Chuan Huang, K., Geller, J.: Naturalness of ontology concepts for rating aspects of the semantic web (2006)
15. Sarraipa, J., Jardim-Gonçalves, R.: Semantics adaptability for systems interoperability (2010)

16. Djedidi, R., Aufaure, M.-A.: Ontological knowledge maintenance methodology. In: Lovrek, I., Howlett, R.J., Jain, L.C. (eds.) KES 2008, Part I. LNCS (LNAI), vol. 5177, pp. 557–564. Springer, Heidelberg (2008)
17. Sarraipa, J.: Semantic adaptability for the systems interoperability. PhD thesis presented at Faculdade de Ciências e Tecnologia da Universidade Nova de Lisboa (2012)
18. Rolls, E.: Memory systems in the brain. Annual Review of Psychology (2000)
19. Patterson, K., Nestor, P.J., Rogers, T.T.: Where do you know what you know? The representation of semantic knowledge in the human brain. Nat. Rev. Neurosci. 8, 976–987 (2007)
20. Shamsfard, M., Abdollahzadeh Barforoush, A.: The state of the art in ontology learning: a framework for comparison. Knowl. Eng. Rev. 18, 293–316 (2003)
21. Gómez-Pérez, A., Manzano-Macho, D.: A survey of ontology learning methods and techniques. Deliverable 1.5, OntoWeb Consortium (2003)
22. Zhou, L.: Ontology learning: state of the art and open issues. Inf. Technol. and Management 8, 241–252 (2007)
23. Sabou, M.: Learning domain ontologies for web service descriptions: An experiment in bioinformatics. In: Intl. World Wide Web Conf., WWW (2005)
24. Meystel, A.M., Albus, J.S.: Intelligent Systems: Architecture, Design, and Control, 1st edn. John Wiley & Sons, Inc., New York (2000)
25. Zadeh, L.A.: Fuzzy logic, neural networks, and soft computing. Commun. ACM 37, 77–84 (1994)
26. Kasabov, N., Filev, D.: Evolving intelligent systems: methods, learning, & applications. Evolving Fuzzy Systems (2006)
27. Xi-Hu, Z., Yan-Fei, L.: Building ontology automatically based on bayesian network and part neural network. In: Proceedings of the 2009 WRI Global Congress on Intelligent Systems, GCIS 2009, vol. 04, pp. 563–566. IEEE Computer Society, Washington, DC (2009)
28. Fu, L.: Rule generation from neural networks. Systems, Man and Cybernetics (1994)
29. Towell, G.G., Shavlik, J.W.: The extraction of refined rules from knowledge-based neural networks. Machine Learning, 71–101 (1993)
30. Nagypál, G., Motik, B.: A fuzzy model for representing uncertain, subjective and vague temporal knowledge in ontologies. In: Meersman, R., Tari, Z., Schmidt, D.C. (eds.) CoopIS/DOA/ODBASE 2003. LNCS, vol. 2888, pp. 906–923. Springer, Heidelberg (2003)
31. Zadjabbari, B., Mohseni, S., Wongthongtham, P.: Fuzzy logic based model to measure knowledge sharing. In: 3rd IEEE International Conference on Digital Ecosystems and Technologies (2009)

Part V
Computational Systems

BrainMap – A Navigation Support System in a Tourism Case Study

Luís F.S. Teixeira[1], Rita A. Ribeiro[2], António Falcão[2], Gabriel P. Lopes[1], and Ricardo Raminhos[3]

[1] DI-FCT/UNL, 2829-516 Caparica, Portugal
[2] CA3-Uninova, Campus – FCT/UNL, 2829-516 Caparica, Portugal
[3] Viatecla - Estrada da Algazarra, 72, 2810-013 Almada, Portugal
{lft13809,gpl}@fct.unl.pt,
{rar,ajf}@uninova.pt,
rraminhos@viatecla.com

Abstract. Presently, the amount of documents in corporations can make searching and browsing for a specific topic or information a very hard task. Therefore, it is important to develop tools to ease the retrieval of specific information and to support the exploration by users on corporate intranets (composed of several hundreds of gigabytes of documents). Although not explicitly identified, many of these documents are related among themselves (directly or implicitly). In this paper we discuss a navigation support system to explore graphs applied to document correlations, using a tourism case study.

Keywords: Document Correlation, Graph Exploration, Tf-Idf Metric, Jaccard Metric, Tourism.

1 Introduction

The purpose of this paper is to discuss the capabilities of a Tourism navigation support system, developed in the scope of project BrainMap [1]. The motivation behind such a system is to provide a unique mechanism to find relations between unstructured documentation and provide the results in an innovative visual format that allows an intuitive navigation and data exploration. An illustrative case study, based on vacation packages of the tourism industry, is used to test the prototype.

Thus, this paper is structured as follows: Section 2 will briefly describe how this work and its components correlate with the conference theme. Section 3 will describe techniques and technologies used to achieve the objectives and some similar studies, as well as a sub-section describing the system design infrastructure. Section 4 will describe a usage example of the prototype. Section 5presents conclusions and future work.

2 Internet of Things

Considering the concept of the Internet of Things [2], and taking into account the idea that everything would be uniquely identified and connected, but not explicitly related,

L.M. Camarinha-Matos, S. Tomic, and P. Graça (Eds.): DoCEIS 2013, IFIP AICT 394, pp. 99–106, 2013.

any mechanism that would allow the discovery and exploration of possible relationships would be very advantageous. In the context of this prototype, we can imagine that any document could have a unique identifier that would enable not only the identification of its content, but also other practical information that could be useful to extract relations to other documents with regard to their contents. A user would also be identified and have an associated profile containing preferences, usage domain, needs and expertise etc. That profile allows the system to better adjust the relations and gives the user the ability to interact with these documents in a more productive way. This relation and correlations would evolve in time with the interaction with documents, making the task of searching more focused, since the results would automatically have filtered themselves. These could be achieved by providing to the documents a unique identifier (usually termed a "URI" – Uniform Resource Identifier), which would further enhance the integration between physical objects and digital contents. The system presented in this paper would also allow navigation with seamless transition between physical objects and the digital contents.

3 Techniques and Technologies and Similar Studies

In this section, we describe the techniques and technologies used in the development of the navigation support system. We also present similar studies made in this scientific area. In the following sub-sections the document representation issue is reported, the extraction of key-terms is described and finally the correlation metrics problematic is described. Considering first similar studies, we have the work done in [3]where term and document correlation is addressed. Similarity metrics were used between the terms and the web documents. In [4] document correlation is established using a self-organizing map (SOM) to cluster documents by cluster of topics. There have been also attempts at visualizing the results retrieved by means of such systems. Considering the above references, one type of visualization is taken in consideration by [3] where the author projects in space a graph of correlated web documents while in [4] the authors represent the relationships between documents as a SOM (self-organizing map), to visually allow the user to observe the documents within the cluster and their neighbours. Our work mainly differs from the above, as we perform searches within a specific corporation intranet being solely based on textual document data, independently of the language or any other tools that would limit the support system. Specifically the authors of this work developed a navigation support [5], which is overviewed in the next sub-sections, and is then discussed in the scope of a customized tourism package case study.

3.1 Main Concepts of the Navigation Support System

While developing this system, one of the first problems was the issue on document representation [5].There are many representations, specifically in the ontologies domain [6], but we followed a simple text-mining approach. Particularly, we use a "Bag of Words" [5] representation, i.e., the textual content is represented as a collection of all words in the text. To achieve this representation, a preprocessing step is necessary, which was accomplished by the use of scripts that removed unnecessary noise, e.g. html tags. The document text is indexed and therefore represented in a "Vector Space Model" [7] using Term Frequency – Inverse Document Frequency (Tf-Idf) [8] for term weighting purposes, as described below.

The first step in the system is to perform a request of information, using a word search query. The system will return a list of documents that contain the search keywords ordered by importance of the keyword. For accomplishing this task we used statistically based extraction methodologies. These are divided into two major types: approaches that use statistical methods and approaches that use alternative methods, which are not statistical or not purely statistical like the usage of grammars [9] or using statistical with linguistic modulation [10]. There are several ways to calculate metrics to give weight to the extracted words [5].As we are interested in the extraction, independent from the languages of the texts, and knowing that the statistical approaches are based mainly in frequencies counting, we have chosen to use a statistical approach, which uses a common term-frequency - inverse document frequency metric because it poorly scores the words that are not relevant, thus making unnecessary the explicit use of stop-lists for each of the languages envisaged[8].

Term Frequency – Inverse Document Frequency(Tf-Idf) [8] is a statistical metric used in information retrieval and text mining. It is used to evaluate how important a word is to a document in a corpus, increasing proportionally to the number of times a word appears in the document but it is offset by its frequency in the corpus. From [5] we use a probability, $p(W, d_j)$, in equation (1), defined in equation (2), instead of using the usual term frequency factor.

$$Tf - Idf(W, d_j) = p(W, d_j) * Idf\ (W, d_j) \tag{1}$$

$$p(W, d_j) = f(W, d_j)\ /\ Nd_j \tag{2}$$

$$Idf\ (W, d_j) = \log(\|D\| / \|\{d_j : W\ \in\ d_j\}\|) \tag{3}$$

Where $f(W, d_j)$ denotes the frequency of a word W in a document d_j and Nd_j stands for the number of words of d_j; $\|D\|$ is the number of documents of the corpus. So, $Tf - Idf(W, d_j)$ will give a measure of the importance of W within the particular documentd_j. By the structure of term Idf we can see that it privileges words occurring in fewer documents. The choice for using the Tf-Idf metric was based on a comparative study performed by the authors of several metrics for the automatic extraction of document topics [5,11].

Correlation Metrics are needed to generate relationships between documents based on the score of their words. The terminology of networks and graph are used interchangeably, in this work we follow the same definition as presented in [12], where a network indicates a relationship between objects (free text documents in our case) and graph indicates relationships generated through an automatic process. According to [13] crisp relations, are relations of dichotomous type, like in binary logic, where a statement can be true or false and nothing else. Other important concept is the notion of similarity or proximity between elements of a corpus, for enabling us to learn how related they are and to construct the weighted graph. In this work we use Jaccard [14] similarity measure because it is flexible, simple to implement and is easily generalized to fuzzy weighted graphs, but any other similarity measure could have been used ([15][16]). Further, by using the minimum (min) and maximum (max) operators from the T-norms [13] the generalization of the Jaccard measure to the fuzzy interval of [0,1] is done with the following formulation [17]:

$$J_{i,j} = \sum_k min(x_{ik}, x_{jk}) \Big/ \sum_k max(x_{ik}, x_{jk}) \tag{4}$$

Where the indexes i and j stand for the rows of the Tf-Idf Matrix (rows stand for documents, see Tf-Idf Matrix in Fig.).The x_{ik} means the column k (a Tf-Idf score of word) from the row i (a document). The definition for x_{jk} is similar but for line j (another document).

3.2 Navigation Support System Architecture

Having presented the background concepts and technologies required to understand the several parts of the navigation support system used in this work, the developed infrastructure is depicted in Fig.. It shows the flow and the four processing steps included in the navigation support system infrastructure

1. Data preparation: A company or corporation documentation has a large variety of formats, ranging from emails, project reports, studies reports, etc. In order to deal with this fact, normalization is required; this usually involves extracting the actual text from html pages (removing html tags). All required documentation is transformed automatically into plain text (txt) files. Several libraries are available (for many programming languages) that allow us to transform documents in a given format into a simple text file. For example [18] presents Apache Tika, a Java based toolkit for extracting content from a variety of document formats

2. Calculate the weights of documents words: For being able to mine any kind of relation between documents, their words (as we are using the bag of words representation) must be in some way weighed, assigning best scores to best words, i.e. best topic descriptors of the document content will have best scoring values. According to the work of [5,11] the best metrics to accomplish this are those based on Tf-Idf (Term Frequency – Inverse Document Frequency) and Phi-Square metrics.

3. Mine document correlation: This step goal is to discover document relations where they do not explicitly exist, just based on the documents textual content, requiring no semantic knowledge, and applying similarity metrics, as exemplified in the previous section example.

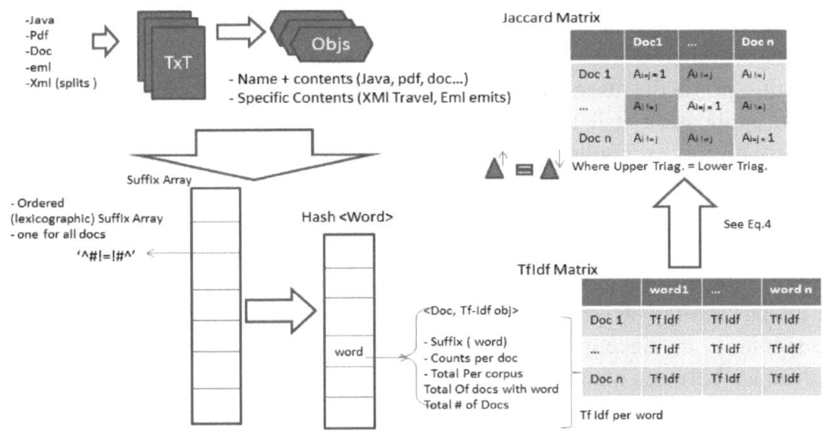

Fig. 1. Overall system design

4 Tourism Case Study

The case study used is based on the Tourism industry as a source of documentation, in the form of textual descriptions from touristic destination packages [19]. These files are characterized by a small destination description, no more than a couple of sentences. Ideally, as more textual contents are available, the more accurate the Tf-Idf metric will be. Some other fields of the source file were also used, on one hand allowing more user interaction, and at the same time enriching the description initially offered. As an example, usually the country of the vacation package was not present in the description, but available in a specific field. This information is automatically appended to the description prior to the statistical treatment. After the system performs all the steps, the user has available a graphic interface application, allowing him to search for a destination, or for some characteristic that he prefers for his vacations, e.g. "beach" or "snow", or "Bali".

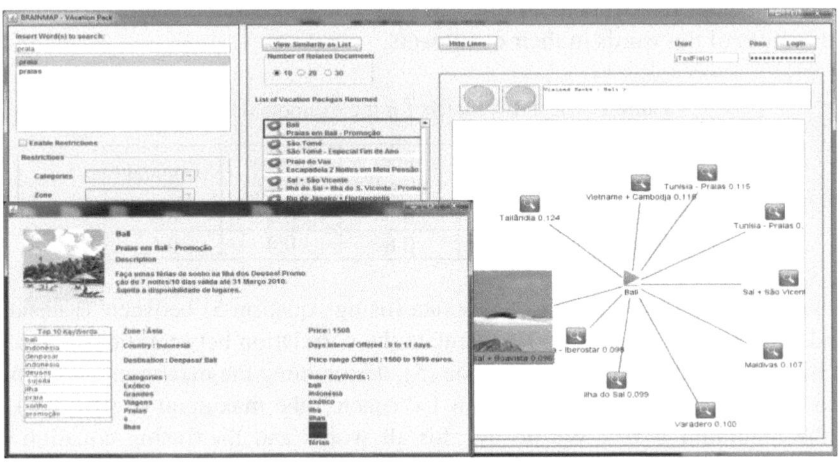

Fig. 2. User Interface overview

The results are presented to the user in a form of ranked list first (according to the score of the searched word in each vacation package).

From this point the user can select one of the vacation packages presented, on which the application will then present him a graph with other vacation packages correlated with the user selection (see Fig. 1). At this point the user can consult the description of each vacation package; the interaction result is presented in Fig. 1. On this graph of packages the user may be presented with results that would not be seen on the ranked list, because it did not contain the exact search word, but will appear because the content of a description correlates to the initial selection of the user. This can be an interesting outcome, while searching for a vacation destination.

The similarity distance given the Jaccard metric value, can be visually represented as the distance between the center node and the related nodes, see Fig. 2

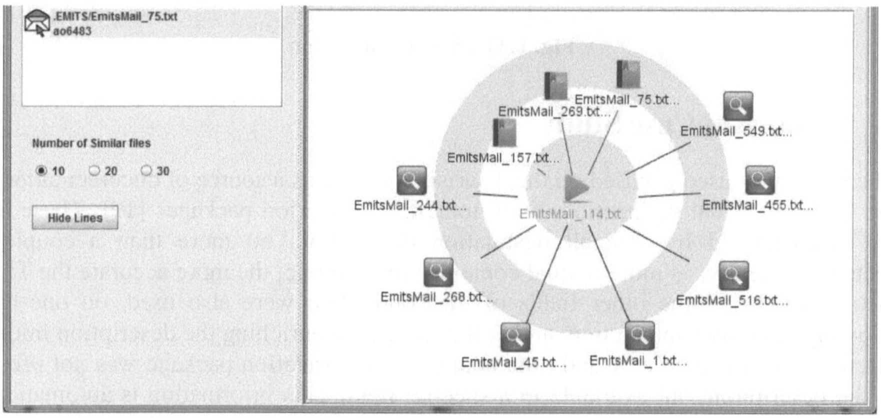

Fig. 3. Correlation graph with Jaccard distances represented by arc lengths

See the illustrative example below. The system would work for a corpus of two documents composed by a bag of 4 words that would generate a matrix containing the Tf-Idf values of the words in their documents.

Table 1. The TfIdf Matrix for the example documents

Docs\words	praia (beach)	ilhas (islands)	exótico (exotic)	tropical
Thailand	0.7	0.3	0.5	0.4
Bali	0.6	0.8	0.4	0.6

Now we want to find the Jaccard Value (using equation 5) between Thailand (Doc 1) and Bali (Doc 2) in Table 1. To calculate the correlation between lines 1 and 2, we start by setting i =1 and j = 2 in equation (5), determining the maximum and minimum for each word. For instance, in column 1 ("beach") the maximum of (0.7, 0.6) is 0.7 and the minimum is 0.6. We do this for all words and then using equation 4, we calculate the Jaccard measure:

$$J_{i=1j=2} = \frac{0.6 + 0.3 + 0.4 + 0.4}{0.7 + 0.8 + 0.5 + 0.6} = 0.65 \quad (5)$$

The result of 0.65 depicts the correlation between Thailand and Bali using the Jaccard metric. This result means that Thailand is similar to Bali in 0.65. Notice that 0 means completely dissimilar and 1 means total similarity (this result only happens when correlating the same document with itself).

5 Conclusions and Future Work

This paper described an application of tourism industry using a novel Navigation Support System. This application demonstrated how the system supports intranet corporate users in the search for either/or keywords and documents, and also enable users to navigate across a network of related documents. Visual inspection of the correlated documents returned by the system, showed similar contents that would allow a user to navigate through a graph and possibly allowing him to reach a document that otherwise would not be found, e.g. does not contain the initial search word. We believe this type of application can be very useful for exploring and finding knowledge and information within the massive databases of corporations. Moreover, it can help novices to learn about the "business" in a faster and user-friendlier way.

As future work we will consider evolving the prototype to the web environment, making it readily accessible from multiple platforms. Other improvements such as considering as keywords of a document not only words but also multi-words, as indicated in the work of [5] will also be considered. Additionally in translations tasks, it is possible to enhance the translation process by finding correlations between already translated documents that are not necessarily parallel to the current document (i.e. not translations of each other).

Acknowledgments. he work on this paper was developed in the context of the project "BrainMap" lead by the company ViaTecla (Portugal) in collaboration with UNINOVA and University of Évora (both in Portugal). Financedbythe Portuguese "QREN- Quadro de Referência Estratégico Nacional; Programa Operacional de Lisboa". This was also supported by the Portuguese Foundation for Science and Technology (FCT/MCTES) through funded research projects ISTRION (ref. PTDC/EIA-EIA/114521/2009) and VIP-ACCESS (ref. PTDC/PLP/71142/2006).

References

1. ViaTecla, BrainMap (2011), `http://www.viatecla.pt/inovacao/brain_map`
2. Van Kranenburg, R., Anzelmo, E., Bassi, A., Caprio, D., Dodson, S., Ratto, M.: The Internet of Things. A critique of ambient technology and the all-seeing network of RFID, Network Notebooks, vol. 2 (2007)
3. Viji, S.: Term and Document Correlation and Visualization for a set of Documents. Stanford University (2002)

4. Klose, A., Nürnberger, A., Kruse, R., Hartmann, G., Richards, M.: Interactive text retrieval based on document similarities. Physics and Chemistry of the Earth, Part A: Solid Earth and Geodesy 25(8), 649–654 (2000)
5. Teixeira, L.F.S., Lopes, G.P., Ribeiro, R.A.: An Extensive Comparison of Metrics for automatic extraction of Key Terms. In: Proceedings of the 4th International Conference on Agents and Artificial Intelligence (ICAART 2012), Algarve, Portugal (2012)
6. Maynard, D., Peters, W., Li, Y.: Metrics for Evaluation of Ontology-based Information Extraction. In: WWW 2006, Edinburgh, UK, May 22-26 (2006)
7. Salton, G., Wong, A.K.C., Yang, C.S.: A vector space model for automatic indexing. Commun. ACM 18(11), 613–620 (1975), doi:10.1145/361219.361220
8. Salton, G., Buckley, C.: Term Weighting Approaches in Automatic Text Retrieval. Cornell University (1987)
9. Afrin, T.: Extraction of Basic Noun Phrases from Natural Language Using Statistical Context-Free Grammar. Virginia Polytechnic Institute and State University (2001)
10. Hulth, A.: Improved Automatic Keyword Extraction Given More Linguistic Knowledge. In: EMNLP 2003 Proceedings of the Conference on Empirical Methods in Natural Language Processing, pp. 216–223. ACL, Stroudsburg (2003)
11. Teixeira, L., Lopes, G., Ribeiro, R.A.: Automatic Extraction of Document Topics. In: Camarinha-Matos, L.M. (ed.) DoCEIS 2011. IFIP AICT, vol. 349, pp. 101–108. Springer, Heidelberg (2011)
12. Mihalcea, R., Radev, D.: Graph-based Natural Language Processing and Information Retrieval. Cambridge University Press, New York (2011)
13. Zimmermann, H.-J.: Fuzzy Set Theory and its Applications, 4th edn. Springer (2001)
14. Miyamoto, S.: Fuzzy Sets in Informational Retrieval and Cluster Analysis, 1st edn. Series D: System Theory, Knowledge Engineering and Problem Solving, vol. 4. Springer (1990)
15. Shyi-Ming, C., Ming-Shiow, Y., Pei-Yung, H.: A comparison of similarity measures of fuzzy values. Fuzzy Sets Syst. 72(1), 79–89 (1995), doi:10.1016/0165-0114(94)00284-e
16. Amit, S., Chris, B., Mandar, M.: Pivoted document length normalization. In: Proceedings of the 19th Annual International ACM SIGIR Conference on Research and Development in Information Retrieval, Zurich, Switzerland (1996)
17. Rocha, L.M., Simas, T., Rechts, A., Di Giacomo, M., Luce, R.E.: MyLibrary at LANL: proximity and semi-metric networks for a collaborative and recommender Web service. In: The 2005 IEEE/WIC/ACM International Conference on Web Intelligence, pp. 565–571. IEEE Press (2005), doi:10.1109/WI.2005.106
18. Mattmann, C.A., Zitting, J.L.: Tika in Action. Manning Publications Co. (2012)
19. Vacation packages database, Viatecla (2011), http://www.viatecla.com

A Scalable Spam Filtering Architecture

Nuno Ferreira[1], Gracinda Carvalho[1], and Paulo Rogério Pereira[2]

[1] Universidade Aberta, Portugal
[2] INESC-ID, Instituto Superior Técnico, Technical University of Lisbon, Portugal
ze@nuno.pro, Gracinda.Carvalho@uab.pt, prbp@inesc.pt

Abstract. The proposed spam filtering architecture for MTA[1] servers is a component based architecture that allows distributed processing and centralized knowledge. This architecture allows heterogeneous systems to coexist and benefit from a centralized knowledge source and filtering rules. MTA servers in the infrastructure contribute to a common knowledge, allowing for a more rational resource usage. The architecture is fully scalable, ranging from all-in-one system with minimal components instances, to multiple components instances distributed across multiple systems. Filtering rules can be implemented as independent modules that can be added, removed or modified without impact on MTA servers operation. A proof-of-concept solution was developed. Most of spam is filtered due to a grey-listing effect from the architecture itself. Using simple filters as Domain Name System black and white lists, and Sender Policy Framework validation, it is possible to guarantee a spam filtering effective, efficient and virtually without false positives.

Keywords: spam filtering, distributed architecture, component based, centralized knowledge, heterogeneous system, scalable deployment, dynamic rules, modular implementation.

1 Introduction

Internet mail spam[2] is a problem for most organizations and individuals. Receiving spam on mobile devices, and on other connected appliances, is yet a bigger problem, as these platforms are not the most appropriate for spam filtering.

Spam can be seen as belonging to one of two major categories: **Fraud** and **Commercial**. In the *fraud* category we include phishing, scams, malware, counterfeit products, and any other criminal activities perpetrated or assisted through Internet mail. In the *commercial* category we include promotional messages – such as newsletters – that we do not want to receive, either being sent legally or illegally[3] from legitimate organizations. We can also classify these two categories as per threat and per volume [1], as shown in Table 1.

---

[1] MTA – Mail Transfer Agent, commonly referred as mail server.

[2] For this paper, spam is defined as every message that most people do not want to receive.

[3] At the European Community countries all commercial mail messages are opt-in, that is to say that it is illegal to send commercial messages without prior consent from the receiver.

L.M. Camarinha-Matos, S. Tomic, and P. Graça (Eds.): DoCEIS 2013, IFIP AICT 394, pp. 107–114, 2013.

Table 1. Category classification

Category	Threat	Volume
Fraud	*High*	*High*
Commercial	*Low*	*Low*

So, we should primarily address the spam messages in the *fraud* category.

What is an efficient – and still effective – solution for this spam filtering?

Can an architecture, that enables filtering based on the source of spam, be both efficient and effective? Designing and implementing a proof-of-concept for such an architecture, should provide us with enough statistical data to find an answer.

2 Relationship to Internet of Things

The usefulness of information platforms, specially of mobile devices and connected appliances, depends largely on the relevance of the information they provide. The spam received in these platforms wastes resources[4], and reduces the overall relevancy of the information provided.

The proposed spam filtering solution is a scalable and distributed architecture that allows the construction of an ecosystem composed by connected heterogeneous systems to collaborate for a common knowledge. This common knowledge provides a more rational use of resources, as it allows a simple and fast decision to be taken at each MTA and, at the same time, it prevents the duplication of complex decisions as these are based on a common source of information. This solution reduces waste and improves the relevancy of the information provided.

3 Traditional Solutions

Traditional and common solutions for spam filtering can be found in two flavours: **client** and **server** side.

Client-side solutions – that reject or obfuscate messages – must be avoided, as in these cases the sender cannot be properly informed, becoming a major problem for any false positives that might occur. And it is inefficient, as the server uses resources to process and store spam, and the client to receive it.

Most common server-side solutions, such as the Spamassassin [2], are monolithic and run a synchronous filtering process. This is highly inefficient, as each connection is kept active until a complex decision process is complete. On busy servers this might end in a Denial of Service (DoS). It is even more inefficient when an incoming connection, from a previously blocked source, triggers the same process again in any MTA of the infrastructure, due to the lack of a common knowledge.

Approaches have been made to spam filtering architectures, yet they focus more on effectiveness of spam identification, rather than on filtering efficiency. Ma, et al. [3]

[4] Such as CPU, bandwidth, connections, storage, and others related to energy consumption.

propose an architecture for content normalization, to improve content analysis. Cottereau [4] proposes a collaborative architecture for client-side filtering. Yih, et al. [5] also propose a method for increasing the effectiveness of spam identification.

4 The Architecture

In order to save resources[5], we need to make a decision as soon as possible, and this means to start our filtering process as soon as a connection is established to an MTA.

Also, we must avoid client-side bounces, as the mail addresses from senders may be forged, and we would be assisting the spammer by spamming innocent victims. In this respect, server rejections are safe, as these are reported to the MTA of the sender and not to any eventually forged mail address. It is vital to inform the sender of a rejection, so that false positives do not go unnoticed, thus giving the chance to a legitimate sender to find a way to bypass any false positive rejection that might occur.

So, the present architectural solution is designed to operate in mail servers (MTA).

4.1 Working Principle

The working principle of the architecture is a quite simple and straightforward one. It is roughly divided into three main areas and five components, as shown in Table 2.

Table 2. Areas and working Components

Areas	Components
Decision	*Adapter + Proxy*
Information	*Knowledge*
Assessment	*Consultant + Agent*

A *decision* process is initiated by the MTA, and its goal is to process each SMTP [6] command and decide to either accept or reject the SMTP session or transaction based on available *information*. When the *information* is insufficient to make a reliable *decision*, the MTA returns a temporary fail, and that fact gets registered.

The *information* is a set of data classified in order to allow a quick and reliable *decision*. Unclassified data is processed by an *assessment* to obtain the proper classification. The type of data to consider should provide reliable identification, such as an IP address of an MTA, or a mail domain. Mail addresses can also be considered for a type of data for identification purposes, although they can easily be forged, as long as this fact is kept in mind. The classification should be easy to interpret as a simple rule to reject (*dark*, *black*), accept (*light*, *white*), or as insufficient to make a decision and thus issue a temporary fail (*grey*).

An *assessment* process runs at regular intervals, checking if insufficient *information* was found by a *decision* process. If there is such an event, an *assessment*

[5] Including human resources, whose time is spent with irrelevant information (spam).

is made, in order to obtain a reliable classification (*dark* or *light*) for what was previously insufficient *information* (*grey*).

4.2 Architecture Components

The components of the architecture allow a distributed environment with a centralized source of information. Fig. 1 shows a possible deployment for the components of the architecture, where the dashed lines denote a tight coupling between components, and full lines denote a loose coupling (network connection).

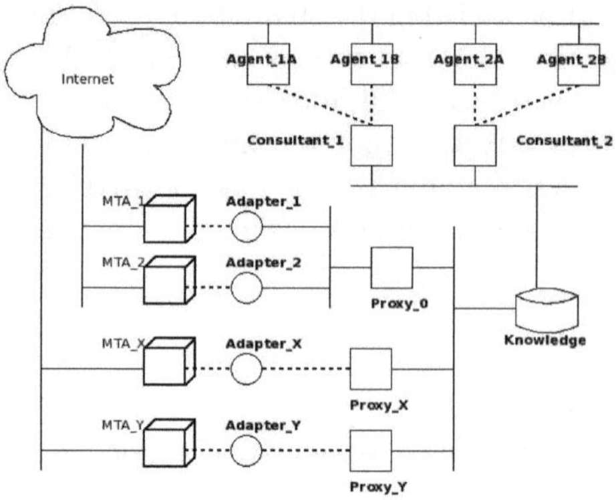

Fig. 1. Diagram of the proposed Architecture

The *Adapter* component allows heterogeneous MTA systems to coexist, and to benefit from a common and central source of information, to which every MTA contributes. This enhances the performance of the system and the rationalization of the resources, as it avoids duplicate assessments. It accesses the solution services through the *Proxy* component. A simple API is provided to enable this access.

The *Proxy* component is the decision centre, based on the *Knowledge*.

The *Knowledge* component gathers data into a database that can be used for informed decision making – by the *Proxy* – and for statistical purposes.

The *Consultant* component is responsible for regularly checking the *Knowledge* for information that requires classification, and to invoke the appropriate *Agents* that are able to properly classify that information.

The *Agent* component acts upon advice from a *Consultant*, and provides the proper classification as a response to an information query asked by the *Consultant*.

5 Results

A proof-of-concept solution was developed to assess the architecture efficiency and effectiveness. The developed *Adapter* was for the Sendmail MTA implementation. The *Proxy* – and associated API – were developed to allow communication only through IPC (Inter-Process Communication), instead of networked communications.

Yet, the developed solution allows for a full analysis and assessment of the architecture on both effectiveness and efficiency.

The whole solution was developed using only Open Source technologies. The test environment was a single CentOS 5 system running a Sendmail MTA server. The *Adapter* and *Proxy* components – as the *Proxy* API – were developed in C. A MySQL relational database was used for the *Knowledge* component. The PERL language was used for the *Consultant* and *Agent* components. Hardware wise, as it is important to assess the efficiency, the Internet connection was assured by an ADSL link of roughly 10 Mbps downstream and 700 Kbps upstream, and the whole software services (plus others like HTTP and DNS) ran on an ATOM D525@1.8GHz processor machine with 2 GB of memory.

Data was collected for two months. For this period of over sixty days[6], at the test server, it was recorded:

- 6,793 distinct source IP addresses
- 14,282 distinct mail addresses (both sender and receiver)
- 23,194 connections, of which:
  - 17,564 reached the envelope phase (sender identification)
  - 2,544 reached the recipient phase

5.1 Effectiveness

Being an implementation of a proof-of-concept, only one filtering *Agent* was developed for the *Consultant*. This developed *Agent* checks DNS lists – both black and white – for an MTA host IP address.

Yet, the architecture has a grey-listing effect, which acts as an additional filter. This happens because, when a source MTA is not yet classified (as to reject or to accept) at the central common *Knowledge* component, the *Proxy* instructs the MTA to return a temporary failure, to delay the reception until the *Consultant* assesses and classifies the required information. The delay time depends on the frequency set for the *Consultant* to run, plus the assessment duration, and the settings of the sender MTA for retry attempts.

In Fig. 2a we can understand why this grey-listing effect is effective to filter *fraud* spam, as most connections are from IP addresses that do not host real MTA servers and so they do not retry, ending with just one single connection established from that source IP address. We can also conclude, from Fig. 2b, that most legitimate messages (*light*) come from a low amount of new IP addresses[7], and that spam (*dark*) come from a large number of different IP addresses. This is consistent with spam being sent from zombie computers of spam botnets.

---

[6] The first couple of days, and the last day of the recorded 66 days of data are incomplete.

[7] Should be even less than the amount shown, as these numbers include – most certainly – IP addresses that are in fact source of spam but were not in any of the DNS lists used.

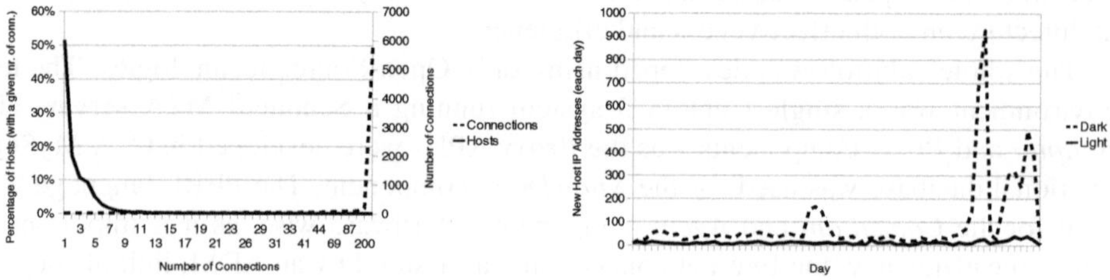

(a) Connections by Host (b) New Hosts per Day

Fig. 2. Distribution of Host Connections

As it can be seen in Fig. 3, one of the DNS black lists (Spamhaus), used by the single implemented *Agent*, was able to identify 5,909 IP addresses as a source of spam from the 6,793 distinct source IP addresses from which connections were established to the testing MTA used. This accounts for over 85% of the sources.

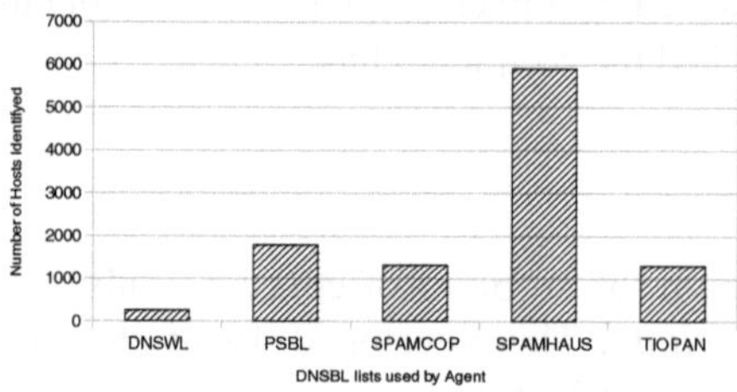

Fig. 3. DNS Lists Effectiveness

As most spam in the *fraud* category has faked mail addresses, adding an *Agent* for SPF [7] validation to the solution would guarantee an even higher effectiveness of the solution.

Considering that over 50% of *fraud* spam is caught by the grey-listing effect, that a single DNS black list can detect 85% of the remaining messages, and that the others DNS black lists can increase this detection ratio, then it is possible to expect a virtual full detection of *fraud* spam just by adding an SPF filtering *Agent*.

False positives (*FP*) are not an issue, up to this point, even with SPF filtering added. Those *FP* that might result from DNS black lists can be reversed by the usage of white lists. The *FP* resulting from SPF and grey-listing are due to poor server configurations, and those systems administrators should properly configure their systems following the RFC directives. Besides, *FP* are always properly addressed, meaning that the sender will always be notified of a rejection, allowing for an alternative form of communication, or by correcting the system configuration.

5.2 Efficiency

The architecture allows spam filtering to occur asynchronously, and this is important to the solution efficiency. Not only it relieves the MTA server from a longer waiting period of validations – spam filtering rules and associated procedures – as it only validates one single instance of each data across all MTA in the ecosystem, being it from several instances at the same MTA or from different MTA.

As the *Proxy* component only makes a decision to accept, reject or temporarily fail a connection based on a simple common local knowledge (as the source MTA IP address, or senders domain and IP as in SPF), the MTA server takes little time and effort to make a decision. This can be seen in Fig. 4a, where almost 50% of the connections took less than one second to process, and almost 95% took less than two seconds.

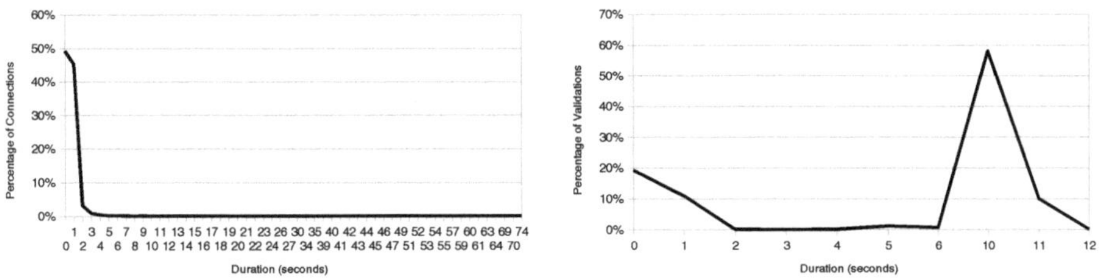

(a) Proxy Performance (b) Consultant Performance

Fig. 4. Performance

On the other hand, the *Consultant* component, as it uses *Agents* that can have complex filtering procedures, or that check external services subject to delays, may take a lot longer to classify a source as spam or ham[8], as it can be seen in Fig. 4b, where almost 60% of the DNS list verifications took 10 seconds to conclude.

So, considering a monolithic solution with a synchronous decision process, where the MTA would need to wait for the filtering process, just to consult DNS black lists would increase each MTA connection duration from an average below one second to one almost 8 seconds long, as it can be seen in Table 3.

Table 3. Processing Times (in seconds)

Component	Minimum	Average	Maximum
Proxy	0	0.7473	74
Consultant	0	7.1089	12

6 Conclusions and Further Developments

It came as a bit of a good surprise that this architecture alone, without any help from a filtering *Agent*, could be responsible for filtering most messages in the *fraud* category. The temporary fail, imposed when no data is available to make an informed decision,

[8] Ham is said to be the opposite of spam, which is to say a wanted message.

has a grey-listing effect that prevents most spam originated from botnets of zombies computers, as most only tries once to send the spam.

The IP addresses DNS lists (both black and white) filtering *Agent*, the only one implemented in the proof-of-concept, was able to identify over 85% of the IP source addresses as being a source of spam. An implementation of an SPF filtering *Agent* would most certainly fill the gap for messages in the *fraud* category.

On the downside, we have a more complex solution due to its many disperse components. Thus, faults due to communication failures are more prone to occur, and security constitutes a major concern. These issues must be addressed with extreme care, as the reliability of the solution depends largely on them.

The final conclusion is that the architecture fulfilled the objectives. It greatly improves spam filtering processing performance, and has an impressive effectiveness record, even with so scarce development of *Agent* components, dubbed as filtering rules and procedures. With a good set of filtering *Agents*, it can achieve an excellent record at spam identification. On top of this, we can run the solution in heterogeneous systems, sharing efforts and information, and scale the distributed system to match the traffic volume of mail messages.

Further developments most certainly will include the development of new *Agents*, such as SPF filtering and trap mail addresses. Yet, the logs recorded will allow a more detailed spam profiling, and this analysis should be used to develop other *Agents* that will increase effectiveness without any loss of efficiency. Also, this server-side filtering can be combined with a client-side filtering that can contribute to the common central *Knowledge*, and with other forms of filtering.

Acknowledgements. I would like express my sincere thanks to professor Gracinda Carvalho and professor Paulo Pereira for all the help in reviewing, guidance, suggestions and incentive provided, whose supervision made this project possible. I also would like to thank my wife Graça, and my parents, for the support, patience and comprehension on my absence of mind and spirit for the time this project lasted.

References

1. Symantec: Internet Security Threat Report, 2011 Trends, vol. 17, p. 31. Symantec Corporation (April 2012)
2. Apache: Spamassassin, `http://en.wikipedia.org/wiki/SpamAssassin`
3. Ma, W., Tran, D., Sharma, D.: On extendable software architecture for spam email filtering. IAENG International Journal of Computer Science 34(1) (2007)
4. Cottereau, L.: A peer-to-peer architecture for collaborative spam filtering. Master's thesis, University of Dublin (2002)
5. Yih, W., McCann, R., Kocz, A.: Improving spam filtering by detecting gray mail. In: CEAS 2007 - Fourth Conference on Email and Anti-Spam. Microsoft (2007)
6. Klensin, J.: Simple mail transfer protocol. RFC 5321 (October 2008)
7. Kitterman, S.: Sender policy framework (SPF) authentication failure reporting using the abuse reporting format. RFC 6652 (June 2012)

Multi-Agent Systems Meet GPU: Deploying Agent-Based Architectures on Graphics Processors

Roman Pavlov and Jörg P. Müller

Clausthal University of Technology, Clausthal-Zellerfeld, Germany
{roman.pavlov,joerg.mueller}@tu-clausthal.de

Abstract. Even given today's rich hardware platforms, computation-intensive algorithms and applications, such as large-scale simulations, are still challenging to run with acceptable response times. One way to increase the performance of these algorithms and applications is by using the computing power of Graphics Processing Units (GPU). However, effectively mapping distributed software models to GPU is a non-trivial endeavor. In this paper, we investigate ways of improving execution performance of multi-agent systems (MAS) models by means of relevant task allocation mechanisms, which are suitable for GPU execution. Several task allocation architecture variants for MAS using GPU are identified and their properties analyzed. In particular, we study three cases: Agents and their runtime environment can be (i) completely on the host (CPU); (ii) partly on host and device (GPU); (iii) completely on the device. For each of these architecture variants, we propose task allocation models that take GPU restrictions into account.

Keywords: Multi-Agent Systems, GPGPU, CUDA.

1 Introduction

Today's software-intensive networked applications, such as e.g. microscopic traffic simulations [1] cause increasing demand for computational resources. Moreover, Ubiquitous Artificial Intelligence (UAI, [5]) applications such as decentralized energy systems control [2] or cooperative traffic management [3], are data-intensive: the amount of data steadily increases, and more powerful algorithms and more computational units are needed, while paying attention to better performance per Watt.

A state-of-the-art approach to this challenge is to make use of recent progress in parallel and distributed computing. Distributed computing helps divide tasks across different computational units. If well coordinated, it can lead to improved performance and scalability performance improvement, as it provides a possibility to decrease computational time by using multiple machines. However, it trades off computation with communication overhead, thus Amdahl's Law [4] and the specifics of the application at hand need to be taken into account.

Multi-Agent Systems (MAS) [8] are a paradigm, which is often used for UAI modeling and exploits both distributed and parallel computing. A MAS represents a system as a set of interacting agents - software units that are able to execute tasks for

L.M. Camarinha-Matos, S. Tomic, and P. Graça (Eds.): DoCEIS 2013, IFIP AICT 394, pp. 115–122, 2013.

reaching local (or system) goals. Agents co-exist in the environment, and they communicate with each other to exchange information and coordinate task execution. MAS can be applied to find solutions for complex problems by decoupling them into less complex subproblems and providing coordinated solution frameworks such as the contract net protocol [25] or simulated trading [28]. MAS are of particular interest when viewed as a modeling and simulation paradigm [27]. The MAS metaphor enables fine-grained modeling of systems with local preferences, motivations, and capabilities, including issues of interaction, coordination, and cooperation. This is of use in many large-scale networked simulation tasks, such as traffic simulation [6] or physical, ambient intelligence simulations. However, a big challenge for agent-based simulation is that they are highly computation-intensive, which to date limits their scalability compared to e.g. approaches based on system dynamics.

Thus, a key objective of this work is to make a contribution to more scalable, high-performing agent-based simulation. In this paper, we focus on a specific aspect of MAS technology, i.e. on the question of how tasks in a MAS are assigned to agents.

Typically, task allocation and scheduling in MAS is achieved by dedicated protocols for cooperative problem solving, such as the contract net protocol [25], which defines how problems are decomposed, assigned to agents, and scheduled for execution. Our hypothesis is that we can exploit information related to task decomposition, allocation, and scheduling contained in MAS models for efficiently mapping these models to models (code) executable on a parallel (e.g., graphics) hardware.

Modern computers are highly parallel. Their Central Processing Units (CPU) have more than one core. In addition, most of the modern computers have Graphics Processing Units (GPU) that can be used as a coprocessor for computational tasks. This becomes possible because of frameworks such as CUDA/OpenCL, which support General-Purpose Computing on Graphics Processing Units (GPGPU) [7].

This paper introduces an approach to tasks scheduling in MAS; our approach couples MAS technology with GPGPU programming. We present three types of task scheduling models for MAS that uses GPU for non-graphical tasks. Our goal is to establish a correspondence between agents (which plan and execute computational tasks) and massively parallel, GPGPU-capable computational resources.

The structure of this paper is as follows: Section 2 describes the relationship to Internet of Things; Section 3 outlines the current state-of-the-art; Section 4 shows the research contribution and innovation of this work. In Section 5 we critically discuss the results; Section 6 gives conclusions and directions for future work.

2 Relationship to Internet of Things

Internet technology is becoming ubiquitous. This concerns the service layer as well as the network layer. At the same time, virtualization is used to provide network users with a homogeneous interface to ubiquitous, decentralized services. Cloud computing is one example of this trend. MAS is a suitable metaphor to describe decentralized environments and applications where large numbers of semi-autonomous, self-interested or cooperative entities (services, objects, devices) interact and solve problems by collaboration and cooperation. Agents are considered as Smart Objects [24] that have their own behavior. MAS have been successfully applied to a range of problems, e.g., simulations (see e.g. [9]). When it comes to simulation, MAS is a very interesting paradigm for the implementation and simulation of Ambient Intelligence scenarios. Here is the link to the Internet of Things: agents can provide semantics, behavior, and autonomy to smart objects created and described with Internet of Things technology. To deal with the complexity and scalability issues that subsist in Internet of Things scenarios, we study mapping MAS models to scalable, massively parallel execution environment and hardware platforms – in this context, task allocation in MAS plays an important role, finally, in performance of the system.

This article contributes to the improvement of task scheduling for MAS based on GPGPU. Our goal is to achieve improved performance and scalability.

3 State-of-the-Art

Agent-Oriented Programming [26] (including agent-based modeling simulation [27]) is a fairly young thread in software development, which extends Object-Oriented modeling / programming by concepts such as beliefs, goals and intentions [5].

Numerous methodologies for MAS development have been proposed (see [11] for an overview). FIPA [10] provides a set of standardized specifications for MAS runtime architecture and interaction protocols. The reason for it is that different problems require specific approaches. GPGPU is represented by a set of technologies, including CUDA [12], OpenCL [13], and DirectCompute [14]. In our work, we use CUDA by NVIDIA. OpenCL is based on similar concepts, so we expect that a solution for CUDA can be rather easily ported to OpenCL.

In [15] the authors present the use of GPU technology for multiagent simulation. They demonstrate a performance improvement of their Multi-Agent Simulator by using a GPU to execute agent-based models, but they don't pay much attention to the agent environment.

In [16] an adaptation of the FIPA standard to GPU is proposed with an example of crowd simulation. It is a continuation of earlier research [17] of the authors, where they compared JADE (Java Agent DEvelopment framework) and GPGPU, while operating with containers as in JADE. Authors claim about the necessity of standardization for agent creation on GPU.

In summary, there is a tendency in attempts to use GPU for agents, and as new GPUs are released, new devices provide more computational resources. In our research we pay great attention to agent communication, collaboration and their environment.

4 Research Contribution and Innovation

A common GPGPU application consists of two parts: host code and device code. Host code is executed by CPU, and device code is executed by GPU. Modern GPUs have a set of Streaming Multiprocessors (SM). Each SM represents a Single Instruction

Multiple Threads (SIMT) architecture [18]. There is a special "task manager" for every SM that groups threads, so they are finally executed in a Single Instruction Multiple Data (SIMD) manner. All threads are organized into blocks, with blocks belonging to grid(s) [19].

In this paper, we consider three architecture variants to implement MASs that exploit GPU. The three variants are illustrated in Figure 1.

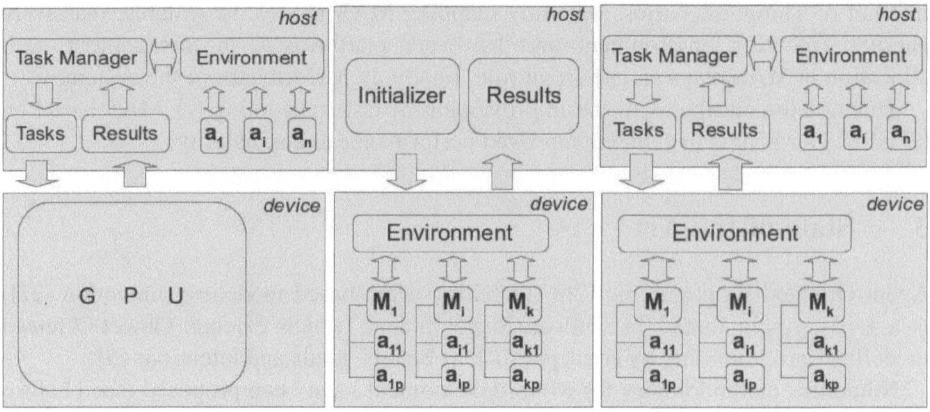

Fig. 1. Architecture variants: a) Agents-on-Host (AoH); b) Agents-on-Device (AoD); c) Hybrid (HA)

The Agents-on-Host (AoH) model shown in Fig. 1a) consists of three top-level components: *environment*, *agents* and *task manager*. We consider the individual components in details. On the host part of the system, agents (denoted by $\mathbf{a_i}$) can communicate with each other only by means of the environment, not directly with each other. An environment here holds all information about the current system state.

The next component is Task Manager (TM) consisting of two modules: Tasks and Results. Module "Tasks" serves for GPU tasks preparation and GPU resource allocation; it also sends data to GPU and calls respective kernel functions. At the beginning, agents – which have their own tasks – make requests for computational resources to accelerate their tasks execution. Then TM starts the GPU task preparation phase, during which data obtained from agents are organized in a way that the kernel function can process them. There are numerous copies of the kernel function, each working on different data. This complies to the needs of similar agents, when they have analogous tasks. The next step is to allocate resources on the GPU and copy the data there. Then computation on the GPU is ready to start. TM sets a number of threads and number of blocks that will be executed on GPU. It means that the total number of copies of our kernel function equals the total number of GPU threads. When execution is completed, the "Results" module takes control. This module finally synchronizes GPU threads and solutions gathered from the kernel functions back to the host. Finally, the TM provides the environment with all these solutions.

This model can be mapped to more than one machine by uniting local environments of model instances. So we will get a global environment, where agents placed on one machine can ask for computational resources on others. (Here it should be noted that characteristics of the network connection and its bandwidth need to be taken into account while evaluating the system performance).

Another model is Agents on the Device (Fig. 1b, AoD), which includes environment on GPU. Here within a GPU we consider its global memory like an environment, and kernel-function call concurrency is considerably helpful here.

The host part has two modules: *Initializer* and *Results*. *Initializer* has initial data and it also allocates GPU resources. Then it copies data to the device. It is also possible to dynamically allocate memory on GPU, but it takes special attention while doing so. For example, as we dynamically allocate GPU global memory - it is the slowest on GPU - we should evaluate the necessary amount of memory and check whether it meets the default size that we can allocate. Otherwise, it should be marked on the host before executing on the device. Another aspect of dynamic memory allocation is the expensiveness of this operation. So a good practice is to allocate memory and then just reuse it by overwriting data. The *Results* module returns the results after computation or simulation on GPU. Depending on next actions it either de-allocates GPU resources and/or passes execution to Initializer.

Components on the device are the environment, managers and agents. Every manager works within single multiprocessor and has access to the environment (here GPU global memory plays this role). As every multiprocessor has its own shared memory, which is a fast memory type on GPU, the manager is going to work with blocks. As threads running in a block can interconnect with each other and have access to shared memory, we can consider one separate thread in a block like a manager which can take part in tasks assignment for the rest threads in this block. But it makes other threads in a block waiting. After tasks assignment, this "manager" thread can also take part in computations. The main role of this manager is to synchronize threads in current block, copy data from global memory to shared one (when necessary). The amount of shared memory is limited and it should be properly used for communications between threads in this block.

After finishing all tasks on the GPU, we synchronize all GPU threads and copy data back to host. And it is Results-module that gets final results (solutions).

As an advantage, by using MAS conception, threads can be considered like agents, and every block has its manager-thread responsible for synchronization in current block and can copy necessary data to faster shared memory.

However, there are some bottlenecks in this model. As we check threads in blocks during execution and conditional expressions (if...then) are evaluated, it influences on GPU's scheduler. It means that our threads for which these conditions are true will be executed, and others have to wait wasting time by standing idle. Also, synchronization is an expensive operation, so we use it either at the end of block execution or during threads communication and using shared memory (to protect variables in shared memory while reading/writing by threads, we use atomic functions).

And the third model – the Hybrid architecture (HA, Fig. 1c) incorporates and integrates both AoH and AoD. In the host part we have also the environment (global, which can be connected or united with remote machines), agents, and the Task Manager TM. Here agents, like in AoH model, make requests for GPU computational resources. Then, the environment passes all requests to TM. TM can act either concurrently or sequentially. It depends on the number and type of the tasks, e.g., TM can organize tasks by agent types. So the module Tasks sets final input representations for the GPU (this module is a part of TM and it also has a concurrent nature).

A device part represents the AoD model: GPU environment, manager-agents and agents (inside blocks). They act in the same manner: managers are responsible for control actions within their block, including synchronization and shared memory use. When a kernel-function completes its execution, module Results synchronizes tasks from the kernel-function and sends results back to the environment with TM help.

So our kernel-functions work independently and solve agents' tasks according to the type of agents or their tasks. TM also plays a role of load-balancer, because it should maximize the usage of both CPU and GPU to get the best performance.

5 Discussion and Critical View

We presented three types of task allocation architectures for MAS that use GPU as an underlying execution hardware. Like the authors in [20], we do not assume a strong definition of an agent, but rather take a "useful-first" approach. Conception of MAS helps to abstract from implementation details, but needs to be adapted with respect to the main restrictions of GPU. We decided to analyze the conceptual model and then implement a prototype.

The AoH model, where agents request their environment for computational resources, can be used for systems with a small number of agents. But agents can be the complex ones. Every agent should have its own tasks (which is, for example, an independent subproblem). In addition, the Task Manager ™ uses several criteria for grouping tasks, e.g. agent type, task type. For instance, if our agents are working with matrices, TM can vary available resources for concrete agents depending on the type of operation (e.g., inversion) and the amount of data.

As modern CPU is also multi-core, and we exploit this parallelism. A TM that operates with kernel-function calls puts them in separate streams and execution is asynchronous. This helps to decrease the amount of wasted computational time, but it takes a thorough analysis of the granularity: number of kernels executing in different streams.

The AoD model – with agents on GPU – suits for such tasks that could be divided into smaller subtasks that can be processed independently. For example, we should find a global minimum of a function in a space of values. We divide our space into subspaces and look for local minima. Here agents communicate when they found local minima and check whether it is less than the current minimum for all agents. After checking the whole space of values, agents will find the global minimum. To accelerate this process, a heuristic approach should be used.

The hybrid model, which is a mixture of AoH and AoD, is the most flexible among these models. Firstly, it can be scaled in two directions: locally and globally. Local scaling includes adding more GPUs to the machine and create a united GPU environment (so we have more GPU computational resources and GPU memory altogether). Global scaling implies extending environment by connecting machine environments with each other and organizing a common space. Technologies such as GPUDirect [21], rCUDA [22], OpenMPI [23] help to implement such environments.

6 Conclusions and Future Work

In this article we presented three types of task allocation architectures for MAS which use GPU. We discussed advantages and disadvantages of every architectural type. Also we mentioned possible bottlenecks. Based on the discussion, we believe that the third (hybrid) type of model is the most promising: it is a general architecture that is aimed at performance improvement of the whole system. Moreover, there is a potential for scaling into two dimensions: local and global.

The main long-term goal of our research, of which this paper is a part, is to create a methodology of multi-agent system development with the help of general purpose computing on graphics processing units. At the end it will be realized as a framework or simulation tool. As there is a wide range of MAS application, we focus on traffic simulations and combinatorial optimization problems. Future versions could be extended to other application domains. Currently there are lots of tools for simulation, even agent-based, but the problem is that within a single machine it is difficult to achieve high performance.

Next steps are to implement all these models and create corresponding prototypes. So we can analyze performance empirically. Firstly, we will consider all models within a context of single machine, and find out the performance and behavior of the system. Then we will investigate ways of scaling, which we mentioned above.

Acknowledgments. Special thanks to Dr. Maksims Fiosins at TU Clausthal for his invaluable help.

References

1. Blatnig, S.: Microscopic Traffic Simulation with Intelligent Agents: Simulation of Human Driving Behaviour. VDM (2009)
2. Rehtanz, C.: Autonomous Systems and Intelligent Agents in Power System Control and Operation. Springer (2003)
3. Adler, J.L., et al.: A multi-agent approach to cooperative traffic management and route guidance. Transportation Research, Part B 39, 297–318 (2004)
4. Amdahl, G.: Validity of the Single Processor Approach to Achieving Large-Scale Computing Capabilities. In: AFIPS Con. Proc., vol. (30), pp. 483–485 (1967)
5. Krumm, J.: Advances in Ubiquitous Computing. Chapman& Hall/CRC (2009)
6. Fiosins, M., Fiosina, J., Müller, J.P., Görmer, J.: Agent-Based Integrated Decision Making for Autonomous Vehicles in Urban Traffic. In: PAAMS 2011, pp. 173–178 (2011)

7. GPGPU, `http://www.gpgpu.org/about`
8. Wooldridge, M.: An Introduction to Multi-Agent Systems, 2nd edn. John Wiley & Sons (2009)
9. Ehmke, J.F., Fiosins, M., Görmer, J., Schmidt, D., Schumacher, H., Tchouankem, H.: Decision Support for Dynamic City Traffic Management Using Vehicular Communication. In: Proc. of SIMULTECH 2011, pp. 327–332. SciTePress Digital Lib. (2011)
10. The Foundation for Intelligent Physical Agents, `http://www.fipa.org`
11. Dastani, M.: Programming MAS. In: 5th Int. Work., ProMAS 2007, USA (2007)
12. What is CUDA? `http://blogs.nvidia.com/2012/09/what-is-cuda-2/`
13. OpenCL, `http://www.khronos.org/opencl`
14. Direct Compute, `https://developer.nvidia.com/directcompute`
15. Laville, G., Mazouzi, K., Lang, C., Marilleau, N., Philippe, L.: Using GPU for Multi-agent Multi-scale Simulations. In: Omatu, S., Paz Santana, J.F., González, S.R., Molina, J.M., Bernardos, A.M., Rodríguez, J.M.C. (eds.) Distributed Computing and Artificial Intelligence. AISC, vol. 151, pp. 197–204. Springer, Heidelberg (2012)
16. dos Santos, L.G.O., Gonzales Clua, E.W., Bernardini, F.C.: A Parallel Fipa Architecture Based on GPU for Games and Real Time Simulations. In: Herrlich, M., Malaka, R., Masuch, M. (eds.) ICEC 2012. LNCS, vol. 7522, pp. 306–317. Springer, Heidelberg (2012)
17. dos Santos, L.G.O., et al.: Mapping Multi-Agent Systems Based on FIPA Specification to GPU Architectures. In: VIDEOJOGOS 2010 (2010)
18. SIMT, `http://docs.nvidia.com/cuda/cuda-c-programming-guide/index.html#simt-architecture`
19. `http://docs.nvidia.com/cuda/cuda-c-programming-guide/index.html#programming-model`
20. Johnson, T., Rankin, J.: Parallel Agent systems on a GPU for use with Simulations and Games. In: 12th WSEAS ACS 2012 (2012)
21. GPUDirect, `http://developer.nvidia.com/gpudirect`
22. rCUDA, `http://www.rcuda.net`
23. OpenMPI, `http://www.open-mpi.org`
24. Kallmann, M., Thalmann, D.: Modeling Objects for Interaction Tasks. In: 9th EGCAS, Lisbon, Portugal, pp. 73–86 (1998)
25. Smith, R.G.: The Contract Net Protocol: High-Level Communication and Control in a Distributed Problem Solver. IEEE Trans. Comput. 29(12), 1104–1113 (1980)
26. Shoham, Y.: Agent-oriented programming. Artif. Intell. 60(1), 51–92 (1993)
27. Klügl, F., Bazzan, A.L.C.: Agent-Based Modeling and Simulation. AI Magazine 33(3), 29–40 (2012)
28. Fischer, K., Müller, J.P., Pischel, M.: Cooperative Transportation Scheduling: an Application Domain for DAI. Journal of Applied Artificial Intelligence 10, 1–33 (1996)

Part VI
Computational Systems Applications

The Architecture of Coupon-Based, Semi-off-Line, Anonymous Micropayment System for Internet of Things

Daniel Wilusz and Jarogniew Rykowski

Department of Information Technology, the Poznań University of Economics
Mansfelda 4, 60-854, Poznań, Poland
{wilusz,rykowski}@kti.ue.poznan.pl

Abstract. In the Internet of Things a lot of business opportunities may be identified. The devices in IoT may create ad-hoc temporary networks to provide services or share some resources. Such services are characterized by a great economical potential, especially while provided at mass-scale and for incidental users. However, the development of paid services or resources in IoT is hampered by relatively big transaction costs of payment operations. To deal with this problem, we propose a novel architecture of coupon-based, semi-off-line, anonymous micropayment system to enable transactions in the scope of Internet of Things. User anonymity and security is assured by the usage of standard cryptographic techniques together with novel architectural design of the payment processes. Utilization of a hash function allows generating and verifying electronic coins in computationally efficient way, so as to be executed even in hardware- and software-restricted environment such as Internet of Things.

Keywords: micropayments, e-money, Internet of Things, Future Internet.

1 Introduction

The Internet of Things (IoT) is the rapidly developing phenomenon, which undoubtedly will influence the society and economy. Originally it was perceived as the intelligent network linking objects, information and people to allow remote coordination of resources by humans and machines [3]. However, nowadays IoT is not only limited to electronic identification of objects, but it is defined as technology integrating physical objects with information network, where these objects may act as active participants in business processes [6].

Multiple applications of Internet of Things have been identified in such economy areas, as manufacturing, supply chains, energy, healthcare, automotive industry and insurance, to mention a few [6]. Moreover, the expansion of Internet of Things outside the internal infrastructures of companies is expected. Such situation should affect the economy by creating human-to-machine (H2M) and machine-to-machine (M2M) markets [14]. In these markets, the machines will match business partners and conclude contracts, according to the preferences of their owners.

In the Internet of Things markets the devices will cooperate by creating temporary networks to exchange data. The devices are expected to provide services on the mass

L.M. Camarinha-Matos, S. Tomic, and P. Graça (Eds.): DoCEIS 2013, IFIP AICT 394, pp. 125–132, 2013.

scale, but the business relations will be temporary, and often formed on an ad-hoc basis. The devices in Future Internet will be used locally, but at mass scale and in many places. Similarly, the particular payments will be made in micro scale, but all transactions will have the global character. One of the features of the services provided by devices from IoT network is activity-based micro-pricing, which means that clients are charged by exact time, amount and sort of services they consume [14]. Therefore, to make the provision of ad-hoc services profitable, an efficient micropayment system is required.

The micropayment system suitable for the transactions in Internet of Things should observe the following requirements. First, it should be possible to implement the system worldwide, by avoiding intermediaries (e.g., a clearing house). Second, the financial institution should bear possible minimal computation costs. Third, a user should be able to make a payment off-line in order to decrease the costs of a network connection, which in particular cases may exceed the value of purchased goods or services (e.g., mobile data transfer under roaming circumstances). Fourth, the vendor should contact a financial institution as seldom as possible. Fifth, the anonymity of the user should be assured by a prevention of payment tracking. Sixth, the security of transaction should be assured by the encryption and the prevention of double spending or forgery.

To deal with the above problems, an anonymous, semi-offline, micropayment system for transactions in Internet of Things is proposed in this paper. The contribution of the proposed solution in the field of anonymous micropayment systems and Internet of Things is described in Section 2. The state of the art of micropayment systems is reviewed in Section 3. In Section 4 overall architecture and main protocols of micropayment system suitable for micro-transactions in Internet of Things are proposed. Finally, Section 5 concludes the paper.

2 Contribution of Proposed System to the Internet of Things

In the paper a new architecture of micropayment system adjusted to the requirements of business transactions in Internet of Things (IoT) is presented. The proposed micropayment system meets the needs of human-to-machine and machine-to-machine markets, where efficient processing of many small-amount transactions is required. Although plenty of anonymous micropayment systems were proposed in the literature, the state of the art reveals that none of them is suitable for transactions in Internet of Things. However, the coupon-based systems, originally proposed to handle pay per view payments on the websites, seem to fit the gap.

A good starting point to realize the idea presented in the paper is related with a micro-payments architecture originally proposed by Wilusz and Rykowski [15]. However, in order to fulfill above-mentioned requirements of IoT transactions, this architecture was significantly modified. First, the clearing house became an unnecessary intermediary and in our proposition a financial institution is sufficient to handle both coin issue and clearing. Second, the coin registration was changed by putting this activity into a vendor task, so that each coin is registered during its first use.

Moreover, this solution strengthened user anonymity, by an elimination of a possibility of user (e.g., his personal device) tracking. Third, the coin blocking procedure was improved by setting the final date for blocking a coin by a vendor. Fourth, the user and vendor specify together in anonymous way the account to be credited, which, by an application of cryptographic methods, improves security of the proposed solution.

According to the long tail concept [2], the Internet of Things has a huge economic potential of services or resources being sold for micro-price in huge amounts to plenty of incidental clients. The potential areas of application of micropayments in IoT are as follows: transportation services (pay per exact distance or number of stops), services concerning augmented reality (virtual guides in a museum or multimedia concerning cultural heritage objects), a reduction of external effects in public spaces (payments for using a car in city centers), and services in public spaces (vending machines, toilets or waiting rooms), to mention a few. However, because of the lack of efficient payment method, high transaction costs strongly limit IoT market. The proposed architecture will allow implementing an efficient micropayment system suitable for payments in IoT. The system, once become de facto standard, will cause the IoT market to flourish on international scale.

3 State of the Art

There were proposed numerous micropayments schemes in the literature. However, to our best knowledge none of them is suitable for the payments in the IoT environment.

Most of already implemented micropayment systems are based on the prepaid account and aggregation of particular payments by single payment system operator (e.g., Amazon Flexible Payment Service [1] or GeldKarte [5]). In this approach vendors and clients are bound to a single payment service provider, which consolidates clients' micropayments into macropayments made to the sellers at the end of specified period. The micropayments aggregation by one entity is not proper for IoT transactions, as purchases happen rather irregularly, incidentally and should be possible worldwide. Moreover, this solution lacks anonymity, as service provider can track each payment [13].

Another sort of micropayment schemes applies probability approach to make micropayments effective (e.g., schemes proposed by Micali and Rivest [8] or Lipton and Ostrovsky [10]). The foundation of these systems is based on an assumption that users make a lot of purchases and vendors have huge volume in sales. Under this assumption, it is enough to provide a coin, which with probability of s will be payable amount of $1/s$ to assure that users will pay and vendors will get amount close to their expenses and earnings. However, there are following drawbacks of such a solution. First, no anonymity is assured, as each user is obliged to sign transaction details. Second, the probabilistic scheme has features of a wager (one may win or lose depending on result of some activity) rather than sale (getting goods or services for an amount of money). Third, both user and vendor are required to have PKI certificates.

The third group of micropayment schemes is coupon-based, which utilizes the idea of hash chains [7] (e.g. PayWord [12], Payeras-Capella et. al. scheme [11] or Wilusz and Rykowski architecture [15]). In coupon-based systems an electronic coin is represented by the last output of a hash function applied iteratively to some seed. This approach enables the electronic coin to be spent in fractions, by presenting the vendor a set of chain nodes. Although, the general scheme presented by Rivest [12] missed the anonymity and allowed user to exceed their bank-account limits, it was improved by Payeras-Capella et al. [11] by an application of Chaum's blind signature protocol [4]. However, in this scheme, a user needs to contact financial institution directly before the payment in order to produce merchant-specific hash chain. Moreover, the Authors do not mention what happens with unspent tokens of merchant specific hash chain, and it appears that unspent value is lost in favor of the financial institution. Wilusz and Rykowski [15] proposed improved, anonymous coupon based micropayment system. In their proposition the user contacts the financial institution only during coin issue phase. During the payment process only the vendor needs to stay on-line. The proposition of locking the coin by the vendor at the clearing house solved the problem of double spending, and in advance to the Payeras-Capella et al. [11] scheme, there is no need to generate merchant-specific coins. Moreover, unspent fraction of the coin may be used in further transactions, provided that the coin was unblocked by the merchant. However, there are still some drawbacks of this system. First, the presence of the clearing house increases transaction costs and hampers the worldwide expansion of the system. Second, there is a serious risk, that the vendor will not unblock the coin after the payment.

The decentralized micropayment systems such as Bitcoin [9] are becoming more popular. However the decentralized systems appear to be of limited trust, as there is no central authority controlling the supply of virtual currency. Moreover, the security of Bitcoin system is preserved under condition that the majority of Bitcoin network nodes are honest. The botnet attack poses a real risk of Bitcoins forgery or doublespending. The above mentioned traits of Bitcoin system cause, that this solution is not suitable for mass scale payments in the Internet of Things.

The analysis of the state of the art in micropayment systems indicate that there is no solution meeting all the requirements of transactions in the Internet of Things. However, the Wilusz and Rykowski solution [15] has a potential, after some modification, to meet the needs of H2M and M2M markets. In this paper, we propose some extensions and improvements to this approach, to be described in the next section.

4 Architecture of Micropayment System for Internet of Things

The architecture and protocols of a micropayment system for Internet of Things, presented in this section, are the results of significant improvements of the proposition of Wilusz and Rykowski [15]. The introduced changes enabled the system to meet the requirements of micro transactions on both H2M and M2M markets.

The proposed system is coupon-based, which means that the hash chain concept is used to produce divisible coin. Such a solution decreases computation costs of a financial institution, as it is enough to sign the coin once and further this coin may be spent in fractions. The anonymity is assured by an application of Chaum's blind signature scheme. In the result, a financial institution is not able to link the coin identifier with a user. Additionally, the vendor stands for an entity to register a coin at financial institution, which strengthens the user anonymity, as no IP tracking of end-user device is possible. The financial institution *F* generates the coin in return for money, so the coin bears obligation of *F* to accept it and pay the holder (e.g. by banking transfer). As a result there is no need to involve a clearing house in the payment process, as the financial institution, that issued a coin, takes care of coin clearing. What is more, a user needs an Internet connection only to get new coins from *F*. In Internet of Things, a user (a person or a device) contacts a vendor (a person or a device) by a short-range radio communication. Moreover, the vendor *V* uses the Internet, to contact *F*, only when blocking the coin or finishing the transaction. The idea of indexing and blocking of coins together with storing the last spent token in financial institution database, prevents the double spending issue. In the system a coin may be blocked only till specified date *T*. Additionally the user *U* specifies who should be credited by using vendor's certificate, which prevents any eavesdropper from gaining benefit.

Figure 1 presents the participants of the system and their interactions. In the next sections the coin generation protocol and payment protocol are described in details.

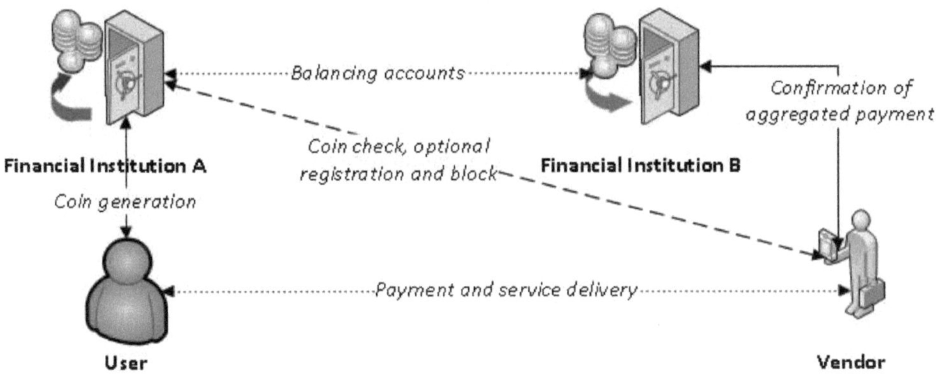

Fig. 1. Overall system architecture

4.1 Coin Generation Protocol

In the proposed system, the coins generated by a financial institution are of fixed value *M* and may be divided into *n* fractions (tokens). Each token has specified value equal to *M/n*. The divisibility of a coin is achieved by an application of hashchain concept, while the anonymity is assured by the use of Chaum's blind signature scheme. The coin generation protocol consists of three phases: blind coin generation, blind coin signing and a removal of a blind factor.

In the first phase, the user *U* generates a random seed *S*, which is kept in secret. Then, hash function *H()* is performed iteratively on the seed. Next, the user generates random blinding factor *r* and finds its reverse modulo *N* (where *N* is modulus of financial institution RSA public key). Finally, blinding factor *r* is encrypted with financial institution RSA public key e and multiplied modulo N by the coin identifier (last value of hash chain $H^n(S)$). As a result of this operation a blinded coin $H^n(S)r^e \ mod \ N$ is produced.

During the second phase, the user is authorized in the payment system and sends the blinded coin to the financial institution, which in turn debits user account and signs blinded coin with RSA decryption key d. Then, the signed blinded coin $H^n(S)^d r^{ed} \ mod \ N$ is returned to the user U.

In the last phase, the user removes the blinding factor. As the $r^{ed} \ mod \ N$ is equal to $r \ mod \ N$ it is sufficient to multiply the blinded coin by the r reverse modulo N ($r^{-1} \ mod \ N$), because the following congruence is met $rr^{-1} \equiv r^0 \equiv 1 \ (mod \ N)$ [4].

Coin generation protocol	
U→F:	user credentials, H^n (S) r^e mod N
F→U:	H^n $(S)^d$ r^{ed} mod N

4.2 Payment Protocol

At the beginning of a transaction, the user U and the vendor V both agree on the date T of finishing transaction (the final date till which the coin will be blocked at a financial institution). Additionally, the vendor provides to the user its certificate C containing credentials and bank account details. The user sends to the vendor the following items: the signed coin, encrypted concatenation of the last unspent token $H^{n-m}(S)$ (where $n>=m$) with T, and encrypted concatenation of $H^{n-m}(S)$ with C. This proves that U (the only entity knowing $H^{n-m}(S)$) authorizes vendor V identified by certificate C to block the coin identified by $H(S)^d mod \ N$ till the date T.

In the next step vendor V sends C, T, the coin $H^n(S)^d \ mod \ N$ and both concatenations $(H^{n-m}(S)||T)^e \ mod \ N$ and $(H^{n-m} \ (S)||C)^e \ mod \ N$ to the financial institution F. Then, the financial institution F uses its encryption key e, in order to obtain coin identifier $H^n(S)$ from signed coin $H(S)^d mod \ N$. Next, the financial institution F decrypts $(H^{n-m}(S)||T)^e mod \ N$ and $(H^{n-m}(S)||C)^e mod \ N$ with its decryption key d and retrieves $H^{n-m}(S)$. Then, the financial institution F checks if the coin was already registered. In the case of a new coin, the financial institution F checks if $H(H^{n-m}(S))$ *is* equal to $H^n(S)$. If this checking succeeded, the coin $H^n(S)$ is registered in the database by storing the coin $H^n(S)^d \ mod \ N$, coin identifier $H^n(S)$, and first token $H^{n-1}(S)$ (in this case $m=1$). If the coin was already registered, the financial institution F checks if the token $H^{n-m}(S)$ is the predecessor of the previously spent token $H^{n-m+1}(S)$ by verification of the following equation $H^{n-m+1}(S)=H(H^{n-m}(S))$. In the next step, the coin is blocked for the period T for the vendor V identified by the certificate C. Then, financial institution F sends to the vendor V the last spent token $H^{n-m}(S)$ and a position (n-m) of this token in the hashchain. The position of the token in hashchain determines the remaining value of the coin ($(n-m)M/n$).

After having validity and blocking of the coin confirmed by financial institution F, the vendor V starts charging the user U the micro prices by requesting tokens representing particular value. Then the user U provides the vendor V the token

$H^{n-m-q}(S)$ of value qM/n and the order number $n\text{-}m\text{-}q$ (where $n>=m+q$). The token is verified by the vendor V by checking if $H^q(H^{n-m-q}(S))= H^{n-m}(S)$. If the checking is positive, the goods or services may be provided. The request and validation of the tokens may be repeated during the consumption of services or goods purchase, in the case when the user U is charged on the pay per use (or unit) basis. For a next service use, the user U is requested for next token $H^{n-m-q-p}(S)$ of value pM/n. The vendor V provides proper token and this process continues till the end of the transaction.

At the end of the payment the vendor V sends to the financial institution F the coin $H^n(S)^d \bmod N$, last received token $H^{n-m-q-p}(S)$, and order number $n\text{-}m\text{-}q\text{-}p$ (where $n>= m+q+p$). The financial institution F locates the coin in its database and stores the last token. Then, the coin is unblocked and vendor's bank account may be credited. If the total value M of the coin is spent (in the case of $n\text{-}m\text{-}p\text{-}q = 0$), it is marked as invalid.

Payment protocol

V→U: T, C
U→V: $H^n(S)^d \bmod N$, $(H^{n-m}(S)||T)^e \bmod N$, $(H^{n-m}(S)||C)^e \bmod N$
V→F: T, C, $H^n(S)^d \bmod N$, $(H^{n-m}(S)||T)^e \bmod N$, $(H^{n-m}(S)||C)^e \bmod N$
F→V: $H^{n-m}(S)$, n-m
V→U: Request for tokens
U→V: $H^{n-m-q}(S)$, n-m-q
V→U: Request for tokens
U→V: $H^{n-m-q-p}(S)$, n-m-q-p
V→F: $H^n(S)^d \bmod N$, $H^{n-m-q-p}(S)$, n-m-q-p

5 Conclusions

In this paper, an anonymous semi-off-line micropayment system for transactions in Internet of Things is proposed. The improvements described in the text allowed simplifying the architecture, strengthening user anonymity and protocols security. The intermediary clearing house became unnecessary and was removed from the architecture, which caused a simplification of communication and transactional overheads of the proposed solution. The original approach of the vendor being the entity who registers the coin at a financial institution has been applied. This solution caused an increase of the user-anonymity level by disabling communication tracking possibilities. Moreover, the coin management in payment process enhanced security of the solution, for at least two reasons. First, encrypted concatenation of token with blocking date prevents the suspension of coin in financial institution database. Second, the concatenation of token with vendor's certificate ensures that the vendor gets the payment, because any malicious intervention in the protocol cannot change the beneficiary.

The proposed system meets the specific needs of financial transactions in Internet of Things by decreasing transaction costs concerning the payment process. What is more, proposed solution may be implemented on wide range of payment instruments concerning smartcards or smartphones. The payment protocol is easy to be executed by even small devices, and users are required only to formulate an intention of a payment. Additionally, a coin management system may be proposed, which would optimize the choice and strategies of coin spending with different vendors. We do hope that the proposed micropayment system will allow several new, commercial innovations to emerge within the scope of Internet of Things.

References

1. Amazon FPS Aggregated Payments Quick Start, https://payments.amazon.com/sdui/sdui/business?sn=devfps/aggregated
2. Anderson, C.: The Long Tail, http://www.wired.com/wired/archive/12.10/tail.html
3. Brock, D.L.: The Electronic Product Code (EPC) A Naming Scheme for Physical Objects, Auto-ID Center, http://www.autoidlabs.org/uploads/media/MIT-AUTOID-WH-002.pdf
4. Chaum, D.: Blind signature for untraceable payments. In: Chaum, D., Rivest, R., Sherman, A. (eds.) Advances in Cryptology: Proceedings of CRYPTO 1982, Santa Barbara, California, USA, August 23-25, 1982, pp. 199–203. Plenum Press, New York (1983)
5. Facts and figures on GeldKarte, http://www.geldkarte.de/_www/en/pub/geldkarte/press/facts_and_figures.php
6. Haller, S., Karnouskos, S., Schroth, C.: The Internet of Things in an Enterprise Context. In: Domingue, J., Fensel, D., Traverso, P. (eds.) FIS 2008. LNCS, vol. 5468, pp. 14–28. Springer, Heidelberg (2009)
7. Lamport, L.: Password Authentication with Insecure Communication. Communication of the ACM 24(11), 770–772 (1981)
8. Micali, S., Rivest, R.: Micropayments Revisited. In: Preneel, B. (ed.) CT-RSA 2002. LNCS, vol. 2271, pp. 149–163. Springer, Heidelberg (2002)
9. Nakamoto, S.: Bitcoin: A Peer-to-Peer Electronic Cash System, http://bitcoin.org/bitcoin.pdf
10. Lipton, R.J., Ostrovsky, R.: Micro-payments via efficient coin-flipping. In: Hirschfeld, R. (ed.) FC 1998. LNCS, vol. 1465, pp. 1–15. Springer, Heidelberg (1998)
11. Payeras-Capella, M., Ferrer-Gomila, J., Huguet-Rotger, L.: An Efficient Anonymous Scheme for Secure Micropayments. In: Cueva-Lovelle, J.M., González-Rodríguez, B.M., Labra Gayo, J.E., del Puerto Paule Ruiz, M., Aguilar, L.J. (eds.) ICWE 2003. LNCS, vol. 2722, pp. 80–83. Springer, Heidelberg (2003)
12. Rivest, R., Shamir, A.: PayWord and MicroMint: Two Simple Micropayment Schemes. In: Crispo, B. (ed.) Security Protocols 1996. LNCS, vol. 1189, pp. 69–87. Springer, Heidelberg (1997)
13. Sherif, M.H.: Protocols for secure electronic commerce. CRC Press, Boca Raton (2004)
14. Yamabe, T., Lehdonvirta, V., Ito, H., Soma, H., Kimura, H., Nakajima, T.: Activity-Based Micro-pricing: Realizing Sustainable Behavior Changes through Economic Incentives. In: Ploug, T., Hasle, P., Oinas-Kukkonen, H. (eds.) PERSUASIVE 2010. LNCS, vol. 6137, pp. 193–204. Springer, Heidelberg (2010)
15. Wilusz, D., Rykowski, J.: Requirements and general architecture of a payment system for Future Internet. Business Informatics (24), 91–103 (2012)

Mechanical Characterization of Ink-Jet Printed Ag Samples on Different Substrates

Dragana Z. Vasiljevic[1], Aleksandar B. Menicanin[2], and Ljiljana D. Zivanov[1]

[1] Faculty of Technical Sciences, University of Novi Sad, Novi Sad, Serbia
[2] Institute for Multidisciplinary Research, University of Belgrade, Belgrade, Serbia
{vdragana,lilaziv}@uns.ac.rs, aleksandar.menicanin@imsi.rs

Abstract. In this paper, the main activity was to investigate how different substrates, temperature of sintering and percentage of silver ink containing silver nanoparticles influence on Young's modulus and hardness of printed silver thin samples. Samples were prepared by low cost ink-jet printing technique using Dimatix Material Printer on polyimide flexible substrate and slide glass. Characterization of these samples was carried out by Nano Indenter using a three sided pyramidal (Berkovich) diamond tip. Measurement results show that the thickness of ink-jet printed silver layer varies for different percent of nanoparticles in silver ink. All measurements were done at same depth of indentation to avoid possibility of perforation of printed layer. The higher temperature of sintering and the higher percent of silver nanoparticles give the bigger Young's modulus and hardness of printed silver sample. This research provides very useful information about mechanical characterization of the silver layers on flexible substrates for printed-electronics.

Keywords: ink-jet printing, silver nanoparticle ink, nanoindentation.

1 Introduction

Ink-jet printing is a particularly attractive technique, for a direct write of patterns and the delivery of precise quantities of materials. It is desirable to fabricate onto polymeric or similar temperature sensitive substrates by solution-based printing process. Development of a solution-based process on a flexible substrate would allow roll-to-roll fabrication, which is an extremely inexpensive way to mass-produce circuits since it eliminates conventional photolithography and complex substrate processing including vapor phase deposition and etching. For these reasons, the ink-jet printing as a convenient and rapid processing technique to fabricate conductive lines has attracted great attention in recent years [1]. In the electronic industry, ink-jet printing tecnology has proved to be more flexible, low cost and more enviroment friendly [2] than the current manufacturing processes. This technology has shorted time of production and this result lower cost of production. The ink-jet printing technology has become an interesting aternative to the current manufacturing methods and has been integrated into the various phases of mass electronics production [3].

Nanomaterials have been of much benefit in improving process procedures in printable electronics, especially in connection with sintering process temperature and

L.M. Camarinha-Matos, S. Tomic, and P. Graça (Eds.): DoCEIS 2013, IFIP AICT 394, pp. 133–141, 2013.

sintering process duration. Silver ink, based on nanoparticles makes it possible to lower the sintering temperature and to shorten sintering time. Evaluation of the mechanical reliability of silver ink-jet printed structures is crucial in reliable electrical design. Some of the analysis of mechanical performance of silver ink-jet printed layer are presented in [4].

For a mechanical characterization of such printed samples nanoindentation has been established as an important tool for characterization on the submicron scale. Such a test is usually conducted using instrumented machines with which indenter load and indenter displacement can be continuously and simultaneously recorded during indenter loading and unloading [5]. The two mechanical properties measured most frequently using load and displacement sensing indentation techniques are the Young's modulus and the hardness [6]. The measured Young's modulus and hardness would be different, even for the same film, depending on the substrate materials because of the substrate effect on the measured properties [7].

In this paper, we present fabrication process of silver layer on different plastic/organic substrate and slide glass in ink-jet printing technique and their mechanical characterization. Relationship to internet of things and state of art/releated literature are shown in Section 2 and 3, respectively. In the Section 4 is present fabrication and characterization processes. Section 5 presents measurement results of mechanical characterization and the discussion. The reached conclusions of the obtained results follow finally in the Section 6.

2 Relationship to Internet of Nano-things

Over the internet, everything is connected through the online system to perform controlled devices and complex systems, for example power plants, industrial equipment, automotive and cellular industry, and even households. This leads to improvements in the quality of production and the saving of time, and all that is connected with the lower prices of products on commercial market.

The internet, there is a lot more data, all the time, growing at 50 percent a year, or more than doubling every two years, estimates IDC (International Data Corporation), a technology research firm. It's not just more streams of data, but entirely new ones. For example, there are now countless digital sensors worldwide in industrial equipment, automobiles, electrical meters and shipping crates. They can measure and communicate location, movement, vibration, temperature, humidity, even chemical changes in the air. Link these communicating sensors to computing intelligence and you see the rise of what is called the Internet of Things or the Industrial Internet [8].

Nanotechnology promises new solutions for many applications in the biomedical, industrial and military fields as well as in consumer and industrial goods. The interconnection of nanoscale devices with existing communication networks and ultimately the Internet defines a new networking paradigm as the Internet of Nano-Things [9]. Many application in microelectronics involve combinations of plastic and metal constituents. Due to the different properties of the materials involved (e.g. Young's modulus and hardness, thermal expansion coefficents, elestic constants yeild strenght) components can be subject to internal stresses during fabrication.

Ink-jet printing technology consist of complex micro- or nano-network of metal structures embedded within a plastic/organic substrate. The low sintering temperature (i.e. below 300°C) can be achieve using inks containing nanoparticles. This makes feasible the use of excelent electronic conductors such us silver or gold with different percentage participation in inks.

During the production of commponents, which is used in connecting things (the internet of things) could be used different materials. The choice of the materials depend on the purpose of the components and the conditions in which it will be used. For this reason, ti is very important to know all the characteristics of the used material. For example, if you want to make a RFID tag to be used in an environment which will significantly affect the material that we used in the preparation of tag, we need to know the materials' mechanical characteristics, like hardness, Young 's modulus, and scratch test. This leads to a better design of components, longer life and higher quality mode.

3 State of the Art / Related Literature

In a last twenty years, nanoindentation was introduced as a method for determining the modulus and hardness of materials by studying nanomechanical response as a function of penetration depth. During the years, this method has been constantly refined. Nanoscale characterization techniques are continuously challenged by the rapid progress in nanostructures and functional materials demanding higher resolutions and advanced measurement techniques for mechanical, chemical, electrical, and thermal characterization. This understanding will be valuable in supporting the impetus to harness multifunctionality of materials to realize nano- and micro-devices [10].

Nanoindentation is widely used to study the displacement of materials under specific applied loads to produce load–displacement curves. This will lead to the aim of presenting its contributions to progress in materials science [11]. One of the most fascinating things about nanoindentation is the number of different applications for which the technique is useful. One of these applications is: the utility of nanoindentation in the geological world, It relates some of the smallest experiments on mechanical properties that can be performed to some of the largest we know of: earthquakes. Other very important areas that have become popular recently are the accurate characterization of viscoelastic materials and polymers, polymers characterization as it relates to nano-imprinting and nano-forming, nanoindentation in the materials science of biomedicine [12].

4 Fabrication and Mechanical Characterization of Ink-Jet Printed Samples

In recent years, there has been tremendous interest in the development of printed electronics components as a means of achieving ultra-low-cost electronic circuits [13]. In the electronic industry, fabrication of conductive tracks is inevitable. Conventional photolithographic and electroless deposition techniques are widely adopted in the printing circuit board (PCB) for manufacturing its conductive circuits. However, this method is not only time consuming but also very complicated and expensive, because many processing steps are required to construct a layer of the circuit. Moreover, the electroplating and etching processes also produce large quantities of chemical waste. Therefore, there is an industrial need for direct digital printing to simplify the processes and to reduce manufacturing costs [1]. For good results it is very important to know all characteristic of materials which are used for ink-jet printing. Because of that we have tested mechanical characteristic of printed silver layers.

4.1 Fabrication of Ink-Jet Printed Sample

The samples for mechanical characterization were prepared by ink-jet printing technique by Dimatix Deposition Material Printer DMP-3000 [14], on polyimide flexible substrates with different thickness and slide glass, shown in Fig. 1.

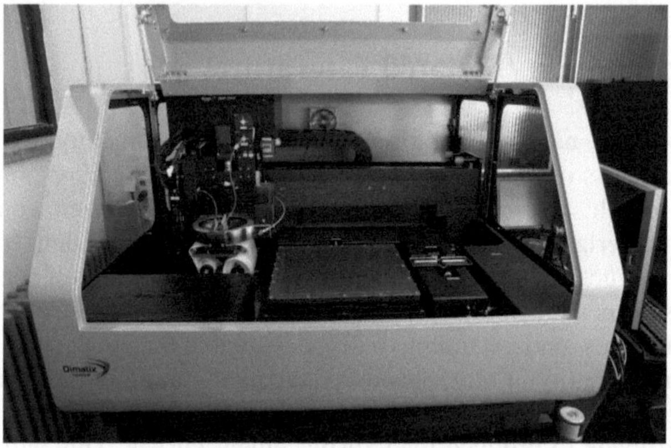

Fig. 1. Dimatix Deposition Material Printer DMP-3000

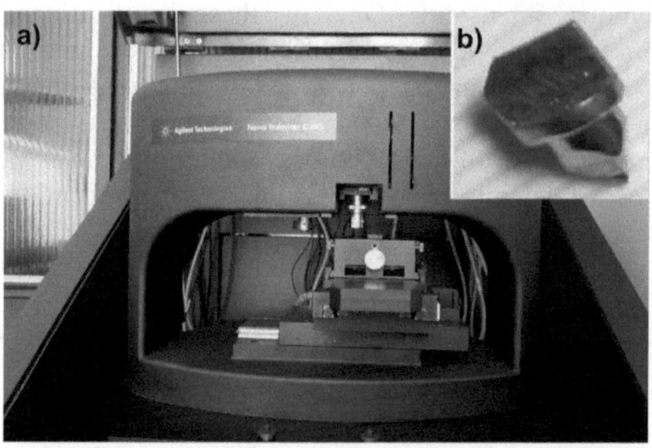

Fig. 2. a) Measurement system - Agilent Nano Indenter G200; b) Berkovich diamond tip

The thicknesses of polyimide film were 50 μm and 75 μm (manufacturer GTS Flexible Materials Ltd [15]) and thickness of slide glass was 1000 μm. As inks have used commercially available SunChemical silver nanoparticle inks with 20 wt% and

40 wt% of silver nanoparticles [16]. During the printing procedure has been used 16-nozzle piezzo cartridge with nozzle diameter of 25.4 μm. The distance between nozzles is 200 μm and drop spacing was 50 μm. After depositing silver nanoparticle ink on polyimide film and slide glass samples are sintered on air atmosphere at three different temperatures (200 °C, 225 °C and 250 °C) for 30 min. Printed samples, after sintering, are presented in Fig. 2.

4.2 Mechanical Characterization of Ink-Jet Printed Samples

After sintering process, characterization of this samples are carried out by measurement system with Agilent Nano Indenter G200 [17] using a three sided pyramidal (Berkovich) diamond tip. Measurement system and Bercovich tip are shown in Fig. 3. Berkovich tip is ideal for most testing purposes. It induces plasticity at very small loads which produces a meaningful measure of hardness. The Berkovich indenter tip has a large included angle which minimizes the influence of friction. Basic principle Nano Indenter G200 testing is employing a high-resolution actuator to force an indenter into a test surface and a high-resolution sensor to continuously measure the resulting penetration. As the indenter is driven into the material, both elastic and plastic deformation cause the formation of a hardness impression conforming to the shape of the indenter to some contact depth. After the indenter is withdrawn, only the elastic portion of the displacement is recovered; this recovery enables one to determine the elastic properties of a material [17]. The first step of measurement is preparing a sample by mounting it on a sample disk. For mounting samples double stick tape is used. Mounted samples are presented in Fig. 4.

Fig. 3. Printed silver layers on different substrate after sintering

Middle position in sample tray is reserved for reference material (Corning 7980, fused silica or Pyrex 7740 [17]). One of these materials are used to indirectly verify the instrument accuracy according to the standards.

After placing sample of silver layer on plastic substrate on tray into the indenter, height of measuring sample must be adjusted to the height of the reference material which is fixed. Adjusting the height of the sample can be done with small thumbwheel for each sample holder; rolling of thumbwheel sets the height of sample. Finally, when samples are in line with reference material, it could be set measurement parameters. All measurements were done at same depth of indentation, around 200 nm to avoid possibility of perforation of printed layer. The Poisson's ratio for silver was set as 0.37. The ten indentations have been made for each sample.

Fig. 4. Samples mounted on holder and prepared for measurement

5 Measurement Results and Discussion

Fig. 5. presents load-displacement curves obtained by nanoindentation for ink-jet printed silver layers on 50 μm polyimide film sintered at 200 °C, 225 °C and 250 °C for 30 min. The observed curves are consistent with intuitive expectations; the largest load is required to silver layer sintered at highest temperature, at 250°C.

Table 1. summarize mean values of the 10 measurement data of Young's modulus, hardness, displacement into surface and load applied to test surface. As it can be seen from the Fig. 5. and Table 1, as it was expected higher temperature of sintering leads to higher hardness and Young's modulus of material.

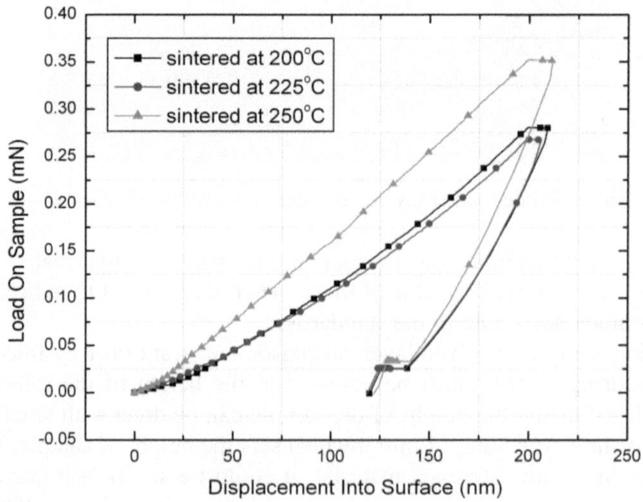

Fig. 5. Typical load-displacemet curves of ink-jet printed silver layers on 50 μm polyimide film at different temperature of sintering

Table 1. Mechanical properties of ink-jet printed silver layers determined by the nanoindentation

	Young's modulus at Max load (GPa)	Hardness at Max Load (GPa)	Displacement at Max Load (nm)	Load at Max Load (mN)
20 %wt on 50 μm polyimide at 200 °C	8.237	1.250	206.773	0.250
20 %wt on 50 μm polyimide at 225 °C	8.658	1.445	204.96	0.240
20 %wt on 50 μm polyimide at 250 °C	10.328	1.737	206.902	0.305
20 %wt on 75 μm polyimide at 200 °C	9.056	1.456	204.69	0.251
40 %wt on 50 μm polyimide at 200 °C	11.968	1.387	207.542	0.383
20 %wt on 1000 μm slide glass at 200 °C	95.506	1.627	219.182	1.697

The load-displacement curves obtained for silver layers printed with 20 wt% and 40 wt% silver nanoparticle ink on 50 μm polyimide substrate and sintered at 200 °C for 30 min are presented in Fig. 6. It can be seen, silver layer printed with 40 wt% silver has larger load during approaching depth of 200 nm of indentation. According to that and to data from Table 1. it also visible that silver layer printed with 40 wt% has a higher value of Young's modulus and hardness that the layer printed with 20 wt% of silver nanoparticles.

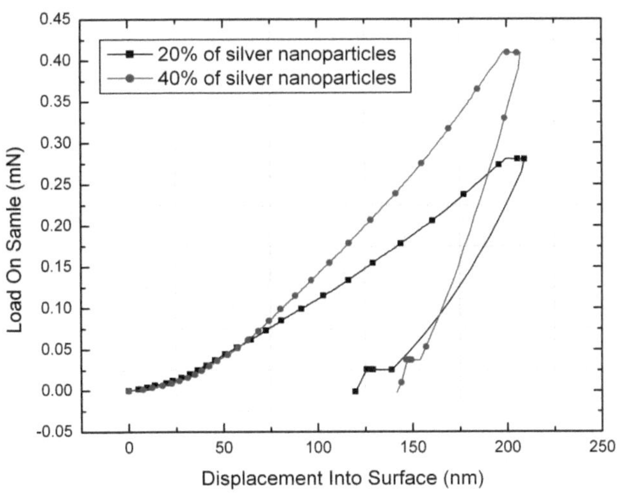

Fig. 6. Typical load-displacemet curves of ink-jet printed silver layers printed on 50 μm polyimide film with different percentage of silver nanoparticles in ink sintered at 200 °C

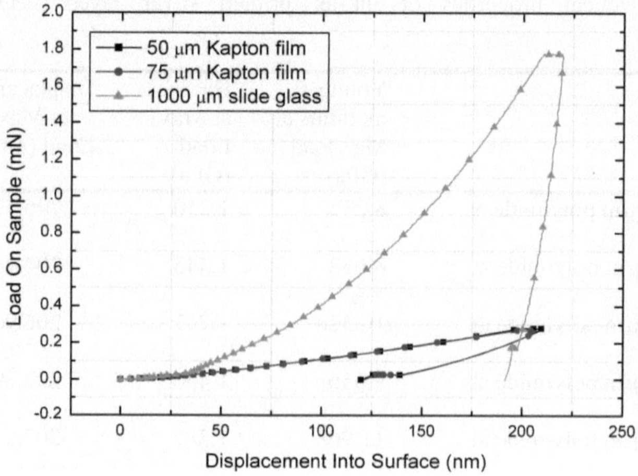

Fig. 7. Typical load-displacemet curves of ink-jet printed silver layers on different substrate sintered at 200 °C

The load-displacement curves for silver layer printed on polyimide film with thicknesses of 50 μm and 75 μm and on slide glass with thickness of 1000 μm are shown in Fig. 7. From Fig. 7. and data from Table 1, the visible is different type of substrate leads to different value of hardness. The silver layer printed on polyimide film with thickness of 50 μm has lowest and silver layer printed on slide glass has the highest value of Young's modulus and hardness. It is also visible that difference in load applied on layer printed on polyimide substrates is almost negligible in comparison with load applied on silver layer printed on slide glass.

6 Conclusion

In the last two decade, much attention has been paid to developing techniques for probing the mechanical properties of materials on the submicron scale. Developing instruments, which can continuously measure load and displacement during the indentation, could see the progress.

Mechanical performance of ink-jet silver layers was considered, suitable for electronic component production. Sample preparation and mechanical test setup were briefly described. Nanoindentation tests were performed to see how different parameters, such as thickness of substrate (50 and 75 μm), the type of substrate (polyimide and glass) and %wt of silver nanoparticles containing in silver ink, influence on Young's modulus and hardness of printed silver layers. According to average value of indentations silver layers, it was found that the higher temperature of sintering and the higher %wt of silver nanoparticles give the bigger Young's modulus and hardness of printed silver sample. It was also found that thickness and type of substrate have low influence on Young's modulus and hardness of ink-jet printed silver layer. This research provides very useful information about mechanical characterization of the silver layers on flexible substrates for different industrial applications on plastic/organic electronics.

Acknowledgments. This work was supported in part by the Ministry of Education and Science, Republic of Serbia, under projects TR-32016 and EC FP7 project APOSTILLE, grant no. 256615.

References

1. Kim, D., Jeong, S., Lee, S., Kyun Park, B., Moon, J.: Organic thin film transistor using silver electrodes by the ink-jet printing technology. Thin Solid Films 515, 7692–7696 (2007)
2. Kunnari, E., Valkama, J., Mäntysalo, M., Mansikkamäki, P.: Enviromental performance evaluation of printed electronics in paralell with prototype development. In: IMAPS 2007, The 40th Int. Symp. on Microelectronics, San Jose, California, USA, November 11-15 (2007)
3. Chason, M., Brazis, P.W., Zhang, J., Kalyanasundaram, K., Gamota, D.R.: Printed Organic Semiconducting Devices. Proc. of the IEEE 93, 1348–1356 (2005)
4. Caglar, U., Kimmo, K., Mansikkamaki, P.: Analysis of mechanical performance of silver inkjet-printed structures. In: 2nd IEEE Int. Nanoelectronics Conf. INEC, pp. 851–856 (2008)
5. Gong, J., Peng, Z., Miao, H.: Analysis of the nanoindentation load–displacement curves measured on high-purity fine-grained alumina. J. Eur. Ceramic Soc. 25, 649–654 (2005)
6. Oliver, W.C., Pharr, G.M.: An improved technique for determining hardness and elastic modulus using load and displacement sensing indentation experiments. J. Mater. Res. 7, 1564–1583 (1992)
7. Li, J., Mei, J., Ni, Y., Lu, H., Jiang, W.: Two-dimensional quasicontinuum analysis of the strengthening and weakening effect of Cu/Ag interface on nanoindentation. J. Appl. Phys. 108, 54309–54317 (2010)
8. Lohr, S.: The Age of Big Data. New Yok Times (February 2012), `http://www.nytimes.com`
9. Akyildiz, I.F., Jornet, J.M.: The Internet of Nano-Things. IEEE Wireless Communication Magazine 17, 58–63 (2010)
10. Albrecht, H.J., Hannach, T., Hase, A., Juritza, A., Muller, K., Muller, W.H.: Can nanoindentation help to determine the local mechanical properties of microelectronic materials? a state-of-the-art review. In: Proc. Electr. Packaging Techn. Conf., pp. 462–467 (2004)
11. Nili, H., Kalantar-zadeh, K., Bhaskaran, M., Sriram, S.: In situ nanoindentation: Probing nanoscale multifunctionality. Progress in Materials Science 58, 1–29 (2013)
12. Oliver, W.C., Pharr, G.M.: Nanoindentation in materials research: past, present, and future. MRS Bull. 35, 897–907 (2010)
13. Jerančе, N., Vasiljević, D., Samardžić, N., Stojanović, G.: A compact inductive position sensor made by inkjet printing technology on a flexible substrate. Sensors 12, 1288–1298 (2012)
14. Dimatix Inc., Dimatix Materials Printer DMP-3000, `http://www.dimatix.com`
15. GTS Flexible Materials Ltd., `http://www.gts-flexible.co.uk`
16. Sun Chemical Corp., SunTronic Jetable Silver, `http://www.sunchemical.com`
17. Agilent Technologies, Nano Indenter G200, `http://www.agilent.com`

Toward a Modeling Framework for Organizational Competency

Reza Vatankhah Barenji[1], Majid Hashemipour[1], and David A. Guerra-Zubiaga[2]

[1] Department of Mechanical Engineering, Eastern Mediterranean University, Famagusta, KKTC, Via 10 Mersin, Turkey
Reza.vatankhah@cc.emu.edu.tr
[2] CIDESI Texas, 1700 Research Parkway, Suite 170, College Station, Texas, 77845, USA

Abstract. Competency modeling framework serves as a; (a) very important basis for the explanation of a generic competency modeling approach, (b) base element in the consolidation of existing knowledge in this area, (c) tool for model developers on selecting appropriate competency models, and (d) basis for competency modeling. This research uses literature review approach to propose a modeling framework for organizational competency. The proposed modeling framework has been developed based on the most relevant well known competency models. The research suggests that organizational competency can be categorized into three groups; individual competency, enterprise competency and collaboration-oriented competency. For modeling each of these groups, it is essential that the modeling process have to be aligned with model developer purpose (Modeling perspective), thus the model developing process will be based on the same segmentation model. Furthermore, competencies have to be model at different levels of abstraction (modeling intent).

Keywords: Competency, Capability, Modeling, Modeling framework.

1 Introduction

In early 1990's [1] described core competences as the "collective learning in the organization, especially how to coordinate diverse production skills and integrate multiple streams of technologies". There exist several definitions for competency. Notable among them, [2] defines competency as "the interaction between three components: the professional situations, the actors, and the resources" and, [3] describes competency as "the smallest autonomous performance unit able to create value, be indivisible and able to exist independently".

Competency model is an information and knowledge model that describes the skills and abilities of a particular organization. Organizations need comprehensive competency model for successful management of internal resources/activities and corresponding their inter-related activities [4]. For an organization to participate in Virtual Organization Breeding Environment (VBE) activities, prior submission of competency model is necessary [5]. On the other hand, competency models are

L.M. Camarinha-Matos, S. Tomic, and P. Graça (Eds.): DoCEIS 2013, IFIP AICT 394, pp. 142–151, 2013.

essential tool for improving organizational core competency [6]. In small and medium size enterprises, the competency models can be developed from oral information while, in a more complex organization , the collection and modeling of competency by human actor is not any more effective [7]. In such cases, computer-based mechanisms are required. Available Literature reviewed emphasizes competency model as a paradigm which depend on modeling purpose that varies from one model to another [8]. Furthermore, collection, analysis and management of competencies for modeling purpose are a complex task involving many aspects of manufacturing and business environment.

These issues motivate the co-authors of this paper to propose a PhD thesis to the author. The main aims of this research thesis are: (a) understand capability and competency concepts (b) introduce an approach to store, manage and maintenance capability and competency of an organization in different levels of abstraction (c) suggest some criteria for using competency as a tool for organization integration. The central research questions which are addressed in this research thesis;

RQ1) how to model an organization with its existing competencies?

RQ2) what are the templates, procedures and methods to store, maintain and manage competency of an organization?

To answer the above research questions, the hypothesis prepared are as follows;

Hypothesis1) if we combine a multi-phase holonic modeling view with associated needed competencies then we can better understand product evolution, customization potential, and retrofitting needs.

Hypothesis1)If we combine competency model paradigm with co-innovation and involvement of local stakeholders then we can have a more effective life-cycle support for complex products, including customization, maintenance, and retrofitting. At the first step of the research thesis, based on existing well-known competency models, a conceptual competency modeling framework for organizational competency is developed. This step will be presented in this paper. For the next step, using a case study, a capability model of a shop is developed. The case study uses Flexible Manufacturing Systems (FMS) Lab of the Eastern Mediterranean University. Finally with the use of another case study, the enterprise competency model is proposed [9]. The case study was a bicycle manufacturing company which is located in Cyprus.

"A modeling framework can be seen as an envelope that includes a number of models, collections of templates, procedures and methods, rules, and even tools" [10]. Competency modeling framework is an important basis towards the development of competency models. Competency modeling framework intend to help model developers to identify what kinds of competency models have to be created for successful decision making [11]. What have not been published in competency modeling context, however, are, a comprehensive modeling framework for organizational competencies and, a generic capability and competency model which can be used by researchers and decision makers. The purpose of this paper is to develop a modeling framework for organizational competency. The modeling framework has been developed using a literature reviewed approach. In essence, the proposed modeling framework has been developed based on the literature review and evaluated with the aid of most relevant well known competency models.

1.1 Contribution of the Paper to "Internet of Things"

Competency modeling is an approach for configuring an organizations data base scheme. In contrast to the other organizational data modeling approaches that covers only the information and knowledge of the organization from one perspective, competency modeling has the capability of providing an organizational data model from different viewpoints. Similar to the other organizational data models, it is accessible through the internet using some specific infrastructures and tools, in which, all the tangible and intangible objects of the organization (such as, machines and processes) can be individually identified and presented in virtual environment. Since, "uniquely identifying objects and their virtual representations in an inter-like structure referred as "Internet of things"", the authors believe is that, competency modeling approach, is a good starting point on applying the "Internet of things" philosophy on organizational management context.

1.2 Research Methodology

To effectively conduct the proposed research investigation and develop a competency modeling framework (CMF) a research methodology developed as below. The proposed methodology illustrates the CMF configuration processes from requirements analysis (literature review approach) to CMF development. In the requirements analysis phase, primary 86 related journal papers were extracted both from computer science and organizational management data bases. After elimination of the papers with less novelty and or fewer citations, 46 papers have been used in the final requirement analysis phase. These requirements are then utilized for generalization phase which proposed the modeling framework perspectives. Ultimately, the generalized perspectives are employed to propose a competency modeling framework.

2 Literature Review

In this section, for discovering competency modeling framework perspectives, the most cited competency models are taken in to consideration. To begin with, the most relevant well known competency models on both organizational and computer science literatures are reviewed. After that, the potential perspectives for competency modeling framework are highlighted.

When attempting to establish a modeling framework, it is important to consider the potential inputs and partial contributions from previous related works for proper system requirements [12]. This contribution uses literature review approach for the requirement analysis phase. In this way, we strived to harness the advantages offered by a number of well-known competency models introduced by other initiatives, and attempted to eliminate some of the common pitfalls which theoretically or empirically, model developers face while establishing a model.

Clearly, this attempt does not argue about the relevance and appropriateness of previous well known models, for the purpose of organizational competency modeling. Rather it argues and represents (i.e. through identifying more elements) that more needs to be done for the purpose of proper organizational competency modeling. As such, instead of starting from scratch to identify the main perspectives and required elements for competency modeling, we have tried as much as possible to reuse some already defined concepts that are more familiar to the users of model, model developers and researchers in this area. As an illustration, table 1 summarizes our analysis results from some well-known competency models for highlighting the potential perspectives for competency modeling framework at intra- enterprise.

Table 1. An attempt to map current competency models applicable to competency modeling framework

Intra	REF.	RESEARCH CONTRIBUTION	MODELING AREA
Managerial sciences	[1]	Core competency notation	Organization competency, definition at concept level
	[13] [14]	Core competency hierarchy	Organization competency, concept model
	[15]	"To assist in core competence management an enabling core competence lens model was presented together with a framework for core competence maintenance"	Enterprise competency, organization competency, at detail level
	[7]	"Investigating the existence and nature of core competency concepts within a section of UK SME manufacturing organizations"	Organization competency, concept and basic level
Computer sciences	[16]	Analyze IS role in raising organizational competencies and prompting the cross-functional integration necessary to achieve scale, scope, and learning curve economies for an enterprise.	Organization competency at concept level
	[17]	Develop a framework for enterprise competency modeling	Enterprise competency, organization competency at concept level

A summary of our main observations are as follow:

- Similar to organization management theory, which hierarchically subdivides organization into four tiers; shop, enterprise, organization and network, the competency models depending on the organizational level differ from one tier to another. Shops as an organization just comprises of capability not competency [13] so there is no competency model for shop level. For enterprise level, there exist three different sets of competencies models namely individual, enterprise and collaboration-oriented. For organization level the models refers to core competencies of the organization and for network level the models deals with sets of competencies which each organization shared within the network.
- The reviewing on competency model literatures emphasis that competency model is a research context from both organizational management and computer science points of view meanwhile, this context is significant at intra-organization, inter-organization and network levels.
- An important task during competency modeling is to provide an abstract representation of the model on which the model is to be used. The structure and level of details depending on the further intentions of using this specific model is a three level architecture: concept level, basic level and detail level.

3 Toward a Comprehensive Competency Modeling Framework

3.1 Competency Inherent Characteristic

Competencies within an organization mainly can be categories in three groups, individual competency; enterprise competency and collaboration-oriented competency. The incorporation of the individual competency, enterprise's competencies and collaboration-oriented competency were referred as organizational competency [18] (see figure 1 from bottom to top), while a portion of the organizational competency which an enterprise decides to share within a network is entitled as sharing competency.

According to the U.S. Department of Labor, employment and training administration (ETA), individual competency, described as: [19] "the ability or capability of the individuals to apply the required set of skills, abilities and related knowledge to perform certain task successfully in a defined work setting". As a result of the ETA definition, Individual competencies denotes from the individuals of the organization. The second class of competencies is enterprise's competencies which refer to a set of competencies that constitute from enterprise facilities beside manufacturing and business knowledge [20]. This class of competencies captures information related to enterprise activists and resources and knowledge correlated to activity and resources [21].The third class of competency is collaboration-oriented competency, which is the organizations ability to cooperate and collaborate by working together towards achievement of a common goal [18]. Interaction and communication are highlighted as establishing the basis for shared understanding, which in turn is considered the basis for the creation of collaborative-oriented competency in a network [22].

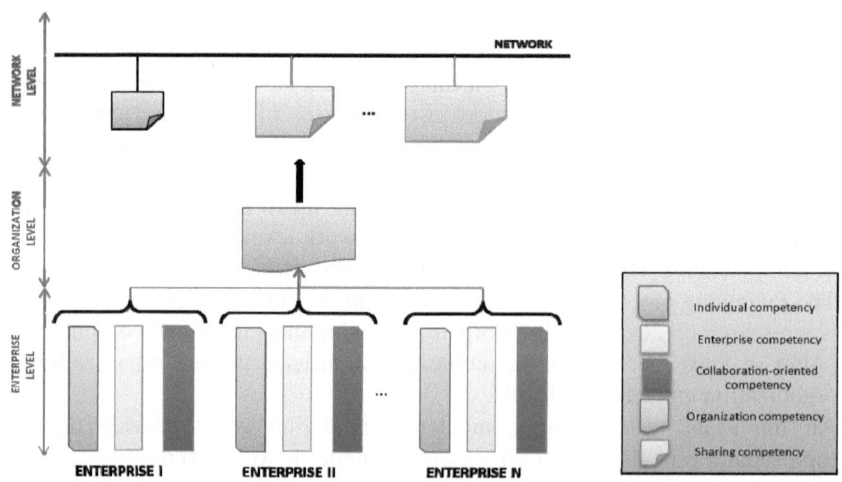

Fig. 1. Competency inherent characteristic

Based on most recent categorization, the first defined dimension for competency modeling framework focuses on capturing competency inherent characteristics is represented by the vertical axis, labeled as "competency inherent characteristic". This perspective further includes three subspaces that comprehensively cover all the inherent characteristics of the organizational competency. The individual competencies of the organization (labeled "Individual competencies"); the enterprise competencies characteristics (labeled "Enterprise competencies competencies"), and collaboration-oriented competencies (labeled" Collective competencies").

3.2 Modeling Viewpoint

From the system theory points of view, competency is complex entities operating in a variety of environments having different purposes and internal/external manufacturing and business processes [23].The existing competency models developed on two spaces namely; (a) managerial sciences and industrial engineering and, (b) information/knowledge managerial sciences (depicted in figure 3). These spaces although are related to each other, have different position in regards to the competency model, and thus the models defines in these views also differ.

On the other hand, the operating of competency models may occur at internal and/or external manufacturing business processes of an enterprise. [5] divide their researches on competency into intra-organization, inter-organization and network levels. Intra-organization oriented studies deal with the competencies within an organization. When the competencies are not bound to a single organization we talk of inter- organization level. Competencies for the static forms of cooperation among organizations that go beyond their boundaries could be subsumed under the term of supplier networks. Network level studies, consider, competencies need for creation of Virtual Enterprises as a network organization as well as competency as a tool for improve the VO performance.

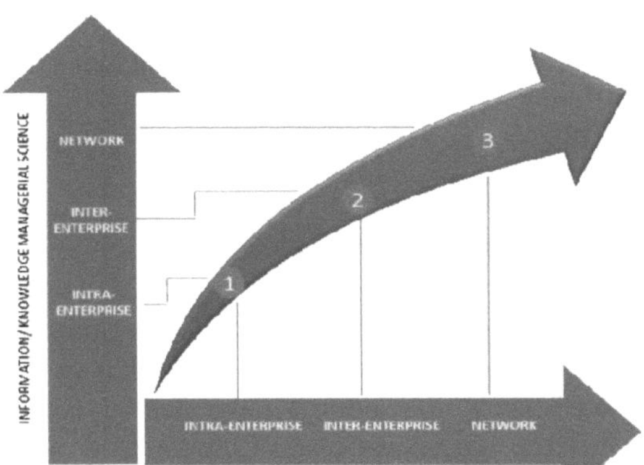

Fig. 2. Modeling perspectives for competency modeling

The modeling perspective capture the modeling requirements and the diversity for internal and external manufacturing and business processes and for different purposes is be represented by the vertical axis, labeled as "modeling perspective".

The second defined modeling perspective addresses the competency modeling viewpoints as represented by horizontal axis on competency modeling framework. These viewpoints include "managerial science and industrial engineering" and "information/knowledge managerial sciences". Accordingly, competency at both viewpoints includes a variety of requirements and these requirements can be categorized into three different groups: intra- organization, competences within an organization; inter- organization, competencies beyond the boundary of the organization but for cooperation and network, competencies for the collaboration networks.

3.3 Modeling Intents

The contemporary organizational theories distinguish three hierarchical levels for organization management: organizational level, infrastructural level, and content level [24]. On the basis of such hierarchical system, it is possible to define three adequate levels in competency modeling processes. The third defined perspective is related to the different intents for the modeling of competency features, will be represented by the diagonal axis, and labeled as "modeling intents". This perspective addresses the three possible modeling stages for competency elements; from the organizational level, to the infrastructure level (e.g. using a specific modeling approach or theory), and finally to the content level.

Following the research practices in modeling, the three layers below are considered as follows:

- Organizational level: includes the most general concepts and related relationships, that is common to all competencies at the highest level, independent of the application domain.
- Infrastructure level: an intermediate level that includes more detailed models, focused on different classes of competencies.
- Content level: that represents models of concrete competencies.

4 Proposed Modeling Frameworks

In this section, the proposed modeling framework for organizational competency is presented. To begin with, the general modeling framework is given.

For the sake of consistency, the authors have named the suggested modeling framework, Comprehensive Organizational Competency Modeling Framework (COCMF). Figure 4 shows the developed modeling framework for the competencies. It can be argued that organizational competency models can be encapsulated into a process of three dimensions: Competency inherent characteristic, Modeling perspective and Modeling intent. COCMF explore the granularity of the competency with the purpose to systemize competency models which will be applicable for the transformation of an organization into a knowledge-based system [25], and its alignment with business goals and the range of other business management functions [26].

In this matrix, for the three subspaces of the individual competency, enterprise competency and collaboration-oriented competencies which forms the "competency inherent characteristic" dimension; their respective dimensions are depicted as different columns. Similarly, for the modeling perspective, each perspective is depicted as one row. The Modeling Intent constitutes the third dimension of the matrix, with its three respective subspaces of organizational level, infrastructure level and content level.

Each of 54 items within the COCMF possesses its own semantics and identifies the definite component of competency model, which integrates three dimensions: modeling perspective, inherent characteristic and modeling intents at the same level of elaboration. E. g., item 111 represents the integration of a competency model concerning three aspects of individual competency, intra-enterprise at managerial science points of view that are used at the organizational level. There are three, two-dimensional subspaces of the COCMF, namely, E1, E2 and E3.

The subspace E1 "modeling perspective- competency inherent characteristic" defines models that are used to support modeling perspective at a definite inherent characteristic (individual, enterprise and collaboration oriented). The subspace E2 "modeling perspective-modeling intents" describes modeling perspectives and their interactions with the modeling intents. The subspace E3 "inherent characterizes-modeling intents" characterizes the competency in the way it is used at each level of competency modeling.

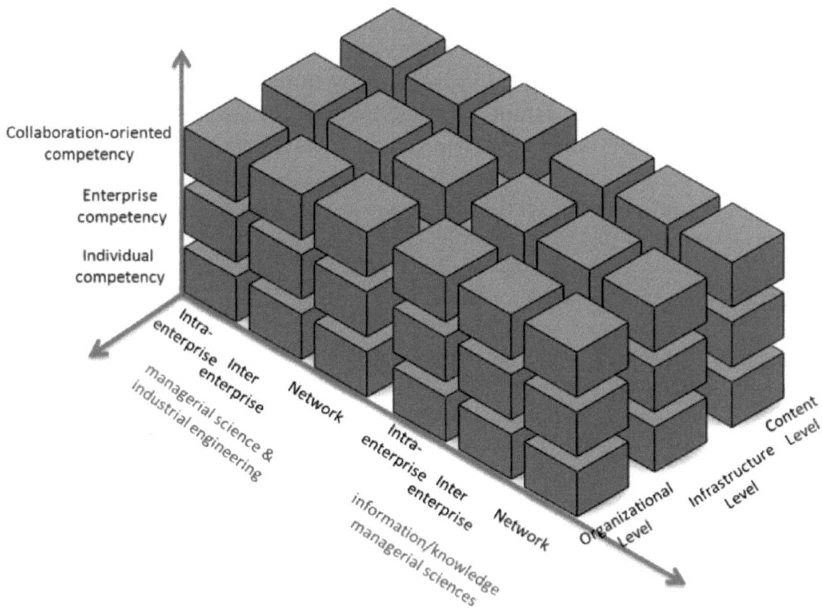

Fig. 3. A modeling framework for enterprise competency

Our result of this study shows that although several related previous works have provided valuable contributions to the understanding of several aspects of this area, they are somewhat limited, when a more holistic modeling is pursued. This research suggests that organizational competency can be categorized into three groups, namely individual competency, enterprise competency and collaboration-oriented competency. For modeling each of these groups of competencies, it is essential that the modeling process have to be aligned with model developer purpose (Modeling perspective), so they exchange information, and operate based on the same segmentation model. Furthermore, competency have be model at different levels of abstraction (modeling intent).The main benefit of a competency modeling framework is enhancing decision making process for model developers. The main research limitation was; since the research is explorative in nature, therefore empirical data from similar and other research settings should be gathered to reinforce the validity of the findings. The major practical limitation of the COCMF is that a generic competency modeling approach is still needed for organizational competency modeling.

5 Conclusions

Establishment of a comprehensive modeling framework for organizational competency is a very important basis for the explanation of a generic competency modeling approach, a base element in the consolidation of existing knowledge in this area, tool for model developers for selecting appropriate competency models and a basis for its consistent further progress. In this way, as a contribution, a modeling framework for organizational competency considering multiple dimensions is proposed. The necessity and detail elements of each of the three dimensions on the proposed modeling framework, i.e. competency inherent characteristic, modeling perspective, and modeling intents are addressed. Finally, to benefit from the knowledge generated by other related research in this area, the most relevant other well-known competency models aspects and domains are evaluated with the three dimensions of the proposed modeling framework.

References

1. Prahalad, C.K., Hamel, G.: The core competence of corporation. Harvard Business Review, 79–91 (1990)
2. Boucher, X., Peillon, S., Burlat, P.: Towards a decision support for a collaborative increase of competencies within networks of firms. Research Report G2IEMSE, 600-609 (2005)
3. Mueller, E.: Production planning and operation in competence cell based networks. Production Planning & Control 17(2), 99–112 (2006)
4. Saulius, G.: Enterprise knowledge modeling: domains and aspects. Baltic Journal on Sustainability 15(2), 281–293 (2009)
5. Ermilova, E., Afsarmanesh, H.: Competency modeling targeted on boosting configuration of virtual organizations. Production Planning & Control: The Management of Operations, 103–118 (2010)

6. Zoiopoulos, I.I., et al.: Identifying organizational competencies in project oriented companies: an evolutionary approach. In: Procs. 24th Annual ARCOM Conference, Cardiff, UK, September 1-3 (2008)
7. Bhamra, R., Dani, S., Bhamra, T.: Competence understanding and use in SMEs: a UK manufacturing perspective. International Journal of Production Research, 2729–2743 (2010)
8. Ermilova, E., Afsarmanesh, H.: Modelling and management of profiles and competencies in VBEs. International Journal of Intelligent Manufacturing 18(5), 561–586 (2007)
9. Vatankhah, R., Hashemipour, M., Guerra, D.: Toward a framework for intra-enterprise competency modeling. In: 2nd IEEE International Conference on Advanced in Computational Tools for Engineering Applications (ACTEA), Beirut (2012)
10. Camarinha-Matos, L.M., Afsarmanesh, H.: Towards a reference model for collaborative networked organizations. In: Shen, W. (ed.) Information Technology for Balanced Manufacturing Systems. IFIP, vol. 220, pp. 193–202. Springer, Boston (2006)
11. Rosemann, M., van der Aalst, W.M.P.: A configurable reference modelling language. Information Systems, 1–23 (2007)
12. Boucher, X., Bonjour, E., Grabot, B.: Formalisation and use of competencies for industrial performance optimisation: A survey. Computers in Industry, 98–117 (2007)
13. Javidan, M.: Core competence: What does it mean in practice? Long Range Planning, 60–71 (1998)
14. Ljungquist, U.: Core competency beyond identification: presentation of a model. Management Decision, 393–402 (2007)
15. Gilgeous, V., Parveen, K.: Core competency requirements for manufacturing effectiveness. Integrated Manufacturing Systems, 217–227 (2001)
16. Zhang, M.J., Lado, A.A.: Information systems and competitive advantage: a competency based view. Technovation, 147–156 (2001)
17. Walsh, S.T., Linton, J.D.: The Competence Pyramid: A Framework for Identifying and Analyzing Firm and Industry Competence. Technology Analysis & Strategic Management, 165–177 (2001)
18. Kokko, N., Vartiainen, M., Lönnblad, J.: Individual and collective competences in virtual project organizations. The Electronic Journal for Virtual Organizations and Networks, 28–52 (2007)
19. Competency, competency framework and competency model on line referencing, http://www.careeronestop.org/competencymodel/default.aspx (accessed February 11, 2011)
20. Kristianto, Y., Helo, P., Takala, J.: Manufacturing capabilities reconfiguration in manufacturing strategy for sustainable competitive advantage. International Journal of Operational Research, 82–101 (2011)
21. Blanch, R., Ferrer, I., Garcia-Romeu, M.L.: A model to build manufacturing process chains during embodiment design phases. International Journal of Advanced Manufacturing Technology, 421–432 (2012)
22. Paszkiewicza, Z., Picard, W.: Modelling Competences for Partner Selection in Service-Oriented Virtual Organization Breeding Environments. Journal of Computers & Industrial Engineering, 1–17 (2011)
23. Vartiainen, M.: Mobile virtual work – concepts, outcomes and challenges. In: Mobile Virtual Work: A New Paradigm?, pp. 13–44. Springer, Heidelberg (2006)
24. Sormaz, D.: Distributed Modeling of Manufacturing Activities using Integrative Manufacturing Process Model. International Journal of Industrial Engineering and Management, 9–18 (2010)
25. Vichare, P., et al.: A Unified Manufacturing Resource Model for representing CNC machining systems. Robotics and Computer-Integrated Manufacturing, 999–1007 (2009)

Part VII
Perceptional Systems

Food Product Traceability by Using Automated Identification Technologies

Ivana Šenk, Gordana Ostojić, Laslo Tarjan,
Stevan Stankovski, and Milovan Lazarević

University of Novi Sad, Faculty of Technical Sciences, Trg Dositeja Obradovića 6, 21000 Novi Sad, Serbia
{ivanas,goca,laci,stevan,laza}@uns.ac.rs

Abstract. Food product traceability from harvesting, through food processing to the final food product and through the retailer to the end consumer is a significant process that has to ensure food quality and safety. The traceability enables the end consumer to get information from all previous stages of the food product, leading back to the food origin. In this way, the consumer can get more information on the specific product, and thus make a decision on buying the product that suits his needs best. In each stage of the food product transformation, important data are generated for the subsequent chain participants. Every participant should have access to certain data of interest to them. This can be achieved by using automated identification technologies, like RFID (Radio Frequency IDentification) and two-dimensional barcode, which allow faster data acquisition, recording and reading processes than the traditional means, and provide up-to-date information in each product stage. Furthermore, these technologies allow the possibility to record large amounts of data for each specific product, and interconnect all the data in a database. This paper discusses the process of providing traceability of food products, recording, transmitting and reading of significant data in specific stages of food product chain, with the application of automated identification technologies, including the possibility of obtaining additional data from a database, according to appropriate access level of each participant in the chain. Advantages and disadvantages of automated identification technologies are discussed, with the proposition for using specific technologies in certain food product stages.

Keywords: Food traceability, RFID, 2D barcode, QR code.

1 Introduction

Great competition within the food market forces the food producers to introduce technologies and principles that make their products more competitive. Studies show that the customers rather chose quality over price in food product, as shown e.g. in [1] where 72% of the respondents said that they are more concerned with quality than the price of food. Similar situation is in Serbia, where an agency conducted research that showed that the main reason for deciding on buying a specific product is its quality, with 61% of interviewed customers [2]. The quality of food products is influenced by

L.M. Camarinha-Matos, S. Tomic, and P. Graça (Eds.): DoCEIS 2013, IFIP AICT 394, pp. 155–163, 2013.

the source and quality of raw food, but also by the processing and production methods, the packaging and transport of semi-products and finished products. This leads to the necessity for a traceability system that would enable tracking the food product through the whole food chain, and also tracing the food end-product back to its origin, which is recognized as forward and backward traceability.

The EU Regulation 178/2002, which took effect on January 1st 2005, set general foundations for traceability requirements in food sector in the EU and at the national level in EU member states [3]. According to this regulation, "traceability means the ability to trace and follow a food, feed, food-producing animal or substance intended to be, or expected to be incorporated into a food or feed, through all stages of production, processing and distribution". This regulation requires traceability of all the abovementioned food products and ingredients, identification of preceding and subsequent participants in the food supply chain, availability of this information to the competent authorities upon demand, and labeling of food products that are ready for the market.

The traceability system is supposed to enable all participants in the food chain to trace food products at all times and to provide full history for the specific product [4], which includes the source of all raw food, treatments to which it was exposed during growing and harvest phase, processing, transport, etc. This system should also protect the consumers from buying out-of-date food, food that contains allergens, religiously prohibited food, etc. In the case of problems with safety and quality of a batch of food products, e.g. appearance of dangerous substances or microorganisms, such a system should enable withdrawal of these products from the market and prevent health disasters.

Traceability systems worldwide have been developed, but they lack standards, companies develop their own systems which are different, and are producing different economic results [5]. Such systems are mainly developed internally within one company or within large food production chains, and they lead to large information gaps between the participants in the food chain. On the other hand, Daives (as cited by [6]) states that 62.2% of European food companies are small and medium enterprises, they produce the largest amount of food products, but cannot expect the potential benefits from implementing traceability systems in comparison to the cost of work needed for implementing such a system, as they are not connected into a unified information system, which would allow efficient and transparent information flow in supply chains.

One of the challenges in traceability systems are means of storing and transferring data between participants in the food chain. Traditional systems include alphanumerical identification of food products, but such systems require manual identification of products, and are time consuming. Automated identification technologies such as barcode and RFID technologies enable automation of traceability systems and faster and less expensive data collection and transfer. These technologies can thus help overcome problems that exist with traditional traceability systems, and enable full integration of data in a food product chain. RFID and barcode technologies have become widely implemented in many areas due to the increasing availability of devices and systems that enable data collection through these technologies. The main challenge in implementing a traceability system based on automated identification technologies is the fact that food products greatly vary from one product to another, regarding the aggregate state, amount of products in a group package, methods of processing, packaging and transportation, storage temperatures, key product data, etc. In order to create a unified system, all of these features have to be taken into account, and the system set up in such a way that it is adaptable to various conditions in food chains.

Taking into account all the abovementioned, several hypotheses can be set: an automated identification system can be developed and applied for food product traceability through different stages; data that are generated in different stages of the food chain can be integrated in one database and thus can be used in subsequent stages; the data generated using automated identification technologies are significant for food safety. Research methodology that is applied for validating the hypotheses includes analysis and synthesis for developing a concept for food traceability based on automated identification technologies, and also experimental validation, that is planned for fulfillment in the next phase of research.

2 Relationship to Internet of Things

In the world of globalization and interconnecting various areas into one integrated whole, a very important aspect is the traceability of objects and linking important information into a unique system that allows access to this information to various users, to the appropriate extent. Such systems include systems for food products traceability, which are specific as they can influence human health. A system for food traceability discussed in this paper precisely allows interconnecting various data about food products, from the means, place and time of the farming of raw food, through its processing, packaging, transport and storing, and to its arrival to retailers and end consumers. Such a system should enable storing of all key data in an adequate database, and allow access to this database to various users and according to their position in the food chain and their privileges to enable viewing and/or updating specific food data. This system should be universal and applicable for different types of food products, and to easily provide incorporation of new participants in the process, in order to help the tendency for integration of all relevant data in a unique system. The main purpose of such a system is the availability of all the key data to the end user, which would enable them to more wisely choose products that they are buying, in accordance with their wants, needs, restrictions (e.g. for allergic, vegetarian, diseased consumer). Other very important aim of such a system is the possibility of fast reactions in the case of contaminated batches of food (e.g. with poisonous chemicals or bacteria), where the food can immediately be traced back to its origin and further spread of the contaminated food can be blocked and the already spread food withdrawn. Moreover, such a system can also help producers, as it enables them to recognize the actual demand for products on the market, the amount of their sold products as well as places and time when they are sold, which would let them create better market strategies.

3 State of the Art

As automated identification technologies had become widely used in various areas for tracking different types of objects, systems based on automated identification were also created for traceability of food products.

In [7], a traceability system with the use of standard linear barcodes is presented, on a case company in the baking industry that started applying barcode labels to the cases and pallets in order to improve warehouse inventory system by recording quantity, location and product identification numbers. This system also allowed them to track the subsequent participants in the food chain, and help in risk situations when batches with bad products were accidentally released. The authors state that improved systems for product traceability can be achieved by using two-dimensional barcodes or RFID technology.

In [6], the authors proposed an information infrastructure that enables traceability in the food supply chain by the application of RFID technology. The presented traceability system is cost effective and applicable in the chain of small and medium enterprises, which have limited financial capabilities. Every participant in the food chain has set up RFID readers. The data from food products is collected via RFID readers and sent to the central database through an IP network. The central database can be accessed through a web interface, and each participant in the food supply chain can view the adequate data. The authors state that the greatest potential for RFID application in a food supply chain is in particular in connecting different participants in the chain.

In the meat processing industry [8], traceability systems have to enable tracing meat along the complete chain, back to the origin, as various diseases can influence the quality of meat. The traceability systems have to be considered the highest priority, and not just legislation, in order to ensure safe and high quality meat for the end consumer. In [9] a case traceability system for chicken meat is presented, that integrates RFID technology with the information system. The system is applied through the complete food chain, from the farm, through slaughter house and processing factory, to the retailer. Food traceability data is gathered and registered through RFID readers, and sent to the central database. At specific places there are devices where a consumer can read data from the central database and get the required information.

In Japan [10], an integrated traceability system was developed for identification of agricultural products and storing the key data from production to consumption stages, by the application of RFID technology, mobile phones and web-based network computing. This system enables farmers to input data for methods and materials used in the production stage through their mobile phones, and store them in the production database. Through the distribution stage, data is inputted to the distribution database, and RFID tags or barcodes are applied to the individual products, and they connect each product to the data in the production and the distribution database. End-consumers can use their internet enabled mobile phones that have integrated RFID or barcode readers to scan data from the RFID tags or barcodes, and access the database with production and distribution data about the particular product.

The same author [11] has developed a unique database system that integrates the previous production and distribution databases, and a visualization tool for traceability of agricultural products in the food chain. This database system allows changing of the ID tags of products through the chain, and concatenating all the available data about the product. The author states that the major drawback in using the RFID technology in a traceability system is the price of tags and devices used for the transfer of data and the time needed for tagging and reading each product.

In [5] a traceability system is presented for a very expensive Italian cheese, Parmigiano Reggiano. This cheese is produced in large cylindrical pieces. In the presented traceability system, each cheese is labeled with one tag, which precisely identifies the specific cheese. The authors state that this system is applicable and effective as the price of the tag is irrelevant considering the price of the whole cheese. They consider RFID a good solution for traceability of high-value products like cheese or wine, while for the low-priced food products they propose systems still based on alphanumerical codes or barcodes.

4 Research Contribution and Innovation

The main objective of this paper is to set up a framework for a unique food traceability system that enables transfer of key product data through the entire product transformation chain, which uses automated identification technologies: RFID and two-dimensional barcodes. This framework should be adaptable to various food chains according to the specific demands for traceability of particular products.

RFID technology has many advantages for implementation in automated identification and traceability systems, such as the amount of data that can be contained in a tag, high reading speed of data, possibility of simultaneous reading of multiple tags, possibility of non-contact reading of data, etc. One of the major disadvantages of RFID technology is the price of its implementation and of single tags. This disadvantage affects the use of RFID technology in food product traceability systems, as the price of an RFID tag would greatly affect the price of a single food product. On the other hand, two-dimensional barcodes can store less but still a significant amount of data, and are not costly like RFID tags. They have other disadvantages, including the need for proximity of readers while reading labels, inability of readers to simultaneously read multiple labels, etc. One of the most often used two-dimensional barcodes is QR code, which can store a sufficient amount of data, has very good readability even on small sized labels, and which also has very good readability in case of physical damage of a part of the code. QR codes are mostly used for recording a certain numeric code or a URL to the website that contains information. In this paper the framework for a traceability system is proposed that uses both the RFID and QR codes for food product traceability, in which both the RFID and QR codes contain information about a particular product, which is therefore immediately readable with the use of an adequate reader.

The traceability system framework presented in this paper includes all the possible stages in a food chain, where the potential participants are: primary producers, processing industries, transport, retail and the end-consumer (Fig. 1). In a specific traceability system, it is not necessary that all of the above mentioned participants be present, but the system can consist of any possible combination thereof. Each participant in the chain represents a specific stage in the transformation of food.

At each of the transformation stages, specific data are generated that describe the current processes, methods used, place and time, etc. Each product gets a unique ID code, which is the primary key data for the product. All the important data at a particular stage are assigned to the unique ID code of the product, and recorded in the database, to which each participant in the food chain can connect via an IP based computer network. If the new product is a combination of several ingredients, its unique ID code is also connected to the ingredients' unique ID codes in the database, which enables traceability back to each raw product that is integrated in the current food product. This way, the database contains all the existing data about a specific product, which is also connected to the data originating from its ingredients. Also, accessibility levels must be set, to define to what extent subsequent users can reach specific product data in the database.

Moreover, at each stage, the key data is picked, which is significant for the subsequent participant in the chain, and it is incorporated in the label for each product in the form of a QR code. The QR code is chosen for this application as it can store a sufficient amount of data, and at the same time it is small, but can be read even if greatly damaged, can be placed on a sticker, and printed on many available printers.

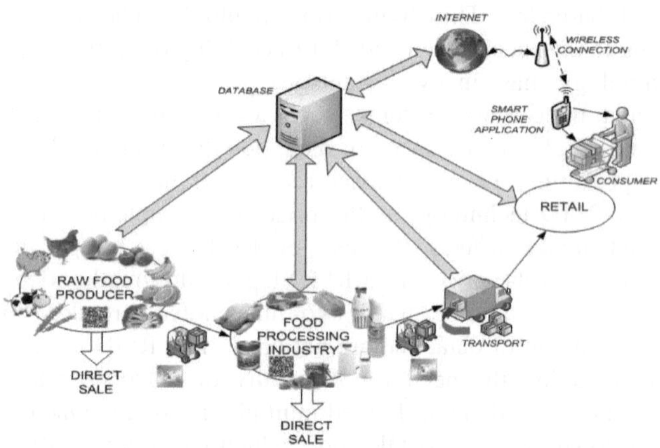

Fig. 1. The concept for food product traceability based on automated identification technologies

An application rarely given to QR codes is the usage of their possibility of storing a sufficient amount of data in a small space, and not just a numeric code or the address of a website. For example, a QR code sized 25x25 mm which consists of 69x69 blocks, can store up to 321 characters of data. The data in the QR code can be accessed by the appropriate reader, such as a dedicated two-dimensional barcode reader or a smart phone with the adequate application. The mandatory data in the food product QR code is a unique ID code that can connect the user to the correct data in the database upon request. The important thing is that here the QR code is used for tagging each single product, with the corresponding information related solely to it. This enables the subsequent participant in the chain or the end consumer to easily access the most important data about the product without the need for internet connectivity, just by using a smart phone with the application for reading two-dimensional barcodes, or a dedicated reader for two-dimensional barcodes. With the increase of everyday usage of smart phones, many users could read such labeling to immediately get the required information about the particular product. If such users install the dedicated application for reading detailed information about products, they can acquire additional information from the database through internet connection, where they can check detailed information about the product or its ingredients at the available accessibility level.

A specific situation occurs in transport, where large amounts of products are simultaneously transferred between the participants in the chain. In the transport stages, products are usually packed in larger group packaging, such as boxes, cases, pallets, barrels, etc. or the products are themselves large. In transport, speed is a very important issue, and if each product or a group packaging should be read with barcode readers, it would require a lot of time. A possible solution to this challenge is to additionally label transport packaging with RFID labels with the required information about the contained products. The RFID technology enables storing of larger amounts of data than two-dimensional barcodes, more RFID tags can be read simultaneously and are readable from a distance with dedicated RFID readers. The disadvantage of RFID technology is its price, which is the reason why it is not yet widely available for identification and traceability systems of individual products. However, this disadvantage is not so significant during the transport stage, as the product packaging is usually large, and contains more products, therefore the use of a single RFID label for the entire packaging would not drastically affect the price of individual products. Furthermore, identification of RFID tags can be done automatically without the need for additional workers for reading labels and registering the products. The implementation of RFID tags during the transport stage would allow immediate registration of sent and received packages, which must be integrated in the whole information system for traceability of particular food products and synchronized between all partners in the chain. This should again be done through the database, where the particular products would get additional data about their new location.

5 Discussion of Results and Critical View

This paper provides a framework for a centralized unified system with the application of automated identification technologies for food product traceability. By reviewing the literature, no such complete system could be found, and on the other hand many regulations, both national and worldwide, require complete traceability of food products. There seems to be no single technology that can provide such a complete system for various kinds of existing food chains, which leads to the conclusion that several identification technologies must be integrated into a system where their individual advantages will be used, and disadvantages suppressed or even avoided. The proposed framework suggests exactly such a system, where two-dimensional QR codes would be used combined with the RFID technology, using their possibilities for easy integration into various information systems, large memory capacities for the required data, easy printing and application to the products etc. This way the proposed framework can be considered general for use in any food product traceability system.

The suggested traceability system based on automated identification technologies must be implemented in various food product chains, and validation of the proposed framework must be carried out in different conditions in order to reach the real impact that the proposed solution potentially has.

6 Conclusions and Further Work

The paper presents a possible conceptual framework for combined application of QR codes and RFID technology in a traceability system for food products. Such traceability systems are very important as they provide the end consumer with information from all the previous stages of the food product, including the information about the origin of the product ingredients, the means and places of food production and processes, the transportation and storage information, etc. By the implementation of the proposed framework in a traceability system, the end consumers can easily get important information about a particular product, and if interested, they can check further information about the product from the database, which enables them to make better decisions when choosing an appropriate product.

In order to validate the proposed framework, further work must include experimental implementations in as many different food product chains as possible, where also primary producers, food processing industries, transporters and retailers must be included and educated to accept such a system upgrade and integrate it with their current information infrastructures. Only by such real implementations, challenges can be revealed that are not obvious in the theoretical analysis, the existing problems can be solved and the system framework improved.

References

1. http://www.adweek.com/news/advertising-branding/consumers-choose-quality-over-price-100447 (accessed on October 31, 2012)
2. http://www.gfk.rs/public_relations/press/articles/005465/index.rs.html (accessed on October 31, 2012)
3. Folinas, D., Manikas, I., Manos, B.: Traceability data management for food chains. British Food Journal 108(8), 622–633 (2006)
4. Abad, E., et al.: RFID smart tag for traceability and cold chain monitoring of foods: Demonstration in an intercontinental fresh fish logistic chain. Journal of Food Engineering 93, 394–399 (2009)
5. Regattieri, A., Gamberi, M., Manzini, R.: Traceability of food products: General framework and experimental evidence. Journal of Food Engineering 81, 347–356 (2007)

6. Kelepouris, T., Pramatari, K., Doukidis, G.: RFID-enabled traceability in the food supply chain. Industrial Management & Data Systems 107(2), 183–200 (2007)
7. Bar Coding and RFID: The Key to Traceability and Safety in the Foodservice Supply Chain. A ZEBRA Black&White Paper (2007)
8. Schwagele, F.: Traceability from a European perspective. Meat Science 71, 164–173 (2005)
9. Chen, R.S., Chen, C.C., Yeh, K.C., Chen, Y.C., Kuo, C.W.: Using RFID technology in food produce traceability. WSEAS Transactions on Information Science and Applications 5(11), 1551–1560 (2008)
10. Sugahara, K.: Traceability System for Agricultural Products based on RFID and Mobile Technology. Computer and Computing Technologies in Agriculture II 3, 2293–2301 (2009)
11. Sugahara, K.: A database system and a visualization tool for agricultural product traceability based on RFID technology. In: Proceedings of EFITA/WCCA, pp. 637–643 (2011)

Implantable Sensor System for Remote Detection of a Restenosis Condition

J.A. Miguel, Y. Lechuga, R. Mozuelos, and M. Martínez

Microelectronics Engineering Group
University of Cantabria Av. de los Castros s/n
39005 Santander, Spain
{joseangel.miguel,yolanda.lechuga,roman.mozuelos,
mar.martinez}@unican.es

Abstract. The increase of life expectancy in the European Union, and the high risk of cardiovascular diseases associated with age, are some of the main factors to contribute to the rise of healthcare costs. An intelligent stent (e-stent), capable of obtaining and transmitting real-time measurements of physiological parameters for its clinical consultation, can be a useful tool for long-term monitoring, diagnostic, and early warning system for arterial blockage without patient hospitalization. In this paper, a behavioural model of capacitive Micro-Electro-Mechanical (MEMS) pressure sensor is proposed and simulated under several restenosis conditions. Special attention has been given to the need of an accurate fault model, obtained from realistic finite-element simulations,to ensure long-term reliability; particularly for those faults whose behavior cannot be easily described by an analytical model.

Keywords: Biomedical Electronics, Implantable Biomedical Devices, Biomedical Transducers, Cardiology, MEMS Testing, Fault Injection.

1 Introduction

Cardiovascular diseases are the leading cause of mortality in the European Union (EU), being responsible of nearly 35% of all deaths in the year 2009 [1]. Percutaneous Coronary Interventions (PCIs) are the most common coronary revascularization procedures, consisting mainly of a balloon angioplasty procedure with stent placement. A stent is a bio-compatible mesh tube designed to be inserted into the body in a collapsed way, mounted at the tip of a catheter with a deflated balloon inside it. At the point of blockage of a blood vessel, the balloon is inflated, so the stent is expanded against the vessel walls, where it is deployed in order to create a durable unobstructed path for blood flow.

The stent has had a strong impact in modern cardiovascular medicine, positioning as a minimally invasive alternative to coronary artery bypass grafting surgery (CABG), with a short recovery time and low major complications rate. Particularly, the Universitary Hospital Marqués de Valdecilla in Santander (Spain) conducts more than 1500 procedures with stent implantation per year.Unfortunately, this procedure is

L.M. Camarinha-Matos, S. Tomic, and P. Graça (Eds.): DoCEIS 2013, IFIP AICT 394, pp. 164–171, 2013.

not exempt of long-term complications as the in-stent restenosis (ISR), that consists of the re-occlusion of the vessel lumen due to physiological repair mechanisms triggered by damage induced to the vessel walls during an angioplasty procedure [2].

The development of MEMS, together with micro-fabrication technologies compatible with CMOS fabrication processes, allow the integration of sensor structures and electronic circuits in the same chip, increasing the capabilities of implantable medical electronic devices. An intelligent stent (e-stent) that incorporates a sensor and integrated electronic circuitry, capable of monitoring and transmitting real-time measurements of physiological parameters related to blood vessel occlusion, such as pressure and flow velocity, can be a useful tool to ISR early detection [3-5].

It is relevant to point out that an implantable sensor must fulfil certain design restrictions, including reduced size, low power consumption, low cost, and above all, long-term reliability and stability. This is why heterogeneous testing, fault modelling and fault injection are critical issues to validate the fabricated device.

In this paper an initial model of a proposed ISR monitoring device is presented and evaluated under different grades of stenosis; focusing on the analysis of MEMS pressure sensors, to estimate the influence of their most common fabrication faults on the system response. Section 2 summarizes the relationship of this work to the Internet of Things (IoT). Section 3 describes the behavioural model of the proposed implantable device under different grades of stenosis. In Section 4 analytical models of a MEMS pressure sensor are validated using FEM simulations. A common fabrication fault error is injected into the FEM model, so that its influence on the sensor response can be quantified. The paper finishes in Section 5 with the conclusions and future work.

2 Relationship to the Internet of Things

The Internet of Things (IoT) was born with the main objective of providing wireless communication capabilities to daily life objects, in order to create a new network concept based on the interaction between an imaginary environment (internet) and the real world (objects and sensors). The application of IoT technology in the healthcare field represents a potential way to minimize, social problems related, among others, to population aging and the associated risk of long-term diseases.

In this sense, a field of interest in the IoT lies in the concept of Medical Body Area Networks (MBAN) [6]. The use of implantable or wearable sensors, together with the addition of network connectivity, short range wireless communication to handheld modules, and remote data transmission via mobile phone or home gateway, can optimize the way medical data is nowadays acquired, stored and analyzed by healthcare providers. This type of sensors offers numerous benefits, such as continuous monitoring of patient's health state and personalized treatment or instantaneous first aid notification in the case of critical condition.

This work proposes a behavioural model of an implantable pressure sensing device for the pulmonary artery, capable of obtaining and transmitting real-time measurements for the early detection of ISR condition without patient hospitalization.

3 Intelligent Stent Model

Blood flow and pressure measurements are some of the most commonly carried out procedures to monitor cardiovascular diseases. In this field, three approaches have been proved to be compatible with e-stent design: electromagnetic, ultrasonic and pressure-based techniques [3-5]. Pressure-based techniques are of special interest, because of their simple implementation, low power requirements, reduced dimensions and integration capabilities of sensors and electronic circuits on the same silicon substrate. Moreover, these approaches allow measuring both blood flow velocity and pressure in the vessel, providing a complete set of data to carry out ISR follow-up. The relationship between the two physiological parameters in an obstructed vessel can be expressed as [7]:

$$\Delta P = R_1 \cdot v + R_2 \cdot v^2 + R_3 \cdot \frac{dv}{dt}. \tag{1}$$

Where ΔP is the pressure gradient between both sides of the stenosis, v is the mean cross-sectional flow velocity in the vessel, and R_1, R_2 and R_3 are coefficients that depend on the geometry of the obstruction and fluid properties. Parameter R_3 is related to fluid inertial effects and has been reported to be despicable under medium to strong stenosis conditions [7].

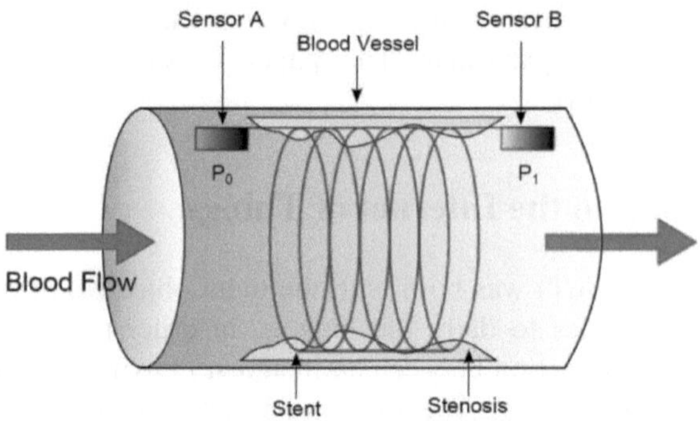

Fig. 1. Sensor placement and measures

The simplest pressure-based e-stent implementation consists of a pair of capacitive MEMS sensors to measure blood pressure and an inductance to form an LC tank for remote power transfer and data transmission through inductive coupling. Fig. 1 shows a simplified representation of the proposed device, where P_0 and P_1 are blood pressure values in the heart and distal sides of the stent respectively.

A behavioural model of the implantable device, implemented in Matlab, is shown in Fig.2. This model has been simulated under different grades of stenosis, by varying R_1 and R_2 parameters, in order to obtain an approximation of its response under real disease conditions.

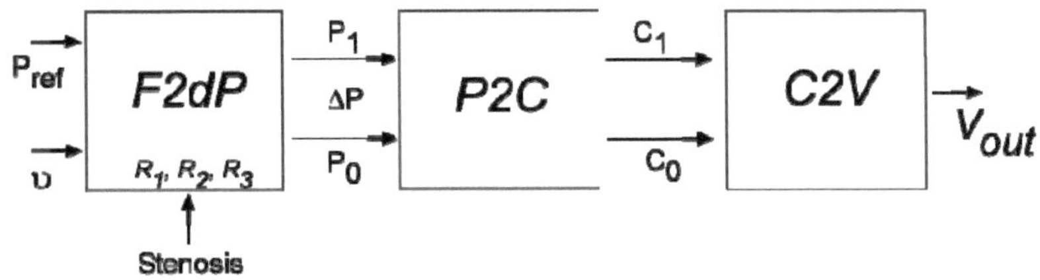

Fig. 2. Simplified model of the electronic system

The first block of the model, called F2dP, converts blood flow velocity to differential blood pressure in the pulmonary artery, using the relationship exposed in (1). The values of R_1 and R_2 have been taken from clinical trials concerning ISR [8], as seen in Table.1. Using pressure and blood flow velocity signals from clinical trials, the F2dP block provides two pressure waveforms to be measured by the proximal (P_0) and distal (P_1) MEMS sensors.

Table 1. Characteristics of the instantaneous flow velocity and pressure gradient relationship

Condition	R1	R2
Normal artery	0.032±0.018	0.00030±0.0049
Intermediate stenosis	0.15±0.11	0.0021±0.0014
Severe stenosis	2.67±1.58	0.0014±0.010

The P2C block contains an analytical model of a capacitive MEMS pressure sensor. This kind of sensor consists of a fully clamped diaphragm suspended over a sealed cavity and a fixed backplate. Once a uniformly distributed pressure is applied to the sensor, the top membrane deforms towards the backplate, reducing the chamber size and causing an increase in the resulting capacitance. A more detailed sensor description will be carried out in section 4.

The C2V block converts the output capacitance value from the pressure sensors to electric digital voltage signals V_0 and V_1. The former voltage is related to the pressure recorded at the proximal location of the stent, while the latter is proportional to the pressure at the distal side. The converter model presents a linear response; and its description has been selected to be similar to common capacitance-to-digital converters (CDC) [9]. The most relevant parameters of the model are: a reference capacitance of 1.5 pF, a conversion range of 1.7 pF and a resolution of 10 bits. Moreover, several undesired effects, such as capacitive input noise, absolute error after calibration, gain error, integral nonlinearity and differential nonlinearity, have been taken into account to obtain a first estimation of their influence over the pressure and flow velocity measurements.

The behaviour of the modelled system for low ISR rate in the pulmonary artery is shown in Fig. 3. It shows the output voltage of the system (right graph), as a function of the blood flow signal for a healthy artery, and pressure waveforms to be measured by the MEMS sensors (left graph).

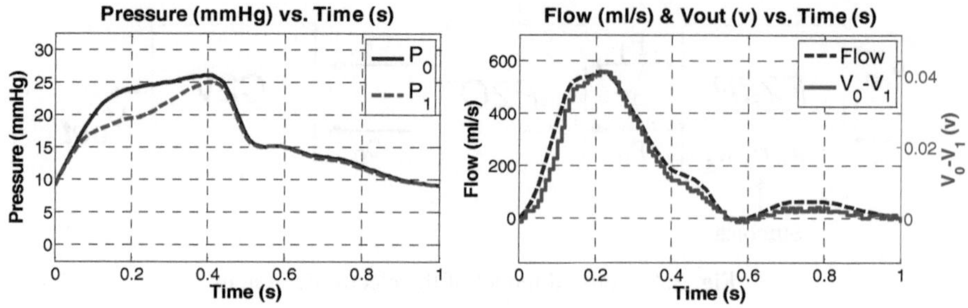

Fig. 3. Output system voltage compared to blood flow (*right*), reflecting the differential pressure measured in a pulmonary artery under low stenosis conditions (*left*)

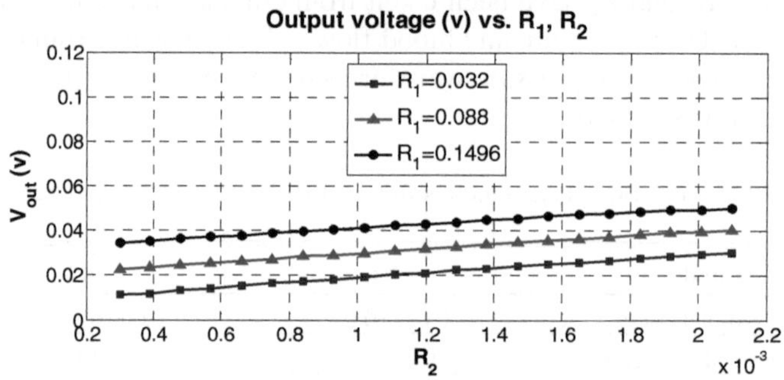

Fig. 4. Output mean voltage for R_1 and R_2 parameter sweep

Several sets of R_1 and R_2 parameters have been introduced in the model, to generate blood flow waveforms under various real ISR conditions. As shown in Fig. 4, worse ISR conditions, reflected in higher R_1 and R_2 values, increase the pressure gradient along the longitudinal axis of the e-stent, as well as the mean output voltage of the system. The simulation results show a similar behaviour to the one expected (1), and have been compared to experimental data obtained from clinical trials [10].

It must be noted that a fault-free analytical model of the sensor has been used to carry out the previous simulations. As will be seen in the next section the accuracy of the mathematical models under faulty conditions must be analyzed, in order to evaluate its influence in the behaviour of the whole device.

4 Capacitive MEMS Pressure Sensor

The principle of operation of MEMS capacitive pressure sensor is based on the concept of a two parallel plate capacitor, as mentioned in Section 3. Fig. 5 shows a typical cross-sectional view of this kind of sensors, where P is a uniformly distributed pressure applied to the sensor, w_0 is the center deflection of the membrane, t_g is the gap between the membrane and the fixed backplate when the external pressure is equal to the one in the cavity, and t_m is the thickness of the diaphragm.

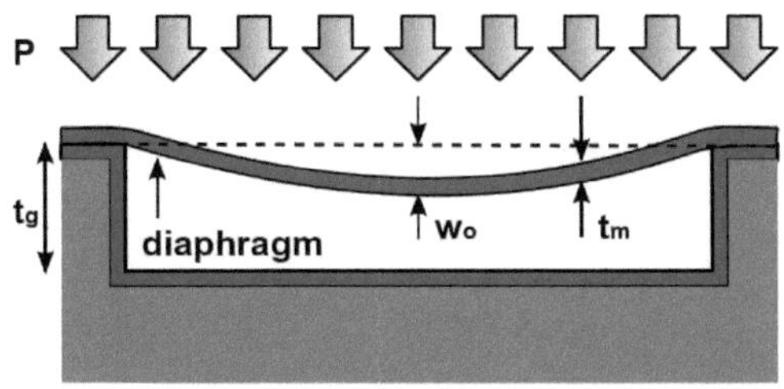

Fig. 5. Output mean voltage for R_1 and R_2 parameter sweep

The deflection of a fully clamped circular diaphragm can be analytically modeled as a function of the radial distance from the center of the plate [11]. To validate this approach, some assumptions must be considered [11]: (a) the material of the diaphragm must have isotropic mechanical characteristics; (b) the thickness of the metallic electrode on the plate has to be neglected; (c) the gap between the plates needs to be small compared to the lateral extents of the plates, so the electric field fringing effects can be depicted; (d) the residual stresses in the diaphragm are not considered. Once fulfilled the previous requirements, the deflection of a circular diaphragm under large deflection conditions ($w_0 > 30\%\ t_m$) can be stated as:

$$w(r) = w_0 \cdot \left[1 - \left(\frac{r}{a}\right)^2\right]^2 . \tag{2}$$

$$w_0 = \frac{3Pa(1-\upsilon^2)}{16Et_m^3} \cdot \frac{1}{1 + \frac{0.488w_0^2}{t_m^2}} . \tag{3}$$

Where r is the distance to the center of the diaphragm, a is the radius of the diaphragm, and w_0 is the maximum center deflection.

Once known the expression relating radial displacement and pressure applied to a fully clamped circular diaphragm, the sensor capacitance can be defined as:

$$C = \iint_A \frac{\epsilon_0 r dr d\theta}{t_g - w(r)} = \frac{\epsilon_0 A}{t_g}\sqrt{\frac{t_g}{w_0}}\tanh^{-1}\left(\sqrt{\frac{w_0}{t_g}}\right) = C_0\sqrt{\frac{t_g}{w_0}}\tanh^{-1}\left(\sqrt{\frac{w_0}{t_g}}\right) . \tag{4}$$

Where C_0 is the capacitance of the undeflected sensor, ϵ_0 is the dielectric permittivity of free space, and A is the common area of the plates.

Finite element analysis (FEA) techniques allow non-linear mechanical and electromagnetic simulations of complex structures. In the particular case of MEMS pressure sensors, FEA analysis enables the validation of the aforementioned analytical models under various environmental conditions, as well as fault injection in the structural model of the sensor. Fig.6 shows the analytical and FEA capacitance and deflection response of a pressure sensor with a polysilicon diaphragm (Young Modulus: E = 169 GPa; Poisson Ratio Coefficient: υ = 0.22) of 4μm thickness, a radius of 350μm and a sealed cavity of 2μm height.

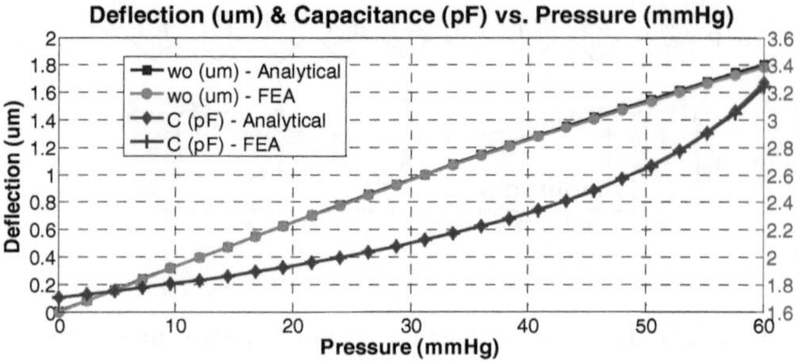

Fig. 6. Output mean voltage for R_1 and R_2 parameter sweep

Because of its small thickness and high deformation degree, the sensor's diaphragm can be considered to be the most vulnerable part to fabrication defects. As the critical element of a capacitive sensor, its analytical model response, obtained from (2), (3) and (4), has been compared to the one from a FEA simulation, in fault injection cases. Fig. 7 contains analytical and FEA results when a defect, in the form of a pyramidal protuberance, appears on the top diaphragm of the sensor. This type of defect has been reported to occur because of the presence of impurities during the anisotropic wet etching of a single crystal silicon; a widely used process to create membranes [12]. Fig. 7 (left) shows that the capacitance response of the sensor, under constant pressure conditions, decreases for higher pyramid base areas at the center of the diaphragm. Fig. 7 (right) contains the capacitance results, under similar pressure conditions, calculated for different locations of a pyramid with a base side of 50μm.

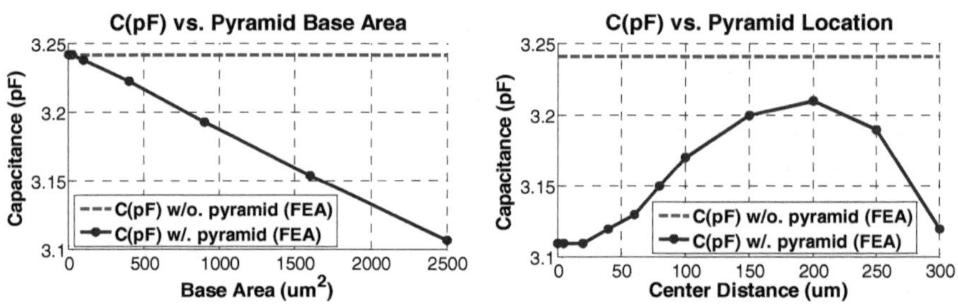

Fig. 7. Capacitance for different pyramid base sizes (*left*) and capacitance for different pyramid locations from the center of the diaphragm (*right*), under a pressure of 8kPa

As shown in Fig. 7, the presence of a pyramid on the top membrane of the sensor reduces its deflection and equivalent output capacitance value, implying a loss of sensor sensitivity which can seriously compromise its reliability.

As shown in Fig. 6, an acceptable analytical solution can be achieved in the fault-free case, based on (2), (3) and (4). However, this mathematical formulation is no longer valid for modeling a faulty membrane. Hence, it is necessary to create additional analytical membrane models under faulty conditions, considering deflections results obtained from FEA simulations; especially for those faulty cases in which the geometry and/or material properties of the membrane are altered.

5 Conclusions and Further Work

In this work, a behavioral model of an implantable device for ISR early detection has been presented. Test related problems for implantable capacitive MEMS pressure sensors have been considered; especially those cases which significantly affect the properties of the diaphragm. The deflection issue of circular membranes has been proved to be solvable for a fault-free case, using an accurate analytical behavioral model. Nevertheless, the aforementioned mathematical model evidences a lack of accuracy when tested under fault-injection conditions. The creation of additional models to precisely describe the behavior of a flexible diaphragm under faulty scenarios, using deflection results from FEA simulations, seems to be an essential requirement to completely characterize implantable MEMS pressure sensors.

Acknowledgments. This work was carried out in collaboration with the cardiology department of the Universitary Hospital Marqués de Valdecilla, Santander (Spain). We would like to acknowledge the funding for this research from Ministerio de Ciencia e Innovación (Spain), Plan Nacional de I+D+i, project TEC2010-19122.

References

1. OECD, Health at a Glance 2011: OECD Indicators, OECD Publishing (2011), http://www.oecd-ilibrary.org/social-issues-migration-health/health-at-a-glance_19991312
2. Hoffmann, R., Mintz, G.S.: Coronary in-stent restenosis - predictors, treatment and prevention. European Heart Journal 21, 1739–1749 (2000)
3. Takahata, K., Gianchandani, Y.B., Wise, K.D.: Micromachined Antenna Stents and Cuffs for Monitoring Intraluminal Pressure and Flow. Journal of Microelectromechanical Systems 15(5), 1289–1298 (2006)
4. Wang, M., Chen, J.: Volumetric Flow Measurement Using an Implantable CMUT Array. IEEE Transactions on Biomedical Circuits and Systems 5(3), 214–222 (2011)
5. Chow, E.Y., Chlebowski, A.L., Chakraborty, S., Chappell, W.J., Irazoqui, P.P.: Fully Wireless Implantable Cardiovascular Pressure Monitor Integrated with a Medical Stent. IEEE Transactions on Biomedical Engineering 57(6), 1487–1496 (2010)
6. Zhao, W., Wang, C., Nakahira, Y.: Medical Application on Internet of Things. In: Proceedings of IET International Conference on Communication Technology and Application (ICCTA 2011), pp. 660–665 (2011)
7. Young, D.F.: Some factors affecting pressure-flow relationships for arterial stenoses. In: ASME Conf. Appl. Mech. Bioeng. Flu. Eng., pp. 87–90 (1983)
8. Marques, K.M.J.: Combined flow and pressure measurements in coronary artery disease. Vrije Universiteit, Amsterdam (2008)
9. Arfah, N., Alam, A.H.M.Z., Khan, S.: Capacitance-to-Voltage Converter for Capacitance Measuring System. In: 4th International Conference on Mechatronics (ICOM), pp. 1–4 (2011)
10. Rothman, A., Perry, S.B., Keane, J.F., Lock, J.E.: Early results and follow-up of balloon angioplasty for branch pulmonary artery stenoses. Journal of the American College of Cardiology 15(5), 1109–1117 (1990)
11. Timoshenko, S.: Theory of Plates and Shells. McGraw-Hill, New York (1940)
12. Landsberger, L.M., Nashed, S., Kahrizi, M., Paranjape, M.: On Hillocks Generated During Anisotropic Etching of Si in TMAH. Journal of Microelectromechanical Systems 5(2), 106–116 (1996)

Internet of Things in Psoriasis Assessment and Treatment

Simona Maria Banu, Laura Mădălina Dascălu, and Gheorghe Toacşe

Transilvania University of Braşov, România,
Department of Electronics and Computers,
Faculty of Electrical Engineering and Computer Science
{simonam.banu}@gmail.com, {madalina.dascalu,toacseg}@unitbv.ro

Abstract. Within the Internet of Things (IoT) paradigm, an everyday object can be transformed into a smart object, able to sense, interpret and react to the environment. IoT is bringing new ways of communicating between people and *things* (objects) to reach common goals, bringing a high impact on everyday-life. The aim of this paper is to present how people with psoriasis and their doctors can benefit from the IoT advantages. There is presented a proposed system for surveillance and treatment plan for patients suffering from psoriasis using assisted IoT and Computer Vision technologies.

Keywords: Internet of Things, psoriasis severity, Expert System.

1 Introduction

The Internet of Things (IoT) is meant to refer to a vision of an Internet where there is a strong connectivity between anyone and anything, and takes place anywhere and anytime. It envisions a future in which physical and digital *things* can be linked to enable and support new classes of applications and services. There are many fields in which IoT has started to bring important contribution: health and e-health, domotics, monitoring systems, logistics [1].

The use of cross-layer communication schemes to offer adaptive solutions for the IoT is motivated by the high heterogeneity in the hardware capabilities and the communication requirements among *things*. IoT permits a global communication on existing layered solutions. It also provides efficient and reliable end-to-end communication. Among the possible IoT domains and corresponding applications, we can consider the following: healthcare (tracking, identification, authentication, data collection), smart environments (smart homes, industrial plants), transportation and logistics (assisted driving, augmented maps), personal and social (social networking, historical queries) [2].

There is a growing demand for more advanced and better healthcare solutions with the focus on: a) decrease of healthcare costs; b) a need for better and more efficient outcomes. Our objective is to bring some preliminary ideas/evidence of improving the quality of healthcare and/or reduce the cost of it.

L.M. Camarinha-Matos, S. Tomic, and P. Graça (Eds.): DoCEIS 2013, IFIP AICT 394, pp. 172–179, 2013.

Psoriasis is a skin condition that develops when the immune system malfunctions. It is estimated that about 5% of the global population have this genetically-determined chronic skin disease [3]. People with psoriasis may have discomfort in the form of pain and itching, and emotional distress (see Fig. 1).

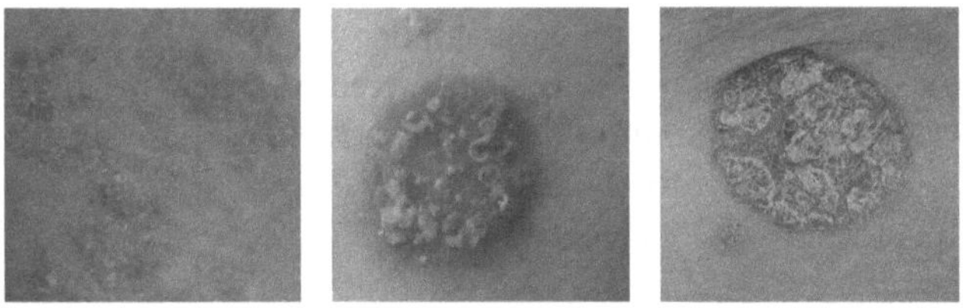

Fig. 1. Different psoriasis lesions photographed under different illumination conditions

In the next sections we will cover the following aspects: (i) technical contribution to the Internet of Things; (ii) related work in both Internet of Things technology and also in Expert Systems for dermatology (psoriasis in particular); (iii) detail presentation of the proposed system; (iv) discussion of preliminary results; and (v) final conclusions and future work.

2 Relationship to Internet of Things

In this paper, we propose a prototype that aims to integrate physical and digital *things* (IoT technology) into the everyday life of patients and their medical doctors. We try to respond to the needs of patients and healthcare professionals by proposing a new expert system that envisions the IoT paradigm. This expert system will objectively assess the erythema (degree of redness) for psoriasis severity evaluation. The IoT solution is meant to give guidance to medical doctors in prescribing the best treatment for patients suffering from psoriasis disease.

The system's architecture is based on smart objects (*things*) and is designed to meet primarily the expectations of patients and doctors. The patients would benefit from a high quality personalized healthcare on an equal access basis, regardless of their social status and location, with full respect of their privacy. Also they can stay at home as much as possible due to the ability of the system to collect data provided by the patients themselves (this data being further evaluated by the doctor who will have access to it through an intelligent web application). On the other hand, the healthcare professional would be able to follow up any development in the patient's health condition at home and also, would have access to the health information of other patients through this healthcare system of sharing information between medical facilities (by having an RFID tag reader present which would be able to read the tag inserted in the patient's smart phone).

3 Related Work

Psoriasis is a long lasting skin disease depicted when patches of skin become inflamed and scales develop. Psoriasis Area and Severity Index (PASI) [4] is an evaluation technique that measures the overall psoriasis severity and coverage (see Fig. 2). PASI is the most used measure in clinical trials when assessing the psoriasis severity and therefore, giving the proper treatment.

The PASI score computation is done in a subjectively manner by evaluating three characteristics (erythema, induration and scaliness) of psoriatic lesions (see Fig. 2). The Abbott PASI meter tries to standardize these psoriasis characteristics into four types (Mild, Moderate, Severe and Very severe).

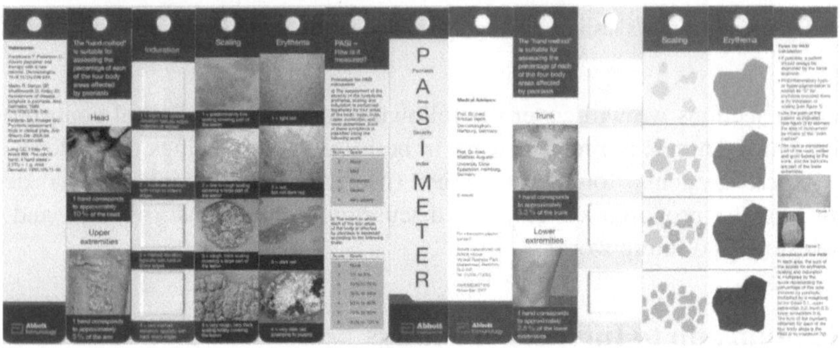

Fig. 2. Abbott PASI meter

There are several papers that give a solution to this subjective evaluation by objectively assess the lesion thickness and erythema for psoriasis severity evaluation [5], [6].

Expert Systems are a sub-type of Artificial Intelligence systems used in clinical diagnosis. They contain medical knowledge and are able to output logical conclusions given as entry point specific data from individual patients. Although there are many variations, the knowledge within an expert system is typically represented by a set of rules or criteria.

Recent research presents different expert systems for differential diagnosis of erythemato-squamous diseases. For example, the paper presented in [7] aimed the implementation of a visual tool for differential diagnosis of skin diseases, which is a difficult problem in dermatology, because several diseases share almost the same clinical features of erythema and scaling.

The main objective of our expert system is not to make a differential diagnosis, but to classify the already known psoriasis disease in order to further help the dermatologist in giving the correct treatment.

The RFID technology (Radio Frequency IDentification) can be used to put identification labels (e,g. tags) into objects (tablets, mobile phones). If the consecrated reader of these tags is in next proximity of the tagged object, the information related to it can be accessed without any physical contact. In this way, the whole process of patient's identification and database access is simplified and the timings are appreciable reduced [8].

There are several medical applications that make use of RFID technology [9], [10], [11].

In the following section we will present our proposed system used for surveillance and treatment plan for patients suffering from psoriasis disease.

4 The Proposed System

In the Romanian healthcare sector, the information related to each individual patient is scattered among several hospitals and medical clinics, with no connection among them. There is a major lack of interoperability due to no standardized content related to patients' medical records. The data provided in each database is sometimes incomplete and there is no way to update it by synchronizing it with different other databases from other cities (or even other medical clinics from within the same city).

Considering the above-mentioned disadvantages we propose in this paper an expert system that can be used for the objective assessment of psoriasis. This system is integrated in a bigger IoT architecture which includes three main layers: a perception (or sensor) layer, a network layer and an application layer. The perception layer gathers information from objects which are transmitted further to the application layer through the network layer.

The basic idea behind IoT is the fact that we are surrounded by many things or objects which interact with each other to reach common goals [12]. Objects or *things* can be considered the following: RFID tags, sensors, actuators, mobile phones. A series of objects that appear in the proposed system are presented in Fig. 3.

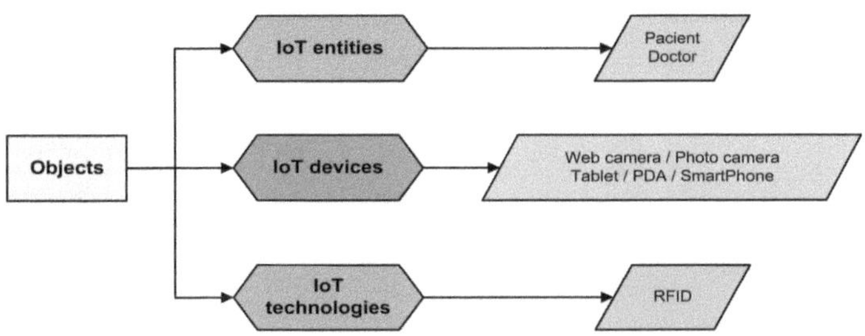

Fig. 3. Objects *(things)* present in the proposed system

The proposed IoT architecture is presented in Fig. 4. The system is based on the RFID technology which includes the following main components:

- RFID readers placed in hospitals, medical clinics etc.
- Smart phones equipped with RFID tags;
- An EMR (Electronic Medical Record) server for storing the medical records of individual patients regardless their location;
- A HL7 server for interoperability with other clinical software.

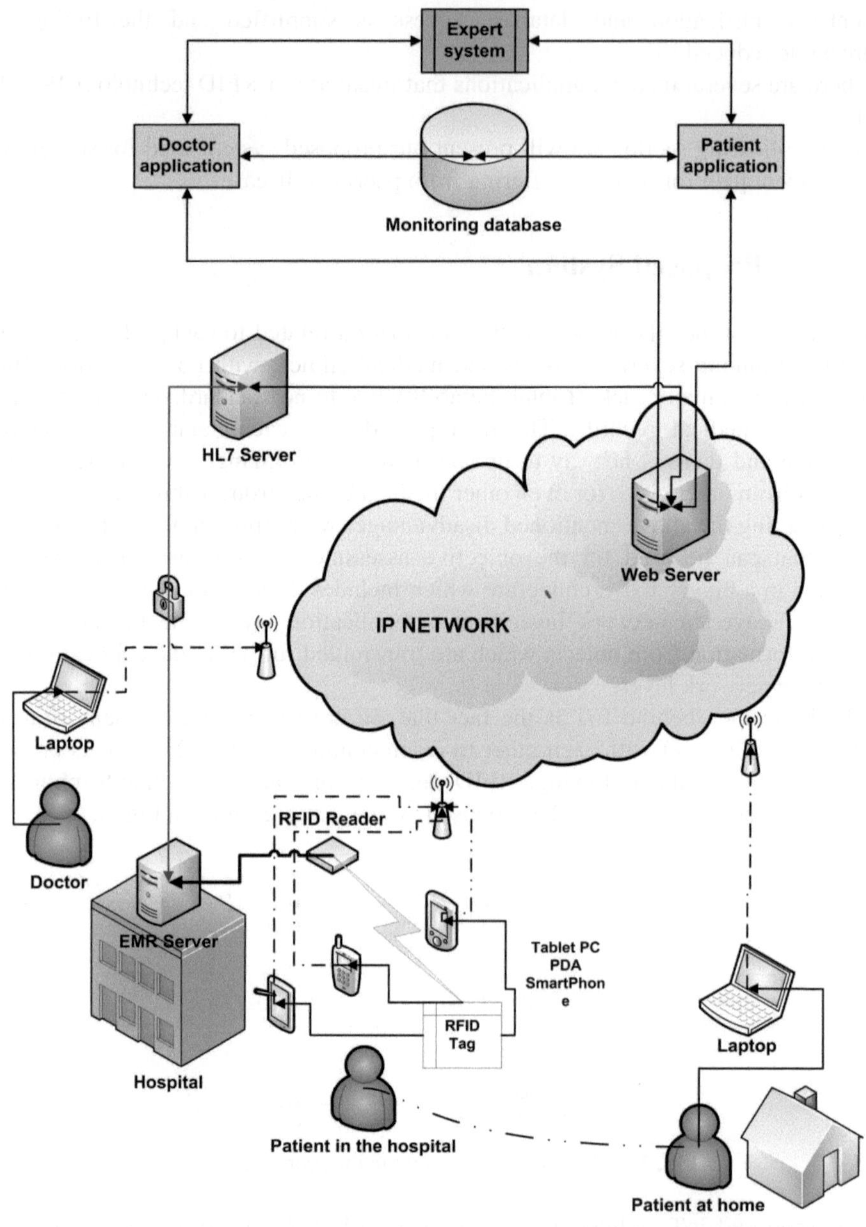

Fig. 4. General view of the IoT architecture for the proposed system

Even though the proposed RFID-based system is similar to the SIMOPAC system described in [13] the application types and the main idea are different. Therefore, our proposed expert system has a more specific task, by objectively evaluating the erythema severity of psoriasis lesions regardless the location of the patient (at a clinical facility or at home) and also by updating the medical records through the intelligent applications which can be accessed by doctors and also by patients.

The ability to integrate and exchange information among similar systems will be achieved through HL7 (Health Level Seven) standards. This type of standard is promoted in the healthcare industry due to its quality, accuracy, and efficiency characteristics [14].

In the application layer the two types of applications (doctor's and patient's application) are present. They are connected to the expert system (see Fig. 5) which returns the degree of erythema for the psoriasis lesions. The psoriasis lesions are first photographed by the patient at home or by the doctor in the hospital. The images are then uploaded through the specific applications onto the system and the expert system is processing this input. From the answer given by the expert system the doctor can give the proper treatment and also, can update the medical data related to each patient.

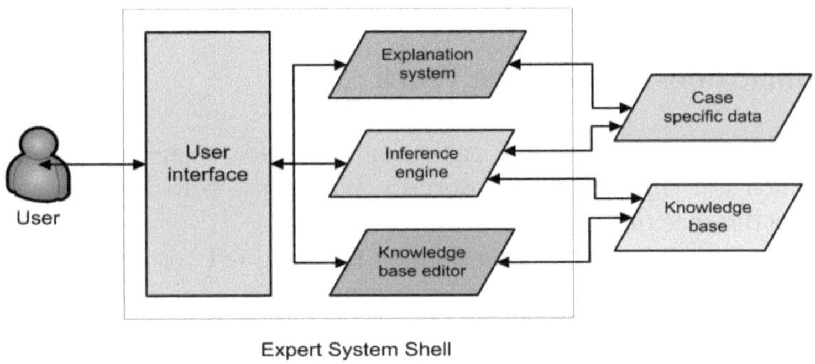

Fig. 5. Expert System architecture

The presented IoT solution enables the system management, respectively the disease pathway. It brings transparency in the responsibility of impact, it reduces the subjective visual analyze of psoriasis by doctors and also gives guidance during treatment plan.

The main benefits of the proposed system consist in: immediate access to data, the possibility of transmitting data when it is available (patients are uploading the pictures whenever they want and doctors can access uploads whenever they want), direct feedback on the treatment efficiency, the time of every day visual diagnosis during the treatment is reduced, all the process of clinical pathway is more efficient and helps the patient to comply with his/her treatment routine.

The main risks that might appear are the information accuracy, which depends mostly on the quality of the image uploaded by the patients and the need of more evidence (the induration or thickness measure for the psoriasis lesion which can be evaluated only by the medical doctor) for treating the disease.

5 Discussions of Results

In order to develop an expert system that can be used in the dermatology field we have to first classify the psoriasis severity for different lesions. The input to our expert system is composed of digital images depicting psoriatic lesions. Due to the different illumination conditions a color correction step is needed.

In the research done in [6] the authors propose an approach for objective assessment of the erythema degree. Their approach comprises of three steps: (i) reference-based color correction, (ii) lesion segmentation and (iii) erythema classification and ranking. Due to the correction step, the color segmentation and classification could be correctly performed, and also, the specialist could assess the erythema severity with less ambiguity when being presented with images of lesion images previously color corrected.

We intend to use a similar approach when developing our psoriasis evaluation expert system. Having a limited set of images containing lesions of psoriasis we are in the process of enlarging our database with different other images (test cases) to get a higher accuracy when objectively assess the erythema severity.

6 Conclusions and Future Work

We strongly believe that the IoT can offer assistance, support and follow-up of patients which are suffering from psoriasis symptoms, both in hospitals/treatment clinics and at distance (home).

We have presented in this paper a solution for using IoT for improving the quality of healthcare in psoriasis diagnosis and treatment. There were presented two possible applications that can be used both by patients and doctors. The applications show how patients can achieve a better quality of life while treating the disease, during the prescript treatment.

Using the proposed Internet platform the costs for transferring patients from one hospital to another, for diagnosis and/or treatment, is reduced. Also, the system can guarantee the privacy and security of patient's records.

In the future, we intend to implement all the necessary *things* for bringing to life the system for surveillance and treatment plan for patients suffering from psoriasis disease.

Acknowledgments. This paper is partially supported by the Sectoral Operational Programme Human Resources Development (SOP HRD), ID76945 financed from the European Social Fund and by the Romanian Government.

References

1. Gusmeroli, S., Haller, S., Harrison, M., Kalaboukas, K., Tomasella, M., Vermesan, O., Vogt, H., Wouters, K.: Vision and challenges for realizing the internet of things. In: Friess, P., Guillemin, P., Sundmaeker, H., Woelfflé, S. (eds.) European Commission (2010)
2. Atzori, L., Iera, A., Morabito, G.: The internet of things: a survey. Computer Networks (Elsevier) Journal 54, 2787–2805 (2010)
3. Raychaudhuri, S., Farber, E.: The prevalence of psoriasis in the world. Journal of European Academy of Dermatology and Venerology 15(1) (January 2001)

4. Langley, R.G., Ellis, C.N.: Evaluating psoriasis with psoriasis area and severity index, psoriasis global assessment, and lattice system physician's global assessment. Journal of the American Academy of Dermatology 51(4), 563–569 (2004)
5. Ahmad Fadzil, M.H., Fitriyah, H., Prakasa, E., Nugroho, H., Hussein, S.H., Affandi, A.M.: Objective assessment of psoriasis lesion thickness for PASI scoring using 3d digital imaging. Proceedings of World Academy of Science, Engineering and Technology 63, 109–115 (2010)
6. Ivanovici, M., Banu, S.M., Irimie, M., Căliman, A., Richard, N.: Objective assessment of erythema for psoriasis severity evaluation. Submitted to International Symposium on Biomedical Imaging: From Nano to Macro (2013)
7. Güvenir, H.A., Emeksiz, N.: An expert system for the differential diagnosis of erythemato-squamous diseases. Expert Systems with Applications 18, 43–49 (2000)
8. Laranjo, I., Macedo, J., Santos, A.: Internet of Things for Medication Control: Service Implementation and Testing. Procedia Technology 5, 777–786 (2012)
9. McCall, C., Maynes, B., Zou, C.C., Zhang, N.J.: RMAIS: RFID-based medication Adherence Intelligence System. In: 2010 Annual International Conference of the IEEE Engineering in Medicine and Biology Society (EMBC), pp. 3768–3771 (2010)
10. Yeung, C.L., Kwok, S.K., Mui, H.C.: An Investigation of an RFID-based Patient-tracking and Mobile Alert System. International Journal of Engineering Business Management (2011)
11. Vilamovska, A.M., Hattziandreu, E., Schindler, R., Van Oranje, C., DeVries, H., Krapelse, J.: RFID Application in Healthcare – Scoping and Identifying Areas for RFID Deployment in Healthcare Delivery. RAND Europe (February 2009)
12. Giusto, D., Iera, A., Morabito, G., Atzori, L.: The Internet of Things. Springer (2010)
13. Turcu, C., Cerlinca, T., Turcu, C., Cerlinca, M., Prodan, R.: An RFID and multi-agent based system for improving efficiency in patient identification and monitoring. WSEAS Transactions on Information Science and Applications 6(11), 1792–1801 (2009)
14. Shaver, D.: HL7 101: A Beginner's Guide. For the Record 19(1), 22 (2007), `http://www.fortherecordmag.com/archives/ftr_01082007p22.shtml`

Part VIII
Robotics and Manufacturing

On Collaborative Aerial and Surface Robots for Environmental Monitoring of Water Bodies

Eduardo Pinto[1], Pedro Santana[1,2], and José Barata[1]

[1] Universidade Nova de Lisboa, Faculdade de Ciências e Tecnologia, Departamento de Engenharia Electrotécnica, 2829-516 Quinta da Torre, Portugal
[2] Instituto Universitário de Lisboa (ISCTE-IUL), Departamento de Ciências e Tecnologias da Informação, Avenida das Forças Armadas, 1649-026 Lisboa, Portugal

Abstract. Remote monitoring is an essential task to help maintaining Earth ecosystems. A notorious example is the monitoring of riverine environments. The solution purposed in this paper is to use an electric boat (ASV - Autonomous Surface Vehicle) operating in symbiosis with a quadrotor (UAV – Unmanned Air Vehicle). We present the architecture and solutions adopted and at the same time compare it with other examples of collaborative robotics systems, in what we expected could be used as a survey for other persons doing collaborative robotics systems. The architecture here purposed will exploit the symbiotic partnership between both robots by covering the perception, navigation, coordination, and integration aspects.

Keywords: Collaborative robotics, ASV, UAV, remote riverine monitoring.

1 Introduction

The monitoring of water in riverine environments is of enormous importance not only in controlling some of the most delicate ecosystems (some of the highest concentration of life forms are in the estuaries) but also for economic reasons, since almost all the potable water come from rivers or lakes making the control of water quality essential for industry, agricultural and human consume. It is an enormous challenge attending to the actual robotics state of the art to build a single robot with all the capabilities to survive and to be autonomous in this kind of environment. Although sensor networks are becoming popular solutions for this kind of problem, they are seriously limited by their static characteristics. Several projects [1], [2] had dealt with this limitation by introducing a very large number of sensors or combining static sensors with sensors with more or less locomotion capabilities or even using ASV - Autonomous Surface Vehicles. The use of small (ASV) are also becoming a reasonable and attractive alternative [3], [4]. However, providing these boats with full autonomous behaviour is not a trivial task, there are many challenges, namely: The low height the sensors equipping these robots are from the water's surface limits their ability to perceive the far-field and consequently to ensure safety by means of a proper path planning strategy. Another related limitation is their inability to inspect

L.M. Camarinha-Matos, S. Tomic, and P. Graça (Eds.): DoCEIS 2013, IFIP AICT 394, pp. 183–191, 2013.

riverbanks, which are key elements in the riverine ecosystem. In RiverWatch we overcome the limitations imposed from observing the environment from a low vantage point, by pairing the ASV with a small unmanned aerial vehicle (UAV). With an aerial perspective, the UAV will extend the visual field of the ASV as well as it will enable the survey of riverbanks. Equipped with solar panels, the ASV will be able to do energy harvesting for itself and for the UAV, in a mechanism known in biology as Trophallaxis. The project will exploit this symbiotic partnership between both robots by covering the perception, navigation, coordination, and integration aspects making the system extremely flexible, reliable and with extended capabilities for adaptation, self-organization and self-development.

We expect the RiverWatch project to impact on scientific (novel perception, navigation, and coordination algorithms), technological (novel autonomous multi-robot integrated platform), and application (novel environmental application scenario) dimensions.

1.1 Progress beyond the State-of-the-Art

In order to overcome some of the limitations of static sensor networks, [5] propose the addition of a small ASV into the monitoring system. Although this approach improves the resolution of the performed analysis, the overall system continues to be spatially constrained. Several solutions have been proposed for dynamic monitoring of riverine ecosystems. A simple solution is to deploy drifters that flow with the river current gathering data along way [6]. This solution is limited to wide and obstacle free river streams, as otherwise drifters may get stuck. A more efficient solution is to use a set of ASVs [7], [8] to navigate along rivers while collecting environmental data. The use of robotic sensors is motivated by the fact that they can achieve sufficient spatial coverage to perform complete measurements, i.e., they are not limited to a set of discrete sampling points. Although many work exists in aquatic robots, some already capable of performing energy harvesting [9] and multi-vessels coordinated behaviour [10], many challenges are still open. One particularly demanding is the ability to robustly segment the river from land and obstacles up to the far-field. The low vantage point afforded by medium size surface vehicles poses additional difficulties to this problem. Previous work [11] capitalized on the benefits of multi-robot systems to handle some problems arising from the low vantage point, by using an helicopter to provide the human operator with improved situation awareness in a hurricane post-disaster situation. In this project we intend to exploit the same idea but in this case to extend the surface vehicle's situation awareness. Hence, we consider the coordination between both robots for autonomous behaviour, instead of the context of tele-operation. Although previous work considered the docking of a UAV on a ground-based platform [12], [13], [14] docking on a moving aquatic platform has not been reported yet. Waves, wind, and displacement are variables that must be taken into account by the model. We will depart from previous work by considering the problem not only under the control perspective. A smart mechanism will be used to grasp the UAV, thus reducing the need to hover close to the landing surface, where chaotic airflow complicates the process. Furthermore, both ASV and UAV will negotiate a rendezvous site where it is more likely to occur successfully, such as by searching a less windy spot. This may require the ASV to actively search for this spot while the UAV is performing its delegated surveying task.

For short range obstacle detection in the ASV, we will mostly recur to 3-D data provided by stereo vision, laser scanner, and underwater sonar scanner. For this purpose we intend to adapt some work on all-terrain obstacle detection [15] [16] to the domain in question. An interesting point of this work is the use of visual attention parallel mechanisms, which we have been shown that, if operating in an intricate way with obstacle avoidance, the amount of processing can be considerably reduced [17], and as a consequence, the computational requirements as well. A key aspect of obstacle avoidance in aquatic environments, yet to be solved, is to filter out false positives induced by waves and large ripples in the water. For this purpose we intend to exploit the fact that false positives in this domain exhibit structured spatio-temporal patterns (propagating waves). These patterns can be determined in volumetric terms using the range image produced by the stereo-vision sensor and in appearance terms as produced by the aerial image provided by the UAV.

To enable path planning in aquatic environments [18], long-range water/land segmentation of the environment is required. This includes the ability to detect sand banks. For this purpose we will recur mostly to appearance models learnt under the self-supervised learning paradigm, similar to work on ground-based vehicles [19]. In these models, near-field range data provided by stereo/laser is used to label patches of the robot's input image as belonging to the obstacle or non-obstacle class. In our case the classes are water and land and the input images are the ones obtained by both ASV and UAV. Then, an on-line machine learning mechanism can be used to learn a classifier in the appearance space given the provided water/land labels. This way the system learns to segment water from land in the appearance-space of the images obtained by both robots. This learning process requires to register the obstacle detection results obtained with range data onto the aerial image produced by the UAV. This in turn will require both robots to localise each other with respect to the other. We expect to solve this problem using their global localisation mechanism (i.e., GPS based) in addition to having the UAV performing visual tracking of the ASV. The transformations required to match the image obtained from an omnidirectional camera mounted on the ASV with the aerial image provided by the UAV may be useful to provide an additional cue on both robots' relative localisation. The proposed method differs considerably from previous work on river [20] and coast-line [21] detection from aerial images, where human intervention was key to assist the learning process. We also expect to exploit maps provided by the Google Maps API, where rivers are already segmented, to score potential river hypotheses generated by the perceptual system against the map.

With the purpose of reducing ambiguities in images classification process, we consider the possibility of having the classifiers as dynamical entities exploring the image to build contextual information [22]. Context is well recognised as a key component of any robust and parsimonious perceptual system [23]. In addition to 3-D information, the sky/water interface can also be useful to generate training labels. This innovative approach can exploit cues regarding the profile of the detected skyline.

Shoreline orientation and position is also an important cue for ASV safe navigation. We intend to use particle swarms to detect and track shorelines. For this purpose we expect to use an omni-directional camera. Particle swarms have recently exhibited interesting properties, even better than particle filters, for object tracking [24]. Moreover, making these particles sensitive to contextual cues, we expect to improve detection rate while reducing computational cost. These techniques will also be used to determine shoreline from the aerial images obtained by the UAV, besides being equipped with a conventional photographic camera, it will be equipped with a near-infrared camera, as it is recognizably useful for water/land segmentation [25]. Our expectation is that the fusion of all these methods in addition to the long-range water/land segmentation mechanism results in a robust system, which in turn reduces the accuracy requirements for each of the techniques alone.

2 Contribution to the Development of the Internet of Things

Remote monitoring is an essential task to help maintaining Earth ecosystems and will be without doubt influenced by the Internet of Things. The use of network sensors for monitoring some environmental aspects is already a reality [26]. Several TIWS – Tsunami International Warning Systems (IOTWS,NEAMTWS,PTWC, etc.) based on fixed DART© stations are actually working all over the world, coordinated by the Intergovernmental Oceanographic Commission of UNESCO integrated within the program GOOS – Global Ocean Observing System. Inspired on these systems and in the IoT the RiverWatch architecture will enable the cooperative robots of the system (UAV and the ASV) to interact and communicate among themselves and with the environment by exchanging data and information 'sensed' about the environment, while reacting autonomously to the 'real/physical world' events with or without direct human intervention depicted as one of the most important aspects in European IoT roadmap. It is our goal that with RiverWatch and by using ROS as the operating system with its capacity of making transparent the communications between several nodes and the recent developments in support to IPv6 to allow those requirements.

3 RiverWatch Architecture

We use the ROS operating system as the background of our system with all the advantages that cross-language inter-process communication system will allow. We´ve decided to adopt as possible open sources solutions as pointed on the following sub-chapters.

3.1 UAV (Multi-rotor, Variable Configuration)

We had evaluated several platforms Comercial: AirRObot, Asctec, Microdones, Draganfly, Cyberquadand several open source: MikroKopter, Mikro'sAeroquad, NG-UAVP, UAVX & UAVP, OpenPilot, Arducopter, Multipilot 32.The price of these commercial solutions ranges from 5000€ to more than 20000€. All of them, uses some patented and closed mechanisms of control what was unacceptable to our project. With the open-source projects, we have access to all the code and in some of them the solutions available are more complex and with more functions than the commercial solutions, and so we decided to build a platform based in open source solutions. For the low level control boards, we´ve adopted the Multipilot 32 solution that is totally open source, compatible with the project Arducopter (a very large active community) and offers the best processor and IMU capabilities. For the mechanical platform we´ve selected the open source Mikrokopter solution that offers reconfigurable frames that allow to build any configuration: Quadrotor, Quadrotor coaxial, X6 , X6 axial, Octo. Power distribution boards that allow a very easy mount with a minimum of connections and wires, very good motors (MK3638) and I2C high speed motor controllers. The batteries selected were the Maxamps 11Ah 4S. For manual and emergency control we will use the system OSRC.

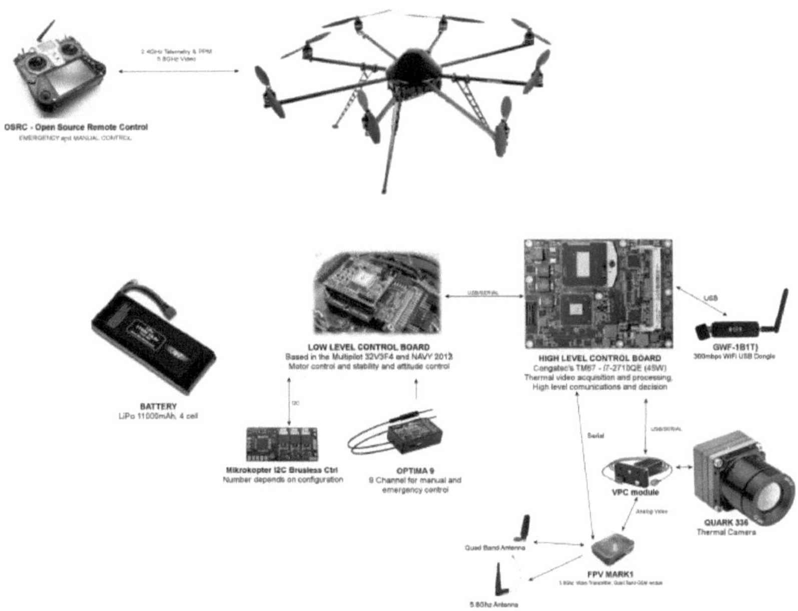

Fig. 1. UAV components and architecture

3.2 USV

In partnership with the company Jetbuster we´ve decided to use the commercial hulls from model JETRIDER XXL and the water turbine propulsion system. We´ve adopted a modular configuration with 4 main components (2 hulls, connection platform and electronics boxes) that will allow the easy setup and transport of the ASV. The propulsion unit based on ducted water jets systems have great advantages in shallow waters when compared with tradition propellers, namely the possibility to navigate with few centimetres of water, more resistance to impacts, etc.

The on board sensors are divided in two categories:"Localization and attitude" and "Obstacle detection". In "Localization and attitude" we have a PB200 weather station and a CS4500 water speed sensor from Airmair and a complete GPS RTK from Ashtech (Proflex 800 base and rover). In "Obstacle detection" we have a multibeam DT837B sonar from Imagenex mounted in a tilt system based on servos EX-106 from Dynamixel, sonar that will allow us to create a 3D map of the bottom of the river to detect and predict what will be the best way in the presence of sand banks or other subaquatic obstacles. For collision detection, the main sensor we will use is the LIDAR LD-LRS2100 system from Sick. For segmentation water algorithms and detection of obstacles, we will use a Ladybug3 360 degrees camera from Pointgrey. For low visibility (nigh, fog, etc) and for segment the ASV from the water using its special thermal signature, we will use thermal cameras. In the ASV the FLIR Quark 640 with wide angle lens, while for the UAV the camera selected was a Quark 336. For special safety precautions we will have on board an active radar deflector, infrared and visible light marks and redundant communications with VHF modem (ADL Vantage radio system), Long range Wifi and GSM modem as we could see in Fig.2.

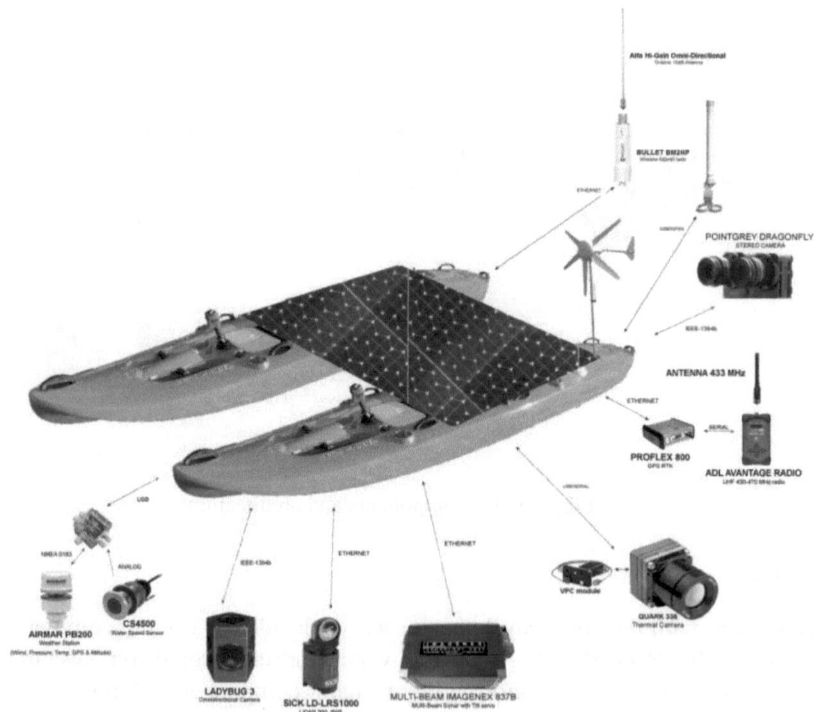

Fig. 2. USV components and architecture

3.3 Operations Centre

In its first version, the control centre has been developed to run as a web-based application. This has the advantage of allowing the use of a simple web-browser for interacting with the robotic system. However, it is also relevant to enable the integration of the control centre software in custom applications. For instance, the control centre can be part of a larger software package that not only allows human-robot interaction but also integration with simulators and information management systems. Bearing this in mind, the control centre has been wrapped as a QT component that can be integrated in any QT application. An important part of this software package will be the GSC – Ground Station Control that will allow to have access to all the attitude information from ASV and from the AGV. Since we adopted the open-source solution based the project Multipilot32, that follow the MAVLINK protocol there are several Mission Planners open source available. In Fig.3 we can see the hardware to run the GSC.

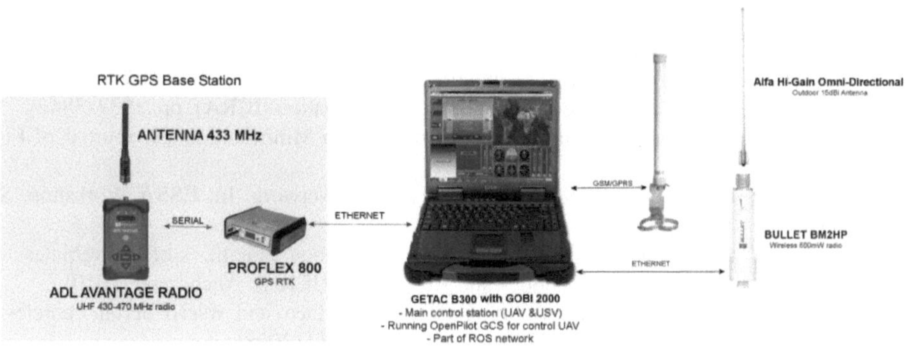

Fig. 3. Control center hardware

4 Conclusions and Further Work

The use of robots working in a symbiotic relation is a solution that shows very promising results, although the topic is still in its infancy. Some results are: Increase of autonomy of the entire system, the increase in flexibility and reliability. There are however drawbacks and problems that must be yet solved. They constitute challenges that could lead to promising research directions. As examples: Intelligent power share mechanisms, intelligent docking, identification of each robot (not only the robot but also its relative orientation) in the system by the others, the possibility of sensorial expansion of the main robot of the system or the whole system itself, etc. With RiverWatch, we hope to contribute with some answers. Many more questions will prevail and it's our hope that this paper contributes to arouse some curiosity in the reader and makes him think about them.

Acknowledgments. This work was developed within the scope of the RIVERWATCH experiment from the European FP7 Large Scale Integrating Project ECHORD and it was also partially supported by CTS multi-annual funding, through the PIDDAC Program funds.

References

1. Bayen, A.: Floating Sensor Network, `http://float.berkeley.edu/`
2. Beacon Institute for Rivers and Estuaries: River and Estuary Observatory Network (REON), `http://www.bire.org/institute/innovation.php`
3. Manley, J.: Unmanned surface vehicles, 15 years of development. In: OCEANS 2008 (2008)
4. Fallon, M.F., Papadopoulos, G., Leonard, J.J.: Cooperative AUV navigation using a single surface craft. In: Howard, A., Iagnemma, K., Kelly, A. (eds.) Field and Service Robotics. STAR, vol. 62, pp. 331–340. Springer, Heidelberg (2010)
5. Sukhatme, G.: Design and Development of a Wireless Robotic Networked Aquatic Microbial Observing System, pp. 205–215 (2007)
6. Emery, L.: Drifting buoy for autonomous measurement of river environment. In: MTS/IEEE OCEANS, pp. 1–8
7. Bhadauria, D.: A Robotic Sensor Network for monitoring carp in Minnesota lakes. In: IEEE International Conference on Robotics and Automation (ICRA), pp. 3837–3842
8. Tokekar, P.: A robotic system for monitoring carp in Minnesota lakes. Journal of Field Robotics (2010)
9. Low, K.: Robot Boats as a Mobile Aquatic Sensor Network. In: ESSA Workshop, San Francisco
10. Arrichiello, F.: Cooperative caging using autonomous aquatic surface vehicles. In: International Conference on Robotics and Automation (ICRA), Anchorage (2010)
11. Murphy, R.: Cooperative use of unmanned sea surface and micro aerial vehicles at Hurricane Wilma. Journal of Field Robotics 25, 164–180 (2008)
12. Herisse, B.: The Landing Problem of a VTOL Unmanned Aerial Vehicle on a Moving Platform Using Optical Flow. In: IEEE/RSJ 2010 International Conference on Intelligent Robots and Systems, IROS (2010)
13. Wenzel, K.E., et al.: Automatic take off, tracking and landing of a miniature UAV on a moving carrier vehicle. Journal of Intelligent & Robotic Systems (2011)
14. Srinivasan, M.V., et al.: How honeybees make grazing landings on flat surfaces. Biological Cybernetics (March 2000)
15. Santana, P.: A Saliency-Based Solution for Robust Off-Road Obstacle Detection. In: IEEE International Conference on Robotics and Automation (ICRA), Anchorage (2010)
16. Santana, P., Guedes, M., Correia, L., Barata, J.: Saliency-Based Obstacle Detection and Ground-Plane Estimation for Off-Road Vehicles. In: Fritz, M., Schiele, B., Piater, J.H. (eds.) ICVS 2009. LNCS, vol. 5815, pp. 275–284. Springer, Heidelberg (2009)
17. Santana, P.: Swarm Cognition on Off-Road Autonomous Robots. Swarm Intelligence (2010)
18. Singh, A.: Efficient informative sensing using multiple robots. J. Artificial Intelligence Research 34, 707–755 (2009)
19. Dahlkamp, H.: Self-supervised monocular road detection in desert terrain. In: Robotics: Science and Systems, RSS (2006)

20. Rathinam, S.: Autonomous Searching and Tracking of a River using an UAV. In: American Control Conference, New York, pp. 359–364
21. Xu, A.: A Vision-Based Boundary Following Framework for Aerial Vehicles. In: IEEE/RSJ Intl. Conf. on Intelligent Robots and Systems, IROS (2008)
22. Santana, P.: Saliency-Based Approach to Boost Trail Detection. In: IEEE International Conference on Robotics and Automation (ICRA), Anchorage (2010)
23. Oliva, A.: The role of context in object recognition. Trends in Cognitive Sciences 11(12), 520–527 (2007)
24. Zhang, X.: Sequential particle swarm optimization for visual tracking. In: IEEE Conference on Computer Vision and Pattern Recognition (CVPR), pp. 1–8 (2008)
25. Rathinam, S.: Autonomous Searching and Tracking of a River using an UAV. In: American Control Conference, New York, pp. 359–364 (2007)
26. Yick, J., et al.: Wireless sensor network survey. Computer Networks 52, 2292–2330 (2008)

An Omnidirectional System for Navigation in Virtual Environments

Răzvan Gabriel Boboc, Mădălina-Ioana Toma, Horaţiu Moga, Alina Ninett Panfir, and Doru Talabă

Transilvania University of Brasov, Romania,
Department of Automotive and Transportation
{razvan.boboc,madalina-ioana.toma,moga.horatiu,alina.panfir, talaba}@unitbv.ro

Abstract. Virtual Reality (VR) is one of the newest technological domains with revolutionary applicability for the tomorrow's Future Internet, including visions of the Internet of Things. The sensation of total immersion in Virtual Environment (VE) is still unresolved. Therefore, our work proposes a new omnidirectional locomotion interface for navigation in VEs. The novel interface was built from an ordinary unidirectional treadmill, a new mechanical device, a motion capturing system to track the human walking and a control method using artificial intelligence techniques. A neural network is used to predict the motion of the new interface based on user's body motion and information about VE. The feasibility of the proposed system is verified through experiments and the preliminary results suggest that the new interface performs very well in a simplest VE based on our control method.

Keywords: Virtual Reality, navigation, Virtual Environment, locomotion interface, neural network.

1 Introduction

Nowadays, Virtual Reality (VR) has become an indispensable technology with practical applications in many areas such as medical, engineering to urban planning, training and education, sports and arts, etc. The VR is defined as the use of a computer-generated 3D environment – called a 'virtual environment' (VE). One of the hardest things to achieve in this area is to provide the immersion of the user, meaning that he can feel in a virtual world like in the real world. To achieve this, navigation is a fundamental task needed for human interaction with the VE.

In our work we deal the navigation into VE by walking because it is the main motor activity of daily life. The human gait is considered for VR field the user's most natural way of exploring a VE [4], [14]. Over the time, locomotion interfaces were built in order to not limit the navigation by walking in a VE. Due to these interfaces, the user can walk freely and his motion was compensated. Ruddle, R. A. and Lessels, S. showed that a walking interface would bring immediate benefits for navigation in virtual environments in a number of VE applications [11]. Therefore, the first

L.M. Camarinha-Matos, S. Tomic, and P. Graça (Eds.): DoCEIS 2013, IFIP AICT 394, pp. 192–199, 2013.

research about omnidirectional (ODT) treadmill was proposed by Darken et al. [2] in 1997 and was named by the authors 'a revolutionary locomotion device that enables bipedal locomotion in *any* direction of travel'. The base principle of the ODT consists of two perpendicular belts, one inside of the other, each of them comprised of an array of freely rotating rollers.

Another similar concept is the CyberWalk platform, one of the largest VR platforms in the world [3]. Over the years, many other locomotion interfaces have been designed and implemented. An example is ATLAS [8], that consists of a remodeled commercial treadmill and a motion platform with three axis. In the first phase the authors control the speed of the belt and further the walking direction, proposing a method than allows user to walk in any direction by cancel the turning motion of the user [7]. In [10] is used a motor-less treadmill resting on a mechanical rotating base. The belt of treadmill contains stripes along the direction of motion and measuring the angle of foot placement, the entire treadmill is rotate accordingly on a rotary base.

To achieve a target in a real environment, the human needs to walk on flat surfaces and also he need to climb or to go down surfaces with hilliness. As we could see in above researches, the currently locomotion interfaces based on ordinary treadmills are controlled by making a user navigation in VEs just on a flat surface. Therefore, in our on-going project we want to build a new locomotion interface by improving the control of an ordinary treadmill in order to allow the user to navigate also on a surface with hilliness in a VE, as in a real environment. Thus, one novelty of our on-going project is to make a complex control of treadmill to fit very well in a VR environment. Besides forward and backward navigation, it provides to the user the possibility to climb to or go down on a virtual surface with hilliness and also to turn left or right during navigation in a natural way. Our complex control of treadmill gives two main opportunities of the new locomotion interface: (i) to move in any direction in virtual worlds, and (ii) to enhance the user's immersion.

Up to the present time, controls of ordinary treadmills were done by using the classical PID controller method and derivatives of this method, such as: PD on/and PI controller [9], [12]. Although these basic controllers are most widely used, they are limited when complex processes are required to perform a task. Our locomotion interface based on an ordinary treadmill involves a complex controller because two motors need to be controlled in the same time, taking into account other two entities: user's actions and the scenarios of the VE. User's actions represent the first entity that involves the modulation through perception-action model. Thus, the user perception activity is given by treadmill motion through a natural gait. Based on 3D virtual scenario users performed the reaction activity upon treadmill.

In order to achieve the intended complex controller of the treadmill, we replaced the classical PID controller with artificial neural networks method. We have chosen to use this method because it is simplified mathematical models of brain-like systems that function as parallel, distributed computing networks [6]. Hence, artificial neural networks method let us to build a flexible cognitive control over the user's behavior and the plan of virtual scenarios.

To build the proposed complex controller of the treadmill, we had to use multiple devices, such as an ordinary treadmill (Energetics Power Run 3000 HRC), an Optitrack Motion Capturing System to track the human walking, and also we design a simplest virtual environment, as see in Fig.1 a). In addition, we had to design and

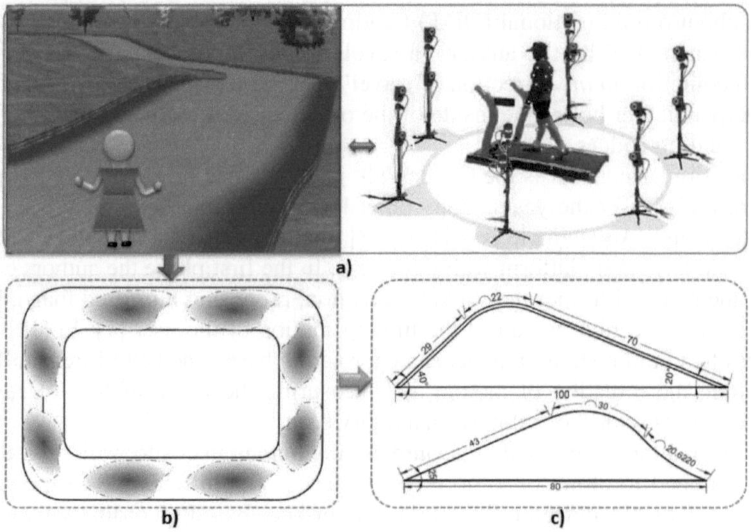

Fig. 1. The 3D model of the omnidirectional interface

build a mechanical support with two degrees of freedom (2-DOF) upon the treadmill was fixed. Therefore we presented preliminary results obtained based on the proposed complex controller of the treadmill and the prototype of the new locomotion interface.

The outline of the paper is as follows. In section 2 is illustrated the contribution for the Internet of Things of this paper, section 3 present our prototype and the used methodologies. Experimental studies, results and a discussion are given in section 4 and finally, in section 5, is given the conclusion and the future work.

2 Contribution to the Internet of Things

Nowadays, the speed of information technologies development is increasing more than ever. Hence, VR is one of these technologies that is often supported by the Internet of Things, by making a natural human-computer interaction for the users. One of the areas where VR found practical application is user's natural navigation by walking in 3D VEs. Thus, human locomotion inside VEs applications is an important concept that can be integrated and supported by novel technologies, promising to expand into most of today's domains and also into the concept of tomorrow's Future Internet, where Internet of Things is one of the core elements of it.

Internet of Things will revolutionize and expand the interaction between persons and objects and object-object. In this context a 'thing' can be defined as a real or virtual entity that exists, moves and can be identified.

According to the definition of Internet of Things from [5]: "*Things having identities and virtual personalities operating in smart spaces using intelligent interfaces to connect and communicate within social, environmental, and user contexts*", our new locomotion interface for navigation in virtual worlds represent a promising step toward the future of the Internet. From this reason, the proposed system with small dimensions manages to reduce the main disadvantage of most active treadmill interfaces – the need for a large walking area [8] – and ensure proper integration in a full immersive CAVE system by making it ideal for future immersion needs.

Based on the experimental study and results presented in this article, we achieved an omnidirectional walking for navigation in 3D VE by implementing a new locomotion interface and building an intelligent control. The preliminary test results will help future development of already existing domains by opening a path to new ideas and trends for more applications which integrated visions about Internet of Things, and also will contribute to VR locomotion interfaces evolution, which represent in our view a huge step making the Internet of Things to become a reality very soon in the next years. Moreover, our preliminary results show that the interaction between user and real/virtual objects is enhanced and it represents the main vision of the Internet of Things. Hence, our new locomotion interface will improve in the same time the quality of humans lives.

3 Proposed Prototype and Methodology

In order to answer the main research question addressed in this paper (*How to build a new locomotion interface with small dimensions by achieving an omnidirectional walking for navigation in 3D VE based on a complex control through AI methods, i.e, neural networks?*) we propose a modular prototype and describe next the functionality of each module. An overview of the proposed prototype was presented in Fig. 1.

The proposed prototype is to allow a user to endlessly walk in a VE in any direction, while he is on a real interface that compensates his motion. Our first scientific challenge in development of the omnidirectional interface was to make a completely immersion of the users and for this purpose a new method to control this interface was implemented.

Simulation Apparatus - We have built the locomotion interface based on a set of devices. The main device of our approach is a walking compensation device, based on an ordinary treadmill adapted for VR purposes. Also we combined other devices for building the prototype, such as: the Optitrack Motion Capturing System to track the human walking. In order to implement the control of the treadmill related to the VE of the application, we designed a simple test scenario that simulates a surfaces with hilliness (see fig. 1, a), left). The display and rendering graphics scenes are done in an immersive CAVE system to provide a whole body immersion and a strong visual feedback. Using this system, the VE scale and the human interaction become equal.

The omnidirectional compensation device – We have adapted an ordinary treadmill with the following dimensions for the surface area: length – 150 cm, width – 50 cm into an omnidirectional device. To achieve an omnidirectional walking for navigation we had to design and to build a mechanical support with two degrees of freedom (2-DOF) that has the following dimensions: the height of mechanical support is 50 cm, the upper surface is by 80 cm (L) and by 50 cm (W) and the upper surface is a cylinder with a diameter of by 50 cm (see Fig. 2, right). Then classical unidirectional treadmill was fixed on it, (see Fig. 2, left) and the new device became an omnidirectional compensation device with three degrees of freedom (3-DOF= 2-DOF of mechanical support + 1-DOF treadmill). The whole prototype was simulated using CAD parts in order to build a good solution. The electrical part of the system was modeled in Matlab Simulink.

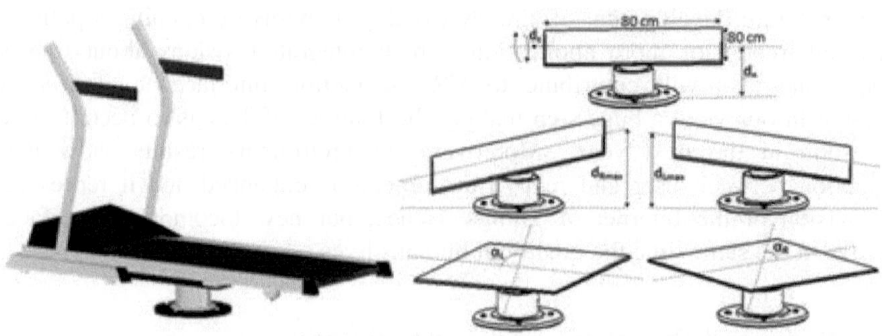

d_a (cm)	d_{Rmax} (cm)	d_{Lmax} (cm)	α_L (°)	α_R (°)
50	20	20	45	45

Fig. 2. The 3D model of the interface and the dimensions of the mechanical base

Three motors were used: one that actuates the treadmill belt and others two for mechanical support of treadmill: one motor to rotate left/right the whole platform and another one motor to tilt up/down the platform. One motor actuate the belt of the treadmill in order to compensate the user walking and another motor actuate the whole base for compensating the change of direction in walking. The first motor is a DC motor of 180 V, 7.5 A, 2 hp, 3000 rpm. Motors of mechanical support are EC motors of 48 V, 9.38 A, 400 W, 5370 rpm. To detect the sense of rotation for the motors are used optical encoders TIRO 1000. This sensor is coupled to the motor and detects the sense and also the speed of rotation by incrementing or decrementing a position number related to a reference value [13]. Hence, the user can navigate by walking on the surfaces of the VE into omnidirectional way on a real omnidirectional treadmill. Our system is able to rotate accordingly to cancel the turning motion when the user intents to change the direction of walking. In order to give user a feeling of safety, the hand supports of the original treadmill were not removed.

The Optitrack Motion Capturing System – A motion tracking system was needed to recognize the motion of the user's feet and maybe also body and then change the view accordingly and the whole platform is rotate. Therefore, we used a passive optical system that use markers coated with a retro-reflective material to reflect light back to the infrared cameras (IR). This system presents the following main advantages: high update rates, low latency and scalable to fairly large areas. In our experiments 12-IR cameras were used for data capture (see Fig. 1, a) right).

VE of the test scenario – We designed a simplest VE using Blender tool, because it is an open source tool. The VE simulates a track with mix surfaces: flat and surfaces with hilliness. The track from scenario has a square shaped, with a length of 4 km and a width of 10 m, with two traffic lanes (Fig. 1, b)). There are two types of hillines, with different angles of inclination (Fig. 1 c)). The human position on the track is given by a simplest virtual avatar (as it can be seen in the Fig. 1, a) left).

4 Experiment and Results

In order to control the sense of motors we use an algorithm implemented with neural networks. A neural network can be designed to perform a specific task and it need to be trained before it can be put in use. Matlab software was preferred to perform the computation, using the Neural Network Toolbox™.

The inputs of the algorithm are extractions of features from a set of data. Data were recorded from six users when they performed a real scenario similar with proposed VE. We tracked the position and orientation of users feet using the motion tracking system. Then we process data in order to extract the angle between the foot and tibia and also feet orientation related to a reference position for right and left rotation of treadmill. Extracted angles and feet orientation were data inputs for a neural network. A feed-forward neural network with one hidden layer of ten neurons is used. Also we modeled the used motors in Matlab Simulink in order to obtain the positions of the motors related to a reference factor and a parameter for the scenario, i.e. used to tilt the platform depending on the inclination angle of hills scenarios.

For choosing the number of neurons, the neural network is trained in Matlab and the number of neurons in hidden layer is reduced until the error can be accepted.

When training multilayer networks, the general practice is to first divide the data into three subsets: training set, validation set and test set [1]. Figure 3 depicts the training process for the proposed neural network, showing the confusion matrices for the training, validation, test and for all phases. For the test phase the accuracy of the neural network is 88,9 %, representing the percentage of correct classifications.

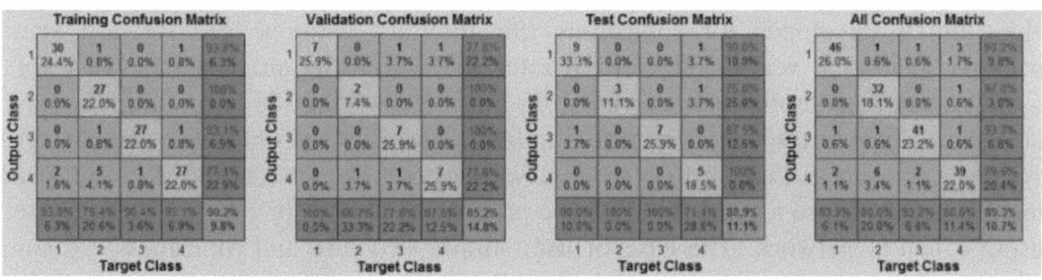

Fig. 3. The confusion matrices for the training process

In Figure 4 (left) it can observe the mean square error during 17 epochs for the phases described above and for all steps the mean square error value decreases, the optimum value for the validation phase being obtained after 11 epochs. In Figure 4 (right) the gradient and the validation checks plots are illustrated. The value of the gradient after 17 epochs is 0.019608. The training process will stop when the validation checks reach 6 (the default value).

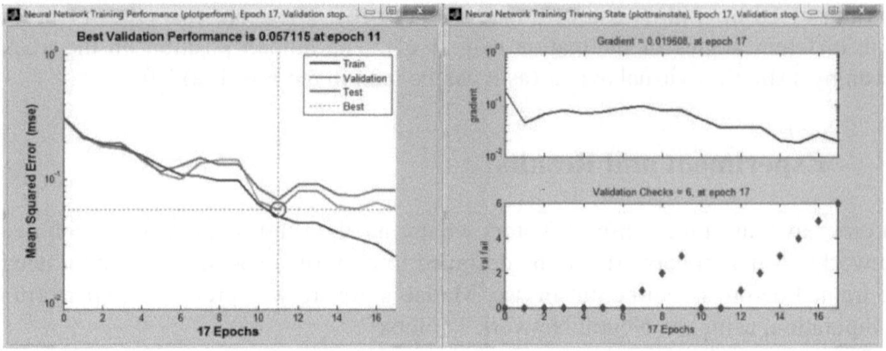

Fig. 4. Validation performance for neural network. Gradient and validation checks

After the training of data we tested the prototype in VEs in offline functionality for checking the solution. We checked the solution with new data recorded from the users. Each user performed three scenarios: (1) normal waking, (2) alert walking, and (3) mixed walking. We tested and checked the prototype in a VE in order to see the avatar behavior and also the control of treadmill (see Fig. 1, a)).

In normal walking good results were achieved with low errors for both the control of treadmill and avatar behavior. The results highlight that the mixed walking scenario presents errors due to fluctuation of waking speed. These errors are caused also of lost markers of the motion tracking system. In alert walking scenario some errors was registered, but less than in the mixed scenario. According to [15] the errors could be reduced if data are filtered previously to use after for testing the prototype.

5 Concluding Remarks and Future Works

In this paper, we present a system for omnidirectional navigation in VE and an algorithm is developed for controlling this interface using a feed forward neural network. Our goal was to develop an interface that allow an endlessly walk for a virtual avatar, while the real user move in a limited workspace in any direction. We designed a simple test scenario that simulate a surfaces with hilliness and a CAVE system was used as display to provide a whole body immersion and a strong visual feedback. In order to test the interface we propose a method for controlling the system using a neural network. The ease of use, simple structure and robustness are some advantages of the neural network controller, overcoming the limitations of PID controllers, like instability, late response and poor control performance. The preliminary results show the feasibility of the system was achieved.

The presented system is an improvement over current implementations, providing a solution for navigation in VEs with a small dimensions interface and enhancing the user immersion.

Future work includes a better control of the interface by implementing a system that recognizes the user intentions of walking. We intend to develop a complete virtual system for navigation with application in actual and future areas of interest.

Acknowledgments. This paper is supported by the Institution Organizing Doctoral Studies (IODS), financed by the Romanian Government under the contract number 1.B.11.1168.

References

1. Beale, M.H., Hagan, M.T., Demuth, H.B.: Neural network toolbox user's guide. The MathWorks Inc., Natick (2012)
2. Darken, R.P., Cockayne, W.R., Carmein, D.: The Omni-Directional Treadmill: A Locomotion Device for Virtual Worlds. In: Proceedings of the 10th Annual ACM Symposium on User Interface Software and Technology, New York, pp. 213–221 (1997)
3. Giordano, P.R., Souman, J.L., Mattone, R., De Luca, A., Ernst, M.O., Bülthoff, H.H.: The CyberWalk Platform: Human-Machine Interaction Enabling Unconstrained Walking through VR. In: First Workshop for Young Researchers on Human-friendly Robotics (2008)
4. Heintz, M.: Real Walking in Virtual Learning Environments: Beyond the Advantage of Naturalness. In: Cress, U., Dimitrova, V., Specht, M. (eds.) EC-TEL 2009. LNCS, vol. 5794, pp. 584–595. Springer, Heidelberg (2009)
5. INFSO D.4 Networked Enterprise & RFID INFSO G.2 Micro &Nanosystems, in Cooperation with the Working Group RFID of the ETP EPOSS, Internet of Things in 2020, Roadmap for the Future, Version 1.1 (May 27, 2008)
6. Nguyen, H.T., Prasad, N.R., Walker, C.R., Walker, E.A.: A First Course in Fuzzy and Neural Control. Chapman and Hall/CRC (2002)
7. Noma, H., Miyasato, T.: A New Approach for Canceling Turning Motion in the Locomotion Interface, ATLAS. Proceedings of ASME-DSC 67, 405–406 (1999)
8. Noma, H., Miyasato, T.: Design for Locomotion Interface in a Large Scale Virtual Environment, ATLAS: ATR Locomotion Interface for Active Self Motion. In: The 7th Annual Symposium on Haptic Interfaces for Virtual Environment and Teleoperator Systems. The Winter Annual Meeting of the ASME, Anaheim, USA, pp. 111–118 (1998)
9. Noma, H., Sugihara, T., Miyasato, T.: Development of Ground Surface Simulator for Tel-E-Merge System. In: IEEE VR 2000, pp. 217–224 (2000)
10. Patel, K.K., Vij, S.K.: Unconstrained Walking Plane to Virtual Environment for Spatial Learning by Visually Impaired. Ubiquitous Computing and Communication Journal, 1–7 (2010)
11. Ruddle, R.A., Lessels, S.: The benefits of using a walking interface to navigate virtual environments. ACM Transactions on Computer-Human Interaction 16(1) (2009)
12. Stavar, A., Dascalu, L.M., Talaba, D.: Design, Test and Experimental Validation of a VR Treadmill Walking Compensation Device. In: Camarinha-Matos, L.M. (ed.) DoCEIS 2011. IFIP AICT, vol. 349, pp. 402–409. Springer, Heidelberg (2011)
13. Stavăr, A.: User's locomotion compensation using haptic systems while navigating through immersive Virtual Environments. Ph. D., Transilvania University of Braşov (2011)
14. Usoh, M., Arthur, K., Whitton, M.C., Bastos, R., Steed, A., Slater, M., Frederick, P., Brooks, J.: Walking > walking-in-place > Flying, in virtual environments. In: Proceedings of the 26th Annual Conference on Computer Graphics and Interactive Techniques, SIGGRAPH 1999, pp. 359–364. ACM Press/Addison-Wesley Publishing Co., New York (1999)
15. Jensenius, A.R., Nymoen, K., Skogstad, S.A., Voldsund, A.: A study of the noise-level in two infrared marker-based Motion capture systems. In: Proceedings of the 9th Sound and Music Computing Conference, Copenhagen, Denmark, July 11-14 (2012)

Indoor Exploration Using a μUAV and a Spherical Geometry Based Visual System

Tiago Caldeira[1], Lakmal Seneviratne[1,2], and Jorge Dias[1,3]

[1] Khalifa University for Science Technology and Research, UAE
[2] Center for Robotics Research from Kings College of London, UK
[3] Institute of Systems and Robotics, University of Coimbra, Portugal
tiago.caldeira@kustar.ac.ae

Abstract. This research presents a new vision system that explores a spherical geometry and will be provide innovative solutions for tracking, surveillance, navigation and mapping with micro Unmanned Aerial Vehicle (μUAV) in unknown indoor environments. The system will be used with μUAV in indoor environment and it is composed by twenty six cameras that are arranged in order to sample different parts of the visual sphere around the μUAV. This configuration allows that some of the cameras will have overlapped field of view. This system has been designed for the purpose of recovering ego-motion and structure from multiple video images, having a distributed omnidirectional field of view. We use the spherical geometry to extend the field of view, from one single direction to a single point of perspective, but with multiple views. This manuscript will prove that spherical geometric configuration has advantages when compared to stereo cameras for the estimation of the system's own motion and consequently the estimation of shape models from each camera. The preliminary field tests presented the theoretical potential of this system and the experimental results with the images acquired by 3 cameras.

Keywords: Vision, μUAV, Spherical Geometry, Indoor, Ego-motion, Trifocal tensor, Motion Flow.

1 Introduction

Mobile robots are important artifacts for the exploration of unknown areas, not only in hazard situations, but also as an extension of human capabilities. Those so called explorer robots can be developed to navigate in unleveled terrain. Both semi-autonomous and completely autonomous machines allowed a real time report of otherwise impossible places to visit. On the last decades researchers expanded their attention to others than Unmanned Ground Vehicles (UGV), for example, μUAV.

The most challenging part of 3D mapping is obtaining the full 6 Degrees of Freedom (DOF) of pose of the robot between each scan [1]. The vehicle position and orientation, is, in most cases, estimated with the combination of inertial sensors and GPS, but in indoor GPS denied environments tends to create an important barrier from the fully autonomous exploration [2]. To complete the autonomous exploration, problems like

L.M. Camarinha-Matos, S. Tomic, and P. Graça (Eds.): DoCEIS 2013, IFIP AICT 394, pp. 200–209, 2013.

localization and mapping unknown environments need to be solved. One way of achieving this is with the use of laser sensors or vision systems. Advances have been made in the use of vision sensors for target tracking and obstacle avoidance or to estimate vehicle pose [3], but the traditional stereo configuration does not provide enough information nor covers the totality of the space. In order to complete the visual information, inertial sensors were added to the system. The inertial sensed gravity [4][5] provides a vertical reference for the spherical vision system, enabling to establish an artificial horizon line and vertical features. Between each pair of cameras there is a rigid transformation that could be determined using a calibration process [6], getting the rectification of the stereo configuration and generate the epipolar constrains, and generating the optical flow [7].

In this paper we intend to demonstrate some of the advantages of the spherical geometry for multiple cameras complemented with inertial data in order to be successfully solving problems of tracking, surveillance, navigation and mapping.

In the next section it will be presented the relation with the Internet of Things (Section 2), followed by the research in UAV field and vision based controls (Section 3). The proposed solution and mathematical support (Section 4) will be complemented with some of the implementation details and the progress towards the full sensor developing (Section 5) that lead us to the conclusions and future work (Section 6).

2 Relationship to Internet of Things

When explaining the definition of “Internet of Things”, Kevin Ashton [8] mention that “people have limited time, attention and accuracy – all of which means they are not very good at capturing data about things in the real world”. This extension to the man capabilities was always present during the development of this project. In the early stages of development of this project, we intended to create a unique capturing device that not only capture video from all of its surroundings but also could be able to identify and track people or objects. By analyzing and categorizing moving objects, it will provide enough information for an organized database which could grow without human interaction. With multiple systems like this and if them are complemented with network connection, multiple nodes will act as smart and autonomous data collection spots, presenting content to the Internet of Things.

3 Related Work

In the last decade, many authors increased their research on the implementation of control algorithms of flying devices, mainly μUAV. Those devices have been used in multiple research goals, but with the goal of autonomous navigation or mapping an unknown environment. The first approaches on the UAV control were only focused on outdoor environments, where there was more space of maneuver and where GPS signals were an important data to close the loop regarding the position and velocity of the UAV.

Samir Bouabdallah and Roland Siegwart [9] proposed an approach for full control of an UAV based on a quadrotor configuration. To achieve autonomy, the UAV needs a navigation system, and many authors proposed several solutions for indoor navigation systems [10][11][12]. Different methods have been used to not only navigate but also for map the environment. Active sensors like lasers or ultrasonic sensors where combined with visual elements [3][13]. The use of visual information for obstacles avoidance and simple navigation [14] was also object of research.

Grzonka et al. [15], already provided a solution to a fully autonomous indoor quadrotor based on LIDAR technology. Bachrach et al. [16] using a similar architecture to explore and map unstructured and unknown indoor environments. The unknown and unstructured environments were also the base for another research by Blöesch, Scaramuzza et al. [17], but using a visual approach to this problem.

4 Research Contribution and Innovation

In this paper we present our progress towards a new vision system that combines multiple cameras in a spherical geometry merged with inertial sensors.

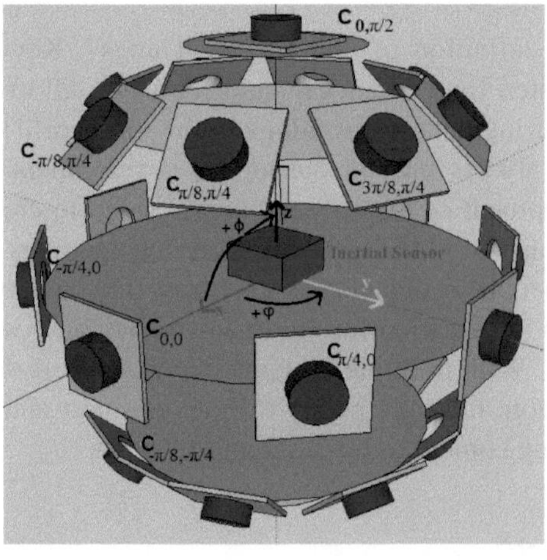

Fig. 1. Three-dimensional representation of the mechanical design with the cameras information and inertial sensor position

This system is composed by twenty six symmetrically distributed cameras with the same intrinsic parameters. As presented in Fig. 1, there are groups of 8 cameras that are equally distributed along the equatorial line and along the tropic lines. This is complemented with one camera in each pole. With this configuration, the angle between any two consecutive cameras is always $\pi/4$ radians. The coordinates of each camera is always $C_{\varphi,\phi}$ where φ is the angle on z-axis and ϕ the angle on x-axis.

In the center of the sphere is a 6 DOF inertial sensor that will provide the system accelerations and orientations.

4.1 Geometry Relationships and Models

Using a spherical geometry allow a unique transformation between each pair of cameras. Even without a complete overlap between all cameras, it is possible to see all of the extension of the exterior space.

The center of projection of the camera $C_{\varphi,\phi}$ could be obtained with a rotation on x-axis $\mathbf{R}_x(\phi)$ and z-axis $\mathbf{R}_z(\varphi)$ from the frontal camera $C_{0,0}$ using (1).

$$ {}^{C_{0,0}}\mathbf{R}_{C_{\varphi,\phi}} = \mathbf{R}_z(\varphi)\ \mathbf{R}_x(\phi) \tag{1} $$

where $\varphi = \frac{m\pi}{4} + \frac{n\pi}{8}$ and $\phi = \frac{n\pi}{4}$, $(\forall m \in M = \{-3,\ldots,0,\ldots,4\} \wedge \forall n \in N = \{-2,-1,0,1,2\})$

That way, the relation between a camera i ($C_{\varphi,\phi}$) with any other camera j ($C_{\varphi',\phi'}$) could be determined with (2).

$$ {}^{C_{\varphi',\phi'}}R_{C_{\varphi,\phi}} = \mathbf{R}_z(\varphi' - \varphi)\ \mathbf{R}_x(\phi' - \phi) \tag{2} $$

By using developments on [18] and assuming that the intrinsic parameters among these two cameras are equal, the homography ${}^{j}\mathbf{H}_{i}$ between the images could be described in (3) where $\mathbf{K}$ represents the intrinsic parameters (the same in both cameras).

$$ {}^{j}\mathbf{H}_{i} = \mathbf{K}\ {}^{j}\mathbf{R}_{i}\mathbf{K}^{-1} \tag{3} $$

That way, it is possible to calculate the epipolar constrain between the cameras (4).

$$ e^{j} = \mathbf{P}' \begin{pmatrix} \vec{0} \\ 1 \end{pmatrix} = \mathbf{K}\ {}^{j}t_{i} \tag{4} $$

The fundamental matrix, F, is obtained in (5).

$$ {}^{j}\mathbf{F}_{i} = [e^{j}] \times \mathbf{K}\ {}^{j}\mathbf{R}_{i}\ \mathbf{K}^{-1} = [e^{j}] \times \mathbf{K}\,\mathbf{R}_z(\varphi' - \varphi)\,\mathbf{R}_x(\phi' - \phi)\ \mathbf{K}^{-1} \tag{5} $$

The unique setup of the geometry between the cameras allows the simplicity of the calculation not only between two, but also with each three cameras. The trifocal tensor [19] could be approached using the relationship of three corresponding lines. The lines $\mathbf{l_i}$, $\mathbf{l_j}$ and $\mathbf{l_k}$ are the projection of L (line in space) on each of the cameras i ($C_{\varphi,\phi}$), j ($C_{\varphi',\phi'}$) and k ($C_{\varphi'',\phi''}$) as represented in Fig. 2.

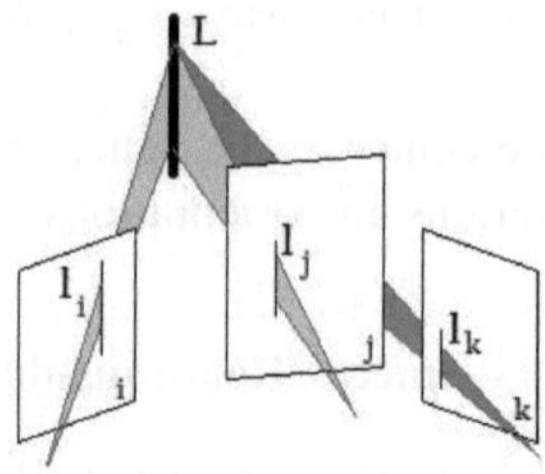

Fig. 2. Symbolic representation of three cameras representing the line L

Assuming the origin at camera I and using development of Hartley et al. [19], the cameras matrices for the three views would be $\mathbf{P}_i=[\mathbf{I}|\vec{0}]$, $\mathbf{P}_j=[\,{}^j\mathbf{H}_i|e^j]$, $\mathbf{P}_k=[\,{}^k\mathbf{H}_i|e^k]$, where ${}^j\mathbf{H}_i$ and ${}^k\mathbf{H}_i$ are the homography matrix between cameras j and i, and k and i respectively (calculated in (3)), while $e^j$and e^k are the epipolar constrain from camera j and k to i, as calculated in (4). Recovering the equation $l_a=\mathbf{l_j}^T\left({}^j\mathbf{H}_{i_a}e^{k^T}\text{-}e^j\,{}^k\mathbf{H}_{i_a}{}^T\right)\mathbf{l_k}$ where l_a represents the a coordinate of $\mathbf{l_i}$. Moreover, by definition, $l_a=\mathbf{l_j}^T\ \mathbf{T}_a\ \ \mathbf{l_k}$, what is equivalent to present the notation in (6).

$$\mathbf{T}_a = \left({}^j\mathbf{H}_{i_a}\ e^{k^T} - \ e^j\ {}^k\mathbf{H}_{i_a}{}^T \right) \tag{6}$$

Using the equations (2) and (3) that refers about the particular cameras configurations, the equation (6) could be rewrite as (7).

$$\mathbf{T}_a = [\mathbf{K}\,\mathbf{R}_z(\varphi'\text{-}\varphi)\,\mathbf{R}_x(\phi'\text{-}\phi)\mathbf{K}^{-1}]_a\, e^{k^T} - \ e^j\,[\mathbf{K}\,\mathbf{R}_z(\varphi''\text{-}\varphi)\,\mathbf{R}_x(\phi''\text{-}\phi)\mathbf{K}^{-1}\,]_a{}^T \tag{7}$$

We could define the trifocal tensor $\mathbf{T}_i^{jk}$ between the cameras i, *j* and k, as being $\mathbf{T}_i^{jk}=[\mathbf{T}_x,\mathbf{T}_y,\mathbf{T}_z]$. This result could be used to extract the fundamental matrix between the cameras i and j in (8).

$${}^j\mathbf{F}_i = [e^j] \times [\mathbf{T}_x, \mathbf{T}_y, \mathbf{T}_z]e^k \tag{8}$$

This equation (8) shows that it is possible to recover the relationship between each subset of three cameras using the Trifocal tensor and the epipolar lines. Taking advantage of the unique relation between each cameras (pure rotations on x-axis and z-axis) the calculation will be simplified.

4.2 Motion Flow Properties in Image Plan

One very important part of our research is related with the movement of the entire system motion flow. In order to simplify the calculations we analyze the system using the camera $C_{0,0}$ as reference.

Assuming a three dimension motion has two components: translation and rotation.

If we define the projection $\vec{p}$=f **P**$/Z$, with p the image point, $\boldsymbol{P} = [X\ Y\ Z]^T$ the scene point with a depth Z, we could define the equation (9), where **V** is the world motion, **T** the translation and **ω** a vector that represents the rotation.

$$\mathbf{V}=-\mathbf{T} - \boldsymbol{\omega}\times\mathbf{P} \tag{9}$$

Assuming that between two consecutives frames the motion is small, we could derive the both sides of the projection equation. Using the previous equation (9) and from the image perspective analyzing its velocity projection components, we obtain the equations in (10).

$$\boldsymbol{v} = f\frac{Z\mathbf{V}-V_Z\mathbf{P}}{Z^2} = \begin{cases} v_x = \frac{T_Zx-T_xf}{Z} - \omega_y f + \omega_z y + \frac{\omega_x xy}{f} - \frac{\omega_y x^2}{f} \\ v_y = \frac{T_Zy-T_yf}{Z} + \omega_x f - \omega_z x - \frac{\omega_y xy}{f} + \frac{\omega_x y^2}{f} \end{cases} \Leftrightarrow$$

$$\Leftrightarrow \boldsymbol{v} = \frac{1}{Z}\begin{bmatrix} -f & 0 & x \\ 0 & -f & y \end{bmatrix}\mathbf{T} + \begin{bmatrix} \frac{xy}{f} & -f-\frac{x^2}{f} & y \\ f+\frac{y^2}{f} & -\frac{xy}{f} & -x \end{bmatrix}\boldsymbol{\omega} \tag{10}$$

From this equation (10) we see that is difficult to decouple the motion from rotations around x-axis or y-axis of the camera. Therefore, we use inertial sensors to confirm those. The relation between inertial sensor and the camera axis is represented on Fig. 3.

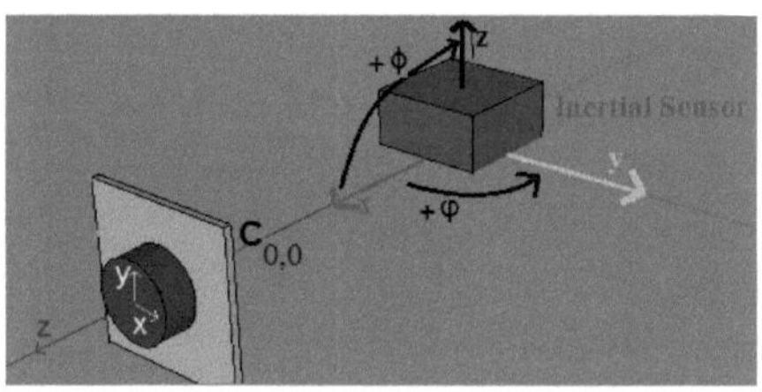

Fig. 3. Inertial and camera $C_{0,0}$ referential

4.3 Visual and Inertial Integration

Inertial information provided by accelerometer and gyroscopes are by themselves very important to obtain the attitude and motion parameters of the vision system.

However, the accelerometer provides the linear acceleration in the 3 axes of $\boldsymbol{a}$, sensing the gravity vector $\boldsymbol{g}$ summed with the visual system acceleration $\boldsymbol{a}_{SYSTEM}$. If the vision system is attached to the UAV ($\boldsymbol{a}_{SYSTEM} \equiv \boldsymbol{a}_{UAV}$), then the acceleration could be defined in (11).

$$\boldsymbol{a} = -\boldsymbol{g} + \boldsymbol{a}_{UAV} \tag{11}$$

This will mean that, if the system is motionless ($\boldsymbol{a}_{UAV} = 0$), only the gravity will be present, allowing the system to have a vertical reference from the gravity, and calculate the orientation n of the UAV using (12).

$$\hat{n} = \begin{bmatrix} n_x \\ n_y \\ n_z \end{bmatrix} = -\frac{\boldsymbol{g}}{\|\boldsymbol{g}\|} = \frac{1}{\sqrt{a_x{}^2+a_y{}^2+a_z{}^2}} \begin{bmatrix} a_x \\ a_y \\ a_z \end{bmatrix} \tag{12}$$

The gyroscope measures the rotational acceleration in the 3 axes, and it is possible to separate the gravity from the sensed acceleration by performing the rotation update [4].

As the inertial sensor is fixed in the center of the sphere as soon as we determine the orientation $\hat{n}$ of the device, we could calculate the orientation $\hat{c}$ of every camera $C_{\varphi,\phi}$, based on the rotation between the inertial system and the $C_{0,0}$ in equation (13).

$$\hat{c} = \begin{bmatrix} c_x \\ c_y \\ c_z \end{bmatrix} = {}^{IMU}\boldsymbol{R}_{C_{0,0}} \quad {}^{C_{0,0}}\boldsymbol{R}_{C_{\varphi,\phi}} \quad \hat{n} \tag{13}$$

The ${}^{IMU}\boldsymbol{R}_{C_{0,0}}$ could be easily determined using calibration [6]. Having the complete pose information of each camera, it is possible to predict location of the horizon line, from where it is also possible to easily identify vertical features on each image.

5 Implementation and Experimental Setup

In order to present the initial proof of concept, only one oct part of the sphere where designed. Using the configuration displayed in Fig. 4 it was possible to test different angles configurations.

Fig. 4. Hardware used on the trials, (a) 3D printed version of part the Sphere(b) UI 1226LE-M camera from IDS-Imaging (c) 6DOF Inertial Measurement Unit

Each camera is connected through a serial connection link (USB) and could display at a 43 frames per second (fps) rate, value that decreases when multiple cameras are connected on the same bus. This frame rate is enough to use vision based tracking algorithms.

5.1 Experiments

The initial experiments proved the system concept and the correspondence between points in cameras were the expected. Images from the three cameras (see Fig. 5) were captured and displayed simultaneous at a 25 fps rate. The inertial data captured was also validated in order to confirm the orientation of the system. The square chess-board was also important in order to calibrate the trifocal tensor.

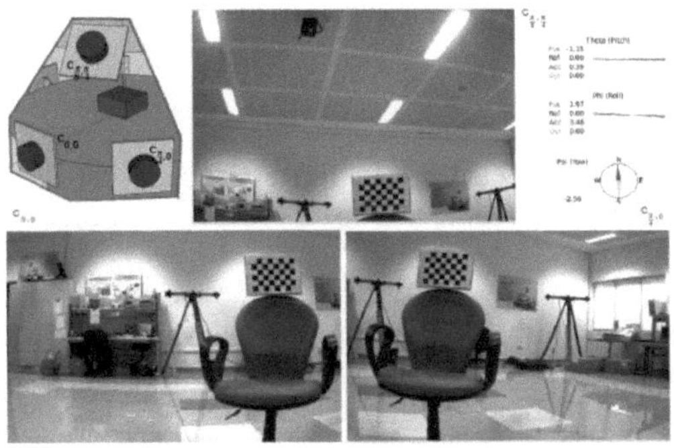

Fig. 5. Cameras Position, Attitude Control and respective Images - $C_{\frac{\pi}{8},\frac{\pi}{4}}$ (top) $C_{0,0}$ (left) and $C_{\frac{\pi}{4},0}$ (right)

After calibration (between each camera and with the inertial system) a set of images were collected from the three cameras (key frames selected and then grouped on Fig. 6) where two people move in opposite directions. With this set of images is possible to recover the motion on space of each person.

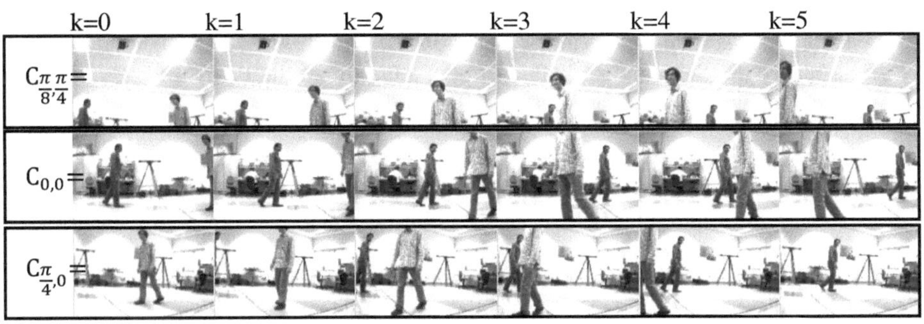

Fig. 6. Sequence with 6 frames from $C_{\frac{\pi}{8},\frac{\pi}{4}}$ (top) $C_{0,0}$ (middle) and $C_{\frac{\pi}{4},0}$ (bottom)

6 Conclusions and Future Work

We introduced a visual system based on multiple cameras with a spherical geometry with a complete field of view. This system simplifies the mathematical equations for correspondence and movement perception. The use of an orthogonal matrix (rotation matrix) in the homography and epipolar calculation will reduce the calculation time.

The first trials provided enough data to test basic vision algorithms and proved the capability and potential future of the system.

Other algorithms could be developed and proved based on this system, eventually aiming an autonomous μUAV to explore, navigate and map indoor environments.

References

1. Morris, W., Dryanovski, I., Xiao, J.: 3D Indoor Mapping for Micro-UAVs Using Hybrid Range Finders and Multi-Volume Occupancy Grids. In: Proc. of Robotics: Science and Systems, RSS (2010)
2. Ahrens, S., Levine, D., Andrews, G., How, J.P.: Vision-based guidance and control of a hovering vehicle in unknown, GPS-denied Environments. In: Proceedings of the IEEE International Conference on Robotics and Automation, pp. 2643–2648 (2009)
3. Achtelik, M., Bachrach, A., He, R., Prentice, S., Roy, N.: Stereo Vision and Laser Odometry for Autonomous Helicopters in GPS-Denied Indoor Environments. In: Proceedings of the SPIE Conference on Unmanned Systems Technology XI, Orlando, FL (2009)
4. Lobo, J.: Integration of Vision and Inertial Sensing. PhD Thesis, U. of Coimbra (2006)
5. Corke, P., Lobo, J., Dias, J.: An introduction to inertial and visual sensing. International Journal of Robotics Research, Special Issue 2nd Workshop on Integration of Vision and Inertial Sensors 26(6), 519–535 (2007)
6. Lobo, J., Dias, J.: Relative Pose Calibration Between Visual and Inertial Sensors. International Journal of Robotics Research, Special Issue 2nd Workshop on Integration of Vision and Inertial Sensors 26(6), 561–575 (2007)
7. Dias, J., Araújo, H., Paredes, C., Batista, J.: Optical Normal Flow Estimation on Log-polar Images. A solution for Real-Time Binocular Vision. Real-Time Imaging Journal 3, 213–228 (1997)
8. Ashton, K.: That 'Internet of Things' Thing. RFID Journal (July 22, 2009)
9. Bouabdallah, S., Siegwart, R.: Full Control of a Quadrotor. In: Proc. of the IEEE/RSJ Int. Conf. on Intelligent Robots and Systems, IROS (2007)
10. Michael Sobers Jr., D., Chowdhary, G., Johnson, E.N.: Indoor Navigation for Unmanned Aerial Vehicles. In: AIAA Guidance, Navigation, and Control Conference (2009)
11. Achtelika, M., Bachrach, A., He, R., Prentice, S., Roy, N.: Autonomous navigation and exploration of a quadrotor helicopter in GPS-denied indoor environments. In: Robotics: Science and Systems (2008)
12. Chowdhary, G., Michael Sobers Jr., D.: Integrated Guidance Navigation and Control for a Fully Autonomous Indoor UAS, Portland, Oregon (2011)
13. Morris, W., Dryanovski, I., Xiao, J.: 3D Indoor Mapping for Micro-UAVs Using Hybrid Range Finders and Multi-Volume Occupancy Grids. In: Proceedings of Robotics: Science and Systems, RSS (2010)
14. Ahrens, S., Levine, D., Andrews, G., How, J.P.: Vision-based guidance and control of a hovering vehicle in unknown, GPS-denied environments. In: Proceedings IEEE International Conference on Robotics and Automation, ICRA 2009, pp. 2643–2648 (2009)
15. Grzonka, S., Grisetti, G., Burgard, W.: A fully autonomous indoor quadrotor. IEEE Transactions on Robotics (99), 1–11 (2012)
16. Bachrach, A., He, R., Roy, N.: Autonomous Flight in Unknown Indoor Environments. Intl. Journal of Micro Air Vehicles, 217–228 (2009)

17. Blöesch, M., Weiss, S., Scaramuzza, D., Siegwart, R.: Vision Based MAV Navigation in Unknown and Unstructured Environments. In: IEEE International Conference on Robotics and Automation (2010)
18. Hartley, R.I., Zisserman, A.: Multiple View Geometry in Computer Vision, 2nd edn. Cambridge University Press (March 2004)
19. Hartley, R.I.: Lines and points in three views and the trifocal tensor. International Journal of Computer Vision 22(2), 125–140 (1996)

Context Awareness for Self-adaptive and Highly Available Production Systems

Sebastian Scholze[1], José Barata[2], and Oliver Kotte[1]

[1] Institut für angewandte Systemtechnik Bremen GmbH, 28359 Bremen, Germany
{scholze,kotte}@atb-bremen.de
[2] CTS – UNINOVA, Dep. De Eng. Electrotecnica, Faculdade de Ciencias e Tecnologica, Universidade Nova de Lisboa, 2829-516 Caparica, Portugal
jab@uninova.pt

Abstract. A new approach for the realization of self-adaptive and highly available production systems based on a context aware approach, allowing self-adaptation of flexible manufacturing processes in production systems and effective knowledge sharing to support maintenance, is presented. The usage of dynamically changing context as basis for adaptation of flexible manufacturing lines/processes and effective knowledge sharing is proposed. The presented solution includes services for context extraction, adaptation and self-learning allowing high adaptation of production systems depending on the identified context. A generic architecture following Service Oriented Principles is presented allowing for integration of the proposed solution into various production systems. A successful initial application of the developed solution in real world manufacturing environment is presented.

Keywords: Context Awareness, Context Extraction, Ontology, SOA.

1 Introduction

The need for highly flexible manufacturing processes is nowadays indispensable. In order to enhance the availability and at the same time the efficiency of modern production systems, two key requirements arise a) self-adaptation to support change of process parameters aiming at increasing flexibility and efficiency and b) effective knowledge acquisition / sharing to support maintenance increasing availability of production systems, as indicated in [1].

In order to realize such production systems, which are able to solve both above mentioned requirements, there is a clear need for powerful ICT infrastructures/solutions able to handle high amount of data, complex models and algorithms. Key challenge is to find a common approach for addressing these two problems:

- To allow for a self-adaptation, production systems need advanced monitoring and control. Thereby changes (e.g. changing process parameters) need to be identified and used to (self) adapt control processes.

L.M. Camarinha-Matos, S. Tomic, and P. Graça (Eds.): DoCEIS 2013, IFIP AICT 394, pp. 210–217, 2013.

- To allow for knowledge sharing, production systems need advanced monitoring to allow enhancement of monitored knowledge in order to facilitate knowledge sharing needed for effective maintenance.

A service oriented approach promises new perspectives for realizing such self-adaptive production systems. It is likely that approaches based on Service Oriented Architecture (SOA) principles, using distributed networked embedded services in device space (sensors, controllers etc.), are the most appropriate for implementation of such production systems in general.

While successful application of SOA based principles in manufacturing lines is emerging, there is still a major challenge:

How to realize reliable, production systems for complex manufacturing processes, which support both, self-adaption of process parameters and knowledge sharing to support improved maintenance of such production systems, while using a common approach.

1.1 Research Question

Taking the above mentioned major challenge and the related sub-challenges into account, a new innovative approach for future production systems is required, which raises the following research question:

Which methods and tools are required, to allow for the realization of reliable production systems that are able to support both, self-adaption of production processes without the need to reprogram or reconfigure such manufacturing processes and knowledge sharing to support effective maintenance, increasing thereby availability and efficiency of complex manufacturing processes?

The research question elaborated above can be addressed by the following two hypotheses:

- Self-adaptive production systems can be achieved if a context aware approach is used to identify the current context of (complex) manufacturing processes and use this extracted context as a basis for self-adaptation of process parameters.
- Effective sharing of knowledge in production systems can be achieved, if a context aware approach is used to identify the current context under which the knowledge is monitored and use this extracted context as foundation to support knowledge sharing, improving effective maintenance.

2 Relationship to Internet of Things

The application of context awareness within production systems to allow the realization of self-adaptive and highly available systems may on one side lead to enormous benefits regarding availability and efficiency of production systems and, on the other side increase the implementation of such possibilities by using a generic solution which can be used in several scenarios (i.e. Control, Maintenance).

To allow for the realization of such solutions, the presented work describes a new approach for such self-adaptive production systems based on a context aware approach, allowing self-adaptation of flexible manufacturing processes in production systems. The usage of dynamically changing context of production lines, as well as changing ambient conditions, for adaptation of flexible manufacturing lines/processes based on the usage of connected sensors and machines, as well as scalable architectures is proposed.

3 Survey of the State of the Art

This section gives a brief overview of the state-of-the-art for RTD topics which are relevant for the presented approach.

In the area of self-learning production systems, the research has demonstrated that the application of machine learning techniques, dynamic self-adaptation and operator's feedback in the loop promises to increase the intelligence of the overall system. In this approach, machine learns whenever it changes its structure, program, or data based on inputs or in response to external information in such a manner that its overall performance is expected to improve [2], [3], [4]. In production systems in particular, these methods have been proven to be especially useful for monitoring/diagnosis [5], [6], [7] and control [8]. However, the applications of self-adaptation of production systems and learning in industrial practice are still in initial phase. In this paper a novel approach in production systems is presented, which is context aware, adaptive to contextual changes at run time and learns from adaptation and operator's action.

Of high interest for the work presented in this paper is Service Oriented Architecture (SOA) in manufacturing i.e. the relation between self-learning production systems and SOA. SOA gained popularity with the advancement of web service technologies in industrial networks. For example, Devices Profile for Web Services (DPWS) allows devices with embedding computing capabilities, sensors, actuators etc. to interact within the industrial network and enabling SOA integration to higher levels [9]. OPC Unified Architecture is the recent OLE (Object Linking and Embedding) for process control (OPC) specification from the OPC Foundation and differs significantly from its predecessors. The Foundation's goal was to provide a path forward from the original OPC communications model (namely COM/DCOM) to a cross-platform SOA for process control, while enhancing security and providing an information model [10]. Scalable SOA holds promise for seamless integration, interoperation and flexibility in manufacturing environment. But, there is a lack of adoption of overall SOA based self-adaptive production systems in discrete manufacturing environment.

Context awareness is widely applied in modern ICT solutions. Different approaches to context modeling are developed, where ontology based context modeling is mostly investigated. Based on the formal description of context information, context can be processed with contextual reasoning mechanisms. Since contextual information has some inherent features (it can be considered incomplete, temporal, and interrelated) context reasoning can exploit reasoning mechanisms to deduce high level, inferred context from low-level raw contextual information. Furthermore, contextual reasoning can be used to verify and possibly solve inconsistent context knowledge due to imperfect input. For example, Luther et al. [11] show the needs for ontology support and reasoning in mobile applications; their case study is conducted with the Protégé knowledge workbench [12] for ontology modeling and OWL editing, and the RACER inference engine [13] for proof checking, ontology validation and classification. A more flexible use of ontological reasoning is presented in [26]. Their framework utilizes context-awareness for service classification.

Application of context awareness for SOA based self-learning production systems has not yet been sufficiently investigated. The approach proposed in this paper applies the ontology based context modeling, and re-use of experience of other projects [14], for context extraction in highly flexible and dynamic self-learning production systems [15].

4 Proposed Approach to Identify Context

The key assumption is that the extraction of current context is the foundation for self-adaptation / self-configuration of service-oriented production systems on one side, as well as for knowledge sharing to support effective maintenance. The application of such a context aware approach based on extraction of dynamically changing current context seems to be an effective way to assure availability and efficiency of modern, flexible production systems.

In a more detail, the assumption is that it is possible to gain more efficient embedded services by using context awareness than by using classical embedded services. The foundation of such context aware services is data, which will be gathered from various sources (e.g. sensors, inputs of the human operator, etc.). This information will be used to identify the current context of the services, which is realized via monitoring services, which are, for example, services for monitoring of sensors or of a user interacting with a system. These Monitoring services will transform all monitored information into a "standardized" matrix/action which is used to extract the current context. The extraction uses the Monitoring/Action Matrix together with context reasoning methods to identify the context. Basis of the context identification is a context model for device spaces. The context model consists of a generic model, to allow for effective use of different application domains, and a specific model which instantiates generic concepts to allow for optimal adjustment to the domain. In order to allow context extraction to produce tangible results, a well-defined context model is crucial.

The identified context will be used to a) allow embedded services to adopt their behavior according to the identified context (e.g. change of process parameters) and/or b) support sharing of knowledge of production processes according to the identified context (e.g. identify necessary maintenance tasks).

The objective of the presented work is to realize a context aware approach which can be used as a common solution for the realization of self-adaptive production systems and effective knowledge sharing to support effective maintenance in such production systems. The aim is to develop a generic solution, which can be used in several scenarios (i.e. Control, Maintenance) and various domains [16], [17], [18].

4.1 Architecture

A context aware approach allows a manufacturing system to be dynamically self-adapted during run time. To achieve this, the solution adapts based on contextual changes and learns from these adaptions. The learning is further enhanced by the operator's feedback.

A generic solution for context based self-adaptation of production systems is proposed which can be applied in various manufacturing systems. This self-learning reference architecture is derived from analyzing three application cases which belong to different industrial sectors. The overall proposed reference architecture, which is following SOA principles, is illustrated in Figure 1.

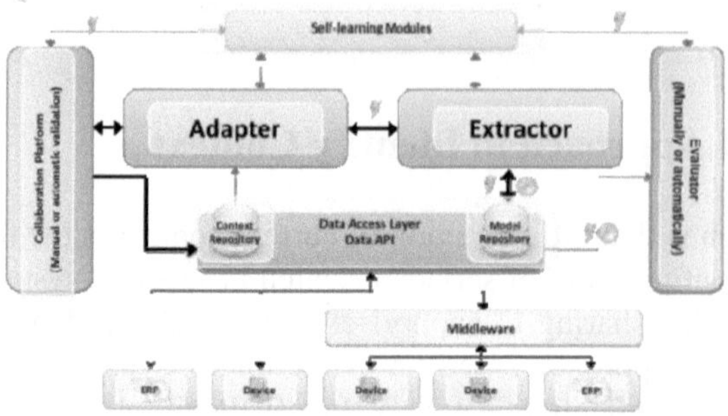

Fig. 1. Proposed Reference Architecture [19],[20]

The modules of the reference architecture are not described in this paper, for a detailed description see also [21].

4.2 Context Extraction

Context Extraction is based on a set of embedded services responsible for identifying changes in the context of the environment (see Figure 2). The current identified context is used to extract available context knowledge. The results of the Context Extraction are used in the Adapter which is responsible for updating the system behavior.

Context Extraction uses all "raw data" provided via the data access layer to derive the machine's current contextual situation [22]. Using the ontology/context model the monitored data is evaluated and the current context extracted [23]. Based on the identified context, situations can be compared to previous ones and stored. A continuous process, which is coordinating with the monitoring and followed by the adaption process to give current contextual meaning to the provided knowledge, is built around the main extraction of current contextual. The core modules of the proposed Context Extractor Architecture are the following:

- **Adapter Interface:** Via this interface the Context Extractor and the Adapter will exchange data and call specific functionality on both sides.
- **Data Access Layer:** Responsible for accessing the device layer.

- **Data Processing:** This module is responsible for the pre-processing of monitored raw data acquired via the data access layer, before the context will be identified. Main functionality will be the normalization of monitored data to transform the raw data in a format which serves as basis for context identification

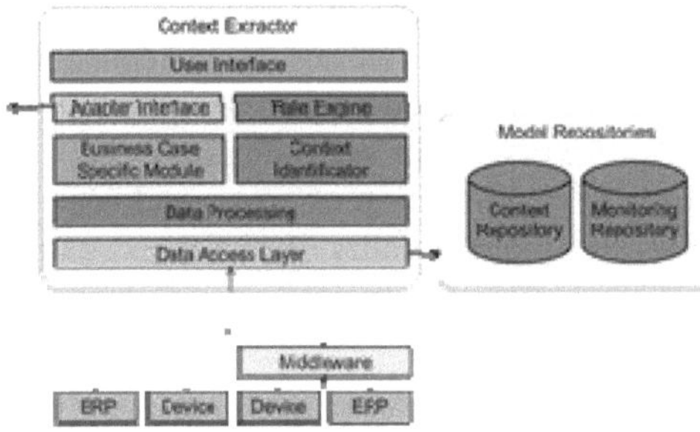

Fig. 2. Context Extractor Architecture

- **Context Identifier:** Main component of the Context Extractor. It is responsible for the identification of the current context, based on monitored raw data, the ontology and historic context information stored in the context repository.
- **Rule Engine:** Responsible for providing rules for the identification of context.
- **User Interfaces:** User interfaces for maintaining and administering the rules and the context repository.
- **Model Repositories:** Inside the Context Repository the identified context will be stored for further processing and reuse. Model Repository is used for the ontology.
- **Business Case Specific Modules:** This module is a placeholder and represents all components and user interfaces which needs to be developed for each business case individually.

5 Application

The presented concept has been developed following SOA principles and initially applied in different application scenarios. These application scenarios belong to different industrial sectors.

One of these application scenarios addresses control systems/machines and automation systems for shoe industry. The overall objective is to enhance machines with self-adaptive functionalities by allowing machines to inspect statistically the condition of products and equipment, report and analyze proactively the gathered statistic values, enabling the machines to decide and adapt the parameters and keep them always inside the "optimized" working range. The objective is to achieve high adaptation of the machines to the changing conditions. The parameter variations are in terms of pressure and temperature, speed frequencies of drives and pumps, proper material mix ratio and filling of materials into shoe forms. For example, one of the scenarios addresses synchronicity of the different valve circuits when dosing of several components, caused by e.g.

different force requirements or different air supply. As the valve synchronization is designed by an electronic system with different valve opening times, and/or due to valve abrasion, it may come to flaws in the product. Currently the synchronization can only be adjusted by a technician during downtime of the machines. By implementation of the proposed self-adaptive solution an automatic adjustment of the valve switching to different conditions is achieved. The advantage of having such an intelligent solution for the valve synchronizations is that it provides a basis for preventive maintenance.

The implementation of the proposed self-adaptive solution for the automatic adjustment of machine parameters based on changing context, for example changing ambient conditions, leads to minimization of errors and keeps the machine utilization high, as well as the overall product quality. The context extraction serves as a basis for identifying machine adjustment parameters.

6 Conclusions and Further Work

A novel approach for the realization of self-adaptive and highly available production systems is presented. The proposed solution addresses the adaptation of process parameters based on a context aware approach.

Further research will focus on advanced algorithms for self-learning based on extracted context to (semi-)automatically update the context model. In addition the context model itself will be addressed by further research to allow better utilization of the presented model for other companies as well as for other application domains. High complexity of data acquisition and real-time data analysis algorithms will be addressed in further research to "fully" utilize the opportunities offered by service-based self-learning systems.

Acknowledgements. This work is partly supported by the Self-Learning project of European Union's 7th Framework Program, under the grant agreement no. NMP-2008-228857 and by the EPES project of European Union's 7th Framework Program, under the grant agreement no. FoF-ICT-2011.7.3-285093. This document does not represent the opinion of the European Community, and the European Community is not responsible for any use that might be made of its content.

References

1. MANUFUTURE Strategic Research Agenda. Report of the High-Level Group (September 2006)
2. Alpaydin, E.: Introduction to machine learning. The MIT Press (2004)
3. Michie, D., Spiegelhalter, D., Taylor, C., Campbell, J.: Machine learning, neural and statistical classification. Ellis Horwood Series in Artificial Intelligence, p. 289 (1995)
4. Carbonell, J., Michalski, R., Mitchell, T.: An overview of machine learning. Machine Learning: An Artificial Intelligence Approach 1, 3–23 (1983)
5. Liu, S.C., Liu, S.Y.: An Efficient Expert System for Machine Fault Diagnosis. The International Journal of Advanced Manufacturing Technology 21, 691–698 (2003)

6. Palluat, N., Racoceanu, D., Zerhouni, N.: A neuro-fuzzy monitoring system application to flexible production systems. Comput. Ind. 57, 528–538 (2006)
7. Jianbo, Y., Lifeng, X., Xiaojun, Z.: Intelligent monitoring and diagnosis of manufacturing processes using an integrated approach of KBANN and GA. Comput. Ind. 59, 489–501 (2008)
8. Monostori, L.: AI and machine learning techniques for managing complexity, changes and uncertainties in manufacturing. Engineering Applications of Artificial Intelligence 16, 277–291 (2003)
9. Ribeiro, L., Barata, J., Colombo, A.W., Jammes, G.: A Generic Communication Interface for DPWS-based Web Services. In: IEEE International Conference on Industrial Informatics, INDIN, Daejeon, Korea. IEEE (2008)
10. Mahnke, W., Leitner, S.-H.: OPC Unified Architecture - The future standard for communication and information modeling in automation. ABB Review 3/2009, pp. 56–61 (2009)
11. Luther, M., et al.: Situational reasoning - a practical OWL use case, Chengdu, Jiuzhaigou, China. IEEE (2005)
12. Noy, F.N., et al.: Creating Semantic Web contents with Protege-2000. IEEE Intelligent Systems 16(2), 60 (2000)
13. Haarslev, V., Moller, R.: RACER system description, Siena, Italy. Springer (2001)
14. Ziplies, S., Scholze, S., Stokic, D., Krone, K.: Service-based Knowledge Monitoring of Collaborative Environments for User-context Sensitive Enhancement. In: ICE 2009 (2009)
15. Sorli, M., Stokic, D.: Innovating in Product/Process Development. Springer, Heidelberg (2009)
16. K-NET Project - Deliverable D1.4 Public Concept (2008)
17. AsKoWi Project (2011), `http://www.askowi.de`
18. EPES Project (2011), `http://www.epes-project.eu`
19. Self-Learning, EU project NMP-2008-228857 "Reliable self-learning production system based on context aware services", Public report (2010)
20. Stokic, D., Scholze, S., Barata, J.: Self-Learning Embedded Services for Integration of Complex, Flexible Production Systems. In: IECON 2011-37th Annual Conference on IEEE Industrial Electronics Society, Melboure, Australia (2011)
21. Scholze, S., Stokic, D., Barata, J., Decker, C.: Context extraction for self-learning production systems. In: 2012 10th IEEE International Conference on Industrial Informatics (INDIN), July 25-27, pp. 809–814 (2012)
22. Kelly, D., Teevan, J.: Implicit Feedback for Inferring User Preferences: A Bibliography. SIGIR Forum 37(2), 18–28 (2003)
23. Oard, D.W., Kim, J.: Modelling information content using observable behaviour. In: Proc. ASIST Annual Meeting, pp. 481–488 (2001)

Part IX
Embedded Systems and Petri Nets

Augmenting High-Level Petri Nets to Support GALS Distributed Embedded Systems Specification

Filipe Moutinho and Luís Gomes

Universidade Nova de Lisboa, Faculdade de Ciências e Tecnologia, Portugal
UNINOVA – CTS, Portugal
fcm@uninova.pt, lugo@fct.unl.pt

Abstract. High-level Petri net classes are suited to specify concurrent processes with emphasis both in control and data processing, making them appropriate to specify distributed embedded systems (DES). Embedded systems components are usually synchronous, which means that DES can be seen as Globally-Asynchronous Locally-Synchronous (GALS) systems. This paper proposes to include in high-level Petri nets a set of concepts already introduced for low-level Petri nets allowing the specification of GALS systems, namely time domains, test arcs and priorities. Additionally, this paper proposes external messages and three types of (high-level) asynchronous communication channels, to specify the interaction between distributed components based on message exchange. With these extensions, GALS-DES can be specified using high-level Petri nets. The resulting models include the specification of each component with well-defined boundaries and interface, and also the explicit specification of the asynchronous interaction between components. These models will be used not only to specify the system behavior, but also to be the input for model-checking tools (supporting its verification) and automatic code generation tools (supporting its implementation in software and hardware platforms), giving a contribution to the model-based development approach and hardware-software co-design of DES based on high-level Petri nets.

Keywords: Distributed embedded systems, GALS systems, model-based development, high-level Petri nets, asynchronous-channels.

1 Introduction

This work gives a contribution to the use of high-level Petri nets in the model-based development approach of distributed embedded systems (DES). Model-based development approaches [1, 2, 3, 4] and hardware-software co-design techniques [5] have been providing several development methods for embedded systems. Often supported by tools, these methods can provide benefits such as the reduction in the development time and in the number of development errors (bugs). DES are composed by a set of embedded systems (components) in interaction, which may be geographically distributed or not (implemented in a single implementation platform or chip). When these components, which may be hardware or software components, are synchronous with a specific clock tick (or clock signal), the DES is globally-asynchronous locally-synchronous (GALS) [6].

L.M. Camarinha-Matos, S. Tomic, and P. Graça (Eds.): DoCEIS 2013, IFIP AICT 394, pp. 221–228, 2013.

Several modeling formalisms, such as State-Diagrams, StateCharts [7] and Petri nets [8], have been used in model-based development approaches to develop embedded systems. Given that Petri nets are appropriate to specify concurrency, they were chosen in this work as the modeling formalism for DES. Low-level Petri net classes are appropriate to specify "control dominated" systems, whereas high-level Petri net classes are appropriate to specify not only control but also "data processing".

Low-level Petri nets were extended and used in [9, 10, 11] to specify GALS systems. Additionally, several works (such as in [12, 13, 14, 15]) addressed the development of distributed systems using low-level Petri nets. But no works are known, where globally-asynchronous locally-synchronous distributed embedded systems (GALS-DES) are explicitly specified through high-level Petri nets.

The work presented in this paper is integrated in a PhD work, and contributes to answer the following research question: *How to extend high-level Petri net classes to allow the explicit specification of GALS-DES?* An explicit specification should include the specification of each component, its interface, and the specification of the asynchronous interaction between components.

This paper presents an extension to high-level Petri nets, which contributes both to the model-based development and to the hardware-software co-design of distributed embedded systems that are also globally-asynchronous locally-synchronous (GALS-DES). The extended high-level Petri net models are intended to support not only the system specification, but also to be used as input in model checking and automatic code generators tools for hardware and software platforms.

Section 2 mentions how this paper contributes to the Internet of Things, section 3 presents the proposed extension for high-level Petri nets, and finally in section 4 conclusions and future work are presented. Due to space limitations it was not possible to present any complete application example in this paper.

2 Relationship to Internet of Things

The spread of embedded systems by devices, machines and infrastructures around us, and the possibility to interconnect via Internet or mobile connections, enabled the creation of many types of distributed embedded systems. The development of distributed systems when compared with the development of centralized systems is a challenging task, due to the interaction of concurrent components and also due to the large size/complexity of the overall system. This paper aims to contribute to the development of such systems using a model-based development approach and high-level Petri net classes, which are suitable to specify concurrent controllers with control and data processing capabilities. High-level Petri nets were augmented in this work to allow the specification of distributed embedded systems, and are also intended to be the input for verification and automatic code generation tools.

3 Extending High-Level Petri Nets for GALS-DES

High-level Petri nets were defined in the international standards ISO/IEC 15909-1 and ISO/IEC 15909-2 [16]. Briefly presented in Fig. 1, high-level Petri nets (*HLCoreStructure*) are an extension of low-level Petri nets (*PNMLCoreModel*). It is important to note that the international standard does not define a concrete syntax for high-level Petri nets, and that in this paper is used a concrete syntax similar to the one used in Colored Petri nets [17].

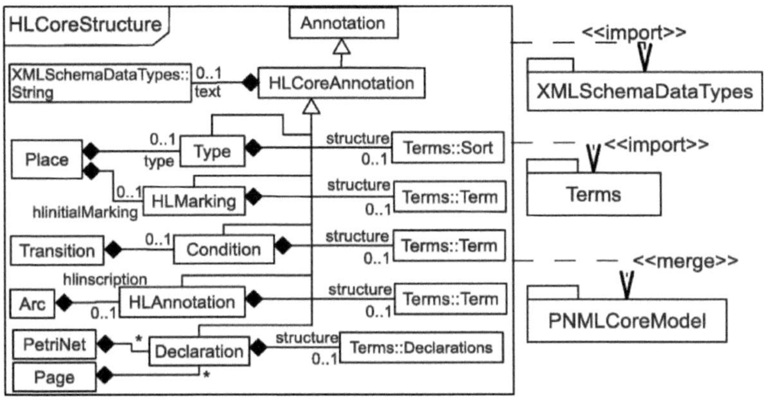

Fig. 1. The UML class diagram of the *HLCoreStructure* package, image adapted from [16]

This paper extends high-level Petri nets with a set of concepts to support the development of GALS-DES with emphasis both on control and data processing. The proposed concepts are presented in Fig. 2, which is represented by an UML class diagram and by a set of constraints expressed in the Object Constraint Language (OCL). Some of the concepts (test arcs, priorities, and time domains) have already been proposed for low-level Petri nets in [18, 11], whereas the other concepts (external messages and three types of asynchronous channels) are new.

Test arcs and priorities proposed here for high-level Petri nets, were proposed in [18] to support conflict arbiters. Test arcs (also known as read arcs) do not remove tokens from places, and are represented in Figs. 3, 4, and 5, by arcs with an arrow in the middle. When two or more transitions are in conflict, the one with higher priority (lower value) will fire. Priorities are represented in Fig. 5, by *p:1* and *p:2*.

3.1 Execution Semantics and Time-Domains

The concept of time-domains was proposed in [11] to extend the IOPT-net class [18], which is a low-level Petri net class. This class without time-domains has (globally) synchronous and maximal-step execution semantics, which means that transitions can only fire at specific time instants (given by a clock tick) and that all transitions that are enable and ready (see [18]) to fire (and also not in conflict) at a specific clock tick,

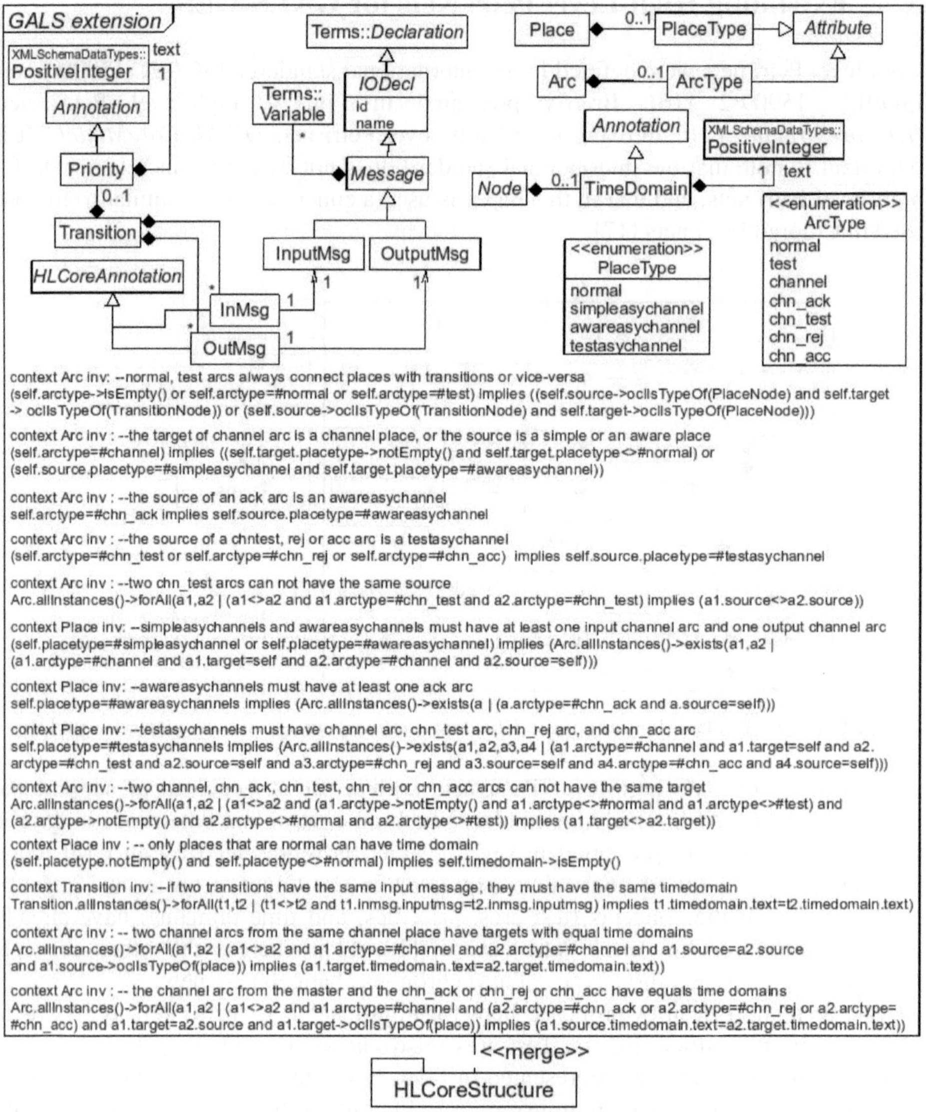

Fig. 2. The proposed *GALS extension* package

will fire simultaneously. The IOPT-net class extended with time-domains allows the specification of globally-asynchronous locally-synchronous systems, because all transitions with a specific time-domain are "locally synchronous" and have a "locally maximal-step" execution semantics. The concept of time-domain associates each Petri net node (transition or place) with a specific component, which is synchronous with a specific clock tick. Transitions with different time-domains belong to different components (synchronous with distinct clock signals).

The concept of time-domain when added into high-level Petri net classes, introduces the globally-asynchronous locally-synchronous execution semantics into high-level Petri nets, allowing the specification locally synchronous components (synchronous with specific clock signals). Fig. 2 presents the concept of time-domain added into high-level Petri nets, where Petri net nodes (transitions and places) can have an associated time-domain (represented as a node annotation). An OCL specifies that if two transitions have the same input message, they must have equal time-domains, because an input message cannot be an input of several components. Fig. 3 presents a high-level Petri net model with two components, each one specified by a specific time-domain (represented by *td:1* and *td:2*).

3.2 Asynchronous-Channels

Time-domains are used to identify components, whereas asynchronous-channels enable their interaction specification (the exchange of data between components). Three types of directed asynchronous communication channels are proposed in this paper: (1) Simple-Asynchronous-Channel (SAC); (2) Aware-Asynchronous-Channel (AAC); and (3) Test-Asynchronous-Channel (TAC). To ensure that messages are delivered by the same order that are sent, asynchronous channels have FIFO (First In, First Out) semantics. All the proposed channels assume that all sent messages eventually arrive (sometime in the future) at the destination component.

The models at the right hand side of Figs. 3, 4, and 5 present the semantics of each channel and the models at the left hand side present their representation (clouds connected to transitions by dashed arcs). Although it is only presented for SAC (in the right model from Fig. 3), the behavioural model of the other two channels (right models from Figs. 4 and 5) also have a *reset* transition to prevent the messages identifier to grow indefinitely. Due to space restrictions the channels proposed in Fig. 2 will be briefly described.

Each SAC sends messages from one (master) transition to a set of (slave) transitions. All slaves have the same time-domain (belong to the same component), which is different from master time-domain. Fig. 3 (left) presents a high-level Petri net model with a SAC, and its behavior is presented in Fig. 3 (right) model.

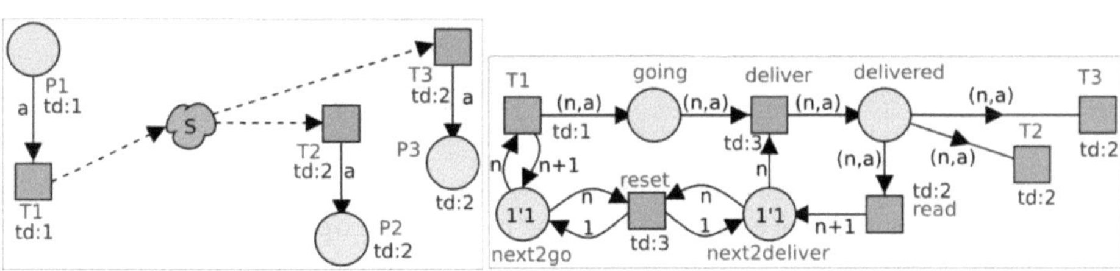

Fig. 3. A high-level Petri net model with a Simple-Asynchronous-Channel (at the left) and its behavioral specification using a high-level Petri net model (at the right)

In an AAC, messages are sent as in a SAC, but an acknowledge is sent when the message is read by destination component (even if the message is ignored). Fig. 4 (left) presents a model with an AAC, and its behavior is presented in Fig. 4 (right).

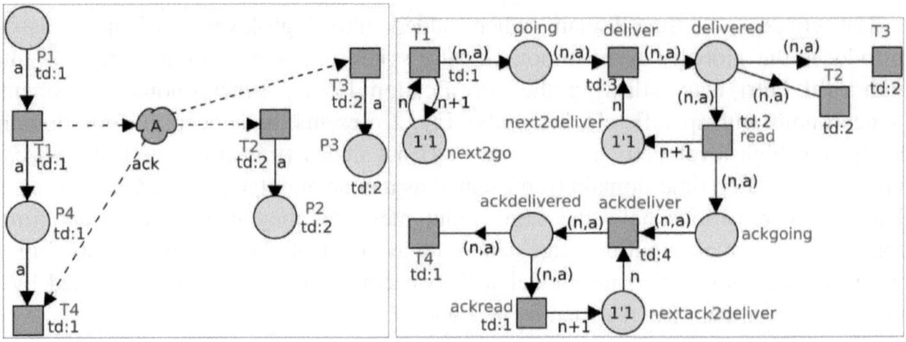

Fig. 4. A high-level Petri net model with an Aware-Asynchronous-Channel (at the left) and its behavioral specification using a high-level Petri net model (at the right)

Finally, in a TAC messages are sent as in a SAC, but when the message is read by destination component, an accepted (*acc*) message (if the message fired the transition) or a rejected (*rej*) message (if the message is ignored) is sent back. Fig. 5 (left) presents a model with a TAC, and the TAC behavior is presented in Fig. 5 (right) model. Each SAC or AAC can have several slave transitions, whereas each TAC can only have one slave transition.

Any Petri net model with these three types of communication channels can be specified by a Petri net model without channels, replacing the channels by their behavioral models (right models from Figs. 3, 4 and 5).

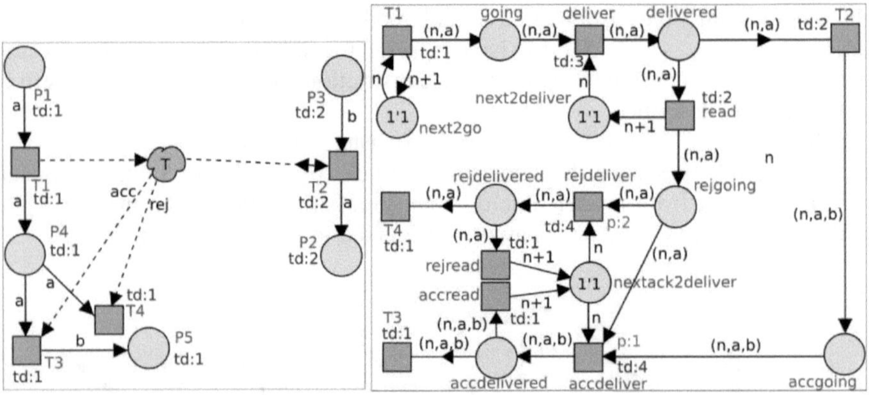

Fig. 5. A high-level Petri net model with a Test-Asynchronous-Channel (at the left) and its behavioral specification using a high-level Petri net model (at the right)

The GALS-DES specification using high-level Petri net classes extended with time-domains and asynchronous-channels have a strong mathematical definition (not presented in this paper) and a well defined execution semantics, allowing its use as input for model-checking tools, to verify the extended high-level Petri net models proprieties.

3.3 Messages

Input and output messages are proposed to specify the components interface (to allow the specification of the interaction between the components and the environment or between the components and the communication channels). After GALS-DES specification and verification, and before components implementation, asynchronous channels must be removed, and input and output messages must be inserted into Petri nets. Inserted input and output messages are associated with master and slave transitions (previously connected with asynchronous channels). The resulting models with input and output messages can be used as input for automatic code generator tools for software and hardware platforms.

Input and output messages are proposed in this paper as presented in Fig. 2. *Messages* are *Petri nets* or *Pages* declarations, and are associated with transitions. Both input and output messages can have associated *Variables*, which defines the carried data. When an input message occurs, the associated transitions fire if enable and ready (the carried data influence transition bindings). Output messages are generated when associated transitions fire. An output message can carry data obtained from place marking.

4 Conclusions and Future Work

This extension gives a contribution to the use of high-level Petri nets in the model-based development approach of distributed embedded systems. The identification of components sub-models is made through the use of time-domains; the interaction between components is specified by asynchronous communication channels; and external messages (input and output messages) are the components interface.

Some of the concepts (such as time-domains) proposed in this paper for high-level Petri nets had already been proposed in the past for low-level Petri nets, whereas other concepts (such as the three types of asynchronous communication channels with FIFO semantics and the external messages) as far as we know, are new.

During our experience using these communication channels, it was possible to specify all encountered communication scenarios between components of globally-asynchronous locally-synchronous distributed embedded systems.

As future work we intend to extend the tool chain framework for a low-level Petri net class (online available at *http://gres.uninova.pt/*), to allow the edition, the model-checking, and the automatic code generation of distributed embedded systems using high-level Petri nets.

Acknowledgments. This work was partially financed by Portuguese Agency "FCT - Fundação para a Ciência e a Tecnologia" in the framework of project PEst-OE/EEI/UI0066/2011. The first author was supported by a FCT grant, ref. SFRH/BD/62171/2009.

References

1. Schätz, B., Pretschner, A., Huber, F., Philipps, J.: Model-based development of embedded systems. In: Bruel, J.-M., Bellahsène, Z. (eds.) OOIS 2002 Workshops. LNCS, vol. 2426, pp. 298–311. Springer, Heidelberg (2002)
2. Rust, C., Kleinjohann, B.: Modeling Intelligent Embedded Real-Time Systems using High-Level Petri Nets. In: Proceedings of the Forum on Design Languages, FDL (2001)
3. De Niz, D., Bhatia, G., Rajkumar, R.: Model-Based Development of Embedded Systems: The SysWeaver Approach. In: Proceedings of the 12th IEEE Real-Time and Embedded Technology and Applications Symposium, Washington, DC, USA (2006)
4. Gomes, L., Fernandes, J. (eds.): Behavioral Modeling for Embedded Systems and Technologies: Applications for Design and Implementation. IGI Global's (2009) ISBN 978-1-60566-750-8
5. Wolf, W.H.: Hardware-software co-design of embedded systems [and prolog]. Proceedings of the IEEE 82(7), 967–989 (1994)
6. Chapiro, D.M.: Globally-Asynchronous Locally-Synchronous Systems. Ph.D. Thesis: Stanford University (1984)
7. Harel, D.: Biting the silver bullet: toward a brighter future for system development. Computer 25(1), 8–20 (1992)
8. Reisig, W.: Petri nets: an introduction. Springer-Verlag New York, Inc., NY (1985)
9. Nielsen, M., Sassone, V., Srba, J.: Towards a Notion of Distributed Time for Petri Nets. In: Colom, J.-M., Koutny, M. (eds.) ICATPN 2001. LNCS, vol. 2075, pp. 23–31. Springer, Heidelberg (2001)
10. Kleijn, H., Koutny, M., Rozenberg, G.: Processes of Petri nets with localities, Technical Report CS-TR-941, School of Computing Science, Newcastle upon Tyne, UK (2006)
11. Moutinho, F., Gomes, L.: Asynchronous-channels and time-domains extending Petri nets for GALS systems. In: Camarinha-Matos, L.M., Shahamatnia, E., Nunes, G. (eds.) DoCEIS 2012. IFIP AICT, vol. 372, pp. 143–150. Springer, Heidelberg (2012)
12. Hopkins, R.: Distributable nets. In: Rozenberg, G. (ed.) APN 1991. LNCS, vol. 524, pp. 161–187. Springer, Heidelberg (1991)
13. Badouel, E., Caillaud, B., Darondeau, P.: Distributing finite automata through Petri net synthesis. Formal Asp. Comput. 13(6), 447–470 (2002)
14. Van Glabbeek, R., Goltz, U., Schicke, J.-W.: On synchronous and asynchronous interaction in distributed systems. CoRR, abs/0901.0048 (2009)
15. Van Glabbeek, R., Goltz, U., Schicke-Uffmann, J.-W.: On distributability of Petri nets. In: Birkedal, L. (ed.) FOSSACS 2012. LNCS, vol. 7213, pp. 331–345. Springer, Heidelberg (2012)
16. Hillah, L., Kindler, E., Kordon, F., Petrucci, L., Treves, N.: A primer on the Petri Net Markup Language and ISO/IEC 15909-2. Petri Net Newsletter 24(76), 9–28 (2009) (Originally Presented at the 10th International Workshop on Practical Use of Colored Petri Nets and the CPN Tools – CPN 2009)
17. Jensen, K., Kristensen, L.M., Wells, L.: Coloured Petri Nets and CPN Tools for Modelling and Validation of Concurrent Systems. International Journal on Software Tools for Technology Transfer (STTT) 9(3-4), 213–254 (2007)
18. Gomes, L., Barros, J., Costa, A., Nunes, R.: The Input-Output Place-Transition Petri Net Class and Associated Tools. In: Proceedings of the 5th IEEE International Conference on Industrial Informatics (INDIN 2007), Vienna, Austria (2007)

On Structuring Events for IOPT Net Models

Rogério Campos-Rebelo[1,2], Anikó Costa[1,2], and Luís Gomes[1,2]

[1] Universidade Nova de Lisboa, Faculdade de Ciências e Tecnologia, Portugal
[2] UNINOVA – Centro de Tecnologias e Sistemas, Portugal
{rcr,akc,lugo}@uninova.pt

Abstract. This paper presents a proposal for structuring events for system models expressed using IOPT nets (Input-Output Place-Transition Petri nets). Currently, a non-autonomous event within an IOPT model is defined based on change of input signals with respect to a specific threshold, when two consecutive execution steps are considered. New types of events are proposed allowing the definition of an event activated not only by crossing a fixed threshold, but also considering a change in associated signal value on a specific amount (belonging to an interval of values). The concept is further extended allowing the definition of an event based on signal values presented on previous execution steps. The proposal results on a classification of several types of events, namely threshold events, momentum events, impetus events, as well as delayed events and logical events. Usage of these types of events allows improvements in terms of expressiveness and compactness of the resulted model.

Keywords: Petri nets, embedded systems, human-system interaction.

1 Introduction

Since its presentation by Carl Adam Petri in 1962 [1], Petri Nets have suffered many changes and developments, which have led to the growth of their use in various areas. Engineering is one of those areas that take advantage of features such as the ability to model synchronization, concurrency, or conflicts [2], or the simultaneous visualization of the structure and behavior of the system [3]. These features allow Petri nets to capture the dynamics of the system, making them useful in simulation [4].

The inclusions of non-autonomous characteristics in Petri nets models have significantly contribute to increase the impact of their use when compared with other modeling formalisms used for embedded systems design [5]. These non-autonomous characteristics allow the connection of the net with other environments, making them very suitable for modeling of system's controller behavior.

There are a set of non-autonomous Petri net classes applicable for this kind of application areas. Some examples are the Interpreted Petri nets[6], synchronized Petri nets [7] or IOPT nets (Input-Output Place-Transition nets) [7].

After working with these formalisms, it is easy to conclude that whenever a complex interface modeling is needed the model tends to become complex and with

L.M. Camarinha-Matos, S. Tomic, and P. Graça (Eds.): DoCEIS 2013, IFIP AICT 394, pp. 229–238, 2013.

hard interpretation, due to limited number of modeling primitives to explicitly handle human-system interaction.

In this paper a set of proposals to define different types of events are presented addressing the effectiveness of Petri nets modeling for human-system interaction systems, having also application within controllers modeling.

On top of the traditional way to define an event (as a signal trespassing a threshold level), new ways to define an event associated with signal changes are proposed based also on concepts of amplitude and changing speed of variation on signal values.

In this sense, starting with the concept of event definition based on a signal threshold when comparing its value on two consecutive execution steps, the new concepts are proposed considering a larger observation window (not only two consecutive execution steps), as well as reasoning on speed and acceleration of the signal values changes. This led to the definition of impetus events and momentum events, as well to the definition of delayed events and logical events, resulting from the pre-processing of signal variations.

An overview on regular and new events is provided in this paper. A simple potential application for this new modeling attitude uses the model of a joystick controlling a plane simulator as inspiring source, and can be used to informally support the rationale around each new type of event, illustrating modeling benefits for their usage and potential compactness of resulting models. Due to lack of space, it is not possible to present detailed models.

2 Relationship to Internet of Things

In this paper a set of additions to the non-autonomous part of Petri nets models allowing a structuring on events is proposed. These additions directly support a more compact modeling of human-system interactions, permitting also a simpler implementation. With the growth of internet and its opening to the general public, rather than just for those who have expertise in the area, brought new problems to solve like how to develop easier and more enjoyable interfaces to facilitate the interaction of the people with the systems.

Petri nets are in a good position to provide solutions to these problems, as tools are available to model distributed systems and to support a good way to answer to these modeling challenges.

The ideas proposed in this paper leads to an evolution in the modeling of the human-system interactions with Petri nets facilitating the modeling of new interfaces within the “internet of things” concept.

3 IOPT Nets

The IOPT nets, proposed in [8], are a low-level and non-autonomous Petri nets class, extending Place Transition nets and allowing modeling of interaction with the environment through inputs and outputs (events and signals). These signals and events are associated with the transition firing and places marking, allowing communication with the environment.

The IO extension adds a minimal set of notations to Place-Transition nets that allow the specification of control of discrete event systems. The IOPT nets support specification of:

- External input signals and external input events constraining transition firing, as well as generation of external output events as a result of transition firing (Mealy style).
- External output signals associated with marking of places (Moore style).

As common in modeling controller systems with discrete events, the evolution of the model is possible only in specific moments, called "ticks", and the period between two "ticks" is called step (which is of paramount importance along this paper). All transitions that are enabled and ready will fire at the same step ("maximal step"). In this sense, the synchronized Petri nets paradigm [9] is adopted, where a transition is enabled to fire whenever the marking of its input places provides adequate marking, and is ready to fire whenever associated conditions on signals and events are evaluated as true.

The IOPT nets may also accommodate mechanisms for model structuring [10] supported by decomposition and composition techniques, as in [11].

Currently there are two types of events defined: autonomous events (in the sense that they come directly from environment), and non-autonomous events (in the sense that they are automatically generated after processing of external signals). Only the later ones are considered for this paper. A non-autonomous input event or simply input event from now on, is associated with one input signal and is associated with a threshold level. It is the starting concept and will be formally presented in next section under the classification of "threshold event".

4 Events Definition

In this section the definitions of the existing threshold events are presented, followed by the definitions of the new types of events. Notation used in this section considers representation of signal X, being x_n the value of signal X at execution step n, and x_{n-1} the value of signal X at previous execution step, and x_{n-p} the value of signal X occurring p execution steps before.

4.1 On Threshold Events

Two types of threshold events are defined: the "up" event and the "down" Event, as proposed in [8].

The event "up", defined against the definition of the threshold level K, occurs when the value of the associated signal in the last execution step was lower or equal to the level K and the current value is higher than the level K. This event is presented and defined in (1). Similarly, the event "down" occurs when the value of associated signal in the last execution step was higher than the level K and the current value is higher or equal to the level K. Expression (2) presents the mathematical definition and a representation of this type of signals.

Table 1. Threshold events definitions

	Definition	
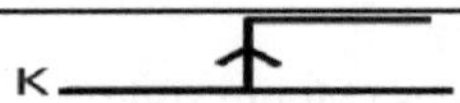	$evx^{+}(k) = (x_n > k) \wedge (x_{n-1} \leq k)$	**(1)**
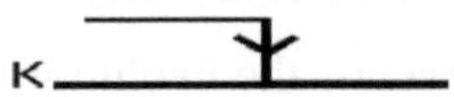	$evx^{-}(k) = (x_n \leq k) \wedge (x_{n-1} > k)$	**(2)**
	$evx^{\pm} = evx^{+} \vee evx^{-}$	**(3)**

A third type of event, an event "up or down" can be composed from the previous two types of events, is defined in expression (3) and reflects a situation where a crossing of the threshold level is observed.

4.2 On Momentum Events

The new type of events (the momentum events) is associated with reasoning on the difference of values of signal X along consecutive execution steps (where the concept of acceleration can be associated with this type of events, if X is related with position representation). So, let dx = $x_n - x_{n-1}$. Starting with equations in Table 1 related with threshold events *evx*, the momentum events *evdx* defined in Table 2 are obtained applying the same reasoning to *dx* as to *x* previously.

Table 2. Momentum events definitions

Definition	
$evdx^{+}(k) = (x_n - x_{n-1}) > k \wedge (x_{n-1} - x_{n-2}) \leq k$	**(4)**
$evdx^{-}(k) = (x_n - x_{n-1}) \leq k \wedge (x_{n-1} - x_{n-2}) > k$	**(5)**
$evdx^{\pm}(k) = evdx^{+}(k) \vee evdx^{-}(k)$	**(6)**

Expression (4) presents the definition of momentum event "up". This event detects that the variation of the difference on the signal values becomes higher than a specific value k. It means that in x_{n-1} the momentum value was lower or equal than k and in x_n the value becomes higher than k.

In the same way, expression (5) presents the momentum event "down". Similarly, this event can detect a variation of the difference of the signal X lower than a specific value k. It means that in x_{n-1} the momentum value was higher than k and in x_n the value becomes lower or equal than k.

Similarly to (3), a momentum event "up or down" is defined allowing the detection of the crossing of the k value, without taking into account the direction. This event is composed by event "up" (4) and "down" (5) and is defined in (6).

4.3 On Impetus Events

Impetus events are events that occur with the analysis of the differences in the signal magnitude. Four types of basic events can be defined, according with the signal magnitude and the variation direction:

- evxdpu event defined in (7), associated with a signal having a large positive increase rate, larger than k.
- evxdpd event defined in (8), associated with a signal having a small positive increase rate, smaller than k.
- evxdnu event defined in (9), associated with a signal having a large negative increase rate, larger than k.
- evxdnd event defined in (10), associated with a signal having a small negative increase rate, smaller than k.

On top of these four types of events, it is possible to define composed impetus events, detecting if the variation is between two levels, k_1 and k_2, and not only below or above a value k. Thus, two additional events are defined:

- evxdp event defined in (11), associated with a variation of the signal between two values (k_1, k_2) and composed by evxdpu and evxdpd events, considering a positive variation on the associated signal.
- evxdn event defined in (12), associated with a variation of the signal between two values (k_1, k_2) and composed by evxdnu and evxdnd events, considering a negative variation on the associated signal.

Note that in the events (11) and (12) the values k_1 *and* k_2 can be equal. If this occurs, the generated event is associated with a specific value k on the difference of values of signal X.

Table 3. Impetus events definitions

Definition	
$evxdpu\ (k) = \|x_n - x_{n-1}\| \geq k \wedge (x_n > x_{n-1})\ ,\ k>0$	**(7)**
$evxdpd\ (k) = \|x_n - x_{n-1}\| \leq k \wedge (x_n > x_{n-1})\ ,\ k>0$	**(8)**
$evxdnu\ (k) = \|x_n - x_{n-1}\| \geq k \wedge (x_n < x_{n-1})\ ,\ k>0$	**(9)**
$evxdnd\ (k) = \|x_n - x_{n-1}\| \leq k \wedge (x_n < x_{n-1})\ ,\ k>0$	**(10)**
$evxdp\ (k_1, k_2) = evxdpu(k_1) \wedge evxdpd(k_2)$	**(11)**
$evxdn\ (k_1, k_2) = evxdnu(k_1) \wedge evxdnd(k_2)$	**(12)**
$evxdmu\ (k) = \|x_n - x_{n-1}\| \geq k \wedge (x_n \neq x_{n-1})\ ,\ k>0$ $\equiv evxdpu\ (k) \vee evxdnu\ (k)$	**(13)**
$evxdmd\ (k) = \|x_n - x_{n-1}\| \leq k \wedge (x_n \neq x_{n-1})\ ,\ k>0$ $\equiv evxdpd\ (k) \vee evxdnd\ (k)$	**(14)**
$evxdmn = (x_n = x_{n-1}) \equiv \neg(evxdmu \vee evxdmd)$	**(15)**

As previously defined for threshold and momentum events, two additional events can be defined based on the magnitude of the difference, namely:

- evxdmu event defined in (13) that detects if the signal had a variation higher that a certain value k; this event can be obtained by composition of the two "up" events, evxdpu and evxdnu.
- evxdmd event defined in (14) that detects if the signal had a variation lower that a certain value k; this event can be obtained by composition of the two "down" events, evxdpd and evxdnd.

Note that for these events to occur there must have some variation on the signal, so it is needed that $x_n \neq x_{n-1}$.

Finally, to conclude on impetus events, an event detecting no variation in the signal is defined in expression (15).

4.4 Delayed Events

In all cases presented in the previous sub-sections, the analysis of the event is made taking into account the signal value at current execution step and the signal value at immediately prior execution step(s).

In order to accommodate adequate modeling flexibility (particularly for human-system interaction modeling), generation of events may depend on the analysis of signal values not so close in time. In these cases, it can be of interest to consider an event which takes into account a larger number of execution steps in its trigger condition, leading to the comparison of the current value x_n with the value p-execution steps before (x_{n-p}).

This modeling strategy, relying on the analysis of a delayed signal, can be applied to all previously proposed events.

Applying this strategy to threshold events as in Table 1, Table 4 is obtained.

Table 4. Threshold delayed events definitions

	Definition	
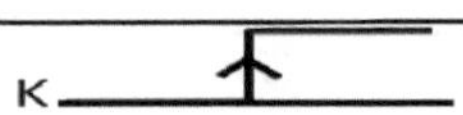	$evx^{+}(k) = (x_n > k) \wedge (x_{n-p} \leq k)$	**(16)**
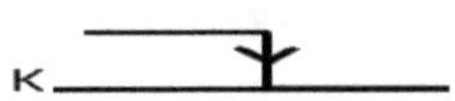	$evx^{-}(k) = (x_n \leq k) \wedge (x_{n-p} > k)$	**(17)**
	$evx^{\pm} = evx^{+} \vee evx^{-}$	**(18)**

Likewise, applying the delayed strategy to momentum events as in Table 1, Table 5 is obtained.

Table 5. Momentum delayed events definitions

Definition	
$evdx^{+}(k) = (x_n - x_{n-1}) > k \wedge (x_{n-p+1} - x_{n-p}) \leq k$	**(19)**
$evdx^{-}(k) = (x_n - x_{n-1}) \leq k \wedge (x_{n-p+1} - x_{n-p}) > k$	**(20)**
$evdx^{\pm} = evdx^{+} \vee evdx^{-}$	**(21)**

Finally, applying the same strategy to impetus events as in Table 3, Table 6 is obtained.

Table 6. Impetus delayed events definitions

Definiton	
$evxdpu\ (k) = \lvert x_n - x_{n-p} \rvert \geq k \wedge (x_n > x_{n-p})\ ,\ k>0$	**(22)**
$evxdpd\ (k) = \lvert x_n - x_{n-p} \rvert \leq k \wedge (x_n > x_{n-p})\ ,\ k>0$	**(23)**
$evxdnu\ (k) = \lvert x_n - x_{n-p} \rvert \geq k \wedge (x_n < x_{n-p})\ ,\ k>0$	**(24)**
$evxdnd\ (k) = \lvert x_n - x_{n-p} \rvert \leq k \wedge (x_n < x_{n-p})\ ,\ k>0$	**(25)**
$evxdp\ (k_1, k_2) = evxdpu(k_1) \wedge evxdpd(k_2)$	**(26)**
$evxdn\ (k_1, k_2) = evxdnu(k_1) \wedge evxdnd(k_2)$	**(27)**
$evxdmu\ (k) = \lvert x_n - x_{n-p} \rvert \geq k \wedge (x_n \neq x_{n-p})\ ,\ k>0$	**(28)**
$\equiv evxdpu\ (k) \vee evxdnu\ (k)$	
$evxdmd\ (k) = \lvert x_n - x_{n-p} \rvert \leq k \wedge (x_n \neq x_{n-p})\ ,\ k>0$	**(29)**
$\equiv evxdpd\ (k) \vee evxdnd\ (k)$	
$evxdmn = (x_n = x_{n-p}) \equiv \neg(evxdmu \vee evxdmd)$	**(30)**

These events have similar characteristics as the single step events but their usage allows a tuning with time constants associated with the specific modeling situations.

4.5 Logical Events

The next proposed type of complex events is based on the preprocessing of signals evolution history and allows the identification of specific sequences on those signals evolution. They are classified as logical events. Due to space restrictions, only two types of logical events are proposed, having two instances, the "up-down" logical event, and the "down-up" logical event, with and without hysteresis.

These events are of big interest to support production of compact models and rely on the detection of a "cycle" on the level of the signal.

An "up-down" event occurs when the signal value starts below or at level *k* and goes above *k*, returning to level k afterwards. The "up-down" event occurs in the execution step when signal returns or cross level *k*. This event is presented and defined in expression (31)

A "down-up" event occurs when the associated signal starts above level *k*, reaches level *k* or below, and returning above level k afterwards. This "down-up" event is defined in expression (33), and occurs when the signal return above level *k* again.

Delayed logical events can naturally be defined as extensions to the above proposed events, as defined in expression (32) and expression (34), respectively, where *p* determines the pulse width (in terms of number of execution steps). A particular case of interest is observed when *p* could hold an arbitrary value, allowing the detection of a complete cycle on signal values.

Table 7. Logical events definitions

	Definition	
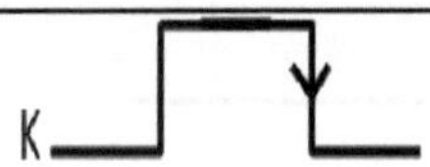	$evx^{+-}(k) = (x_n \leq k) \wedge (x_{n-1} > k) \wedge (x_{n-2} \leq k)$	**(31)**
	$evx_p^{+-}(k) = (x_n \leq k) \wedge (x_{n-1} > k) \wedge (x_{n-1-p} \leq k) \wedge (\forall_{q \in [1,p]} : x_{n-1-q} > k)$	**(32)**
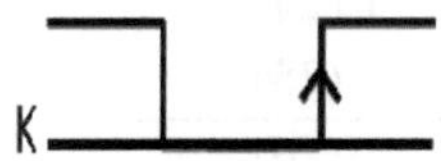	$evx^{-+}(k) = (x_n > k) \wedge (x_{n-1} \leq k) \wedge (x_{n-2} > k)$	**(33)**
	$evx_p^{-+}(k) = (x_n > k) \wedge (x_{n-1} \leq k) \wedge (x_{n-1-p} > k) \wedge (\forall_{q \in [1,p]} : x_{n-1-q} \leq k)$	**(34)**

A second extension to the referred events, with particular interest for non-Boolean signals, is the definition of a hysteresis on associated signal.

In this case, the events are defined using two levels (k_1 and k_2), corresponding to the hysteresis window.

The "up-down" event with hysteresis, presented in expression (35), is generated when associated signal starts at or below k_1, goes above k_2, and return at or below k_1. Similarly, the "down-up" event with hysteresis, presented in expression (36), is generated when associated signal starts above k_2, goes at or below k_1, and return above k_2.

Table 8. Logical events with hysteresis definitions

	Definition	
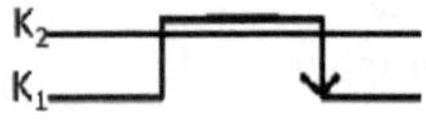	$evx_p^{+-h}(k_1,k_2) = (x_n \leq k_1) \wedge (x_{n-1} > k_1) \wedge (x_{n-1-p} \leq k_1) \wedge (\forall_{q \in [1,p]} : x_{n-1-q} > k_1) \wedge (\therefore_{l \in [1,p]} : x_{n-1-l} \geq k_2)$	**(35)**
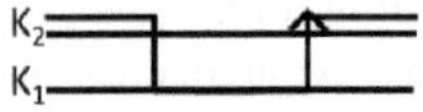	$evx_p^{-+h}(k_1,k_2) = (x_n > k_2) \wedge (x_{n-1} \leq k_2) \wedge (x_{n-1-p} \leq k_2) \wedge (\forall_{q \in [1,p]} : x_{n-1-q} \leq k_2) \wedge (\therefore_{l \in [1,p]} : x_{n-1-l} \leq k_1)$	**(36)**

5 Brief Discussion

The initial type of events, based on a threshold level, only allow modeling of very simple dependencies in terms of analysis of input signals evolution. This means that one event is associated with a specific signal threshold value. All of the other modeling possibilities need to be explicitly represented through sub-models augmented with annotations integrating signal dependencies. This solution allows the modeling of these situations, but leads to a complex model.

Returning to the inspiring example of a joystick controlling an airplane simulator, situations as quick changes in the controls that should lead to different simulation results are difficult to model if only the threshold event is available. On the other hand, the set of proposed new events will adequately support a compact and expressive modeling of the different reactions of the user when using the joystick. In the referred example some examples to of use of each type of event can be found.

For example, whenever the speed at which the joystick is moving is above a certain value, a momentum event can be used considering a K value equals to the value that should not be crossed. When the signal changes from a lower value to a higher value of K, an event is generated.

On the other hand, whenever one wants to identify variations in the joystick speed, it is possible to create an impetus event with the values that wants to detect. The event is generated whenever the speed variation is higher than the set value regardless of the previous value.

Additionally, whenever detection of activation-deactivation sequence of a pushbutton is required, an up-down logical event associated with the pushbutton signal can be used. This event only occurs when the button has just been released.

Any of these previous events could be a delayed event as well. It would be used if the analysis is being done at a rate slower than the analysis rate of the system.

The introduction of the concept of analysis of signal evolution in more than one execution step allows the possibility of introducing some filtering on signal variation.

The proposed logical event paves the way to complex pre-processing of signals, allowing embedded detection of special sequences on signal evolution.

These characteristics make the model easier to produce, understand, and maintain, being easily integrated in automatic code generation tools, as the ones available at http://gres.uninova.pt applicable to IOPT nets.

6 Conclusions and Further Work

The use of these new types of events provides effective modeling capabilities when using IOPT nets models. It allows the creation of smaller models for the same system, making them easier to read and easy to interpret.

Another advantage of the use of such systems lies in the possibility to increase the processing speed of the implemented system using code generation tools.

These event updates are proposed intrinsically matched with IOPT net models, but can also be applied and useful in other classes of Petri nets, both high-level and low-level Petri nets classes.

Other possibility that can be interesting to be further analyzed is the possibility of create events depending on more than one signal, allowing a new level of complex events.

Acknowledgments. This work was partially financed by Portuguese Agency "FCT - Fundação para a Ciência e a Tecnologia" in the framework of project PEst-OE/EEI/UI0066/2011.

The first author was supported by a Grant from project ARTEMIS/FCT-100261 SIMPLE project, co-funded by Portuguese FCT (Fundação para a Ciência e a Tecnologia) and EU ARTEMIS Joint Undertaking.

References

1. Petri, C.A.: Kommunikation mit Automaten. Technische Hochschule, Darmstadt (1962)
2. Peterson, J.L.: Petri Nets. ACM Comput. Surv. 9(3), 223–252 (1977)
3. Zurawski, R., MengChu, Z.: Petri nets and industrial applications: A tutorial. IEEE Transactions on Industrial Electronics 41(6), 567–583 (1994)
4. Murata, T.: Petri nets: Properties, analysis and applications. Proceedings of the IEEE 77(4), 541–580 (1989)
5. Gomes, L., Barros, J.P., Costa, A.: Modeling Formalisms for Embedded Systems Design. In: Zurawski, R. (ed.) Embedded Systems Handbook, ch. 5, pp. 5–1, 5–34. CRC (2005) ISBN 0849328241
6. Ramirez-Trevino, A., Rivera-Rangel, I., Lopez-Mellado, E.: Observability of discrete event systems modeled by interpreted Petri nets. IEEE Transactions on Robotics and Automation 19(4), 557–565 (2003)
7. Giua, A., Di Cesare, F.: Easy synchronized Petri nets as discrete event models. In: Proceedings of the 29th IEEE Conference on Decision and Control (1990)
8. Gomes, L., Barros, J.P., Costa, A., Nunes, R.: The Input-Output Place-Transition Petri Net Class and Associated Tools. In: 2007 5th IEEE International Conference on Industrial Informatics (2007)
9. David, R., Alla, H.: Petri Nets and Grafcet: Tools for Modeling Discrete Event Systems. Prentice Hall (1992)
10. Gomes, L., Barros, J.P.: On structuring mechanisms for Petri nets based system design. In: Proceedings of the IEEE Conference on Emerging Technologies and Factory Automation, ETFA 2003 (2003)
11. Jensen, K.: A brief introduction to coloured Petri Nets. In: Brinksma, E. (ed.) TACAS 1997. LNCS, vol. 1217, pp. 203–208. Springer, Heidelberg (1997)

Minimalist Architecture to Generate Embedded System Web User Interfaces

Fernando Pereira[1,2,3] and Luís Gomes[1,3]

[1] Universidade Nova de Lisboa - Faculdade de Ciências e Tecnologia – Portugal
[2] ISEL, Instituto Superior de Engenharia de Lisboa, – Portugal
[3] UNINOVA - CTS – Portugal
fjp@deea.isel.ipl.pt, lugo@fct.unl.pt

Abstract. This paper presents a new architecture to semi-automatically generate Web user interfaces for Embedded Systems designed using IOPT Petri Net models. The user interfaces can be used to remotely control, monitor and debug embedded systems using a standard Web Browser. The proposed architecture takes advantage of the distributed nature of the Internet to store all static user interface data and software on third-party Web services (the Cloud), and execute the user-interface code on the user's Web Browser. A simplified protocol is proposed to enable remote control, status-monitoring, debugging and step-by-step execution, minimizing resource consumption on the physical embedded devices, including processing load, memory and communication bandwidth. As the user interface data and code are kept on third-party Web services, these resources can be shared among multiple embedded device units, and the hardware requirements to implement the devices can be simplified, leading to reduced cost solutions. To prevent down-time due to network problems or server failures, a fault-tolerant topology is suggested. The distributed architecture is transparent to end-users, observing just a Web interface for an embedded device on the other side of an Internet URL.

Keywords: Embedded-systems, Petri-Nets, Web User Interface.

1 Introduction

The widespread dissemination of the Internet and the mass production of telecommunication technology has significantly reduced the cost barrier to add Internet connectivity to the most inexpensive embedded-system devices, ranging from home appliances, medical and health monitoring devices, surveillance and security equipment, in-vehicle systems, to industrial machinery. However, the traditional Internet connectivity implementation strategies generally lead to increased product complexity, longer development time and increased hardware requirements, including more memory to store images and multimedia files and higher processing power to execute user interface code, with the corresponding implications on power-consumption and battery life.

L.M. Camarinha-Matos, S. Tomic, and P. Graça (Eds.): DoCEIS 2013, IFIP AICT 394, pp. 239–249, 2013.

This paper proposes a new architecture to overcome these limitations, taking advantage of the distributed nature of the Internet and the processing capabilities offered by modern Web browsers. All static data files are stored on external Web servers and the user-interface code is executed directly on the user's personal computer Web browser, greatly reducing the embedded devices hardware requirements and contributing to minimize bandwidth consumption, as many user interactions are dealt directly by the user interface code running on the browser.

To implement the new architecture, a new protocol is proposed to establish the communication between the user-interface code running on the browser and remote embedded devices, taking advantage of AJAX (Asynchronous Javascript and XML) principles. The same protocol used to implement the Web user interfaces can also be used to enable remote monitoring, debug and step-by-step execution of embedded system controllers running on physical devices, or simulated on personal computers.

The proposed architecture is applied to embedded systems designed using the IOPT Petri-net modeling framework [1] and the associated Web-based IOPT-tools [2] tool-chain (http://gres.uninova.pt), including an editor to design IOPT models, a model-checking framework based on a state-space generator and a query system, and automatic "C" and VHDL code generators to produce the controller implementations.

The user interfaces are designed using an Animator tool [3] that permits the interactive rule-based definition of Animated Graphical User interfaces for embedded-systems. This tool has been previously used to design and generate simulation control-panels and graphical user interfaces for systems running on FPGA (Field Programmable Gate Array) reconfigurable devices. Using the Animator tool, the user-interface designer creates a set of screen background images, used to implement application dialogs, and sets of static or animated icons and images. Each dialog contains a table of rules associating internal state variables (IOPT Marking) and Input/Output Signals with the appearance and screen location of the selected icons and images, creating an animated SCADA-like user interface that reflects the system-state in real-time. To control the system from the Web interface, Input Signals can be associated with icons and images activated using mouse clicks to implement bidirectional user-feedback.

The new architecture proposes the execution of Animator-generated user interfaces inside a Web browser, converting the Animator rules into equivalent Javascript code and establishing the communication with the physical systems using remote procedure calls over a protocol based on AJAX *XmlHttpRequests*.

The creation of Debug and monitoring interfaces has been presented in [4], with the automatic generation of Animator screens that directly depict the corresponding IOPT models and the automatic generation of rules that display the system status in real-time, including place marking, transition firing readiness and the state of Input and Output Signals. However, in order to fully support remote debug sessions over the Internet, the communication protocol will be extended with additional remote procedure calls implementing step-by-step execution, continuous execution, system reset, force input signal values and the definition of breakpoints associated with Transition firings.

Finally, this paper proposes a new type of Transition, called the Test-transition, used during the debug and simulation development phases. Test-transitions differ from standard IOPT Transitions because they are not allowed to change system behavior in any form. This way, the new transitions can be freely added to existing models without the risk of accidentally introducing behavioral modifications, to define breakpoints associated with new conditions that were not verified in the original models.

2 Related Work and Research Innovation

Over the past decade, Internet enabled embedded devices with Web interfaces have been offered by commercial systems and were implemented on many research prototypes. The traditional architectures used to implement these solutions have resorted to full-featured Web servers running over embedded operation systems, as embedded Linux [5] and QNX [6], using standard interface technologies like common-gateway-interface (CGI) to control the physical embedded devices.

However, solutions based on complex operating systems require advanced microprocessors and occupy large amounts of memory, including many megabytes of RAM (Random Access Memory) and mass storage devices to store operating system files. The Web interfaces are generally created using standard Web page authoring tools and the connection to the physical embedded systems are manually programmed. All files used by the Web interface, including images and scripts are usually stored in the device. Due to these requirements, Internet connectivity has usually been skipped from the least expensive devices.

The new solution presented in this paper has many advantages over traditional technologies. To start, the Web user interfaces are semi-automatically generated using the rule-based Animator tool [3] and the Debug interfaces are fully automatically generated [4] without the need to manually write any code. As the static files are stored in external Web servers and the user interface code (Animator rules) is executed by the user's Web browser, the computational requirements of the embedded controller are largely reduced and the need to employ complex operating systems is avoided, allowing the addition of internet connectivity and Web interfaces to the most inexpensive devices without a significant cost increase. Instead of requiring 32-bit microprocessors, the proposed minimalist architecture can be implemented with simpler 8-bit embedded micro-processors using small TCP/IP stacks as uIP and LwIP [7] that require just tens of Kilobytes of RAM and can entirely fit inside the memory blocks offered by FPGA devices without external RAM. The usage of 8-bit micro-processors also contributes to reduce FPGA resource consumption, enabling the choice of smaller and less expensive reconfigurable devices.

In addition to the automatic generation of Web user interfaces, the proposed communication protocol also permits remote debugging and step-by-step execution of embedded-system controllers running on real hardware devices, allowing long distance troubleshoot and maintenance operations over the internet. As the graphical debug interfaces automatically generated by the PNML2Anim4Dbg [4] tool can be directly presented in the new architecture, it is possible to monitor the state of remote embedded-systems in real time, observing the graphical evolution of the underlying IOPT Petri net model. Contrary to traditional remote administration tools, this solution offers a high degree of intuitiveness and user friendliness, as the system designer operates directly on the Petri net model used to design the original system.

Other Petri net tools, including CPN tools [8], CPN-Ami [9], Renew [10] and others have implemented debug and step-by-step execution tools, but as these classes of Petri nets are autonomous, the scope of these tools is generally restricted to simulations running on personal computers and not for final implementations. Inside these simulation tools, some authors have also defined the concept of breakpoints associated with transitions and changes on place-markings [11], but the new concept of Test-transition presented in this paper offers many advantages over these solutions as it enables the definition of generic breakpoint conditions.

Finally, the new architecture was built on top of previous work, starting with the definition of the IOPT Petri Net class [1], IOPT design tools [2], automatic code generators and the Animator Tool [3]. The proposed architecture enables the porting of previous Animator-designed user interfaces that were executed inside simulations or on FPGA hardware [12] [13], to a distributed Web/Internet platform.

3 IOPT Petri Nets

The characteristics of the IOPT Petri net class [1] were selected to support the design of embedded-system controllers. Beyond the Places and Transitions inherited from classical Petri nets [14], IOPT nets also contain a set of non-autonomous properties used to specify the interface between the controllers and the external world. This interface comprehends Input and Output Signals, that can hold Boolean logic values or Integer range values, and Input and Output Events associated with changes in Signal values. Figure 1 displays an example IOPT Model implementing a UART transmitter hardware module, edited with the IOPT-Tools model editor.

The model presented in Fig 1, has three Places (yellow circles), four Transitions (cyan rectangles), three Input Signals (cyan circles), five Output Signals (green circles), and one Output Event (green triangle). Associated with Places there are Output Expressions that assign values to Output Signals whenever these Places are marked. Guard conditions, associated with Transitions, inhibit the firing according to the value of Input/Output Signals, Literals and Place marking. Transition firing is also triggered by Input Events, and can produce Output Events that perform changes in Output Signals. For example, transition ***TCount*** raises a ***Cntr*** Output Event, used to count the number of bits transmitted by incrementing the value of the ***Cnt*** Output.

In order to ensure determinist execution, IOPT nets employ maximal-step execution semantics, where every transition ready to fire will immediately fire on the next execution step. Firing conflicts, when more than one transition is simultaneously ready to fire, but the number of available tokens is not enough to fire all of them, can be solved by assigning different priorities to each conflicting transition. A state-space generation tool offered by IOPT-Tools can be used to automatically detect conflicts, deadlocks and to calculate the maximal and minimal bound of each Place.

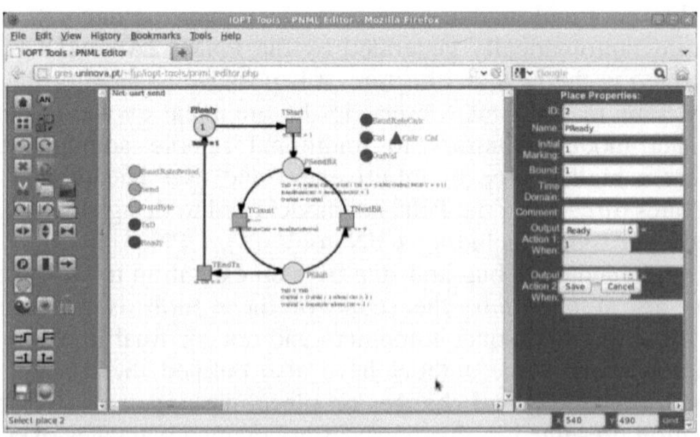

Fig. 1. IOPT-Tools Editor (UART Transmitter IOPT Model)

4 Distributed Web User-Interface Architecture

Figure 2 displays the distributed topology of the proposed embedded-system's Web user interface architecture, employing the Web browser running on the user's personal computer, an external Web server and the embedded-system.

The user interfaces are designed with the Animator tool [3], producing a set of screen dialogs, image files and animation rules. Image files hold the background pictures of each screen and graphical representations of icons, buttons and animation frames. Animation rules define when images are show or hidden according to the instantaneous embedded-system state, including the Place marking, Input signals and Output signals.

As the Animator tool stores the animation rules under XML files, these rules can be easily translated into equivalent Javascript code using XSL (Extensible Style Sheet Transformation) transformations or other XML processing technologies. The resulting Javascript code is executed inside the Web Browser, removing the computational load of the user interface code from the physical embedded device.

All static files, including images and Javascript code, are stored in external servers, eliminating the need to store large files on the embedded device. By moving the user interface code and file storage to external computers, the embedded-system controller can be implemented using minimal hardware specifications. The embedded system controller only needs to implement a micro HTTP server, used to answer remote procedure call from the Web browser and transmit information about the current system state (almost) in real time.

5 Internet-Enabled Embedded-System Internal Architecture

The main goal of the architecture proposed in this paper is the ability to add Internet support to embedded-system devices with minimalist hardware specifications. Figure 3 presents a possible hardware architecture that fulfills these goals.

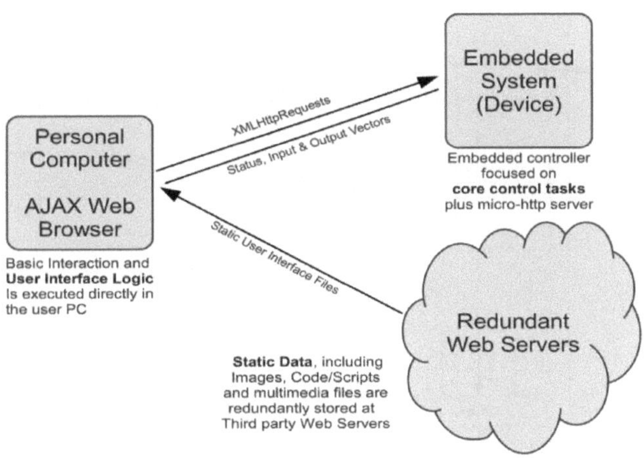

Fig. 2. Distributed Web GUI Topology

The proposed hardware architecture requires just a minimal set of components to implement the TCP/IP communication protocol over Ethernet networks. The entire system can be implemented in a single FPGA (or ASIC), with the addition of an external Ethernet Physical Layer Integrated Circuit (PHY) and the respective magnetics and connector, or equivalent components to support wireless networking.

Resource consumption inside the FPGA/ASIC device to implement TCP/IP over Ethernet requires a standard Ethernet MII/MAC (Media Independent Interface / Media Access Controller) module and an embedded microprocessor, used to execute the TCP/IP stack and a mini Web-server. There are many microprocessor cores offered by FPGA vendors and many open-source processor cores publicly available. As the open-source LwIP (Light Weight Internet Protocol) is coded using the "C" Programming Language, the only processor choice requirements are the availability of a "C" compiler and a 16 bit memory address space (or larger), in order to store data packets, plus the 40Kb LwIP code and a small HTTP command interpreter. These requirements cover a wide range of microprocessors, ranging from 8-bit Zilog Z80 clones to more advanced 32 bit MicroBlaze cores.

A clock management unit is also employed to allow debugging and step-by-step execution of the embedded-system controllers, by disabling the controller clock signal or sending individual clock pulses to run single execution steps.

The embedded microprocessor I/O bus is connected to the embedded controller core and is able to read signals containing the instantaneous values of Input Signals, Output Signals and Place marking. It can also define the value of Signals associated with Graphical User Interface objects. For example, an Icon or a Button inside an Animator screen can be associated with an Input Signal in the embedded system controller and when the user manipulates that button, a HTTP request is sent to the microprocessor to update the corresponding Signal value.

A set of optional internal signals is used to support remote debugging and maintenance operations, including a breakpoint mask, a forced-input-signal mask, a reset signal and three signals to control the clock management unit: stop, run and step.

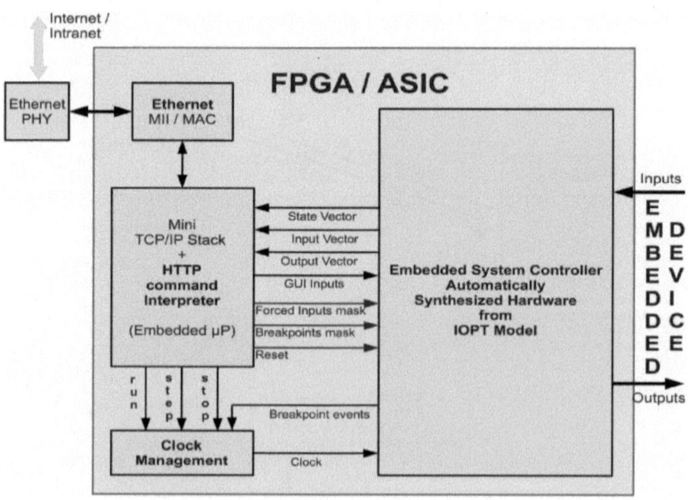

Fig. 3. Internal Controller Architecture

The breakpoint mask is used to enable breakpoints on selected IOPT Transitions. When a selected Transition fires, it will trigger a breakpoint event that stops the clock management unit. The forced-input-signal mask can be used to remotely force the value of Input signals read by the embedded controller, in order to bypass mechanical problems in the physical embedded devices or to easily force error situations and test the model behavior during these conditions.

6 Information flow over the Internet

Figure 4 presents the flow of information over the Internet between the Browser, the embedded-system and an external Web server. A typical interaction session starts when the user accesses the embedded-system URL (Web address) using a Web Browser. The embedded system answers with a Top/Start HTML page containing just a reference to a main script file stored in an external server. As soon as the main script is loaded and the user is successfully authenticated, it immediately starts downloading all Animator files required to execute the graphical user interface, including images and the scripts that execute animation rules. To obtain redundancy and prevent service failures in case the external Web server is not reachable, the Start HTML page might include multiple references to copies of the main script located in different servers that will load only if the first server does not respond during a predefined time interval.

After the start-up sequence is finished, a Web user interface main loop starts running, continuously requesting updated state information from the embedded-system, including Place marking and Input and Output Signal values. Using these values, the animation scripts will update the graphical user interface, almost in real-time, with an update rate configurable according to the available bandwidth and the requirements of each application.

As the static files stored in the external server are loaded in background, the distributed topology is transparent to end user that only sees the address (URL) of the embedded device.

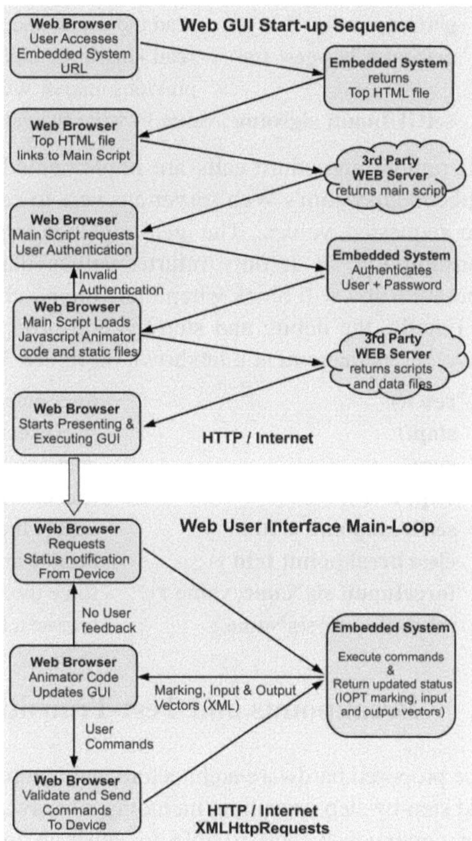

Fig. 4. Internet-IOPT Information Flow

7 The IOPT-Internet Protocol

The proposed Internet-IOPT protocol defines a set of commands, viewed as remote procedure calls over HTTP, used to establish the communication between the user interface code running on the Web Browser and the embedded-system HTTP command interpreter. It comprehends three types of procedures calls: 1) front-page and authentication; 2) status-monitoring and GUI interaction and 3) debug and step-by-step execution.

Start-page requests and user authentication are implemented with the standard HTTP commands used to serve static HTML files. User sessions can be implemented using a temporary HTTP cookie holding a random number that identifies each user after the authentication phase is successfully passed.

The second group of commands implements remote procedure calls that produce XML results, implementing the following methods:

getMarking()	- read the instantaneous IOPT net marking vector
getInputs()	- read the instantaneous input signal vector
getOutputs()	- read the instantaneous output signal vector
getStateChanges(tm)	- read «marking, inputs and outputs» changed since the previous read or wait «tm» seconds if there are no changes
setGUInput(sigName, value)	- write the value of a GUI input signal.

All remote procedure calls are implemented using AJAX *xmlHttpRequests*, and the embedded-system's Web server answers to each command with XML files containing the requested values. The *getStateChanges()* method provides reduced bandwidth consumption, as it only returns values that suffered changes since the last read operation and will block when there are no changes during a specified timeout period.

Finally, the debug and step-by-step group of commands, used to directly control the clock management unit shown on figure 3, implement the following methods:

reset()	- reset embedded-system controller state
stop()	- stop execution
run()	- continue execution
step(n=1)	- execute «n» execution steps (by default n=1)
setBreakpoint(trId)	- enable breakpoints on transition «trId»
clearBreakpoint(trId)	- disable breakpoints on transition «trId»
forceInput(sigName, value)	- force the value of a non GUI input signal
releaseInput(sigName)	- release a forced input signal

8 Breakpoints and Test-Transitions

The proposed hardware architecture and communication protocol supports remote debug and step-by-step execution, including a hardware clock management unit that can stop or run continuously and is able to generate individual clock pulses to perform single execution steps. However, as complex embedded-systems often execute thousands of execution steps before reaching a critical situation being tested, step-by-step execution may not be practical. Other embedded-systems simply cannot be run step-by-step because the controllers must respond to external input changes during very fast time intervals in order to prevent malfunctions and mechanical damages. For example, an automatic door controller must immediately turn off the door motor when the door is being closed and a presence sensor detects a person inside the door limits.

To solve this problem, the concept of breakpoints, usually employed in software debugging systems, was extended to the IOPT Petri net modeling framework, adding the possibility to assign breakpoints to Transitions. When a Transition fires, a Breakpoint event will be raised and execution is stopped before the Transition is actually fired. Observing figure 3, breakpoint events are connected to the clock management unit and will immediately stop the clock signal. This concept can also be easily ported to software implementations and may be applied on simulations or on embedded devices running on microprocessors.

Finally, the test conditions corresponding to the error situations being debugged may not directly correspond to the firing of any Transition existing in the model. As a consequence, it might be necessary to add additional new Transitions to the model, containing Arcs, Input events and Guard conditions that detect the (un)desired situations. However, adding new Transitions to an existing model will usually introduce changes to the behavior of the original model, potentially invalidating the debug conclusions, as the new model may behave differently from the original one.

To solve this problem, a new concept of Test-transition is proposed. A Test-transition has a set of restrictions that does not allow behavioral changes and can be safely used to define test/debug conditions associated with breakpoints. In order to achieve this effect, Test-transitions can only be connected to Test Arcs, cannot be connected to Normal Arcs and may not be associated with Output events. As a result, Test-transitions can only have input Test Arcs and cannot have output Arcs.

The concept of Test-transition has many advantages. First, Test-transitions can be safely added to any model in any configuration without the risk of introducing any behavioral changes. Second, Test-transitions can continue to be viewed as regular Transitions by all IOPT Tools, including the automatic software and hardware code generators, simulators and validation tools, without requiring additional development. Finally, Test-transitions can also be safely removed from models using automated filters, to generate final controller implementations without Debug code.

9 Conclusions and Future Work

The goal of this paper is to propose a new architecture to automatically add Web User Interfaces to embedded-system controller designed with IOPT Petri net models. This architecture employs a distributed topology to take advantage of the Web browser processing power and the storage capacity offered by external Web servers, in order to minimize the hardware requirements on the physical embedded-systems, implementing Web awareness without almost no additional cost.

This work is an extension to previous work, where the Animator [3] concept and tools were introduced. The proposed architecture can be used to automatically convert existing user interfaces designed with the Animator tool, producing equivalent Web interfaces.

In addition to automatic Web interface generation, the same communication protocol was also extended to support remote debugging and step-by-step execution, offering the possibility to perform maintenance and diagnose problems over the Internet. The association of breakpoints to Transition firing and the new concept of Test-transitions also contribute to reduce test and debugging effort, simplifying the debug of systems where step-by-step execution would be impractical.

Although the proposed architecture was not yet implemented, all the components used in the architecture are currently disseminated technologies, including the suggested hardware platforms, the embedded TCP/IP protocol stack and the AJAX technologies used to execute the user interface code in the Browser. All the components employed are readily available, leading to the conclusion that the implementation of the proposed ideas is feasible. Future work may lead to the creation of prototypes based on the new architecture, with possible improvements to overcome technical difficulties that may occur during development.

Acknowledgments. This work was partially financed by Portuguese Agency "FCT – Fundação para a Ciência e Tecnologia", in the framework of project Pest-OE/EEI/UI0066/2011.

References

1. Gomes, L., Barros, J., Costa, A., Nunes, R.: The Input-Output Place-Transition Petri Net Class and Associated Tools. In: Proceedings of the 5th IEEE International Conference on Industrial Informatics (INDIN 2007), Vienna, Austria (July 2007)
2. Pereira, F., Moutinho, F., Gomes, L.: Model-checking framework for embedded systems controllers development using IOPT Petri nets. In: 2012 IEEE International Symposium on Industrial Electronics (ISIE), May 28-31, pp. 1399–1404 (2012), doi:10.1109/ISIE.2012.6237295
3. Gomes, L., Lourenco, J.: Rapid Prototyping of Graphical User Interfaces for Petri-Net-Based Controllers. IEEE Transactions on Industrial Electronics 57, 1806–1813 (2010)
4. Pereira, F., Gomes, L., Moutinho, F.: Automatic generation of run-time monitoring capabilities to Petri nets based Controllers with Graphical User Interfaces. In: Camarinha-Matos, L.M. (ed.) DoCEIS 2011. IFIP AICT, vol. 349, pp. 246–255. Springer, Heidelberg (2011)
5. de Souza, R.N., Muniz, D.N., da Silva Fidalgo, A.V.: Ethernet communication platform for synthesized devices in Xilinx FPGA. In: 2011 IEEE International Conference on Computer as a Tool (EUROCON), April 27-29, pp. 1–4 (2011), doi:10.1109/EUROCON.2011.5929377
6. QNX website, `http://www.qnx.com/developers/docs/6.3.0SP3/neutrino/user_guide/embedded_web_server.html` (accessed January 6, 2013)
7. Dunkels, A.: Full TCP/IP for 8 Bit Architectures. In: Proceedings of the First ACM/Usenix International Conference on Mobile Systems, Applications and Services (MobiSys 2003), San Francisco (May 2003)

8. Jensen, K.: Coloured Petri Nets. Basic Concepts, Analysis Methods and Pratical Use. Basic Concepts, vol. 1. Springer, Berlin (1997)
9. Hamez, A., Hillah, L., Kordon, F., Linard, A., Paviot-Adet, E., Renault, X., Thierry-Mieg, Y.: New features in CPN-AMI 3: focusing on the analysis of complex distributed systems. In: Sixth International Conference on Application of Concurrency to System Design, ACSD 2006, June 28-30, pp. 273–275 (2006), doi:10.1109/ACSD.2006.15
10. Kummer, O., Wienberg, F., Duvigneau, M., Cabac, L.: Renew – User Guide, University of Hamburg, Department for Informatics, Theoretical Foundations Group, Release 2.2, (August 28, 2009)
11. Sadilek, D.A., Wachsmuth, G.: Prototyping Visual Interpreters and Debuggers for Domain-Specific Modelling Languages. In: Schieferdecker, I., Hartman, A. (eds.) ECMDA-FA 2008. LNCS, vol. 5095, pp. 63–78. Springer, Heidelberg (2008)
12. Moutinho, F., Gomes, L.: From models to controllers integrating graphical animation in FPGA through automatic code generation. In: IEEE International Symposium on Industrial Electronics (ISIE 2009), Seoul Olympic Parktel, Seoul, Korea, July 5-8 (2009)
13. Moutinho, F., Pereira, F., Gomes, L.: Automatic Generation of Graphical User Interfaces for VHDL based Controllers. In: ISIE 2011 – 20th IEEE International Symposium on Industrial Electronics, Gdansk, Poland, June 27-30, pp. 1491–1496 (2011), doi:10.1109/ISIE.2011.5984381, ISBN: 978-1-4244-9312-8
14. Reisig, W.: Petri nets: an introduction. Springer Verlag New York, New York (1985)

Application of an Exact Transversal Hypergraph in Selection of SM-Components

Łukasz Stefanowicz, Marian Adamski, and Remigiusz Wisniewski

University of Zielona Góra, Institute of Computer Engineering and Electronics,
ul. Licealna 9,65-417 Zielona Góra
L.Stefanowicz@weit.uz.zgora.pl,
{M.Adamski,R.Wisniewski}@iie.uz.zgora.pl

Abstract. The paper deals with the application of the hypergraph theory in selection of State Machine Components (SM-Components) of Petri nets [1,2].As it is known, Petri nets are widely used for modeling of concurrency processes. However, in order to implement the concurrent automaton, an initial Petri net ought to be decomposed into sequential automata (SM-Components), which can be easily designed as an Finite-State-Machine (FSM) or Microprogrammed Controller [3]. The last step of the decomposition process of the Petri nets is selection of SM-Components. This stage is especially important because it determines the final number of sequential automata. In the article we propose a new idea of SM-Components selection. The aim of the method is reduction of the computational complexity from exponential to polynomial. Such a reduction can be done if the selection hypergraph belongs to the exact transversal hypergraphs (xt-hypergraphs) class. Since the recognition and generation of the first transversal in the xt-hypergraphs are both polynomial, the complete selection process can be performed in polynomial time. The proposed ideas are an extension of the concept presented in [1]. The proposed method has been verified experimentally. The conducted investigations have shown that for more than 85% of examined Petri nets the selection process can be done via xt-hypergraphs.

Keywords: hypergraph, exact transversal hypergraphs, SM-Components, Petri nets, graphs, transversal, exact cover, exact transversal, concurrency hypergraphs, selection hypergraph, sequentiality hypergraph.

1 Introduction

Discrete systems model discrete processes: sequential and concurrent, so they are mostly applied to model and represent controlling systems described by Petri nets [2]. In some cases the designer may encounter a problem: size of system exceeds the frames imposed by the size of prototyped systems thus the system needs to be decomposed into a series of collaborating subsystems. The most frequently applied synthesis method is decomposition of a concurrent automation into sequential automata [4,5].

Petri nets [2,4] are widely used for designing and modeling of the concurrency processes [5,6]. Analysis of the relations (such as concurrency or sequentiality) and

L.M. Camarinha-Matos, S. Tomic, and P. Graça (Eds.): DoCEIS 2013, IFIP AICT 394, pp. 250–257, 2013.

decomposition of a Petri net are very often hard to perform because of exponential dependence between number of states and size of the initial net [2,4,6,7].

The decomposition process can be divided into several stages. At the beginning the concurrency relation is obtained. The operation can be executed via concurrency graph or concurrency hypergraph. The detailed description about those structures can be found in [5,8]. Next, the set of SM-Components is obtained. There are several methods that permit to calculate this stage [1]. Finally, the selection process should be executed to eliminate the redundant elements. The main problem is computational complexity of such an operation. In a general case the selection can be done in an exponential time. Since for some Petri net classes there are polynomial methods for selection of the concurrency graph/hypergraph and generation of SM-Components [1, 9], the selection process is a bottleneck of the whole decomposition process.

In the article we propose anapplication of xt-hypergraphs in the selection process. Initially, the selection hypergraph is formed. Such a structure is determined from the set of SM-Components. Next, if the selection hypergraph is an xt-hypergraph, the first exact transversal is computed. It directly indicates the proper SM-Components that ought to be selected to reach the optimal solution. The proposed idea will be described later in more details.

2 Relationship to Internet of Things

Petri nets are widely used for modeling of concurrency processes. Thus, they are very useful in designing and prototyping of concurrent automata. Decomposition of a Petri net permits to split the device into independent finite-state-machines (FSMs) [1]. SM-Components that are obtained during the decomposition process refer to FSMs. Each FSM can be analyzed and designed as a separate module. Since SM-Components ought to be synchronized [1,6], usually they are implemented in the same programmable device (like an FPGA) [1]. However, Internet of Things may revolutionize such an approach. Imagine that SM-Components can be synchronized via Internet. Then, different devices may be applied for implementation of modules. Moreover, each module can be designed with different technology! Since the distance is not important, SM-Components can cooperate all over the World. Therefore, decomposition of Petri nets maybe especially fruitful in relation to Internet of Things.

3 Main Definitions

3.1 Hypergraph

Hypergraph H is defined by a pair [9]:

$$H=(V,E), \tag{1}$$

where:

$V=\{v_1,\ldots,v_n\}$ is arbitrary, nonempty set of vertices;
$E=\{E_1,\ldots,E_n\}$ is set of hypergraph edges, i.e., subset of set P(V) of all the possible non-empty sets, elements of which belong to V [9].

3.2 Transversal

Transversal (hitting set, vertex cover) of hypergraph H is set [9,10]:

$$T \subseteq V, \tag{2}$$

containing vertices incident to each edge of hypergraph [9].
A *minimal transversal* is such a transversal which contains no other transversal of hypergraph H.

3.3 The Exact Transversal

The exact transversal D of hypergraph H is set [1,10]:

$$D \subseteq V, \tag{3}$$

of vertices of hypergraph H, which is incident to all egdes of hypergraph H, where each edge is incident to exactly one vertex of set D.

3.4 Exact Transversal Hypergraph (xt-hypergraph)

Exact transversal hypergraph H_{xt} is hypergraph in which all minimal transversal are also exact transversals [10]. In the other words, hypergraph H_{xt} belongs to class xt.

4 Problem Formulation

The decomposition process of a Petri net can be divided into three main steps [1]:

1) Formulation of concurrency graph or concurrency hypergraph.
2) Formulation of SM-Components.
3) Selection of SM-Components.

In the first step of the decomposition process, the concurrency relations are determined. They can be obtained in two ways: via concurrency graph or concurrency hypergraph. The concurrency graph is computed during the analysis of the structure and relations in a Petri net. The graph contains concurrency relations between any *two* places of a net. It was proved that the concurrency graph can be calculated in a polynomial time [7].

The alternative decomposition method bases on the concurrency hypergraph computation. In opposite to the concurrency graph, hypergraph can hold the information between more than two places in a net. Its vertices represent the places of an initial Petri net, while hyperedges are related to the concurrency relations. As it was shown in [5], the computation of the concurrency hypergraph may be polynomial, however the main problem is the number of hyperedges, which in some cases can be exponential.

The SM-Components are generated from the concurrency graph or hypergraph. There are several methods; like concurrency graph/hypergraph coloring, application of the graph-search algorithms or computation of exact transversals [1,12,13]. Depending of the structure of an initial Petri net, this process is either polynomial or exponential [1,13].

Most of algorithms generate more SM-Components, that are required to cover the whole Petri net. Then, regardless of the initial structure (concurrency graph or hypergraph), the process of selection ought to be applied.

In the article we propose a new idea of SM-Components selection. The aim of the proposed method is to apply an algorithm that can be performed in polynomial time in most of cases.

5 Idea of the Proposed Method

The proposed idea of selection of SM-Components can be divided into the following steps (Fig.1):

1) Formulation of the *sequentiality hypergraph* that represents the set of achieved SM-Components [1]. Vertices refer to the places of a Petri net while each hyperedge represents one of SM-Components.
2) Calculation of the dual hypergraph to the sequentiality hypergraph obtained in the previous stage. In practice this operation can be done via transposition of an incidence matrix of the sequentiality hypergraph [1].
3) Reduction of dominant edges and dominated vertices and formulation of the *selection hypergraph*. The dominant edges refer to the two or more places that are covered by the same SM-Components. Analogically, the dominated vertices are also reduced. They refer to the SM-Components that are covered by other SM-Components. Therefore, they can be removed from the further analysis. The concept of presented reduction was taken from [11] and [1], where it was proved that this operation has polynomial computational complexity. Finally, the selection hypergraph is formed which is the result of the reduction process.
4) Examination of the selection hypergraph. If the hypergraph belongs to the xt-hypergraphs class, it means that the selection process can be done via xt-hypergraphs. Otherwise, the selection ought to be performed in the other way (like calculation of transversals, coloring, etc.) [1].
5) If the selection hypergraph belongs to the xt-class, the first exact transversal has to be calculated. Obtained result indicates SM-Components that form the final solution (the selected set of SM-Components). Since the examination and selection of the subsequent exact transversals in the xt-hypergraph are polynomial, the whole selection process is polynomial.

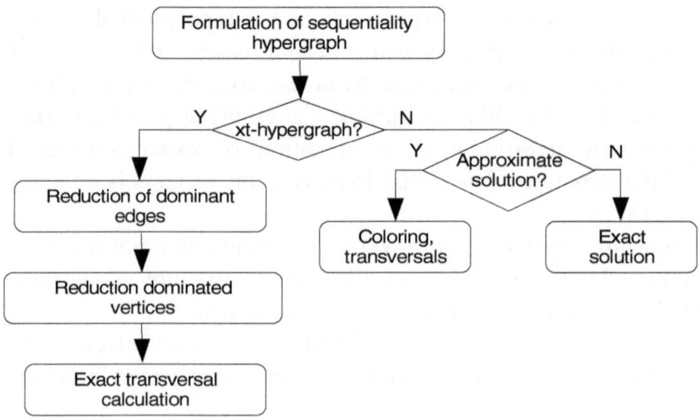

Fig. 1. Idea of the proposed method

6 Example of the Proposed Method

Figure 2 illustrates Petri net P_1that will be used as an example. It is a simplified version of a controlling system of the preparation to a single heat and race during a speedway tournament [1].

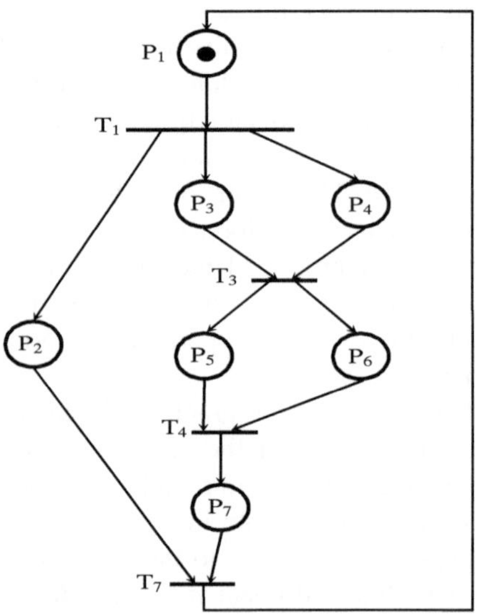

Fig. 2. Exemplary Petri net P_1

There can be totally five SM-Components found in the presented net: $S_1=\{P_1,P_2\}$, $S_2=\{P_1,P_3,P_5,P_7\}$, $S_3=\{P_1,P_3,P_6,P_7\}$, $S_4=\{P_1,P_4,P_5,P_7\}$, $S_5=\{P_1,P_4,P_6,P_7\}$. According to the conception presented in the previous section, at the beginning, sequentiality hypergraph H_C is formed. An incidence matrix of H_C is presented in fig. 3.

$$A_C = \begin{array}{c} \begin{array}{ccccccc} P_1 & P_2 & P_3 & P_4 & P_5 & P_6 & P_7 \end{array} \\ \begin{bmatrix} 1 & 1 & 0 & 0 & 0 & 0 & 0 \\ 1 & 0 & 1 & 0 & 1 & 0 & 1 \\ 1 & 0 & 1 & 0 & 0 & 1 & 1 \\ 1 & 0 & 0 & 1 & 1 & 0 & 1 \\ 1 & 0 & 0 & 1 & 0 & 1 & 1 \end{bmatrix} \end{array} \begin{array}{c} \\ S_1 \\ S_2 \\ S_3 \\ S_4 \\ S_5 \end{array}$$

Fig. 3. Incidence matrix of sequentiality hypergraph H_C

Next, the duality hypergraph to the H_S is calculated. Such an operation can be easily done by transposition of incidence matrix A_C. After the reduction of the dominant edges and dominated vertices, the selection hypergraph $H_S$is reached. In our example the selection hypergraph belongs to the xt-class, thus the proposed idea have sense. Incidence matrix $A_S$of selection hypergraph H_S is shown in the fig. 4.

$$A_S = \begin{array}{c} \begin{array}{ccccc} S_1 & S_2 & S_3 & S_4 & S_5 \end{array} \\ \begin{bmatrix} 1 & 0 & 0 & 0 & 0 \\ 0 & 1 & 1 & 0 & 0 \\ 0 & 0 & 0 & 1 & 1 \\ 0 & 1 & 0 & 1 & 0 \\ 0 & 0 & 1 & 0 & 1 \end{bmatrix} \end{array} \begin{array}{c} \\ P_2 \\ P_3 \\ P_4 \\ P_5 \\ P_6 \end{array}$$

Fig. 4. Incidence matrix of selection hypergraph H_S

There are two exact transversals in the selection hypergraph. The first one includes components S_1, S_2 and S_5, while the second is realized by S_1, S_3 and S_4. Since both transversals have the same number of SMCs, it does not matter which one will be selected. For further analysis we chose the first one. Finally, the initial Petri net can be decomposed. The proper SMCs are indicated by the selection process and in our case P_1 is decomposed into three SMCs, represented by S_1, S_2 and S_5. Figure 5 illustrates the result of the decomposition process. The initial concurrent automaton has been decomposed into three sequential automata: S_1(a), S_2 (b) and S_5 (c).The detailed description about formation of the particular automaton can be found in [1].

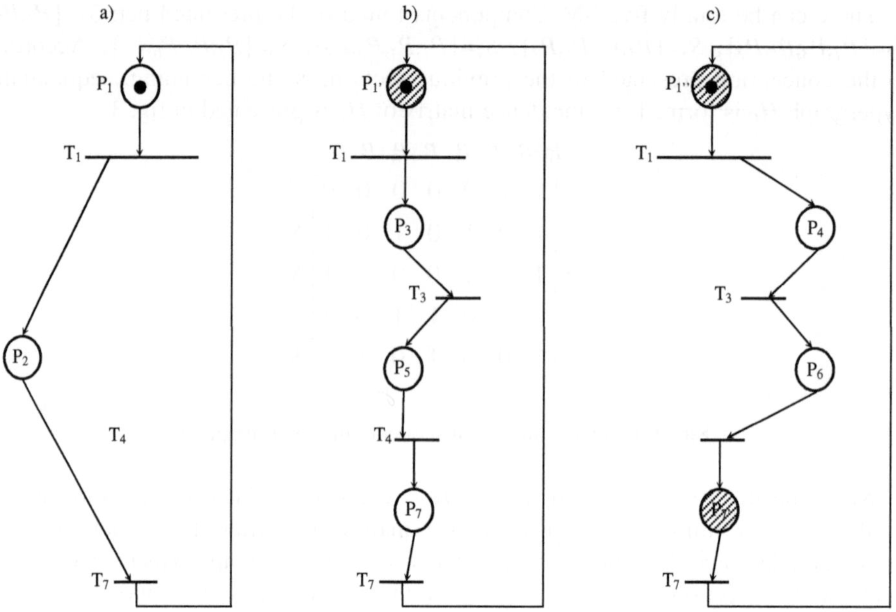

Fig. 5. The result of the decomposition process of Petri net P_1

7 Results of Experiments

The proposed method has been verified experimentally. The library of test modules consist of over than 100 Petri nets that belong to the different classes (such as *Marked Graphs*, *Free-Choice*, *Extended Free-Choice* and *Simple Nets*[2,4]). Most of the nets describe real digital controllers and were taken from literature [1,2,3,4,5,6,14].

The main criterion of the selection was to obtain the minimal (optimal) number of SMCsthat are required in order to decompose the initial Petri net. Obtained average results of experiments are shown in Tab. 1. From the table we can see, that 85% of examined Petri nets have xt-selection hypergraphs, thus the selection process for those nets can be done in polynomial time.

Table 1. Average results of experiments

Average number of places	Average number of transitions	Average number of SM-Components	Number of achieved xt-selection hypergraphs
12	10	4	85%

8 Conclusions

In the article the application of the xt-hypergraphs in the selection of SM-components was presented. The main advantage of the proposed idea is availability (in some cases) to perform the whole selection process in polynomial time. The initial experiments have proved the effectiveness of the presented method. According to the results of conducted researches showed that over 85% of examined nets have the selection hypergraph that belongs to the xt-class. It means that the selection process can be executed in polynomial time if hypergraph belongs to xt-class.

At the moment the essence of research was to determine if the selection hypergraph belonged to the xt-class. Further work will include comparison of achieved results with other selection methods (based on the coloring, transversals, etc.)

Acknowledgment. The research was partially financed from budget resources intended for science in 2010-2013 as an own research project No. N N516 513939.

References

1. Wisniewska, M.: Application of hypergraphs in decomposition of discrete systems. Lecture Notes in Control and Computer Science, vol. 23. Univ. of Ziel, Gora Press (2012)
2. Murata, T.: Petri Nets: Properties, Analysis and Applications. Proceedings of IEEE 77(4), 541–580 (1989)
3. Barkalov, A., Titarenko, L., Hebda, O.: Synthesis of Moore finite state machine with nonstandard presentation of state codes. Electrical Review 9, 134–136 (2010)
4. Karatkevich, A.: Dynamic analysis of Petri net-based discrete systems. Springer, Berlin (2011)
5. Wisniewski, R., Adamski, M., Wisniewska, M.: A polynomial algorithm to compute the concurrency hypergraph in Petri nets. Measurement Automation and Monitoring 07, 650–652 (2012) (in Polish)
6. Karatkevich, A.: SM-Components problem reductions of Petri nets. Telecommunication Review 6 (2008) (in Polish)
7. Kovalyov, A.V.: Concurrency Relations and the Safety Problem for Petri Nets. In: Jensen, K. (ed.) ICATPN 1992. LNCS, vol. 616, pp. 299–309. Springer, Heidelberg (1992)
8. Karatkevich, A., Wisniewski, R.: Computation of Petri nets covering by SM-components based on the graph theory. Electrical Review 8, 141–144 (2012) (in Polish)
9. Berge, C.: Graphs and Hypergraphs. American Elsevier Pub. Co., New York (1989)
10. Eiter, T.: Exact Transversal Hypergraphs and Application to Boolean μ-Functions. Journal of Symbolic Computations 17, 215–225 (1994)
11. Rudell, R.L.: Logic Synthesis for VLSI Design. PhD Thesis, EECS Department, University of California, Berkeley (1989)
12. Kubale, M., Obszarski, P., Piwakowski, K.: Hypergraphs coloring. Technical Reports of the Silesian Technical University (2006) (in Polish)
13. Knuth, D.: Dancing Links. In: Millennial Perspectives in Computer Science, pp. 187–214. Palgrave (2000)
14. Blanchard, M.: Comprendre, Maîtriser Et Appliquer Le Grafcet Cepadues, Tolouse (1979) (in French)

Part X
Control and Decision

Safety in Supervisory Control for Critical Systems

Reinaldo Squillante Jr., Diolino J. Santos Fo, Jeferson A. L.de Souza, Fabrício Junqueira, and Paulo E. Miyagi

University of São Paulo, São Paulo, Brazil
{reinaldo.squillante,diolinos,jeferson.souza, fabri,pemiyagi}@usp.br

Abstract. Recent studies show the designs of automated systems are becoming increasingly complex to meet the global competitive market. Additionally, organizations have focused on policies to achieve people's safety and health, environmental management system, and controlling of risks, based on standards. In this context, any industrial system in the event of a fault that is not diagnosed and treated correctly could be considered to pose a serious risk to people's health, to the environment and to the industrial equipment. According to experts, the concept of Safety Instrumented Systems (SIS) is a practical solution to these types of issues. They strongly recommend layers for risk reduction based on control systems organized hierarchically in order to manage risks, preventing or mitigating faults, or to bringing the process to a safe state. Additionally, the concept of Risk and Hazard Control can be applied to accomplish the required functionalities. It is based on problem solving components and considers a cooperative way to find a control solution. In this context, the software architecture can be based on a service-oriented architecture (SOA) approach. This paper initially proposes a new architecture for design of safety control systems for critical systems, based on Safety Supervisory Control Architecture, in accordance with standards IEC 61508 and IEC 61511. Furthermore, a method is also proposed for design the control layer of risk prevention within Safety Supervisory Control Architecture.

Keywords: Safety Supervisory Control Architecture, Safety Instrumented System, Critical Fault diagnosis, Critical Fault Treatment, Service-oriented architecture.

1 Introduction

Recent studies show that automation is highly influenced by the advance of technologies such as mechatronics and Internet, moreover, the design of automated systems are becoming increasingly complex to meet the global competitive market. Industrial processes must considerer an increasing number of functionalities associated with customized products and concepts such as design for manufacturing, design for quality, etc. and their control solutions must frequently consider contradictory specifications. Additionally, organizations have focused on policies to achieve people's safety and health, environmental management system, and controlling of risks, based on standards

L.M. Camarinha-Matos, S. Tomic, and P. Graça (Eds.): DoCEIS 2013, IFIP AICT 394, pp. 261–270, 2013.

like Occupational Health and Safety Assessment Services – OSHAS 18001[1], and ISO14001[2], respectively.

In this context, any industrial system in the event of a fault that is not diagnosed and treated correctly could be considered to pose a serious risk to people's health, to the environment and to the industrial equipment [3].Thus, several approaches for fault-tolerant reconfigurable control system have been proposed [4].However, although the development of techniques for diagnosis and treatment of faults exists, accidents still occur. These issues are fully explained because there is no zero risk in industrial processes since: (i) physical devices do not have zero risk of failure; (ii) human operators do not have zero risk of error; and (iii) there is no control programs developed that can predict all the possibilities. Thus, studies that aim to diagnose and treat faults rely on restricting its state space for the control and treatment of a particular class of faults.

According to experts, the concept of Safety Instrumented Systems (SIS) is a practical solution to these types of issues. They strongly recommend layers for risk reduction based on control systems organized hierarchically in order to manage risks, preventing or mitigating failures, or bringing the process to a safe state. In this sense, some safety standards such as IEC 61508 [5], IEC 61511 [6], among others, guide different activities related to a SIS Safety Life Cycle (SLC),such as design, installation, operation, maintenance, tests and others [7].

The term "risk" defines a metric for quantifying injury, environmental damage and economic losses; in reference to both probability of a fault occurrence and magnitude of the injury or loss [8]. According to IEC 61508 [5], the term "fault" is defined as an abnormal condition that can cause a reduction or loss of the ability of a functional unit. In this work, faults are classified into two groups: (a) non-critical faults that define risks to be tolerated and therefore automatically recovered by the Basic Process Control System (BPCS); and (b) critical faults that define unacceptable magnitude of risks and must be either prevented or mitigated in order to avoid a catastrophic scenario. The principal is that industrial processes should always be placed into a safe state via the degeneration of the processes by layers of risk reduction of SIS.

One of the challenges is the development of safety control systems, based on SIS. The initials basic questions are: (a) how to design safety control systems based on Safety Supervisory Control Architecture (SSCA) in accordance with standards IEC 61508 and IEC 61511; (b) how to design a layer of risk prevention based on control system, which will be incorporated in the SSCA, according to standards. Therefore, this paper initially proposes the design of safety control systems, based on Safety Supervisory Control Architecture (SSCA), as shown in Fig. 1, in accordance with standards IEC 61508 and IEC 61511. The functions for each layer of both risk prevention and mitigation control systems consider safety programmable controllers with their respective safety sensors and actuators, safety programs based on analysis, validation and verification of mathematical models for diagnosis, treatment and mitigation of critical faults and integration of these layers in a cooperative way by a Risk and Hazard Supervisory Control using service-oriented architecture (SOA).

Fig. 1. Safety Supervisory Control Architecture (SSCA)

Additionally it is proposed a method for modeling and validating the control layer of risk prevention within SSCA, in compliance to IEC 61508 and IEC 61511 standards. This method considers diagnostic and treatment for each Safety Instrumented Function (SIF) including hazard and operability (HAZOP) studies of the equipment or system under control. For modeling critical faults diagnosis, approaches based on Artificial Intelligent (AI), such as Fuzzy Logic, Neural networks, and Bayesian Network (BN) has the potential to solve them [3, 12]. In this method BN and Interpreted Petri Net(IPN) are used.

IPN is also used for modeling critical faults treatment, since this technique is proper for treat dynamic system behavior oriented by the occurrence of instantaneous events and discrete states, i.e., dynamic discrete event systems (DEDS). Additionally, a coordination model based on IPN is used to link each fault treatment model to a corresponding diagnostic model. The mathematical models generated enable the validation of the control program by using a computational resource ensuring that the Safety Integration Level (SIL) is achieved. Finally, these models can be translated to a control program in any language defined by IEC 61131-3 in accordance to IEC 61511 standard [5] and implemented in a safety Programmable Logic Controller (PLC) as a layer of prevention control within SSCA.

This paper is organized as follows: Section 1 presents the introduction and proposes the design of safety control systems, based on Safety Supervisory Control Architecture (SSCA). Section 2 presents its relationship to Internet of Things. Section 3 presents the fundamental concepts for BN, and IPN. Section 4, presents the proposal of a risk prevention control system layer. Section 5 presents the example of application. Section 6 presents the conclusion. Finally references are presented.

2 Relationship to Internet of Things

According to [9], the Internet infrastructure will retain its vital role as global backbone for worldwide information sharing and diffusion, interconnecting a wide range of services and technologies, such as RFIDs, sensors/actuators, machine-to-machine communication devices, etc. In the context of service-oriented architecture (SOA)-based automation systems, the communication among components in a cooperative way is completely based on open standards, and the components can be easily rearranged. This saves down-times and costs when a system needs to be reconfigured or extended [10]. Furthermore, it is possible the distribution of the control logic as independent software-blocks is possible. Therefore, the concept of SSCA can be applied considering a cooperative way to find a control solution by using SOA-based automation system. Works adopted SOA, in which Web Service (WS) is a popular instance of this architecture [11].Finally, SOA architecture approach will be used to allow Risk and Hazard Supervisory Control to obtain information from risk prevention and mitigation control systems in order to generate a data repository that will be used as input for Artificial Intelligence (AI) algorithms and also for continuous improvement of safety control systems [12].

3 Fundamental Concepts

This section introduces fundamental concepts of Bayesian Network (BN) and interpreted Petri net (IPN) for critical faults diagnosis. Moreover, it introduces IPN for coordination and treatment of critical faults.

3.1 Bayesian Network (BN)

The BN provides a method to represent partial beliefs under conditions of uncertainty [13]. Its graphical structure models relationships of probabilistic dependence of cause-effect considering a group of variables. Bayesian networks have been extensively applied for fault diagnosis [14]. BN allow the combination of human expert knowledge of the process under observation and probability theory for building a diagnostic structure, based on algorithms, like *K2* [15].

3.2 Interpreted Petri Net (IPN)

Presented in 1962 by Carl A. Petri, the Petri net (PN) is a powerful tool for modeling, analysis and design of DEDS. PN can represent processes with synchronism, concurrent, causality, conflict, share resources and normal situations in DEDS. It is especially useful in applications in which security is a relevant factor. As mentioned in section 1, the proposal of this work is to use PN as a tool for modeling coordination and treatment of critical faults in a SSCA design. Interpreted Petri net (IPN) is defined as a tool which is associated with either an interpretation or meaning to their places and transitions; representing something real which aims to modeling (i.e.: safety sensors and safety actuators).

4 Proposal of Risk Prevention Control System Layer

The proposal of a method for modeling and validating control programs is based on BN, and IPN. The initial idea was introduced in [16], and is presented in Fig.2.

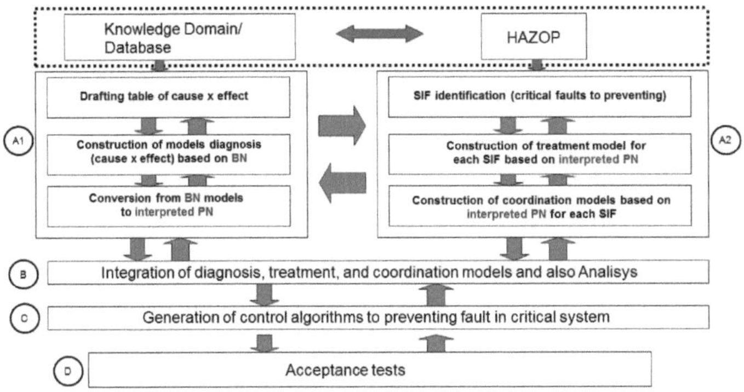

Fig. 2. Method for modeling and validating control programs for preventing risks in SSCA projects

This method defines four steps: (A) modeling, (B) integration and analysis of models, (C) generation of control programs for preventing faults in critical system and (D) acceptance tests. The modeling step is divided in the stages: (A1) diagnosis of critical fault and (A2) critical fault treatment and coordination. The Step (B) is performed to: (i) verify if some PN properties are met and; (ii) validate integrated models in compliance with specification. The Step (C) is performed to convert verified models into a language recommended by standard IEC 61131-3 [17] and accepted by IEC 61511 [6] for implementation in a Safety PLC. Finally, Step (D) is performed for validating if a control program for each SIF complies with the specifications.

4.1 Description of the Proposed Mathematical Method

A1 - Critical Fault Diagnostic modeling

Step 1 - Drafting table of cause x effect:
In this step, it is build a table that lists the causes (critical faults) defined for each SIF, and effects that are observed by sensors when these critical faults occur. The criteria for the building of this table can be: (a) based on domain knowledge about the system and / or (b) based on database obtained from either field experiments or historical data.

Step 2 - Construction of models for diagnosis (cause x effect) based on BN
The cause x effect table obtained is used for building the BN model for diagnosis of faults. Additionally, it is strongly recommended to apply some restrictions from knowledge to improve the network structured.

Step 3 - Conversion from BN models to IPN
In the diagnostic reasoning, causes should be diagnosed based on the monitored effects, in other words, the relationship *effect x cause* must be established. Furthermore, these diagnosis models must be converted into IPN to make its possible. Finally, IPN model should be transcribed into IEC languages [18] to be implemented into a Safety PLC.

A2 - Critical Fault Treatment and Coordination modeling

Step 1 - SIF identification
From HAZOP studies, SIF and SIL are identified. For each SIF, that represent critical fault to prevent, important data are obtained as SIL; that includes initialization events (sensors) and actions(actuators) to be performed by SSCA to prevent such critical faults.

Step 2 - Construction of Treatment Model for each SIF
An IPN model is built based on information obtained from each SIF, as shown in the previous step. It ensures that the Safety PLC takes appropriate actions to prevent undesirable risks in the critical system. The aim of each SIF is verified based on dynamic behavior of IPN models at the end of this step.

Step 3 - Construction of Coordination models
A coordination model for each SIF is built based on IPN. Once a critical fault is identified by the diagnose model, the coordinator model is responsible for calling their respective treatment model to be run, taking actions to prevent risks. These models should be designed to be robust against the occurrence of spurious faults that may unduly de-energize final elements and produce unwanted system downtimes.

B – Integration and Analysis of models
In this step, the models of critical fault diagnosis, treatment, and coordination are integrated to compose the SIF general model. The integration is made from logical connections, since no flow of tokens should occur among IPN models. After integration, SIF general model should be simulated based on computational tools (e.g.: HPSim [19]) for validating control programs for each SIF complies with the specifications. For the proceeding of simulation of SIF general models, the models of devices for sensing and actuation are considered to close the control loop.

C – Generation of control programs to prevent fault in critical system
The SIF general model obtained in the previous step should be converted into control program based on the IEC 61131-3 [16] language and accepted by IEC 61511 [6] such as (a) Ladder Diagram, (b) Function Block Diagram and (c) SFC (Sequential Function Chart). Many works have been published about methods for converting PN models into algorithms based on IEC 61131-3 languages. An example is showed in [18].

D – Acceptance tests

In according with the IEC 61508 / IEC 61511 standards, one of the Safety Life Cycle (SLC) steps is related to final testing for commissioning and start-up procedures.

5 Example of Application

A natural gas compression station is presented to illustrate the proposed method. To evaluate this approach, it is considered one SIF, identified as SIF-01; obtained from HAZOP.

5.1 Process Description

The natural gas compression station has at least a natural gas supply line, called suction, from a gas pipeline which transports this natural gas. At the station entrance, natural gas goes through filters before it being compressed by the turbo-compressor machine. A portion of this gas is directed to the utility unit. The utility unit accounts for controlling the gas temperature and pressure for use in the compression station, such as fuel gas for the turbo-compressor machine, gas heaters, and gas power generators. Then it is sent back to the gas pipeline through discharge lines.

5.2 Application of the Proposed Method

A1 - Critical Fault Diagnostic modeling
Step 1 - Drafting table of cause x effect
This step was based on the knowledge about the system. The cause x effect table is shown in Fig. 3a. All values are binary (e.g.: 0=*off*, 1=*on*); except the first column, which defines the number of knowledge cases for the specific SIF-01. The second column defines the critical fault (e.g.: Very High Pressure on Discharge Header) considered for the SIF-01 under study and the remaining columns represent the states values from sensors observed when the critical fault occurred. PSHH-006A, PSHH-006B, and PSHH-006C states are based on thresholds for very high pressure observed via sensors installed in the discharge lines in the natural gas compression station.

CASE	VERY HIGH PRESSURE ON DISCHARGE HEADER	PSHH-006A	PHSS-006B	PSHH-006C
1	*1*	*0*	*1*	*1*
2	*1*	*1*	*0*	*1*
3	*1*	*1*	*1*	*0*
4	*1*	*1*	*1*	*1*

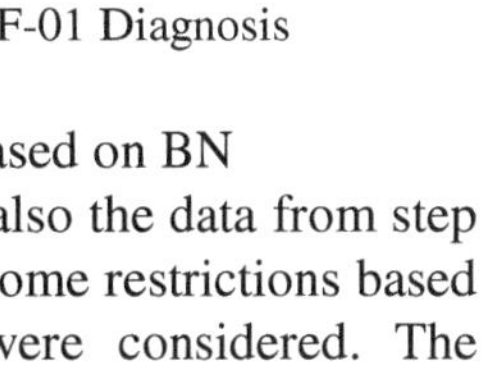

Fig. 3. (a) Cause x effect table (b) Resulting BN model for SIF-01 Diagnosis

Step 2 - Construction of models for diagnosis (cause x effect) based on BN
In this work, the learning algorithm K2 (search and score) and also the data from step 1 were used to obtain the initial structure of BN. Furthermore, some restrictions based on human knowledge of relationships between variables were considered. The resulting BN model for SIF-01 diagnosis is shown in Fig.3b.
Step 3 - Conversion from BN models to IPN
The relationship *cause x effect* is:

Fault 1→ (PSHH-06A ^ PSHH-06B) v (PSHH-06A ^ PSHH-06C) v (PSHH-06B ^ PSHH-06C)

Finally, the model of Fig.4awas constructed including the representation of the additional elements necessary to reset the model after no further fault detection and also considering the possibility of a failure be spurious(eg.: the possibility of the diagnosis do not be effected). The places F_{AB}, F_{AC} and F_{BC} represent transient states that may correspond to spurious failures. If the transitions t_4 or t_5 or t_6 are enabled for firing, it means that the signal from sensor is no longer active and therefore, the fault diagnosis should be aborted, returning the system to the initial state (place READY_D1 marked). BesidesFig.4b describes the interpretations given to the model elements.

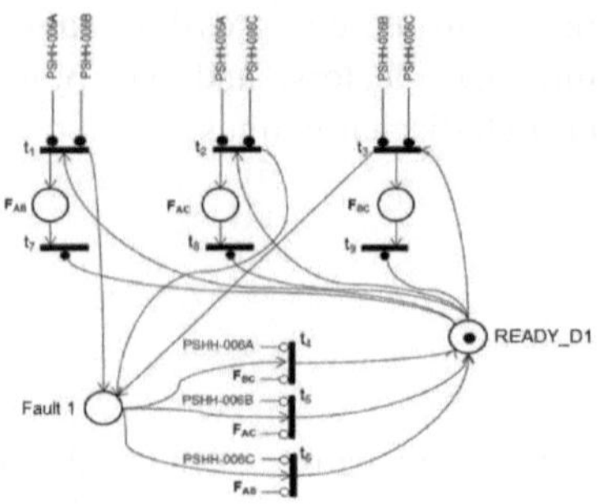

Element	Description
Enable Arc associated with "PSHH-006A"	*Very High Pressure Deviation in the discharge detect by sensor PIT-006A.*
Enable Arc associated with "PSHH-006B"	*Very High Pressure Deviation in the discharge detect by sensor PIT-006B.*
"Enable Arc associated with "PSHH-006C"	*Very High Pressure Deviation in the discharge detect by sensor PIT-006C.*
Place "F_{AB}"	*Auxiliary memory that transition t1 was fired.*
Place "F_{AC}"	*Auxiliary memory that transition t2 was fired.*
Place "F_{BC}"	*Auxiliary memory that transition t3 was fired.*
Place "READY_D1"	*Model is ready to re-start in case of pressure deviation in the compression station discharge.*
Place "Fault 1"	*Fault Diagnosis due High Pressure Deviation in the compression station discharge.*

Fig. 4. (a) SIF-01Diagnosis model based on IPN (b) Description of model Elements

A2 - Critical Fault Treatment and Coordination modeling

Step 1 - SIF identification (critical faults to prevent)

From the risk analysis report in a real situation, 13 SIF's were obtained [15]. In this work, it is only considered SIF-01.

Step 2 - Construction of Treatment Model for each SIF based on IPN

In this step, the treatment model should represent the action to be taken by the SSCA when the SIF-01 is diagnosed. The action should be close valves XV-001/017/019/020 (suction line), close valves XV-003 / 018 (discharge header),and send the shutdown command for the turbo-compressor.

Step 3 - Construction of Coordination models based on IPN for each SIF, according to [15].

B – Integration and Analysis of models

In this step, the models of critical fault diagnosis, treatment, and coordination were integrated to compose the SIF general model of SSCA. After the integration, the following activities were accomplished: (a) The SIF general model was simulated using HPSim to validate safety requirements for preventing faults and it was also verified that the SIF general model is restartable; (b) The properties of liveness and safety were verified for SIF general model based on PIPE2[20].

C – Generation of control programs to prevent fault in critical system

The models for fault critical diagnosis, coordination, and treatment were converted into a control program based on IEC 61131-3. The language used in this example of application was the Ladder Diagram, although, other languages based on the IEC 61131-3 and accepted by IEC 61511 such as Function Block Diagram and, SFC (Sequential Function Chart) also could be used.

D – Acceptance tests
In the first instance, the control programs were tested on-line in a simulation tool based on Siemens PLC technology (e.g.: Simatic S7-300F, where "F" means Fail Safe), and then they were validated in compliance with technical requirements.

6 Conclusions

In the first, a new architecture for the design of safety control system, based on Supervisory Safety Control Architecture (SSCA), was proposed in accordance with standards IEC 61508 and IEC 61511, concerning cooperative and hierarchical layers of control prevention and mitigation of critical faults. In the second, a method for design the layer of risk prevention control system was presented and validated to an application example of a gas compression station, showing to be an efficient method. Furthermore, SSCA will use SOA [12] to work with the new Risk and Hazard control module responsible for the acquisition and maintenance of data for synthesis needs in safety programmable controllers. Finally, some issues must be solved: (a) how to design a layer of risk mitigation based on control system that will be incorporated in the SSCA; (b) how to proceed the dynamic commissioning to ensure effective tests of safety devices in accordance with SIS SLC of IEC 61508. Research is being developed to address these issues.

Acknowledgments. The authors would like to thank the Brazilian governmental agencies CNPq, FAPESP, and CAPES for their financial support to this work.

References

1. OSHAS18001. International standard of occupational health and safety assessment services (2007)
2. ISO14001. International standard for environmental management systems (2004)
3. Sallak, M., Simon, C., Aubry, J.: A fuzzy probabilistic approach for determining safety integrity level. IEEE Transaction on Fuzzy Systems 16(1), 239–248 (2008)
4. Zhang, Y., Jiang, J.: Bibliographical review on reconfigurable fault-tolerant control systems. Annual Reviews in Control 32, 229–252 (2008)
5. IEC. Functional safety of electrical/electronic/programmable electronic safety-related systems (IEC 61508) (2010)
6. IEC. Functional safety - safety instrumented systems for the process industry sector - part 1 (IEC 61511) (2003)
7. Lundteigen, M.-A., Rausand, M.: Architectural constraints in IEC 61508: Do they have the intended effect? Reliability Engineering and System Safety, 520–525 (2009)
8. Bell, R.: Introduction to IEC 61508. In: Proceedings of ACS Workshop on Tools and Standards, Sydney, Australia (2005)
9. Miorandi, D., Sicari, S., De Pellegrini, F., Chlamtac, I.: Internet of things: vision, applications and research challenges. Ad Hoc Networks (2012)
10. Feldhorst, S., Libert, S., Hompel, M.T., Krumm, H.: Integration of a Legacy Automation System into a SOA for Devices. In: IEEE Conference on Emerging Technologies & Factory Automation, ETFA, pp. 1–8 (2009) ISSN 1946-0759

11. Garcia Melo, J.I., Junqueira, F., Morales, R.A.G., Miyagi, P.E.: A procedure for modeling and analysis of service-oriented and distributed productive systems. In: Proceedings of 4th IEEE Conf. on Automation Science and Engineering (CASE), Washington, DC, USA, pp. 941–946 (2008)
12. Florea, G., Ocheana, L., Popescu, D., Rohat, O.: Emerging technologies - the base for the next goal of process control - risk and hazard control. In: Proceedings of Recent Advances in Signal Processing, Computational Geometry and Systems Theory, Bucharest (2011) ISBN: 978-1-61804-027-5
13. Pearl, J.: Causality: Models Reasoning and Inference. Cambridge University Press (2000)
14. Chien, C.F., Chen, S.L., Lin, Y.S.: Using Bayesian network for fault location on distribution feeder. IEEE Transactions Power Deliv. 17, 785–793 (2002)
15. Cooper, G.F., Herskovitz, E.: A Bayesian method for the induction of probabilistic networks from data. Machine Learning 9, 309–347
16. Squillante Jr., R., Santos Filho, D.J., Riascos, L.A.M., Junqueira, F., Miyagi, P.E.: Mathematical method for modeling and validating of safety instrumented system designed according to IEC 61508 and IEC 61511. In: Proceedings of International Congress of Mechanical Engineering (COBEM), Natal, RN, Brazil (2011)
17. IEC, Programmable controllers IEC 61131- part 3: Programming languages (2003)
18. Mello, A.T.F., Barbosa, M.C., Santos Filho, D.J., Miyagi, P.E., Junqueira, F.: A Transcription Tool from Petri Net to PLC Programming Languages. In: Proceedings of the 21st International Congress of Mechanical Engineering. ABCM, Rio de Janeiro (2011)
19. Anschuetz, H.: HpSim, http://www.winpesim.de (accessed January 12, 2012)
20. Bonet, P., Llado, C.M., Puijaner, R., Knottenbelt, W.J.: PIPE2, http://pipe2.sourceforge.net/ (accessed January 12, 2012)

ARMA Modelling for Sleep Disorders Diagnose

João Caldas da Costa[1], Manuel Duarte Ortigueira[2], Arnaldo Batista[2], and Teresa Paiva[3]

[1] DSI, Escola Superior de Tecnologia de Setúbal, Instituto Politécnico de Setúbal, Portugal
[2] UNINOVA and Department of Electrical Engineering, University Nova, Lisbon, Portugal
[3] Faculty of Medicine University of Lisbon, Lisbon, Portugal

Abstract. Differences in EEG sleep spindles constitute a promising indicator of sleep disorders. In this paper Sleep Spindles are extracted from real EEG data using a triple (Short Time Fourier Transform-STFT; Wavelet Transform-WT; Wave Morphology for Spindle Detection-WMSD) algorithm. After the detection, an Autoregressive–moving-average (ARMA) model is applied to each Spindle and finally the ARMA's coefficients' mean is computed in order to find a model for each patient. Regarding only the position of real poles and zeros, it is possible to distinguish normal from Parasomnia REM subjects.

Keywords: Sleep Spindles, ARMA, EEG, Parasomnia REM.

1 Introduction

Sleep spindles are particular EEG patterns which occur during the sleep cycle with center frequency in the band 11.5 to 15 Hz. They are used as one of the features to classify the sleep stages [1]. Sleep spindles are promising objective indicators in sleep disorders. In order to interpret then, their structure needs to be clarified or a suitable model needs to be found. The correct detection of human sleep spindles and posterior characterization can lead to early detection of changes in brain and prevent or, at least, mitigate the influence of certain diseases [2].

In [2] automated spindle characterization by using autoregressive moving average (ARMA) was proposed by the authors to distinguish between normal, elderly and dementia patients. In this work, ARMA model for sleep spindles is used to detect meaningful differences when applied to spindles from different types of people, in this case to distinguish normal from Parasomnia REM subjects.

2 Relationship to Internet of Things

As one of the main ideas of the Internet of Things (IoT) is that all objects and people in daily life will be equipped with radio tags and they could be identified and inventoried by computers, it will bring great advantages in the bio-signal processing fields. It will come a time where people are constantly monitored in their "biological values". Signals like body heat, heart rate, ECG and EEG amongst others will be real-time monitored in order to rapidly and efficiently detect certain deceases.

L.M. Camarinha-Matos, S. Tomic, and P. Graça (Eds.): DoCEIS 2013, IFIP AICT 394, pp. 271–278, 2013.

3 Sleep Spindles

It is commonly referred in literature that sleep spindles are the most interesting hallmark of stage 2 sleep electroencephalograms (EEG) [1]. A sleep spindle is a burst of brain activity visible on an EEG and it consists of 11-15 Hz waves with duration between 0.5s and 2s in healthy adults, they are bilateral and synchronous in their appearance, with amplitude up to 30 μV (Fig. 1 and Fig. 2).

The spindle is characterized by progressively increasing, then gradually decreasing amplitude, which gives the waveform its characteristic name. It is now reliable that sleep spindles are originated in the thalamus and can be recorded as potential changes at the cortical surface [3].

Sleep spindles were first described in human EEG by Loomis in 1935, but the first commonly accepted definition of sleep spindle was given by [4]:

"The presence of a sleep spindle should not be defined unless it is of at least 0.5sec duration, i.e., one should be able to count 6 or 7 distinct waves within the half-second period. Because the term "sleep spindle" has been widely used in sleep research, this term will be retained. The term should be used only to describe activity between 12 and 14 cps."

4 ARMA Models and Sleep Spindle Detection

4.1 ARMA Model

In signal processing, autoregressive moving average (ARMA) models are typically applied to correlated time series data. Given a time series, we can consider it as the output of an ARMA system driven by white noise. The ARMA model is a tool for understanding and, whenever necessary, predicting future values in time series. The model consists of two parts, an autoregressive (AR) part and a moving average (MA) part. The model is usually referred to as ARMA(p,q) where p is the order of the autoregressive part and q is the order of the moving average part .

Compared with the pure MA or AR models, ARMA models more suitable for describing the characteristics of a given process with minimum number of parameters using both poles and zeros; rather than just poles or zeros [5].

As referred, a stationary ARMA process of order (p,q) is considered as the output of a linear time-invariant(LTI) digital filter driven by white noise. The transfer function of the system is given by:

$$H(z) = \frac{\sum_{m=0}^{q} b_m \, z^{-m}}{\sum_{k=0}^{p} a_k \, z^{-k}}, \tag{1}$$

with a_0=1. The process corresponding to this model satisfies the difference equation:

$$x(n) = -\sum_{k=1}^{p} a_k \; x(n-k) + \sum_{m=0}^{q} b_m \; w(n-m) \,, \tag{2}$$

where w(n) is the input sequence, a zero-mean white noise and x(n) is the output sequence. The main task in the modeling can be formulated as:

Given a segment of a time series, x(n), n=0,1,2 …, L-1, estimate the p+q+1 ARMA parameters.

4.2 Sleep Spindle Detection

In this paper a combination of three different approaches is used for the automatic detection of sleep spindles: Short Time Fourier Transform, Wavelet Transform and Wave Morphology for Spindle Detection.

The individual detection algorithms are explained from section 4.3.1 trough 4.3.3. In order to improve the results, the three detectors are mixed together using the procedure presented in 4.3.4. The best performance obtained resulted in a sensitivity and specificity of 94% when compared to human expert scorers. The algorithms were previously implemented, tested and evaluated by the authors in manual human scored signals in [6].

4.3.1 Sleep Short Time Fourier Transform (STFT)

The use of STFT is used to determine the sinusoidal frequency and phase content of local sections of a signal as it changes over time and it is commonly used in signal processing [7].

The STFT of a discrete signal is:

$$STFT\{x[n]\} = X(m,\omega) = \sum_{n=-\infty}^{\infty} x[n]\,\omega[n-m]e^{-j\omega n} \,. \tag{3}$$

The magnitude squared of the STFT yields the spectrogram of the signal:

$$spectrogram\{x[n]\} = \left|X(\tau,\omega)\right|^2 . \tag{4}$$

The SS detection is based on the spectrogram. A segment is marked as SS when a peak (above a pre-specified threshold) with duration between 0.5s and 2s occurs in the SS frequency range.

In the STFT SS detection used, the threshold value used corresponds to the cumulative value of peaks in the spectrogram

4.3.2 Wavelet Transform (WT)

In this method, the detection of sleep spindles employ the continuous wavelet transform of EEG signal x(t):

$$CWT\ \psi\ x(a,b) = \frac{1}{\sqrt{|a|}} \int_{-\infty}^{\infty} x(t)\ \psi * (\frac{t-b}{a}) dt\ , \tag{5}$$

where Ψ(t) is called the 'mother wavelet', the asterisk denotes complex conjugate, whereas a and b are scaling parameters [8]. The corresponding normalized wavelet power is defined by:

$$w(a,b) = W^2(a,b) / \sigma^2\ , \tag{6}$$

and σ is the standard deviation of the EEG segment used.

Complex Morlet WT is defined as

$$\psi(x) = \frac{1}{\sqrt{\pi fb}}\ e^{2\pi fcx}\ e^{x2/fb}\ dt\ , \tag{7}$$

where fc is the center frequency and fb the bandwidth frequency. In order to find SS using the WT, the normalized WT power was determined and when a peak (greater then a determined threshold) with a duration between 0.5s and 2s occurred a SS was marked.

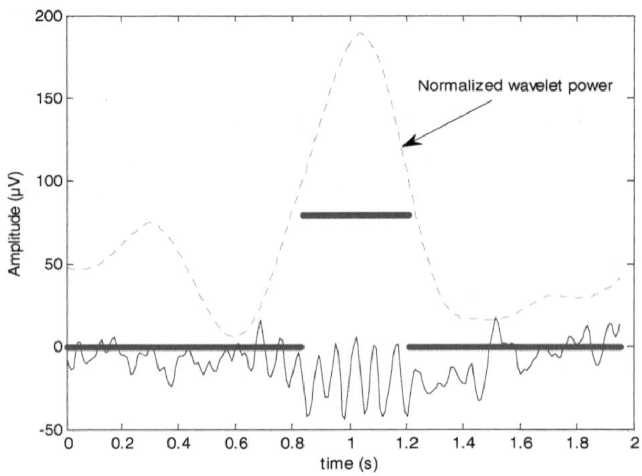

Fig. 1. Example of SS detection using WT

4.3.3 Wave Morphology for Spindle Detection (WMSD)

The WMSD algorithm proposed in this paper is based on the definition of Sleep Spindle by Rechtschaffen and Kales [4]

The WMSD algorithm was for the first time published by the authors in [6].The implemented algorithm consists of:

a) Detection of peaks in the signal (maxima and minima), based on a defined threshold, thus, eliminating small peaks;

b) Determination of extreme to extreme time distance and conversion to frequency:

$$f = \frac{1}{T}, \tag{8}$$

c) Verification if the determined frequencies lie in the SS range (11-15 Hz);
d) If there are more than 12 consecutive peaks (6 maxima and 6 minima) in the SS frequency band a spindle is marked.

The whole process mimics the visual detection mechanism.

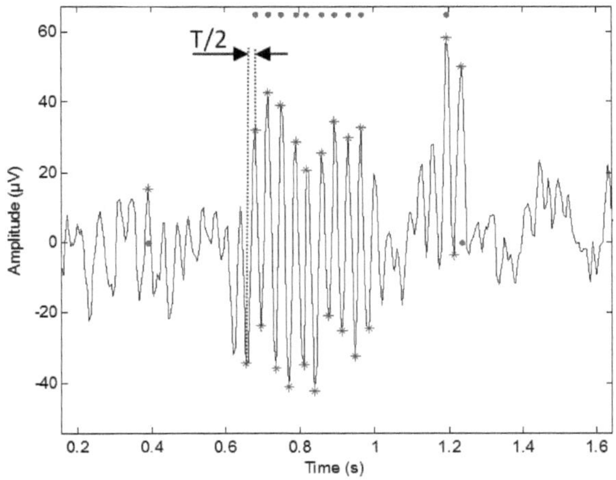

Fig. 2. Example of SS detection using WMSD

4.3.4 Mixed Detection Using WT, STFT and WMSD

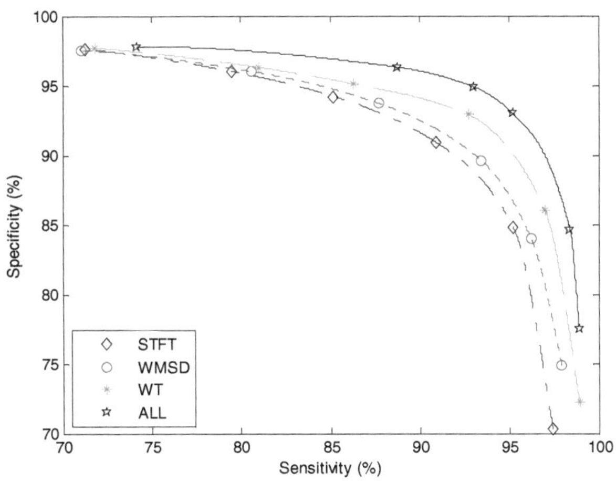

Fig. 3. Sensitivity x Specificity curves

In this work, a mixed algorithm using WT, STFT and WMSD algorithms, was used.

In this approach, we use a vector to characterize the signal (same length as the sampled signal).This vector defines each point as belonging to a SS or not. The mixed result is computed, i.e., a point is considered belonging to a SS if it is marked as SS in WT, STFT and WMSD algorithms. Finally, if there are not enough consecutive points marked as belonging to a SS, in order to last at least 0.5 seconds, they are considered as non-spindle. The best performance obtained resulted in a sensitivity and specificity of 94% when compared to human expert scorers. In Fig. 3, Sensitivity and Specificity curves are shown for the individual algorithms together with the combination of the 3.

5 Experimental Results

This study makes use of a sample representative of human sleep, obtained from 23 volunteers, males and females with ages between 35 and 87 years old. Briefly, all polysomnograms were obtained by a Nicolet EEG 1A97 18-channel polygraph with a sampling rate of 256Hz. From the group, 8 subjects were completely healthy and the remaining 15 had some kind of REM Parasomnia: REM sleep behavior disorder (RBD), Recurrent Isolated Sleep Paralysis or Catathrenia. The signals were unclassified and the whole night signal of C3-A2 channel was used. At this stage our objective was only to distinguish healthy from Parasomnia subjects.

The detection methods were applied with a combination of threshold parameters for the STFT, WMSD and WT algorithm. In the STFT case, the threshold value used corresponds to the cumulative value of peaks in the spectrogram. In the WMSD algorithm, a point is considered a maximum peak if it has the maximal value, and was preceded (to the left) by a value lower than the threshold defined. The Normalized Wavelet Power amplitude is used as threshold in the WT case.

After the algorithms were applied to the signals and SS identified, ARMA modelling was performed for all the SS from all subjects. After this, computation of the arithmetic means of the coefficients from the ARMA transfer functions was performed. This gave us a typical SS transfer function for each subject. 4 groups have then been created, comprising subjects with similar pole/zero distributions. Only the zero and real pole position have been taken into care. The characteristics of each group are as follows:

- Group 1: pole on the left complex plane and zero on the right complex plane (see Fig. 4. Zeros and poles from Group 1.);
- Group 2: pole on the right complex plane close to 1 and zero on the left complex plane (left from -0.1) (see Fig. 5. Zeros and poles from Group 2.);
- Group 3: pole on the right complex plane close to 1 and zero near the origin (right from -0.1 and left from +0.1) (see Fig. 6. Zeros and poles from Group 3.);
- Group 4: pole on the right complex plane close to 1 and zero on the right complex plane (right from +0.1) (see Fig. 7. Zeros and poles from Group 4.).

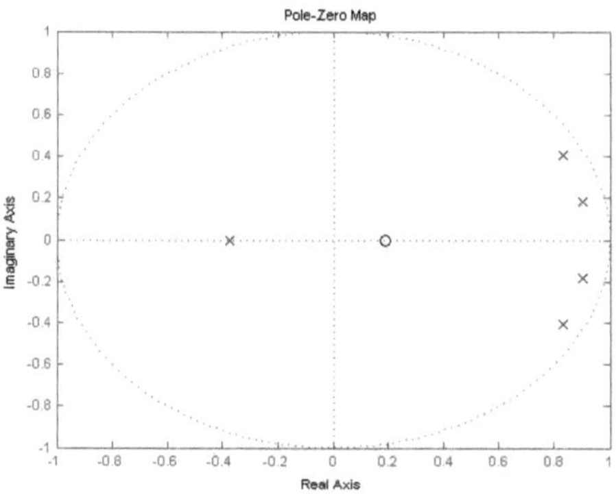

Fig. 4. Zeros and poles from Group 1

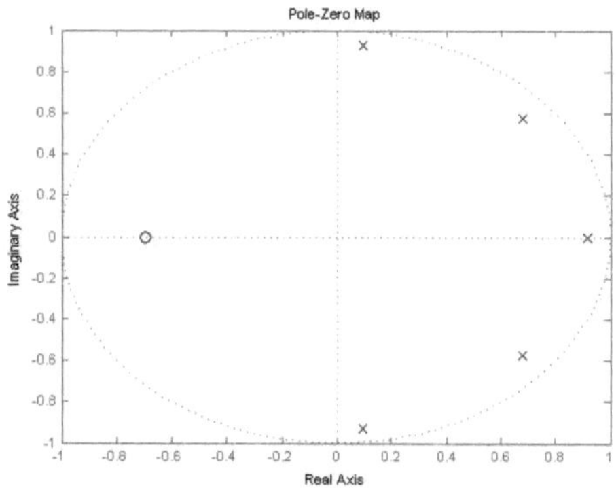

Fig. 5. Zeros and poles from Group 2

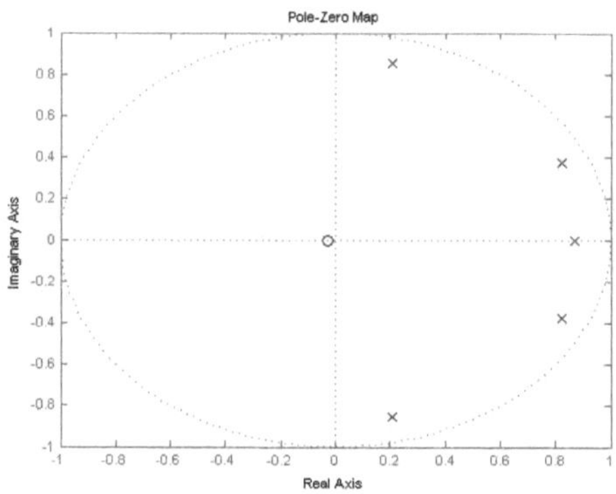

Fig. 6. Zeros and poles from Group 3

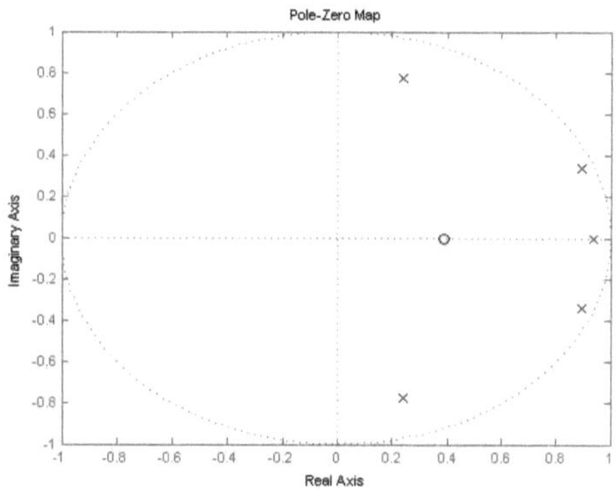

Fig. 7. Zeros and poles from Group 4

The subjects' allocation to the groups resulted as follows:

- Group 1: 3 subjects, all suffering from REM Parasomnia;
- Group 2: 12 subjects, all suffering from REM Parasomnia;
- Group 3: 3 healthy subjects;
- Group 4: 5 healthy subjects.

It is word notice that all the healthy subjects were classified as belonging to Groups 3 or 4 and all the Parasomnia REM patients were allocated in Groups 1 or 2. A further in depth analysis of the subjects' deceases and the deceases grades should be taken into account to understand the existence of four groups instead of 2. This can also lead to the conclusion that for some reason there are significant differences between the healthy subjects.

6 Conclusions

ARMA modeling seems a promising indicator for Sleep disorders. In this work subjects suffering from several Parasomnia disorders were automatically identified based only on the zeros and poles position of their Sleep Spindle model. The work to follow will be to distinguish pathologies from each other, that is, once it is known that a patient suffers from a sleep disorder, how to automatically diagnose its condition based on the Sleep Spindle model.

Acknowledgements. The authors would like to acknowledge sleep laboratory CENC – Centro de Electroencefalografia e Neurofisiologia Clinica for providing the data used for this work. This work was funded by Instituto Politécnico de Setubal, IPS UAII&DE and by Portuguese National Funds through the FCT – Foundation for Science and Technology under the project PEst-OE/EEI/UI0066/2011.

References

1. De Gennaro, L., Ferrara, M.: Sleep spindles: an overview. Sleep Med. Rev. 7, 423–440 (2003)
2. da Costa, J.C., Ortigueira, M.D., Batista, A.: ARMA Modelling of Sleep Spindles. In: Camarinha-Matos, L.M. (ed.) DoCEIS 2011. IFIP AICT, vol. 349, pp. 341–348. Springer, Heidelberg (2011)
3. Steriade, M., Jones, E.G., Llinas: Thalamic Oscillations and Signaling. Neuroscience Institute Publications, John Wiley & Sons, New York (1990)
4. Rechtschaffen, A., Kales, A.: A manual of standardised terminology, techniques and scoring system for sleep stages of human subjects. Public Health Service, U.S. Government Printing Office, Washington, DC (1968)
5. Kizilkaya, A., Kayran, A.H.: ARMA model parameter estimation based on the equivalent MA approach. Digital Signal Processing 16(6) (2006)
6. Costa, J.C., Ortigueira, M.D., Batista, A., Paiva, T.: An Automatic Sleep Spindle detector based on WT, STFT and WMSD. International Journal of the World Academy of Science, Engineering and Technology 68, 1298–1301 (2012)
7. Proakis, J., Manolakis, D.: Digital Signal Processing, 4th edn. Prentice-Hall (2006)
8. Omerhodzic, I., Avdakovic, S., Nuhanovic, A., Dizdarevic, K., Rotim, K.: Energy Distribution of EEG Signal Components by Wavelet Transform, pp. 45–60. InTech Publishing (2012)

Fault Detection and Diagnosis in Induction Machines: A Case Study

Miguel Marques[1], João Martins[2], V. Fernão Pires[3,4], Rui Dias Jorge[5], and Luís Filipe Mendes[5]

[1] Department of Electrical and Computer Engineering, Faculty of Sciences and Technology – University New of Lisbon, Quinta da Torre, 2829-516 Caparica, Portugal
[2] CTS/UNINOVA, Faculty of Sciences and Technology – University New of Lisbon, Quinta da Torre, 2829-516 Caparica, Portugal
[3] Instituto Politécnico de Setúbal, Campus do IPS, Estefanilha, 2914-508 Setúbal, Portugal
[4] Center for Innovation in Electrical and Energy Engineering (CIEEE), Lisbon, Portugal
[5] EFACEC, Rua da Garagem, nº1, Carnaxide, 2790-078, Oeiras, Portugal

Abstract. Nowadays in industry there many processes where human intervention is replaced by electrical machines, especially induction machines due to his robustness and performance. Although, induction machines are a high reliable device, they are also susceptible to faults. Therefore, the study of induction machine state is essential to reduce human and financial costs. It is presented in this paper an on-line system for detection and diagnosis of electrical faults in induction machines based on computer-aided monitoring of the supply currents. The main objective is to detect and identify the presence of broken rotor bars and stator short-circuits in the induction motor. The presence of faults in the machine causes different disturbances in the supply currents. Through a stationary reference frame, such as αβ-vector transform it is possible to extract and manipulate the results obtained from the supply currents using Principal Component Analysis (PCA).

Keywords: induction motor, diagnosis, fault detection, principal component analysis, PCA, eigenvalue, eigenvector.

1 Introduction

Three-phase induction machines perform critical functions as part of industrial processes. It is estimated that induction motors typically consume 40 % to 50 % of all electrical energy produced in a country [1]. Therefore, induction motors have a special role in the economy of industrialized countries.

Due to its importance, is urgent to develop intelligent systems that detect the presence of faults in the machines in order to reduce maintenance costs. These systems will allow the possibility of scheduled maintenance and predict the need for maintenance before serious deterioration or fault occurs, making it possible to increase the reliability of equipment, the improvement of his behavior and performance [2].

L.M. Camarinha-Matos, S. Tomic, and P. Graça (Eds.): DoCEIS 2013, IFIP AICT 394, pp. 279–286, 2013.

Concerning to maintenance of induction machines, preventive maintenance is widely used currently. It consists in periodic inspections with the objective of replacing parts that are supposed to break after a certain number of hours. However, even with periodic inspections this type of maintenance presents disadvantages such as, unneeded maintenance costs and in some situations catastrophic failures still likely to occur. Thus, the concept of condition-based maintenance (CBM) has emerged as an alternative to preventive maintenance. This type of maintenance is defined as the process of monitoring characteristics or parameters of a machine, in order to verify early changes and trends that can be used to indicate a fault situation or the need for maintenance.

In the last 20 years, FDD in induction machines is a research area that had a great evolution, as seen by the number of proposed methodologies, such as neural networks [3], current space patterns [4], fuzzy logic [5], spectral analysis [6] and vibration monitoring [7].

2 Contribution to Internet of Things

The paper presents the development of a commercial application for fault detection and diagnosis of electrical faults in induction machines. This analysis, based on PCA methodology [8,9], allows to conclude the practical feasibility of on-line monitoring through current space pattern analysis using an industrial product, such as the Terminal Protection Unit (TPU) developed by EFACEC. In the future the use of condition-based systems, based on non-invasive measurements, will contribute to connect the induction machines with cyberspace making it possible to know the state of the machine remotely. Thus, since TPU allow the use of multiple communication standards it will be possible to make a remote monitoring and diagnosis of the devices connected to TPU.

3 Principal Components Analysis (PCA)

Principal Component Analysis is a non-parametric statistical method used to reduce the number of original variables, which are correlated, in a set of new uncorrelated variables referred as Principal Components (PC). The first public descriptions of this method were given in 1901 by Pearson [10] and latter developed in 1933 by Hotelling [11].

PCA can be obtained through several ways, such as eigenvalue decomposition of a matrix or single value decomposition (SVD) of a matrix [12]. In the case of eigenvalue decomposition it consists in the representation of matrix in terms of its eigenvalues and eigenvectors. Through the definition of eigenvectors, this technique is able to obtain the main directions of the data sample on a space-vector. It also possible to measure the weight of the sampled data spread through the main directions defined by the eigenvectors. These metric values are defined as eigenvalues [12].

Let $X \in \Re^{n\times m}$ represents a data matrix, where n denotes the number of measurements and m denotes the number of physical variables. The $X^T \in \Re^{m\times n}$

represents the transposed matrix of X, where m and n have the same meaning as in the *X* matrix. From the product of the two matrices X and X^T is obtained a square matrix $E \in \Re^{n \times n}$ called correlation matrix.

$$E = X^T X \tag{1}$$

After establishing the correlation matrix the eigenvectors and the respective eigenvalues, of E are calculated. There are several ways to define eigenvectors and eigenvalues, the most common approach defines an eigenvector of the matrix E as a vector that satisfies the following equation:

$$(E - \lambda I)u = 0 \tag{2}$$

In fact, this concept of reducing the number of variables through PCA is useful in energy systems, particularly three-phase systems, such as three-phase induction machines. In three-phase energy systems without neutral connection it is usual to use the αβ-vector transformation to reduce the number of original variables. This transformation converts the three-phase currents or voltages into an equivalent two-phase system. So in ideal conditions, the three-phase currents lead to αβ-vector with the following components:

$$\begin{cases} i_\alpha = \sqrt{\dfrac{2}{3}}\, i_a - \dfrac{1}{\sqrt{6}}\, i_b - \dfrac{1}{\sqrt{6}}\, i_c \\ i_\beta = \dfrac{1}{\sqrt{2}}\, i_b - \dfrac{1}{\sqrt{2}}\, i_c \end{cases} \Leftrightarrow \begin{cases} i_\alpha = \dfrac{\sqrt{6}}{2}\, I \sin \omega t \\ i_\beta = \dfrac{\sqrt{6}}{2}\, I \sin\left(\omega t - \dfrac{\pi}{2}\right) \end{cases} \tag{3}$$

Under normal conditions and with a balanced and constant frequency power supply, a pure sinusoidal signal makes a circular pattern centered at the origin of the αβ coordinates. In Fig. 1 there is the representation of a healthy motor input current in the αβ-vector pattern.

However under abnormal conditions and considering a constant frequency power supply the previous conditions are no longer valid and the αβ pattern loses its circular

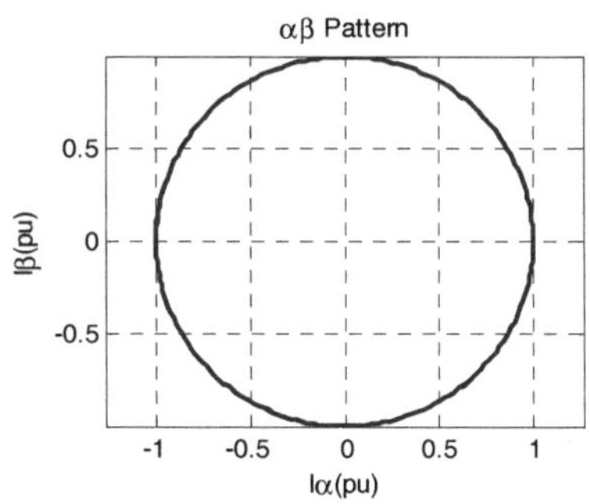

Fig. 1. Healthy motor input current αβ –vector pattern

shape. For a situation where occurs a stator winding fault the input current αβ pattern becomes an ellipse because there is an amplitude variation in the current of the winding that is in a fault situation. The patterns related to a stator winding fault are presented in the Fig. 2.

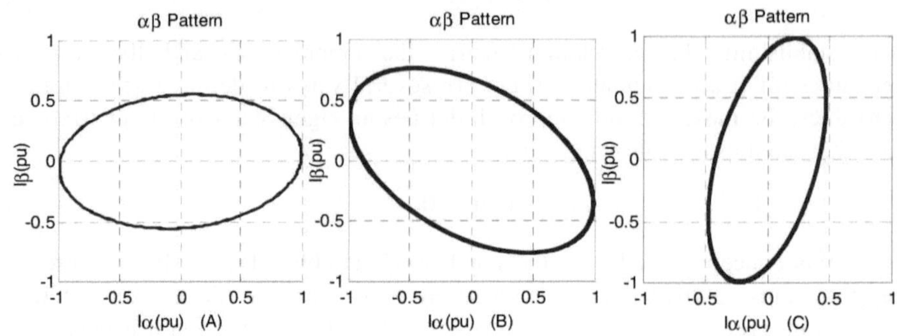

Fig. 2. Stator fault input current αβ-vector patterns. (A) stator fault in phase A (B) stator fault in phase B (C) stator fault in phase C

When the motor presents a rotor fault situation the αβ-vector pattern presents a circular shape but the eigenvalues are not constant. It is possible to observe (Fig. 3) the appearance of a thick ring and the thickness of the ring increases with the severity of the fault. Cardoso *et al.* [13] concluded that the severity of the fault is proportional to the number of the rotor bars, but there is a moment where severity of the factor decreases as the number of broken bars increases.

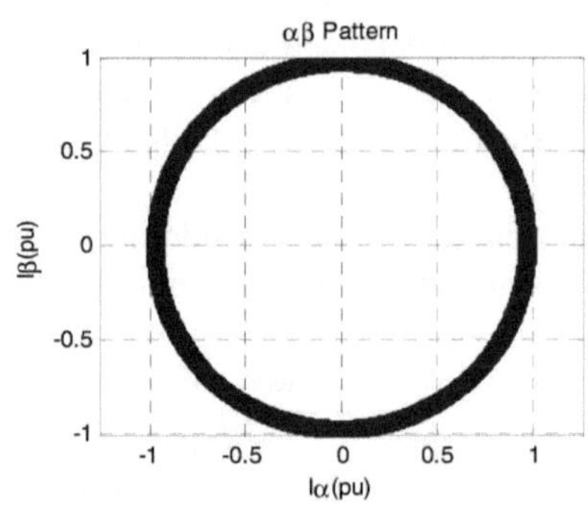

Fig. 3. Rotor fault input current αβ-vector pattern

4 Experimental Results

4.1 Experimental Set Up

The experimental set up used is depicted in Fig. 4. A series of tests were conducted on three squirrel-cage induction motors with a mechanical power (P_{mec}) of 2 Hp, 230/400

V nominal voltage (V_{nom}), a rated speed (N) of 3000 rpm, all with same parameters. One motor was considered a healthy motor and tested. The other two motors were tested with stator short-circuits and broken rotor bars faults.

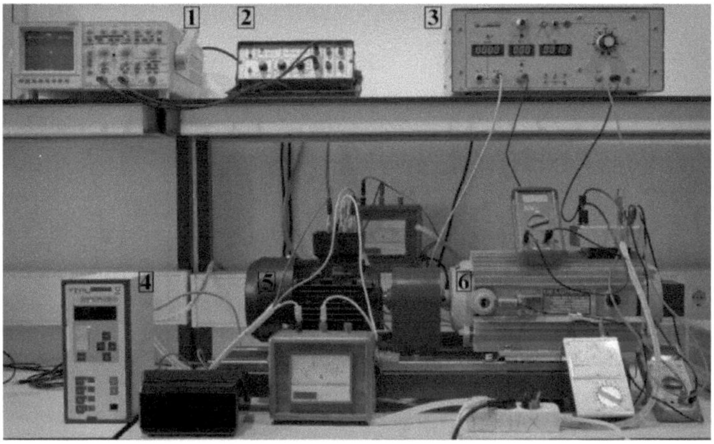

Fig. 4. Experimental apparatus used in this work

In the performed tests the motors {5} were fed with a three-phase auto-transformer with an apparent power (S) of 4 kVA and 0-400 V_{lnl} (line-to-line). The mechanical load was applied to the induction motor by connecting the shaft to a DC generator {6} (controlled by a dc current source{3}) of 0.75 kW rated power (P_{el}). The output of the DC generator was connected to a variable resistive load. In order to allow tests to be performed at different load levels, the DC excitation current and the load resistor were both adjustable.

The data acquisition, signal conditioning and data processing are performed by the TPU S220 {4} developed by EFACEC. For the laboratory tests, a broken rotor bar fault was introduced by drilling a hole into a bar, the hole diameter is slightly larger than the bar width. In the case of short-circuits in the stator windings they were applied by introducing an external variable resistor in series with the windings of each phase.

4.2 Experimental Results

The motor was initially tested in a healthy situation, with the stator windings and its cage intact, in order to verify the current αβ-vector reference pattern. In the conducted tests were applied various levels torque, in order to verify the robustness of the algorithm at different load levels. The start-up condition was discarded since it is not considered by the algorithm.

The temporal evolution of the motor stator currents is represented in the Fig. 5 (A). The Fig. 5 (B) does not present the αβ-vector with a circular shape due to experimental limitation, such as harmonic distortion in the supply voltage and the magnetic field distribution in the machine is not perfectly sinusoidal. However the machine was considered in a healthy condition. For stator faults were used three variable resistors with 11.2 Ω/ 5A in series with the impedance of each phase of the machine (Z = 4.8 Ω). In the case of a fault severity factor of 60% the resistance value is 1.2 Ω. The αβ-vector (Fig. 6) no longer presents a "circular" shape and exhibit an elliptical shape.

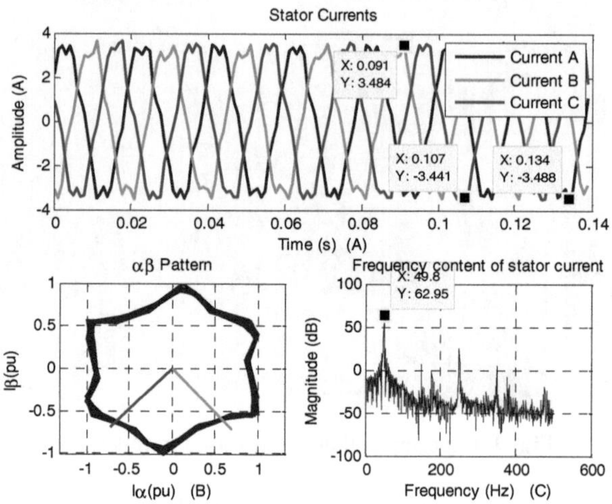

Fig. 5. (A) Stator currents of the machine in nominal operation (B) Experimental αβ-vector (C) Current A spectrum

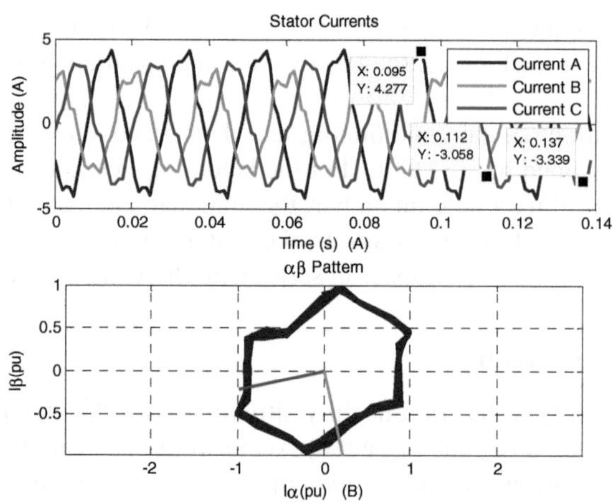

Fig. 6. Experimental results obtained for a stator fault situation in nominal operation with a SF = 60 % in the phase A (A) Stator currents of the machine (B) Experimental αβ-vector

The results obtained for 6 broken rotor bars (major fault) are shown in the Fig. 8. The appearance of a thick ring (Fig. 8 (A)) in a rotor fault situation can be detected through the variation of eigenvalues (Fig. 8 (B)). In the experimental results the eigenvalues does not present a sinusoidal behavior, but show a periodic variation over time.

The results obtained for 6 broken rotor bars (major fault) are shown in the Fig. 8. The appearance of a thick ring (Fig. 8 (A)) in a rotor fault situation can be detected through the variation of eigenvalues (Fig. 8 (B)). In the experimental results the eigenvalues does not present a sinusoidal behavior, but show a periodic variation over time.

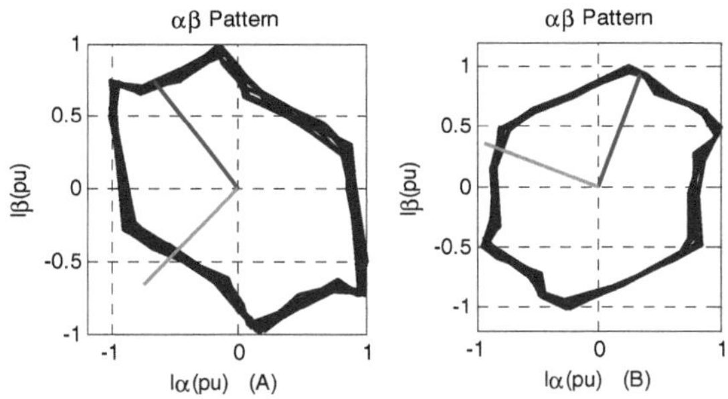

Fig. 7. Experimental αβ-vector pattern for a SF = 60 % (A) Phase B (B) Phase C

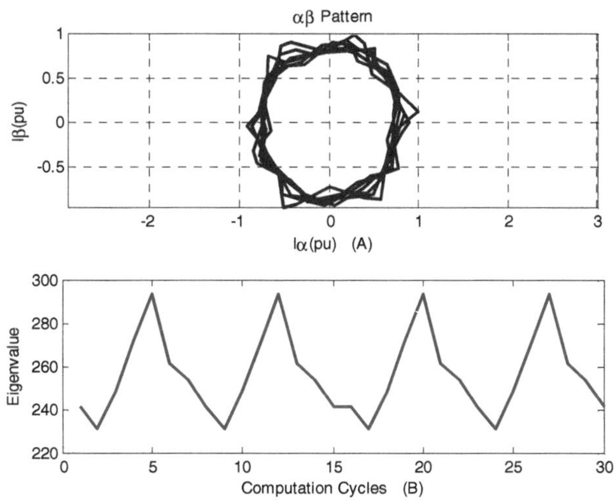

Fig. 8. Experimental results obtained for the machine with 6 broken rotor bars (A) Experimental αβ-vector pattern (B) Variation of the eigenvalues over the computation cycles

5 Conclusions and Future Work

The research in the field of fault detection and diagnosis could provide a reduction in operational costs and the increase of the machine's reliability that associated with remote systems will enable the interconnection of the induction machines with the cyberspace.

In this paper is presented an on-line fault detection and diagnosis system for three-phase induction motors based on PCA. The proposed system uses the eigenvalue decomposition to detect changes in the αβ-vector pattern.

The experimental results indicate that it is possible to detect the presence of short-circuits in the stator windings and broken rotor bars in induction machines with this FDD method. Through the analysis of the eigenvalues from the αβ–vector components it is possible to infer if the machine is in a fault situation or not. For stator fault situations using the eigenvectors, it is also possible to identify in which phase the fault occurred.

Concerning to future works the stator faults were applied through the addition of variable resistors in series with the stator windings. This set up can be considered as an approximation to the fault situation. In the future should be short-circuited some stator windings and test the algorithm to observe his behavior.

Acknowledgements. Authors would like to thank to CTS of UNINOVA for the financial support and to EFACEC for providing its facilities and equipment for this work.

References

1. Ahmed, I., Supangat, R., Grieger, J., Ertugrul, N., Soong, W.L.: A Baseline Study for On-Line Condition Monitoring of Induction Machines. In: Australasian Universities Power Engineering Conference, pp. 26–29 (2004)
2. Siyambalapitiya, D., Mclaren, P.: Reliability Improvement and Economic Benefits of On-Line Monitoring Systems for Large Induction Machines. IEEE Transactions on Industry Applications 26(6), 1018–1025 (1990)
3. Chow, M., Mangum, P.: Neural Network Approach to Real-Time Condition Monitoring of Induction Motors. IEEE Transactions on Industry Electronics 38(6), 448–453 (1991)
4. Cardoso, A., Saraiva, E.: Computer Aided Detection of Airgap Eccentricity in Operating Three-phase Induction Motors, by Park's Vector Approach. IEEE Transactions on Industry Applications, 897–901 (1993)
5. Nejjari, H., Benbouzid, M.: Application of Fuzzy Logic to Induction Motors Condition Monitoring. IEEE Power Engineering Review, 52–54 (1999)
6. Thomson, W.: On-line MCSA to Diagnose Shorted Turns in Low Voltage Stator Windings of 3-Phase Induction Motors prior to Failure. In: IEEE International Electric Machines and Drives Conference, pp. 891–898 (2001)
7. Tsypkin, M.: Induction Motor Condition Monitoring: Vibration Analysis Technique - a Practical Implementation. In: IEEE International Electric Machines & Drives Conference, pp. 406–411 (2011)
8. Martins, J.F., Pires, V.F., Amaral, T.: Induction motor fault detection and diagnosis using a current state space pattern recognition. Pattern Recognition Letters 32(2), 321–328 (2011)
9. Pires, V., Martins, J., Pires, A.: On-Line Diagnosis of Three-Phase Induction Motor Using an Eigenvalue αβ-vector approach. In: IEEE Conference on Power Electronics and Drive Systems, pp. 689-693 (2007)
10. Pearson, K.: On Lines and Planes of Closest Fit to Systems of Points in Space. Philosophical Maganize, 559–572 (1901)
11. Hotelling, H.: Analysis of a complex of statistical variables into principal components. Journal of Educational Psychology 24, 417–441 (1933)
12. Jolliffe, I.: Principal Component Analysis. Springer, New York (2002)
13. Cardoso, A., Cruz, S., Carvalho, J., Saraiva, E.: Rotor Cage Fault Diagnosis in Three-Phase Induction Motors, by Park's Vector Approach. In: IEEE Industry Applications Conference, vol. 1, pp. 642–646 (1995)

Part XI
Integration of Power Electronics Systems with ICT - I

Multilevel Inverter for Grid-Connected Photovoltaic Systems with Active Filtering Function

Kleber C. Oliveira[1], João L. Afonso[2], and Marcelo C. Cavalcanti[3]

[1] Department of Electrical Engineering, Federal University of Paraíba, João Pessoa, Brazil
[2] Centro Algoritmi – University of Minho – Guimarães, Portugal
[3] Department of Electrical Engineering and Power Systems, Federal University of Pernambuco, Recife, Brazil
kleber_ufpe@yahoo.com.br, joao.l.afonso@algoritmi.uminho.pt, marcelo.cavalcanti@ufpe.br

Abstract. This paper presents a single-phase inverter with multilevel topology adopted to interface photovoltaic systems with the electrical power grid, and at the same time, to eliminate harmonics currents and to compensate reactive power, operating as shunt active power filter. Multilevel inverters have been attracting increasing attention in the past few years as power converters of choice in high voltage and high power applications. Various topologies to implement these inverters have been introduced and studied recently. The theory of instantaneous reactive power applied to compensation of single-phase circuits is also presented in this paper. This theory is a particular case of the well-known p-q theory, which was originally developed for three-phase circuits. Simulation results demonstrate this control algorithm applied to multilevel single-phase topologies.

Keywords: Photovoltaic System, Active Power Filter, Multilevel Inverters, p-q Theory.

1 Introduction

There has been increasing interest in electrical power generation from renewable-energy sources, such as photovoltaic (PV). In the last decades, solar energy has been one of the most active research areas among renewable-energy sources. Photovoltaic modules can be used in stand-alone or grid-connected applications [1]. While photovoltaic systems connected to the electrical power grid are gaining more and more visibility, demand for energy is increasing worldwide.

To interface photovoltaic modules with the electrical power grid it is necessary to use a power electronics inverter, which has to perform two tasks: to inject a sinusoidal current into the grid and to ensure that the PV modules operate at the maximum power point (MPP).

Most of the topologies for PV systems are multi-stage, having a high frequency transformer or line-frequency transformer that adjusts the inverter DC voltage and isolates the PV modules from the grid. However, the conversion stages decrease the efficiency and make the system more complex [2].

L.M. Camarinha-Matos, S. Tomic, and P. Graça (Eds.): DoCEIS 2013, IFIP AICT 394, pp. 289–298, 2013.

2 Contribution to Internet of Things

The use of multilevel inverters for photovoltaic applications is becoming popular in recent years. By using multiple levels on the DC bus, the stress on each switching device is reduced. Researches related to multilevel inverters have been reported in applications about FACTS (flexible AC transmission systems), active filters and grid interface for non-conventional energy sources.

By synthesizing the AC output terminal voltage from several levels of voltages, staircase waveforms can be produced, which approach the sinusoidal waveform with low harmonic distortion, thus reducing filter requirements.

The fact that multilevel converters need several DC sources in the DC side makes them attractive for photovoltaic applications.

In the last few decades, the proliferation of nonlinear loads, like switching power supplies, adjustable speed drives and commercial lighting, has resulted in serious degradation of power quality in the electrical grids, mainly because of harmonics problems. Reactive power consumption by the loads is another significant problem to the electrical power grids. Both harmonics and reactive power generate extra power losses in the distribution system transformers and cables, reducing the overall efficiency of the power system. Therefore, compensation of harmonics and reactive power is very important in order to reduce costs due to power losses and bad functioning of equipment in the consumers and in the electrical distribution system.

This paper presents an inverter with multilevel topology, which interfaces photovoltaic systems with the electrical power grid, and that at the same time compensates harmonics currents and reactive power. The presented solution can be an interesting option to the future Smart Grids, contributing to the Internet of Things.

3 p-q Theory

The p-q theory was first introduced by Akagi [3], and is also known as the instantaneous power theory. It is an interesting tool to apply to the control of active power filters (APF), or even to analyze three-phase power systems in order to detect problems related to harmonics, reactive power and unbalance.

The p-q theory applies a Clack transformation to the system voltages and currents, which is represented by the matrices in (1). This transformation converts the voltages and currents from the phases *a-b-c* to the α-β-0 coordinates. It has the advantage of allowing separating the zero sequence components of the voltage (v_0) and current(i_0).

With the voltage and currents determined in the α-β-0 coordinates, can be calculated the instantaneous real power (p), the instantaneous imaginary power (q), and the zero sequence power (p_0), according to (2).

The power components p and q depend only on the α and β coordinates of the voltages and currents, and the power component p_0 depends only on the 0 coordinate of the voltages and currents.

In order to compensate reactive power, harmonics currents and current unbalance (eliminating the neutral wire current), it is necessary to compensate the power components q and p_0, and the oscillating value of p ($\tilde{p}$). The compensation reference currents in the α-β-0 coordinates are obtained from (3) and (4), which come from the inverse of (2), and where $q_x = q$, and p_x is given by (6).

$$\begin{bmatrix} v_\alpha \\ v_\beta \\ v_0 \end{bmatrix} = \sqrt{\frac{2}{3}} \begin{bmatrix} 1 & -\frac{1}{2} & -\frac{1}{2} \\ 0 & \frac{\sqrt{3}}{2} & -\frac{\sqrt{3}}{2} \\ \frac{1}{\sqrt{2}} & \frac{1}{\sqrt{2}} & \frac{1}{\sqrt{2}} \end{bmatrix} \cdot \begin{bmatrix} v_a \\ v_b \\ v_c \end{bmatrix} \begin{bmatrix} i_\alpha \\ i_\beta \\ i_0 \end{bmatrix} = \sqrt{\frac{2}{3}} \begin{bmatrix} 1 & -\frac{1}{2} & -\frac{1}{2} \\ 0 & \frac{\sqrt{3}}{2} & -\frac{\sqrt{3}}{2} \\ \frac{1}{\sqrt{2}} & \frac{1}{\sqrt{2}} & \frac{1}{\sqrt{2}} \end{bmatrix} \cdot \begin{bmatrix} i_a \\ i_b \\ i_c \end{bmatrix} \quad (1)$$

$$\begin{bmatrix} p \\ q \\ p_0 \end{bmatrix} = \begin{bmatrix} v_\alpha & v_\beta & 0 \\ -v_\beta & v_\alpha & 0 \\ 0 & 0 & v_0 \end{bmatrix} \cdot \begin{bmatrix} i_\alpha \\ i_\beta \\ i_0 \end{bmatrix} \quad (2)$$

In order to compensate reactive power, harmonics currents and current unbalance (eliminating the neutral wire current), it is necessary to compensate the power components q and p_0, and the oscillating value of p ($\tilde{p}$). The compensation reference currents in the α-β-0 coordinates are obtained from (3) and (4), which come from the inverse of (2), and where $q_x = q$, and p_x is given by (6).

$$i_{c0}^* = \frac{p_0}{v_0} \quad (3)$$

$$\begin{bmatrix} i_\alpha^* \\ i_\beta^* \end{bmatrix} = \frac{1}{v_\alpha^2 + v_\beta^2} \begin{bmatrix} v_\alpha & v_\beta \\ -v_\beta & v_\alpha \end{bmatrix} \cdot \begin{bmatrix} p_x \\ q_x \end{bmatrix} \quad (4)$$

An active power filter with a control system that presents a fast response, like the one based on the p-q theory, has basically two advantages [4]:

- It presents a good dynamic response, producing the correct values of compensating current in a short time after variations of the power system operating conditions;
- If the active filter has a fast control system, its energy storage element will suffer less to compensate the power system parameters variation.

The calculations of the p-q theory also allow a simple method to adjust and regulate the DC link voltage. For that it is necessary to calculate a regulation power (p_{reg}), which can be obtained through a proportional controller, K_p. Thus:

$$p_{reg} = K_p(v_{dc}^* - v_{dc}) \quad (5)$$

Where:

K_p – proportional controller;
v_{dc}^* – DC link reference voltage;
v_{dc} – DC link voltage.

The regulation power, besides regulating the DC link voltage, can be used to set the value of power to be injected in the electrical grid, produced by the solar photovoltaic system. The regulation power is included in the instantaneous real power value to be compensated, so that p_x is given by:

$$p_x = \tilde{p} - \bar{p}_0 + p_{reg} \tag{6}$$

The p-q theory is a theory for three-phase electrical systems. An easy way to apply this theory to single-phase systems consists in creating virtual voltages and currents for the two non-existing phases, delaying them by +120º and +240º in relation to the existing phase voltage and current.

In Fig. 1 is presented a block diagram of the proposed control system.

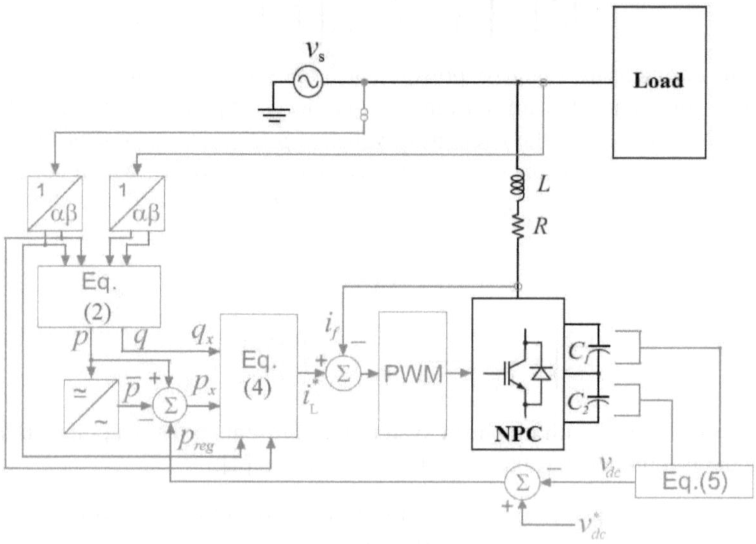

Fig. 1. Block diagram of the control system

4 Multilevel Inverters

The term multilevel was introduced by Nabae et al. [5], and their structures were developed from three-level inverters. These inverters have the ability to operate at high voltage levels using devices with low power voltages, producing low harmonic content at the output waveforms, when compared with the waveforms obtained with two-level inverters.

In the Multilevel Inverters topologies, when the number of levels increases, the rate of total harmonic distortion (THD) of the output voltage decreases. However, the number of power devices also increases, increasing the complexity and cost of the inverter.

The main topologies of Multilevel Inverters are the Neutral Point Clamped (NPC) Inverter, the Flying Capacitor Inverter and the Cascaded H-Bridge Inverter.

Neutral Point Clamped Inverter

The output voltages of this topology have a lower harmonic content, enabling the use of semiconductor devices in half the threshold voltage of the devices used in the two-level topologies.

The three-phase three-level NPC inverter has a function of transferring energy from the source at the DC side to the AC side, or acting as a rectifier, transferring energy from the AC side to the DC side. This property of reversibility, usually present in other multilevel converters, requires two circuits connected to the inverter that are able to provide and/or receive this energy.

Flying Capacitor Inverter

Although the Flying Capacitor Inverter (FCI) topology is not as common as other structures, it has some distinct advantages over the diode-clamped topology, including the absence of clamping diodes and the ability to regulate the flying-capacitor voltages through redundant state selection, even if the number of voltage levels is greater than three.

Unlike the NPC inverter topology, the Flying Capacitor Inverter has redundant switching states for synthesizing the phase voltage, so that some phase voltage vectors can be synthesized by more than one switching state [6].

Cascaded H-Bridge Inverter

This topology consists of two or more single-phase full bridge converters with their outputs connected in series at the AC side. Thus, each converter creates three different levels of voltage. The major advantages of the Cascaded H-Bridge converters are scalable power rating, modularity, and cost effectiveness. The output voltage of the Cascaded H-Bridge converter is the summation of the output voltage of the individual H-bridges. By increasing the number of series H-bridge converters, the output voltage of the converter can be increased, while the switching frequency of the individual H-bridge converter can be decreased to achieve the same output waveform quality [7].

After studying the three basic topologies of the multilevel inverters (Neutral Point Clamped - NPC, Cascaded and Flying Capacitor), the NPC topology was chosen mainly due to its characteristics of easy control, especially when compared to the Flying Capacitor topology that has a difficult voltage control of the clamping capacitors [8][9]. The Cascaded H-Bridge inverter needs separate DC sources for real power conversions, and thus its applications are somewhat limited. The Flying Capacitor Inverter control would be very complicated, and the switching frequency and switching losses would be high for real power transmission [10].

5 Simulations Results

The proposed control system, with function of active power filter using the p–q theory, is applied to a single-phase NPC inverter connected to the electrical power grid. Simulations were performed in Matlab/Simulink for different loads: linear and non-linear. The linear load is composed by one inductance (1 mH) in series with one resistance (5 Ω). The non-linear load is composed by a single-phase rectifier with C = 640 μF in parallel with R = 5 Ω at the DC side.

The simulations conditions are the following: switching frequency T_s= 4 kHz, source voltage v_S= 230 V (RMS value), source frequency f_s= 50 Hz, source inductance L_s = 1mH, link DC voltage V_{dc} = 700V, and DC link capacitors $C_1 = C_2$ = 4.7 mF.

In order to compare the behavior of the output filter currents and the balance of the DC link voltages of the NPC inverter, simulations were performed with two different PWM techniques:

- In-Phase Disposition PWM (IPD PWM);
- Phase Opposition Disposition PWM (POD PWM).

Booth techniques require each of the 2 carrier waveforms for 3-level phase waveform. In the POD PWM the carrier waveforms are all in phase above and below the zero reference value, however, there is 180 degrees phase shift between the ones above and below zero respectively.

The IPD PWM technique, generally, is widely used in multilevel inverters. This technique is similar to the previously described one, except that all carriers are in phase. The zero reference is placed in the middle of the carrier set.

The Flying Capacitor inverter presents difficulties in controlling the balancing of the voltage of the clamped capacitors and of the flying capacitor, too, for booth PWM techniques. One problem is that an excessive number of storage capacitors are

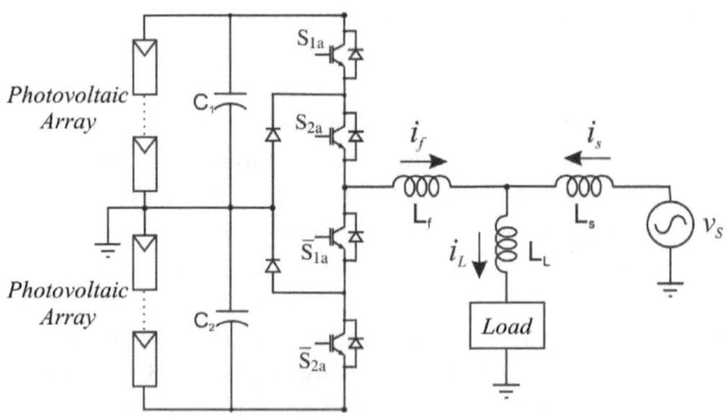

Fig. 2. Single-phase NPC inverter connecting the PV system to the electrical power grid

required when the number of converter levels is high. The Cascaded H-Bridge inverter and the NPC inverter present good results in balancing the DC link voltage. But the Cascaded inverter needs separate DC sources for real power conversions, and thus its applications are somewhat limited. Reactive power flow can be controlled more easily in the NPC inverter.

The NPC inverter presents the best behavior among the multilevel inverters. Using the IPD PWM, the NPC inverter has a lower THD grid current. Therefore, the NPC inverter with the IPD PWM was chosen to be simulated with the two loads.

In order to validate the control system for the NPC topology, the single-phase NPC leg connected to the grid was tested as an inverter. In fact, a three-phase NPC topology can be obtained using three independent single-phase inverters, like the one shown in Fig. 2, connected through the common neutral.

The grid voltage and current, the reference current, the filter current and the load current are shown in the Fig. 3 and Fig. 4, respectively for a linear and non-linear load. For booth loads the NPC inverter presents the best current control among the studied multilevel topologies. The control of the DC link presents good results for the NPC and the Cascaded inverters.

6 Experimental Results

In order to obtain results for the single-phase NPC inverter topology with active power filter function, it was developed an experimental laboratory prototype. The complete system control was executed in discrete time using the TMS320F2812 DSP from *Texas Instruments*, working with sampling and switching frequencies of 10 kHz. The conditions of the experimental results are the same of the simulation results, except that the adopted experimental grid voltage was 115 V. The two loads conditions are the same, too.

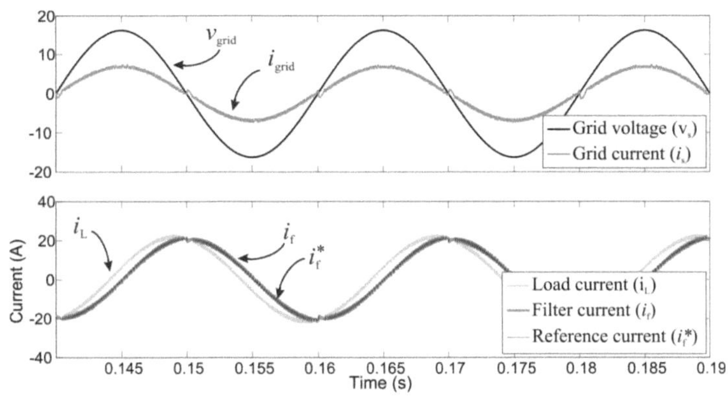

Fig. 3. NPC inverter with linear load using a PWM technique IPD

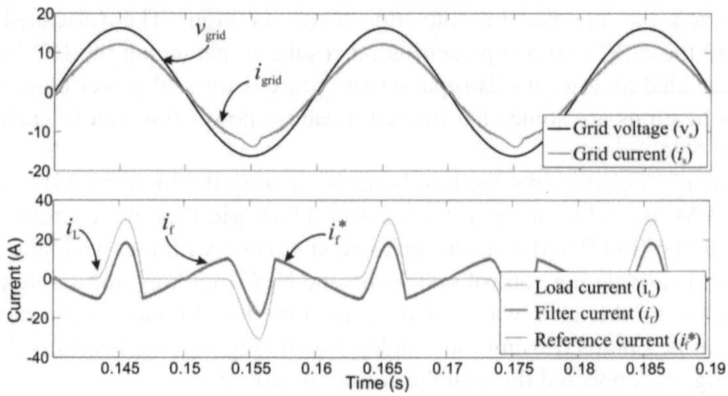

Fig. 4. NPC inverter with non-linear load using a PWM technique IPD

Fig. 5 shows the results for operation with the linear load, without and with active power filter function, in which, for this case, the inverter compensates reactive power. In Fig. 5(a) the active filter function is off, and the grid current (i_s) is delayed almost 90° in relation to the grid voltage (v_s). The amplitude of the reference current (i^*_f) is close to the amplitude of the load current (i_c), since the load current should be almost completely compensated. However, the filtering current is zero, because the active filter function is off. In Fig. 5(b) the active filter function is on, and the filtering current (i_f) follows its reference current. As result, the grid current (i_s) has its amplitude reduced, and it becomes in phase with the grid voltage.

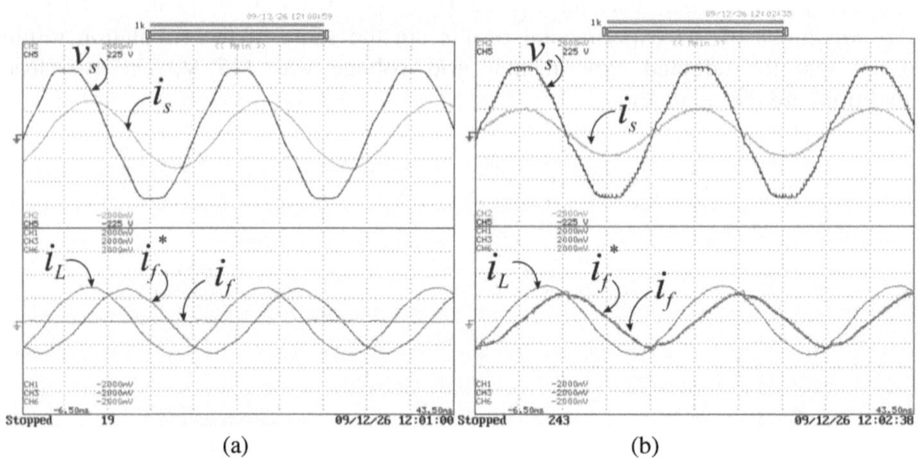

Fig. 5. Experimental results of the single-phase NPC inverter with linear load: (a)Without active filtering; (b)With active filtering

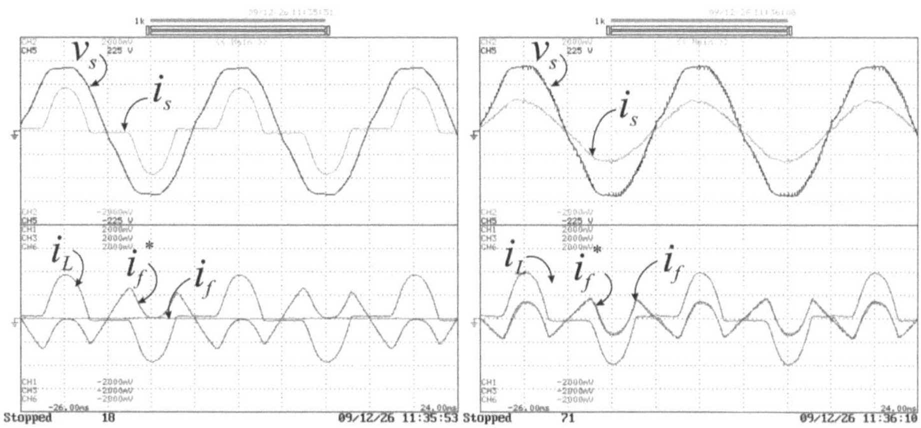

(a) (b)

Fig. 6. Experimental results of the single-phase NPC inverter with non-linear load: (a)Without active filtering; (b)With active filtering

Fig. 6 shows the experimental results for operation with the non-linear load (single-phase rectifier with RC parallel load), without and with active power filter function, in which, for this case, the inverter compensates mainly harmonics currents. Fig. 6(a) shows the waveforms when the active filter function off. The grid current (i_s) is distorted but almost in phase with the grid voltage (v_s). In Fig. 6(b) the active filter function is on, the filtering current (i_f) follows the reference current (i^*_f), compensating the harmonics currents, and the grid current becomes sinusoidal.

7 Conclusions

This paper presents a single-phase multilevel NPC (Neutral Point Clamped) inverter with active filtering function, to be used to interface photovoltaic systems with the electrical power grid, and at the same time, to eliminate harmonics currents and to compensate reactive power, operating as shunt active power filter.

The control system is based on the p-q theory, proposed by Akagi, adapted to work in single-phase electrical systems. This control theory permits the compensation of harmonics currents and reactive power, and the setting of the power produced by the photovoltaic system, to be injected by the inverter in the electrical power grid.

Three multilevel inverters topologies are presented, and the choice of the NPC topology is explained.

The single-phase multilevel NPC inverter with active filtering function is tested, and are presented both simulation and experimental results, obtained with a laboratory prototype. Simulations and experiments were performed with two types of load, corresponding to different active filtering conditions: linear load, with reactive power compensation; non-linear load, with harmonics currents compensation.

The studied multilevel inverter presented satisfactory performance for the cases here evaluated, showing that the combined operation of active filtering and interfacing of a photovoltaic system with the electrical power grid can be an interesting possibility.

References

1. Oliveira, K.C.: Avaliação da Conversão da Energia Fotovoltaica em Sistemas Isolados. Master Thesis, UFPE (March 2007)
2. Gubía, E., Sanchis, P., Ursúa, A., López, J., Marroyo, L.: Ground currents in single-phase transformerless photovoltaic systems. Progress in Photovoltaics: Research and Applications 15(7), 629–650 (2007)
3. Akagi, H., Kanazawa, Y., Nabae, A.: Instantaneous Reactive Power Compensators Comprising Switching Devices without Energy Storage Components. IEEE Transactions on Industry Applications (3), 625–630 (1984)
4. Afonso, J.L., Freitas, M.J.S., Martins, J.S.: p-q Theory power components calculations. In: ISIE 2003 - IEEE Int. Symposium Ind. Electronics, Rio de Janeiro, Brazil, June 9-11 (2003)
5. Nabae, A., Takahashi, I., Akagi, H.: A new neutral-point-clamped PWM inverter. IEEE Transactions on Industry Applications (5), 518–523 (1981)
6. Rech, C., Gründling, H.A., Hey, H.L., Pinheiro, H., Pinheiro, J.R.: Uma metologia de projeto generalizada para inversores multiníveis híbridos. Revista Controle & Automação 15(2), 190–201 (2004)
7. Anuradha, K., Muni, B.P., Kumar, A.D.R.: Cascaded Multilevel Converter based DSTATCOM using p-q theory with DC link voltage balancing. In: International Conference on Power and Energy Systems, ICPS, December 22-24, pp. 1–6 (2011)
8. Oliveira, K.C., Cavalcanti, M.C., Afonso, J.L., Farias, A.M., Neves, F.A.S.: Transformerless photovoltaic systems using neutral point clamped multilevel inverters. In: IEEE Int. Symposium on Industrial Electronics (ISIE), July 4-7, pp. 1131–1136 (2010)
9. Cavalcanti, M.C., Farias, A.M., Oliveira, K.C., Neves, F.A.S., Afonso, J.L.: Eliminating Leakage Currents in Neutral Point Clamped Inverters for Photovoltaic Systems. IEEE Trans. Industrial Electronics 59(1), 435–443 (2012), doi:10.1109/TIE.2011.2138671, ISSN: 0278-0046
10. Lai, J.-S., Peng, F.Z.: Multilevel converters-a new breed of power converters. In: Thirtieth IAS Annual Meeting Industry Applications Conference (IAS), Conference Record of the 1995 IEEE, October 8-12, vol. 3, pp. 2348–2356 (1995)

Analysis of the Features of a UPQC to Improve Power Quality in Smart Grids

J.G. Pinto, Carlos Couto, and João L. Afonso

Centro Algoritmi - University of Minho – Guimarães, Portugal
{gabriel.pinto,joao.l.afonso}@algoritmi.uminho.pt,
ccouto@dei.uminho.pt

Abstract. An UPQC (Unified Power Quality Conditioner) is an equipment composed by two active conditioners operating in a combined way. One of the active conditioners is connected in series with the electrical power system, allowing the compensation of problems in the system voltages. The other active conditioner is connected in parallel with the electrical system, and allows the compensation of current harmonics, current unbalances and power factor. In three-phase four-wire systems the parallel connected active conditioner also compensates the zero sequence current components, eliminating the neutral wire current. The UPQC operates in an automatic way, adjusting itself dynamically to the variations of the load and of the electrical system, keeping high levels of power quality in the voltages delivered to the load. At the same time, it only consumes from the electrical system the active power necessary to the load operation, in a balanced way through the three phases, and with sinusoidal currents. The compensation capabilities of the UPQC can be very useful to ensure high levels of power quality in the future Smart Grids, which are not characterized as a single technology or device, but rather as a vision of a distributed electrical system, supported by reference technologies, as Power Electronics Devices, Renewable Energy Resources, Energy Storage Systems (ESS), Advanced Metering Infrastructures (AMI), and Information and Communication Technologies (ICT).

Keywords: UPQC - Unified Power Quality Conditioner, Active Conditioners, Power Quality, Smart Grids, p-q Theory.

1 Introduction

The effective research in the area of semiconductor technologies in recent decades allowed the refinement and development of faster power semiconductors with lower operating losses [1-3]. The use of these electronic components enabled the optimization of production processes in terms of performance, controllability and cost, allowing the execution of tasks that were previously impossible. Due to the massive production, the price of power semiconductors decreased and the proliferation of these components has been so high that they are currently present in almost all electrical equipment. However, this technological evolution not only brought advantages. The equipment based on

L.M. Camarinha-Matos, S. Tomic, and P. Graça (Eds.): DoCEIS 2013, IFIP AICT 394, pp. 299–306, 2013.

power semiconductors is responsible for causing many problems in the electrical power systems. The IEEE 1159 Standard classifies various electromagnetic phenomena in power systems that cause malfunction of equipment which is sensitive to disturbances in the power supply voltage, namely: oscillations, sags, swells, interruptions, undervoltages, overvoltages, harmonics, interharmonics, notches, noise, flicker, and frequency variation [4-8]. Some of these problems are caused by nonlinear loads and can be mitigated by specific equipment. Passive filters have been widely used as a solution to mitigate harmonic problems, but they present several limitations, namely: they only filter the frequencies for which they were previously tuned, and their operation cannot be limited to a certain load or installation. Furthermore, the interaction between passive filters and other loads may result in resonances with unpredictable results [9].In order to improve the mitigation achieved with passive filters and to minimize the impact of the power quality problems, in 1976 Gyugi and Strycula introduced the concepts of Active Filters [10]. Active filters, differently from passive filters, have the capability to dynamically adjust to the conditions of the electrical system, which consists in a great advantage. Starting from this new concept, in the last years various solutions based in power electronics to compensate power quality problems were investigated [11-18]. These types of equipment are usually designated in the literature as Active Conditioners. Some of these active conditioners are connected in parallel with the electrical grid and are designated as Shunt Active Conditioners. Other conditioners are connected in series with the electrical grid and are designated as Series Active Conditioners. The conjunct operation of shunt and series active conditioners originated a new concept of equipment called Unified Power Quality Conditioner (UPQC).

2 Contribution to Internet of Things

Studies conducted by several international organizations show that the economic losses caused by power quality problems are, nowadays, very high, and defends that little investment in equipment and procedures to increase power quality can result in an effective reduction in these losses [19]. According to EPRI (Electric Power Research Institute) problems related to power quality and supply interruptions cost to the U.S. economy more than 100 billion euros per year[20]. The European COPPER Institute - Leonardo Energy Initiative, estimates that the cost of electric quality problems in Europe is superior of 150 billion euros per year [21], [22]. Under this scenario, the development of equipment to improve power quality, such as the UPQC presented here, is a matter of utmost importance to ensure a favorable environment for the proper functioning of businesses, contributing to higher productivity and for the reduction of economic losses derived from power quality problems. In addition to the ability to mitigate power quality problems, the UPQC performs the measurement of several electrical signals that may be very useful for other devices connected to the Smart Grid. These measurements can be made available in an open way, through a bidirectional interface contributing to the Internet of Things.

3 UPQC Working Principle

The UPQC is basically constituted by two conditioners that share the same DC Link. One is connected in series and the other is connected in parallel with the electrical system. The shunt (parallel connected)active conditioner works as a controlled current source and drains from the electrical system the undesired current components produced by the load. By the action of this active conditioner the phase currents upstream of the installation point (source currents) become sinusoidal, balanced, and in phase with the fundamental positive sequence component of the electrical system voltages. The series active conditioner works as a voltage source connected in series with the electrical system, allowing the compensating of voltage harmonics, sags, swells, and flicker (a cyclic variation of light intensity of lamps caused by fluctuation of the system voltage).The ability of the UPQC to transfer active power between the two active conditioners also allows the compensation of long term undervoltages and overvoltages, delivering to the load voltages with the desired amplitude, and with high levels of power quality. Fig. 1 shows a block diagram that explains in a simplified way the working principle of the UPQC. As it is represented in this figure, the voltage (v_F) and the current (i_F) produced by the UPQC make sinusoidal the current upstream and the voltage downstream of the connection point.

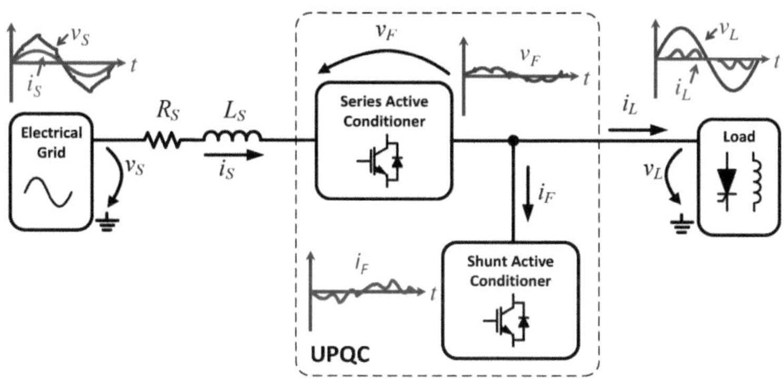

Fig. 1. UPQC (Unified Power Quality Conditioner) block diagram

4 UPQC Simulation

In order to validate the topology and the control algorithms it was developed a simulation model of the UPQC using the PSIM 9.0 software tool. Fig. 2 presents the schematic of the UPQC power circuit. In this figure it is possible two see the power inverters of the two active conditioners (sharing the same DC link capacitors C_1 and C_2) and the coupling inductors (L_1 to L_6). For the proper operation of the UPQC, and to prevent short-circuits between the phases, it is necessary to galvanic isolate one of the active conditioners from the electrical system. In the presented UPQC topology the galvanic isolation is done by three isolating transformers used to connect the series active conditioner with the electrical system.

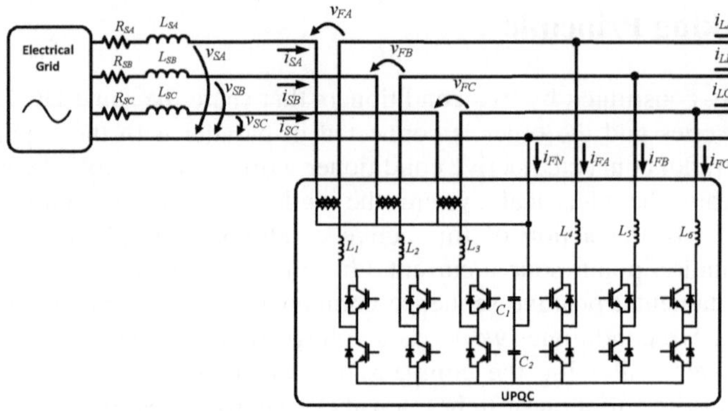

Fig. 2. Schematic of the UPQC power circuit

Several algorithms were developed to control the UPQC. The first algorithm consists in a digital Phase Locked Loop (PLL) and is responsible for the synchronization of the controller with the positive sequence of the fundamental component of the source voltages. The PLL receives the three source voltages and returns two sine waves with unitary amplitude that are used as synchronizing signals. These synchronizing signals are used to calculate the compensation currents by applying the concepts of the p-q theory, and also to generate the compensation voltages of the series active conditioner, by calculating the difference between the source voltages and the ideal desired load voltages.

Fig. 3 shows the voltage and current waveforms in the source, in the UPQC and in the load. Fig. 3 (a) shows the source voltages (v_{SA}, v_{SB}, v_{SC}), which are distorted, unbalanced, and with low amplitude. Fig. 3 (b) shows the voltages produced by the series active conditioner (v_{FA}, v_{FB}, v_{FC}) to compensate the source voltages. As it can be seen in the Fig. 3 (c), the load voltages (v_{LA}, v_{LB}, v_{LC}) present sinusoidal waveforms with the same aptitude and with a constant phase shift of 120º. This means that, even with the voltages of the electrical system distorted and unbalanced, in a system with this UPQC in operation, the loads are always fed by a three phase balanced voltage system with constant and nominal amplitude. Fig. 3 (d) shows the waveforms of the phase currents consumed by the loads (i_{LA}, i_{LB}, i_{LC}). It is possible to see that these currents present high distortions and that the current in the phase *B* is significantly greater than the currents in the other two phases, which have similar amplitudes. With the currents produced by the shunt active conditioner (i_{FA}, i_{FB}, i_{FC}),which are presented in Fig. 3 (e), the source phase currents (i_{SA}, i_{SB}, i_{SC}) become sinusoidal, balanced and in phase with the positive sequence of the fundamental component of the system voltages, as can be seen in the Fig. 3 (f). By the analysis of the Fig. 3 (f) it is possible to see that the shunt active conditioner produces a current (i_{FN}), exactly equal to the current in the neutral wire of the load (i_{LN}), which eliminates the current in the neutral wire upstream of the connection point (i_{SN}).

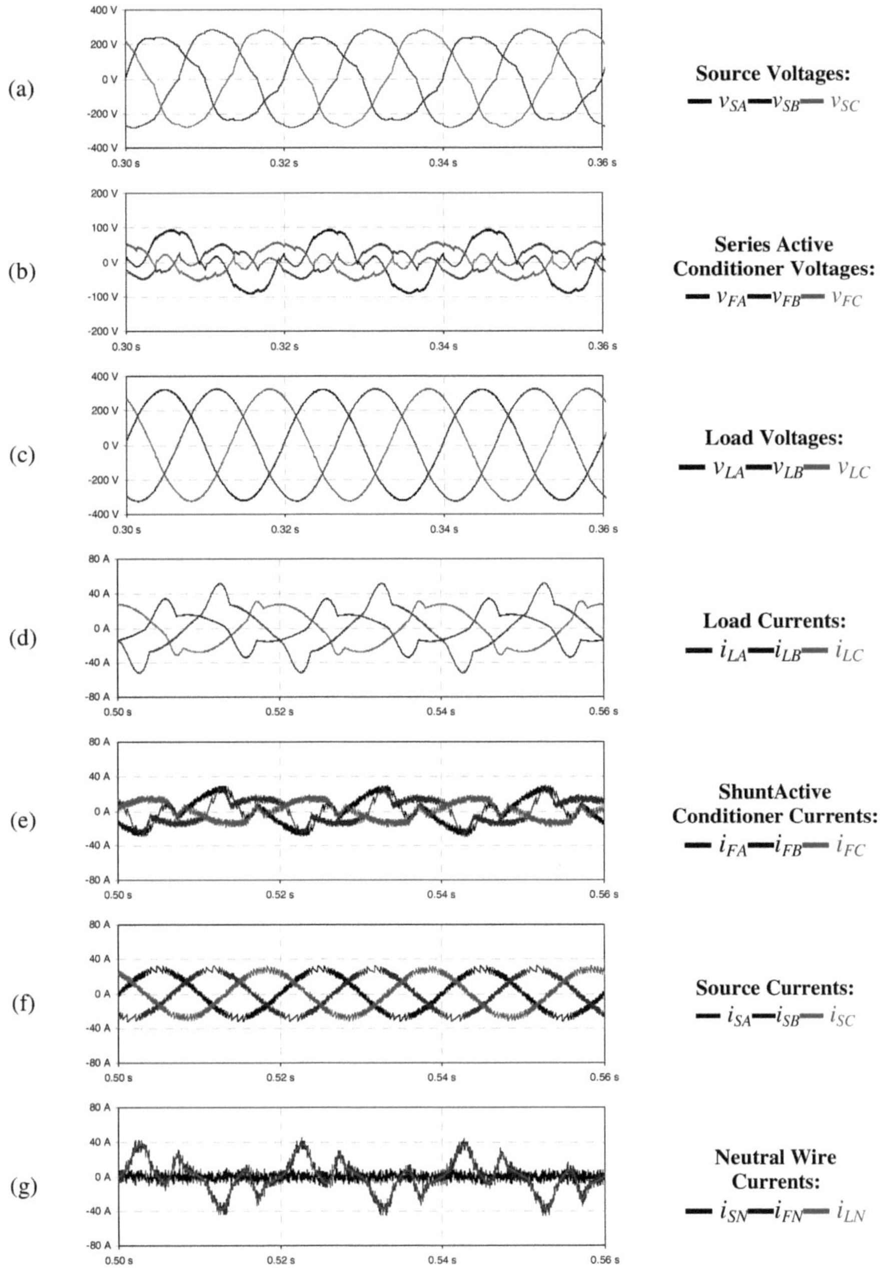

Fig. 3. Voltage and current waveforms in the source, in the UPQC and in the load: (a) Source voltages (v_{SA}, v_{SB}, v_{SC}); (b) Series active conditioner voltages (v_{FA}, v_{FB}, v_{FC}); (c) Load voltages (v_{LA}, v_{LB}, v_{LC});(d) Load currents (i_{LA}, i_{LB}, i_{LC}); (e) Shunt active conditioner currents (i_{FA}, i_{FB}, i_{FC}); (f) Source currents (i_{SA}, i_{SB}, i_{SC}); (g) Neutral wire currents (i_{SN}, i_{FN}, i_{LN})

5 UPQC Experimental Results

In order to obtain experimental results, it was developed a laboratory prototype of the UPQC with the topology presented in Fig. 2. Fig. 4 shows some experimental results obtained with the UPQC laboratory prototype. Fig. 4 (a) shows the waveforms of the source voltages (v_{SA}, v_{SB}, v_{SC}) and load currents (i_{LA}, i_{LB}, i_{LC}, i_{LN}). Fig. 4 (b) shows the waveforms of the compensated load voltages (v_{LA}, v_{LB}, v_{LC}) and source currents (i_{SA}, i_{SB}, i_{SC}, i_{SN}).The source voltages are slightly distorted and with amplitude lower than desired. The load currents are highly distorted and unbalanced. The neutral wire current is high, not only due to the load unbalancing, but also to the high zero sequence harmonic contents of the load currents.

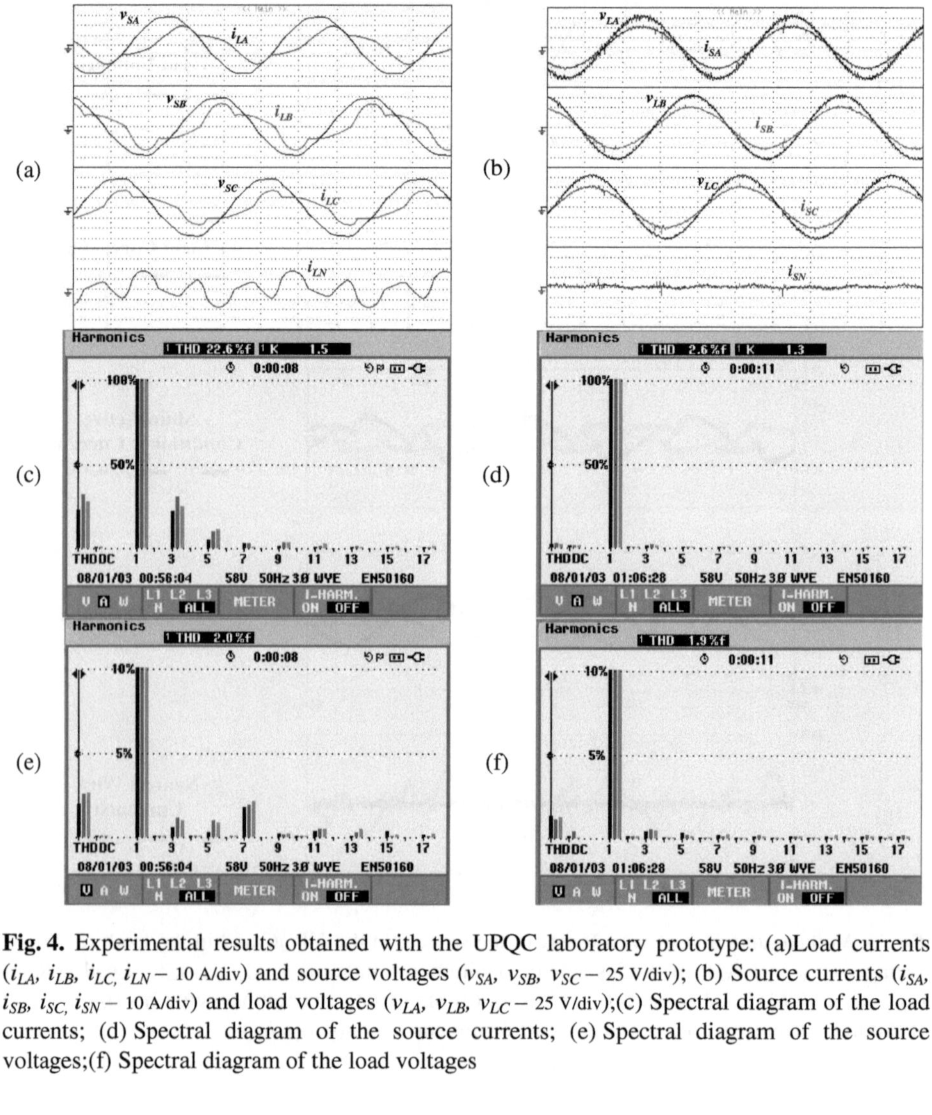

Fig. 4. Experimental results obtained with the UPQC laboratory prototype: (a)Load currents (i_{LA}, i_{LB}, i_{LC}, i_{LN} – 10 A/div) and source voltages (v_{SA}, v_{SB}, v_{SC} – 25 V/div); (b) Source currents (i_{SA}, i_{SB}, i_{SC}, i_{SN} – 10 A/div) and load voltages (v_{LA}, v_{LB}, v_{LC} – 25 V/div);(c) Spectral diagram of the load currents; (d) Spectral diagram of the source currents; (e) Spectral diagram of the source voltages;(f) Spectral diagram of the load voltages

With the UPQC in operation the load voltages became sinusoidal and with the desired amplitude. The source currents became sinusoidal, balanced and in phase with the positive sequence of the fundamental component of the source voltages. The neutral wire current is practically eliminated. In terms of spectral analyses, as it can be seen in the Fig. 4 (c), the load currents present harmonics of various orders. In all the three phases the 3rd order harmonic is the more relevant, however the 5th is also present with a significant amplitude, especially in phase *C*. The shunt active conditioner of the UPQC compensates practically all the harmonic content, resulting in three source currents composed only by the fundamental, as it can be seen in Fig. 4 (d). As it can be seen in Fig. 4 (e), the source voltages present some harmonic contents, where the 7th order harmonic is the most significant. The series active conditioner compensates almost all of the harmonics contents, as shown in Fig. 4 (f).

6 Conclusions

In this paper was presented an UPQC and it was demonstrated its capacity to compensate various problems related to the currents and voltages in the electrical power system. The topology and the control algorithms of the presented UPQC were validated trough computer simulations, and some simulation results are shown. The high-level validation of the introduced concepts and simulation model was guaranteed by experimental results, obtained with a prototype of the UPQC, which was developed in this PhD work. The unique features of the UPQC to compensate the load currents and the system voltages, and its capability to dynamically and instantaneously adapt to changes in the loads or in the power system, makes it a very interesting equipment to ensure high levels of power quality in future Smart Grids.

Acknowledgments. This work is financed by FEDER Funds, through the Operational Program for Competitiveness Factors – COMPETE, and by National Funds through the Foundation for Science and Technology of Portugal, under projects: PTDC/EEA-EEL/104569/2008 and FCOMP-01-0124-FEDER-022674.

References

1. Adler, M.S., Owyang, K.W., Baliga, B.J., Kokosa, R.A.: The evolution of power device technology. IEEE Transactions on Electron Devices 31(11), 1570–1591 (1984)
2. Baliga, B.J.: Trends in power semiconductor devices. IEEE Transactions on Electron Devices 43(10), 1717–1731 (1996)
3. Hower, P.L.: Power semiconductor devices: an overview. Proceedings of the IEEE 76(4), 335–342 (1988)
4. Wagner, V.E., et al.: Effects of harmonics on equipment. IEEE Transactions on Power Delivery 8(2), 672–680 (1993)
5. Fuchs, E.F., Roesler, D.J., Kovacs, K.P.: Sensitivity of Electrical Appliances to Harmonics and Fractional Harmonics of the Power System's Voltage. Part II: Television Sets, Induction Watthour Meters and Universal Machines. IEEE Transactions on Power Delivery 2(2), 445–453 (1987)

6. Fuchs, E.F., Roesler, D.J., Alashhab, F.S.: Sensitivity of Electrical Appliances to Harmonics and Fractional Harmonics of the Power SYSTEM's Voltage. Part I: Transformers and Induction Machines. IEEE Transactions on Power Delivery 2(2), 437–444 (1987)
7. Bachry, A., Styczynski, Z.A.: An Analysis of Distribution System Power Quality Problems Resulting from Load Unbalance and Harmonics. In: IEEE PES, September 7-12, vol. 2, pp. 763–766 (2003)
8. Fuchs, E., Masoum, M.: Power Quality in Power Systems and Electrical Machines. Elsevier (2008)
9. Afonso, J.L., Couto, C., Martins, J.: Active Filters with Control Based on the p-q Theory. IEEE Industrial Electronics Society Newsletter 47, 5–10 (2000)
10. Gyugyi, L., Strycula, E.C.: Active AC Power Filters. In: Proc. IEEE Ind. Appl. Ann. Meeting, vol. 19-C, pp. 529–535 (1976)
11. Akagi, H.: New trends in active filters for improving power quality. In: Proceedings of the 1996 International Conference on Power Electronics, Drives and Energy Systems for Industrial Growth, January 8-11, vol. 11, pp. 417–425 (1996)
12. Fujita, H., Akagi, H.: The unified power quality conditioner: The integration of series active filters and shunt active filters. In: 27th Annual IEEE Power Electronics Specialists Conference, June 23-27, vol. 1, pp. 494–501 (1996)
13. Pinto, J.G., Neves, P., Gonçalves, D., Afonso, J.L.: Field Results on Developed Three-Phase Four-Wire Shunt Active Power Filters. In: IECON 2009, Porto, Portugal, November 3-5 (2009)
14. Watanabe, E., Afonso, J., Pinto, J.G., Monteiro, L., Aredes, M., Akagi, H.: Instantaneous p-q Power Theory for Control of Compensators in Micro-Grids. In: IEEE ISNCC, Łagów, Poland, June 15-18 (2010)
15. Pinto, J.G., Exposto, B., Monteiro, V., Monteiro, L., Couto, C., Afonso, J.L.: Comparison of Current-Source and Voltage-Source Shunt Active Power Filters for Harmonic Compensation and Reactive Power Control. In: IECON 2012, Québec, Canada, October 25-28 (2012)
16. Pinto, J.G., Carneiro, H., Exposto, B., Couto, C., Afonso, J.L.: Transformerless Series Active Power Filter to Compensate Voltage Disturbances. In: Proceedings EPE 2011, Birmingham, United Kingdom, August 30-September 1, pp. 1–6 (2011)
17. Carneiro, H., Pinto, J.G., Afonso, J.L.: Single-Phase Series Active Conditioner for the Compensation of Voltage Harmonics, Sags, Swell and Flicker. In: ISIE 2011, Gdansk, Poland, June 27-30, pp. 384–389 (2011) ISBN: 978-1-4244-9310-4
18. Afonso, J.L., Pinto, J.G., Gonçalves, H.: Active Power Conditioners to Mitigate Power Quality Problems in Industrial Facilities. In: Zobaa, A. (ed.) Power Quality. InTech (2012) ISBN 980-953-307-532-2
19. Chapman, D.: The Cost of Poor Power Quality. Leonardo Power Quality Initiative - European Copper Institute (2001)
20. Lineweber, D., McNulty, S.: The Cost of Power Disturbances to Industrial & Digital Economy Companies. EPRI- Electric Power Research Institute (2001)
21. Poor Power Quality Costs European Business More than € 150 Billion a Year. Leonardo Power Quality Initiative - European Copper Institute (2008)
22. Manson, J., Targosz, R.: European Power Quality Survey Report. Leonardo Power Quality Initiative - European Copper Institute (November 2008)

Intelligent Energy Management System for Residential and Community Applications

Vicente Botón-Fernández[1], Máximo Pérez Romero[2], Adolfo Lozano-Tello[1], and Enrique Romero-Cadaval[2]

[1] Quercus Software Engineering Group
University of Extremadura, Escuela Politécnica, Campus Universitario, s/n
10071 Cáceres, Spain
[2] Power Electrical and Electronics R+D group
University of Extremadura, Avda. De Elvas, s/n
06006 Badajoz, Spain
{vboton,mperez,alozano,eromero}@unex.es

Abstract. This paper presents a Smart Storage System able to manage the energy and the smart home devices of a house for optimizing the local consumption of energy, even if there is not renewable generation. The proposed system is composed by two main systems. On the one hand, the Local Energy Management units that will be located in the houses, which are able to maintain the power consumption under a maximum reference value and to switch on/off the devices in the house by using domotic protocols. On the other hand, the Central Energy Management and Intelligent System, that receives operation data from each local unit, analyzes them using behavioral and optimization algorithms and determines the best way in which each local unit has to operate, communicating the operation references back to them.

Keywords: Smart Energy Storage, Intelligent Systems, Pattern Recognition, Ontologies, SWRL.

1 Introduction

The availability of having energy storage systems (ESS) [1, 2] is a key factor for establishing new operation strategies in electric distribution grids related with the hot topic of Smart Grids. This ESS will do easy to integrate distributed generation systems based in renewable energies placed in houses, residential and commercial building and also will contribute to have some control on the load profile, and so achieving a better overall performance and utilization of the distribution grid.

Although this kind of systems has been widely used associated to distributed generation plants, mainly with photovoltaic installations [3, 4], they could be used without any energy generation systems with the aim of smoothing the load curve or even control the consumption of energy depending on its price-by-hour [5].

An important feature that these environments need to possess is the ability to adapt themselves to the residents' consumption habits and have the versatility to make

L.M. Camarinha-Matos, S. Tomic, and P. Graça (Eds.): DoCEIS 2013, IFIP AICT 394, pp. 307–314, 2013.

decisions in a variety of situations. In this sense, finding consumption patterns in a sequence of events in order to predict future peaks on the load curve, can lead us to the energy optimization we are looking for. Therefore, the system will be able to recognize power consumption patterns and determine the best way to smooth future peaks. Obviously, discovering these patterns requires a previous task of learning. In a smart grid environment, learning means that the system has to gain knowledge about the use of energy and common behavior of the electric distribution grids in an unobtrusive and transparent way.

Using ontologies [6] to classify the types of appliances and sensors and their functionality can be an appropriate way to understand the load profile of a house or residential building. Ontologies and SWRL (Semantic Web Rule Language) rules provide a precise definition of smart grid taxonomy and are reusable, so other systems can get them to classify their own components and to build rules that will allow inferring new information.

In this paper an Intelligent Energy Management System (INTELEM) is presented. This system is composed by Local Energy Management Units (LEMU) and a central system that receives data from a group of LEMUs by conventional connections. This central system analyses that information using behavioral algorithms in order to decide their actions and also to be communicated with the distribution grid operator, so that it can manage efficiently the energy. In this way, the system becomes a Central Energy Management and Intelligent System (CEMIS).

2 Contribution to Internet of Things

This work merges two areas: energy storage and Information and Communications Technology (ICT), providing an autonomous power consumption data capture and event transfer through TCP/IP. A central server analyzes the information transmitted over the network using smart algorithms to optimize the load curves in residential or community environments.

3 Existent Work about Intelligent Energy Management Systems

Nowadays, the development of data mining techniques to recognize habitual behavior and provide feedback in context constitutes a key factor to empower individuals to take control over residential electricity consumption. There are several factors that drive household electricity consumption behavior, such as energy-related attitudes, socio-demographic factors, building characteristics, energy prices, etc. van Raaij and Verhallen [7] recognize this problem and suggest that habits can become alternative predictors of electricity consumption, because routines or habits may resist the cognitive and financial drive and still prevail over rational alternatives.

In this sense, Joana M. Abreu et al. [8] propose a methodology which demonstrates that it is possible to use pattern recognition methodologies to identify habitual electricity consumption behavior given the intrinsic characteristics of the residents.

The present work aims to achieve similar objectives but with a different approach. It presents an intelligent system that recognizes electricity consumption patterns to automate the grid behavior in an efficient way, providing electricity supply when needed and saving as much energy as possible. The algorithms are based on the analysis of temporal energy consumption data and the use of the building resources. In spite of the limit literature about ontologies and production rules in smart grid systems, this work uses both of them to automate the behavior of the building.

4 INTELEM

The Intelligent Energy Management System presented in this paper, is built upon an existing project called IntelliDomo [6]. It is a smart system able to control the devices of a home automation system automatically and in real time using SWRL rules. Its main feature is the ability of reasoning and responding in view of the continuous changes in domotic environments.

Each event captured from sensors and actuators is recorded. To manage all this knowledge about the environment, this system uses an ontology called Domo OWL, whose concepts have been modeled to be in sync with the physic devices that constitute a home automation environment, so that it can be stored its outstanding values and properties. Furthermore, the ontology works together with a set of established SWRL rules. The users can manage these rules or create new ones through IntelliDomoRules tool, automating the system behavior as well.

There also exists a domotic database (QDS_DB) where IntelliDomo updates the state values of the physic devices in real time. This way, IntelliDomo is in charge of transmitting the changes instantly from physical (devices) to logical (database) levels and vice versa, maintaining the data integrity.

Finally, there is a learning module which incorporates algorithms in order to acquire users' habits and automatically generate production rules for behavior patterns to anticipate the user's periodic actions. As a result, the system can adapt to changes in the discovered patterns based on the user implicit and explicit feedback.

INTELEM uses some of the above modules to achieve its objectives. Its architecture is presented in Fig. 1. As it can be seen, the CEMIS can take control of

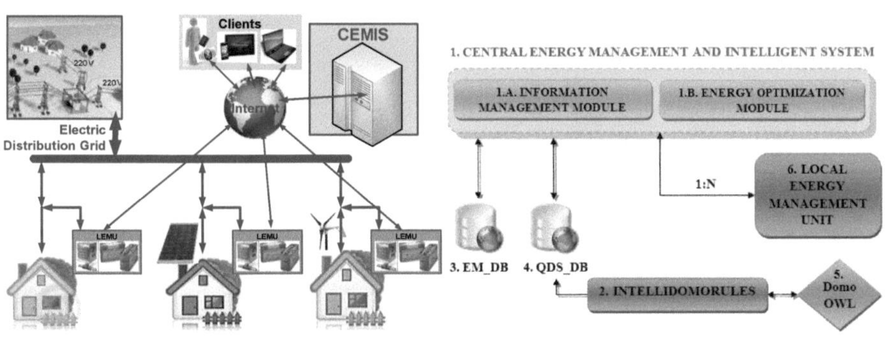

(a) (b)

Fig. 1. Main components of the INTELEM architecture (a) and their relations (b)

several (n) LEMUs and is composed by two main modules: the Information Management Module (IMM) and the Energy Optimization Module (EOM). It's important to remark that these LEMUs are distributed over different houses or residential buildings whereas the CEMIS is located at a central server, using a conventional connection via TCP/IP to exchange information with all these LEMUs.

On the one hand, the CEMIS receives data (energy consumption, battery state, etc) from a group of LEMUs and stores it into a central database called EM_DB. On the other hand, the CEMIS analyzes periodically the data logs with different behavioral algorithms in order to recognize energy consumption patterns and use this information to generate production rules which can optimize the use of the grid and save energy. These rules are stored into the QDS_DB database and managed through the IntelliDomoRules tool. Finally, once the rules are fired, the corresponding changes are transmitted, via control signals, to the affected LEMUs.

In the following sections we're going to check in depth each of these modules, with special emphasis on the CEMIS structure.

4.1 Local Energy Management Unit

This unit is designed to be placed inside houses or residential buildings with the aim of controlling their energy consumption, and it corresponds to the general scheme shown in Fig.2, having the objectives described below.

In the first place the LEMU is able to limit the power consumption peaks, which can take place due to start-up transients or high power devices operating during short-time intervals, by supplying extra energy from the ESS. As a result, the user could contract with the electric power company to reduce one or two steps the level of contracted power that he would normally have without this device. As this contracted power is usually included as a fixed cost in the electricity invoice, the user could have direct economic benefits.

In the second place the LEMU can decouple the consumed energy from the energy demanded to the grid. As far as time-of-use (TOU) rates are concerned, it would be possible to charge the ESS from the grid during the off-peak hours (cheaper prices)

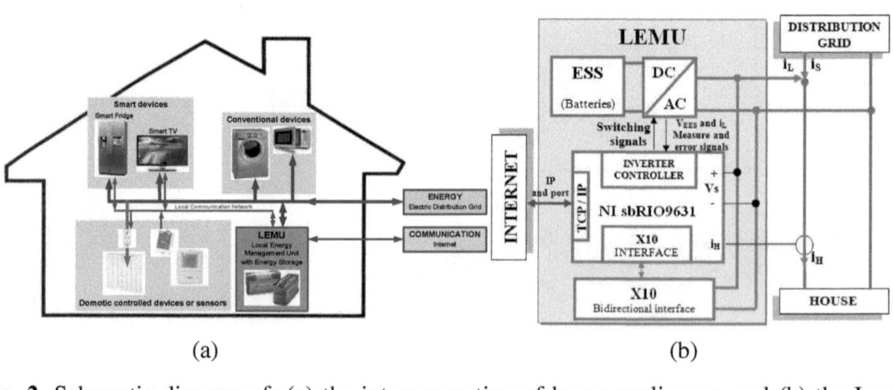

(a) (b)

Fig. 2. Schematic diagram of: (a) the interconnection of home appliances, and (b) the Local Energy Management Unit (with Energy Storage System)

and then supply its energy to the devices during the peak hours (higher prices). In our ongoing work, we plan to let the LEMU work as a constant power sink, by using the ESS to put the difference of energy between the energy demanded from the grid and current level of energy consumed by the devices into the house or building.

Otherwise, the LEMU can switch on or switch off the devices by using the X10 protocol to maintain the power consumption under an established limit, and it can also receive the X10 events sent by X10 sensors such as motion sensors.

Finally, the LEMU has been designed to be integrated into an energy management network, and it has the possibility of being connected to a central system in order to send local data, such as electrical magnitudes (like smart meters do), the state of charge of its ESS or X10 events. The LEMU can also receive operation references from the central system such as power consumption limits, or some commands to switch on/offX10 actuators.

4.2 Central Energy Management and Intelligent System

This central system is in charge of controlling the group of LEMUs which are distributed over the energy management network. It's designed to be hosted in a remote server and its main objective is to efficiently manage the energy consumption of a house or building through data mining algorithms.

Basically, the LEMUs gather power consumption data through the sensors that are connected to the grid, sending the information to the CEMIS. Once the CEMIS receives that information, it can analyze it and then use it to send back control signals to the LEMUs to control their behavior and save energy. As it was mentioned before, the CEMIS has two different modules to achieve those objectives: the IMM and the EOM. Next, we describe the operation of these modules and the relations between them.

4.2.1 Information Management Module

This module is continuously reading and storing into a database the information provided by the LEMUs. The Fig. 3 describes the general operation of this module. The data sent by the LEMUs can be values of electrical magnitudes, X10 events or the state of charge of an ESS. The electrical magnitudes are often sent every 5 seconds, whereas the other values are sent when a certain event takes place (e.g. an X10 sensor detects motion).In order to better identify each type of information, it has been defined three different message format, all of them with the following structure: header field plus data field.

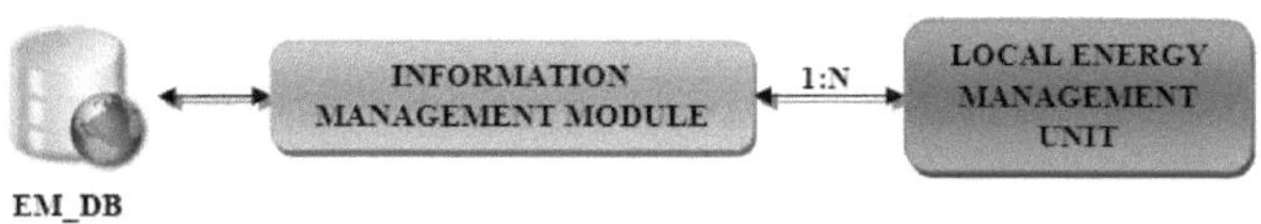

Fig. 3. Information Management Module schema

The X10 message format is as follows: header = "X10", data = {house code, unit code, value}. The data field uses one character to represent the house code (A-P), two characters to describe the unit code (1-16) and three more characters to identify the value. The ESS message format has the following fields: header = "BAT", data = state of charge (%).Finally, the power consumption message format uses the data field to store the value of electrical magnitudes (kW) and its header is "CON".

Otherwise, these messages are transmitted through a communication channel(or socket) between each LEMU and the CEMIS. In this case, it's the LEMU who takes the client role and the one who initiates the communication session with the CEMIS, which is waiting for incoming requests. Once the socket is established, the LEMU will send data packets continuously, and then the IMM will process them and save the information into the EM_DB. This way, the database keeps a log of everything that happens inside the houses or residential buildings. Obviously, after each packet reception the IMM will response with a message indicating if the data was correctly saved into the database or not. If there is any error, the packet will be send once again, but there is a limit number of attempts.

4.2.2 Energy Optimization Module

The main objective of this module, which is still in development, is to coordinate a group of LEMUs, optimizing the use of the grid and saving energy. To that end, this module will incorporate data mining algorithms with the purpose of controlling the ESS charging time. These algorithms will aim to discover the resident power consumption patterns and try to have the ESS prepared (full charge) for those moments where the consumption is considerably high (peak periods). With this information, the system could predict future demands on the grid and manage them with the best strategy. Fig. 4 sums up the operation of this module.

First of all, the EOM reads the log entries from the EM_DB database. These entries are composed by three fields: energy consumption value, date and time of the event and IP of the associated LEMU. Then, the EOM selects a specific LEMU and sends the corresponding entries, based on the LEMU's IP, to the data mining algorithms.

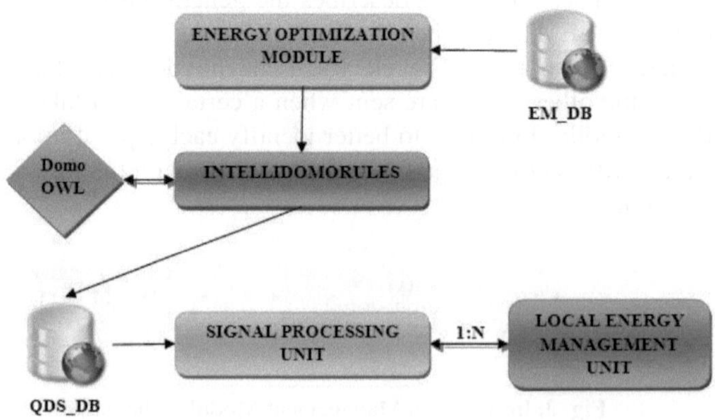

Fig. 4. General view of the *Energy Optimization Module*

Then, the algorithms will analyze this data set and look for frequent peaks in the load curves (see Fig. 5). Once they recognize the patterns, they will design a strategy to coordinate the behavior of the LEMU and optimize the use of the resources. The way to carry out this strategy involves generating SWRL rules that automate the corresponding actions. Therefore, the EOM will send the results of the data mining process to the IntelliDomoRules tool. This one can build the rules and save them into the ontology (Domo OWL) and the database (QDS_DB).

The Signal Processing Unit (SPU) monitors the rules of the QDS_DB database. When a rule affecting a LEMU is fired, it sends a control signal to the associated LEMU. There are two types of control signal: one to control the charge of the ESS and another one to manage the X10 devices. The control signals for X10 devices follow the same format as that used in the IMM-LEMU communication, but now these signals can remotely modify the physical state of an X10 device. This type of signal is usually used to switch non-priority X10 devices off when the consumption curve reaches a peak and there's no enough electricity supply in the ESS. Otherwise, the first type of signals has a header value of "BAT" and it uses the data field to store the ESS command, which would be "on" if the EOM wants to start the charge of the ESS or "off" if it wants it to stop the charge. To communicate the SPU with a LEMU it's necessary to create another socket. However, this time it's the CEMIS who takes the role of the client and each LEMU will work as a server, listening for new requests. The SPU sends a control signal and then receives an acknowledgement response from the LEMU, which can be a message like "OK", "ERROR", etc. Once the control signal has been correctly transmitted, the socket is closed and the LEMU waits again for new requests.

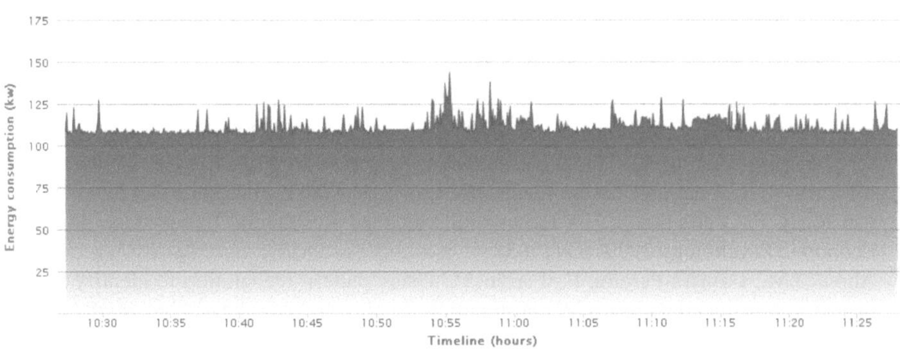

Fig. 5. Example of the power consumed over time in our lab. This data can be checked at anytime at http://158.49.245.68:8810/ecosaver/app/events/events.jsp.

5 Conclusions and Future Works

There are many opportunities to use INTELEM (Intelligent Energy Management System) in Smart Houses or Buildings to achieve economic benefits and an optimized use of the distribution grid. These opportunities increase if the house or building has a distributed generation system, which is another factor that must be taken into account when defining the operation strategy. In this sense, this paper presents the bases of an intelligent energy management system that aims to achieve these objectives, describing the relations and the main features and functions of each component.

At present, INTELEM can control several LEMUs (Local Energy Management Units) to optimize the power energy consumption and manage smart home devices remotely. In our ongoing work, we plan to design data mining algorithms in order to coordinate the behavior of a set of LEMUs in an efficient way, so that we can optimize the use of the distribution grid in different houses at the same time while saving energy altogether. We will test these behavioral algorithms in different scenarios to better understand their strengths and weaknesses.

Acknowledgements. This work has been developed under support of Telefónica Chair of the University of Extremadura, FEDER and Junta de Extremadura.

References

1. Pieper, C., Rubel, H.: Revisiting Energy Storage: There Is a Business Case. The Boston Consulting Group (2011)
2. Guerrero, M., Romero, E., Barrero, F., Milanés, M., González, E.: Supercapacitors: Alternative Energy Storage Systems. Przegląd Elektrotechniczny (2009)
3. Smart Energy Storage System, `http://panasonic.net/energy/storage_battery/index.html`
4. Towards a "Lifestyle with Virtually Zero CO2 Emissions throughout the Entire House", `http://panasonic.net/eco/zero-co2/`
5. Auväärt, A., Rosin, A., Belonogova, N., Lebedev, D.: NoordPoolSpot Price Pattern Analysis for Households Energy Management. In: 7th International Conference-Workshop Compatibility and Power Electronics, pp. 103–106. Tallinn Univ. of Technol., Tallinn (2011)
6. Botón-Fernández, V., Redondo-García, J.L., Lozano-Tello, A.: Learning Action Sequences for Decision-Making in Home Automation Systems. In: 7th Iberian Conference on Information Systems and Technologies, pp. 126–131. AISTI, Madrid (2012)
7. van Raaij, F., Verhallen, T.: A Behavioral Model of Residential Energy Use. Journal of Economic Psychology 3, 39–63 (1983)
8. Abreu, J.M., Câmara Pereira, F., Ferrão, P.: Using pattern recognition to identify habitual behavior in residential electricity consumption. Energy and Buildings 49, 479–487 (2012)

Community and Residential Energy Storage in Smart Grids

Máximo Pérez-Romero[1,2], Adolfo Lozano-Tello[2], Enrique Romero-Cadaval[1], and Joao Martins[3]

[1] Power Electrical and Electronics R+D group
University of Extremadura, Avda. De Elvas, s/n
06006 Badajoz, Spain
[2] Querqus Software Engineering Group
University of Extremadura, Escuela Politécnica, Campus Universitario, s/n
10071 Cáceres, Spain
[3] Universidad de Nova de Lisboa-FCT-DEE and UNINOVA-CTS
P-2829-516 Monte de Caparica, Portugal
mperez@peandes.unex.es, {alozano,eromero}@unex.es,
jf.martins@fct.unl.pt

Abstract. This paper reviews the most important energy storage systems for applications in residential environments. Normally, these systems are associated with renewable energy in order to achieve specific characteristics, although new possibilities and challenges are implemented over them in any residential places. Nowadays, the development of such kind of energy systems represents a technical challenge in the market. In residential and small building applications, energy storage solutions supply the demanded power by consumers of residential areas or small stand-alone buildings placed on isolated zones. In context of smart grids, energy storage management systems are not only concerned about energy storage, also inefficiency optimization and power consume. This work presents INTELEM, a new smart storage architecture which integrates the energy storage and the intelligent energy management in communities and in residential applications.

Keywords: Renewable energies, energy storage systems, residential and community applications, smart storage, smart grids.

1 Introduction

In coming years, the energy sector will have several challenges due to the demanded primary energy in the world producing an increase of the emissions and greenhouse effect. The International Energy Agency predicts that world primary energy consumption will increase from 2008 to 2035 by 36% [1]. The international community is supporting the use of renewables and efficient energy sources, and one important consequence of the European 20-20-20 targets [2] (The European Union's climate change package, including 20% reduced in emissions, 20% improvement in

L.M. Camarinha-Matos, S. Tomic, and P. Graça (Eds.): DoCEIS 2013, IFIP AICT 394, pp. 315–322, 2013.

energy efficiency and 20% increase in renewable energy by 2020) is that thermal power generation must be reduced to boost the renewable power generation, ensuring the environmental sustainability [3].

The most popular renewable resources are photovoltaic cells and wind turbines as they do not yield pollution and inexpensive primary energy are widely available in many areas. Actually, renewable energy resources are becoming cost competitive and are used instead of traditional fossil fuels as a second option to produce electric power to the local loads in remote areas.

The principal problems of renewable energies are the seasonal, daily or instantaneous variations due to dependence of weather conditions, so the power generation is stochastic. Therefore, the solution is the use of energy storage systems. These devices must feed the residential loads maintaining a low cost. They must also be so fast to produce a continuous supply to the loads. The advantages of energy storage devices are improved stability, power quality and reliability of power supply.

Renewable resources are integrated into the electrical grid as distributed generation taking advantages such as high reliability, high power quality, modularity, efficiency, low emissions, security and load management [4]. It has also some disadvantages, for example complexity of protections and variable power generation. Distributed generation allows the development of micro-grids, needing smart grids for an intelligent and efficient integration on the electrical systems.

In the same context, Virtual Power Plants (VPPs) are gaining a significant momentum within the energy industry. They have the ability to tap resources in real time, and with enough details, to control the load profiles of customers and to aggregate these resources hosting a trader's desk.

In this paper, it will be shown different energy storage systems used for residential or community applications (community of residential energy storage, CRES) (Fig. 1). The main targets of this research work are a higher efficiency, lower cost and longer life.

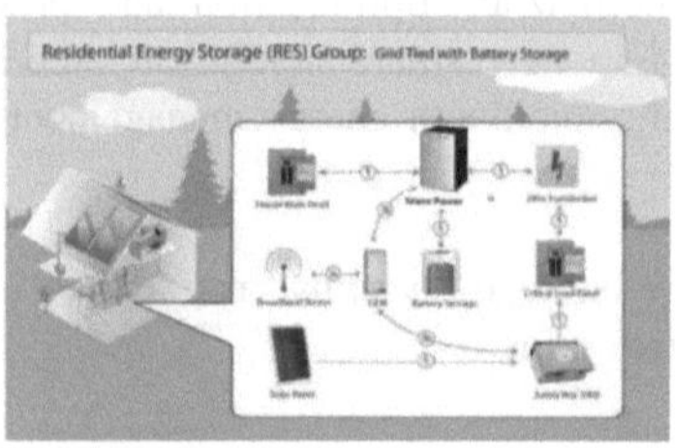

(a)

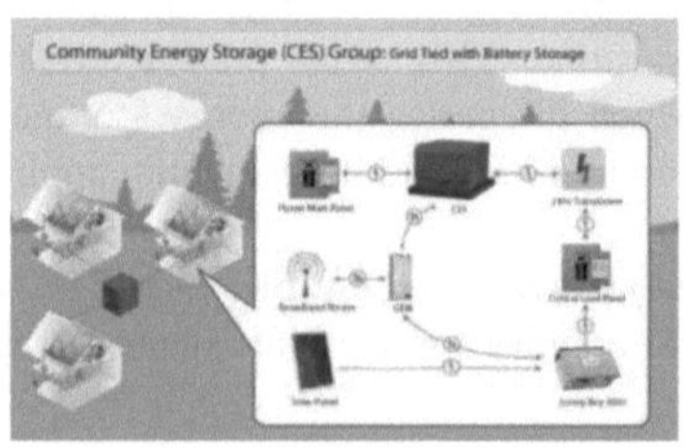

(b)

Fig. 1. Residential (a) and Community (b) Energy Storage systems

2 Contribution to Internet of Things

Many different battery technologies can be used in an intelligent device taking part of the future project INTELEM using it both concepts: energy storage and information and communications technology (ICT). It provides a new function to Internet of Things. This device is able to receive domestic power consumption data and send them through TCP to a central intelligent server. This server can manage the energy in an smart way, so this unit can control the demanded energy in a residential or community environment.

3 CRES on the Market

The market of batteries and other kinds of energy storage will gradually increase because the use of renewable energy and technologies mature will reach approximately € 10 billion annually in2020.

According to Pike Research, solar power systems will have3 giga watts in 2020 (Fig. 2a) [5]. If residential energy storage devices cover just 10% of this fact, it would reach annual 300 megawatt for energy storage. If this were accompanied by a decrease in price of lithium ion batteries until 345 $/kW, this market could represent a $ 100 million annual opportunity for lithium ion batteries by 2020.In coming decade the top technology in CRES systems will be lithium ion (Li-ion) batteries [6]. Nowadays, these batteries are the most important technology for energy storage in utility demonstration projects.

Some energy experts identify the CRES sector as one of the newest and least understood application areas for energy storage systems, where the market is still in situation of technical demonstration, and vendors are developing products. The implication for the CRES market is that purpose-built technologies are not expected to be commercialized in next two years [7] (Fig. 2 b).

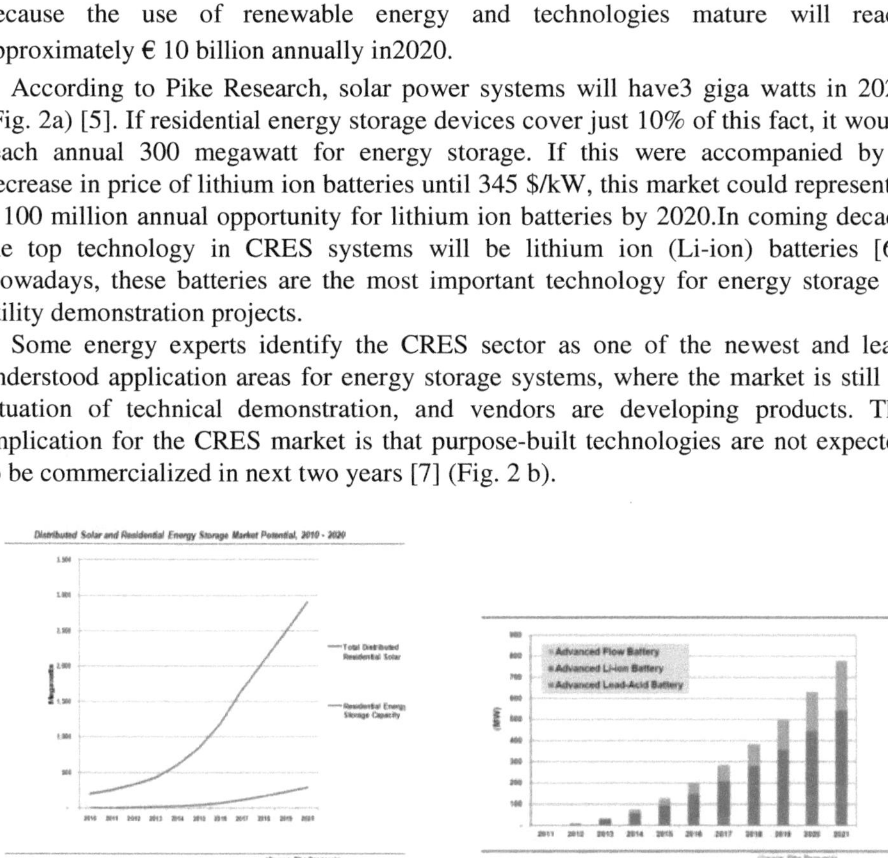

(a) (b)

Fig. 2. (a)Distributed Solar and Residential Energy Storage Market Potential, 2010 – 2020 (Courtesy of Pike Research) (b)CRES Installed Capacity by Technology. World Markets: 2012-2022. (Courtesy of Pike Research).

Moreover, the company NEC is developing the Smart House [8], which is a house that uses information control technologies (ICT) to manage energy supply and demand. This makes houses more comfortable, saving energy and reducing the electricity receipt. This enterprise is one of a few vendors in the world that has both energy storage technology and ICT.

4 Energy Storage Technologies

At present, there are many incipient energy storage technologies, such as pumped hydropower, batteries, flywheels, fuel-cells, superconducting magnetic energy storage and electrochemical capacitors.

It should make a distinction between short and long-time responses in the energy storage technologies. Energy storage systems with long-time response can produce energy during minutes or hours and, therefore, they are used in many tasks, as for energy management, frequency regulation and grid congestion management. Short-term energy storage systems work during transient state from a few second to minutes to improve the power quality and maintain the voltage stability of the electrical grid [9].

The considered technologies in this paper for long-time response are pumping-hydro, batteries (electrochemical batteries and redox flow), compressed air and fuel-cells.

A. Pumping-hydro energy storage technology

In this case the energy is stored as potential energy, pumping water in tanks located in high places whenever electricity is cheap and there is low demand. If power consumption is high, the stored water is released to rotate hydroelectric turbines. Pumping-hydro energy storage technology has the great problem of requiring two tanks separated by a short distance and located at different heights. As a result, infrastructures are quite expensive and also depending on the availability of the water resources and the topography of the place.

B. Batteries technology

Battery is the most popular storage system. However, it is not ideal having the following physical properties as was analyzed in [10]. The first feature is the lifetime that depends on the number of times it is charged or discharged and the depth of discharge. Secondly, there are losses due to energy conversion during the charging or discharging process. This fact is characterized by the measure of charging/discharging efficiency. Finally, the battery will leak energy while is working.

There are many important factors to consider in batteries for distributed energy storage applications, such as energy density, energy capability, efficiency, cycling capability, life time, initial cost and modularity.

In general, there are two types of batteries, some are the electrochemical system and the others are redox flow system.

The first type, electrochemical batteries, is based on chemical reactions between anode and cathode to create ions electrically charged. In the charging process, a direct current is converted in chemical energy. In the reverse process, when the battery discharges, the chemical energy is converted back into a flow of electrons in form of direct current. Most popular material used in electrochemical batteries are lead-acid, nickel cadmium, sodium sulphur and lithium ion.

Lead-acid batteries are widely used as storage element in vehicles and stationary equipment, so that this technology is mature and stable. For residential environments, the lead-acid technology is the best option, in particular the type deep-discharge specially designed for stationary solar electric systems. Regarding to the aging processes in lead-acid batteries are anodic corrosion, positive active mass degradation, irreversible formation of lead sulfate in the active mass, short-circuits and loss of water [11].

However, there are better storage systems than lead-acid batteries for stationary applications that has higher energy density capability, but they are still expensive for higher power applications. As examples, may be named to nickel-cadmium and lithium ion batteries being its application in electric vehicles. In these cases, the cost is balanced by high energy density capability.

The second type, vanadium redox flow batteries (VRB), is an electrochemical system that converts chemical energy to electrical energy, and vice versa.

It uses the redox active substance which is stored in two tanks separately. The process consists of a solution flowing through the battery by an electric pump, then the oxidation-reduction occurs at the electrodes in both sides of an ion exchange membrane producing electricity. Many stacks connected in series through bipolar plates, resulting in batteries.

A VRB has three main parts: a stack where the conversion of electric energy and chemical energy takes place, tanks which hold liquid electrolyte and a power conversion system.

The VRB energy storage has four main features [12]. First, the arrangement and design of VRB are very flexible, allowing higher storage capacity. Second, the battery system can work at a high speed and output high power. Third, the battery system is safe, stable and easy to maintain. Finally, this storage system is clean because the electrolytic liquid of the battery can be reused and materials such as carbon dioxide, etc. do not appear.

C. Hydrogen fuel cells

A fuel cell is a solid state DC power generator that converts chemical energy into electricity with no pollution. There is a chemical reaction which is composed of two elements, hydrogen and oxygen. Due to fuel cells do not have moving parts and do not rely on combustion, they are easy to maintain, very efficient and quiet.

Fuel cells can be classified according to two features: operation temperature and used electrolyte. In the high temperature range (65ºC to 1050ºC), there are two technologies, molten carbonate fuel cell (MCFC) and solid oxide fuel cell (SOFC). For low temperature (65ºC to 250ºC), it can be distinguished some types, proton exchange membrane fuel cell (PEMFC), alkaline fuel cell (AFC) and membrane phosphoric acid fuel cell (PAFC) [13].

The used fuel is mainly hydrogen in the high and low temperature technologies, although high temperature fuel cells use carbon monoxide and some hydrocarbons as well. One company that offers this technology is Idatech [14].

D. Other energy storage technologies

There are other energy storage technologies that they are mainly used for short-term response energy storage systems:

- Flywheels.
- Supercapacitors or ultracapacitors.
- Magnetic superconductors.

Flywheel energy storage systems operate mechanically in a rotating flywheel. Electrical energy is stored by using a motor which rotates the flywheel, thus converting electrical energy into mechanical energy. The energy that can be stored depends on its angular velocity and inertia. More energy can be stored if the flywheel rotates at higher angular speed. The main advantages of flywheel energy storage system are fast charge/discharge, high energy storage density, high power density, low risk of overcharge or over-discharge, wide range of operation temperature, very long life cycle and environmental behaviour [15]. A problem with flywheel systems is large vibration of the rotor when it is rotating, so it is difficult to increasing the angular speed. Companies that offer this technology are EATON [16] and Beacon POWER [17].

Supercapacitors, also called ultracapacitors, are a new technology employed for energy storage systems. These devices store energy in form of electric charge between two metallic or conductive plates, separated by a dielectric medium. Supercapacitors operate in DC allowing enough power and high energy density. They have a long life cycle and a short charging-discharging time [18]. Nowadays, the company Maxwell Technologies manufactures and sells energy storage and power delivery solutions through supercapacitors [19].

Superconducting magnetic energy storage (SMES) is a large superconducting coil capable of storing electric energy in the magnetic field generated by DC current flowing through it. System converts the AC current supplied by power system into DC current [20]. The superconductor operates at low temperatures (cryogenic) without resistive losses, so that the energy can be stored all the necessary time. The main advantage of this system is that the charging-discharging time is very short.

5 Conclusions

In residential applications there are many energy storage technologies and some are available in the market being the main problem the charge-discharge efficiency. Lead-acid batteries are devices chosen for any new project because they are cheap and having high availability. These batteries are widely used as storage element in vehicles and stationary equipment, so this technology is mature and stable. For residential environments, the lead-acid technology is the best option, in particular the deep-discharge ones specially designed for stationary solar electric systems.

Normally, energy storage systems are related to renewable energy applications to improve the stability, power quality and reliability of power supply. Moreover, the word ICT is used by smart home more and more to control energy supply and demand. In other words, it refers to a residential building equipped with embedded intelligence and communication infrastructure.

If the two previous concepts, energy storage systems and ICT, are mixed appears the term “Smart Storage” that would enable store energy when it was cheaper and deliver energy when it is needed, thus reducing the electricity receipt. To achieve this, the consumers must become active customers and they must understand and accept this. Therefore, active customer participation plays a key role.

6 Future Work

In the context of smart grids, future smart houses will have DC loads, pv systems, evs and energy storages (mainly based in batteries)connected with power and control systems by means of power lines and communication infrastructures. The control system has to send control commands to the smart-house according to system conditions and the smart house determines the operation of controllable loads.

In this scenario is defined a new concept named smart storage which are concerned not only about the energy storage (that has been discussed previously) but also about energy management systems (EMS).

EMS will be able to adapt to the environment, grid and user requirements. In this context, it can be referred the project INTELEM (Fig.3) that implements ideas related with the smart storage. In this project the smart storage is based in two systems implementing a master/slave configuration that integrated electric grids and communication networks.

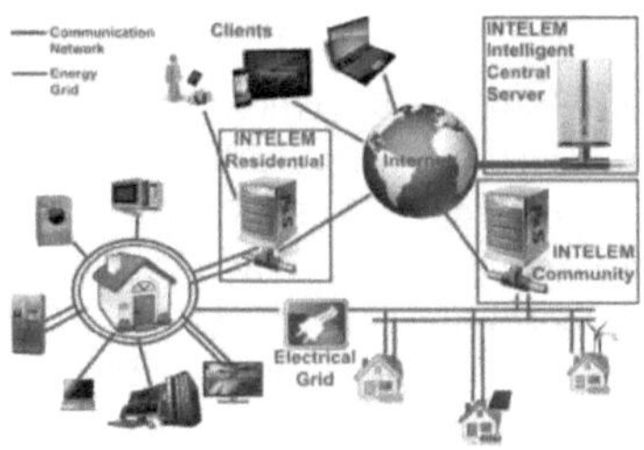

Fig. 3. Schematic diagram of INTELEM Project

One system is a device placed in a house or in a community of houses, able to store energy and to connect or disconnect different loads in order to control the power consumption, implementing in this way energy saving functions. This device will operate take into consideration the guidelines received from the central server.

The other system is an Artificial Intelligent Central Server that receives data from different INTELEM devices (consume profiles, load turned on/off …). The central server optimizes the overall performance of all devices, and is sensible to environment (temperature, irradiance, energy prices by time period, forecast of DG generation if exists, forecast of consume at controlled houses …) using commands. The central server will send information back to INTELEM devices, that could include the priority for connecting or disconnecting loads in houses if it is necessary, the maximum power that the house could demand from the grid, and INTELEM devices will operate locally according to these commands.

References

1. International Energy Agency. World Energy Outlook, Executive summary (2010)
2. European Commission. Climate action, http://ec.europa.eu/clima/policies/package/index_en.htm
3. Bedir, A., Ozpineci, B., Christian, J.E.: The impact of plug-in hybrid electric vehicle interaction with energy storage and solar panels on the grid for a zero energy house. In: 2010 IEEE PES Transmission and Distribution Conference and Exposition, pp. 1–6 (2010)
4. Roggia, L., Rech, C., Schuch, L., Baggio, J.E., Hey, H.L., Pinheiro, J.R.: Design of a sustainable residential microgrid system including PHEV and energy storage device. In: Proceedings of the 2011-14th European Conference on Power Electronics and Applications (EPE 2011), pp. 1–9 (2011)
5. PikeResearch. Residential Energy Storage, http://www.pikeresearch.com/blog/articles/residential-energy-storage
6. PikeResearch. Investment in Community and Residential Energy Storage Systems to Total $4.2 Billion through 2022, http://www.pikeresearch.com/newsroom/investment-in-community-and-residential-energy-storage-systems-to-total-4-2-billion-through-2022-2
7. Globe-Net. Community and Residential Energy Storage Systems Investment to Soar, http://www.globe-net.com/articles/2012/february/29/community-and-residential-energy-storage-systems-investment-to-soar/
8. NEC. Smart Houses, http://www.nec.com/en/global/environment/energy/house.html
9. Angheliţă, P., Chefneux, M., Balaban, R., Trocan, L.: Energy storage systems for buildings equipped with photovoltaic cells. In: 2010 3rd International Symposium on Electrical and Electronics Engineering (ISEEE), pp. 332–335 (2010)
10. Urgaonkar, B., Neely, M.J., Sivasubramaniam, A., Urgaonkar, R.: Optimal power cost management using stored energy in data centers. In: ACM International Conference on Measurement and Modeling of Computer Systems, SIGMETRICS 2011, San Jose (2011)
11. Sun, Y.-H., Jou, H.-L., Wu, J.-C.: Aging Estimation Method for Lead-Acid Battery. IEEE Transactions on Energy Conversion, 264–271 (2011)
12. Zhang, X., Wei, Y., Xu, S., Zhang, Y.: Optimizition design of all-vanadium redox flow battery energy storage system. In: 2010 China International Conference on Electricity Distribution (CICED), pp. 1–4 (2010)
13. Weickgenannt, M., Sawodny, O.: Efficient Hydrogen Production for a Stationary Fuel Cell System: Two-Degree-of-Freedom Control of a Steam Reformer Unit. Control Systems, 54–69 (2012)
14. Idatech. Power for the long run, http://www.idatech.com/products-and-services-electragen-h2-i-system.asp
15. Nguyen, T.D., Tseng, K.J., Zhang, S., Nguyen, H.T.: On the modeling and control of a novel flywheel energy storage system
16. EATON, http://www.eaton.com/Electrical/USA/index.htm
17. Beacon Power, http://www.beaconpower.com/
18. Guerrero, M., Romero, E., Barrero, F., Milanés, M., González, E.: Supercapacitors: Alternative Energy Storage Systems. Przegląd Elektrotechniczny (Electrical Review) (2009)
19. Maxwell Technologies, http://www.maxwell.com/
20. Ali, M.H., Dougal, R.A.: Comparison of SMES and SFCL for transient stability enhancement of wind generator system. In: Energy Conversion Congress and Exposition (ECCE), pp. 3382–3387 (2010)

Part XII
Integration of Power Electronics Systems with ICT - II

Development of a Photovoltaic Array Emulator in a Real Time Control Environment Using xPC Target

S. Polo-Gallego, Carlos Roncero-Clemente, Enrique Romero-Cadaval, V. Miñambres-Marcos, and M.A. Guerrero-Martínez

Power Electrical and Electronic Systems (PE&ES),
School of Industrial Engineering (University of Extremadura)
http://peandes.unex.es

Abstract. This paper is devoted to the design and construction of a photovoltaic array emulator for high power applications in order to test all kind of photovoltaic inverters. To develop such device, a rapid prototyping tool based on xPC Target of Matlab/Simulink has been used, providing a real-time testing environment. PV array emulator can be used to evaluate the performance of photovoltaic inverters as any test conditions can be programmed. The proposed emulator operates as a distributed control system taking advantage of the TCP/IP protocol features.

Keywords: photovoltaic array, inverter performance, rapid prototyping tool, real-time, xPC Target, TCP/IP protocol.

1 Introduction

Nowadays, the power electrical system scenario is very different in comparison with the traditional configuration. Several factors, such as an electrical consumption increase, the electrical market liberalization, the need to reduce CO_2 emissions, and the new technological development, are boosting the distributed generation (DG).

Photovoltaic solar energy is one of the most relevant distributed energy resources taking an important part in this new scenario [1]. Due to the increased use of this technology, several regulations [2] have been established in order to manage the inverters of photovoltaic plants. The main deal of these rules says that the inverters must work providing support and stability in the electrical grid. This fact obliges that every inverter must be tested under different conditions. For high power testing applications a DC power supply is hard to find and certainly a photovoltaic array could not be available.

The well-known software Matlab/Simulink [3] has been widely used in system modeling and simulation of control algorithms. Since years ago, efforts have been made to control the physical systems by means of Matlab and its toolboxes. The xPC Target Matlab toolbox provides a rapid prototyping host-target environment by using TCP/IP [4] to construct the real time control system.

This paper is devoted to the designed and construction of a photovoltaic array emulator for high power applications in order to test all kind of photovoltaic inverters.

L.M. Camarinha-Matos, S. Tomic, and P. Graça (Eds.): DoCEIS 2013, IFIP AICT 394, pp. 325–333, 2013.

xPC Target from Matlab/Simulink has been used as rapid control prototyping tool. The different characteristics and configuration of this control platform will be explained in detail.

2 Relationship to Internet of Things

A photovoltaic array emulator is a useful tool for testing an inverter. The non-availability of high DC power supplies is solved by means of the proposed device. This work shows the developed platform for evaluating the performance of PV inverters. As a result an optimum use of natural resources can be achieved. At the same time, xPC Target has been chosen to implement the real time embedded controller, to interact with the real system, to support control and functionality and to allow communication between the user and the system by means of internet connection. It operates as a distributed control system taking advantage of the TCP/IP protocol features as it can be seen in Fig. 1.The proposed emulator could be accessible by Internet for programming the test, monitoring the execution and acquiring data in a remote way.

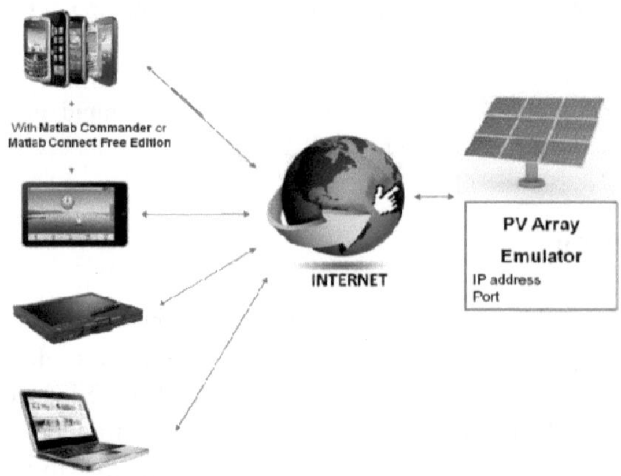

Fig. 1. Global communication scheme of the PV array emulator

3 Creating a Real-Time Testing Environment with xPC Target

The real time environment built for rapid prototyping is shown in Fig. 2.Controllersand algorithms are designed in Simulink on the host PC (which runs Matlab/Simulink, Real-Time Workshop and xPC Target toolbox). xPC Target operates with the code generated by a C compiler from the Simulink models in order to develop the real-time target application. Target application can be executed in real time once it is downloaded in the target PC from the host PC through TCP/IP.

The host-target communication can be obtained by means of TCP/IP communication protocol. Once the target application has been downloaded to the target PC, it can be controlled and changed from the host PC with the external control mode. It also allows tuning the parameters before, during and after real-time execution and viewing, logging and acquiring signal data.

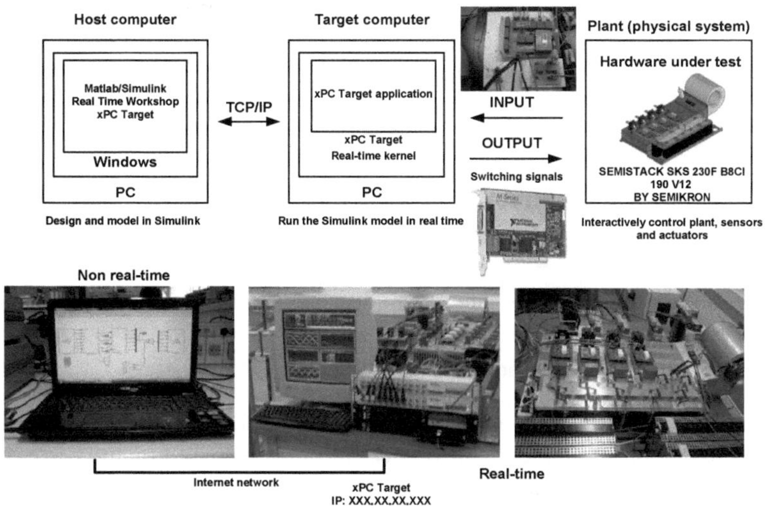

Fig. 2. Rapid prototyping environment based on xPC Target

4 Plant Under Test. Physical System

In order to achieve PV array emulation, an electronic converter working as a rectifier has been used.

AC/DC converters constitute the interface circuit between the electrical grid and the DC loads. With the ever increase of power quality requirements at the point of common coupling (PCC), these converters are nowadays required to achieve different task such as: provide high input power factor, low current distortion [5] and fixed output voltage.

Synchronous rectifier allows demanding sinusoidal currents from the grid which are synchronized with the voltage at the PCC. Fig.3 shows the topology used in this work. The rectifier is connected to the grid by means of an autotransformer.

The real system is the Semikron SEMISTACK SKS 230F B8CI 190 V12 inverter shown in Fig. 2.

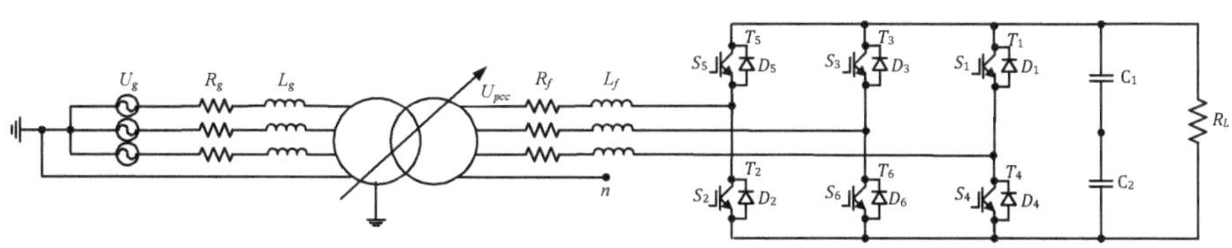

Fig. 3. Three-phase two-level rectifier electronic converter topology connected to the grid

5 Host PC. Photovoltaic Array Emulator Models and Controllers

In this section, the explanation of the designed models will be explained. Models have been implemented in Simulink on the host PC in order to be executed in the target PC in real time after its compilation. This PC has the hardware characteristics shown in Table 1.

Table 1. Host PC characteristics

Equipment	CPU	RAM
Host PC	Pentium Dual-Core T4500 at 2.3 GHz	4 GB

5.1 Photovoltaic Panel Model

In order to reproduce the typical curve of a photovoltaic (PV) panel, a PV model has been employed. This model, explained in detail in [6], is based on the I-V exponential curve defined by the information provided by the manufacturers. I-V curves obtained in the simulation of the panel Shell SP 150-P in different conditions are shown in Fig. 4 a) and 4 b), as it is the panel that has been used to develop the proposed PV array emulator.

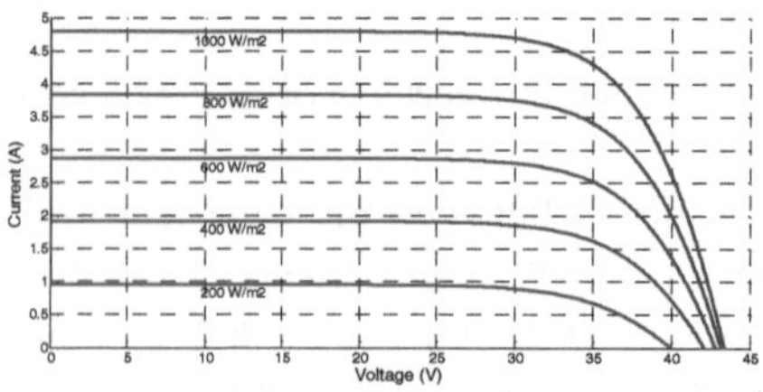

Fig. 4a. I-V curves (differents W)

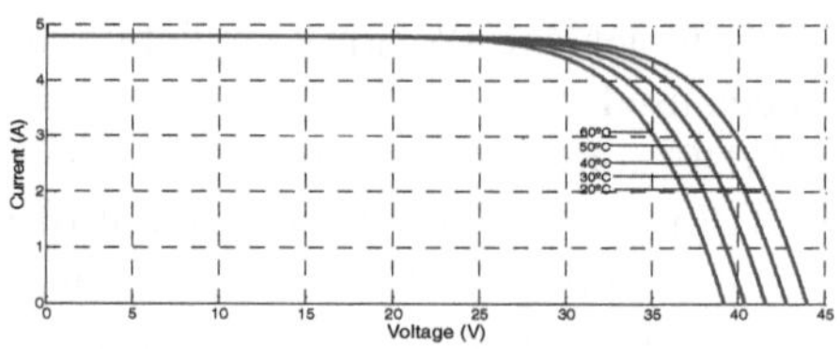

Fig. 4b. I-V curves (differents T)

5.2 Photovoltaic Array Emulator Control Strategy

In Fig. 5, the control strategy of the photovoltaic array emulator is shown. The measured DC load current is the input of the photovoltaic array model. By this way, the reference DC link voltage is generated by the model according to the I-V curve.

In order to generate the reference current demanded from the grid, a proportional-integral (PI) controller produces its RMS value (control variable) depending on the DC voltage error. The amplitude of the reference current is multiplied by three unitary wave forms synchronized with the voltage (v_{pcc}) at the PCC. These waveforms are provided by an adjustable synchronous reference frame (ASRF) [7].

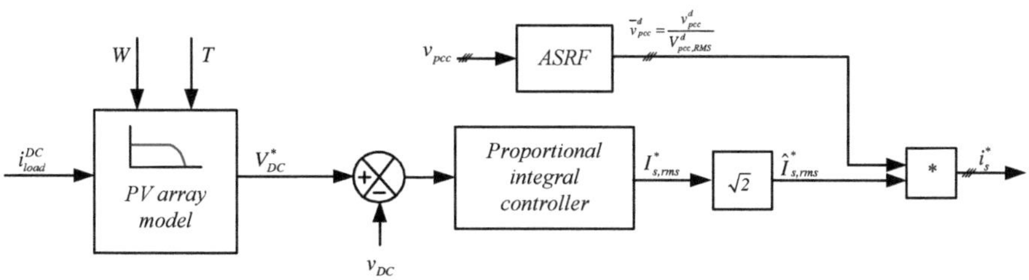

Fig. 5. Control strategy in PV array emulator

5.3 Reference Tracking

In order to track the reference current, a hysteresis band has been used [8]. In Fig. 6 a) one can see a block diagram of this current controller. The reference current and the measured current are compared to generate the switching signals that control the system (Fig. 6 b).

The switching signals are generated by the next way. The voltage on the filter inductance is given by (1) and (2):

$$v_L = L\frac{di(t)}{dt}, \tag{1}$$

$$v_L = v_{sn} - v_{xn}, \tag{2}$$

wherev_{sn} is the line to neutral voltage and v_{xn} is the voltage between the middle point of any leg and the neutral point.

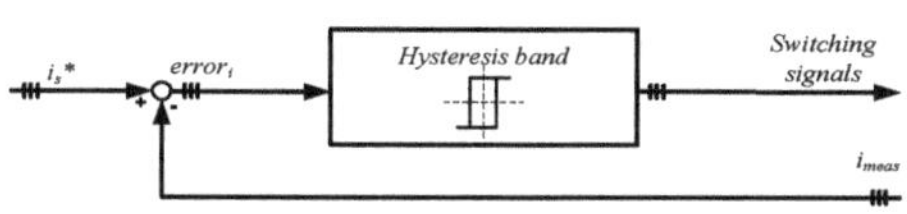

Fig. 6a. Hysteresis band diagram

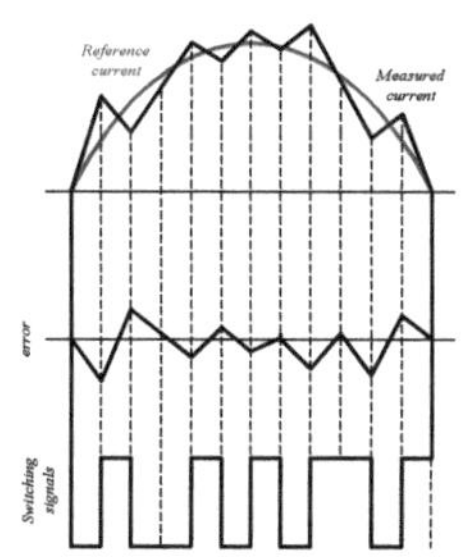

Fig. 6b. Switching signals generation

This value could take the next limits:

$$v_{xn}: \frac{2v_{DC}}{3} \text{ or } -\frac{2v_{DC}}{3}. \tag{3}$$

Depending on the obtained error signal between the reference current (i^*_s) and the measured current, the slope (di/dt) of the demanded current must be changed. This is achieved by controlling u_{xn} through the appropriate switching states defined in Table 2.

Table 2. Different states of the switching signals

Error	Scheme	Situation
is*-imeas<0	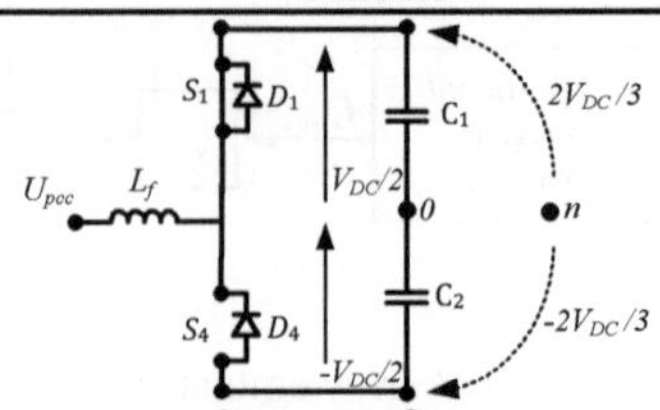	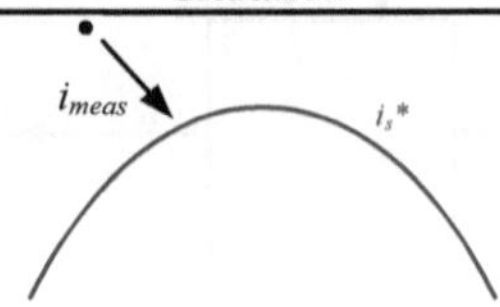
is*-imeas>0	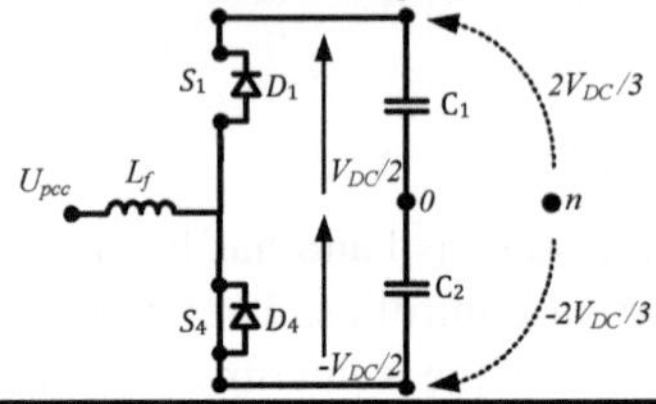	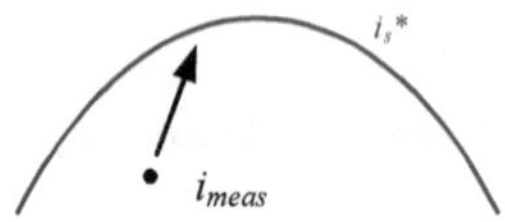

6 Target PC. Data Acquisition, Measurement and Control Signal

An industrial PC has been used as target. Its characteristics are shown in Table 3.

Table 3. Target PC characteristics

Equipment	CPU	RAM
Target PC	Pentium Core 2 T4500 at 2.66 GHz	2 GB

The executable code is generated in the host PC and the target PC runs it in real-time to control the plant. This computer is equipped with a data acquisition (DAQ) board PCI-6259 by National Instruments. This board has the necessary I/O channels according to the designed control strategy and tracking technique. In this case, eight analogical input channels (required measurements) and six digital output channels (switching signals) are used.

6.1 Measurement Board

A measurement board with the necessary sensors has been built (Fig. 7.a). A total of eight sensors have been used(Table 4).

Table 4. Sensor characteristics

Magnitude	Sensor	Number of sensors
Voltage	LV 25-P by LEM	4 (v_{pcc} (v_{an}, v_{bn}, v_{cn}) and v_{DC})
Current	LA 25-NP by LEM	4 (i_{meas}(i_a, i_b, i_c) and i_{DC})

7 Full System. Experimental Results

The proposed photovoltaic array emulator has been studied experimentally in real-time environment (Fig. 7.b.).The experimental test parameters are shown in Table 5. Resistance value correspondswith the MPP for the chosen panel configurations under standard conditions. The PV array user interface created for the host PC is shown in Fig.8.

Experimental results are displayed in Fig. 9 a) and b). One can see how the system produces the MPP voltage and current according to the parameters of the panel. Experimental results with different load conditions are shown in Fig. 10. Critical points have been emphasized. The test scheme can be seen in Fig. 11.

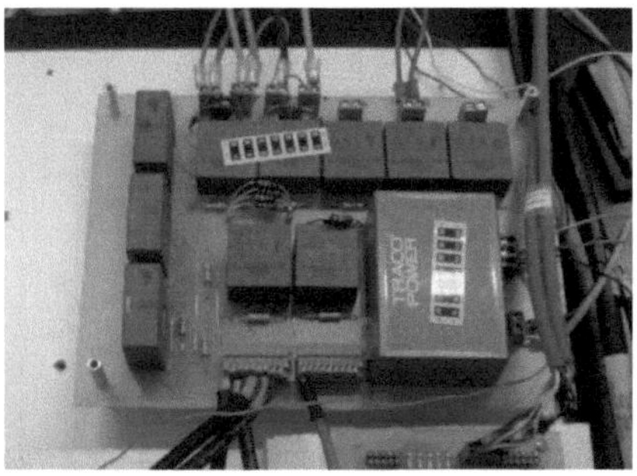

Fig. 7a. Built measurement board

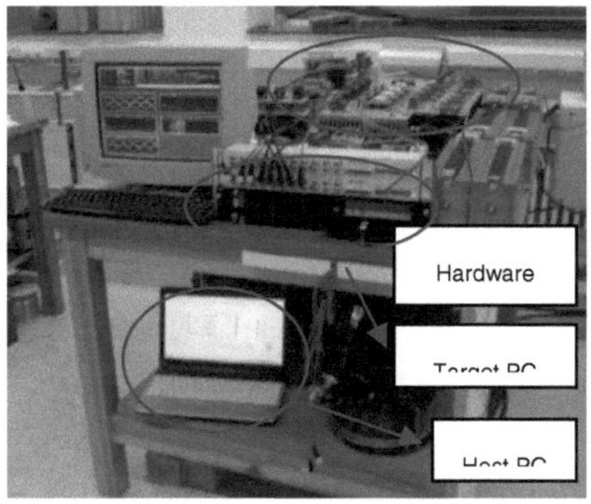

Fig. 7b. PV array emulator (full system)

Table 5. Used parameters in experimental tests

Parameter	Description	Value
n_p	Number of panel connected in parallel	1
n_s	Number of panel connected in series	6
V_s(V)	v_{pcc} RMS voltage	50
L_f (mH)	Filter inductance	6.2
C_1, C_2(mF)	DC-Link capacitors	28.2
R_L (Ω)	Load	48.22
f_w(KHz)	Switching frequency	8

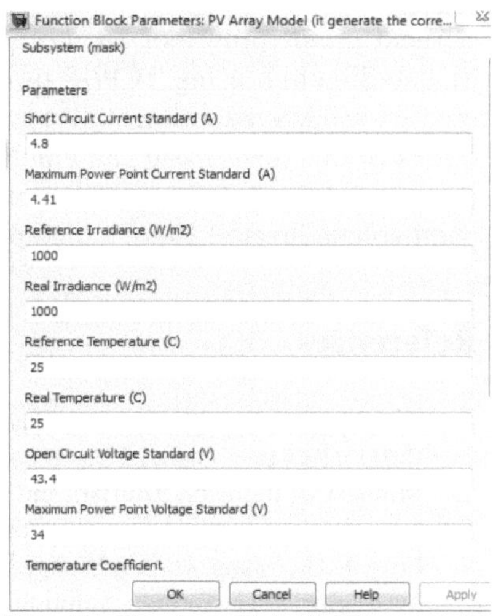

Fig. 8. PV array model interface

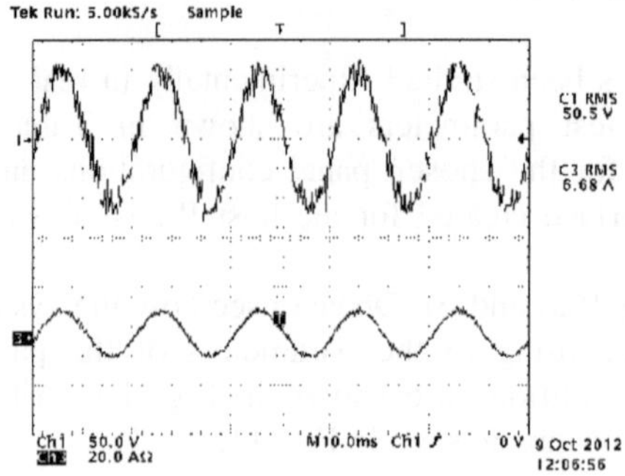

Fig. 9a. v_{pcc} and demanded current (phase a)

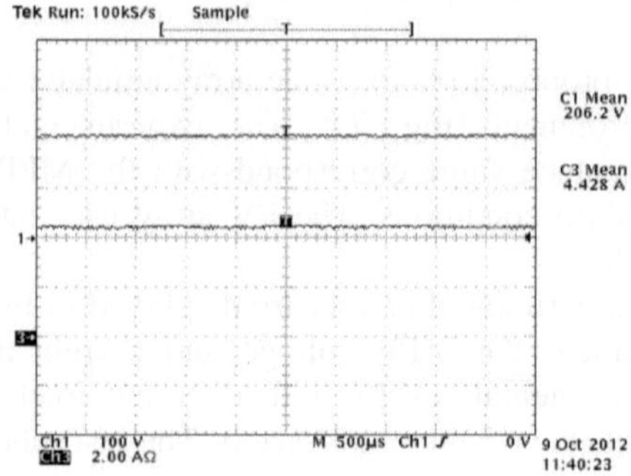

Fig. 9b. PV voltage and PV current

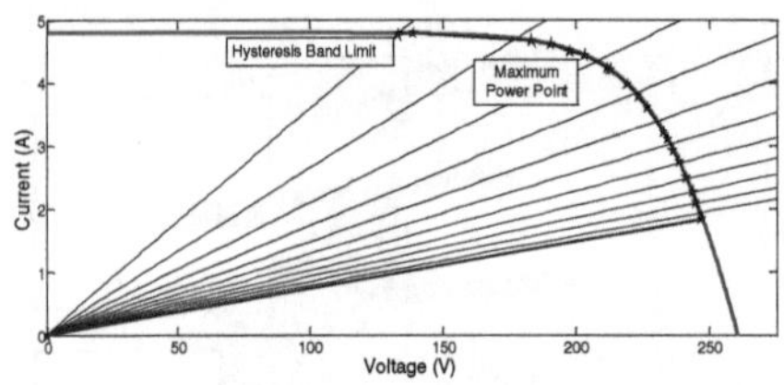

Fig. 10. Experimental results (stars)

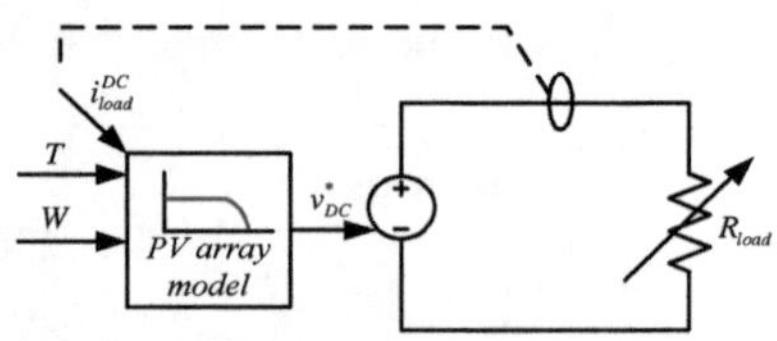

Fig. 11. Equivalent test scheme

8 Conclusion and Future Works

A rapid prototyping tool has been used and configured based on xPC Target from Matlab/Simulink using TCP/IP protocol to communicate host and target. It has been used to validate experimentally, and in real-time, a photovoltaic array emulator. By means of this device one can emulate any PV panel and any array configuration s in every condition. This platform will be used in the future to test any kind of photovoltaic inverter [9] in a high power level.

References

1. International Energy Agency. Photovoltaic Power Systems Programme. Statistic Reports (2011), `http://www.iea-pvps.org`
2. Ministry of Industry, Tourism and Trade of Spain, R.D 1565/2010 Spain (2010)
3. `http://www.mathworks.com`
4. Low, K.H., Wang, H., Yu Wang, M.: On the Development of a Real Time Control System by Using xPC Target: Solution to Robotic System Control. In: IEEE International Conference on Automation Science and Engineering, pp. 345–350 (2005)
5. Wilamowski, B.M., Irwin, J.D.: The Industrial Electronics Handbook. Power Electronics and Motor Drives, 2nd edn. (2011)

6. Roncero-Clemente, C., Romero-Cadaval, E., Roncero-Sánchez, P., González-Romera, E.: Comparison of Two Power Flow Control Strategies for Photovoltaic Inverters. In: 38th Annual Conference of the IEEE Industrial Electronics Society (IECON 2012), Montreál, Canadá (2012) (accepted for publication)
7. Milanés-Montero, M.I., Romero-Cadaval, E., Miñambres, V.M., Barrero-González, F.: Novel Method for Synchronization to Disturbed Three-phase and Single-phase Systems. In: IEEE International Symposium on Industrial Electronics, pp. 860–865 (2007)
8. Roncero-Clemente, C., Milanes-Montero, M.I., Minambres-Marcos, V.M., Romero-Cadaval, E.: Three-phase regenerative electronic load to test shunt power conditioners. In: 7th International Conference-Workshop on Compatibility and Power Electronics (CPE), pp. 178–183 (2011)
9. Valentini, M., Raducu, A., Sera, D., Teodorescu, R.: PV inverter test setup for European efficiency, static and dynamic MPPT efficiency evaluation. In: 11th International Conference on Optimization of Electrical and Electronic Equipment, OPTIM 2008, pp. 433–438 (2008)

Three-Level Neutral-Point-Clamped Quasi-Z-Source Inverter with Maximum Power Point Tracking for Photovoltaic Systems

Carlos Roncero-Clemente[1], Serhii Stepenko[2], Oleksandr Husev[3], Víctor Miñambres-Marcos[1], Enrique Romero-Cadaval[1], and Dmitri Vinnikov[3]

[1] Power Electrical & Electronic System (PE&ES), University of Extremadura, Campus Universitario, Avda. de Elvas s/n, Escuela de Ingenierías Industriales, Lab. C2.7, 06006 Badajoz, Spain
croncero@peandes.unex.es

[2] Department of Industrial Electronics, Chernihiv State Technological University, Shevchenko street 95, 14027 Chernihiv, Ukraine
stepenko.sergey@gmail.com

[3] Department of Electrical Drives and Power Electronics, Tallinn University of Technology, Ehitajate tee 5, 19086 Tallinn, Estonia
oleksandr.husev@ieee.org

Abstract. This article is focused on a photovoltaic system based on the three-level neutral-point-clamped quasi-z-source inverter. The maximum power point tracking (MPPT) algorithm based on dP/dV feedback was used in the photovoltaic system to adjust the duration of the shoot-through states of power switches and achieve a maximum power. Proper system operation in the case of irradiance step is demonstrated by simulation in Matlab/Simulink software.

Keywords: three-level inverter, neutral-point-clamped inverter, quasi-z-source inverter, shoot-through, maximum power point tracking, photovoltaic system, dP/dV feedback.

1 Introduction

Many studies propose the quasi-z-source inverter (QZSI) for use in photovoltaic (PV) systems. Some module architecture designs provide a high efficiency up to 98% [3]. Moreover, the output current of a QZSI based system has good quality (THD up to 3%) [4]. The QZSI also has necessary voltage boost properties and can be used for the Maximum Power Point Tracking (MPPT) in PV systems [5].

Since approaches for the MPPT implementation using QZSI have been required, a lot of new modifications of the MPPT have appeared. Some methods track the MPP without the current sensor [6]. Most of the studies suggest tracking by adjusting the shoot-through of the QZSI [7]. The methods to control shoot-through by artificial intelligence techniques are known as well. In [8] a method based on an adaptive neuro-fuzzy inference system (ANFIS) is proposed to harness the maximum power of the PV system based on the QZSI. It offers a very fast dynamic response with high accuracy.

L.M. Camarinha-Matos, S. Tomic, and P. Graça (Eds.): DoCEIS 2013, IFIP AICT 394, pp. 334–342, 2013.

A new QZSI topology was proposed and described in [9]. It is a combination of the QZSI and the three-level neutral-point-clamped (3L-NPC) inverter. The three-level neutral-point-clamped quasi-z-source inverter (3L-NPC QZSI) has advantages of both of these topologies. It can buck and boost the input voltage, it has short circuit immunity and due to the multilevel topology, high energy density is attainable. The 3L-NPC QZSI is especially suitable for renewable energy sources.

Since the mentioned topology is rather new, in all previous studies the 3L-NPC qZSI was considered as an isolated system [9], [10], [13], [14]. The aim of the current work is to develop and study the capabilities of this topology assuming it as an integrated system. A wide range of problems should be considered (MPPT, anti-islanding methods, reactive power control). This article is focused on the MPPT.

2 Contribution to the Development of the Internet of Things

Continual focus of DoCEIS is on general topics of wide-scale importance that create opportunities for researchers to discuss their contribution to the challenges involved in their proposed subjects [1]. A current topic "Technological Innovation for the Internet of Things" is closely connected to the smart grid.

In some studies the application of the Internet of Things in the smart grid is gravely emphasized. But we should keep in mind that the Internet of Things is a rather broad concept. It may concern not only to the smart grid and therefore is not limited by it.

However discussing the contribution of this work to the development of the Internet of Things we will consider it in terms of the smart grid, since this paper is focused on the photovoltaic system as an element of the smart grid.

Smart grid can solve the problem of an energy alternative and compatible use, which integrates system data on the basis of building an open system and a shared information model to optimize the operation and management of the grid. The principal characteristics of the smart grid include self-healing, mutual operation and participation of the users, perfect electricity quality, distributed generations and demand response, sophisticated market and effective asset management [2].

Our conviction is that the greatest development of the mentioned concept can be achieved if attention is paid first to the objects of control, and then to the ways and methods of the management. Thus, renewable energy sources (fuel cells, photovoltaic arrays, wind turbines etc.) can be used as electrical sources for a smart grid supply. This work is devoted to the photovoltaic system based on the new topology of a quasi-z-source inverter (QZSI).

The control system for the 3L-NPC qZSI can be built on the FPGA (Field Programmable Gate Array), as described in [10]. The FPGA makes it easier to implement a shoot-through mode. Using FPGA resources, it is sufficient to integrate the MPPT based on the same chip. Furthermore, we can use the FPGA to integrate a PV system in the smart grid and use the remote control of power converters. Sensor technology of the Internet of Things can form an interactive real-time network connection between the users, the corporation and power equipment, which will improve the overall efficiency of the integrated power grid [2]. Such control basis makes an excellent contribution to the furtherance of the Internet of Things.

3 Description of the System

The general structure of the studied system is depicted in Fig. 1. Different models of the system were developed and implemented using Matlab/Simulink as a simulation tool.

The power stage of the system is composed of a photovoltaic array and a three-level neutral-point-clamped quasi-z-source inverter with an output LC filter. The MPPT algorithm and a shoot-through pulse-width modulation (SPWM) technique compose the control stage of the system.

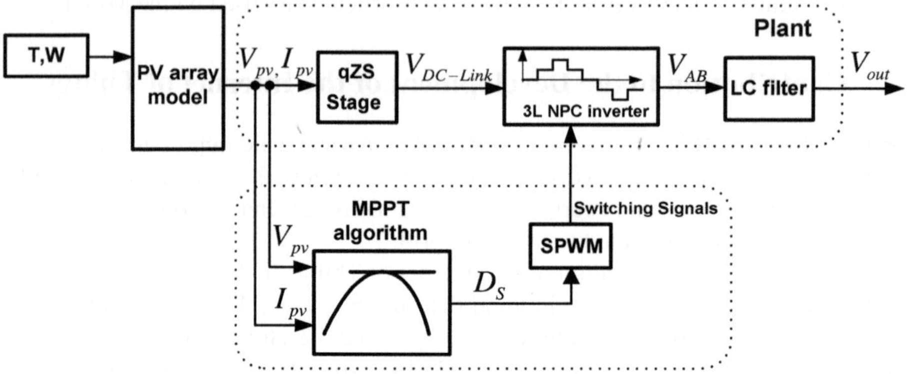

Fig. 1. General block-diagram of the analyzed photovoltaic system

All the mentioned components of the PV system will be described in detail in the following sections.

3.1 Photovoltaic Array Model

One of the principal components of the system is the PV array model. It is modelling the real photovoltaic module Shell SP150-P [11].

Several models for solar panel simulation are reported in the literature. Most of them are modelling the solar cell as an electrical equivalent circuit. Due to such representation some electrical parameters are required, such as junction resistance, dark current, effective cell area etc. Usually these parameters are not provided in the solar panel manufacturer datasheet. It makes such model use difficult for engineers and users.

In this work the model based on I-V exponential curve has been used for simulation. This model was proposed in [12], where detailed description of the model could be found. The influence of temperature and irradiance was also considered.

The simulated I-V curves in different irradiance conditions for the mentioned PV panel are shown in Fig. 2a. Proper work of the model was validated by the comparison of these curves with those provided by manufacturer for the solar panel Shell SP150-P (Fig. 2b).

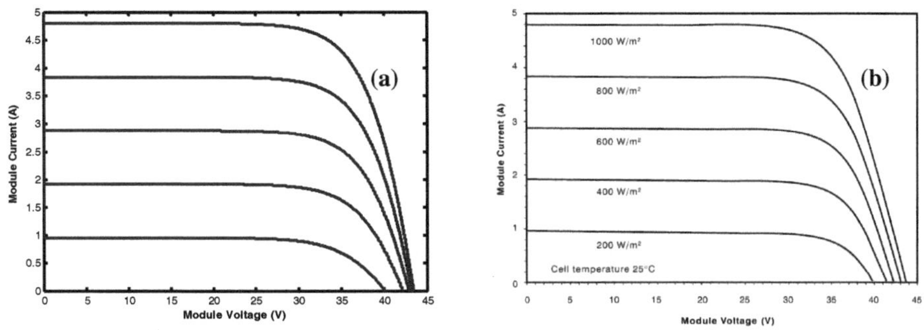

Fig. 2. I-V curves family of the photovoltaic array obtained using Matlab/Simulink model (a) and I-V curves family from the Shell SP150-P Product Information Sheet (b)

Suitable series-parallel associations of the single photovoltaic panels enable the necessary DC input voltage and the PV system power to be provided.

3.2 Three-Level Neutral-Point-Clamped Quasi-Z-Source Inverter

DC/AC conversion is based on the 3L-NPC qZSI proposed and described in [9]. It is depicted in Fig. 3.

Each leg of the converter consists of two complementary switching pairs of transistors and four anti-parallel diodes (Fig. 3). The required parameters of the passive elements of the QZS-stage were calculated according to the methodology proposed in [13].

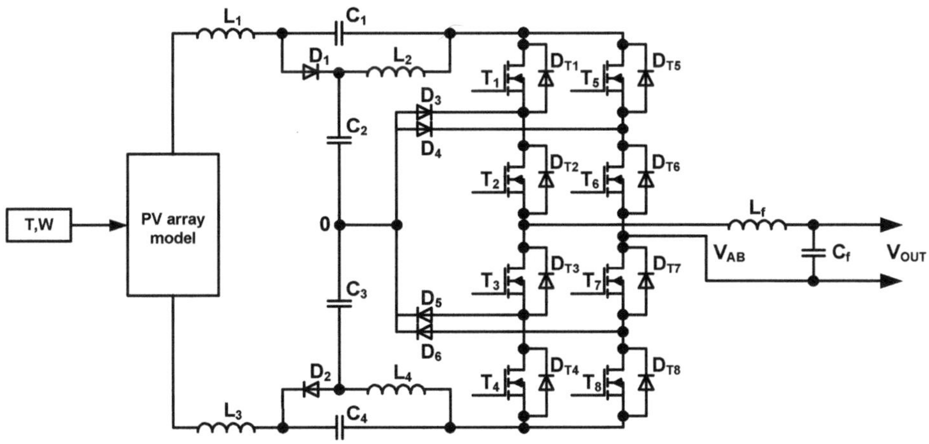

Fig. 3. Three-level neutral-point-clamped quasi-z-source inverter fed by a photovoltaic array

This topology has such advantages as continuous input current, the possibility to use shoot-through, lower switching losses and balanced neutral-point voltage in comparison with the traditional two-level voltage source inverter. Due to the above features the neutral-point-clamped quasi-Z-source inverter is especially suitable for the PV systems.

3.3 Maximum Power Point Tracking Algorithm

The structure of the proposed MPPT algorithm is depicted in Fig. 4.

The input parameters of this stage are values of the voltage and current of the photovoltaic array. As can be seen from Fig. 4, this MPPT algorithm is based on dP/dV.

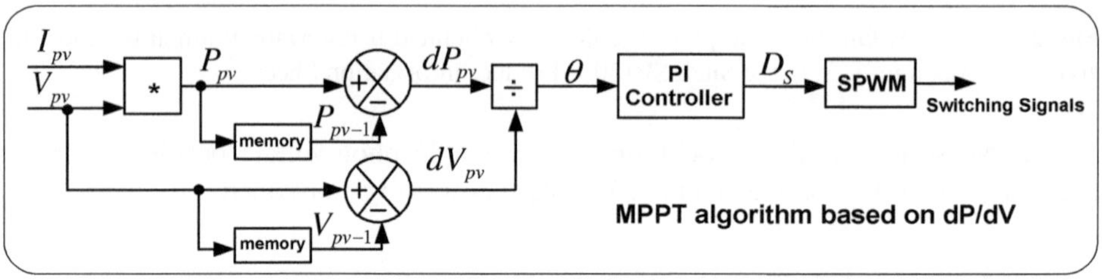

Fig. 4. The structure of the maximum power point tracking algorithm

The novelty of this approach lies in the usage of the shoot-through state of power switches for maintenance of the maximum power. Using the value of the dP/dV, denoted as θ in Fig. 4, PI controller forms the necessary shoot-through duty cycle Ds to achieve the maximum power. Ds is used by the SPWM block to generate the control signals for all switches. Thus, the MPPT algorithm is adapted for the 3L-NPC qZSI in the PV system. In Figs. 5a and 5b, I-V and P-V curves show the MPPT.

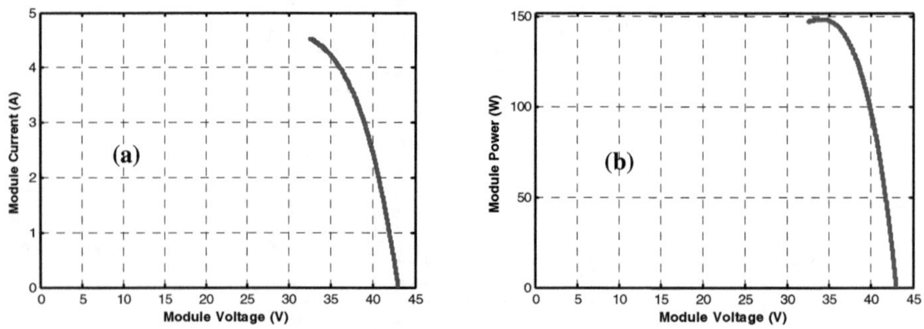

Fig. 5. Modelled I-V (a) and P-V (b) curves with maximum power point tracking

3.4 Modulation Technique

A special shoot-through pulse-width modulation (SPWM) technique was implemented for the 3L-NPC qZSI. This modulation technique as well as the switching strategy are described in detail in [14].

The modulation technique differs from the classical modulation techniques for such converters. It has shoot-through states of the power switches distributed during the whole period of the output voltage, while in the classical approaches the shoot-through states are usually applied when the output voltage is equal to zero.

This modulation technique has several benefits. It significantly reduces the DC voltage pulsations and output voltage distortion. Thus it allows a decrease in the values of the passive elements used for the quasi-z-source stage and neutral point clamping. It also influences the output filter parameters.

In addition, the proposed modulation technique allows the boost factor that is needed to be easily achieved only by changes of the shoot-through duty cycle. Thus, it combines the necessary boost factor with acceptable quality of the output power.

4 Simulation Results

The photovoltaic system was simulated using MATLAB Simulink. All the parameters of the PV array model are presented in Table 1 under standard conditions (irradiance 1000 W/m^2 and temperature 25^{0}C).

Table 1. Simulation parameters

	Symbol	Description	Values
Parameters of the used photovoltaic array model	V_{OC}	Open circuit voltage	43.4 V
	I_{SC}	Short circuit current	4.80 A
	V_{MPP}	Maximum power point voltage	34.0 V
	I_{MPP}	Maximum power point current	4.41 A
	N_S	Number of panels connected in series	7
	N_P	Number of panels connected in parallel	1
	P_{MAX}	Maximum power	1050 W
Parameters of the used passive elements	L_1, L_2, L_3, L_4	Inductors of the quasi-Z-stage	160 μH
	C_1, C_4	Capacitors of the quasi-Z-stage	1180 μF
	C_2, C_3	Capacitors of the neutral point	940 μF
	L_F	Inductor of the output filter	2.2 mH
	C_F	Capacitor of the output filter	0.47 μF
	R_F	Resistor of the output filter	0.25 Ohm
	R_L	Resistor of the load	15 Ohm

In the PV system study it was necessary to validate its ability to track the maximum power point under nonstandard conditions. Thus, using the changes of irradiance the response of the system was observed and results are depicted in Fig 6.

At the initial time, irradiance was 1000 W/m^2 (Fig. 6a). The module voltage is about 240 V (Fig. 6b). So, the voltage of each of 7 panels was 34 V. Such panel voltage according to the I-V curves in Fig. 2a corresponds to the current 4.4 A in Fig. 6c. The PV module power was 1050 W (Fig. 6d).

At second 8, irradiance was gradually decreased to 900 W/m^2 (Fig. 6a). It caused a sharp decline of the module power (Fig. 6d). The MPPT block by changes of the shoot-through duty cycle (Fig. 6e) is looking for the new maximum power point in such conditions. Thus, when the transient process is finished, the voltage is near to 240 V (Fig. 6b). The voltage of a single panel is about 34 V. For such conditions, the current is 4.0 A (Fig. 6c). The power in a steady state is 960 W (Fig. 6d).

After second 16, the irradiance returns to its initial level. MPPT block by changes of shoot-through duty cycle (Fig. 6e) establishes the initial power. It allows all the variables get their original values.

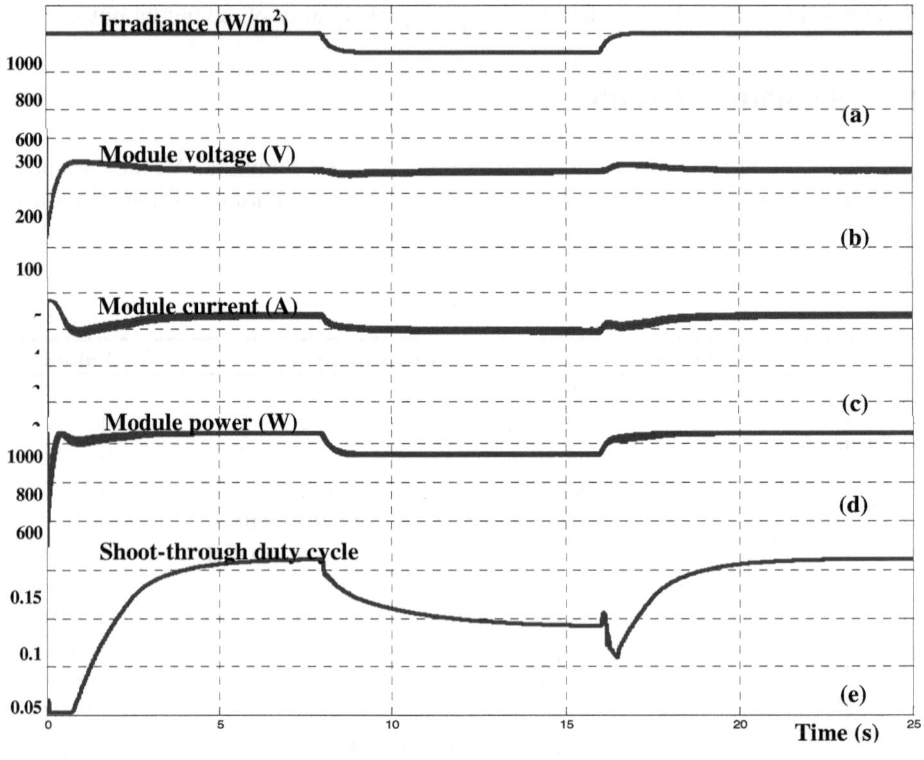

Fig. 6. Simulation results: irradiance (a), PV module average voltage (b), PV module average current (c), PV module average power (d), shoot-through duty cycle (e)

5 Conclusions and Further Work

A photovoltaic system based on three-level neutral-point-clamped quasi-z-source inverter with the maximum power point tracking has been modelled and described in this work. The MPPT algorithm based on dP/dV is featured by the shoot-through duty cycle to achieve the maximum power. Simulation results proved the ability of the proposed system to track the maximum power point under irradiance changes. The results of simulation show the average values of investigated variables. It is connected with perceptible high-frequency pulsations which exist in module current, module voltage and module power, however continuous current mode has been obtained. It should be mentioned as well that PI controller that drives the shoot-through duty cycle was customized manually. Thus the opportunity to achieve better quality of the transient processes by essential adjusting of PI controller will be studied in further work. Another important thing that will be investigated particularly is the input current influence on the DC-link voltage that results in the output voltage quality.

Acknowledgment. This research work was supported by MOBILITAS Postdoctoral Research Grant (MJD391), by Estonian Ministry of Education and Research (project SF0140016s11) and partially supported by Spanish institutions "Ministerio de Economía y Competitividad" and "Junta de Extremadura" and the funding of "Fondos FEDER".

References

1. Camarinha-Matos, L.M., Goes, J., Gomes, L., Martins, J.: Raising Awareness for Value Creation Potential in Engineering Research. In: Camarinha-Matos, L.M., Shahamatnia, E., Nunes, G. (eds.) DoCEIS 2012. IFIP AICT, vol. 372, pp. 3–6. Springer, Heidelberg (2012)
2. Yun, M., Yuxin, B.: Research on the architecture and key technology of Internet of Things applied on smart grid. In: 2010 International Conference on Advances in Energy Engineering, pp. 69–72 (2010)
3. Zhou, Y., Liu, L., Li, H.: A High Performance Photovoltaic Module-Integrated Converter (MIC) Based on Cascaded Quasi-Z-Source Inverters (qZSI) using eGaN FETs. IEEE Transactions on Power Electronics 28(6), 2727–2738 (2013)
4. Shahparasti, M., Sadeghi Larijani, A., Fatemi, A., Yazdian Varjani, A., Mohammadian, M.: Quasi Z-source inverter for photovoltaic system connected to single phase AC grid. In: 1st Power Electronic & Drive Systems & Technologies Conference, pp. 456–460 (2010)
5. Zakis, J., Vinnikov, D.: Study of Simple MPPT Converter Topologies for Grid Integration of Photovoltaic Systems. Sc. J. of RTU, Power and Electrical Engineering 29, 67–72 (2011)
6. Liu, Y., Abu-Rub, H., Ge, B., Peng, F.Z., Almeida, A.T., Ferreira, F.J.T.E.: An improved MPPT method for quasi-Z-source inverter based grid-connected photovoltaic power system. In: IEEE International Symposium on Industrial Electronics, pp. 1754–1758 (2012)
7. Chun, T.W., Lee, H.H., Kim, H.G., Nho, E.C.: Power control for a PV generation system using a single-phase grid-connected quasi Z-source inverter. In: IEEE 8th International Conference on Power Electronics and ECCE Asia, pp. 889–893 (2011)
8. Abu-Rub, H., Iqbal, A., Moin Ahmed, S., Peng, F.Z., Li, Y., Baoming, G.: Quasi-Z-Source Inverter-Based Photovoltaic Generation System With Maximum Power Tracking Control Using ANFIS. IEEE Transactions on Sustainable Energy 4(1), 11–20 (2013)
9. Ott, S., Roasto, I., Vinnikov, D., Lehtla, T.: Analytical and Experimental Investigation of Neutral Point Clamped Quasi-Impedance-Source Inverter. Sc. J. of RTU, Power and Electrical Engineering 29, 113–118 (2011)
10. Stepenko, S., Husev, O., Vinnikov, D., Ivanets, S.: FPGA Control of the Neutral Point Clamped Quasi-Z-Source Inverter. In: 13th Biennial Baltic Electronics Conference, pp. 263–266 (2012)

11. Shell SP150-P. Photovoltaic Solar Module. Product Information Sheet, http://www.meet-egypt.com/Downloads/SOLARM/MONO/SHELLsp150p.PDF
12. Roncero-Clemente, C., Romero-Cadaval, E., Roncero-Sanchez, P., Gonzalez-Romera, E.: Comparison of Two Power Flow Control Strategies for Photovoltaic Inverters. In: 38th Annual Conference of the IEEE Industrial Electronics Society, IECON 2012, pp. 5149–5155 (2012)
13. Husev, O., Roncero-Clemente, C., Stepenko, S., Vinnikov, D., Romero-Cadaval, E.: CCM Operation Analysis of the Single-Phase Three-Level Quasi-Z-Source Inverter. In: The 15th International Power Electronics and Motion Control Conference and Exposition, EPE-PEMC 2012 ECCE Europe, pp. DS1b.21-1–DS1b.21-6 (2012)
14. Roncero-Clemente, C., Romero-Cadaval, E., Husev, O., Vinnikov, D.: Simulation Study of Different Modulation Techniques for Three-Level Quasi-Z-Source Inverter. Sc. J. of RTU, Electrical, Control and Communication Engineering 1, 11–17 (2012)

Instantaneous, Short-Term and Predictive Long-Term Power Balancing Techniques in Intelligent Distribution Grids

Alexander Suzdalenko[1,2] and Ilya Galkin[1]

[1] Riga Technical University, Riga, Latvia
aleksandrs.suzdalenko@rtu.lv, gia@eef.rtu.lv
[2] Tallinn University of Technology, Tallinn, Estonia

Abstract. An increased number of distributed small generators connected to the power grid allows higher total efficiency and higher stability of electrical power supply by exporting energy to the grid to be achieved during peak demand hours. On the other hand, it poses new challenges in structuring and developing the control approaches for these distributed energy resources. This paper proposes an improved method of real-time power balancing targeted to reaching long-term energy management objectives. The novel long-term energy management technique is proposed, that is based on load categorization and regulation of energy consumption by regulating electricity price function estimated with the proposed mathematical model. The method was evaluated by a LabVIEW model by simulating various types of loads. The price function for the defined energy generation pattern from renewable energy sources was obtained.

Keywords: Intelligent Distribution Grid, Nanogrid, Energy Management, Instantaneous Power Balancing, Short-Term Energy Management, Pricing.

1 Introduction

The dynamic of worldwide installed power utilizing renewable energy sources (RES) is increasing year by year [1], showing global awareness of climate change and the footprint of human behavior on the nature: like mining activities that change the landscape and damages caused by oil plants. Another problem is related to the total effectiveness of energy from the primary fuel. It relates to the total system efficiency of energy delivery to the end-user from a mining site, which includes energy losses at middle stages, like use of energy at the mining stage, energy conversion losses, transmission (mechanical and electrical) losses, as well as end-device efficiency. Consequently, locally generated energy (especially from renewable energy sources) is preferable because it excludes most of the mentioned losses.

As more distributed small generators are connected to the power grid, higher total efficiency and higher stability of electrical power supply by exporting energy to the grid during peak demand hours. On the other hand, it poses new challenges in structuring and developing the control approaches for these distributed energy resources.

L.M. Camarinha-Matos, S. Tomic, and P. Graça (Eds.): DoCEIS 2013, IFIP AICT 394, pp. 343–350, 2013.

Thus, it can be concluded that the centralized generation is moving towards distributed, plug-and-play type of generation that requires a special control approach. Implementation of centralized control for such a dispersed grid is unacceptable, thus some kind of distributed control is to be realized.

The advantages of Multi-Agent System (MAS) use in control applications described in [2] demonstrate the effectiveness of this approach in the organization and management of the intelligent distribution grid in large-scale applications.

This paper discusses different energy management approaches for an intelligent grid that consists of RES, energy storage devices and controllable loads. An improved instantaneous energy management method is described targeted to obtaining long-term objectives and a novel long-term energy management approach is proposed based on load categorization and regulation of energy consumption by regulating electricity price function.

2 Relationship to Internet of Things

Concerning growing interest in intelligent systems that acquire data from network sensors, much higher predictive operation of household appliances is expected [3]. The concept named Internet of Things delivers the idea of connecting each smart appliance to the Internet, helping to communicate with the data center and providing advanced functionality (such as predictive connection of loads) that could help to solve the problem of peak demand, by scheduling of distributed generators and shedding the end-user's load or in other words, enabling demand-side management [4] [5].

The idea of Internet of Things covers a large functionality of the smart nodes, but this article uses a simplified model to test the long-term scheduling approach by using a bidirectional communication between the sources and loads on the one side and the data center on the other side.

3 State of the Art in Energy Management

Currently a major effort is focused on the approaches that keep energy balance in small-scale networks, such as nanogrids [6][7], microgrids [8]. The energy management approaches are divided into groups which differ by the period of management: instantaneous (real-time) energy management, keeping balance in the period of milliseconds or seconds, short-term energy management, with periods of seconds to hours and long-term energy management, which defines energy management strategy to fulfil objectives for weeks and months.

3.1 Instantaneous Energy Management

The operation in islanded mode requires special control techniques for energy generation units as they should regulate the output power with the demand of the local grid. For this reason the droop control approach is mostly utilized both in AC and in DC power distribution grids, which defines the method of power sharing between various sources by means of definition of different droops according to their nominal power [9].

The droop control differs between AC and DC grids, because in a DC grid only supply bus voltage should be regulated, but in an AC grid – voltage and frequency. AC droop control is realized in the following way – the amount of active power is tight with the frequency, but reactive power with voltage reference. When a generator's output power becomes more capacitive, it increases the voltage reference, which increases the output power of the generator, at the same time decreasing the output frequency, thus sharing power with other generators.

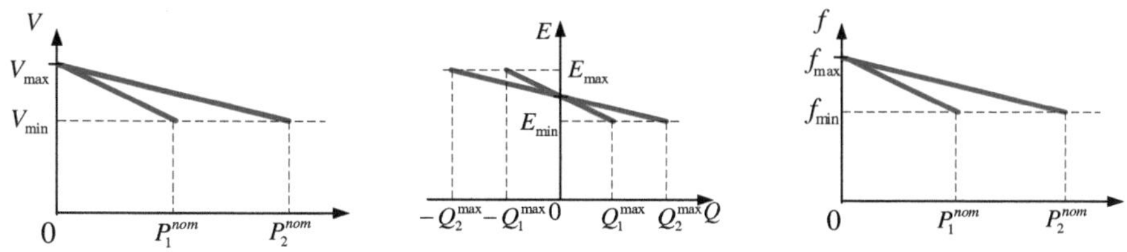

Fig. 1. Droop control method – V-droop for DC grid, E*f*-droop for AC grid

From the point of view of energy management, DC distribution grids provide less problems, as only single parameter should be regulated within the defined levels. In contrast, energy management in an AC grid requires more attention due to inherent problems of reactive power fluctuations, harmonics, unbalances, as well as certain difficulties for generators to operate in no load conditions. These problems are compared in the table below.

Table 1. Comparison of instantaneous power balancing control challenges in AC and DC grids

Problem	AC grid	DC grid
P – active power balance	•	•
Q – reactive power balance	•	
F – frequency control; synchronization;	•	
Harmonics, 3-phase unbalance	•	
R=∞ – no load condition	•	
Imax - Inrush currents	•	•

3.2 Short-Term Energy Management

Instantaneous power balancing provides power sharing approach between various of sources. However it does not provide information about the resources of the grid, for instance, islanded DC grid consisted of 1 kW source and 500 W load will operate at the same voltage level as second islanded DC grid, which consists of 10 kW source and 5 kW load. It means, that in one case there is no possibility to supply additional 1 kW load, but in other grid – there is possibility. Thus, special control approach is needed to overcome this issue.

In [10] the DC bus signalling (DBS) approach is applied to this problem. The control technique is based on different disconnection delays for different priority loads when DC bus undervoltage condition is occurred, thus enabling to lowest priority loads with shorter delays to disconnect faster, releasing power for higher-priority loads. The drawbacks of described approach are the following: the delays are set constantly, meaning that the controllability of such approach is limited; there is always trade-off between the delay periods (with longer periods it is possible to define more steps of load prioritization) and the quality of DC bus voltage, as the longer are delays, the bigger are voltage dips.

In our opinion, the communication hardware should be used to provide better short-term energy management that would eliminate unwanted voltage dips.

3.3 Long-Term Energy Management

According to literature [11], with proper control of microgrids it is possible to realize long-term objectives, such as the reduction of CO2 emissions by a certain percentage or export a defined amount of energy to the grid during some period (week, month or even year), which could be required by contract statements or transmission network operator's tasks.

It is recommended to solve that problem with artificial neural networks [12], which will successfully predict the amount of generated power by RESs. Another solution is to use genetic algorithms [13] that help to choose an optimized operation schedule for energy storage components providing reliable power supply and keeping lifetime of energy storages longer due to full charge and discharge cycles. An additional solution is provided by a pricing approach [14] based on changing the price per electrical energy. Different loads and sources behave according to an internal logic – how to operate under certain price limits. This solution results in additional load when the price is low and vice versa, load shedding when the price is high. However, it is not possible to predict amount of switched on or off load with this solution.

4 Contribution of Research

Instantaneous energy management approach based on the droop control method can be used to obtain long-term objectives in energy management by regulating the reference values of DC bus voltage for a bidirectional AC/DC converter that will influence the consumed energy in the local grid.

The proposed method of long-term energy management is based on the mathematical model of price function estimation. It can be interpreted as a graphical approach of predictive energy generation surface intersection with acquired operation schedule of categorized loads, which can also be drawn as a 3D surface, thus enabling the control of the intelligent grid by providing only price function information.

4.1 Improved Instantaneous Power Balancing

Our proposal is to use the well known droop control method for keeping power balance in the grid, but with variable reference values of a common point interface (CPI) converter. It is especially suitable for DC supply grids that always operate under the droop control method, as these are not connected to the utility grid directly, but through an AC/DC interface converter, which imports or exports energy to or from the grid, depending on the DC bus voltage. We propose to use variable reference value of the DC voltage grid for an interface converter that will influence the operation of local grid elements, for instance by decreasing the DC bus voltage reference value to few per-cents below the nominal value, the unnecessary loads will not operate or will operate at the highest energy saving mode and as a result, more energy will be exported to the grid.

In conclusion, it is possible to fulfil long-term objectives with a CPI converter which regulates its reference value for the regulation of a local grid operation mode, as other grid elements choose their operation mode in accordance with the grid voltage – the parameter available for all elements of the grid.

4.2 Novel Long-Term Energy Management Approach

Our proposal is to define categories of household appliances which switch and operate at different conditions:

1. Time-triggered load (see Fig.2a) – pumps, general illumination etc. – the load which basically switches on at a defined time;
2. Price-triggered load (see Fig.2b) – energy storage element, washing machines etc. – the load which turns on when the price is within an acceptable level;
3. Hybrid-triggered load – heater, boiler, refrigerators etc.– the load which operates according to the daily schedule (depending on consumption), but it can also switch on a little earlier when the price per electricity is economically profitable;
4. Price-responsive load (see Fig.2c) - controllable illumination – the load which changes the consumed power depending on the price per electricity.

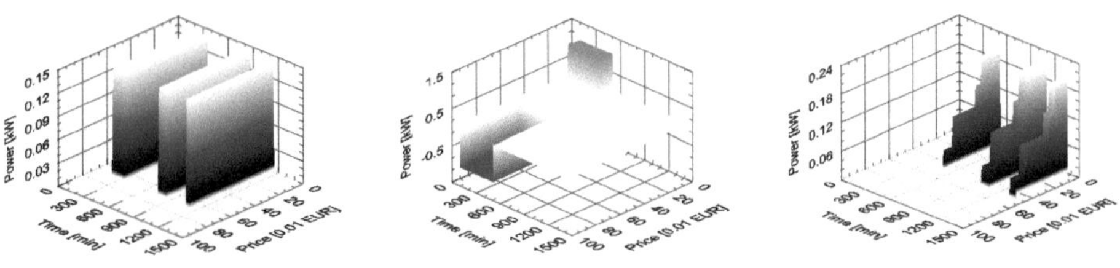

a) Time-triggered load

b) Price-triggered storage element

c) Price-responsive load

Fig. 2. Load pattern graphs

The proposed algorithm is realized in the following way. Each load based on the statistics of use and on the changes of electrical price provides an energy management system (EMS) with a schedule of operation (for time-triggered and hybrid-triggered loads), condition of operation (for price-triggered and hybrid-triggered loads):

$$\Psi_{load}^{(1)}(C,t)=\sum_i\left(P_{T-trig}^i(C,t)\right)+\sum_j\left(P_{C-resp}^j(C,t)\right), \quad (1)$$

$$\Psi_{source}(C,t)=\sum_i\left(P_{source}^i(C,t)\right)\cdot \quad (2)$$

After having aggregated all data from loads ($P_{T-trig}^i(C,t)$ for time-triggered loads and $P_{C-resp}^j(C,t)$ for price-responsive loads), EMS subtracts the 3D surface of demand $\Psi_{load}^{(1)}(C,t)$ (to which the power reserve is added $\Psi_{delta}(C,t)$), from the 3D surface of generation $\Psi_{source}(C,t)$ (see Fig. 3a) to extract the intersection line, the projection of which to Time-Cost axes is used to find the price function $C^{(1)}(t)$ (see Fig. 3b):

$$\Psi_{source}(C,t)-\Psi_{load}^{(1)}(C,t)\geq\Psi_{delta}(C,t), \quad (3)$$

$$C^{(1)}(t)=\Psi_{source}(C,t)\cap\left(\Psi_{load}^{(1)}(C,t)+\Psi_{delta}(C,t)\right). \quad (4)$$

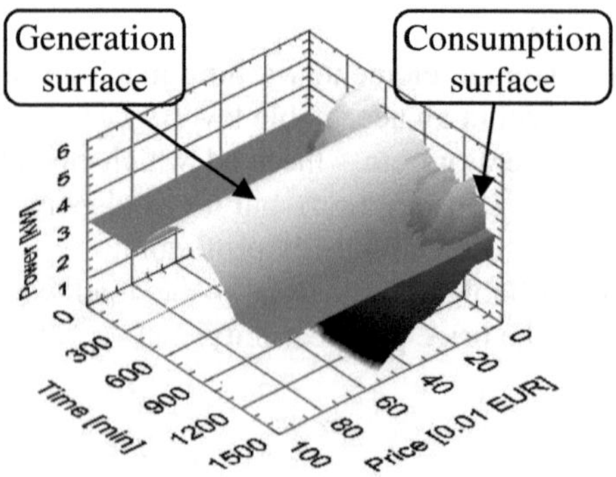

a) Primary energy balancing with time-triggered and price-responsive loads

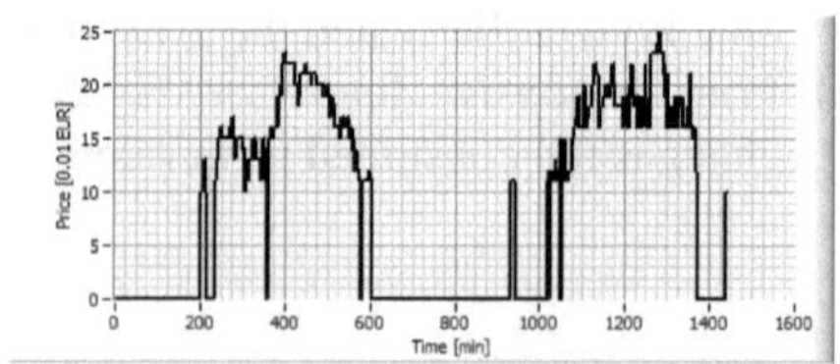

b) Preliminary price function

Fig. 3. Primary equalization of price function

The scheduling of price-triggered loads is also used for the determination of the price function. EMS combines the price-triggered loads in accordance with energy excess, which is calculated using (5) (see Fig.4a):

$$\Psi_{excess}(C,t)=\Psi_{source}(C,t)-\left(\Psi_{load}^{(1)}(C,t)+\Psi_{delta}(C,t)\right), \quad (5)$$

$$\Psi_{excess}(C,t)\geq\sum_i P_{C-trig}^i\left(C^i,t^i\Big|_{t_{start}}^{t_{end}}\right),\begin{cases}C^i<C^{i-1}\\ t_{satrt}^i\geq t_{start}^{i-1}\\ t_{start}^i+t_{\min}^i\leq t_{end}^i\leq t_{start}^i+t_{\max}^i\end{cases}. \quad (6)$$

As a result, the price function $C^{(1)}(t)$ is changed in accordance with the switching schedule of price-triggered loads, keeping the price at a desirable level, to regulate the amount of switched price-triggered loads and the power balance (see Fig.3b).

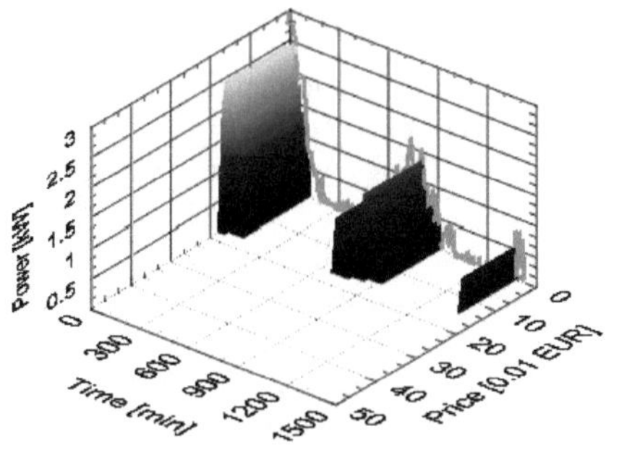

a) Secondary balancing (energy excess is projected to Time-Power axis;3D surface - scheduled price-triggered loads)

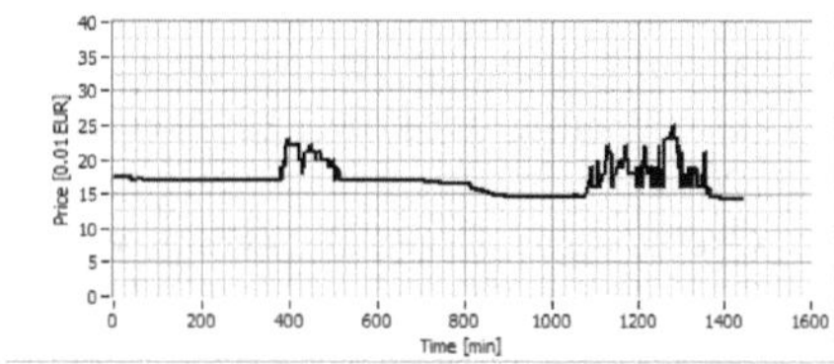

b) Resulting price per electrical energy

Fig. 4. Proposed algorithm of finding the price function

5 Conclusions and Future Work

This paper proposes an improved real-time power balancing approach for DC distribution grids. It is based on the possibility of changing the reference parameter of the DC voltage for a bidirectional AC/DC interface converter influencing the DC bus voltage. When the voltage is decreased, it will lead to the change of operation of grid elements controlled by the droop control method, unnecessary loads will be switched off during the period of low DC voltage. Consequently, a larger amount of the energy will be exported to the utility grid. Thus, it is possible to fulfil long-term objectives just by controlling the voltage reference value of a bidirectional interface converter.

The novel approach of long-term energy management is based on pricing technique. It uses the load categorization that defines different conditions of operation for different loads. Each grid element can communicate with the energy management system (EMS), which aggregates time scheduling and condition of operation of all elements and finds the intersection line between the 3D surface of the load and the 3D surface of generation. As a result, a certain price function is defined that can be used to realize predictive energy management of an intelligent distribution grid.

Our future plan is to conduct experiments to test the proposed methods on a small-scale network to acquire real evaluation of long-term and short-term energy management approaches taking into account typical summer, winter and spring day consumption patterns, as well as appropriate forecast generation patterns.

Acknowledgements. This research is co-financed by the European Regional Development Fund within the project "Intellectual Hybrid Uninterruptible Power Systems and Component Development and Research to Improve Energy Efficiency". Project agreement No. 2010/0225/ 2DP/2.1.1.1.0/10/APIA/VIAA/160.

References

1. Masson, G., Latour, M., Biancardi, D.: Global market outlook for photovoltaic until 2016, Brussels (2012)
2. Jimeno, J., Anduaga, J.: Architecture of a microgrid energy management system. European Transactions on Electrical Power 21, 1142–1158 (2011)
3. Vermesan, O., Friess, P.: Internet of things strategic research roadmap (2009)
4. Smith, I.G.: Internet of Things 2012 New Horizons, Halifax (2012)
5. Atzori, L., Iera, A., Morabito, G.: The Internet of Things: A survey. Computer Networks 54, 2787–2805 (2010)
6. Bryan, J., Duke, R., Round, S.: Decentralized generator scheduling in a nanogrid using DC bus signaling. In: Power Engineering Society General Meeting, vol. 1, pp. 977–982. IEEE (2004)
7. Cvetkovic, I., Dong, D., Zhang, W., Jiang, L., Boroyevich, D., Lee, F.C., Mattavelli, P.: A Testbed for Experimental Validation of a Low-voltage DC Nanogrid for Buildings. In: 2012 15th International Power Electronics and Motion Control Conference, EPE/PEMC, pp. LS7c.51–LS7c.58 (2012)
8. Piagi, P., Lasseter, R.H.: Autonomous control of microgrids. In: Power Engineering Society General Meeting, p. 8. IEEE (2006)
9. Guerrero, J.M., Vasquez, J.C., Teodorescu, R.: Hierarchical control of droop-controlled AC and DC microgrids—a general approach toward standardization. In: Industrial Electronics Conference, pp. 4341–4346 (2011)
10. Schonberger, J., Round, S., Duke, R.: Autonomous Load Shedding in a Nanogrid using DC Bus Signalling. In: 32nd Annual Conference on IEEE Industrial Electronics, IECON 2006, pp. 5155–5160 (2006)
11. Kanchev, H., Lu, D., Colas, F., Lazarov, V., Francois, B.: Energy Management and Operational Planning of a Microgrid With a PV-Based Active Generator for Smart Grid Applications. IEEE Transactions on Industrial Electronics 58, 4583–4592 (2011)
12. Reddy, Y.J., Kumar, Y.V.P., Kumar, V.S., Raju, K.P.: Distributed ANNs in a Layered Architecture for Energy Management and Maintenance Scheduling of Renewable Energy HPS Microgrids (2012)
13. Logenthiran, T., Srinivasan, D.: Short term generation scheduling of a Microgrid. In: TENCON 2009 - 2009 IEEE Region 10 Conference, pp. 1–6 (2009)
14. Ogimoto, K.: Optimum Operation Scheduling Model of Domestic Electric Appliances for Balancing Power Supply and Demand (2010)

Electric Vehicles On-Board Battery Charger for the Future Smart Grids

Vítor Monteiro[1], João C. Ferreira[1],
Andrés A. Nogueiras Meléndez[2], and João L. Afonso[1]

[1] Centro Algoritmi – University of Minho – Guimarães, Portugal
[2] Departamento de Tecnología Electrónica, University of Vigo – Vigo, Spain
{vitor.monteiro,joao.ferreira,joao.l.afonso}@algoritmi.uminho.pt,
aaugusto@uvigo.es

Abstract. The recent and massive investments in Electric Vehicles (EVs) reveal a change of paradigm in the transports sector and the proliferation of EVs will contribute to an effective reduction in the emissions of greenhouse gases. Nevertheless, for the electrical power grids EVs will be extra loads, which will require the demand energy to charge their batteries. With the advent of the Smart Grids, besides the usual battery charging mode (Grid-to-Vehicle – G2V), where the batteries receives energy from the power grid, arises a new concept for the users of EVs and for the power grid market, denominated as Vehicle-to-Grid (V2G).In the V2G operation mode, EVs return to the power grid part of the energy stored in their batteries. The V2G concept requires the use of battery chargers for the EVs with bidirectional power flow capability and bidirectional communication with the Smart Grids through Information and Communication Technology (ICT) applications. It is important to highlight that the proliferation of EVs and the impact of their battery chargers on the power grid quality is a matter of concern, since conventional chargers present current harmonics and power factor problems. In this paper it is presented the preliminary studies resulting from a PhD work about a bidirectional battery charger for EVs, which was designed to operate in collaboration with the power grid as G2V and V2G through an ICT application. In this way, it is expectable to contribute to the technological innovation of the electric mobility in Smart Grids. To assess the behavior of the proposed battery charger under different scenarios of operation, a prototype has been developed, and some simulation and experimental results of the battery charger are presented.

Keywords: Battery Charger, Communication System, Electric Vehicles, Smart Grids, Grid-to-Vehicle (G2V), Vehicle-to-Grid (V2G).

1 Introduction

The upcoming reality of Smart Grids and energy markets will raise a diversity of advantages to the end-user. However, it will require several technologic developments aiming the interaction of the users as active players. The recent increase in the utilization of Electric Vehicles (EVs) with different architectures represents a real contribution to a

L.M. Camarinha-Matos, S. Tomic, and P. Graça (Eds.): DoCEIS 2013, IFIP AICT 394, pp. 351–358, 2013.

new transports paradigm, a gain for independence of the cost of the oil, and also an effective fight against the climatic changes. Effectively, the EVs are seen as one of the most promising means in order to improve the sustainability of the transportation and energy sectors in near-term [1][2].

The EVs in conjunction with Information and Communication Technology (ICT) applications will play an important role to achieve a sustainable balance between production and consumption, and increase the power quality and the reliability of the power grids. The future Smart Grids are the consequence of this new paradigm for the power grids. Smart Grids are not characterized as a particular technology or device, but rather as a vision of a distributed electrical system, supported by reference technologies, as example, the aforementioned ICT, Advanced Metering Infrastructures (AMI), Energy Storage Systems (ESS), Micro Generation (MG), and Power Electronics Systems (PES).Thereby, the actual power grids will be transformed on a coordinated, collaborative [3], and automatic infrastructure [4][5][6].

Nowadays, with the recent bet in EVs around the world, the actual electrical power grids are facing new challenges, forcing to a wide efforts of investigation in different directions [7], mainly taking into account their integration [8][9][10]. Consequently, a considerable amount of energy is stored in the batteries of these vehicles. Thus, besides the charging process (Grid-to-Vehicle – G2V), the energy stored in EVs' batteries may be suitable for providing regulation services, spinning reserves and peak power demand. This interactivity between the vehicles and the power grid is expected to be one of the key technologies in the future of the Smart Grids, denominated Vehicle-to-Grid (V2G) [11][12]. Fig. 1shows a draft of a scenario for a Smart Grid, where the integration of EVs represents the core of the context for this paper.

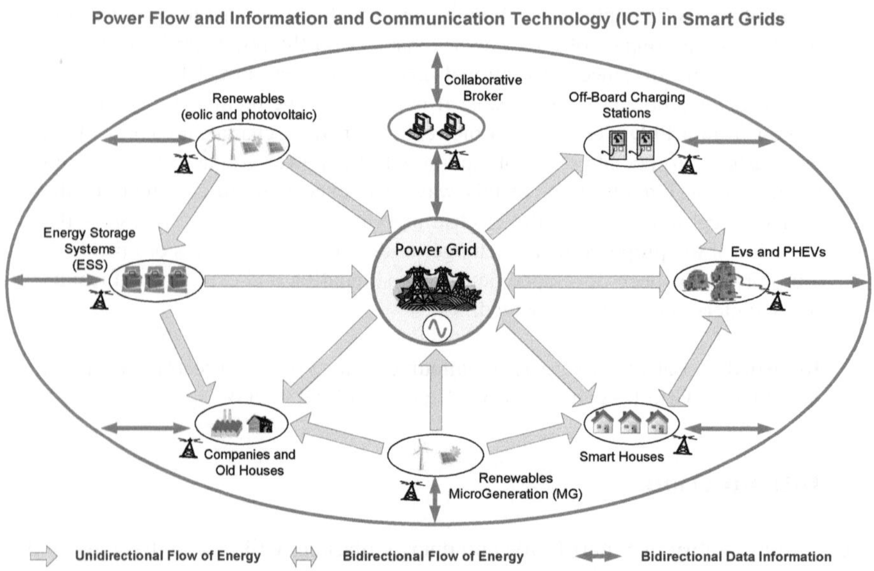

Fig. 1. Scenario for a Smart Grid, where the integration of EVs with bidirectional power flow and communication capabilities represents the core of the context for this paper

This paper results from a PhD work that is yet in the first year, and presents a bidirectional battery charger for EVs based on PES with ICT capabilities. This equipment allows mitigate problems associated with power quality (as distorted current consumption and low power factor), enables the operation of EVs as local ESS, and represents an important contribute to the technological innovation, and to the efficiency and reliability of the electric mobility in Smart Grids.

2 Contribution to Internet of Things

The Functional Areas in Smart Grids represents all the systems related with the production, transmission and distribution [13][14], and considering Advanced Metering Infrastructure (AMI) and ICT [15][16]. In the electric mobility it is important provide the EVs with ICT systems in order to establish a bidirectional flux of information to ensure that the batteries are charged when the electricity is cheapest and the impact on the grid is smallest. With a coordinated charging process of EVs the energy losses in the distribution system are minimized [17]. To reach this goal, the batteries charging process needs assistance of an intelligent process in order to find the periods with cheaper prices to charge the batteries, to identify the available charging slots in public and private areas, and to provide other useful information to the drivers, as their historic use [18][19]. This theme, including supervision, control and communication applications, will be the final topic that will be addressed in the PhD work. The main goal is establish a bidirectional flux of information between the EVs and the power grids aiming to define strategies of cooperation to the G2V and V2G modes of operation [20].

In order to contribute to the technological innovation of the electric mobility in Smart Grids, the bidirectional battery charger for EVs presented in this paper was designed to operate in collaboration with the power grid as G2V and V2G through an ICT application [21]. This way, it will be possible exchange information through the internet, contributing to strengthening the Internet of Things and the impact of the electric mobility in the Smart Grids. As aforementioned, besides the bidirectional flow of active power, through the power factor control, it is also possible control the reactive power in accordance with the capabilities of the battery charger [22]. In Fig.2 is shown the integration of EVs in the power grid considering the active and reactive bidirectional power flow of the presented battery charger.

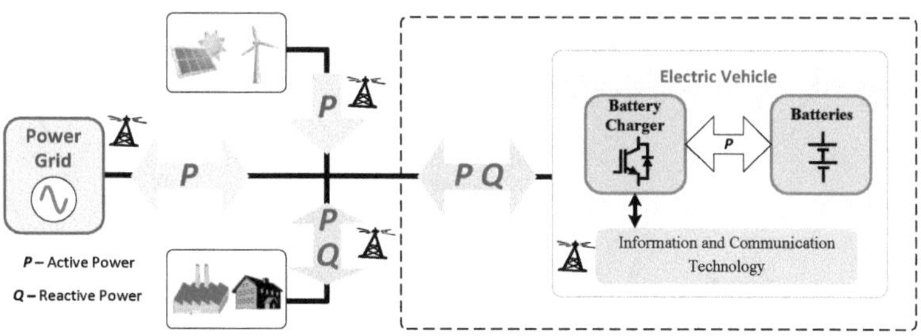

Fig. 2. Smart Grid considering the integration of EVs with bidirectional power flow

3 On-Board Battery Charger

Electric Vehicles are becoming a part of the electric power grid and consequently the battery chargers of these vehicles should have the ability to avoid power quality problems [23][24]. Even more, the EVs power stages can also assume, if needed, the role of the active power quality filters. The battery charger presented in this paper is a device that is composed by two power electronics converters used to adapt the AC electrical energy into DC (AC-DC converter) and to control the output voltage or currents levels (DC-DC converter). To contribute to the power quality in the future Smart Grids, the battery chargers should consume sinusoidal current with controlled power factor. In Fig.3 is presented the schematic of the first prototype of battery charger developed during the PhD work.

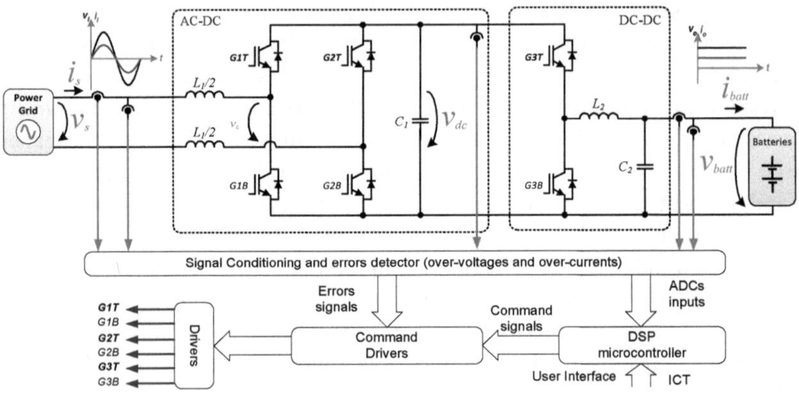

Fig. 3. Schematic of the presented bidirectional power converter

3.1 Modes of Operation

In an early stage, in order to analyses the operation of the converters in both modes of operation was developed a simulation model using the simulation tool PSIM, and were realized different computer simulations. In Fig.4 are shown some simulations results obtained during the operation of the bidirectional power converters.

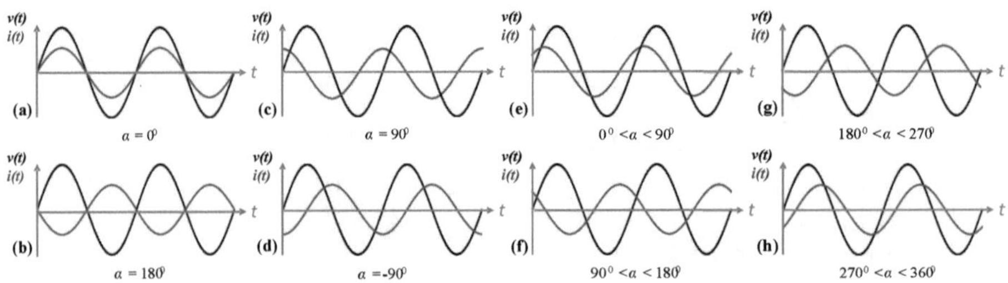

Fig. 4. Summary of some simulation results considering the voltage and the current in the electrical power grid and adjusting the reactive power

3.2 Digital Control System

In a first stage of the PhD work, in this prototype, the operation of the battery charger occur in accordance with the orders given, according to the demands of the user of the EV, which define when it works as G2V and V2G, and the value of reactive power to be produced. The communication with the power grid through an ICT application, instead the user, will be developed along the PhD work where this paper is encompassed. In Fig.5are presented the control algorithms for the AC-DC and DC-DC converters.

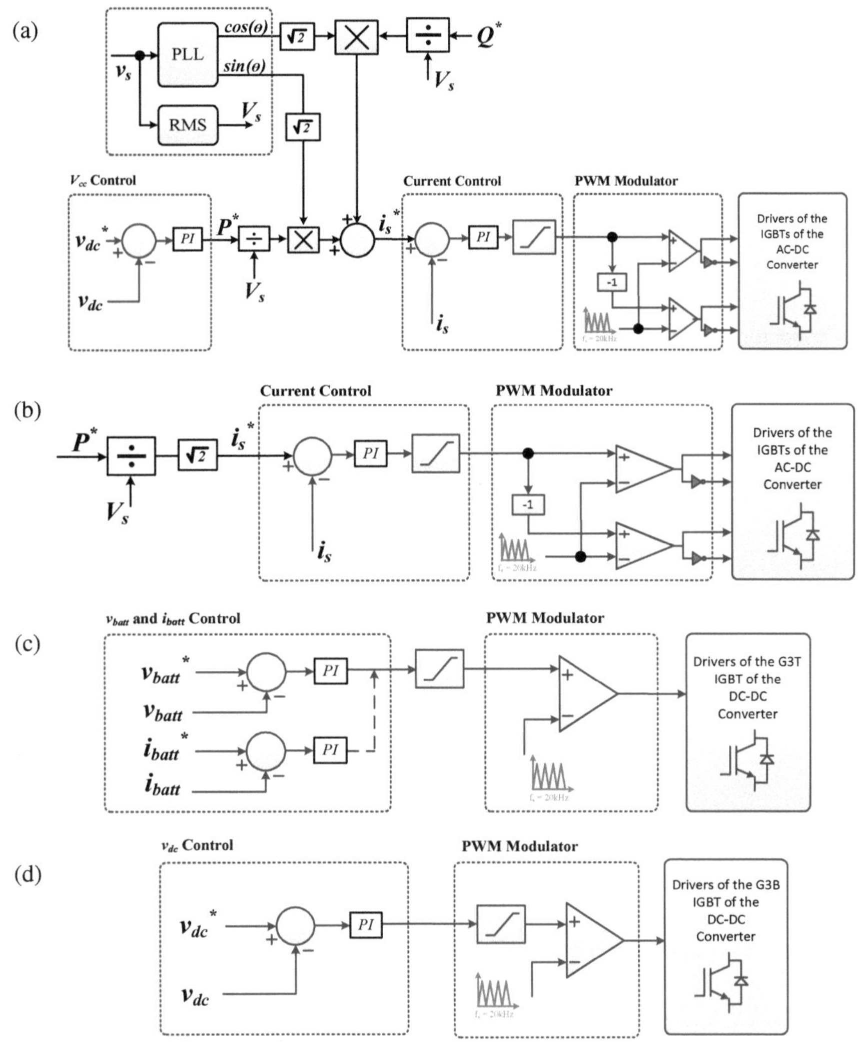

Fig. 5. Control algorithms: AC-DC converter operating as G2V (a) and V2G (b); DC-DC converter operating as G2V (c) and V2G (d)

4 Experimental Results

In order to assess the operation of the on-board battery charger under different modes of operation was developed the prototype presented in Fig.6, aiming to be integrated in an EV with Absorbed Glass Mat batteries with nominal voltage 96 V and nominal capacity 33 Ah. Preliminary experimental results are shown in Fig.7.

Fig. 6. On-board battery charger prototype developed to operate as G2V and V2G controlled through an ICT application, to be implemented in an Electric Vehicle

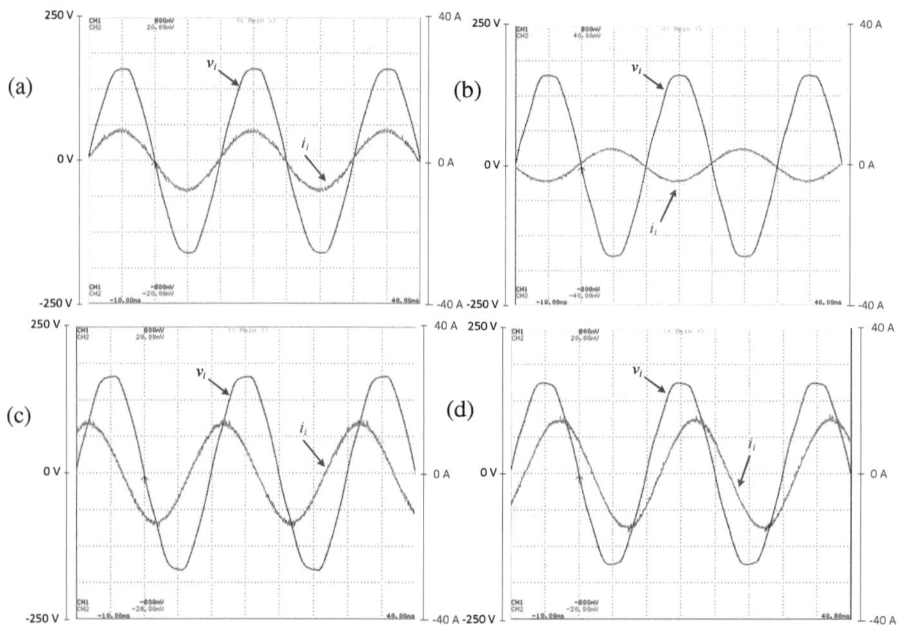

Fig. 7. Experimental results with controlling of the active and reactive power: (a) G2V mode with unitary power factor; (b) V2G mode; (c) G2V mode with capacitive power factor; (d) G2V mode with inductive power factor

5 Conclusions and Further Work

In this paper was presented an on-board battery charger for Electric Vehicles (EVs) aiming their integration in Smart Grids. This battery charger allows the interaction with the electrical power grid to charge the batteries (Grid-to-Vehicle - G2V mode),or to deliver part of the stored energy in the batteries back to the power grid (Vehicle-to-Grid - V2G mode). The modes of operation and the reactive power control are performed considering the user interface of the EVs. However, in the final of the PhD work where this paper is encompassed, the goal is to control the on-board battery charger of the EV through an ICT application aiming to allow the communication with a collaborative broker of the electrical power grid.

In a first stage, the behavior of the bidirectional power converter was evaluated under different scenarios through computer simulations. Then the behavior of the bidirectional power converter was evaluated with a prototype, which was developed aiming to reduce its volume and weight, in order to be integrated in an Electric Vehicle with Absorbed Glass Mat (AGM) batteries. In this paper were presented the simulations and experimental results obtained.

Acknowledgments. This work is financed by FEDER Funds, through the Operational Programme for Competitiveness Factors – COMPETE, and by National Funds through FCT – Foundation for Science and Technology of Portugal, under the projects: FCOMP-01-0124-FEDER-022674 and PTDC/EEA-EEL/104569/2008.

References

1. Bradley, T., Frank, A.: Design, demonstrations and sustainability impact assessments for plug-in hybrid electric vehicles. Elsevier, Renewable and Sustainable Energy Reviews 13, 115–128 (2009)
2. Chan, C., Bouscayrol, A., Chen, K.: Electric, Hybrid, and Fuel-Cell Vehicles: Architectures and Modeling. IEEE Transactions on Vehicular Technology 59, 589–598 (2010)
3. Ferreira, J.C., Silva, A.R., Monteiro, V., Afonso, J.L.: Collaborative Broker for Distributed Energy Resources. In: International Symposium on Computational Intelligence for Engineering Systems, Portugal (2011)
4. Moslehi, K., Kumar, R.: A Reliability Perspective of the Smart Grid. IEEE Transactions on Smart Grid 1, 57–64 (2010)
5. Medina, J., Muller, N., Roytelman, I.: Demand Response and Distribution Grid Operations Opportunities and Challenges. IEEE Transactions on Smart Grid 1, 193–198 (2010)
6. Molderink, A., Bakker, V., Bosman, M., Hurink, J., Smit, G.: Management and Control of Domestic Smart Grid Technology. IEEE Transactions on Smart Grid 1, 109–119 (2010)
7. Khaligh, A., Zhihao, L.: Battery, Ultracapacitor, Fuel Cell, and Hybrid Energy Storage Systems for Electric, Hybrid Electric, Fuel Cell, and Plug-In Hybrid Electric Vehicles: State of the Art. IEEE Transactions on Vehicular Technology 59, 2806–2814 (2010)
8. Monteiro, V., Gonçalves, H., Afonso, J.L.: Impact of Electric Vehicles on Power Quality in a Smart Grid Context. In: 11th IEEE EPQU International Conference on Electrical Power Quality and Utilization, Portugal (2011)

9. Gomez, J., Morcos, M.: Impact of EV battery chargers on the power quality of distribution systems. IEEE Transactions on Power Delivery 18, 975–981 (2003)
10. Lopes, J., Soares, F., Almeida, P.: Integration of Electric Vehicles in the Electric Power System. Proceedings of the IEEE 99, 168–183 (2011)
11. Han, S., Han, S., Sezaki, K.: Development of an Optimal Vehicle-to-Grid Aggregator for Frequency Regulation. IEEE Transactions on Smart Grid 1, 65–72 (2010)
12. Sortomme, E., El-Sharkawi, M.: Optimal Charging Strategies for Unidirectional Vehicle-to-Grid. IEEE Transactions on Smart Grid 2, 131–138 (2011)
13. Zhang, P., Li, F., Bhatt, N.: Next-Generation Monitoring, Analysis, and Control for the Future Smart Control Center. IEEE Transactions on Smart Grid 1, 186–192 (2010)
14. Heydt, G.T.: The Next Generation of Power Distribution Systems. IEEE Transactions on Smart Grid 1, 225–235 (2010)
15. Sauter, T., Lobashov, M.: End-to-End Communication Architecture for Smart Grids. IEEE Transactions on Industrial Electronics 58, 1218–1228 (2011)
16. Yi, P., Iwayemi, A., Zhou, C.: Developing ZigBee Deployment Guideline Under WiFi Interference for Smart Grid Applications. IEEE Transactions on Smart Grid 2, 110–120 (2011)
17. Sortomme, E., Hindi, M., MacPherson, S., Venkata, S.: Coordinated Charging of Plug-In Hybrid Electric Vehicles to Minimize Distribution System Losses. IEEE Transactions on Smart Grid 2, 198–205 (2011)
18. Monteiro, V., Afonso, J.L., Ferreira, J.C.: An Agent Model for the Simulation of Electrical Vehicle Charging Management. In: IEEE CISTI Conference on Information Systems and Technologies, Portugal, pp. 1–4 (2011)
19. Ferreira, J.C., Silva, A.R., Monteiro, V., Afonso, J.L.: Collaborative Broker for Distributed Energy Resources. In: International Symposium on Computational Intelligence for Engineering Systems, ISEC, Portugal (2011)
20. Bilgin, B., Emadi, A., Krishnamurthy, M.: Design considerations for a universal input battery charger circuit for PHEV applications. In: IEEE ISIE International Symposium on Industrial Electronics, pp. 3407–3412 (2010)
21. Monteiro, V., Gonçalves, H., Ferreira, J.C., Afonso, J.L.: Batteries Charging Systems for Electric and Plug-In Hybrid Electric Vehicles. In: Carmo, J.P., Ribeiro, J.E. (eds.) New Advances in Vehicular Technology and Automotive Engineering, 1st edn., pp. 149–168. InTech (2012)
22. Kisacikoglu, M., Ozpineci, B., Tolbert, L.: Examination of a PHEV Bidirectional Charger System for V2G Reactive Power Compensation. In: IEEE APEC Applied Power Electronics Conference and Exposition, pp. 458–465 (2010)
23. Lee, Y.-J., Khaligh, A., Emadi, A.: Advanced Integrated Bidirectional AC/DC and DC/DC Converter for Plug-In Hybrid Electric Vehicles. IEEE Transactions on Vehicular Technology 58, 3970–3980 (2009)
24. Kurohane, K., Senjyu, T., Yona, A., Urasaki, N., Goya, T., Funabashi, T.: A Hybrid Smart AC/DC Power System. IEEE Transactions on Smart Grid 1, 199–204 (2010)

Part XIII
Energy Generation

Schedule of Thermal Units with Emissions in a Spot Electricity Market

R. Laia[1], H.M.I. Pousinho[2], R. Melício[1], V.M.F. Mendes[3], and A.H. Reis[1]

[1] University of Évora, Évora, Portugal
rui.j.laia@gmail.com, ruimelicio@uevora.pt, ahr@uevora.pt
[2] Instituto Superior Técnico and CIEEE, Lisbon, Portugal
hpousinho@gmail.com
[3] Instituto Superior of Engenharia de Lisboa, Lisbon, Portugal
vfmendes@deea.isel.pt

Abstract. A bi-objective optimization approach is presented for solving a generation company short-term thermal schedule problem with a few units, considering the goodness of being schedule, but with emission concern. The startup and shutdown for each unit throughout the time horizon is derived from Pareto-optimal solutions, using a method merging dynamic programming and nonlinear programming to provide schedule of the units. A case study is presented to prove the effectiveness of the approach.

Keywords: Short-term schedule, thermal units, emissions, dynamic programming, Pareto-optimal solution.

1 Introduction

Over the last years, ambitious policy targets and concrete actions have been proposed in order to encourage mitigation of greenhouse anthropogenic gases emissions and ensure environmental sustainability worldwide [1]. A significant share on emission of greenhouse gases into the atmosphere is through the burning fossil fuel on thermal power plants [2]. In 2011, thermal power plants played a dominant role in the mixed power generation, accounting for 54.2 % of total generation in the EU-27 and 57.1 % in Portugal [3]. The expressive share of thermal power plants in a competitive framework forces as a fundamental tackling the schedule problem of thermal units during a short-term horizon in order to conveniently appraise favorable economic conditions for the Generation Companies (GENCO). Within the past 40 years, mathematical programming techniques have been used aiming at the optimization of thermal unit commitment. In early studies, still without the availability of enough computational power to support a full modeling of the problem, the unit commitment was based on priority lists [4]. The development of computational power allowed to model the complexity of the problem, allowing over the years the inclusion of new constraints not only related to operational costs, technical boundaries, but also reserve, minimum up/down time, ramp rate power constraints and supply limits. Furthermore, in nowadays the electricity market framework provides a trading

L.M. Camarinha-Matos, S. Tomic, and P. Graça (Eds.): DoCEIS 2013, IFIP AICT 394, pp. 361–370, 2013.

mechanism based on bilateral contracts suitable for thermal power producers. This kind of mechanism is used in order to allow hedging price volatility. Bilateral contracts are agreements between power producers and consumers to provide a given amount of energy at a predefined price along with a delivering period [5]. Hence, additional modeling must be included in the schedule problem in order to conveniently model electricity market framework influence on the schedule. This paper proposes a bi-objective optimization approach, requiring the use of a method based on Dynamic Programming (DP) and nonlinear programming. A case study is presented for a schedule over a time horizon of 168 hours with hourly periods.

2 Relationship to Internet of Things

Electric power systems are tagged by evolution of computing technologies, allowing the development of powerful optimization approaches able to process decisions under the operational planning. Consequently, every GENCO in the new competitive market paradigm are envisaged in order to ensure convenient management conditions to be a thing in the Internet of Things.

The connection to the Internet of Things conveys adequate decisions for system operators. The interfacing with the internet allows bidirectional communication between the power producers and the remaining decision makers, for instance, regarding real time market information and bilateral contracts. This paper deals with an application contributing to take decision in real time on market information and bilateral contracts.

3 State of the Art

Technical literature presents several optimization methods for solving the unit commitment and the economic dispatch problem: methods have been reported since the old priorities list method [6] to the classical mathematical programming methods until the more recently reported artificial intelligence methods [7]. Although, easy to implement and requiring a small computation time, the priority list method does not ensures an economic convenient solution near a global optimal one, implying a higher operation cost [8]. Within the classical methods are included DP, linear programming, nonlinear programming and Lagrangian relaxation-based techniques [9]. DP methods are flexible but suffer from the "curse of dimensionality", due to the increase in the problem size related with the number of thermal units to be committed and the number of states considered for modeling the thermal behavior of each unit during the time horizon, implying an eventually huge use of computation memory and processing time. Although the Lagrangian relaxation [10] can overcome the previous limitation, does not always lead to a conveniently feasible solution, requiring in order to set a feasible solution the satisfaction of some violated constraints using heuristics, undermining the optimality. Artificial intelligence (AI) methods based on artificial neural networks [11], genetic algorithms [12], evolutionary algorithms [13] and simulating annealing [14] have also been applied. However, the major limitation of the AI methods is the likelihood to obtain a convenient solution near global optimum, especially with a few thermal units. These problems are the ones faced by some GENCO due to the new paradigm, admitting companies with a small number of units to go into business.

Emissions concerns in the literature are mostly addressed for the economic dispatch problem [15], stating only the power output level of each unit but not the on/off status and availability for generation at each hour. The short-term thermal unit schedule problem has to take into account two assessments, i) the thermal units online at each hour and ii) the power output level for each online unit at each hour, so as to suit economic and/or environmental targets on time horizon of one day to one week [9]. The effects of the short-term thermal unit schedule are significant, in sense that a better operation achieves not only a reduction in the fuel consumption [16], but meets environmental concerns: once electric power generation is obtained from fuel-fossil power plants, the anthropogenic emission cannot be overlooked in nowadays. The emission modeling in the short-term schedule problem [17] has not been deeply tackled as in the case of the economic dispatch problem. Hence, in this paper the goodness of being schedule and the emission concern are included in the short-term schedule of thermal units.

4 Problem Formulation

The short-term thermal unit schedule problem can be stated as to find the schedule on status and the power generated for each thermal unit i at each time period t that optimizes performance criterions involving costs, emissions and market trading subject to a set of constraints on the operation of the units.

4.1 Objective Function

The proposed framework considers two different objectives in which the first one expresses the total fuel cost of thermal units and the second one expresses the total emissions of pollutants into the atmosphere. Mathematically, the bi-objective vector to be minimized is given by:

$$\{\sum_{t=1}^{T}\sum_{i=1}^{I} C_{it}(u_{it}, p_{it}), \sum_{t=1}^{T}\sum_{i=1}^{I} \omega E_{it}(u_{it}, p_{it})\}. \tag{1}$$

I is the number of thermal units; T is number of periods in the time horizon; for unit i in t period, $C_{it}(u_{it}, p_{it})$, u_{it}, p_{it}, $E_i(u_{it}, p_i)$ are respectively the fuel cost function, the commitment state (on/off) , the power generated and is the emission function. Since the total fuel cost and total emissions are expected to be conflicting objectives, implying the impossibility to find out a single optimal solution that simultaneously satisfies both objectives. Thus, the set of best compromise solution, known as the Pareto front, can be determine using the weight-sum method, converting the bi-objective vector (1) into a family of single objective function given by a convex combination of the objectives, i.e., given by:

$$(1-\lambda)\sum_{t=1}^{T}\sum_{i=1}^{I}C_{it}(u_{it},p_{it})+\lambda\sum_{t=1}^{T}\sum_{i=1}^{I}\omega E_{it}(u_{it},p_{it}). \tag{2}$$

ω is a unit price penalty factor associated with the emission and λ is a weighting factor, which should obey $0\le\lambda\le 1$. If $\lambda=0$, the solution correspond to minimum cost, the usual unit commitment, and if $\lambda=1$, the solution is minimum emissions commitment. The fuel cost and emissions functions for the operation of each thermal unit depend on the power generated by that unit and can be modeled as a second order quadratic functions respectively given by:

$$C_{itop}(u_{it},p_{it})=(a_{it}+b_{it}p_{it}+\frac{1}{2}c_{it}p_{it}^{2})u_{it}. \tag{3}$$

$$E_{itop}(u_{it},p_{it})=(\alpha_{it}+\beta_{it}p_{it}+\frac{1}{2}\gamma_{it}p_{it}^{2})u_{it}. \tag{4}$$

a_i, b_i, c_i and α_i, β_i, γ_i are respectively the cost coefficients and the emission coefficients for thermal unit i. A GENCO can trade energy via bilateral agreements. Hence, an augmenting term must be included into (2). The augmented objective function is given by:

$$(1-\lambda)\sum_{t=1}^{T}\sum_{i=1}^{I}C_{it}(u_{it},p_{it})+\lambda\sum_{t=1}^{T}\sum_{i=1}^{I}\omega E_{it}(u_{it},p_{it})-\sum_{t=1}^{T}\sum_{i=1}^{I}\pi_t(p_{it}-d_t). \tag{5}$$

π_t is electricity price at period t and d_t is the power contracted with bilateral agreements associated with energy that have to be delivered at period t. This objective function admits the possibility of buying energy in the market if production is not enough to satisfy agreements. The objective function (5) can be seen as the application of the weighted-sum method for a bi-objective optimization problem with the objective vector given by:

$$\{\sum_{t=1}^{T}\sum_{i=1}^{I}C_{it}(u_{it};p_{it})-\pi_t(p_{it}-d_t),\sum_{t=1}^{T}\sum_{i=1}^{I}\omega E_{it}(u_{it},p_{it})-\pi_t(p_{it}-d_t)\}. \tag{6}$$

In other words (5) can be seen as an objective vector with coordinates respectively given by the objective function associated with the minimum emission and minimum cost commitment both with bilateral contract agreements. The start-up cost of thermal units is a term to be added the fuel operation cost in (3) and depends upon the number of periods x_t the unit has been offline prior to startup. The start-up cost is given by:

$$SU_{it}=uccool_{i0}(1-e^{-\frac{5x_t}{utcool_i}}). \tag{7}$$

$uccool_{i0}$ is the cold start-up cost and $utcool_i$ is the cooling time constant. Also, the start-up emission can be included into (4) in the same way as given by (7).

4.2 Constraints

The optimization problem is subject to a set of constraints due to operation conditions that can include the following:

a) Availability of thermal units: The sum of online units cannot exceed the maximum number of thermal units allowed online in period t.

$$\sum_{i \in I} u_{it} \leq NMV_t \ . \tag{8}$$

b) Power balance constraint: The power generated by the thermal units must meet at least a demand D_t in each period t, ignoring transmission losses in the system.

$$\sum_{i \in I} p_{it} \geq D_t \ . \tag{9}$$

c) Spinning reserve constraint: The spinning reserves R_t in period t are necessary in order to ensure reliability.

$$\sum_{i \in J \subset I} \overline{p}_{it} \geq D_t + R_t \ . \tag{10}$$

d) Operating ramp rate constraints: The power generated over any two consecutive online periods is restricted by the ramp-down DR_i and ramp-up UR_i limits.

$$DR_i \leq p_{it+1} - p_{it} \leq UR_i \ . \tag{11}$$

e) Generator Capacity Constraints: Each thermal unit is restricted by its minimum and maximum limits on power generation.

$$u_{it} \underline{p}_{it} \leq p_{it} \leq u_{it} \overline{p}_{it} \ . \tag{12}$$

f) Minimum up time constraint: The minimum up time UD_i imposes that unit i have to be on by at least the minimum up time before shutdown.

$$u_{it}(1 - u_{it+1}) = 1 \ , \text{ if equal to 1, the shutdown occurs at period } t+1 \ . \tag{13}$$

g) Minimum down time constraint: The minimum down time DD_i imposes that unit i have to be down by at least the minimum down time before startup.

$$u_{it+1}(1 - u_{it}) = 1 \ , \text{ if equal to 1, the start-up occurs at period } t+1 \ . \tag{14}$$

h) Bilateral agreement constraint: If necessary in order to ensure that a bilateral agreement is fulfilled the difference between generated power and contracted power with bilateral agreements can be imposed as a non-negative constraint.

$$\sum_{i \in I} p_{it} - d_t \geq 0 \quad \text{for} \quad t = 1, 2, \ldots, T \ . \tag{15}$$

The minimum up/down time constraints are used to avoid thermal stress, implying future augmenting maintenance. More constraints are possible to into the problem, but the complexity will be augmented, implying an augmented processing time. Particularly, instead of (15) is possible to have bilateral agreement only strong fulfilled at particularly periods, leaving to other periods the possibility of buying energy in the market to fulfill agreements.

5 Case Study

The proposed bi-objective optimization approach is tested on GENCO with three thermal units in a competitive electricity market with bilateral contracts. Simulation studies are carried out on hourly basis over a scheduling time horizon of 168 hours. The fuel cost and emission coefficients are given in Table 1.

Table 1. Fuel cost and emission coefficients for thermal units

	U1	U2	U3
a_i	277	300	320
b_i	26.500	26.300	26.100
c_i	0.045	0.051	0.031
α_i	40	43	38
β_i	-1.6×10^{-3}	-1.3×10^{-3}	-1.9×10^{-3}
γ_i	0.008	0.009	0.006
$\underline{p}_i$ (MW)	40	120	240
$\overline{p}_i$ (MW)	120	400	700
UR_i (MW)	40	120	240
DR_i (MW)	20	80	160
UD_i (h)	4	5	7
DD_i (h)	3	3	5
$utcool_i$ (h)	4	5	7
$uccool_{i0}$ (Eur)	2200	2400	3000

Based on [18], the price penalty factor used for emissions is $\varpi = 12.3\,\text{Eur/Mg}$. The numerical computing testing has been performed on a 1.9-GHz-based processor with 2 GB of RAM using as a computing language the VBA for Microsoft Excel platform. Nonlinear and DP methods was used in order to solve the proposed short-term thermal unit schedule problem, being the first step to determine the feasible solutions taking into account the minimum up/down time constraints and total cooling time of each unit. The generated power was computed for each thermal unit schedule. The total number of states is 1120 to be processed at each hour. The energy price profile and the energy contracted at each hour are shown in Fig. 1.

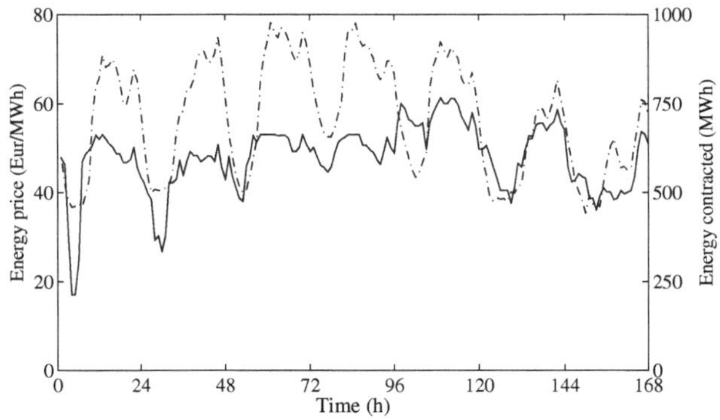

Fig. 1. Price profile *(solid line)* and energy contracted *(dashed–dotted line)*

The computed energy over the time horizon as function of λ is given in Table 2.

Table 2. Energy contracted and supplied, as function of the weighting factor λ

λ	Total costs (Eur)	Energy contracted (MWh)	Energy supplied (MWh)
0.0	4117780	119096	138084
0.2	4092690	119096	120329
0.4	3923691	119096	119135
0.6	3729017	119096	119096
0.8	3568313	119096	119096
1.0	3408946	119096	119096

The major surplus between the supplied and contracted energy happens for with $\lambda = 0$. The power contracted, the power supplied and the hourly units committed are respectively shown for $\lambda = 0$ in Fig. 2 and Fig. 3.

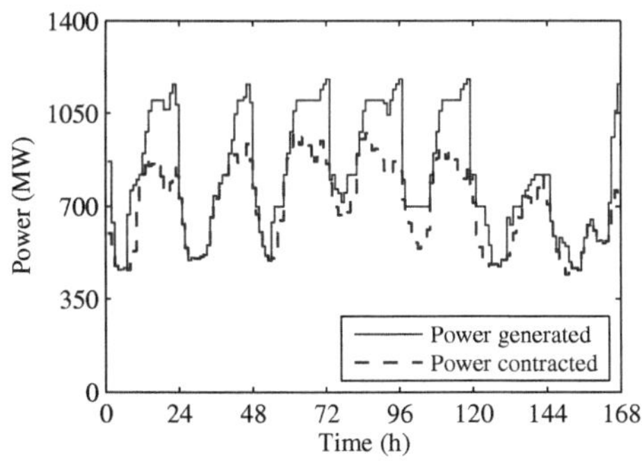

Fig. 2. Power generated and contracted, $\lambda = 0$

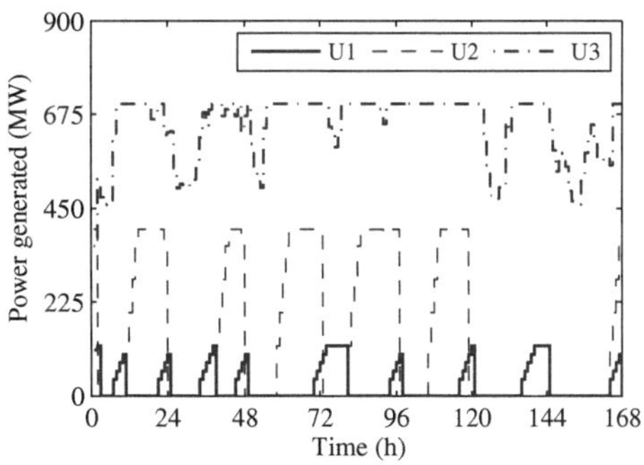

Fig. 3. Unit power generated, $\lambda = 0$

Fig. 2 shows that the power generated is greater than the contracted power at some hours. This is due to higher values of energy prices at these hours. The values of power contracted are closer to the ones obtain for lower energy prices, because is not profitable to generate more energy than the necessary to comply with bilateral agreement. Fig. 4 shows different behaviours for each thermal unit due to differences in startup costs. Both thermal units U1 and U2 are off when the energy price is lower, while U3 is on in order to avoid incur if shutdown occurred on the higher startup cost.

The power contracted, the power supplied and the hourly units committed are respectively shown for $\lambda = 1$ in Fig. 4 and Fig. 5.

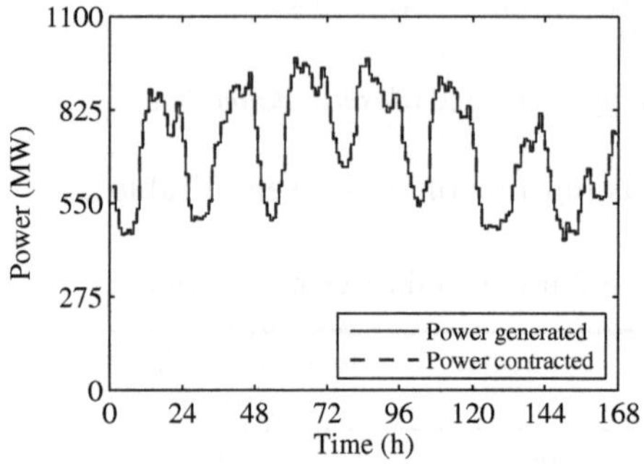

Fig. 4. Power generated and contracted, $\lambda = 1$

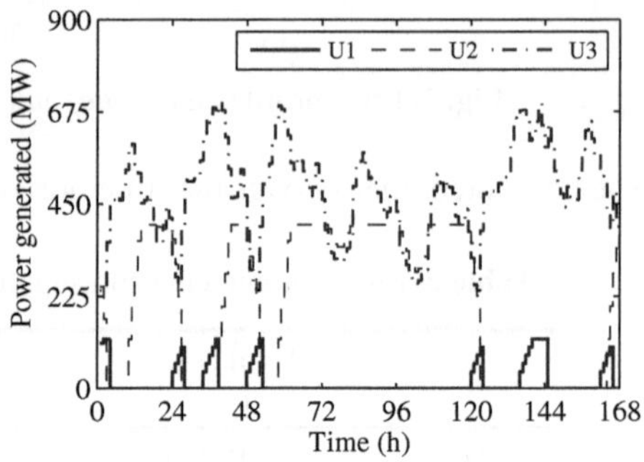

Fig. 5. Unit power generated, $\lambda = 1$

Fig. 4 shows that the power generated trajectory equals the power contracted, comparing Fig. 3 with Fig. 5 the power generated by each thermal unit is different: the schedule of the thermal units U2 and U3 are different due to the coefficients of the fuel cost and emission cost functions. Unit U1 comes into operation for the lower levels of the power contracted either with $\lambda = 0$ or $\lambda = 1$.

The computed trade-off curve between the total fuel cost and emission cost, the Pareto front, drown with the Pareto optimal solutions in shown Fig. 6.

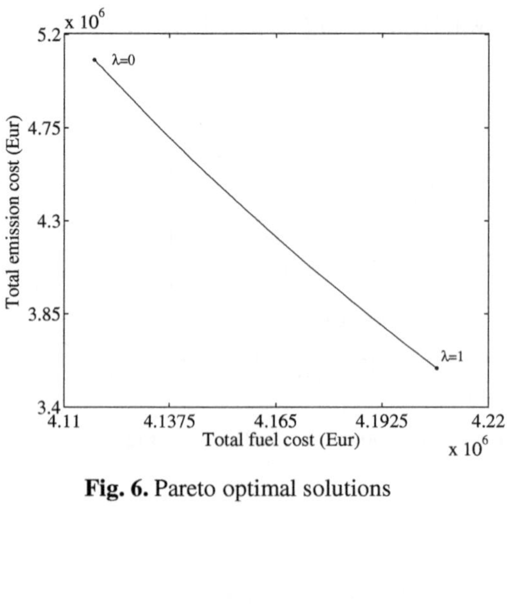

Fig. 6. Pareto optimal solutions

Fig. 6. shows that the objective function are conflicting ones due to the need to compromise between units with significant less cost, but with significant levels of emission, and others with greater cost but lesser emission. The first are committed regarding small costs, the second regarding small emissions.

6 Conclusions

A short-term generation scheduling approach for thermal units in small GENCO's is proposed. The approach considers the goodness of being schedule with the emission concern. Operation constraints of thermal units like ramp rate constraints and minimum up/down time constraints are taken into account, using DP and nonlinear programming. Numerical results allow concluding that the proposed approach is suitable to obtain the trade-off curve to assist on the generation scheduling decisions: according to the trade-off curve, a significant reduction on emissions can be achieved by rescheduling the thermal units, but implying an increase in operational cost. Although, DP suffers from the "curse of dimensionality", therefore impracticable to be used on large-scale systems, on small GENCO's the use of the DP is reasonable. Further work will be carried out including mixed integer programming in order to improve the computation performance and introduce new model considerations.

References

1. Yamashita, D., Niimura, T., Yokoyama, R., Marmiroli, M.: Pareto-optimal solutions for trade-off analysis of CO_2 vs. cost based on DP unit commitment. In: Proc. 2010 Int. Conf. on Power System Technology, Zhejiang, Hangzhou, China, pp. 1–6 (2010)
2. Catalão, J.P.S., Mendes, V.M.F.: Influence of environmental constraints on profit-based short-term thermal scheduling. IEEE Trans. Sustainable Energy 2, 131–138 (2011)
3. Eurostat, Archive agriculture, environment, energy and transport statistics. Eurostat – Statistics Explained 4, 1–1042 (2012)
4. Xie, J., Zhong, J., Li, Z., Gan, D.: Environmental-economic unit commitment using mixed-integer linear programming. Euro. Trans. Electr. Power 21, 772–786 (2011)
5. Heredia, F.J., Rider, M.J., Corchero, C.: A stochastic programming model for the optimal electricity market bid problem with bilateral contracts for thermal and combined cycle units. Annals of Oper. Res. 193, 107–127 (2012)
6. Burns, R.M., Gibson, C.A.: Optimization of priority lists for a unit commitment program. In: Proc. Conf. of IEEE/Power Engineering Society Summer Meeting, pp. 453–456 (1975)
7. Mantawy, A.H., Abdel-Magid, Y.L., Seliin, S.Z.: A simulated annealing algorithm for unit commitment. IEEE Trans. Power Syst. 13, 197–204 (1998)
8. Senjyu, T., Shimabukuro, K., Uezato, K., Funabashi, T.: A fast technique for unit commitment problem by extended priority list. IEEE Trans. Power Syst. 18, 882–888 (2003)
9. Chandrasekaran, K., Hemamalini, S., Simon, S.P., Padhy, N.P.: Thermal unit commitment using binary/real coded artificial bee colony algorithm. Electr. Power Syst. Res. 84, 109–119 (2012)
10. Zhuang, F., Galiana, F.D.: Towards a more rigorous and practical unit commitment by Lagrangian relaxation. IEEE Trans. Power Apparatus Syst. 102, 1218–1225 (1983)

11. Ouyang, Z., Shahidehpour, M.: A hybrid artificial neural network-dynamic programming approach to unit commitment. IEEE Trans. Power Syst. 7, 236–242 (1992)
12. Kazarlis, S.A., Bakirtzis, A.G., Petridis, V.: A genetic algorithm solution to the unit commitment problem. IEEE Trans. Power Syst. 11, 83–92 (1996)
13. Dhillon, J.S., Kothari, D.P.: Economic-emission load dispatch using binary successive approximation-based evolutionary search. IET Gener. Trans. Distrib. 3, 1–16 (2009)
14. Wong, S.Y.W.: An enhanced simulated annealing approach to unit commitment. Int. J. Electr. Power Energy Syst. 20, 359–368 (1998)
15. Abido, M.A.: Multiobjective particle swarm optimization for environmental/economic dispatch problem. Elect. Power Syst. Res. 79, 1105–1113 (2009)
16. Wood, A.J., Wollenberg, B.F.: Power Generation, Operation and Control. Wiley, New York (1996)
17. Gjengedal, T.: Emission constrained unit-commitment (ECUC). IEEE Trans. Energy Convers. 11, 132–138 (1996)
18. Sistema Electrónico de Negociação de Direitos de Emissão de CO_2, `http://www.sendeco2.com`

An Advanced LMI-Based-LQR Design for Load Frequency Control of an Autonomous Hybrid Generation System

S.K. Pandey[1], S.R. Mohanty[1,2], N. Kishor[1,3], and J.P.S. Catalão[2,4]

[1] Motilal Nehru National Institute of Technology, Allahabad, India
skp1111.1969@rediffmail.com, soumyaigit@gmail.com
[2] University of Beira Interior, Covilhã, Portugal
catalao@ubi.pt
[3] Aalto University, Finland
nand_research@yahoo.co.in
[4] CIEEE, Instituto Superior Técnico, Tech. Univ. Lisbon, Portugal

Abstract. This paper proposes a load frequency control scheme for an autonomous hybrid generation system consisting of wind turbine generator (WTG), diesel engine generator (DEG), fuel cell (FC), aquaelectrolyzer (AE) and battery energy storage system (BESS). In wind power generation systems, operating conditions are changing continually due to wind speed and load changes, having an effect on system frequency. Therefore, a robust controller is required for load frequency control. The control scheme is based on Linear Matrix Inequality (LMI)-Linear Quadratic Regulator (LQR). The control optimization problem is obtained in terms of a system of LMI constraints and matrix equations that are simultaneously solved. The proposed load frequency control scheme with the advanced LMI-based-LQR (ALQR) design is applied for the autonomous hybrid generation system. The effectiveness and robustness of the proposed controller is demonstrated for different load and wind power perturbations. The results suggest superior performance of the proposed ALQR controller against an optimal output state feedback controller. The integrated control could be realized though the web by applying Internet of Things technologies within the future smart grid.

Keywords: Load frequency control, distribution generation, optimal control, linear matrix inequalities, Internet of Things.

1 Introduction

In remote areas around the world an autonomous hybrid generation system is useful to supply the demand of electrical energy. A hybrid power system consists of a combination of two or more power generation technologies to enhance their operating characteristics and efficiencies than that could be obtained from a single power source [1]. To meet increasing load demands for an isolated community, expansion of the distributed generation systems is required. The resulting power system may provide

L.M. Camarinha-Matos, S. Tomic, and P. Graça (Eds.): DoCEIS 2013, IFIP AICT 394, pp. 371–381, 2013.

good quality service to the consumer load, which depends mostly on the type and action of the generation controller.

The aim of load frequency control (LFC) is to maintain the power balance in the system such that the frequency deviates from its nominal value within specified bounds and according to practically acceptable dynamic performance of the system.

Several controllers have been presented for LFC problems in order to achieve a better dynamic performance, where the most employed one is the conventional fixed gain controller like proportional-integral (PI) and proportional-integral-derivative (PID) controllers, because of their functional simplicity and simple structure, being widely used in industrial applications. However, their parameters are often tuned using experiences or trial and error methods and may no longer be suitable in all operating conditions. Different methods of tuning of PI/PID controllers are given in [2]. The use of optimal control theory to power systems has shown that an optimal load frequency controller can improve the dynamic stability of a power system. Calovic [3] used output feedback techniques to design a decentralized controller.

Usually a linear model around a nominal operating point is used in the load frequency controller design. However, power system components are inherently nonlinear, so the operating conditions of power systems are continuously changing. Accordingly, the real plant usually differs from the model. Therefore robustness becomes a main issue in the attempt to design a controller to satisfy the basic requirements for zero steady state and acceptable transient frequency deviations.

Various robust control approaches have been used in LFC, such as Riccati equation approach [4], H∞-control [5], μ-synthesis approach [6], robust pole assignment approach [7], Linear Matrix Inequality (LMI) based H∞-loop shaping control [8],loop shaping control [9] and inverse additive perturbation [10].The LMI approach based LFC has been mostly applied in conventional power systems, being very less used yetin hybrid energy systems for LFC.

In this paper, the formulation of LFC problem is based on LMI-Linear Quadratic Regulator (LQR). The basics of LMI-based-LQR control scheme are illustrated in [11]. The control optimization problem is obtained in terms of a system of LMI constraints and matrix equations that are simultaneously solved. The proposed LFC scheme with the advanced LMI-based-LQR (ALQR) is applied for an autonomous hybrid generation system. Results show that the robustness of the proposed controller is much superior to that of the optimal output state feedback controller, called as simple closed-LQR (CLQR), for different load and wind power perturbations.

2 Relationship to Internet of Things

The smart grids are expected to spread the intelligence of the energy distribution and control system from some central core to many peripheral nodes, thus enabling more precise control and adaptation. The concept and objectives of such intelligent nodes are similar to those of the Internet of Things (IoT).

IoT means extending the web paradigm to the connection, monitoring, and control of the objects of everyday life. IoT technologies can provide a viable solution toward a web-enabledsmart gridsystem [12].

The future developments of the IoT related to energy-aware systems and distributed decision will be especially relevant for autonomous hybrid generation systems, making their control ubiquitous.Decision makers will be able to base their actions on real-world, real-time data [13].IoT will make it possible for future energy systems to be self-managing, self-sustaining and robust. Therefore, the IoT based infrastructure will be tightly coupled with the energy domain [14].

3 System Configuration and Description

The interconnected conventional power source and distributed generation (DG) system is shown in Fig. 1. The power-frequency balance is maintained by a proper control of the different power generation resources. In actual systems, WT, DEG, FC generator, AE and BESS are high order models and have nonlinearity. The controller design for higher order non-linear system would be complicated. Thus, the first order linearized model is taken into consideration for frequency control. The first order representation for the LFC is sufficient without sacrificing accuracy as the frequency control loop acts very fast.

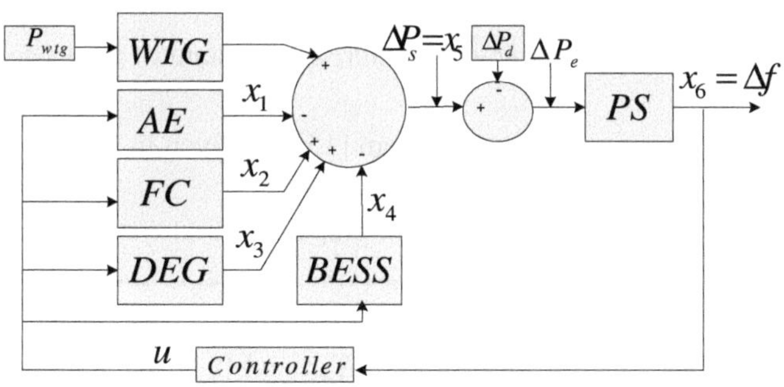

Fig. 1. Block diagram of the hybrid power system

The generated power of the WTG depends upon the wind speed. The transfer function of the WTGis represented by a first-order lag [15], as:

$$G_{WTG}(s) = \frac{K_{wtg}}{1 + sT_{wtg}} \tag{1}$$

Part of the wind power generation is to be used by the AE for hydrogen production. The transfer function model of the AE [15] can be expressed by:

$$G_{AE} = \frac{K_{ae}}{1 + sT_{ae}} \tag{2}$$

The FC converts the chemical energy of the fuel (hydrogen) into electrical energy. The FC is represented by a first-order lag transfer function model [15], given as:

$$G_{FC}(s) = \frac{K_{fc}}{1 + sT_{fc}} \tag{3}$$

The DEG works autonomously to supply the deficit power to the hybrid system in order to meet the supply-demand balance condition. The DEG is modeled by a first-order transfer function [15], given by:

$$G_{DEG}(s) = \frac{K_{\deg}}{1 + sT_{\deg}} \tag{4}$$

The BESS can be utilized to provide additional damping to power system swings, improving both transient and dynamic stability. The transfer function model of BESS is expressed by first-order [15], as:

$$G_{BESS}(s) = \frac{K_{bess}}{1 + sT_{bess}} \tag{5}$$

The power balance is achieved by the equation:

$$\Delta P_e = \Delta P_s - \Delta P_d \tag{6}$$

where ΔP_e is the error in supply power and ΔP_d is the load change.

The transfer function for the system variation to p.u. deviation in power is expressed by:

$$G_{sys} = \frac{\Delta f}{\Delta P_e} = \frac{1}{K_{sys}(1 + sT_{sys})} = \frac{1}{D + sM}. \tag{7}$$

where K_{sys} is the system frequency characteristic constant of the system and T_{sys} is gain of the system.

The parameters of proposed hybrid system [15] are given in Table 1.

Table 1. Parameters of the proposed hybrid system

Components	**Gains (p.u.)**	**Time constants (sec)**
Wind turbine generator	$K_{WTG} = 1.0$	$T_{WTG} = 1.5$
Aqua-electrolizer	$K_{AE} = 0.002$	$T_{AE} = 0.5$
Fuel cell	$K_{FC} = 0.01$	$T_{FC} = 4.0$
Diesel generator	$K_{DEG} = 0.003$	$T_{DEG} = 2.0$
Battery energy storage system	$K_{BESS} = -0.003$	$T_{BESS} = 0.1$
Damping constant of the hybrid system	$D = 0.012$	
Inertia constant of the hybrid system	$M = 0.2$	

4 LMI Representation of LQR

For a linear and robust control framework, the following linear time invariant multiple systems are considered:

$$\dot{x} = Ax + By + \Gamma w \tag{8}$$

$$y = Cx + Du + \eta \tag{9}$$

where A is a $n\times n$ state matrix, B is a $n\times p$ control matrix, Γ is a $m\times n$ output matrix, x is a $n\times 1$ state vector, u is a $p\times 1$ control signal vector, y is a $m\times 1$ output vector, w is a noisy signal vector, assumed to be white.

Using the H_2 representation of the LQR problem [16], [17], we obtain the state-feedback gain matrix k that minimizes the following cost function in terms of output $\tilde{y}$ as:

$$J = \min_{(k)}\{E\left[\tilde{y}^T \quad \tilde{y}\right]\} \tag{10}$$

subject to:

$$(A - Bk)^T P + P(A - Bk) + Q + k^T Rk < 0 \tag{11}$$

where $P \in \Re^{n\times n} > 0$ is a common Lyapunov matrix.

The advanced LMI representation of H_2 LQR optimization problem can be obtained as the minimization of the cost function:

$$\tilde{J} = \min_{(\Lambda, Y, \hat{X})} [tr(QY + \hat{X}) - tr(N\Lambda + \Lambda^T N^T)] \tag{12}$$

subject to:

$$\hat{X} - (R^{1/2}\Lambda)Y^{-1}(\Lambda^{-1}R^{1/2}) > 0 = \begin{bmatrix} \hat{X} & R^{1/2}\Lambda \\ \Lambda^T R^{1/2} & Y \end{bmatrix} > 0 \tag{13}$$

$$\begin{bmatrix} (AY + Y^T A^T - B\Lambda - \Lambda^T B^T & \Lambda^T & Y^T \\ \Lambda & -R^{-1} & 0 \\ Y & 0 & -Q^{-1} \end{bmatrix} < 0 \tag{14}$$

5 Control Formulation

Consider a linear time invariant model for the proposed hybrid power system with the following state space model:

$$\dot{x} = Ax + Bu + \Gamma d \tag{15}$$

$$y = Cx \tag{16}$$

$$u = -Kx \tag{17}$$

where A is system matrix, B is the input distribution matrix, Γ is the disturbance distribution matrix, C is the control output distribution matrix, x is the state vector, u is the control vector and d is the disturbance vector consisting of load change and wind power change.

The state and other variables of the proposed hybrid power system are $x = [x_1 \quad x_2 \quad x_3 \quad x_4 \quad x_5 \quad x_6]^T$ and $y = x_6$. The design of the optimal controller based on output feedback control approach is discussed in detail in [3].

6 Results

6.1 Load Changes

The proposed controllers were applied to the hybrid power system shown in Fig. 1. The parameters of the hybrid power system are considered to be the same as in [15]. In this section, the performance of the ALQR controller is compared with the CLQR controller for different load scenarios.

Case 1: Step Load Change

In this case, a load disturbance of $\Delta P_d = 0.01pu$ and wind power variation of $\Delta P_{wtg} = 0.5pu$ is applied to the system.

The frequency deviation is shown in Fig. 2for ALQR and CLQR. It can be seen that using the ALQR controller the frequency deviation of the system is quickly driven back to zero with a smaller overshoot.

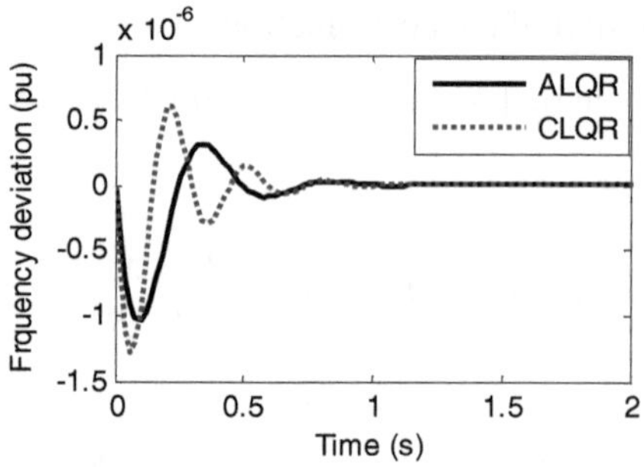

Fig. 2. Frequency deviation for case 1

Fig. 3(a) and 3(b) show the frequency deviation with different load disturbances for ALQR and CLQR controllers, respectively.

It can be seen that the robustness of the ALQR controller is superior comparatively to the CLQR controller.

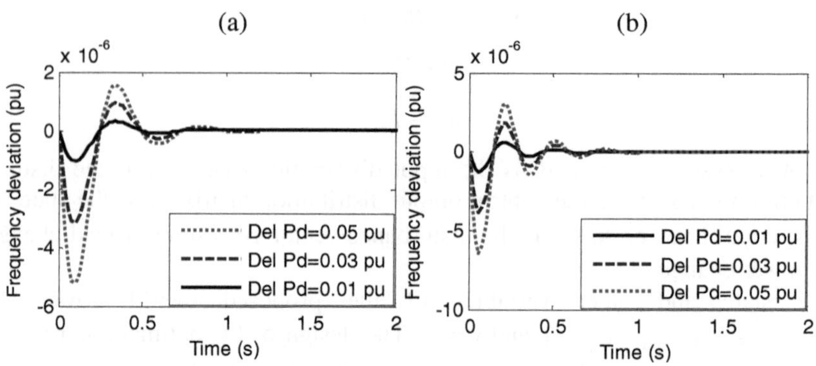

Fig. 3. Frequency deviation for ALQR (a) and CLQR (b) with load disturbances

Case 2: Random Step Load Change

In this case, it is considered that the load disturbance varies with time in step form, as shown in Fig. 4, and $P_{wtg} = 0.5\,pu$ is also applied to the system.

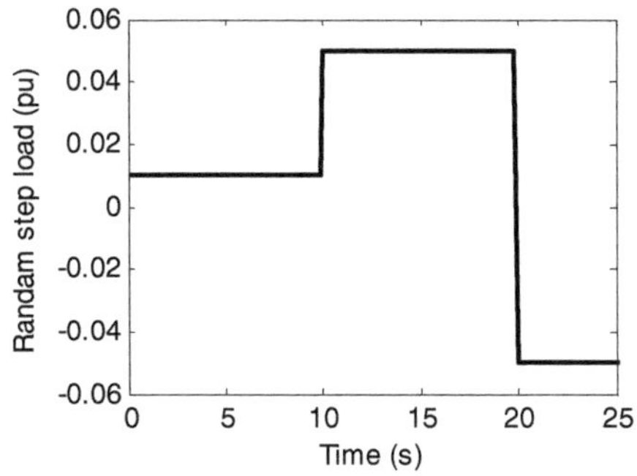

Fig. 4. Random load change in step form

The frequency deviation is shown in Fig. 5(a). The control effect of ALQR is better than CLQR. For clarity, Figs. 5(b) to 5(d) show the frequency deviation for the random step load change considering different time durations.

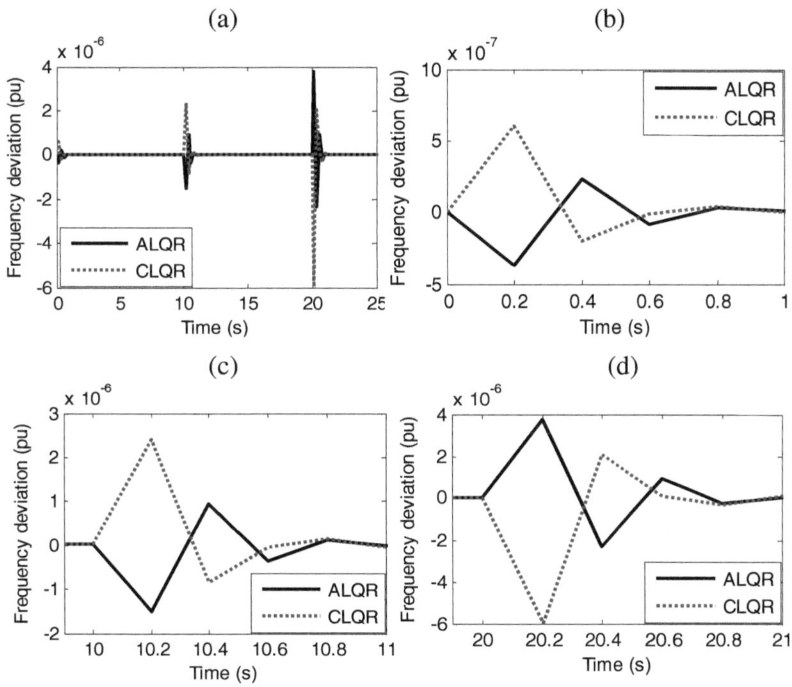

Fig. 5. Frequency deviation for case 2 (a); frequency deviation for case 2from t=0s to 1s (b); frequency deviation for case 2 from t=9s to 11s (c); frequency deviation for case 2 from t=19s to 21s.

Case 3: Random Load Change

In this case, the random load change shown in Fig. 6 (a)is applied to the system, also with $P_{wtg} = 0.5 pu$.

Fig. 6 (b) shows the response of frequency deviation under the random load change. It can be seen that the response with the ALQRcontroller is better than with CLQR.

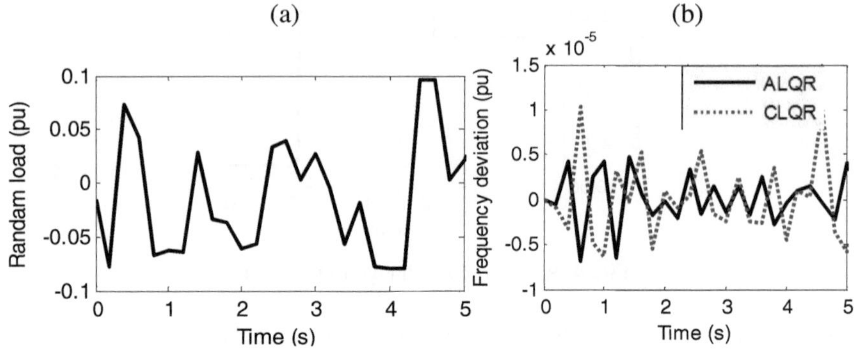

Fig. 6. Random load change (a) frequency deviation for case 3 (b)

6.2 Random Wind Power Input

In this case, the random wind power input shown in Fig. 7(a) is applied to the system, alongsidea load disturbance of $\Delta P_d = 0.01 pu$. Fig. 7(b) shows the frequency deviation for this case.

It can be seen that the response with the ALQRcontroller is again better than with CLQR.

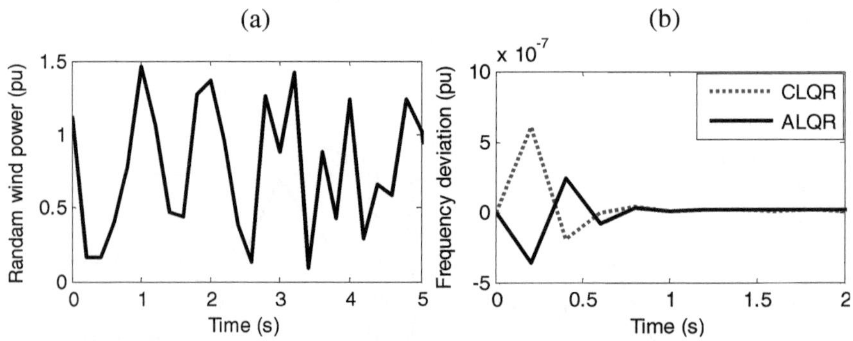

Fig. 7. Random wind power input signal (a); frequency deviation with random wind powerfrom t=0s to 2s (b)

6.3 Robustness with Parameter Variations

Any controller designed for fixed plant parameter or tuned to nominal conditions may not perform satisfactory on variation in their values. Also, in some cases, the value of these parameters is not accurately determined, either experimentally or mathematically. Thus, they are said to belong in an interval range.

A controller design is desired to be robust enough against parameter variations of the system model.

The robustness of theCLQRand ALQR controllers is evaluated by computation of a quantitative performance index, integral square error (ISE) under variation in system parameters.

In order to be realistic, the simulations were done for a limited time. Moreover, the integral values are multiplied by some suitable constants, so that the outcomes of the performance indices attain comparable values.

Thisindexis modified as:

$$\Delta f_{ISE} = 10^5 \int_0^{25} (\Delta f)^2 dt \tag{18}$$

The values of ISE against parameter changes for theCLQRand ALQRcontrollers are shown in Fig. 8.

As can be observed with the parameters change over a large percentage range, the corresponding value of ISE varies significantly and increasingly in the case of the CLQR controller. However, the value of ISE with ALQR controller is much lower and change slightly.

These results confirm the high robustness of the ALQR controller against the random load change, and system parameter variations.

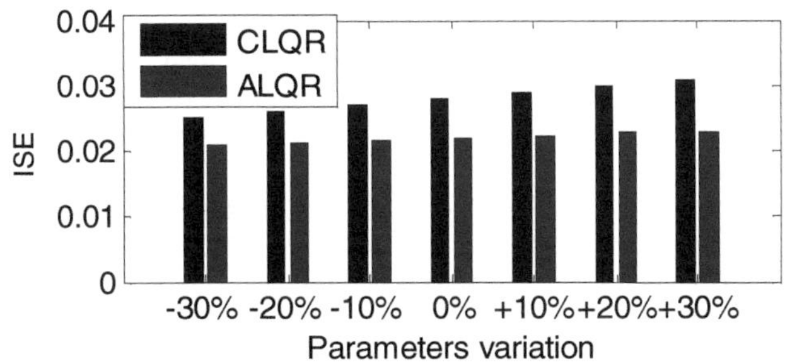

Fig. 8. Robustness evaluation by computing performance index ISE for all parameters variation

7 Conclusions

This paper presented a robust control design technique based on LMI-LQR approach. The proposed ALQR controller is applied for an autonomous hybrid generation system. A comparative study between the proposed ALQR controller and optimal output state feedback controller, called as CLQR controller, is provided in study.

Simulation results using Matlab 2009b version demonstrated that the proposed controller is capable to guarantee robust stability and robust performance, such as precise reference frequency tracking and disturbance attenuation. The results suggest superior performance of ALQR against optimal output state feedback controller, in the case of load changes,parameter variationsand random wind power input. The incorporation of IoT technologies will foster the web-enabled smart grid concept, where all devices will be interconnected and able to interact through the internet, leading to the deployment of a ubiquitous energy controlsystem.

Acknowledgment. This work was supported by FEDER funds (European Union) through the Operational Program for Competitiveness Factors—COMPETE, and by Portuguese funds through the Fundação para a Ciência e a Tecnologia (FCT), under Project FCOMP-01-0124-FEDER-020282 (Ref. FCT PTDC/EEAEEL/118519/2010).

References

1. Ahmad, N.A., Miyarake, M., Al-Othman, A.K.: Power fluctuation suppression of stand-alone hybrid generation combining solar photovoltaic/wind turbine and fuel cell systems. Energy Conversion Manage. 49, 2711–2719 (2008)
2. Cominos, P., Munro, N.: PID controllers: Recent tuning methods and design to specification. In: IEE Proc. Control Theory Appl., vol. 149, pp. 46–53 (2002)
3. Calovic, M.S.: Automatic generation control: Decentralized area-wise optimal solution. Electric Power Systems Research 7, 115–139 (1984)
4. Wang, Y., Zhou, R., Wen, C.: Robust load-frequency controller design for power systems. In: IEE Proc. Part C, vol. 140, pp. 11–16 (1993)
5. Rerkpreedapong, D., Hasanovic, A., Feliachi, A.: Robust load frequency control using genetic algorithms and linear matrix inequalities. IEEE Trans. on Power Systems 18, 855–861 (2003)
6. Wang, Z.Q., Sznaier, M.: Robust control design for load frequency control using μ-synthesis. In: Southco/94, Conference Record, Orlando, FL, USA, pp. 186–190 (1994)
7. Azzam, M.: Robust automatic generation control. Energy Conversion and Management 40, 1413–1421 (1999)
8. Patra, S., Sen, S., Ray, G.: Design of static H∞ loop shaping controller in four-block framework using LMI approach. Automatica 44, 2214–2220 (2008)
9. Majumder, R., Chaudhuri, B., El-Zobaidi, H., Pal, B.C., Jaimoukha, I.M.: LMI approach to normalized H_∞ loop-shaping design of power system damping controllers. In: IEE Proceedings-Generation, Transmission and Distribution, vol. 152, pp. 952–960 (2005)
10. Cuk Supriyadi, A.N., Ngamroo, I., Kunakorn, A., Dechanupaprittha, S., Watanabe, M., Mitani, Y., Hashiguchi, T., Goda, T.: Inverse additive perturbation-based optimization of robust PSS in an interconnected power system with wind farms. In: Proceedings of SICE Annual Conference 2008, Tokyo, Japan, pp. 237–240 (2008)
11. Ko, H.-S., Jatskevich, J., Dumont, G., Yoon, G.-G.: An advanced LMI-based-LQR design for voltage control of grid-connected wind farm. Electric Power Systems Research 78, 539–546 (2008)
12. Bui, N., Castellani, A.P., Casari, P., Zorzi, M.: The internet of energy: a web-enabled smart grid system. IEEE Network, 39–45 (2012)

13. Karnouskos, S., Terzidis, O.: Towards an information infrastructure for the future Internet of energy. In: ITG-GI Conference, Germany (2007)
14. Yu, X., Cecati, C., Dillon, T., Simões, M.G.: The new frontier of smart grids. Industrial Electronics Magazine, 49–63 (2011)
15. Das, D.C., Roy, A.K., Sinha, N.: GA based frequency controller for solar thermal-diesel-wind hybrid energy generation/energy storage system. Electrical Power and Energy Systems 43, 262–279 (2012)
16. Anderson, B.D.O., Moor, J.B.: Optimal Control: Linear Quadratic Methods. New Jersey. Prentice-Hall, Englewood Cliffs (1998)
17. Boyd, S., Ghaoui, L.E., Feron, E., Balkrishanan, V.: Linear Matrix Inequalities in system and control theory, Philadelphia. SIAM Studies in Applied Mathematics, vol. 15 (1994)

Darrieus Wind Turbine Performance Prediction: Computational Modeling

N.C. Batista[1], R. Melício[1], V.M.F. Mendes[2], J. Figueiredo[1], and A.H. Reis[1]

[1] University of Évora, Évora, Portugal
nelson.batista@gmail.com, {rmelicio,jf,ahr}@uevora.pt
[2] Instituto Superior de Engenharia de Lisboa, Lisbon, Portugal
vfmendes@deea.isel.pt

Abstract. The Vertical Axis Wind Turbines Darrieus is facing a rapid installations growth due to the interest for a decentralizing generation. The Darrieus aerodynamic behavior analysis is a hard computational task. This paper starts by offering an insight into the Darrieus wind turbines performance prediction and proposes an algorithm based on the Multiple Streamtube modeling as a prediction approach in order to have an admissible handing in an Internet of things environment.

Keywords: Darrieus, Multiple Streamtube model, performance prediction.

1 Introduction

The eolic energy source is one of the most cost effective of renewable sources. Electric demand has been satisfied by mixed energy sources and in nowadays the eolic energy source is one of those sources in current industrial use. Investments in eolic energy exploitation are reported as increasing ones [1]. Studies have been conducted [2], [3] regarding the use of eolic energy in order to research behaviors and to justify the approval on the usage of eolic energy in the market, offering tools to help and ease the enterprise R&D. Also, the eolic energy source is acknowledged as a favorable energy sources for scenarios regarding urban areas [4]. Where, the Vertical Axis Wind Turbine (VAWT) has advantages over the Horizontal Axis Wind Turbines (HAWT) [5], [6]. The high penetration of eolic energy is increasingly affecting the overall performance of the electric grid, due to characteristics of intermittency and high variability in space and time. One of the concerns related to the high penetration of wind turbines is the impact on the system stability and power quality [7]. Hence, the accurate performance prediction of scattered urban wind turbines is crucial in a Smart Grid (SG) due to the fact that a fast and flexible adjustment to the requirements of all players especially the consumer needs of energy usage is an important issue in the SG. The performance prediction of VAWT wind turbines especially in urban areas is a complex task. The Darrieus type VAWT performance prediction is usually a time consuming task, requiring high processing times. The Darrieus VAWT performance prediction can reside in cloud services [8] or be predefined in the wind turbine

L.M. Camarinha-Matos, S. Tomic, and P. Graça (Eds.): DoCEIS 2013, IFIP AICT 394, pp. 382–391, 2013.

instrumentation in a SG connected to a data network with the possibility for centralized data processing capabilities offered by cloud infrastructures.

The pursuit for the most desirable design for the urban Darrieus VAWT performance prediction in a SG and the relation to the involved data interexchange network on an Internet of Things (IoT) motivated the work presented in this paper. The presented computation algorithm brings the ability to fast predict the Darrieus VAWT aerodynamic performance, contributing to make possible the implementation of services for the interaction between assets and urban wind turbines.

The rest of the paper is organized as follows: Section 2 presents the contribution to innovation: the ability to fast predict the Darrieus VAWT aerodynamic performance, allowing the design of sophisticated services to be implemented on the SG in an IoT environment. Section 3 presents features about models for the Darrieus aerodynamic performance. Section 4 presents the formulation main assumptions for the prediction algorithm. Section 5 presents results from the proposed algorithm and anticipates further work. Finally, concluding remarks are given in Section 6.

2 Relationship to Internet of Things

The expected way to interconnect all the assets, players and services in a SG is over the internet infrastructure. Hence, the urban Darrieus VAWT will be an object in the IoT [9] involved in the SG operation [10], [11]. This object requires an adequate modeling to be conveniently processed as a thing in IoT.

This paper computational model is a contribution offering the ability to fast predict the Darrieus VAWT aerodynamic performance, allowing the design of sophisticated services to be implemented on the SG [12]. The computational model can be implemented over an infrastructure interacting not only with the SG but with other services offered over the IoT.

3 Darrieus Aerodynamic Models

Despite the complexity of the Darrieus VAWT aerodynamic behavior, mathematical models have been developed [13], [14] in order to predict the performance. These mathematical models have strengths and weaknesses with a more or less accurate prediction, depending on the wind turbine configuration, wind behavior and prediction time consumed. The most commonly used models can be divided in three categories: Vortex model, Cascade model and Blade Element Momentum (BEM) model.

The Vortex model is characterized by: the vortices in the wake of the blades are calculated; the vortex filaments substitute the blades and have the strengths determined by the blade profile coefficients, relative flow and angle of attack. The Vortex model has a good accuracy. But as a main disadvantage, this model requires a high computational time.

The Cascade model arranges the blades in vanes called cascade, placed in equal interspaces. The blades aerodynamic behavior are calculated independently, taking in consideration the angle of attack, the position in the upwind and downwind side of the rotor and the local Reynolds number. Although no convergence problems are reported and good predictions results are presented, the cascade model takes a high computational time.

The BEM model is divided in three submodels: Single Streamtube, Multiple Streamtube and Double-Multiple Streamtube.

The BEM modeling combines momentum theory with blade element theory in order to study the behavior of the air flow on the blades and related forces. The turbine in the Single Streamtube model is placed inside a single streamtube and the blades revolution is translated in an actuator disk. The effects outside the streamtube are assumed negligible and the wind speed in the upstream and downstream sides of the turbine are considered constant. This model as a consequence of all the assumptions made does not deliver good prediction accuracy, giving in most cases higher estimate values, but has a fast processing time. The Multiple Streamtube modeling is a variation of the Single Streamtube modeling, dividing the single streamtube in several parallel adjacent streamtubes that are independent from each other and have own undisrupted, wake and induced velocities. Although not accurate, the predicted performance is close to field test values, tending to give values a little higher for high Tip Speed Ratios (TSR) and having a fast processing time. The Double Multiple Streamtube (DMS) modeling is a variation of the Multiple Streamtube modeling, dividing the actuator disc is in two half cycles in tandem, representing the upstream and downstream sides of the rotor. This model presented by Paraschivoiu [15] has suffered several improvements over time and offers a good performance prediction, but presents convergence problems for high solidity turbines, giving high power prediction for high TSR and having a high processing time.

4 Prediction Algorithm

The most suitable Darrieus wind turbine performance prediction modeling is the Multiple Streamtube modeling in a SG environment with the assets connected over an internet of thing. This is due to simplicity, reasonable accuracy and low computational processing time, permiting the real time interexchange of data between several SG assets and the wind turbine over an internet of things platform.

The Darrieus wind turbine performance computation divides the wind turbine space horizontally and vertically in several parallel and adjacent streamtubes considered to be independent from each other, having own wake and induced velocities. The streamtube width is given by $c\sin(\theta)$, with c the chord of the blade profile and θ the blade azimuthal angle. The streamtube width when θ is equal to 0º and 180º is considered to be equal to the blade profile height. A variation in the blade azimuth angle $\Delta\theta$ is chosen for the horizontal streamtube division. The number of Horizontal Streamtubes (HST) is equal to $180°/\Delta\theta$.

The streamtube height h is equal to the wind turbine height H divided by the number of predefined vertical divisions, *VertDiv*. Each set of horizontal streamtubes has the same undisrupted wind velocity V_∞ that varies with the height. Several forces must be considered in order to study the blade behavior in each streamtube [13], [15]. These forces are computed in function of the airfoil velocities. The airfoil velocities diagram is shown in Fig. 1.

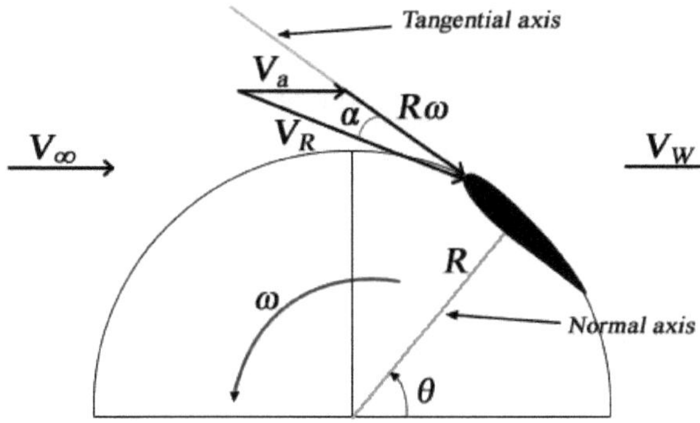

Fig. 1. Airfoil velocities diagram

The relative velocity V_R at the blade is given by:

$$V_R = \sqrt{(V_a \cos\theta + \omega R)^2 + (V_a \sin\theta)^2} \ . \tag{1}$$

ω is the rotor angular speed, R is the radius of the wind turbine and V_a is the local wind speed at rotor. The Froude's Momentum theory for an actuator disk [15] gives a relation between the undisruptive wind velocity V_∞ and the wind speed at the rotor. This relation is given by an induction factor U given by:

$$V_a = V_\infty(1-U) \, . \tag{2}$$

Normalizing the relative velocity V_R by V_∞, (1) is given by:

$$V_R / V_\infty = \sqrt{(V_a / V_\infty \cos\theta + \omega R / V_\infty)^2 + (V_a / V_\infty \sin\theta)^2} \ . \tag{3}$$

Introducing (2) into (3) and considering λ as the TSR, (3) is given by:

$$V_R / V_\infty = \sqrt{[(1-U)\cos\theta + \lambda]^2 + [(1-U)\sin\theta]^2} \ . \tag{4}$$

The blade angle of attach α is given by:

$$\alpha = \tan^{-1}\left(\frac{V_a \sin\theta}{V_a \cos\theta + \omega R}\right) = \tan^{-1}\left[\frac{(1-U)\sin\theta}{(1-U)\cos\theta + \lambda}\right] \ . \tag{5}$$

The Tangential coefficient C_T and normal coefficient C_N [13] are given by:

$$\begin{aligned} C_T &= C_L \sin\alpha - C_D \cos\alpha \\ C_N &= C_L \cos\alpha + C_D \sin\alpha \end{aligned} \ . \tag{6}$$

C_L and C_D are the lift and drag coefficients of the airfoil for the angle of attack α. The aerodynamic trust coefficient C_{AT} is given by:

$$C_{AT} = \frac{2}{\pi}(\frac{Nc}{D})(C_N - \frac{C_T}{\tan\theta})(\frac{V_R}{V_\infty})^2 . \tag{7}$$

N is the number of blades and D the diameter of the wind turbine. Applying the momentum theory, the momentum loss trust coefficient C_{MLT} is given by [15]:

$$C_{MLT} = 4U(1-U) . \tag{8}$$

A Multiple Streamtube model diagram is shown in Fig. 2.

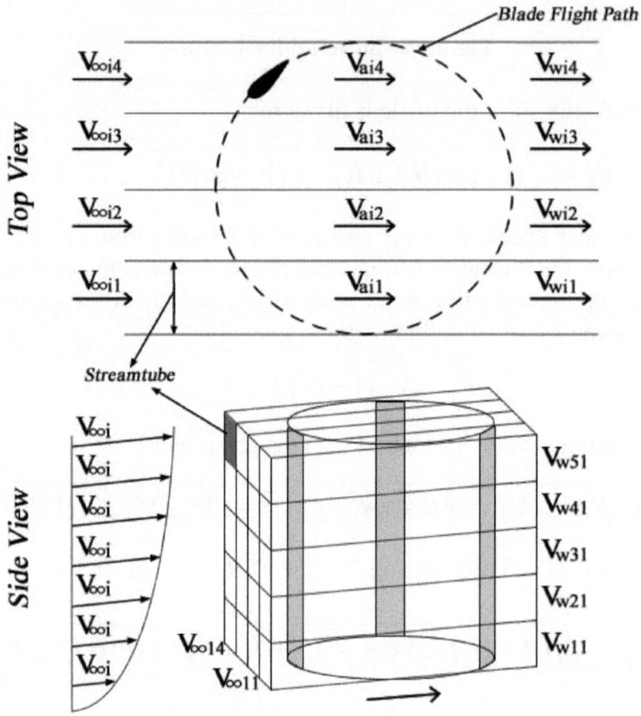

Fig. 2. Multiple Streamtube model diagram

The aerodynamic behavior of the wind turbine in all the streamtubes is calculated in order to compute the Darrieus type VAWT wind turbine performance prediction. A starting interference factor U is chosen for each streamtube. The local wind speed Va, relative wind speed VR, local angle of attack α and Reynolds number are calculated. Hence, the aerodynamic trust coefficient CAT can be calculating, using lift and drag coefficients that can be previously calculated for the blade airfoil and saved in a file or database or calculated in real-time.

The momentum loss trust coefficient C_{MLT} is calculated using the existing interference factor U. Both C_{AT} and C_{MLT} are compared and a difference between the two trust coefficients is an error. If this error is outside an error tolerance, a new interference factor U is chosen and all the process starts again, until the error between the C_{AT} and C_{MLT} are within the error tolerance. The Multiple Streamtube computational model flowchart is shown in Fig. 3.

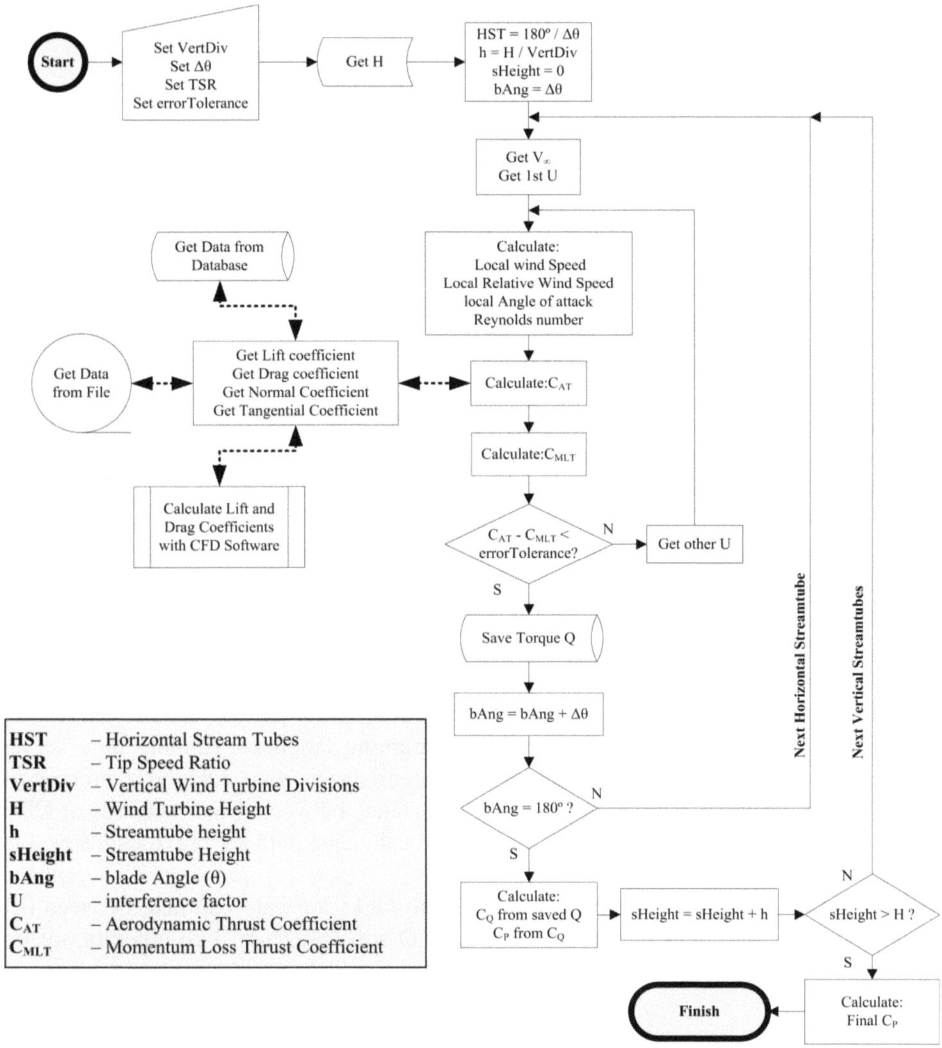

Fig. 3. Multiple Streamtube computational model flowchart

Once an interference factor for a streamtube is found, the torque Q is then calculated and saved.

The final torque coefficient C_Q is calculated when all the horizontal and vertical streamtubes are processed. The final torque coefficient C_Q is computed by the expression given by:

$$C_Q = (\frac{Nc}{D})\frac{\sum_{i=1}^{s}[(\frac{V_R}{V_\infty})^2 C_T]}{s}. \tag{9}$$

The final Power coefficient C_P is given in function of C_Q (9) and the TSR by the expression given by:

$$C_P = C_Q\,\lambda. \tag{10}$$

The incorporation of the wind turbines behavior in an internet of things platform, requiring fast interexchange of data, is simplified by using this model to compute the aerodynamic performance prediction of Darrieus VAWT.

5 Results and Further Work

The aerodynamic behaviors of the Darrieus VAWT blade profiles are found to be the most time consuming task. The lift and drag coefficients of the Darrieus VAWT blades can be analyzed in field tests or computed with Computational Fluid Dynamic (CFD) software.

The lift and drag coefficients can be calculated on demand with CFD software, with the disadvantage of increasing the computational time spent in the Darrieus VAWT performance prediction. Hence the use of previously saved data in file or database format is advisable, decreasing the time consumption performance prediction to a few seconds.

The Darrieus VAWT performance prediction computation of the previously presented algorithm developed in ABAP programing language, running on a server with Windows 2008 R2 64-bit Operating System server, two 2.53 GHz processors and 8 GB of Random Access Memory (RAM), takes between 1 to 2 seconds of CPU time to compute, if all the airfoil lift and drag coefficients data are previously saved in a Microsoft SQL Server 2008.

The lift and drag coefficients calculated with CFD software can take between 0.1 seconds up to 5 minutes depending on the CFD software, airfoil complexity, airfoil number of points, blade angle of attack, Reynolds number and convergence criteria for the iterations.

A comparison between the several mathematical performance prediction models elaborated by Sandia National Laboratories is shown in Fig. 4 between the Single Streamtube model and the Multiple Streamtube model (DART computer code) [16]; in Fig. 5 between the Multiple Streamtube model and the Double-Multiple Streamtube model [15].

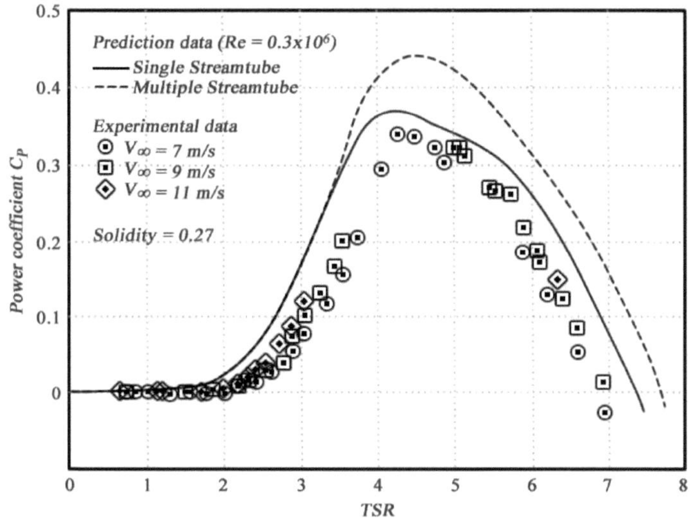

Fig. 4. Single Streamtube model and Multiple Streamtube model comparison with experimental data [16]

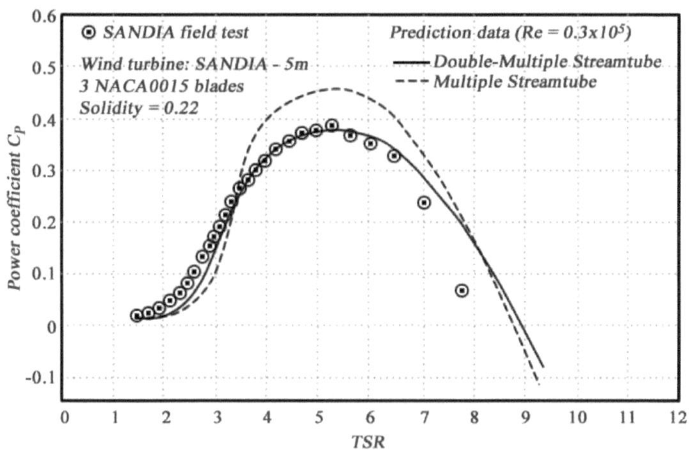

Fig. 5. Multiple Streamtube model and Double-Multiple Streamtube model comparison with experimental data [15]

Although, advantages and disadvantages are intimately related with the Darrieus VAWT design and installation conditions, several most used mathematical performance prediction models advantages and disadvantages are reported. A resume of those reports are presented in [13], and summarized in Table 1.

Table 1. Darrieus VAWT mathematical performance prediction models advantages and disadvantages [13]

Prediction Model	Advantages	Disadvantages
Single Streamtube	Predicts the overall performance of a lightly loaded wind turbine. Very fast computational prediction.	Does not predict the wind variations across the rotor presenting disparities compared with experimental data.
Multiple Streamtube	Predicts the overall performance reasonably, especially when the rotor is lightly loaded. Fast computational prediction.	Some convergence problems, increasing computational time. Slight disparities with field tests, depending on the wind turbine solidity.
Double-Multiple Streamtube	Offers a good correlation between the performance prediction and the experimental data.	Presents convergence problems that increase computational time. Over prediction power for high solidity wind turbine and higher TSR.
Vortex	With the latest improvements the model presents a high correlation between the performance prediction and the experimental data.	Takes the highest computational time of all the prediction models.
Cascade	Reasonable overall prediction for both low and high solidity wind turbine. No convergence problems.	Takes a reasonable computation time.

Further work will target the validation with field tests of the presented solution for the Darrieus VAWT performance prediction in the SG included in an IoT. Also, the developments of optimized computational solutions to other Darrieus VAWT performance prediction models are envisaged in the near future.

6 Conclusions

Aerodynamic behavior analysis models of Darrieus wind turbines are discussed. The Multiple Streamtube modeling is further analyzed, regarding methodology and formulation. The computation approach is presented in a flowchart and is designed in the most suitable approach for a SG included in an IoT environment, achieving a fast simulation of the aerodynamic performance of a Darrieus VAWT.

References

1. Snyder, B., Kaiser, M.J.: A comparison of offshore wind power development in Europe and the U.S.: patterns and drivers of development. Appl. Energy 86(10), 1845–1856 (2009)
2. Melício, R., Mendes, V.M.F., Catalão, J.P.S.: Dynamic stability of wind turbines with permanent magnet machines and power-electronic converters. In: Proc. Power Eng. 2009, Lisbon, Portugal, pp. 484–489 (2009)
3. Faria, D.L., Castro, R., Philippart, C., Gusmão, A.: Wavelets Pre-Filtering in Wind Speed Prediction. In: Proc. Power Eng. 2009, Lisbon, Portugal, pp. 168–173 (2009)
4. Mañana, M.: Small wind energy systems, state of the art and new challenges. In: Proc. ICREPQ 2011, Gran Canaria, Spain (2011)
5. Batista, N.C., Melício, R., Matias, J.C.O., Catalão, J.P.S.: Self-start performance evaluation in Darrieus-type vertical axis wind turbines: methodology and computational tool applied to symmetrical airfoils. In: Proc. ICREPQ 2011, Gran Canaria, Spain (2011)
6. Batista, N.C., Melício, R., Matias, J.C.O., Catalão, J.P.S.: New blade profile for Darrieus wind turbines capable to self-start. In: Proc. IET RPG 2011, Edinburgh, United Kingdom (2011)
7. Melício, R., Mendes, V.M.F., Catalão, J.P.S.: Power converter topologies for wind energy conversion systems: integrated modeling, control strategy and performance simulation. Renewable Energy 35(10), 2165–2174 (2010)
8. Miorandia, D., Sicarib, S., De Pellegrinia, F., Chlamtaca, I.: Internet of things: vision, applications and research challenges. Ad Hoc Networks 10(7), 1497–1516 (2012)
9. Atzoria, L., Ierab, A., Morabitoc, G.: The internet of things: a survey. Computer Networks 54(15), 2787–2805 (2010)
10. Ou, Q., Zhen, Y., Li, X., Zhang, Y., Zeng, L.: Application of internet of things in smart grid power transmission. In: Proc. 3rd FTRA Int. Conf. on Mobile, Ubiquitous, and Int. Comp.- MUSIC, Beijing, China, pp. 96–100 (2012)
11. Bui, N., Castellani, A.P., Casari, P., Zorzi, M.: The internet of energy: a web-enabled smart grid system. IEEE Network 26(4), 39–45 (2012)
12. Atzoria, L., Ierab, A., Morabitoc, M., Nittia, N.: The social internet of things (SIoT) – when social networks meet the internet of things: concept, architecture and network characterization. Computer Networks 56(16), 3594–3608 (2012)
13. Islam, M., Ting, D.S.-K., Fartaj, A.: Aerodynamic models for Darrieus-type straight-bladed vertical axis wind turbines. Renewable and Sustainable Energy Reviews 12(4), 1087–1109 (2008)
14. Melo, R.R.S., Neto, A.S.: Integral analysis of rotors of a wind generator. Renewable and Sustainable Energy Reviews 16(7), 4809–4817 (2012)
15. Paraschivoiu, I.: Wind Turbine Design: with emphasis on Darrieus concept, 1st edn. Polytechnic International Press, Canada (2002)
16. Strickland, J.H.: The Darrieus turbine: a performance prediction model using multiple streamtube. Sandia Laboratories Report SAND 75-0431, United States of America (1975)

Part XIV
Energy Distribution

The Electric Vehicle Integration into the Power System: An Application to the Portuguese Case

Ezequiel Carvalho[1], Jorge Sousa[1,3], and M. Ventim Neves[2]

[1] ISEL - Lisbon Engineering Superior Institute, Rua Conselheiro Emídio Navarro, 1, 1959-007 Lisboa, Portugal
{ecarvalho,jsousa}@deea.isel.ipl.pt
[2] Centre of Technology and Systems, Faculty of Science and Technology, Quinta da Torre, 2829-516 Caparica, Portugal
ventim@uninova.pt
[3] Cie3 - Center for Innovation in Electrical and Energy Engineering, Av. RoviscoPais, 1, 1049-001 Lisboa, Portugal

Abstract. Electric vehicles (EV) offer a great potential to address the integration of renewable energy sources (RES) in the power grid, and thus reduce the dependence on oil as well as the greenhouse gases (GHG) emissions. The high share of wind energy in the Portuguese energy mix expected for 2020 can led to eventual curtailment, especially during the winter when high levels of hydro generation occur. In this paper a methodology based on a unit commitment and economic dispatch is implemented, and a hydro-thermal dispatch is performed in order to evaluate the impact of the EVs integration into the grid. Results show that the considered 10 % penetration of EVs in the Portuguese fleet would increase load in 3 % and would not integrate a significant amount of wind energy because curtailment is already reduced in the absence of EVs. According to the results, the EV is charged mostly with thermal generation and the associated emissions are much higher than if they were calculated based on the generation mix.

Keywords: CO2 emissions, Electric vehicle, Renewable integration, Power grids, Wind energy, Economic dispatch.

1 Introduction

Nowadays, worldwide a large share of mobility relies on passenger car use, and this share is expected to increase even further, especially in developing and emerging countries [1]. Transportation sector is largely dominated by internal combustion engines, relying almost entirely on oil as primary energy source (94 % of the world energy used for transportation in 2007) [2], and contributes largely to GHG emissions.

In the last years, energy and environmental concerns led to a fast growing of the renewable energies, motivated efforts for increasing energy efficiency, and promoted the search for new transportation solutions.[1] Considering the efficiency of the EV as

[1] Light-duty vehicles (LDV) account for most of the world energy use and GHG emissions in the transportation sector [2].

L.M. Camarinha-Matos, S. Tomic, and P. Graça (Eds.): DoCEIS 2013, IFIP AICT 394, pp. 395–402, 2013.

well as their potential to address the integration of RES, the use of electricity as fuel in the electric vehicles has a great potential to reduce the dependence on oil and the GHG emissions, as well as to provide a low cost alternative to liquid carbon fuels.

Portugal is highly dependent on external energy resources (81 % in 2009) [3], however the Portuguese territory presents a large potential in terms of RES with a special focus on hydro, wind, solar and biomass. Great efforts have been made in the last years, in terms of RES installation and Portugal in one of the leading countries in the EU in terms of electricity generation from renewable sources [4].[2] In Portugal, accordingly with the "EU 20-20-20 target", the strategic energy guidelines established by the Portuguese government point that in 2020, 31 % of the final energy and 60 % of the electricity produced must be from renewable sources [5]. Accordingly, targets for installed hydro, wind and solar power capacities were defined as 8600 MW, 6875 MW (revised) and 1500 MW respectively [6].

As one of the major challenges in integrating large amounts of renewable energy in the power systems, is dealing with the intermittent nature of these sources [7-8], storage is a crucial issue. Thus, the distributed storage capacity provided by the integration of the EV in the transportation sector may play a significant role to overcome renewable variability and contribute to mitigate potential unbalance between the electricity generation and demand [9], as well as contribute to reduce GHG emissions. However, when substituting ICEs by EVs, the potential gains achieved by this integration, both in terms of renewable sources integration and emissions, are highly dependent on the generation mix and time of charge. In [10] a hypothetical penetration of EVs into the present (2011) power grid was already studied. In this work the impact of the EV integration into the Portuguese power system in terms of additional power generation, CO2 emissions, renewable energy integration and thermal generation costs, will be assessed under a specific night charge scenario with a 10 % of EV penetration, for the year of 2020.

2 Relationship to Internet of Things

As the power grid integrates increasing amounts of distributed generation, the need to improve power quality, transmission efficiency and renewable energy integration increases the urge for smart grids implementation. Furthermore, in a growing renewable generation environment, the integration of the electric vehicle into the power grid will rely on networks to charge/discharge in a controlled manner (Vehicle to Grid) and thus improve renewable generation integration. In such environment the EV and the Smart Grids will be integrating and essential elements of the Internet of Things (IoT).

3 Objectives and Methodology

In 2010, the Portuguese government estimated a 10 % penetration of EVs in the Portuguese vehicle fleet by the year of 2020 [5]. The main purpose of this work is to

[2] Portuguese figures are well above the European Union objective for 2010.

study the impact of those EVs on the renewable integration in the power system as well as the environmental impacts. In particular it is intended to answer the following questions: Would the integration of the EVs into the grid help to integrate wind power and reduce potential wind curtailments? What would be the benefit in terms of CO_2 reduction that could be achieved from that integration?

A methodology based on a unit commitment and economic dispatch is implemented using the General Algebraic Modeling System (GAMS), in order to evaluate the impact of the EVs integration into the grid. The methodology proposed is an evolution of the model used in [11]. The optimization procedure here proposed performs the economic dispatch of the thermal, storage hydro and storage pumped-hydro technologies, in order to minimize the variable generation costs. Data concerning the run-of-the-river hydro power plants, the renewable generation and the load is given as input data, previously forecasted based on the historic 2011 time-series data. From this methodology, the thermal generation costs, the CO_2 emissions and the potential need for wind generation curtailment is computed. The additional EVs load is then added to the baseline time-series and the impact on the thermal generation, CO_2 emissions and renewable integration is assessed. The objective function considered for the economic dispatch includes the generation and start-up costs of the thermal units, assigns a penalty to wind curtailment, and is expressed as:

$$\min \sum_{t=1}^{T} \left\{ \left[C_{RC} . P_{RC}(t) \right] + \sum_{j=1}^{J} \left[C_G^j . P^j(t) . \delta t . u^j(t) + C_{SU}^j . y^j(t) \right] \right\} . \tag{1}$$

where $P^j(t)$ is the power generated by the thermal unit j at moment t, in MW, $P_{RC}(t)$ is the renewable generation curtailment at moment t, in MW, $y^j(t)$ and $u^j(t)$ are a binary variables which indicate if the thermal power unit j starts-up/is running at moment t, δt is the time interval between t and $t + 1$, T is the number of time-series periods and J is the number of thermal units. Also C_G^j is the power generation cost of the unit j, in €/MWh, C_{SU}^j is the start-up cost of the unit j, in € and C_{RC} is the renewable generation curtailment penalty, in €/MW.

Subjected to the following constraints:

3.1 Thermal Units Constraints

Thermal units are subjected to restrictions concerning maximum and minimum generation limits and maximum ramp-down and ramp-up rates [12], which are expressed by (2), (3) and (4):

$$P_{Min}^j \le P^j(t) \le P_{Max}^j . \tag{2}$$

$$P^j(t) - P^j(t-1) \le P_{up}^j . \tag{3}$$

$$P^j(t-1) - P^j(t) \le P_{down}^j . \tag{4}$$

In (2), (3) and (4), P^j_{up} and P^j_{down} correspond to the ramp-up and ramp-down power rates, respectively.

3.2 Reservoir Hydro Constraints

Limits to the maximum discharge, stored energy and final energy of reservoir hydro unit are imposed by (5), (6), (7) and (8), where $E_H(t)$ is the stored energy in the unit at the moment t, in MWh, $P_H(t)$ is the power of the hydro unit at the moment t, in MW and $E_{HInf}(t)$ corresponds to the given inflow expressed in energy units (MWh).

$$0 \leq P_H(t) \leq P_{H\,Max} \; . \tag{5}$$

$$E_H(t) = E_H(t-1) - P_H(t) \,.\, \delta t + E_{H\,Inf}(t) \; . \tag{6}$$

$$E_{H\,Min} \leq E_H(t) \leq E_{H\,Max} \; . \tag{7}$$

$$E_{H\,Fin} - E_{H\,Init} = \sum_{t=1}^{T} E_{H\,Inf}(t) - \sum_{t=1}^{T} P_H(t).\delta t \; . \tag{8}$$

3.3 Pumped-Hydro Constraints

For the pumped-hydro unit the following constraints apply:

$$0 \leq P_{PH}(t) \leq P_{PH\,Max} \; . \tag{9}$$

$$-P_{PH\,Max} \leq P_{PHp}(t) \leq 0 \; . \tag{10}$$

$$E_{PH\,Min} \leq E_{PH}(t) \leq E_{PH\,Max} \; . \tag{11}$$

$$E_{PH}(t) = E_{PH}(t-1) - \lfloor P_{PHp}(t) \,.\, \eta_{PH} + P_{PH}(t) \rfloor . \delta t + E_{PH\,Inf}(t). \tag{12}$$

$$E_{PH\,Fin} - E_{PH\,Init} = \sum_{t=1}^{T} E_{PH\,Inf}(t) - \sum_{t=1}^{T} \lfloor P_{PHp}(t) \,.\, \eta_{PH} + P_{PH}(t) \rfloor . \delta t \; . \tag{13}$$

$$P_{PH}(t) + P_{PHp}(t) \leq P_{PH\,Max} \; . \tag{14}$$

In (9), (10), (11), (12), (13) and (14) $E_{PH}(t)$ is the stored energy in the unit at the moment t, in MWh, $P_{PHp}(t)$ is the power of the pumped-hydro unit at the moment t, when pumping, $P_{PH}(t)$ is the power of the pumped-hydro unit at the moment t when generating, η_{PH} is the pumping efficiency and E_{PHInf} corresponds to the given inflow expressed in energy units (MWh).

3.4 System Constraints

There are constraints which are applied to the overall system and do not consider a specific technology. The system balance and the reserve power are expressed by (15) and (16), where $P_L(t)$ is the load at time t, $P_R(t)$ is the sum of the renewable powers (PRE) with the run-of-the river hydro and l represents the transmission losses.

$$\sum_{j=1}^{J} P^{j}(t) + P_{H}(t) + P_{PH}(t) = (1+l)\,P_{L}(t) - P_{R}(t) + P_{RC}(t) + P_{PHp}(t)\,. \tag{15}$$

$$\sum_{j=1}^{J} P_{Max}^{j}\,.\,u^{j}(t) + P_{H\,Max}\,.\,u^{j}(t) + P_{PH\,Max}\,.\,u^{j}(t) = (1+l)\,.\,P_{L}(t) + P_{RES}(t)\,. \tag{16}$$

4 EV Modeling and Charging Scenario

To evaluate the impact of the EV, both in terms of wind integration and CO_2 emissions, a night charge scenario is considered. This off-peak charging scenario is assumed as the most likely to occur in the next years. In this scenario, most consumers are expected to delay the starting of EVs charge until 10 p.m. to benefit from the off-peak low electricity prices as well as due to eventual power constraints at the residential level.[3]

It is assumed that the EVs are equipped with a medium 24 kWh lithium battery, and drive 38 km a day with a fuel consumption of 0.167 kWh/km. Charging power is considered to be 3.3 kW with an 85 % of charge efficiency, which includes distribution grid losses. Based on previous assumptions the EVs are modeled as electric loads and a normalized vehicle charge profile is created, according to the scenario considered. To simulate the beginning of EVs charge, a normal distribution ($\mu = 10\,p.m.$, $\sigma = 1h$) is used.

5 Case Study

5.1 Framework

In this work, the proposed methodology is applied to the Portuguese case study for the year of 2020, taking into account the main characteristics of the power system and the estimated vehicle fleet.

---

[3] Most of Portuguese households have a contract capacity lower than 7 kW [13].

The EV fleet is assumed to be 600 thousand vehicles which correspond to about 10 % of the Portuguese LDV fleet estimated for the year 2020, with base on 2010 registration data [14]. Considering the mean occupation of light passenger vehicles referred in [4] and [15] and passenger kilometers data [4], EVs are assumed to drive 38 km a day. Also EVs are assumed to charge according to the off-peak scenario referred previously.

Concerning the power system the values for 2020 installed capacity are considered according to the planned additions and decommissioning of power plants [5-6], [16]. By this year the overall installed capacity in the Portuguese system is expected to reach 28.8 GW, from which 9.5 GW (33 %) correspond to large hydro (4GW of pumped-hydro), 7.2 GW (25 %) correspond to the thermal technologies (Coal and CCGT) and 12.1 GW (42 %) correspond to the RES technologies.[4] According to the estimated values, by this year the energy consumption will be 56.3 TWh.

Coal, natural gas and CO_2 prices were based on 2011 data and were considered respectively, 87.6 €/ton for coal, 22.5 €/MWh$_t$ for natural gas and 10 €/ton for the CO_2 emissions allowances. Efficiencies considered for coal and fuel-oil thermal units ranged from 37 % to 38 % and a 55 % efficiency was considered for the CCGT units [17].

5.2 Simulation Results

Based on the model and previously referred assumptions, simulations were carried out for the Portuguese power system for one entire year of operation, in an hourly basis. In Fig. 1 the generation profiles corresponding to winter and summer typical periods are presented.

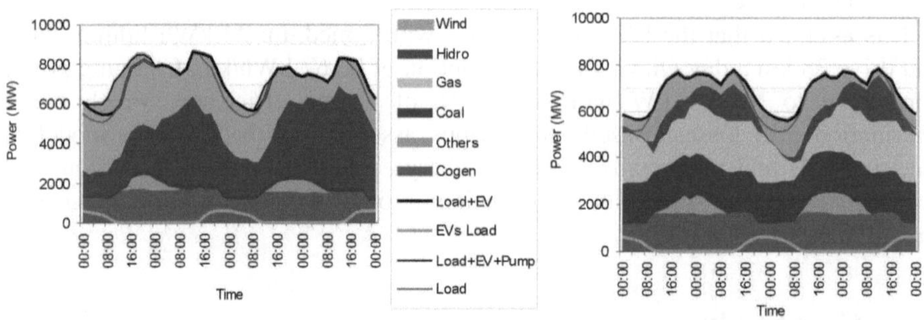

Fig. 1. Generation by technology - typical wet (left) and dry (right) periods

Table 1 presents the simulation results obtained concerning load, power generation, curtailment, costs and emissions for the cases of the load without and with the EV charging.

The EV column of table 1 show the impact of the EV on the power system considering the marginal methodology used. The results show that the considered

[4] Wind power will correspond to 23.7 % of the overall installed capacity.

10 % penetration of electric vehicles in the Portuguese fleet would lead to an increase of 3 % in the electrical load but would not integrate a significant amount of curtailed energy. This results from the fact that the curtailment in the base case is already reduced, due the high values of hydro power installed, which comprises a significant share of pumped-hydro units.

Table 1. Simulation results

	Load	Load+EV	EV
Load (GWh)	56281	57912	1631
Generation (GWh)	57215	58783	1568
Wind Curtail. (GWh)	89.4	1.9	-87.5
Hydro Gen. (GWh)	16121	16458	337
Thermal Gen. (GWh)	10798	12028	1230
CCGT	1308	1823	515
Coal	9490	10206	716
Emissions (ton CO_2)	9038	9872	834
Total Cost (M€)	474	528	55
Average Cost (€/kWh)	0.0084	0.0091	0.0334
Average Thermal Cost (€/kWh)	0.0445	0.0446	0.0450
Average Emissions (g CO_2 /kWh)	160.6	170.5	511.5
Avg Thermal Emissions (g CO_2 /kWh)	849.5	833.1	688.5

The additional generation required to supply the electric vehicles is mostly thermal (45.6 % coal and 32.8 % CCGT) as can be verified in table 1.

Also in table 1 one can see that the marginal generation mix has a specific emission of 511.5 gCO_2/kWh, which leads to the EV emissions of 85.3 g CO_2/km. This compares to a 33.4 g CO_2/km of emissions that would result from the total generation mix emissions of 170.5 g CO_2/kWh.

6 Conclusions

The electric vehicle, seen as a distributed storage system, has a potential to reduce eventual wind curtailments and, therefore, increase the wind integration in the power system. To evaluate this contribution, a methodology based on hydrothermal coordination, unit commitment and economic dispatch was applied to the Portuguese power system for the simulation of an entire year of operation. Results show that the considered 10 % penetration of electric vehicles in the Portuguese fleet would lead to an increase of 3 % in the electrical load and would not integrate a significant amount of wind energy. This is because curtailment is already low in the base case due to the significant pumped-hydro power installed. As a consequence, electric vehicles are charged mostly with thermal generation and the associated emissions are much higher than those that would be obtained considering the generation mix emissions.

In fact, 85 g CO_2/km were computed for the electric vehicle in the marginal approach used in this work, which compares to 33 g CO_2/km of emissions resulting from the generation mix emissions approach.

References

1. Hacker, F., Harthan, R., Matthes, F., Zimmer, W.: Environmental impacts and impact on the electricity market of a large scale introduction of electric cars in Europe - critical review of literature. Technical Report 4/2009, European Tropic Centre on Air and Climate Change, ETC/ACC (2009)
2. IEA-International Energy Agency: CO2 emissions from fuel combustion-highlights, 2009 edition. Technical report (2009)
3. DGEG-Direcção Geral de Geologia e Energia: Caracterização energética nacional (2011)
4. DG-TREN - Directorate-General for Energy and Transport - European Commission: EU energy and transport in figures-Statistical pocketbook 2010 (2010)
5. Governo Português: Estratégia nacional para a energia 2020 (ENE 2020), resolução do conselho de ministros no 29/2010 (2010)
6. REN-Rede Eléctrica Nacional: Plano de desenvolvimento e investimento da rede de transporte 2012-2017 (2022). Relatório técnico (2011)
7. Burke, D., O'Malley, M.: Factors influencing wind energy curtailment. IEEE Transactions on Sustainable Energy 2(2), 185–193 (2011)
8. Ummels, B.C., Pelgrum, E., Gibescu, M., Kling, W.L.: Comparison of integration solutions for wind power in the Netherlands. IET Renewable Power Generation 3(3), 279–292 (2009)
9. Lund, H., Kempton, W.: Integration of renewable energy into the transport and electricity sectors through V2G. Energy Policy 36(9), 3578–3587 (2008)
10. Carvalho, E., de Sousa, J., Ventim Neves, M., Faias, S.: Is the electric vehicle a solution for the wind power integration in the Portuguese power system? In: 2012 9th International Conference on the European Energy Market (EEM), pp. 1–6 (2012)
11. Faias, S., de Sousa, J., Reis, F., Castro, R.: Assessment and optimization of wind energy integration into the power systems: Application to the Portuguese system. IEEE Transactions on Sustainable Energy 3(4), 627–635 (2012)
12. Wang, C., Shahidehpour, S.M.: Effects of ramp-rate limits on unit commitment and economic dispatch. IEEE Transactions on Power Systems 8(3), 1341–1350 (1993)
13. ERSE-Entidade Reguladora do Sistema Eléctrico Nacional: Caracterização da procura de energia eléctrica em 2011. Relatório técnico (2010)
14. ACAP-Associação Automóvel de Portugal: Estatisticas do sector automóvel. Technical report (2010)
15. INE-Instituto Nacional de Estatistica: Mobilidade casa - trabalho da população empregada residente na AMP - 2000. Relatório técnico (2003)
16. REN-Rede Eléctrica Nacional: Caracterização da rede nacional de transporte para efeitos de acesso à rede em 31 de dezembro de 2011. Relatório técnico (2012)
17. Faias, S., de Sousa, J., Castro, R.: Environmental dispatch of the Portuguese power system for CO2 emissions reduction. In: 2011 8th International Conference on the European Energy Market (EEM), pp. 389–394 (2011)

Demand Response Analysis in Smart Grids Using Fuzzy Clustering Model

R. Pereira[1,2], A. Fagundes[2], R. Melício[1],V.M.F. Mendes[2], J. Figueiredo[1], J. Martins[3], and J.C. Quadrado[2]

[1] University of Évora, Évora, Portugal
ruimelicio@uevora.pt
[2] Lisbon Superior Engineering Institute - ISEL, Lisbon, Portugal
rpereira@deea.isel.pt
[3] Center of Technology and Systems, University Nova de Lisboa, Lisbon, Portugal

Abstract. This paper focuses on an analysis of demand response in a smart grid context, presenting the model considerations and architecture. Domestic consumption is divided into groups in order to cover the adequate modeling. A fuzzy subtractive clustering method is applied to demand response on several domestic consumption scenarios and results analyses are presented. The demand response developed model aims to support consumers decisions regarding their consumption needs and possible economical benefits.

Keywords: Demand response, smart grid, fuzzy subtractive clustering.

1 Introduction

The smart grid engages consumer active role. This is crucial for grid management in order to efficiently ensure electric energy generation and usage. For instance, time-scheduling or shedding of load requirements is more favorable when consumer has an active role. Demand Side Management (DSM) such as energy efficiency, energy conservation and Demand Response (DR) programs [1] are required in order to adequate the profile of energy load diagram to generation. DR aims at shaping energy usage in some specified time periods with opportunely economic advantage due to adequate balance between load and generation. The shaping is attained by a modus operandi efficiently taking into account the need to apply time-scheduling or shedding of the load. The modus operandi is intended not only at lowering expenditures, such as the ones resulting from the need to call utilities expensive peaking power plants in short-time or to avoid building new power plant in order to satisfy future forecasted augmented energy consumption needs; but also in case of impossibility to avoid new power plants building, then chiefly discarding the ones with anthropogenic greenhouse gases emission [2]. Already, a European Union program started in 2008 for decreasing the end-user consumption is on effect, aiming to achieve 1 % reduction in energy consumption in the following nine years [3]. Consumers have to be persuaded to adhere to a smart grid environment and be an active element in the smart

L.M. Camarinha-Matos, S. Tomic, and P. Graça (Eds.): DoCEIS 2013, IFIP AICT 394, pp. 403–412, 2013.

grid management, implying the necessity to furnish adequate conveniences to take advantage of the capabilities given by smart grid. DR techniques allows the accomplishment of shifting or decreasing the energy usage in several economic activity sectors, bringing benefits not only to consumers, but also to grid operators due to a reduction in system operation costs [4].

In this paper the consumer´s active role in grid management is supported by a DR method developed using a fuzzy clustering model. The developed methodology provides the consumer an intended efficient tool which allows to support consumer decisions on load management regarding their consumption needs and consumption priorities along one day, taking into consideration the associated costs and possible economical benefits. The desirable massive consumer adherence to DR actions justifies a consumption pattern definition for the model implementation. This consumption pattern justifies the fuzzy clustering techniques applied.

This paper focuses on DR actions and modeling, giving details and results analysis. Domestic consumers are the target of the paper and are divided accordingly to their consumption profiles into three groups. Each group has an equipment priority list of Controllable Loads (CL) to be accomplished, considering for the decision on load satisfaction the available generation capacity and the energy price. Also, consumer non-controllable loads are considered.

The rest of the paper is organized as follows: Section 2 presents the contribution to innovation on DR involving research and customization of programming techniques. Section 3 presents aspects about DSM and DR. Section 4 presents DR model case study main assumptions. Section 5 presents the fuzzy subtractive clustering method to be applied to consumption scenarios. Section 6 presents results analysis based on the scenarios. Finally, concluding remarks are given in Section 7.

2 Relationship to Internet of Things

Internet of Things (IoT) is a name conceived to accommodate an extension of the web paradigm in order to consider the connection, monitoring and control of devices: things of everyday life [5].

The smart grid characteristics [6], the expectation to spread intelligence and control from some central core over peripheral nodes on electric energy usage entails the need to research, test and customize computing techniques for DR analysis, enabling opportune interfacing management of energy usage [5].

The DR analysis in this paper is on a line of technological contributions tied with the IoT aptitude to: (i) collect data on energy usage, (ii) provide this data to other participants on local utility grid, and (iii) get energy prices information [7].

3 Demand Side Management and Demand Response

The oil shocks of the 1970s brought a new research concerning policies and measures targeting energy demand. Since then development of DR and energy efficiency policies are on the way, aiming to influence quantities or patterns of energy [8]. A discussion and analyze DSM appeared in 1985 with alternatives related to load shape benefits derived from techniques concerning, for instance: valley filling of utility's loads, clipping of peaks, shifting of loads to off peak hours, strategic conservation to reduce demand [9]. DSM became more ambitious with smart metering in domestic buildings and services and bidirectional communication provided in smart grids.

DR allows load adaptation to the generation; this adaptation is important due to the intermittent characteristic of renewable energy sources integrated into the electric grid. DR programs are mainly conditioned by grids requests and consumers have to adapt the consumption in order to take advantage of economical benefits. DR is defined as any program which communicates with the consumer and either enables or encourages the consumer to lower or shift energy consumption during periods of unfavorable energy prices to favorable periods [10]. DR programs can be divided into: Incentive-Based Programs (IBP) and Time-Based Programs (TBP) [11].

4 Demand Response Model Case Study

The proposed model is a TBP one, aiming to support consumer's decision and actions on load management also referred as CL management, taking into consideration the available power capacity and energy price. This TBP model intends to give consumers flexibility in order to take advantage of economical benefits allowing the load management that best fits consumer's profiles or life-styles. Consumers load management is possible throughout methods of load time-scheduling and load shedding. The economical benefits mainly come from valley filling of utilities' loads and shifting of loads to off peak hours.

The model assumes that grid power dispatch provides hourly information to consumers about available power and energy price. The assumptions considered in the model are the following: available power is considered non-constant along the time horizon to emulate available distributed generation behavior and consequently allowing a convenient DR modeling; consumer can deselect priority on a CL in order to allow available power to the next priority CL; once the in progress power sum of CL reaches the total available power, no further CL can be connected. These assumptions are intended to give to the consumer an opportunity to adapt the consumption diagram to what the consumer is willing to pay for energy price.

The consumers' profiles and operation modes were obtained from analysis of consumption behavior that allowed a consumption pattern definition. Three consumer profiles were set to ensure the coverage of DR generalization on the model and help consumers cope with price changes over one day period. These three consumer profiles were designated as follows: economic, moderate and extravagant. Two operation modes designated by cleaning and comfort are set for each consumer profile. An example of a priority list is shown in Table 1. In table 1 the CL priority list is the same for the three profiles because it guarantees that consumer decision is obeyed independently of the selected profile, regarding the energy price. For example, if economical profile is selected but for one time period the energy price is superior to the one allowed for the economical profile, consumer can change his profile to moderate, keeping the same CL priority list but with increased costs. Also, a common CL priority list to the three profiles allows a DR model comparison analysis for the scenarios considered in this study.

Table 1. Controllable load priority list for cleaning mode

Economic profile	Moderate profile	Extravagant profile
	1 Dishwasher	
	2 Washing machine	
	3 Dry Machine	
	4 Air Conditioner	
	5 Thermo ventilation	
Low price	Low or medium price	Low, medium or high price

The economic profile is set to allow only the use a CL when the energy price is low. The moderate profile is set to allow the use a CL when energy price is low or medium and in extravagant profile any CL could be used independently of energy price. The necessity of pattern recognition associated with a control which supports consumers' decisions for DR model design is fulfilled using fuzzy clustering method. The controller is implemented in the Matlab-Simulink® software resorting to the Fuzzy Logic Toolbox.

The demand response model is shown in Fig. 1.

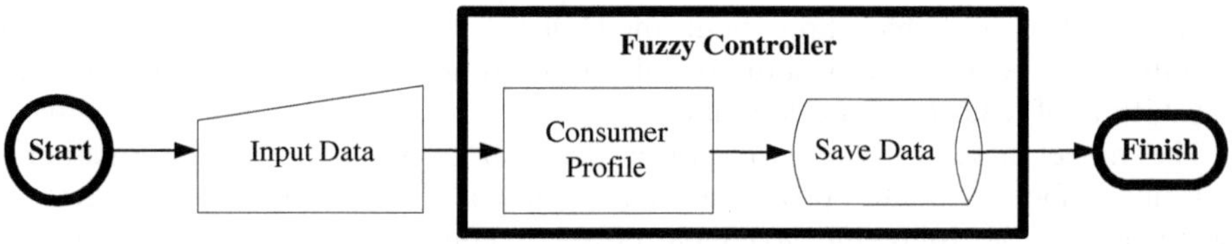

Fig. 1. Demand response model

The input data was set considering one residence with 6.9 kVA of installed power and the CL are home appliances and air conditioner with nominal reference power given in [12]. The model input data are chosen considering CL consumption stages, available power and energy price. The reactive power is neglected.

The consumption stages assumed are the null power or the reference nominal power, i.e., for the air conditioner the consumption stages are off power or full power, which is 1.8 kW. Nevertheless, the model is valid for any other consumption stages introduced as input data. The model is applied considering a power range between 1.8 kW and 6.9 kW, respectively given by the CL lower value and the installed electric power value.

5 Fuzzy Clustering Model

The required DR behavior is standardized in an Excel sheet by creating relations between the input and output parameters desired for the DR model. In order to represent the DR behavior in a fuzzy controller, data clustering technique was used.

Given an input, available in the Excel data sheet, data clustering apprehends the pattern in it and gives the outputs. The outputs are cluster centers corresponding to the pattern, which can be utilized in a fuzzy controller. Each cluster center is composed by twelve elements. A total of 7 inputs are considered. These inputs are given by: 5 CL consumption states required by the consumer, the energy price and the available power. And 5 outputs given by the 5 CL states.

Data clustering techniques are necessary due to the fact that otherwise the required number of rules is a large number in order to acceptably define the fuzzy controller. Fuzzy clustering consists on a strategic division of the data space into fuzzy clusters. Each fuzzy cluster is introduced in order to stand for the convenient representation of one specific part of system behavior. After projecting the clusters onto the input space, the antecedent parts of the fuzzy rules can be found [13].

The subtractive clustering technique is an extension of the mountain clustering method based on data sets evaluated using mountain function resulting in cluster centers. Accordingly to this technique, given a collection of n data points in an m-dimensional space, the technique starts by setting each data point x_i as a potential cluster center with a defined potential in function of the Euclidian distances of all data points. This function, specifying the potential at x_i, is given by

$$P_i = \sum_{j=1}^{n} e^{-\alpha \| x_i - x_j \|^2} . \tag{1}$$

α is a parameter given by,

$$\alpha = 4 / r_a^2 . \tag{2}$$

r_a is the cluster radius, setting a hypersphere of data points with significant influence on the cluster center potential in order to be acknowledged as neighbors.

The r_a value has a strong effect on the numbers of the generated cluster. Three observation about this value are in order: a data point outside a neighborhood of radius r_a has little influence on the potential of the neighborhood center data point; a higher value for r_a generally results in a generation of few clusters and therefore a model too generalized; while a low value can lead to an excessive generation of clusters and therefore a model that has not enough generalization [14]. The r_a values were chosen in order to introduce an adequate amount of clusters which are related to the resulting number of fuzzy rules. The r_a parameter can be adjusted based on the required model complexity and generalization ability. In this paper the r_a values determined resulted in a rules number which allowed an adequate control action with an adequate model time response. Hence, it was considered that no major advantages arrived with improved mathematical methods or optimization methods applied on the r_a determination. The values for r_a are shown in Table 2.

The subtractive clustering technique after computing all the data point potentials, using the expression (1), identifies the first cluster center as the data point with the highest potential.

Table 2. Fuzzy rules number and r_a value

Profile	Operation mode	Rules number	r_a
Moderate	Cleaning	14	1.4
	Comfort	10	1.5

Then, the potential of the data point is revised using the assignment given by

$$P_i \Leftarrow P_i - P_1^* e^{-\beta \| x_i - x_1^* \|^2}. \quad (3)$$

In (3) P_1^* and x_1^* are respectively the first cluster center potential value and center location, β is a parameter given by

$$\beta = 4 / (\eta \, r_a)^2. \quad (4)$$

In (4) η is the squash factor defining the neighborhood of data points that will have significant measurable reduction in the potential value. Normally, a value of $\eta = 1.5$ is a good choice. After the first cluster center has been obtained and all the potential of the data points have been revised using the procedure given by assignment (3), the data point with the highest potential is selected as second cluster center. In general, after k^{th} cluster center has been obtained, the potential of each data point is revised using the assignment given by

$$P_i \Leftarrow P_i - P_k^* e^{-\beta \| x_i - x_k^* \|^2}. \quad (5)$$

In (5) P_k^* and x_k^* are respectively the kth cluster center potential value and center location. A description of further details about the subtractive clustering technique employed in this paper can be seen in [14]. Fuzzy rules are defined based on valid cluster centers. The assumptions considered in the model presented in section 4 and the data sets to be submitted to the subtractive clustering technique were implemented in Excel®. Each profile and respective operation modes have a proper data set. A file with Matlab® extension .m was created, where the function *genfis2* perform the subtractive clustering technique on the data sets. Then the Matlab® function *fuzzy* is applied, which allows articulating the created clusters into if-then rules, establishing the fuzzy control saved in a file with Matlab® extension .fis.

Other approach, such as neural networks (NN) can be applied to the model controller. However, for the model implementation in NN the required data sets characteristics in association with the controllable loads power consumption initial value definition, revealed a controller design complexity which have leaded to considerably more time dispended in this model implementation then with the fuzzy clustering model developed.

6 Results Analysis

Two scenario analyses are performed with and without the DR model, considering a residence with a limited power of 6.9 kVA and with a base power consumption shown in Fig. 3. The base power consumption represents the consumption of all loads excluding the CL.

The first scenario, shown in Fig. 2, refers to a consumption diagram where no planning or restriction to CL connection exists, i.e., no DR model is on consideration. This scenario considers that, when consumers are at home and awake the consumption is high and limited by the 6.9 kVA, otherwise the consumption values are low and limited by the base power consumption. This scenario is used as a comparison basis between the consumption diagram where no planning or restriction to CL connection exists and a consumption diagram resulting from the DR model.

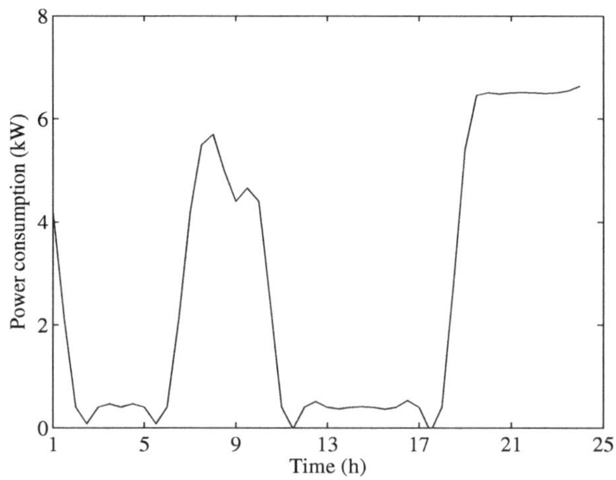

Fig. 2. Power consumption of consumer in first scenario

The second scenario has data information, shown in Fig. 3, about power and energy prices available in one day period on an hourly basis and consumer selects operation mode accordingly with – when at home and awake, comfort mode is selected, otherwise cleaning mode is selected, considering two hypotheses: i) consumer sets an economic profile; ii) consumer sets a moderate profile.

The input data to simulate the DR model for moderate profile is shown in Fig. 3.

A comparison between Fig. 2 and both DR model responses shown in Fig. 4 and Fig. 5, shows the influence of price on consumer consumption. On Fig. 2 the consumer is not influenced by the price because it is always the same, so the consumer consumes where is more convenient. On the other hand, accordingly to Fig. 4 and Fig. 5, the consumer is concerned with the price because it changes along one day, so the consumer reschedules the CL in a way to be connected on hours where the price is lower. A comparison of responses in the second scenario from hypotheses i), Fig. 4 and ii), Fig. 5, shows for the moderate profile that CL has more favorable conditions to be connected in one day period, than for the economic profile. Hence, the consumer is subjected to a lesser restricted DR due to the higher acceptance of the energy price in moderate profile than in economic profile. The extravagant profile allows the consumer a DR only depending on available power. It can be observed in

Fig. 4 and Fig. 5 that DR model guarantees that the sum of base power consumption with CL consumption power is never superior to the available power. The individual CL power consumption for moderate profile in one day period is presented in Fig. 6.

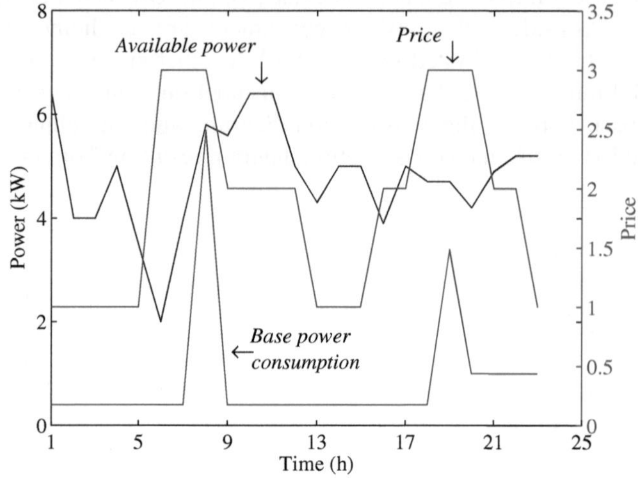

Fig. 3. Base power consumption, available power and price evolution

Together, both Fig. 3, Fig. 4, Fig. 5 and Fig. 6 allow to confirm that the energy price is satisfactory associated with CL power consumption.

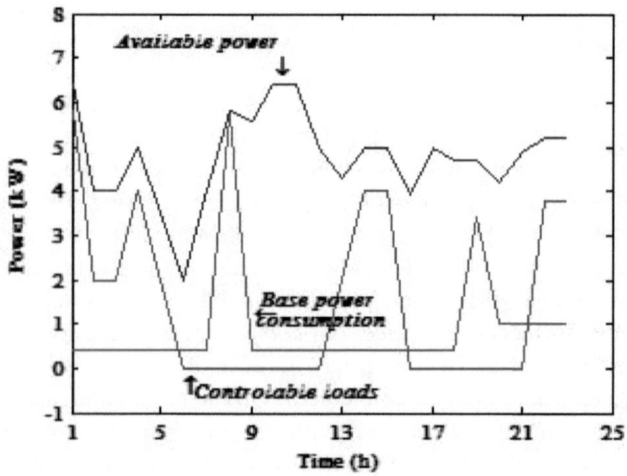

Fig. 4. Power evolution for economic profile

Both Fig. 3 and Fig. 6 allow to conclude that in the time period between 6 a.m and 8 a.m there is no CL available because the energy price is higher than the energy price assigned to moderate profile. The analysis of Fig. 5 and Fig. 6 allows to conclude that only at 8 a.m there is no sufficient power available to connect any CL. However, regarding the energy price shown in Fig. 3 and taking into account the defined CL priority list shown in Table 1, no CL is connected in the period between 6 a.m and 8 a.m as shown in Fig. 6.

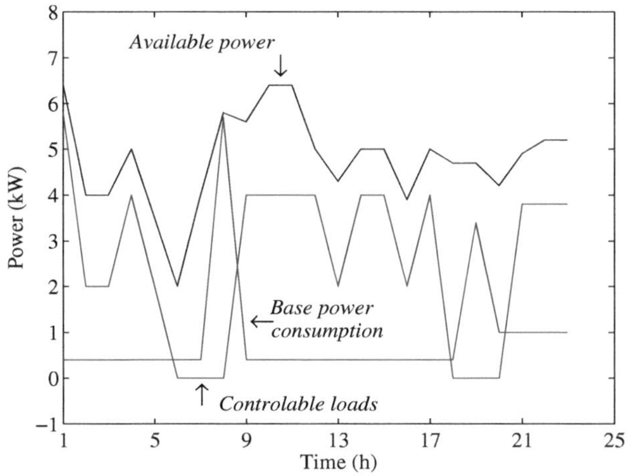

Fig. 5. Power evolution for moderate profile

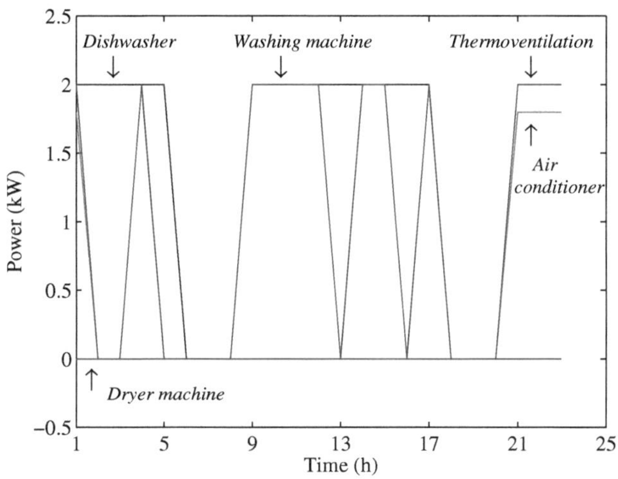

Fig. 6. Controllable loads evolution stage for moderate profile

7 Conclusion

The proposed model ensures that power consumption is never superior to the available power and that no CL can be connected if energy price is superior to the correspondent energy price assigned to the selected consumption profile.

Consumer is subjected to a lesser restricted DR in moderate and extravagant profiles than in economic profile due to a higher acceptance of the energy price.

The consumption diagram can be shaped by the price changes along one day, which can lead to a more efficient grid because the consumption diagram can be conveniently adapted as it can be seen by comparison between the first and second scenarios, where a flatten diagram is desirable.

Consumer evidences a flexibility on consumption schedule when data on available power and on energy price are to be taken into account in due time. Hence, collecting and providing data on energy usage to participants on local grid, having energy prices information and model development, as the one proposed in this paper, are crucial for IoT to conveniently assist on a successful DR.

References

1. Boshell, F., Veloza, O.P.: Review of developed demand side management programs including different concepts and their results. In: Proc. on Trans. Dist. Conf. Exp.: Latin America, Bogota, Colombia, pp. 1–7 (2008)
2. Johnson, K., Thomas, E.: The fundamentals of liking demand side management strategies with program implementation tactics (June 27, 2012), `http://www.utilityexchange.org`
3. Feenstra, C.F.J., Backhaus, J., Heiskanand, E.: How to change consumers' energy-related behaviour? (August 20, 2012), `http://www.energychange.info`
4. Ferreira, J.: Demand Side Management, Igenium, pp. 1–16 (August 15, 2012), `http://www.jesusferreira.com.pt`
5. Bui, N., Castellani, A.P., Casari, P., Zorzi, M.: The internet of energy: a web-enabled smart grid system. Network 26, 39–45 (2012)
6. Yun, M., Yuxin, B.: Research on the architecture and key technology of internet of things (IoT). In: Proc. IEEE Int. Conf. on Adv. in Energy Eng., Beijing, China, pp. 69–72 (2010)
7. Ciuciu, I., Meersman, R., Dillon, T.: Social network of smart-metered homes and SMEs for grid-based renewable energy exchange. In: Proc 6th IEEE Int. Conf. Digital Ecosystems Tech. – DEST, Campione d'Italia, Italy, pp. 1–6 (2012)
8. Haney, A.B., Jamasb, T., Platchkov, L.M., Pollitt, M.G.: Demand-side management strategies and the residential sector: lessons from international experience. EPRG, University of Cambridge (2010)
9. Delgado, R.M.: Demand-side management alternatives. Proceedings of IEEE 73, 1471–1488 (1985)
10. Stromback, J.: The development of demand response in Europe. In: Proc. Smart Grid Tech. Conf. and Exhib., San Diego, USA (2010)
11. Aalami, H., Yousefi, G.R., Moghadam, M.: Demand response model considering EDRP and TOU programs. In: Proc. IEEE Trans. Distri. Conf. and Exp., Chicago, USA, pp. 1–6 (2008)
12. ERSE (July 23, 2012), `http://www.erse.pt`
13. Priyono, A., Ridwan, M., Alias, A.J., Atiq, R., Rahmat, O.K., Hassan, A., Ali, M.A.M.: Generation of fuzzy rules with subtractive clustering. Jurnal Teknologi 43, 143–153 (2005)
14. Chiu, S.L.: Fuzzy model identification based on cluster estimation. Journal of Intelligent and Fuzzy Systems 2, 267–278 (1994)

Modeling the Input Variables and Setting on the Static System Model at Using the Genetic Algorithm for Fault Location in the Power Transmission Grid

Tonka Sharenkova[1] and Rosen Rusev[2]

[1] PhD Student, Electrical Network TU-Sofia, Kl. Ohridski. 8,
1756 Sofia, Bulgarien
[2] PhD, Elektrik plant and Station, TU-Sofia, IPF-Sliven, Burgasko shose 59,
8800 Sliven, Bulgarien
rstamatov@yandex.ru

Abstract. In the paper is presented a method for fault location in the power grid through waveform matching of the recorded wave from failure with simulation from the static system model wave failure. The basis of the approach is comparing of the phase of the waves. The search process to find the best waveform match is actually an optimization problem. The genetic algorithm is used to find the optimal solution. The proposed method is suitable in cases where data from digital fault recorders are scarce. In these circumstances, the proposed approach provides more accurate results compared to the other known techniques. But for the correct operation of this method for fault locating in the system exercise influence both the form of the acquired form from digital fault recorders input data thus the correlation between the power transmission system and the static system model. Namely these issues are the subject of this paper.

Keywords: Genetic Algorithm, Fault Location, Digital Fault Recorders, Simulation.

1 Introduction

Rapid and accurate identification of the site of accident in the transmission system is crucial for its reliability and is the first step towards rapid recovery system. The accuracy of evaluating of the fault essentially depends on the available information. While there are some successful algorithms for the fault location, using data from two or three substations, it is difficult to achieve satisfactory results if the information is from a limited number of substations [1, 2].

To improve the accuracy for fault location when only limited recorded data are available, the “waveform matching" based approach may be used. In this approach, simulation studies are carried out to obtain simulated waveforms under specified fault conditions. The simulated waveforms are then compared with the recorded ones. By iteratively posing faults in the system, running simulations, and comparing the simulated waveforms with the recorded ones, an optimal estimate of the fault location

L.M. Camarinha-Matos, S. Tomic, and P. Graça (Eds.): DoCEIS 2013, IFIP AICT 394, pp. 413–420, 2013.

may be obtained. It may be determined as the one specified in the simulation studies that allows simulating the waveforms that best match the recorded ones. The matching is made at the phasor level presently.

In this research, the fault location estimation is mathematically formulated as an optimization problem of which the fault location and fault resistances are unknown variables. An efficient genetic algorithm (GA) based searching scheme is developed for obtaining the solution that is globally optimal [3].

The article explains the key concepts of "sparse data" and "waveform matching". Below are presented the requirements towards the input data to the GA and an algorithm for tuning the static model of the system. The algorithm is tested and the results of before and after tuning the Static System Model are presented in tabular form.

2 Contributions to Internet of Things

The paper presents a method for fault location in the power grid through waveform matching of the recorded wave from failure based on simulation by using a static system model. For solving this optimization problem used genetic algorithm. The accuracy of the proposed method depends on the actuality of static system model with actual power grid in the moment. The contribution of this publication is to develop an algorithm for setting the static model of the system. The proposed method is suitable in cases where data from digital recorders are scarce.

3 Motivating Problem

The proposed approach uses the method "waveform matching". There are presented two key concepts, namely "sparse data" and "waveform matching" as follows.

3.1 Sparse Data

Sparse data, in the work, is referred to the data obtained from recording devices sparsely located at various substation locations. The recording devices may include digital fault recorders (DFRs), digital relays, or other intelligent electronic devices (IED). The data captured by the recording devices may include analog quantities such as voltage and current waveforms and digital quantities such as breaker status and relay operation status. The both analog and digital quantities may be useful for locating the fault.

If only sparse data are available for fault location, in many cases, none of the one-end, two-end and three-end algorithms may be applicable for locating the faults with satisfactory accuracy [1-2]. To solve the fault location problem utilizing sparse data, the “waveform matching” based approach may be used as illustrated next.

3.2 Waveform Matching

In the "waveform matching" approach, a power system model is used to carry out simulation study. To simulate the waves in the model need to ask fault. Then are compare the simulated and recorded waves and on the degree of match is determined the fault. Theoretically, the simulated fault waveform will match completely with the recorded fault waveform if the assumed fault location and fault resistance correspond to the real fault condition.

The process to determine the fault location is iterative because several lines in the system and variety of possible fault resistances should be searched to obtain the optimal matching. When searching for the appropriate fault resistance first selects the most probable locations of the accident. Changing the fault resistance according to a specific increment, fault locations are searched thoroughly. The process will proceed until the selected sections in power system and possible fault resistance range are exhausted.

After completing the search, the fault is determined based on the optimal matching scheme. There are two possible schemes - phasor matching and transient matching. In the test is used the phasor for matching. The degree of similarity can be expressed by the following criterion:

$$f_c\left(x, R_f\right) = \sum_{k=1}^{N_V} r_{kV}\left|V_{ks} - V_{kr}\right| + \sum_{k=1}^{N_i} r_{ki}\left|I_{ks} - Ikr\right| \quad (1)$$

where: $f_c(x,R_f)$ – the cost function using phasors for matching, it is a non-negative number.; x, R_f – the fault location and fault resistance.; r_{kv}, r_{ki} – weights for the errors of the voltages and currents respectively.; V_{ks}, V_{ky} – the during-fault voltage phasors obtained from the short-circuit studies and from recorded waveforms respectively.; I_{ks}, I_{ky} – the during-fault current phasors obtained from the short-circuit studies and from recorded waveforms respectively.; N_v, N_i – the number of the selected voltage and current phasors respectively.; k – the index of the voltage or current phasors.;

The best estimate of the fault will be achieved when the value of the function is the smallest. Therefore, the problem of assessing the fault actually is optimization problem.

The genetic algorithm (GA) based on the optimization approach is a good choice to search for the global optimal solution. To use GA has to convert the minimization problem to maximization problem in order to utilize the GA. That requires us to convert the cost function to a fitness function of GA. The fitness function is as follows:

$$f\left(x, R_f\right) = C_{\max} - f_c\left(x, R_f\right) \quad (2)$$

where:

$f(x,R_f)$ – is fitness function

C_{max} – is the maximum fitness value in the current population.

In equation (2), the location of the short circuit and resistance were selected as two variables. They are represented as binary lines in GA. Usually in GA are used three operators: selection, crossover and mutation. The process of searching through GA is as follows: at the beginning the initial population is generated randomly. Then posing the fault according to the initial population the short circuit study is carried out to obtain the simulated during-fault phasors and further calculate the fitness value for each individual. The next generation is produced by applying the three steps as described above. The process is repeated until the best match is found.

4 Implementation of the Algorithm for Setting Parameters of the Static System

4.1 Data Requirement

The data inputted by the user includes necessary information for the fault, matching options and selected fault data. The necessary information for the fault refers to the estimated fault type and faulted circuit that can help limit the GA range search. The matching options are used for specifying currents through the circuits or voltages at buses used for waveform matching. Selected fault data refers to a choice in the use of different DFR combinations under the situation where multiple DFRs are triggered.

The static system model refers to the saved case of PSS/E. It should contain the raw data for power flow, data for sequence impedance and system topology. The model is static since it may not reflect the prevailing system conditions. This may affect the accuracy of the algorithm and some measures overcoming the shortcoming should be taken.

The fault data refers for the data captured through Digital Fault Recorders (DFRs). The current software reads fault data provided in COMTRADE format [6]. The COMTRADE file should include two files: COMTRADE configuration file, which contains information for interpreting the data file, and COMTRADE data file, which contains analog (current and voltage) and digital values (breaker contacts and relay status) for all input channels for a specific substation.

The substation interpretation data contains information that relates to the channel numbers to the monitored signals. It also represents the correspondence between the designations used in the DFR files and those used in the PSS/E file. Each substation should have one interpretation file and the interpretation file needs to be modified to reflect the DFR configuration or the system model changes. The information should be provided by the user in advance.

The data inputted by the user includes necessary information for the fault, matching options and selected fault data. The necessary information for the fault refers to the estimated fault type and faulted circuit that can help limit the GA range search. The matching options are used for specifying currents through the circuits or voltages at buses used for waveform matching. Selected fault data refers to a choice in the use of different DFR combinations under the situation where multiple DFRs are triggered.

4.2 Synchronizing the Phasors from DFR Recordings

By applying Fourier algorithm using equation (2) are calculate the fitness values the voltage and currents phasor for different substations. The DFR data from different channels in the same substation or in different substation may lack synchronization.

In order to reduce the error of matching, the phasor calculated from DFR must be synchronized. Fig. 1 shows the relationship between the phasors obtained from the load flow study and from the recorded waveforms. S_{na}, S_{nb}, S_{nc} represent pre-fault phasor of phases A, B, C respectively are obtained by the load flow study. R_{na}, R_{nb}, R_{nc} represent pre-fault phasors of phases A, B, C respectively are obtained from the recorded waveform. R_{fa}, R_{fb}, R_{fc} represent during-fault phasors of phases A, B, C respectively are obtained from the recorded waveform. α is the angle difference between the pre-fault phasor obtained by the load flow study and the pre-fault phasor obtained from the recorded waveform.

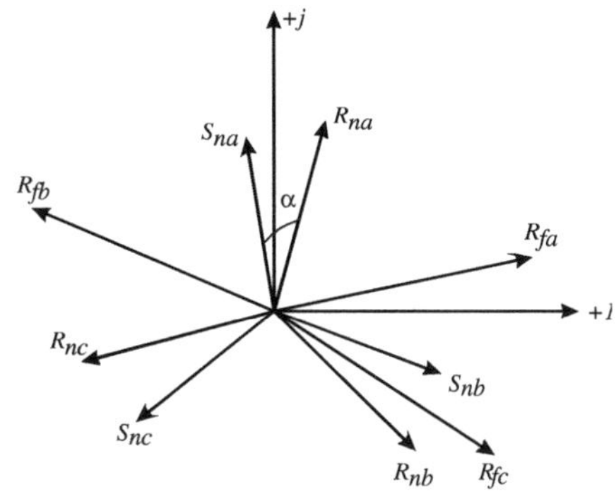

Fig. 1. The relationship between the phasors obtained by load flow study and by recorded waveform

The synchronization is done by rotating counterclockwise the pre-fault phasors R_{na}, R_{nb}, R_{nc} by an angle of α. Consider that the angle difference between pre-fault phasor and during fault phasor for a specific phase and current (voltage) is fixed; during-fault phasor R_{fa}, R_{fb}, R_{fc} is also rotated by an angle of α.

4.3 Tuning the Static System Model

As mentioned before, the given static system model, used in the simulation studies, may not reflect the prevailing operation conditions of the system when the fault occurs. The generator power output and load power may not always keep the same value and may vary with time. To match exactly the phasor extracted from DFRs and those obtained from simulation studies, it may be beneficial that the system model used in simulation studies is updated by utilizing the information captured close to the moment before the fault occurs. The process of updating the system is called "tuning of the static system model".

Tuning the static system model includes two aspects [5]: tuning the topology of the system model and tuning the static parameters such as generator and load data.

The updating of the system topology relates to updating the service status of the circuits in the system. The pre-fault status of breaker in the circuit or the current through the circuit is used to update the system topology. There are two possible situations: 1) In the first situation, where both the circuit's breaker status and currents (or only the currents) are monitored by DFR, the current magnitude will be utilized to update the service status of the circuit. If the current magnitude is smaller than a pre-specified threshold, the circuit will be designated as being out of service. Otherwise, the circuit will be designated as being in service. 2) In the second situation, where only the breaker status of the circuits is monitored, the pre-fault breaker status will be utilized to determine the service status of the circuit. If the pre-fault status breaker indicates an open circuit, then the circuit will be determined as out of service. Otherwise, the circuit will be determined as being in service.

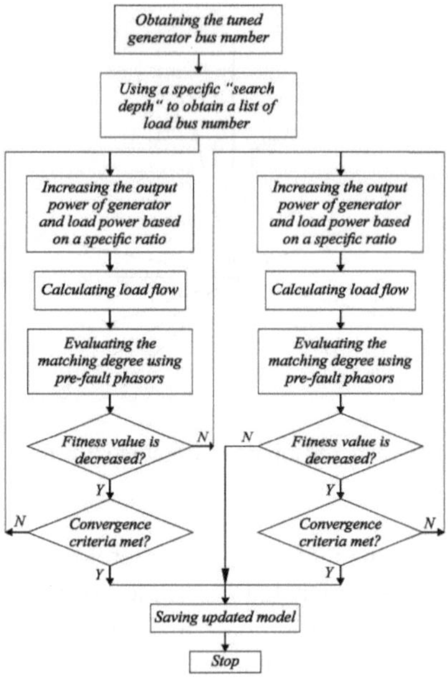

Fig. 2. Block diagram of the static parameter settings

The goal of updating the generator output power and the power load is to bring the static system model closer to the real life system. To reach the purpose, again is utilized the waveform-matching. The matching is made between the voltage and current waveforms obtained by DFRs and those generated by load flow studies. The equation (1) is applied as the objective function to evaluate the matching degree of the simulated and recorded waveforms. Here, the corresponding V_{ks}, V_{kr} and I_{ks}, I_{kr} have different meaning. They are the pre-fault voltage or current phasors obtained from the load flow study and recorded waveforms respectively. The static parameters that provide the best match are the ones that minimize the objective function. The flowchart for tuning the static system parameters is shown in Fig. 2.

5 Test Algorithms

The fault case is selected for which two DFRs located in substations Granitovo and Podves locations respectively, were triggered. Table 1 shows the effect of tuning the static parameters. The first column represents the different combinations of DFR data. The second column shows the fitness value calculated using the pre-fault phasor. The fitness value decreases significantly after tuning the system based on the strategy mentioned above. The updated fitness value is shown in the third column. These results prove the tuning strategy is effective. The more accurate tuning depends on the region of tuning and more real life data.

Table 1. The change of fitness value using pre-fault phasor before and after tuning

DFR UTILIZED	FITNESS VALUE USING PRE-FAULT PHASOR BEFORE MODEL TUNING	FITNESS VALUE USING PRE-FAULT PHASOR AFTER MODEL TUNING	QUANTITIES MATCHED
Granitovo	14,8	8,87	All monitored currents
Granitovo Podves	6,21	0,62	Currents on affected Ckt.25
Granitovo Podves	37,26	27,75	All currents
Granitovo Podves	7,45	0,92	Currents on affected Ckt.25, Ckt.27
Granitovo	17,21	19,41	All currents and voltages
Granitovo Podves	37,45	27,97	All currents and voltages

6 Conclusion

After tuning up of the system based on the strategy mentioned above, compatible values decreases significantly. These results demonstrate that the strategy is effective. The more accurate tuning depends on the region of tuning and more real life data.

The paper offers requirements toward input data for GA and tuning the static system model. The results presented in Table 1 are been used to test the GA [3].

The biggest advantage is using limited data to find the fault and using data from other digital recording devices. It is suitable for situations where conventional algorithms cannot be applied. The approach does not refer to a specific section or line; it is based on a system view. However, the known system model including the static parameters and topology is assumed.

The algorithm has been tested on four emergencies [3]. The test results show that the approach is very promising.

References

1. Girgis, A.A., Hart, D.G., Peterson, W.L.: A new fault location technique for two- and three-terminal lines. IEEE Transactions on Power Delivery 7, 98–107 (1992)
2. Waikar, D.L., Elangovan, S., Liew, A.C.: Fault impedance estimation algorithm for digital distance relaying. IEEE Transactions on Power Delivery 9, 1375–1383 (1994)
3. Goldberg, D.E.: Genetic Algorithms in Search, Optimization and Machine Learning. Addison Wesley, Reading (1989)
4. Power Technologies, Inc., PSS/E 25 Program Operation and Application Manuals (1997)
5. Yehsakul, P.D., Dabbaghchi, I.: A topology-based algorithm for tracking network connectivity. IEEE Trans. on Power Delivery 10(1), 339–346 (1995)
6. IEEE Std C37.111-1999, IEEE Standard Common Format for Transient Data Exchange (COMTRADE) for Power Systems (approved March 18, 1999)

Part XV
Energy Transformation

Modeling and Characterization of Leakage Inductances for Transformer Winding Fault Studies

Luís M.R. Oliveira[1,3] and A.J. Marques Cardoso[2,3]

[1] Instituto Superior de Engenharia, Universidade do Algarve, Faro, Portugal
[2] University of Beira Interior, Department of Electromechanical Engineering, Portugal
[3] Instituto de Telecomunicações, Department of Electrical and Computer Engineering, University of Coimbra, Pole II, P – 3030-290 Coimbra, Portugal
lolivei@ualg.pt, ajmcardoso@ieee.org

Abstract. This paper presents an analytical method to compute the leakage inductances of power transformers with a turn-to-turn winding fault. A leakage inductance model to represent the transformer with faulty turns is also proposed. The results obtained from the application of the analytical method are validated by using data obtained from finite-element analysis and experimental tests.

Keywords: Transformers, winding faults, leakage inductances, modeling.

1 Introduction

Power transformers are key elements of the electric generation, transmission and distribution network and constitute one of the most capital-intensive investments made by power system utilities. The unexpected failure of a power transformer can generate substantial costs for repair and financial loss due to unscheduled electrical outage. Therefore, it is of crucial importance to detect internal defects in their incipient stage, so that the faulted unit can be immediately disconnected, avoiding the progression of the defective condition into a catastrophic failure, minimizing the damages in the transformer and other expensive neighboring equipment, and thus reducing downtime and total outage costs.

The development of new techniques for transformer condition monitoring and fault protection requires a detailed characterization of the transformer behavior during the occurrence of turn-to-turn winding short-circuits. The experimental study of these incipient internal faults presents some difficulties, mainly due to the high magnitudes of the faulty currents involved, which can damage the test transformer. Therefore, a detailed analysis of these phenomena can be better investigated by the use of a suitable digital simulation transformer model [1].

Several circuit-based transformer models were presented in the last few years for winding fault studies [1]-[4]. One of the major difficulties in these approaches is to model the leakage inductances when the turn-to-turn short-circuit is present. The difficulty arises because the distribution of the magnetic flux is substantially modified

L.M. Camarinha-Matos, S. Tomic, and P. Graça (Eds.): DoCEIS 2013, IFIP AICT 394, pp. 423–430, 2013.

when such a fault occurs [5]. In [2] the method to determine the leakage inductances relies on correction factors, in order to take into account the radial component of the leakage flux. In [3] an analytical formula for the leakage inductance of the faulty winding is developed, but only for the simpler case of transformer no-load conditions. Nevertheless, the derivation of the models is not straightforward and difficult to be implemented. More recently a simplified method was proposed [4], in which the leakage inductances are computed from the nameplate short-circuit reactance. Errors up to 68% are reported (as compared with experimental results) and only a model based on finite elements method would yield good results.

This work presents an analytical method to compute the leakage inductances of power transformers with a turn-to-turn winding fault. The influence of the fault position and fault progression is also characterized. Additionally, a leakage inductance equivalent circuit to represent the transformer with faulty turns is proposed. The results obtained from the application of the analytical method are validated by using data obtained from finite element analysis and experimental tests.

2 Relationship to Internet of Things

The smart grid can be seen as one of the Internet of Things application domains, in which intelligent electronic devices and their communication capabilities can be used to provide unprecedented reliability levels in the power network. However, for this to occur it is necessary the development of new fault detection methods, which must be integrated with the smart grid technologies. This in turn requires detailed and extensive modeling and simulation of the power system network. A power transformer model for winding fault studies is essential for these purposes.

3 Winding Fault Characterization

For the experimental investigation a three-phase, two winding, three limb transformer, of 10.3 kVA, 230/132 V, was used. The primary and the secondary windings have 152 and 90 turns, respectively. In each transformer winding there are five additional tappings connected to the coils, allowing for the introduction of shorted turns at several locations in the winding, as shown in Fig. 1(a), for one phase of the primary and the secondary windings. Only the coils of the center limb were used to perform the short-circuit single-phase tests, which results in a shell-type core design.

When a fault occurs in the primary-side, the short-circuited turns act as an autotransformer load on the winding, as shown in Fig. 1(b). The initial effect of the inter-turn short circuit is limited to a slight increase in the primary current. However, the insulation failure can lead to a high circulating current in the shorted turns, even if only a small number of turns is affected.

Fig. 2(a) presents the current waveforms in the transformer windings for the case of a load-test during the occurrence of a turn-to-turn fault in the primary winding. To protect the transformer from complete failure when the fault was introduced the current in the shorted turns is limited by an auxiliary resistor, which represents the fault contact resistance, R_{sh}. It can be seen that the current in the shorted turns (i_b) and the current in the secondary winding (i_s) are in phase opposition with the primary-side current (i.e., they are opposing the primary magnetomotive force (MMF)).

The leakage inductance is usually obtained by performing the short-circuit test. It might be thought that the effect of the inter-turn short-circuit could be analyzed by performing a short-circuit test where both N_s and N_b turns were individually shorted, Fig. 1(c). However, as explained next, this test is far from representing the true leakage flux distribution. The resultant current waveforms obtained for this case are shown in Fig. 2(b). It can be seen that the current i_b presents a completely different behavior than the one for the load test: it has a very small magnitude and is opposing the secondary MMF, instead of opposing the primary MMF. The main reason is that the mutual flux that links the three coils (N_a, N_b and N_s) is significantly different in these two tests, and, therefore, the induced currents do not follow the same pattern. As a result the flux distribution obtained by this short-circuit test is very different from the leakage flux component in the load test and it is not representative of this latter condition.

A better solution to study these phenomena can be obtained by performing the short-circuit test of Fig. 1(d), where the short is applied to N_s and N_b connected in series-addition. With this approach both currents in these coils are opposing the primary magnetomotive force, Fig. 2(c), which is much more similar to the results of the load-test of Fig. 2(a).

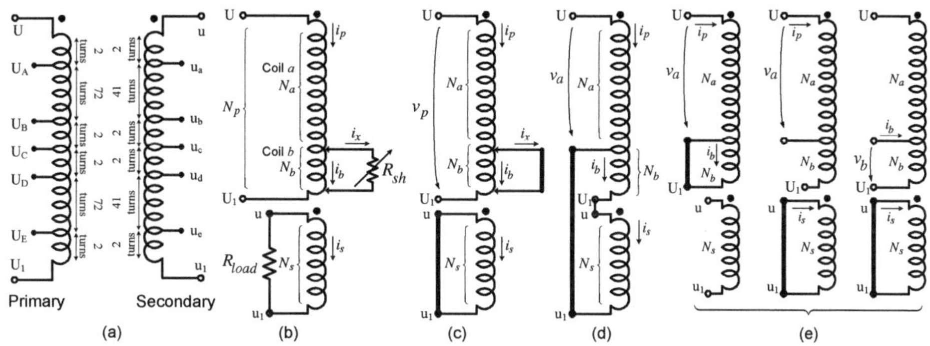

Fig. 1. (a) Location of the tappings of the windings; (b) equivalent circuit for a fault occurring in the primary winding; (c), (d) and (e) short-circuit test schematics

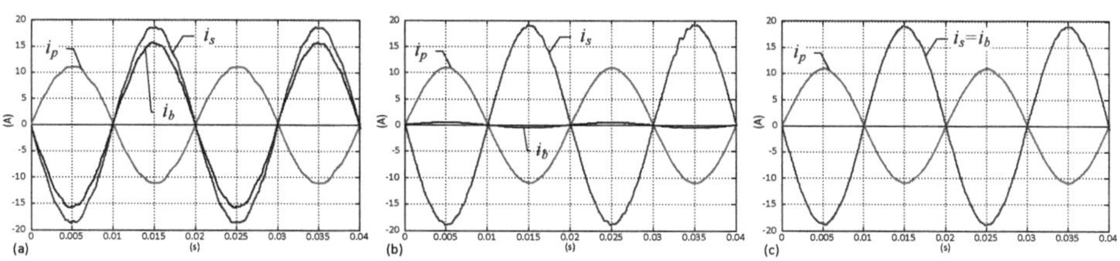

Fig. 2. Current waveforms for the case of: (a) load-test of Fig.1(b); (b) short-circuit test of Fig. 1(c); (c) short-circuit test of Fig. 1(d). ($N_b = 2$ turns.)

Alternatively, the leakage inductances can be obtained by performing the traditional short-circuit test were one winding is shorted at a time, Fig. 1(e). With this approach the three-winding transformer theory can be applied to obtain the leakage inductance transformer model, which must be consistent with the conditions of the non-standard short-circuit test of Fig. 1(d).

4 Leakage Inductance Analytical Computation

Several formulas for the analytical computation of the leakage inductances have been proposed in the past. For the case of concentric transformer windings, with the same height h, the leakage flux that flows due to the load current is virtually parallel to the axis (except in the ends of the windings). Under this assumption, Fig. 3(a) shows the per unit MMF distribution (m_k) obtained for the windings geometric structure of the test transformer. It is considered that the MMF's of the two windings are equal and opposite, which is valid for normal operating conditions (no fault). The leakage inductance can be computed by using (notation as per Fig. 3) [6]:

$$L_{\sigma(ax)} = K_\sigma \frac{2\pi N_1^2 \mu_0}{h} \sum_{k=1}^{n} \left[m_k^2 \cdot g_k \cdot r_{gk} + \left(m_k^2 + m_{k-1}^2 + m_k \cdot m_{k-1} \right) \cdot w_k \cdot r_{wk} / 3 \right] \quad (1)$$

where N_1 is the number of turns of the excited winding, μ_0 is the permeability of free space, n is the number of vertical layers (7 in this case), r_{wk} is the mean radio of the coil k, r_{gk} is the mean radio of the gap between coils k and k+1, and m_k is the per unit MMF acting on gap between coils k and k+1, see Fig. 3(a).The Rogowski correction factor, K_σ, is used to take into account the flux fringing at the top and bottom of the windings and the effect of the iron core. For the case of a shell-type design:

$$K_\sigma = 1 - \frac{1 - e^{-\frac{2\pi E}{T}}}{2\pi E/T} \left[1 - \frac{1 - e^{-\frac{2\pi E}{T}}}{2e^{\frac{4\pi C}{T}}} \left[\frac{L_2}{L} + \frac{L_1}{L} \left(1 + e^{-\frac{2\pi (C_1 - C)}{T}} \right) \right] - \frac{L_1}{L} e^{-\frac{2\pi (E + 2C + 2C_1)}{T}} \right] \quad (2)$$

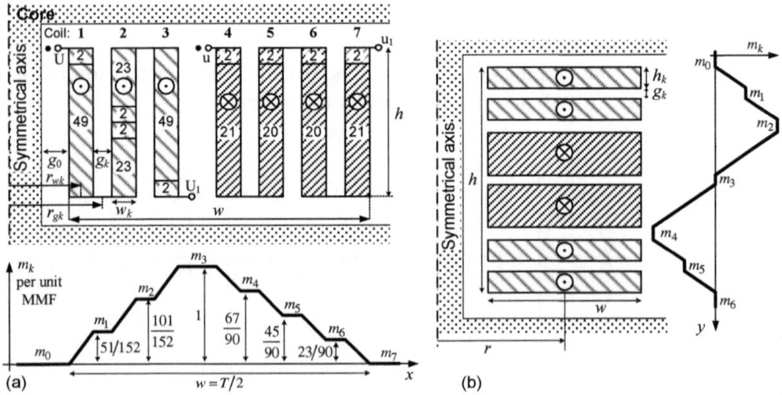

Fig. 3. Geometric structure and per unit leakage MMF distribution for the case of: (a) axial configuration; (b) radial configuration

being E the height of the winding, T the wave length of the MMF wave, L the mean coil perimeter, L_1 that part of the perimeter which has iron on both sides, L_2 the rest of the coil perimeter, and C and C_1 the distances from iron to coil on the two sides [6].

Equation (1) can also be used, with proper adaptations, for the computation of the leakage inductance of a disk-type winding configuration. Fig. 3(b) presents a generic disk-type winding arrangement. In this case it can be assumed that the leakage flux has only one component in the radial direction and the radial leakage inductance becomes:

$$L_{\sigma(rad)} = K_{\sigma} \frac{2\pi N_1^2 \mu_0 r}{w} \sum_{k=1}^{n} \left[m_k^2 \cdot g_k + \left(m_k^2 + m_{k-1}^2 + m_k \cdot m_{k-1} \right) \cdot h_k / 3 \right] \tag{3}$$

When there are irregularities in the concentric windings, such as a fault, the leakage flux is no longer parallel to the axis, but has significant components of radial flux, depending on the amount of asymmetry [7]. The formulas given by (1) and (3) are no longer valid under these asymmetrical conditions and a direct computation of the leakage inductance for these winding arrangements can be very complicated and extremely laborious. A very ingenious and useful way of dealing with these situations was proposed by Stephens [7] (and generalized in [6]), in which the leakage inductance is divided into two components, one axial and the other radial. Each component can be computed separately and then added together to give, very nearly, the value of the total leakage inductance. Fig. 4 illustrates the basic principle of the method with a simplified diagram, assuming a fault in the middle of the primary winding. The axial component is obtained by dividing the affected winding into two coils, one referring to the healthy portion of the winding and the other to the faulty part. These two parts are uniformly distributed along the axis, resulting in a concentric design configuration, which can be computed by (1).

The procedure to obtain the leakage inductance radial component is shown in Fig. 4(b). First, the secondary-side current is converted to the primary-side. Next, the secondary winding is divided into three segments, with the same dimensions of those in the primary winding. The MMF of each segment is then computed assuming an uniform distribution of the ampere·turns in the windings. Finally, the MMF's of the corresponding segments of each winding are summed up and the radial component of the leakage inductance can be computed by using (3).

The total leakage inductance is then obtained:

$$L_{\sigma} = L_{\sigma(ax)} + L_{\sigma(rad)} \tag{4}$$

A finite elements method (FEM) based transformer model [8] is also used to investigate the adequacy of the analytical calculations. The energy method was used to compute the leakage inductance from the FEM results [9].

Fig. 5(a) presents the results obtained when the position of the faulty turns (N_b) are moved along the vertical axis of coil 1, from top to bottom, using the series-addition short-circuit test of Fig. 1(d). The leakage inductance takes greater values when the N_b turns are located at the coil ends, because the radial component is higher in this situation. Obviously, the asymmetry increases with the number of the faulty turns, and, as a consequence, there is also an increase in the variation of the leakage inductance when the N_b turns are moved along the winding. Only one experimental result can be obtained for this specific test conditions, due to the fixed tap positions in the coils.

Fig. 5(b) presents the variation of the leakage inductance when the fault progresses vertically, involving additional turns, first from the top to bottom in coil 1, and then affecting the neighboring vertical layer (coil 2). The leakage inductance initially grows, due the increasing values of the radial component. It is interesting to note that this behavior is opposing the fault progression, since it tends to limit the faulty current. The radial component reaches its highest level when the fault has extended to just about the vertical center of the coil and then decreases more or less symmetrically. This pattern is repeated in the other vertical layers (only shown for coils 1 and 2). The axial component of the leakage inductance drops, since the number of primary turns is effectively decreasing as the fault evolves.

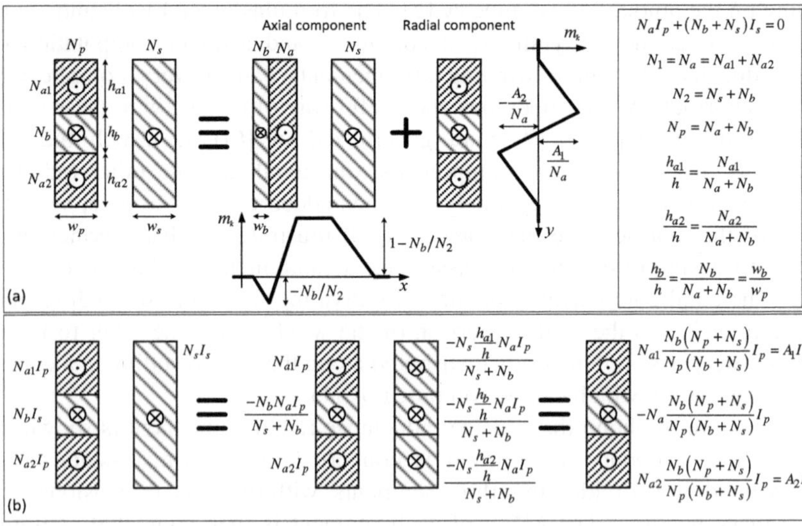

Fig. 4. (a) Equivalent representation of a fault in terms of two components, axial and radial; (b) diagram illustrating the procedure to obtain the radial component

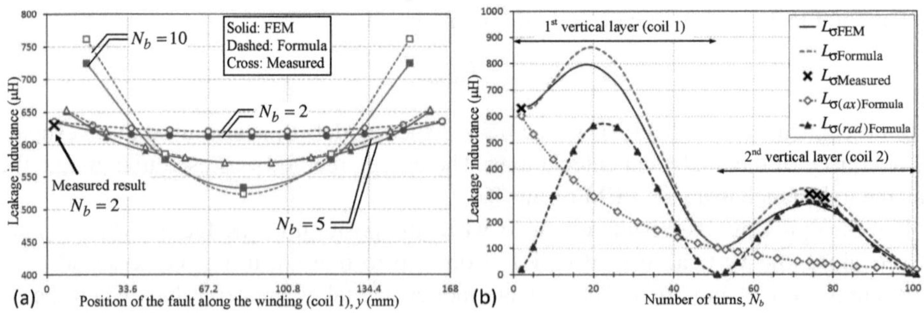

Fig. 5. Leakage inductance as a function of the: (a) number of N_b turns and their relative position along the winding (coil 1); (b) fault progression. (Short-circuit test of Fig. 1(d).)

The results obtained by the analytical calculation, the FEM analysis, and the experimental tests are in relatively good agreement.

The aforementioned method is also valid for computing the leakage inductances between pairs of windings for the case of the standard short-circuit tests of Fig. 1(e). Fig. 6(a) presents the corresponding results as a function of the N_b turns, which are also in good agreement with the FEM analysis and the measured values ($L_{\sigma(ij)} \equiv$ leakage inductance when the N_i turns are excited and the N_j turns are shorted).

5 Leakage Inductance Equivalent Circuit Model

The equivalent circuit for the leakage inductance of three-winding transformers proposed in [10] is adapted here to represent the transformer with faulty turns. The equivalent circuit is shown in Fig. 7 and its parameters are computed from the short-circuit inductances between the pairs of windings obtained by the tests of Fig. 1(e):

$$L_{\sigma 1} = L_{\sigma(ab)} \tag{5}$$

$$L_{\sigma 2} = L_{\sigma(bs)} \left(N_a / N_b \right)^2 \tag{6}$$

$$M_\sigma = \left[L_{\sigma(as)} - L_{\sigma(ab)} - L_{\sigma(bs)} \left(N_a / N_b \right)^2 \right] / 2 \, . \tag{7}$$

The leakage inductance equivalent circuit model can be used to simulate the short-circuit test of Fig. 1(d). By analyzing the circuit of Fig. 7, with a short applied to the series-connected N_b and N_s turns, the equivalent inductance becomes:

$$L_{\sigma(\mathrm{a,\,b+s})} = L_{\sigma 1} + L_{\sigma 2} \left[N_s / \left(N_s + N_b \right) \right]^2 + 2 M_\sigma N_s / \left(N_s + N_b \right) \tag{8}$$

Fig. 6(b) compares the results obtained by applying (8) and the ones previously presented in Fig. 5(b). Globally, the proposed leakage inductance model yields good and consistent results. The FEM results are almost coincident. Minor differences can be detected in the analytical computed results, mainly due to approximations in the calculation of the mean radius and the Rogowski correction factors.

The leakage inductance network of Fig. 7 can be integrated with other transformer models [1], [10], in order to take into account the core and loss components.

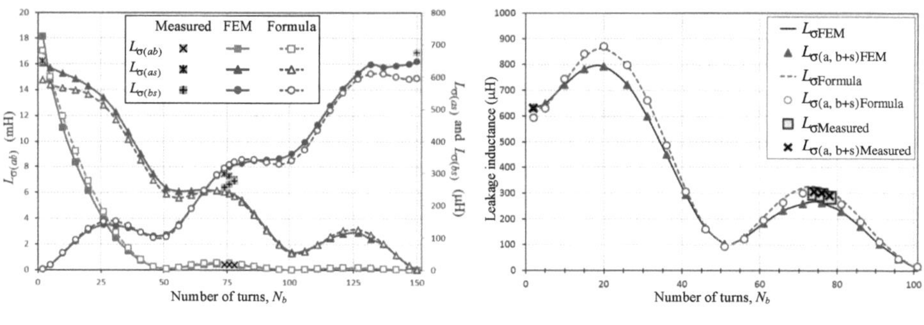

Fig. 6. (a) Leakage inductance as a function of the number of N_b turns for the case of the short-circuit tests of: (a) Fig. 1(e); (b) Fig. 1(d)

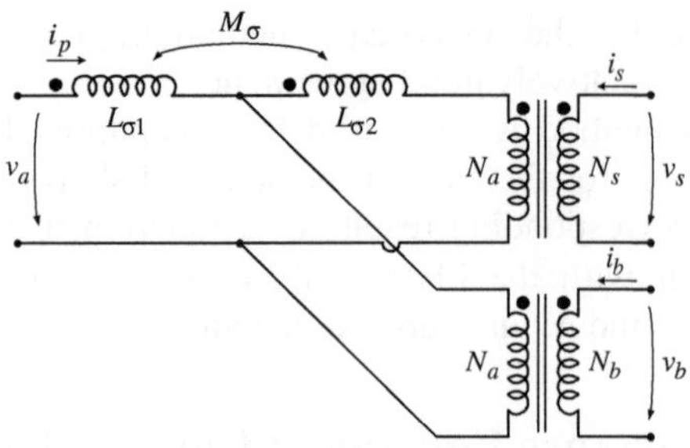

Fig. 7. Leakage inductance equivalent circuit

6 Conclusions

This paper has presented a method for the analytical determination of the leakage inductances of transformers with winding interturn short-circuits. An equivalent circuit for the representation of the leakage inductance of transformers with winding faults is also proposed. The experimental and FEM analysis results confirm the adequacy of the proposed analytical calculation method.

Work is currently in progress to further simplify the leakage inductances computation method, in order to allow their determination from nameplate data and core window dimensions.

References

1. Oliveira, L.M.R., Cardoso, A.J.M.: A Permeance-Based Transformer Model and its Application to Winding Interturn Arcing Fault Studies. IEEE Trans. Power Delivery 25, 1589–1598 (2010)
2. Bastard, P., Bertrand, P., Meunier, M.: A Transformer Model for Winding Fault Studies. IEEE Trans. Power Delivery 9, 690–699 (1994)
3. Jablonski, M., Napieralska-Juszczak, E.: Internal Faults in Power Transformers. IET Electric Power Applications 1, 105–111 (2007)
4. Avendaño, A., Mork, B.A., Høidalen, H.K.: Transformer Internal Fault Modeling in ATP. In: Int. Conf. Power Systems Transients (2011)
5. Billig, E.: Mechanical Stresses in Transformer Windings. Journal IEE, Part II 93, 227–243 (1946)
6. Blume, L.F. (ed.): Transformer Engineering, 2nd edn. John Wiley & Sons (1951)
7. Stephens, H.O.: Transformer Reactance and Losses with Nonuniform Windings. AIEE Trans. 53, 346–349 (1934)
8. Meeker, D.C.: Finite Element Method Magnetics, Version 4.2, User's Manual (2010)
9. Kulkarni, S.V., Khaparde, S.A.: Transformer Engineering: Design and Practice. Marcel Dekker (2004)
10. León, F., Martinez, J.A.: Dual Three-Winding Transformer Equivalent Circuit Matching Leakage Measurements. IEEE Trans. Power Delivery 24, 160–168 (2009)

Preliminary Studies and Test Results of a Superconducting Hysteresis Motor with Multiphase Windings and Variable Number of Magnetic Poles

Raul Dionísio[1], João Murta Pina[2], David Inácio[2], and Amadeu Leão Rodrigues[2]

[1] Electrical Engineering Department
[2] Centre of Technology and Systems
Faculdade de Ciências e Tecnologia, Universidade Nova de Lisboa
Monte de Caparica, 2829-516 Caparica, Portugal
raulmmdionisio@gmail.com, {jmmp,ddi}@fct.unl.pt,
leao@uninova.pt

Abstract. In this paper a procedure for determining the number of different synchronous speeds that can be obtained from the stator of a drum motor as a function of the number of slots is presented. This preliminary study is foreseen for a hysteresis high-temperature superconducting motor, but the approach is directly applied in conventional motors. The targeted device has multiphase windings, in order to achieve full flexibility in torque-speed space through electronic variation of magnetic poles. Simulations are performed in order to achieve a qualitative understanding of the behaviour of the motor, namely in what concerns to torque and settling times from initial to synchronous speed. A prototype with eighteen slots in the stator and a bulk YBCO rotor is described and built, and experimental values of torque are obtained.

Keywords: Superconducting hysteresis motor, Electronic pole variation, multi-phase motor.

1 Introduction

A variety of techniques for changing and control the speed of electric motors have been proposed for decades, including virtual gearboxes, i.e., electronically changing the number of magnetic poles of stator windings during motor operation in multiphase motors [1,2]. On the other hand, the exploitation of hysteresis in high temperature superconducting materials, when this material builds the rotors of electric motors, results in higher torque values than those obtained in hysteresis motors with ferromagnetic rotors, i.e., higher power density is obtained in comparison with conventional hysteresis motor [3,4,5,6,7]. This higher power density allows considering applications like transportation or variable load winch drives, where the changing of power and torque are the normal operating conditions.

L.M. Camarinha-Matos, S. Tomic, and P. Graça (Eds.): DoCEIS 2013, IFIP AICT 394, pp. 431–440, 2013.

The synchronous speed of an electric motor depends on the number of magnetic poles of the stator winding and supply voltage frequency. Changing synchronous speed is commonly achieved by acting on the latter, trough power electronic inverters, which is a well established process [10]. Synchronous variation through changing the number of magnetic poles is a more complex process, and hardly ever used. This can be implemented mechanically [11], typically with ratio 2:1, by the following approaches: a stator with a single set of windings divided in sectors, where number of poles is set by commutation of those sectors; a stator with two independent sets of windings; a stator with two independent sets of windings, each divided in sectors, which allows other ratios. Besides maintenance and reliability issues, mechanical commutation limits flexibility, as all possible pole configurations are established after the machine is built.

Techniques for electronic commutation of poles have already been defined, such as Pole Amplitude Modulation [12] or Pole Phase Modulation [13]. However, these still raise some difficulties, as limited pole changing ratios, namely p: p-1, where p is the number of pairs of poles; discontinuous stators with asymmetrical windings, which are hard to build; or inefficient use of stator slots.

The technique foreseen in this paper was described elsewhere [2], and solves some of the previous problems. It uses as much windings as slots, allowing operation in a large number of poles configurations. Power electronics is used to generate supply voltages with controlled frequency and phase shift, allowing obtaining different configurations of stator poles.

In this paper, preliminary design of multiphase motor, aiming the previous strategy is described. Since the number of armature and rotor poles must be the same, only asynchronous (conventional) or hysteresis (either conventional or superconducting) motors are eligible. A superconducting hysteresis motor is selected.

Considerations about the possible number of different synchronous speeds and configuration of a multiphase motor are detailed in next section. Dynamic simulations of different topologies under a qualitative approach are subsequently presented, followed by the details of the prototype manufactured. First experimental results are then shown, and conclusions and future work are drawn in the last section.

2 Contribution to the Internet of Things

Electronically changing the configuration of the windings, through the number of phases, poles and different frequencies of the electrical currents, allows the adjustment of speeds and torques to the application context. These variables are key factors for the knowledge enhancement of the hysteresis motor in transport applications, for example, where the adaptation of torques and speeds to the context demands are very important features. Adequate interface devices and software for communication and control allows the abstract representation of the motor, aiming the control of its variables through the internet communication infrastructure. Data may be obtainable through different variables in a number to be defined and developed accordingly to the research work.

3 Theory of Variable Poles and Phases

Power converters allow generating multiphase voltages, with controlled phase angles and equal amplitudes. As in conventional three-phase stators, symmetry must prevail, arising from equal spacing or equal number of stator slots between active conductors of a phase, from equal values of currents and frequencies in all phases and a constant relationship between geometrical slot position of a phase and the time difference between phase currents. The result is a rotating magnetic field with constant magnitude where its speed depends on the value of supply frequency and number of magnetic poles. Changing phase angles allows changing the number of magnetic poles on the stator [2]. This imposes not only different motor speeds, but also different torque and power. It is important to mention that the number of different voltages generated by the power converter depends on the value of the phase angles of each voltage, e.g. 24 voltages with phasors displaced by 30º builds a 12-phase system.

To improve magnetic field paths within the motor, electric phase coils fill stator slots and they are thus determinant in the number of phases and the number of poles in each phase. Table 1 shows the relationship between the number of slots, poles and phases in a stator with 18 slots and a single layer of conductors in each slot. There are thus three possible synchronous speeds with this configuration, corresponding to 2, 6 and 18 poles, although the theoretical single phase corresponding to 18 poles is meaningless, as it corresponds to a pulsating field, thus with zero speed. Figure 1 illustrates the particular case of three phases and six poles, within a stator of 18 slots.

Table 1. Number of different speeds achieved with a motor with 18 slots, single layer windings, and different number of phases. There are thus three different possible speeds, although one corresponds to nil speed.

Number of poles	Number of different supply voltages (phases)
2	3, 9
4	-
6	3
8	-
10	-
12	-
14	-
16	-
18	1

The number of different synchronous speeds in Table 1 is obtained by counting the number of integer divisions of the number of slots and the even numbers of poles, $C/2p$, where C is the number of slots and p the number of pairs of poles. With 18 slots, two different synchronous speeds are thus possible.

The relationship between the number of slots and possible different speeds is illustrated in Figure 2, for stators up to 200 slots. It can be observed in also that the number of different speeds is roughly limited by the root of the number of slots.

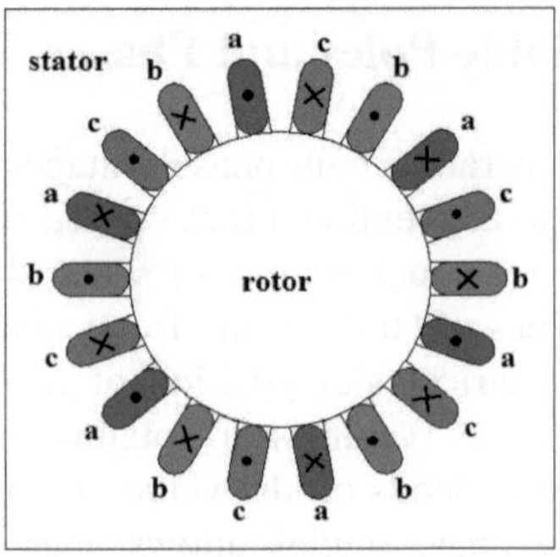

Fig. 1. Winding configuration for a stator with 18 slots, in a three phase and six poles configuration

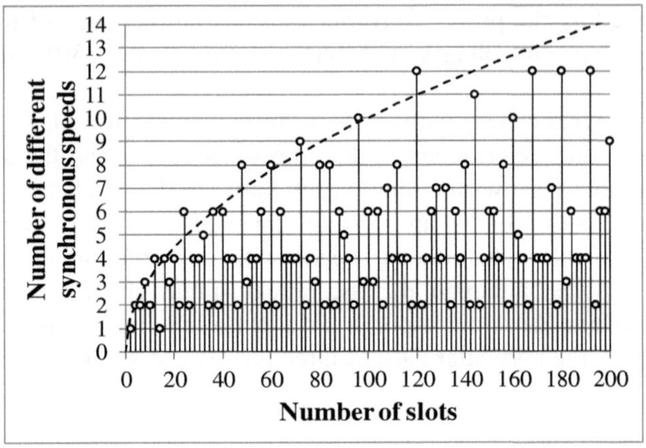

Fig. 2. Number of different speeds as function of the number of stator slots. There is an envelope on the number of speeds given roughly by $\sqrt{C}$, where C is the number of slots available in the stator.

The magnetomotive force (MMF) produced by an m-phase symmetrical system, F_{tot} , where each phase is geometrically displaced by $2\pi/m$ rad, with $2p$ poles per phase, and current in phase j is given by $i_j(t) = I_{\max}\cos(\omega t + \theta_j)$, where I_{max} is the amplitude, $\omega = 2\pi f$ is the angular speed, f is the frequency, and θ_j is the phase angle, is

$$|F_{tot}| = \sum_{i=1}^{m} |F_i| = \frac{m}{2}\frac{4}{\pi}\left(\frac{N}{2} I_{\max}\right)\cos(\omega t - p\theta) \quad (1)$$

where N is the number of conductors in each winding, $4/\pi$ corresponds to the fundamental amplitude of each squared MMF and θ is the geometric angle over stator surface.

The slots occupied by each phase are related with the number of poles in the same phase. After the previously explained calculation, it is now possible to regroup the slots in each phase, thus changing the number of poles but not the number of phases. As an example, the six poles/three-phase stator illustrated in Figure 1 can be rearranged in a two poles/three-phase system, as shown in Figure 3.

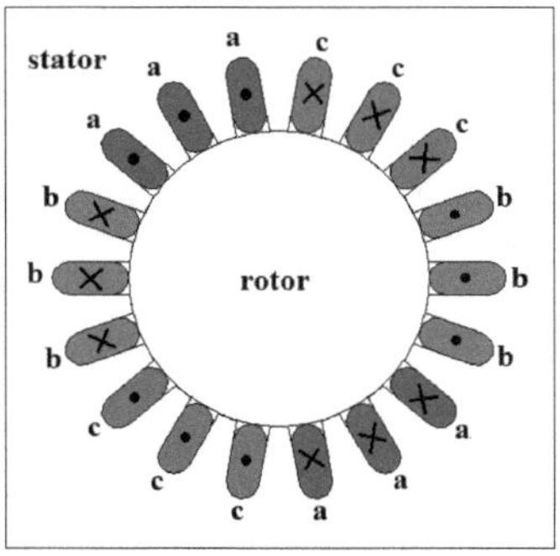

Fig. 3. Regrouping slots allows changing a three-phase hexapolar motor into a three-phase bipolar configuration

It is possible to regroup the slots per each phase if the condition $N_p = 2p \times G$ applies, where N_p is the number of slots occupied by each phase and G the number of adjacent slots. In Figure 1, $N_p = 6$, $p=3$ and $G=1$. In Figure 3, N_p remains equal to 6, $p=2$ and $G=3$. The total number of different speeds calculated according to Table 1 is not altered.

Despite the unchanged total number of different speeds, when regrouping phase slots the total MMF force amplitude is changed as result of

$$\left|F_{tot}\right| = \sum_{i=1}^{m}\left|F_i\right| = G\frac{m}{2}\frac{4}{\pi}\left(\frac{N}{2}I_{\max}\right)\cos\left(\omega t - p\theta\right) \tag{2}$$

Different torque values are possible with the same speed, which improves the number of different values of available power. In (2), the effects of geometrical displacement of adjacent coils in a regrouped phase are not taken in account. Adjacent coils are considered to be a lumped parameter.

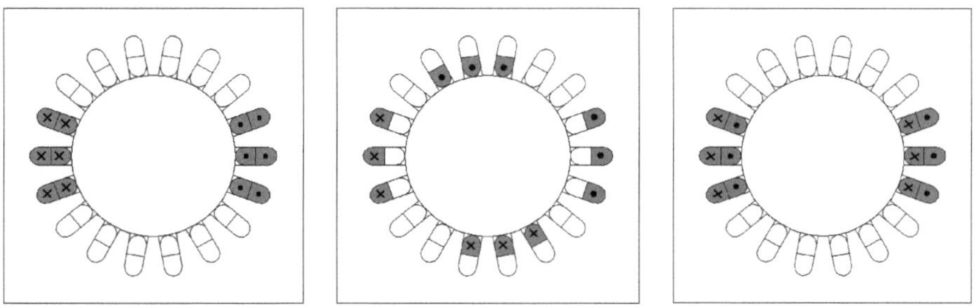

a) b) c)

Fig. 4. Geometrical displacement of two windings of the same phase, in a double layer stator

In Figure 1 each phase occupies the whole volume of the slot, in a single layer configuration. In a double layer configuration, (1) still applies. The number of phases remains the same but there are more than one set of coils for each phase electrically connected in parallel. Each set of coils of each phase may be geometrically displaced throughout the slots and each MMF interacts with the other subsets of coils of the same phase. Figure 4 illustrates the geometric displacement and correspondent interaction between the two sets of coils of the same phase, when each coil occupies half of a slot. In Figure 4, from left to right, there is a reinforcement of MMF in the left (Figure 4.a)) and a full opposition on the right (Figure 4.c)). The total MMF produced from the N coils of the same phase is given by

$$F_{tot} = F\sum_{i=1}^{N}\cos(\theta_i) \tag{3}$$

where F is the MMF of each individual coil, and θ_i is the geometric angle between the phase windings subsets. Despite the MMF combination the number of poles remains unchanged no matter the geometrical displacements between the sets of coils within a phase.

4 Qualitative Simulation of Bulk Superconductor Rotor Dynamics

The concept of changing number of poles (either electronically, either mechanically) is only possible in devices where stator poles correspond to rotor poles, which is achieved with induction or hysteresis motors. In the present work, a HTS hysteresis motor with 18 slots and 18 independent windings in the stator is foreseen. Coils are assembled in double layer configuration, each one spanning by four slots. In practice, as shown later, all the terminals are available outside the motor, allowing maximum freedom in connections amongst coils. Simulations with Flux2D from Cedrat Company were first carried out in order to qualitatively analyse the dynamic behaviour of the device. Three-phase superconducting hysteresis motors have been extensively described elsewhere [5,6].

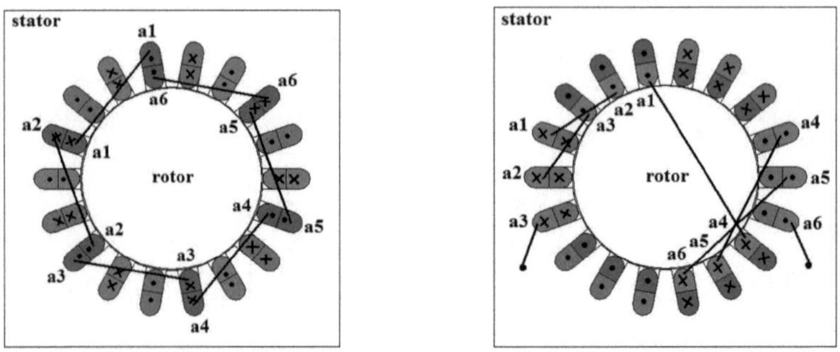

a) Six poles configuration, T_1. b) Two poles configuration, T_2.

Fig. 5. Different configurations in a stator with 18 slots and 18 independent coils, allowing maximum degrees of freedom in connections between windings. The stator is supplied by a three-phase system.

In this study, two three-phase topologies are studied, namely T_1, corresponding to six poles, and T_2, corresponding to two poles. Winding arrangements are represented in Figure 5.a) for topology T_1 and in Figure 5.b) for topology T_2. The superconducting rotor is 52 mm height, 52 mm external diameter and 30 mm internal diameter. In these simulations only qualitative and relative results are foreseen, for which superconducting parameters (critical current density and n parameter) are not determinant.

4.1 Simulation of Topology T1

For MMF forces with amplitudes bellow the full penetration threshold (840 A.t in the simulations, naturally dependent on the characteristics of the HTS material), only partial magnetization in the rotor is achieved. This situation is represented in Figure 6.a). In these regimes the developed torques are considerably lower (in this case, bellow 10^{-4} N·m). Only when flux fully penetrates the rotor, see Figure 6.b), significant torques are developed.

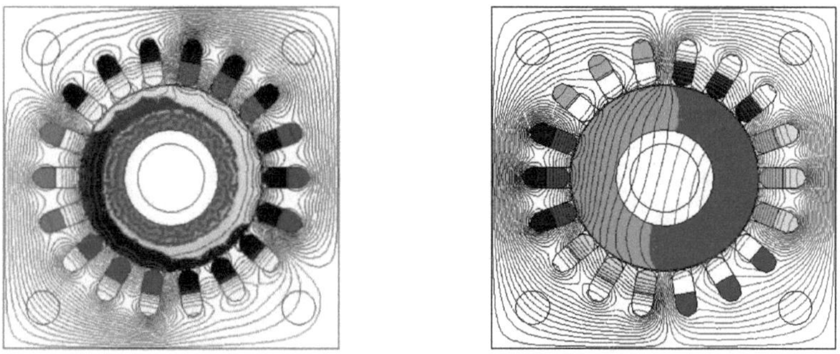

(a) **(b)**

Fig. 6. Profiles of current density in the rotor, for different values of MMF amplitude, corresponding to (a) partial penetration (170 A·t) and (b) full penetration (840 A·t). Superconducting properties where not a concern in simulations, only qualitative results.

Settling times (time needed to reach synchronous speed) were obtained by simulations for different starting speeds and different MMF. For c (3000 rpm synchronous speed) settling times are illustrated in Figure 7.

Fig. 7. Settling times for topology T_1 (bipolar), as a function of (left) MMF and initial speed and (right) as a function of MMF for -3,000 rpm initial speed. Synchronous speed is 3,000 rpm.

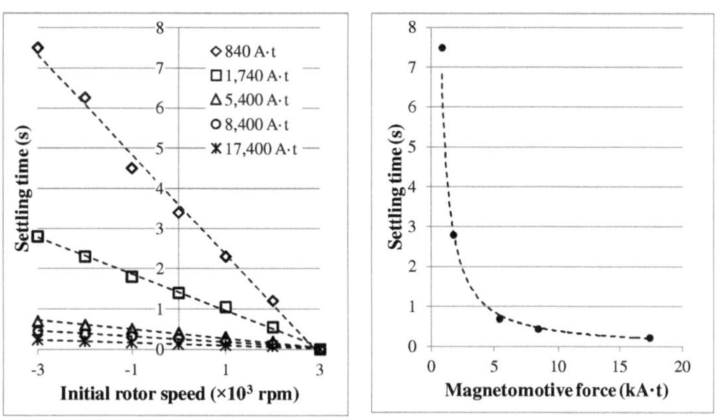

Considerably longer settling times were registered when the external magnetic flux was not able to fully penetrate the rotor material. In this region there is a high nonlinearity between settling time and MMF, thus conditioning the performance of the motor and demanding a higher control complexity. Nevertheless, torque is linear with MMF, as shown later.

4.2 Simulation of Topology T_2

Due to lower magnetic fields produced in topology T_1 (bipolar) when compared with T_2 (hexapolar), settling times are higher on the later for the same MMF produced by each coil, as illustrated in Figure 8.

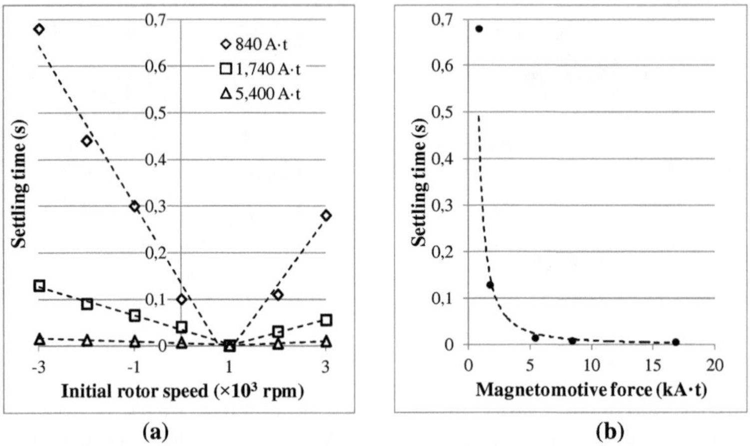

Fig. 8. Settling times for topology T_2 (hexapolar), as a function of (a) MMF and initial speed, where results above 5,400 A·t are not shown as they overlap, and (b) as a function of MMF for -3,000 rpm initial speed. Synchronous speed is 1,000 rpm.

4.3 Comparison of Topologies

Time responses observed in the previous subsections are related to torque differences as illustrated in Figure 9. As expected, torque produced with an hexapolar topology is considerably higher than with bipolar.

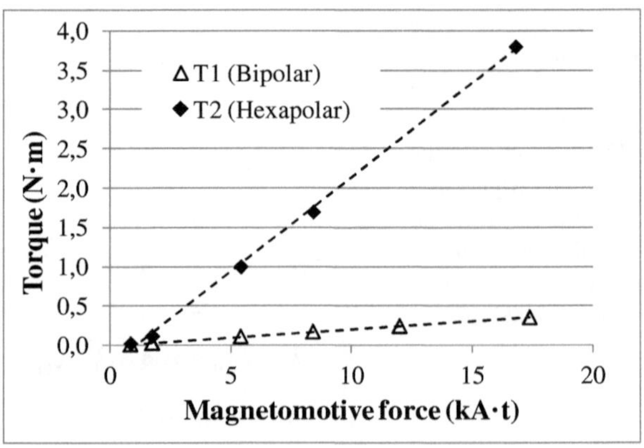

Fig. 9. Comparison of developed static torques for topologies T_1 and T_2. Torque is linear with MMF.

5 Motor Construction

In order to investigate pole changing techniques applied to superconducting hysteresis motors, and to validate the previous results, a motor was built, with a HTS YBCO rotor supplied by ATZ GmbH Company. This rotor, with a stainless steel shaft applied, is shown in Figure 10.a). The cylinder diameter, with outside protection bandage is 54 mm. The stator has 18 slots, 18 independent coils, each coil spanning for four slots, was built with stacked laminations (52 mm height, the same dimension of the rotor) of non-oriented silicon iron (M15), see Figure 10.b). Each lamination is 0.5 mm thick, 10 cm quadrangular shape. Airgap is 0.3 mm.

(a) **(b)** **(c)**

Fig. 10. Constitutive parts of the assembled motor, namely (a) HTS rotor from ATZ GmbH Company with stainless steel shaft inserted; (b) stator built by stacked laminations of non-oriented silicon iron; and (c) view of the motor with terminals of 18 independent coils

6 First Experimental Results

First experiments on the motor, at liquid nitrogen temperature, were performed, using a 1 kW asynchronous split phase motor as load. This machine was used to impose a mechanical load to the superconducting motor, and torque was controlled through stator current. The experimental results for the motor working in bipolar configuration (topology T_1) are shown in Figure 11. As expected, there is a linear relation between torque and stator MMF.

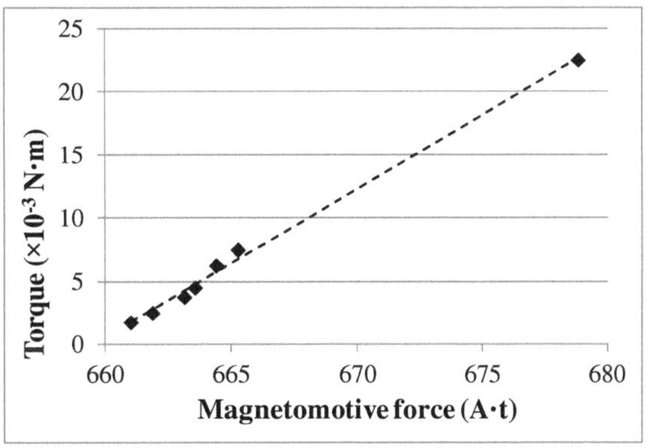

Fig. 11. Comparison of developed static torques for topologies T_1 and T_2. Torque is linear with MMF.

7 Conclusions and Future Work

In this paper, a systematic and recommended procedure for planning stator windings and consequently determining the number of synchronous speeds available in a motor is presented. To be effective the process has a standard approach sequence: number of different phases; possible regrouping; coil slot occupation with eventual geometrical offsets. The performances of different topologies of a multiphase hysteresis motor were simulated, in order to compare them in a qualitative approach. A motor was assembled and first experiments were carried out. Future work includes measuring characteristics of the HTS material (critical current and *n* parameter) in order to perform quantitative simulations. Future work includes also building a power electronic converter that can supply 18 independent phases, in order to investigate the performance of the motor with any configuration of poles.

Acknowledgements. This work was supported by National funds through FCT (Fundação para a Ciência e Tecnologia) under the framework of PEst-OE/EEI/UI0066/2011 project.

References

1. Inácio, S., Inácio, D., Murta Pina, J., Valtchev, S., Ventim Neves, M., Rodrigues, A.: An electrical gearbox by means of pole variation for induction and superconducting disc motors. Journal of Physics: Conference Series 97, 012221 (2007)
2. Murta Pina, J.: Proof-of-concept of a variable poles multiphase motor. In: Proceedings of the Flux Users Conference, France (2009)
3. Máquez, I., Granados, X., Obradors, X., Pallarés, J., Bosch, R.: Radial and Axial Flux Superconducting Motors in a Levitating Rotor Configuration. IEEE Transactions on Applied Superconductiviy 9(2) (1999)
4. Barnes, G.J., McCulloch, M.D., Dew-Hughes, D.: Torque from hysteresis machines with type-II superconducting segmented rotors. Physica - Elsevier Science C331, 133–140 (2000)
5. Kovalev, L.K., Ilushin, K.V., Penkin, V.T., Poltavets, V.N., Koneev, S.M.A., Akimov, I.I., Gawalek, W., Oswald, B., Krabbes, G.: High output power electric motors with bulk HTS elements. Physica C 386, 419–423 (2003)
6. Kovalev, K.L., Kovalev, L.K., Ilushin, K.V., Ladonov, S.A., Poltavets, V.N., Gawdek, W.: Experimental and theoretical study of electrical machines with bulk HTS elements. In: Power Electronics and Motion Control Conference, pp. 568–570 (2004)
7. Inácio, D., Inácio, S., Pina, J., Gonçalves, A., Ventim Neves, M., Leão Rodrigues, A.: Numerical and experimental comparison of electromechanical properties and efficiency of HTS and ferromagnetic hysteresis motors. Journal of Physics Conference Series, 1742-6596 97 1 012218 (2008)
8. Jackson, R.D.: Digital simulation of the hysteresis motor. Proceedings of the IEEE 120(12), 1533–1537 (1973)
9. Kim, H.-K., Hong, S.-K., Jung, H.-K.: Analysis of hysteresis motor using finite element method and magnetization-dependent model. IEEE Transactions on Magnetics 36(4), 685–688 (2000)
10. El-Sharkawi, M.A.: Fundamentals of Electric Drives. Brooks/Cole Publishing Company (2000)
11. Kostenko, M., Piotrovsky, L., Chernukhin, A.: Electrical Machines. Central Books Ltd. (1980)
12. Rawcliffe, G.H., Fong, W.: Slip-Ring P. A. M. Induction Motors for Two Synchronous Speeds. IEEE Transactions on Power Apparatus and Systems PAS-90, 495–499 (1971)
13. Miller, J.M., Stefanovic, V., Ostovic, V., Kelly, J.: Design considerations for an automotive integrated starter generator with pole-phase modulation. In: Conference Record of the 2001 IEEE Industry Applications Conference, vol. 4, pp. 2366–2373 (2001)

A Voltage Limiter Circuit for Indoor Light Energy Harvesting Applications

Carlos Carvalho[1,2] and Nuno Paulino[2,3]

[1] Instituto Superior de Engenharia de Lisboa (ISEL – ADEETC), Instituto Politécnico de Lisboa (IPL), Rua Conselheiro Emídio Navarro, nº1, 1949-014 Lisboa, Portugal
cfc@isel.ipl.pt

[2] UNINOVA/CTS, Instituto de Desenvolvimento de Novas Tecnologias, Campus FCT/UNL, 2829-516 Caparica, Portugal

[3] Departamento de Engenharia Electrotécnica, Faculdade de Ciências e Tecnologia, Universidade Nova de Lisboa, Campus FCT/UNL, 2829-516 Caparica, Portugal
nunop@uninova.pt

Abstract. A voltage limiter circuit for indoor light energy harvesting applications is presented.This circuit is a part of a bigger system, whose function is to harvest indoor light energy, process it and store it, so that it can be used at a later time. This processing consists on maximum power point tracking (MPPT) and stepping-up, of the voltage from the photovoltaic (PV) harvester cell. The circuit here described, ensures that even under strong illumination, the generated voltage will not exceed the limit allowed by the technology, avoiding the degradation, or destruction, of the integrated die. A prototype of the limiter circuit was designed in a 130 nm CMOS technology. The layout of the circuit has a total area of 23414 μm^2. Simulation results, using Spectre, are presented.

Keywords: CMOS integrated circuits, Energy harvesting, Voltage limiters, Voltage reference circuits.

1 Introduction

To achieve indefinite operation in any location, sensor nodes must obtain their power directly from the environment. In a wireless sensor network, it may be difficult to power each node, either using batteries or the power grid. Sensor networks supplied by grid connections are limited to a small range of applications, as they can never be too far from the power outlet. Although the use of batteries allows for more freedom of sensor distribution, its replacement can become burdensome and costly, if a large number of sensors are deployed. As such, energy harvesting systems tend to take over the powering paradigm for pervasive operation [1], being also ecological and thrifty. The sources that can be harvested are diverse [2] and, amongst those, light is the one with the highest energy density per unit of volume [2]. Energy can also be obtained from indoor light. In this ambient, powering an electronic application is an increased challenge, as the levels of available light power inside buildings are much lower than those obtained outside. At the most, the Sun can provide about 1 kW/m^2, but in indoor environments, the Sun and the artificial lighting provide a much lower energy density.

L.M. Camarinha-Matos, S. Tomic, and P. Graça (Eds.): DoCEIS 2013, IFIP AICT 394, pp. 441–448, 2013.

2 Relationship to Internet of Things

The circuit presented in this paper is an essential part of an indoor light energy harvesting system. This energy harvesting will allow a variety of electronic systems to operate indoors without needing batteries or a connection to the power grid, adding an enormous degree of freedom, regarding their location. These electronic systems can communicate with other devices and, ultimately, connect to the Internet. The light energy source, when compared to other sources, has the highest energy density, making the harvester discreet and small. Light energy harvesters are simple and cheap to build. As long as light is available, any device supplied by energy harvested from ambient light can be deployed to anywhere, making it a part of the Internet of Things.

3 Motivation and Background

The unpredictable nature of light energy, results in a variable amount of available power. Since the PV cells must be sized in order for the energy harvesting system to be able to receive enough energy even when the light intensity is weak, this can result in large power, available from the same PV cells, when the light intensity is strong. The circuit described in this paper, fits in a system designed to deal with very low levels of light energy, typically found in indoor environments. However, when this system is illuminated by direct sunlight or generally, by high levels of light intensity, the energy harvested by the PV cells is much higher than the value that was expected under the normal indoor illumination conditions. In this situation, the energy harvesting system must be able to manage the excess of energy that is obtaining.

The architecture of the system is depicted in Fig.1, being constituted by PV cells followed by a voltage step-up converter with MPPT capability.

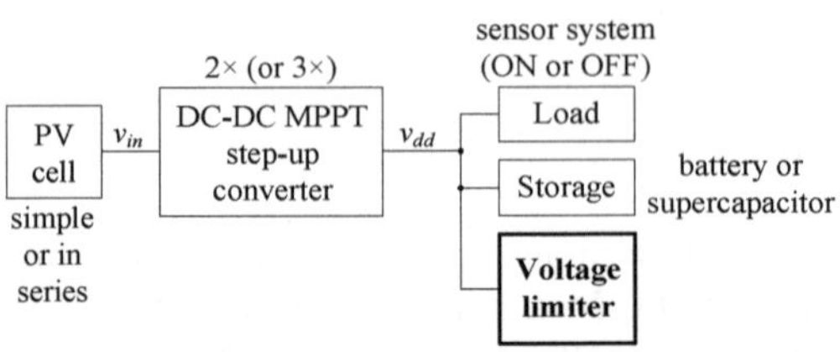

Fig. 1. An energy harvesting architecture that hosts the voltage limiter described in this paper

The output voltage of the converter (v_{dd}) is the one that will be limited by the voltage limiter. The v_{dd} voltage also supplies the inner circuits of the step-up converter. The system can work with one or two PV cells in series and step-up the input voltage (v_{in}) by two or three, depending on the type of PV cells and on the expected light intensities. The power from the PV cell is dependent on its area and on the illumination level, as shown in Fig.2. According to a set of light power intensity measurements, the lowest illumination obtained indoors had an ambient irradiance of 0.1 W/m^2 and the typical irradiance using artificial lighting, was about 0.7 W/m^2.

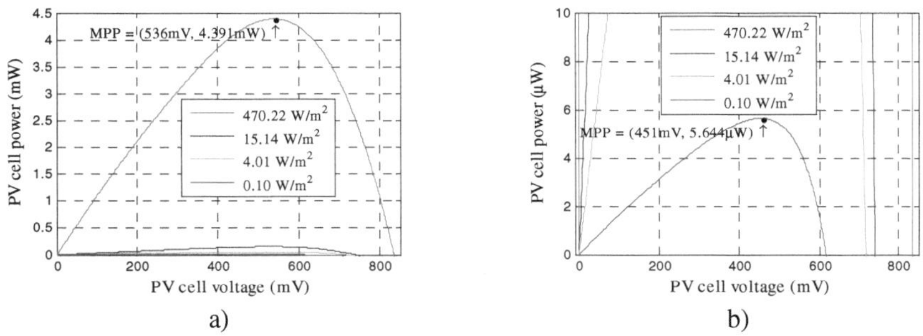

Fig. 2. a) Available power from the PV cell with 1 cm^2, for different irradiance levels. b) Detail showing the power function that corresponds to the lowest irradiance level.

The system must be designed for a worst case scenario (low illumination), meaning that when the light is stronger, the system will produce a large voltage at its output.

The system was designed in a 130 nm CMOS technology, having a maximum voltage of about 1.4V. If this value is exceeded, the devices in the die are stressed, reducing their operating lifetime. With a larger voltage, they can even be destroyed.

The v_{dd} voltage can be used to power any application, for example, a sensor node. Since the available power from the PV cells is low, this application typically will work using an ON-OFF regime with a low duty-cycle. While the application is OFF, the harvested energy is stored in a supercapacitor. When the application is ON, it works using this stored energy. In the case of a high light intensity, after the supercapacitor is charged to a desired voltage value, the step-up converter continues to supply a current to the output node (v_{dd}). This results in the output voltage to increase to a value that can be dangerous, being necessary to add a circuit to limit it, by absorbing the excess current supplied by the step-up converter.

In literature, some work can be found concerning voltage limiters. In [3], the voltage limitation is achieved by opening a switch to a capacitor, whenever its voltage exceeds a certain limit, on a sample-and-hold basis. There are other systems that also use voltage limiter circuits, like in [4] and [5], where the voltage limitation problem appears under the context of RFID applications. The amount of available power to a RFID tag is dependent upon the distance to the reader device, resulting in a similar problem to the one in a light energy harvesting system. In [4], the voltage limitation is done by performing a comparison to a desired limit. In [5], a band gap reference circuit is used to generate a reference voltage to which the desired voltage is compared to. This approach is similar to the one proposed in this paper.

4 Voltage Limiter Circuit Architecture

The architecture of the voltage limiter is shown in Fig.3.

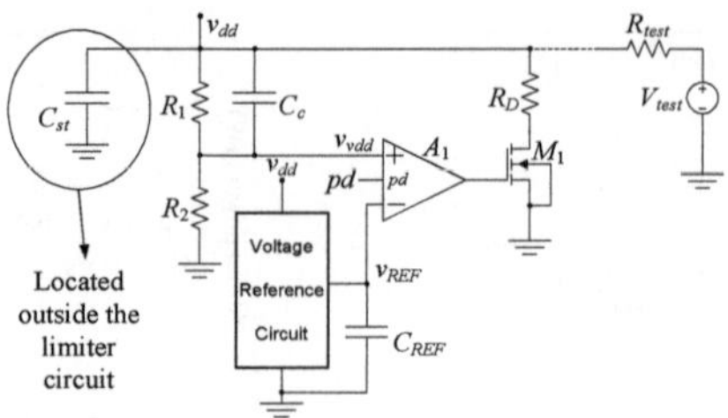

Fig. 3. Architecture of the voltage limiter described in this paper

The Thévenin equivalent, at the right hand side of the dashed line, represents the voltage step-up converter. The aim of the circuit is to limit the v_{dd} voltage, to be lower than 1.4 V – 5% = 1.33 V. The voltage limiter operates by drawing current through M_1. This current causes a voltage drop across R_{test}, such that v_{dd} remains constant. The value, at which v_{dd} is to be limited, is controlled by a Voltage Reference Circuit (VRC), providing a stabilized and temperature compensated voltage reference.

In normal operation, the voltage divider formed by R_1 and R_2, provides at v_{vdd}, a voltage close to v_{REF}. The amplifier A_1 amplifies the error between v_{vdd} and v_{REF}, providing, at its output, a voltage that directly controls the v_{gs} voltage of M_1, and thus the current drawn from the v_{dd} node, through R_D.

4.1 Voltage Reference Circuit

The VRC was adapted from [6], and its schematic is shown in Fig.4 a).The generated reference voltage (v_{REF}), versus the supplying voltage (v_{dd}), and the supply current of the VRC are depicted in Fig.4 b).

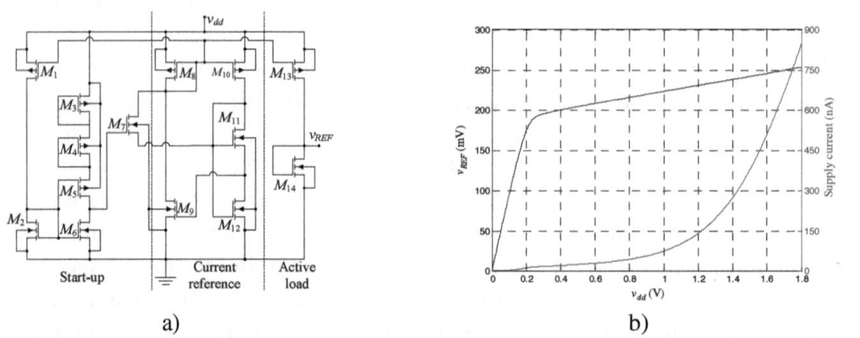

Fig. 4. a) CMOS Voltage Reference Circuit. b) Output reference voltage (v_{REF}), and supply current, as a function of v_{dd}.

M_2 and M_{12} are high voltage transistors (3.3 V), while the others are low voltage ones (1.2 V). The main difference between the original circuit, and the one that was built, is the fact that the one in [6] was built in a 180 nm CMOS technology, whereas the present one uses a 130 nm technology. The transistors were sized based on the sizes presented in [6], making only adjustments due to the different technologies.

The layout of the VRC circuit occupied 15647 μm^2. In normal operation, considering that v_{dd} = 1.2 V and a temperature of 27 °C, the nominal output voltage of the VRC is 230.9 mV. The supply current, in this case, is 139.5 nA. These data are depictable in Fig.4 b). In Fig.5, the supply current and the reference voltage values, as a function of temperature, for different supplying voltages, are shown.

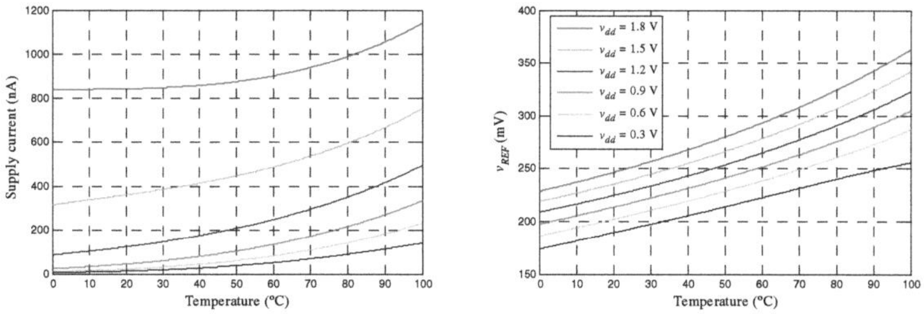

Fig. 5. Temperature dependence of the supply current, and generated voltage reference, for different supplying voltages, for the VRC that is used

4.2 Differential Voltage Amplifier

The topology of the amplifier is shown in Fig.6. This amplifier has a power down feature (*pd*) to disable it, in order to turn off the voltage limiter circuit. The layout of the amplifier occupied 7767 μm^2.The input stage of this amplifier uses PMOS devices, as the typical values of the input voltage are around 200 mV to 300 mV.

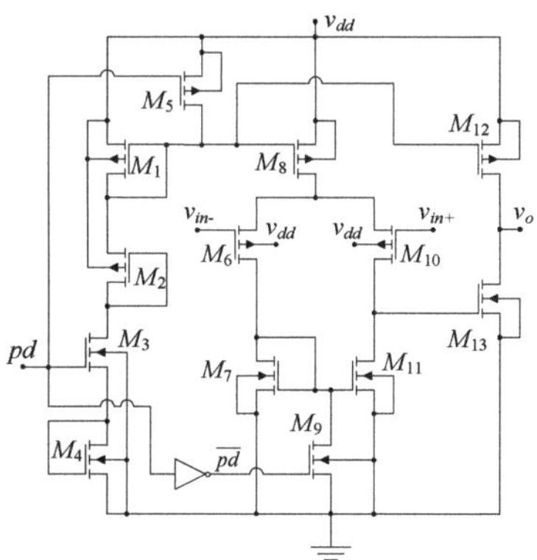

Fig. 6. Differential amplifier circuit

5 Stability Analysis

Since this is a closed loop system (Fig.3), it is important to analyze its stability. The feedback network is constituted by R_1, R_2, C_c and C_p (parasitic capacitance at the amplifier input). The feedback factor, β, is given by

$$\beta(s) = \frac{v_{vdd}}{v_{dd}} = \frac{R_2 + sC_c R_1 R_2}{R_1 + R_2 + sR_1 R_2 (C_c + C_p)}. \tag{1}$$

The impedance, Z_{22}, of the feedback network from v_{dd} to ground, is

$$Z_{22}(s) = \frac{R_1 + R_2 + sR_1 R_2 (C_c + C_p)}{(1 + sR_1 C_c)(1 + sR_2 C_p)}. \tag{2}$$

The small-signal model of the amplifier, including the output transistor M_1, is depicted in Fig.7. As the output variable, i_{out}, is the current drawn from the v_{dd} node, and the input variable is the input differential voltage of the amplifier (v_x), there will be a transconductance function $G_M(s)$ given by

$$G_M(s) = \frac{i_{out}}{v_x} = \frac{g_{m1} g_{m2} g_{m3}}{(g_{ds1} + sC_{out1})(g_{ds2} + sC_{out2})} \tag{3}$$

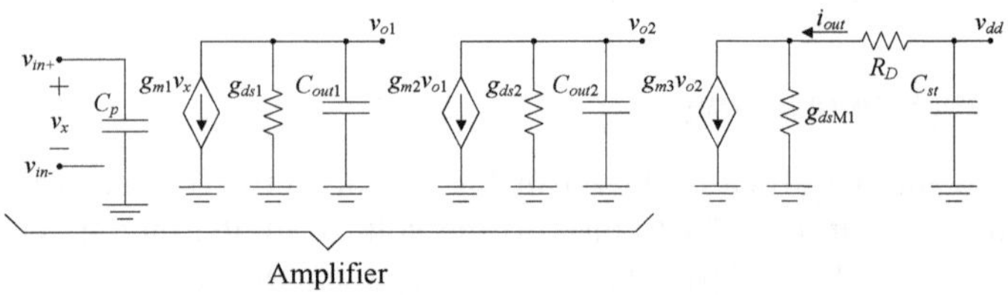

$g_{ds1} = g_{ds10} + g_{ds11}$ $\quad C_{out1} = C_{gd10} + C_{db10} + C_{db11} + C_{gs13}$ $\quad (g_{dsM1})^{-1} >> R_D \Rightarrow i_{out} \cong g_{m3} v_{o2}$

$g_{ds2} = g_{ds12} + g_{ds13}$ $\quad C_{out2} = C_{gd12} + C_{db12} + C_{gd13} + C_{db13} + C_{gsM1}$

Fig. 7. Small-signal model of the amplifier and the output M_1 transistor

The impedance seen from v_{dd} to ground, defining $R_o = (g_{dsM1})^{-1} + R_D$, is

$$Z_{out}(s) = Z_{22}(s) // \left(\frac{1}{sC_{st}} \right) // R_o. \tag{4}$$

The loop of the system in Fig.3 was open before the "+" terminal of A_1. By injecting signal in the "–" terminal, and computing the ratio to the signal obtained at the output of the feedback network, the loop gain is

$$G_L(s) = G_M(s) Z_{out}(s) \beta(s). \tag{5}$$

This function has four poles and one zero. Their expressions are as follows, considering that C_{st}, connected to the output node, is much higher than C_c and that C_p:

$$f_{p1} = \frac{g_{ds1}}{2\pi C_{out1}}; \quad f_{p2} = \frac{g_{ds2}}{2\pi C_{out2}}; \quad f_{p3} = \frac{1}{2\pi (R_o // (R_1 + R_2)) C_{st}};$$
$$f_{p4} = \frac{1}{2\pi (R_1 // R_2)(C_c + C_p)}; \quad f_z = \frac{1}{2\pi R_1 C_c}. \tag{6}$$

The values of f_{p4} and f_z can be approximately equal, by selecting an appropriate value for C_c, thus canceling out each other, making the system equivalent to having only the other three poles. This is useful in improving the phase margin of the system.

Assuming that $R_1 + R_2 >> R_o$, which is the case, the DC loop gain is given by

$$G_L(0) = \frac{g_{m1} g_{m2} g_{m3} R_2 R_o}{g_{ds1} g_{ds2} (R_1 + R_2)}. \tag{7}$$

The magnitude and phase of the open loop gain are determined through an AC sweep analysis and shown next in Fig.8a) and b), respectively. C_c is swept from 0.8 pF to 2.0 pF, to check what would be the consequence on the phase margin of the system. In these simulations, the value of the storage capacitor, C_{st}, was 1.8 μF.

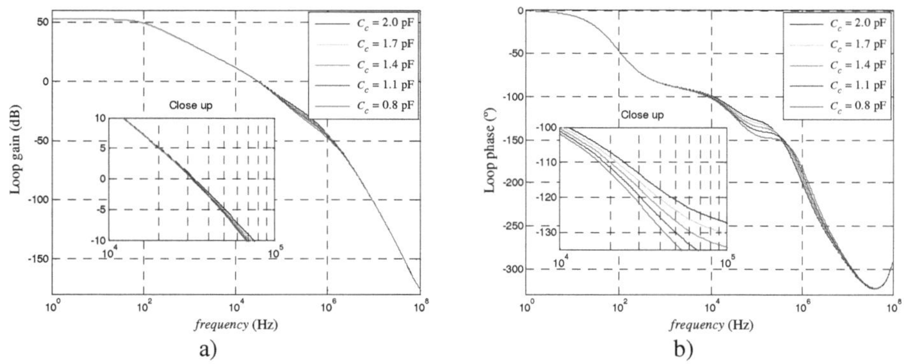

Fig. 8. Magnitude and phase functions of the feedback loop circuit of Fig.3

It can be seen, in the close up inset in Fig.8 a), that regardless of the value of C_c, the zero crossing of the loop gain function occurs at about 32 kHz. As seen by the close up inset in Fig.8 b), at the previous frequency, the spreading of phase values according to the value of C_c, results in a minimum phase margin value of 57°. In order to have an acceptable phase margin value of 60°, the value of C_c was selected to be 1.4 pF. It is important to note that, if the value of C_{st} is increased, the phase margin will also increase. The value of 1.8 μF is still a relatively small value for a storage capacitor, as it can get to units of Farads, or more, thus corresponding to a worst case.

6 Simulation Results

To check the behaviour of the circuit, it was simulated in Spectre. In Fig.9 a) and b), there are depicted the results of a DC sweep and a transient response, respectively, to show the correct operation in different ways. The function in Fig.9 a) was obtained by running a DC sweep, using V_{test}, on the right hand side of the dashed line in Fig.3. The function in Fig.9b) was obtained by using a square wave voltage generator for V_{test}. As seen in both Fig.9a) and b), the voltage limiter comes into action whenever the voltage in the v_{dd} node tends to be higher than the limit. The voltage at which the

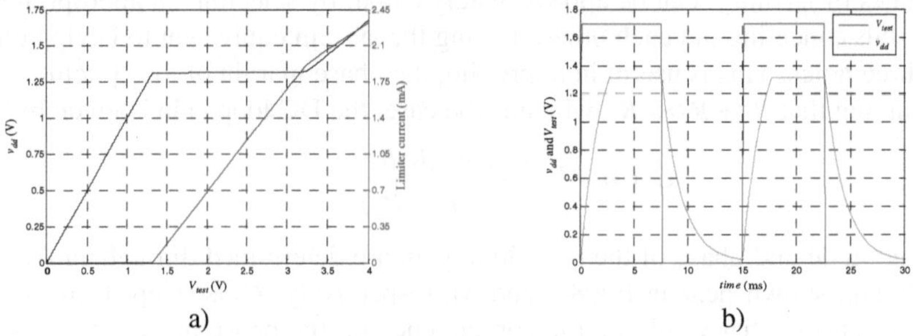

Fig. 9. a) DC sweep response. b) Dynamic transient response, with a period of 15 ms

limitation in v_{dd} is achieved, is about 1.31 V, which is inside the desired limit. In these tests, the supply current of the amplifier was, typically, about 900 nA.

7 Conclusions

A voltage limiter for indoor light energy harvesting applications has been presented. This circuit is a part of a bigger system, whose function is to harvest indoor light energy, process it and store it so that it can be used at a later time. This system can be used as the supplying module for a node of the Internet of Things.

The circuit that was described, ensured that even under a strong illumination, the generated voltage would not exceed the maximum value allowed by the technology, to avoid stressing, or destroying the devices in the die. Simulation results, using Spectre, were presented and showed that the proposed circuit can successfully limit the output voltage (v_{dd}) to the desired ceiling.

References

1. Paradiso, J.A., Starner, T.: Energy scavenging for mobile and wireless electronics. Pervasive Computing 4(1), 18–27 (2005)
2. Chalasani, S., Conrad, J.M.: A survey of energy harvesting sources for embedded systems. In: Southeastcon, April 3-6, pp. 442–447. IEEE, Los Alamitos (2008)
3. Silva-Martinez, J.: Adjustable CMOS voltage limiters for low-voltage applications. In: Proc. IEEE International Symposium on Circuits and Systems (ISCAS 1996) - Connecting the World, May 12-15, pp. 465–468 (1996)
4. Kim, J., Nam, C., Lee, K.-Y.: A design of transceiver for 13.56 MHz RFID reader using the peak detector with automatic reference voltage generator and voltage limiter. In: Proc. International SoC Design Conference (ISOCC 2010), November 22-23, pp. 287–289 (2010)
5. Fernandez, E., Beriain, A., Solar, H., Garcia-Alonso, A., Berenguer, R., Sosa, J., Monzon, J.M., Garcia-Alonso, S., Montiel-Nelson, J.A.: Low power voltage limiter design for a full passive UHF RFID sensor. In: Proc. IEEE 54th International Midwest Symposium on Circuits and Systems (MWSCAS 2011), August 7-10, pp. 1–4 (2011)
6. Magnelli, L., Crupi, F., Corsonello, P., Pace, C., Iannaccone, G.: A 2.6 nW, 0.45 V Temperature-Compensated Sub threshold CMOS Voltage Reference. IEEE Journal of Solid-State Circuits 46(2), 465–474 (2011)

A Study on Superconducting Coils for Superconducting Magnetic Energy Storage (SMES) Applications

Nuno Amaro[1], João Murta Pina[1], João Martins[1], and José Maria Ceballos[2]

[1] Centre of Technology and Systems, Faculdade de Ciências e Tecnologia, Universidade Nova de Lisboa, 2829-516 Caparica, Portugal
[2] "Benito Mahedero" Group of electrical Applications of Superconductors, Escuela de Ingenierías Industriales, Universidad de Extremadura, Avenida de Elvas s/n, 06006, Badajoz, Spain
nma19730@campus.fct.unl.pt, {jmmp,jf.martins}@fct.unl.pt, jmceba@unex.es

Abstract. Superconducting coils (SC) are the core elements of Superconducting Magnetic Energy Storage (SMES) systems. It is thus fundamental to model and implement SC elements in a way that they assure the proper operation of the system, while complying with design specifications. As a part of a larger model, a coil design model is here presented and verified with tests made in a laboratory prototype. The limitations of the superconducting tape used, namely the negative effect of magnetic field components on its critical current value, are also verified and a possible solution to avoid that effect is studied.

Keywords: Superconducting Magnetic Energy Storage, SMES, High temperature superconductivity, HTS coil, HTS coil modeling.

1 Introduction

SMES system is a superconducting device with several possible applications in electric grids, and specially in the future's Smart Grids [1]. With such device it is possible to overcome several power quality issues, as voltage dips and swells, frequency oscillations or harmonic distortion [2]. Power quality is a very important topic in power grids, considering the huge dissemination of microprocessors and electronic devices. Electric energy is usually required to have high quality, i.e. with limited disturbances [3], making SMES a potential candidate in this field [1].

The operation principle of the SMES is based on storing energy in the magnetic field of a SC coil according to

$$E = \frac{1}{2} L I^2 \tag{1}$$

where E is the stored energy, L is the inductance of the SC coil and I is the current flowing in it. This energy is discharged into the grid when necessary. Since current

L.M. Camarinha-Matos, S. Tomic, and P. Graça (Eds.): DoCEIS 2013, IFIP AICT 394, pp. 449–456, 2013.

flowing in the coil is DC and usually grids operate in AC, a bidirectional AC/DC converter and proper control strategies are necessary to assure energy exchanges with grid. This type of storage is only possible with superconducting materials, with virtually zero DC losses, as current decay due to e.g. copper resistivity would made it unfeasible.

The main goal of this work relies on deriving models for the SC coil that can be used in the design stage. These models must include not only electromagnetic effects but also mechanical and thermal characteristics of the coil. As a first approach, it was created a model to obtain the inductance of a coil given its geometric parameters. This model was then validated by comparing its results with measurements made in a real coil.

The coils used in experiments were implemented using first generation (1G) SC tape, built by BiSrCaCuO ceramic on its phase 2223. This material shows a significant drop in its critical current value in the presence of external magnetic field [4]. This degradation increases dramatically with components of magnetic field perpendicular to the tape. The influence of magnetic field in the value of the critical current is verified by measurements of critical current. In order to verify such dependence and improve the critical current a possible solution is also presented in the paper.

2 Contribution to the Internet of Things

Unlike existing unidirectional power grids, in electric grids of the future, energy is foreseen to flow up- and downstream. Smart Grids concept involves both energy and information flows on the grid [5]. To add information flow several standards and protocols must be created or adapted [6]. This paradigm change transforms the power grid in a completely new grid, much more similar to information networks like internet. In the last years several efforts are being made to create the concept of Internet of Things (IoT) in which every devices are connected to the Web. Smart Grids represent a particular case of IoT, where reliability and security of supply are requirements of utmost importance, considering energy transmission and distribution. Devices like SMES are foreseen to address these challenges, as well as contributing to integration of distributed generation, in a context of Smart Grids and IoT.

3 Superconducting Coil Modeling

3.1 Geometry Based Model

According to (1), stored energy in SMES depends on the current flowing through the SC coil and on its inductance. Since current depends on the superconducting material itself (critical current of the SC tape limits the admissible operation current) it is fundamental to maximize the inductance of the coil, so that the stored energy can be maximized while optimizing the required volume of material. Several approaches can be followed to determine the maximum inductance of a coil with specific geometric characteristics. In this case, a mathematical method was used. This method is based in classical formulas to determine self and mutual inductances of air core pancake coils [7, 8].

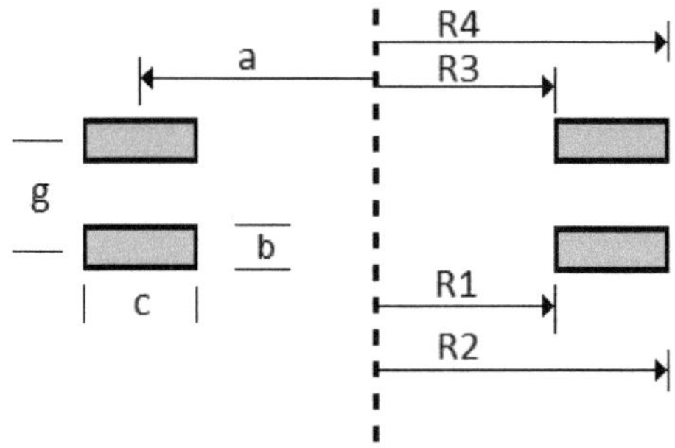

Fig. 1. Geometric variables of a set of air core pancake coils

Considering the geometric variables defined in Fig. 1, considering that N is the number of turns of a single coil, then its self inductance L in mH, with dimensions in cm, is approximated by

$$L = 4\pi a N^2 (\lambda + \mu), \tag{2}$$

where parameters λ, μ and x are defined by

$$\lambda = \log\left(\frac{8a}{c}\right) + \frac{1}{12} - \frac{\pi x}{3} - \frac{\log(1+x^2)}{2} + \frac{\log(1+x^2)}{12x^2} + \\ + \frac{x^2 \log\left(1+\frac{1}{x^2}\right)}{12} + \frac{2}{3}\left(x - \frac{1}{x}\right)\tan^{-1}(x), \tag{3}$$

$$\mu = \frac{c^2}{96a^2}\left\{\left[\log\left(\frac{8a}{c}\right) - \frac{\log(1+x^2)}{2}\right](1+3x^2) + 3.45x^2 + \frac{221}{60} - \\ -1.6\pi x^3 + 3.2x^3 \tan^{-1} x - \frac{\log(1+x^2)}{10x^2} + \frac{x^4 \log\left(1+\frac{1}{x^2}\right)}{2}\right\}, \tag{4}$$

$$x = b/c \tag{5}$$

In order to compute mutual inductance M, also in mH, between two coils with N_1 and N_2 turns, the following expression is used, with dimensions in cm,

$$M = \frac{\mu_0 N_1 N_2}{(R_2 - R_1)(R_4 - R_3)} \int_0^{\pi} \int_{R_1}^{R_2} \int_{R_3}^{R_4} \frac{\cos(\theta) r_1 r_2}{\sqrt{g^2 + r_1^2 + r_2^2 - 2 r_1 r_2 \cos(\theta)}} dr_1 dr_2 d\theta. \tag{6}$$

If the SC coil is built only by one pancake, then inductance is given by (2)-(5). If more than one pancake coil is used, then mutual inductances between each pair of pancakes must be calculated and total inductance of the SC coil is given as follows

$$L_T = L_i + 2\sum_{i=1}^{N_C - 1} \sum_{j=2}^{N_C} M_{ij} \quad , i \neq j, \tag{7}$$

where N_C is the number of pancakes in the set.

One common task consists on maximizing inductance of the coil given available length of superconducting tape. An optimization algorithm derived elsewhere [9] uses (2)-(5) to obtain internal radius and number of turns for a single coil that maximizes self inductance value. After this optimization, given the number of pancakes used, total inductance of the system is derived by (6)-(7).

3.2 Validation of the Model

To validate the results given by the model represented by equations (2)-(7), i.e. to verify that the values of mutual and self inductance given by equations are accurate, two small pancake coils were implemented using two similar 1G tapes (InnoST Bi-2223 HTS insulated wire), named A and B, whose characteristics are shown in Table 1.

Table 1. Characteristics of the SC tapes, according to supplier

Characteristic	Tape A	Tape B
Critical current @ 77 K, self-field (A)	90	85
Minimum bending radius (mm)	30	
Width (mm)	4.2	
Thickness (mm)	0.25	
Available length for each coil (m)	37.5	

The specified length of tape for each coil was 37.5 m, and the optimization algorithm returned an internal radius of 32 mm. After building coils with required dimensions, their inductance was measured using a digital RLC meter (LTD edc1620) and the results of such measurements are presented in Table 2. The implemented coils are shown in Fig. 2. As can be seen in Table 2, the results obtained with measurements are within a 10% margin of the calculated value, except for Coil A. Such deviation is due either to uncertainty in dimensions of the prototype due to shrinking of materials in liquid nitrogen (77 K) experimental measurements environment.

Table 2. Comparison between model outputs and experimental measurements

		Characteristic	Value (mH)
Model outputs	Coil A	Self inductance	1.75
	Coil B	Self inductance	1.75
	Set	Mutual inductance	2.5
		Total inductance	**6**
Experimental measurements	Coil A	Self inductance	1.45
	Coil B	Self inductance	1.8
	Set	Mutual inductance	2.7
		Total inductance	**5.95**

4 Critical Current of the Superconducting Set of Coils

Critical current measurement is needed as operating currents (and stored energy) are limited by this value. Manufacturers specify this value measured in a single portion of tape under self- or external field. In the present case, the effect of magnetic field due to adjacent stacked tapes is expected to degrade critical current. This value is often defined as the current that generates an electric field of 1 μV/cm in the tape.

4.1 Critical Current Measurement

Using the four point method, critical current of the two pancakes operating both separated and in series was measured and results are presented in Fig. 3. In both cases voltage drops at the terminals of each pancake were measured and added to obtain the total voltage drops in the SC set. Using this method, the voltage drop in the resistive connection between the two pancake coils does not influence results.

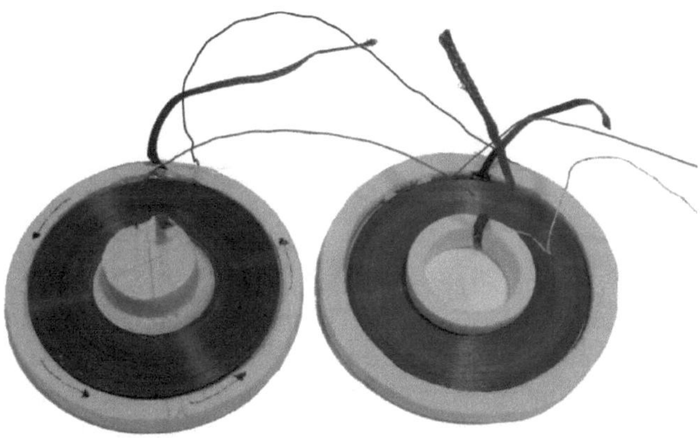

Fig. 2. Implemented coils: coil A (*left*) and coil B (*right*)

As can be seen in Fig. 3, critical current of both coils is similar and approximately 30 A, which is only one third of the value given by manufacturer. Such value is coherent with results previously obtained by other research groups [10], and confirms that in 1G tapes the critical current value decreases rapidly with the increase of the magnetic field. It is also shown that when the two coils are operating in series, critical current value decreases and the losses on the tape increase. This also validates the fact that the increase of magnetic field decreases critical current value, since a set of two coils generate higher magnetic field than a single coil, for the same current.

4.2 Improvement in Critical Current

The degradation of critical current value measured in the SC coils is due to the existence of self field. As mentioned, this effect is accentuated for field components perpendicular to the tape surface, which is more relevant in the internal and external turns of the coils. One approach to minimize this effect consists on using ferromagnetic flux diverters on the extremities of the SC coil [11]. If the air gap between the SC coil and the ferromagnetic material is small enough then flux lines close through the ferromagnetic material and are diverted from the SC tape.

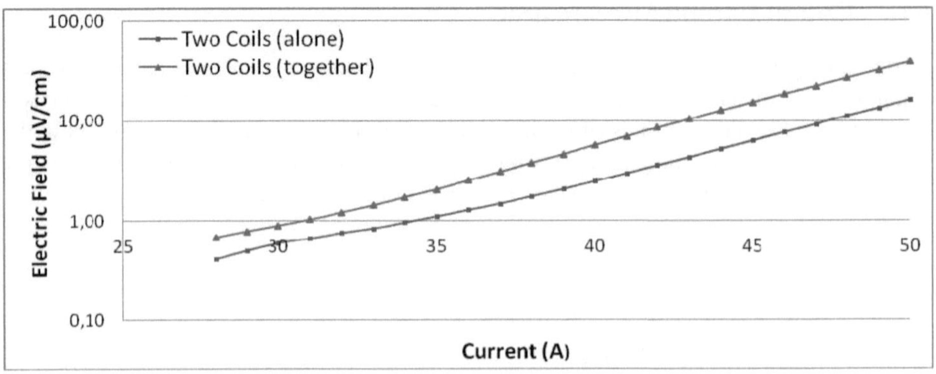

Fig. 3. Critical current measurement

To verify the previous concept, the SC coil built by two series connected pancakes, was tested again with the ferromagnetic material on top and beneath. Fig. 4 shows a comparison between the results obtained with and without diverters. According to that figure, critical current increased with diverters, which is accomplished by a decrease in resistive losses in the SC tape. This demonstrates that ferromagnetic diverters have a positive effect in the field distribution.

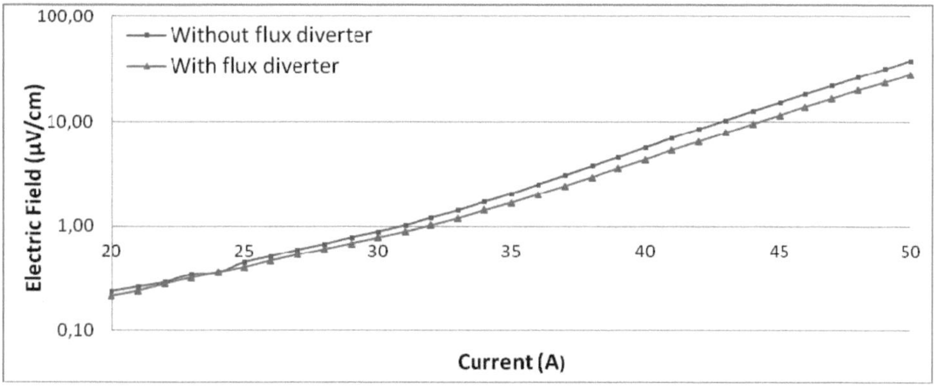

Fig. 4. Critical current measurement with flux diverter

5 Conclusions

In the design of a SMES system, several steps must be followed. The superconducting coil must be properly modeled as it represents the core element of the device. As a first approach, a mathematical model to determine the inductance of superconducting coils was applied. This model is based on classical mathematical expressions, usually applied to conventional materials. To validate the model, two superconducting pancake coils were built using first generation tape and the inductance of the set was measured. The critical current of the set was measured and the influence of the magnetic field in the critical current value was verified. This is also very important on SMES because the stored energy depends on the squared value of current that flows through the coil. As a possible solution to minimize the effect of the magnetic field on the critical current value, a ferromagnetic diverter was added to the system leading to an increase in the current.

As future work, a complete model of the whole system, including mechanical and thermal effects is foreseen.

Acknowledgments. This work was supported by Fundação para a Ciência e a Tecnologia (Centre of Technology and Systems multiannual funding) under the framework of project PEst-OE/EEI/UI0066/2011.

References

1. Amaro, N., Murta Pina, J., Martins, J., Ceballos, J.M.: Superconducting Magnetic Energy Storage - A Technological Contribute to Smart Grid Concept Implementation. In: Proceedings of the 1st International Conference on Smart Grids and Green IT Systems, pp. 113–120. SciTePress - Science and Technology Publications (2012)
2. Eurelectric: Power Quality in European Electricity Supply Networks, Brussels (2003)
3. Benysek, G.: Improvement in the Quality of Delivery of Electrical Energy using Power Electronics Systems. Springer (2007)

4. Matsushita, T., Himeda, Y., Kiuchi, M., Fujikami, J., Hayashi, K.: Characterization of Critical Current Density in Silver-Sheathed Bi-2223 Tape. IEEE Transactions on Applied Superconductivity 15, 2518–2521 (2005)
5. E.U., E. union: European SmartGrids Technology Platform - Vision and Strategy for Europe's Electricity Networks of the Future. Office for Official Publications of the European Communities, Brussels (2006)
6. Gungor, V.C., Sahin, D., Kocak, T., Ergut, S., Buccella, C., Cecati, C., Hancke, G.P.: Smart Grid Technologies: Communication Technologies and Standards. IEEE Transactions on Industrial Informatics 7, 529–539 (2011)
7. Rosa, E.B., Grover, F.W.: Formulas and Tables for the calculation of mutual and self-inductance. United States Government Printing Office, Washington (1948)
8. Babic, S., Salon, S., Akyel, C.: The Mutual Inductance of Two Thin Coaxial Disk Coils in Air. IEEE Transactions on Magnetics 40, 822–825 (2004)
9. Amaro, N., Murta Pina, J., Martins, J., Ceballos, J.M., Álvarez, A.: A fast algorithm for initial design of HTS coils for SMES applications. IEEE Transactions on Applied Superconductivity (preprint version) (2012), doi: 10.1109/TASC.2012.2231912
10. Jiang, Z.Q., Jin, J.X.: Critical current measurement and experimental comparison of 1 G and 2 G HTS tapes. In: 2011 International Conference on Applied Superconductivity and Electromagnetic Devices, pp. 145–149 (2011)
11. Pardo, E., Šouc, J., Vojenčiak, M.: AC loss measurement and simulation of a coated conductor pancake coil with ferromagnetic parts. Superconductor Science and Technology 22, 075007 (2009)

Part XVI
Optimization Techniques in Energy

A Multi-objective Simulation Based Tool: Application to the Design of High Performance LC-VCOs

Amin Sallem[1], Pedro Pereira[2], Mourad Fakhfakh[1], and Helena Fino[2]

[1] LETI-ENIS, University of Sfax, Tunisia
[2] CTS, Uninova, Departamento de Engenharia Electrotécnica, Faculdade de Ciências e Tecnologia, FCT, Universidade Nova de Lisboa, 2829-516 Caparica, Portugal
{sallem.amin,pmrp,hfino}@ieee.org,
fakhfakhmourad@gmail.com

Abstract. The continuing size reduction of electronic devices imposes design challenges to optimize the performances of modern electronic systems, such as: wireless services, telecom and mobile computing. Fortunately, those design challenges can be overcome thanks to the development of Electronic Design Automation (EDA) tools. In the analog, mixed signal and radio-frequency (AMS/RF) domains, circuit optimization tools have demonstrated their usefulness in addressing design problems taking into account downscaling technological aspects. Recent advances in EDA have shown that the simulation-based sizing technique is a very interesting solution to the 'complex' modelling task in the circuit design optimization problem. In this paper we propose a multi-objective simulation-based optimization tool. A CMOS LC-VCO circuit is presented to show the viability of this tool. The tool is used to generate the Pareto front linking two conflicting objectives, namely the VCO Phase Noise and Power Consumption. The accuracy of the results is checked against HSPICE/RF simulations.

Keywords: Multi-Objective Optimization, Metaheuristic, Pareto Front, Simulation-based optimization, LC-VCO, Inductor 2-π model.

1 Introduction

Analog, mixed signal and radio-frequency (AMS/RF) circuit design are becoming more and more complex; there is a pressing need for Electronic Design Automation (EDA) to meet the time to market constraints. Indeed, automation in circuit design has successfully demonstrated its usefulness, from circuit level design, see for instance [1–4], to system level designs, see for example [5–8]. The current computers features have favoured a transition from the hand-calculation-based design, to the simulation-based design. The first, which is known as the knowledge-based design, is one of the earlier approaches. Its basic idea is to have a predefined design plan for sizing circuits to meet the performance specifications [9–11]. The second one, i.e. the simulation-based approach, is based on the use of a circuit simulator such as SPICE, which evaluates the circuit's performance(s) and constraints, as well [12–14]. The key of

L.M. Camarinha-Matos, S. Tomic, and P. Graça (Eds.): DoCEIS 2013, IFIP AICT 394, pp. 459–468, 2013.

this transition is the feasibility to include SPICE like simulators within the loop of any circuit design optimization problem.

The simulation-based approach is adopted in this work. This technique uses a standard circuit simulator in the optimization loop to evaluate the circuit performance. In this way, this approach can handle a large variety of AMS/RF circuits. Among the advantages of this method, it offers a great accuracy at the expense of design time while an analytical approach is faster but suffers from accuracy limitations.

Actually, AMS/RF circuit optimization problems generally involve more than one objective. These objective functions are conflicting and non-commensurable ones. A multi-objective metaheuristic is used in this work to efficiently find the optimum design of an LC voltage controlled oscillator (LC-VCO).

The main concerns in this design are the passive elements, principally the inductor.

When implementing an on-chip inductor, the influence on circuit behaviour that come up from inductor parasitics are usually ignored and the inductor is considered as an 'ideal' element. Yet, this is a wrong approach, since at high frequencies the parasitic capacitances that appear between metal layers and oxide/substrate can have significant weight and jeopardize the expected inductance value.

Despite the fact that the simulation-based technique is accurate and that the obtained results are precisely those of the simulator (since the component simulator models are adopted), at high frequencies, parasitics effects of passive components (mainly inductors) can not be neglected: using the ideal models of these components is a common mistake.

In this work we deal with inductor parasitics and use the proposed tool to size the circuit and the inductor, as well, since classical sizing approaches cannot be used for such design due to its complexity.

For the inductor the 2-π model [15-16] is used in the in-loop optimisation simulation based tool in order to highlight potentialities and performances of the proposed tool and to show the real effect of parasitics of inductors. An LC-VCO circuit is considered as an application example.

The rest of the paper is structured as follows. In section 2, we clearly show the main contributions of this work. In section 3, we put the light on the proposed tool and highlight its capabilities. In section 4, we present the multi-objective optimization. In section 5, the LC-VCO circuit, which includes the inductor equivalent circuit, is used to validate the optimization tool. Finally, conclusions are offered in section 6.

2 Relationship to Internet of Things

Internet of things relies on the interconnections of a large number of heterogeneous cooperating devices. The development of these devices has been made possible due to the rapid evolution of electronic technologies, which enable the implementation of ever more complex functions, in smaller and more rapid circuits. To cope with the necessity of minimizing the power consumption of such systems, new design methodologies must be adopted so that the ever more stringent specifications may be attained. In the particular case of communications services, e.g. wireless communications devices, Voltage Controlled Oscillators (VCOs) are fundamental building blocks where optimization design methodologies must be adopted to cope with such multi-objective design goals, e.g. low power consumption and low phase noise.

The contributions of this paper are as follows:

- A multi-objective simulation based tool that allows generation the Pareto front of conflicting objectives of analog and RF circuits, is proposed;
- The use of the 2-π model for the electrical simulation of the inductor in the loop-optimization. By using a lumped element model, an efficient process is obtained, thus overcoming the need for using lengthy electromagnetic simulations of the inductor.

3 The Simulation-Based Sizing Tool

As introduced in Section 1, the simulation-based optimization technique is adopted in order to be able to deal with complex circuits. Its basic idea consists of 'by-passing' the modelling stage in the sizing problem and to directly use an electrical simulator to evaluate the objective functions, and to check the circuit's constraints, as well. In each optimization iteration the circuit simulator is called to evaluate the circuit's performance(s) for a set of design parameters. A pictorial diagram of this approach is presented in Figure 1 [14].

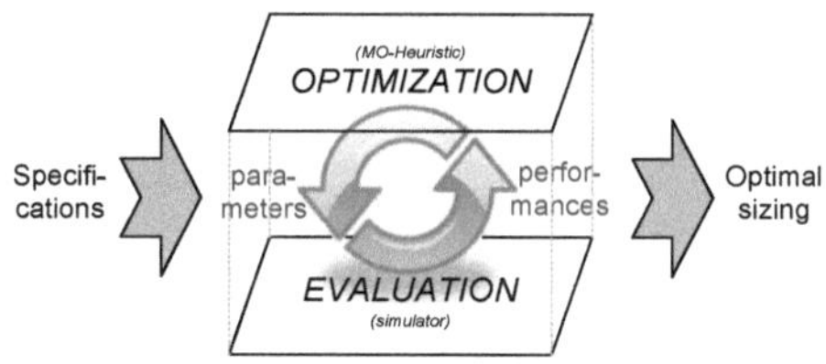

Fig. 1. The simulation-based sizing/optimizing technique [14]

One of the key components in this technique is the optimization block, whose purpose is to find the best circuit's sizing that will lead to the best performance while satisfying a set of constraints. In this work, we use a multi-objective metaheuristic called MOHA [17]. This algorithm is based on a communication tool between HSPICE/RF and C++ software.

In short, this tool works as follows. After generating the circuit's netlist, the parameters' values are introduced in this netlist. Then, the C++ program calls HSPICE/RF thanks to which, constraints are checked and performances are evaluated. These performances are then introduced in the optimization algorithm and 'new' sizes are generated, etc.

In order to deal with multi-objective problems, an external archive was added to the optimization routine [17] where non-dominated solutions are stored and updated at each iteration. A dominance sorting routine [18] was integrated into the optimization program.

This external memory encompasses the 'best' non-dominated solutions obtained so far. It is to be mentioned that these performance evaluations are performed using the simulator .MEAS statement which prints user-defined electrical specifications of a circuit, and the results could be manipulated in a post-processing step. Figure 2 shows the corresponding algorithm flowchart. Further details regarding MOHA can be found in [17].

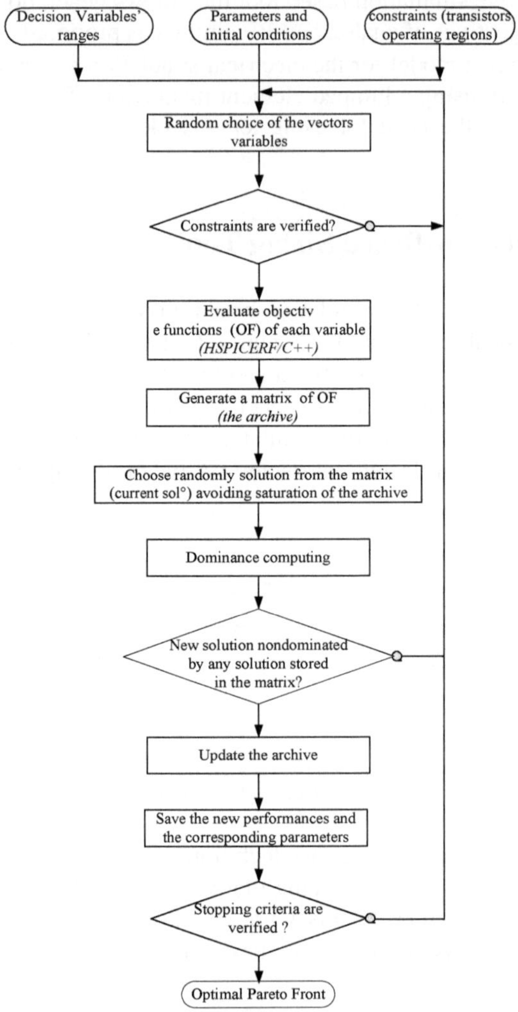

Fig. 2. Flowchart of MOHA

4 The Multi-Objective Optimization

Actually, circuit optimization problems involve more than one objective function that meets all the performance functions and imposed/inherent constraints. In most cases, the considered sizing problems are multi-objective ones. Equation (1) illustrates this multi-objective optimization (MOO) problem:

$$\left| \begin{array}{ll} Minimize\ \vec{f}(\vec{x}); & \vec{f}(\vec{x}) \in R^{k} \\ subject\ to: & \\ \vec{g}(\vec{x}) \leq 0; & \vec{g}(\vec{x}) \in R^{m} \\ and\ \vec{h}(\vec{x}) = 0; & \vec{h}(\vec{x}) \in R^{n} \\ where\ x_{Li} \leq x_i \leq x_{Ui}, & i \in [1, p] \end{array} \right. \quad (1)$$

where, k, m, n and p are the numbers of objectives ($k \geq 2$), inequality constraints to satisfy, equality constraints to assure and parameters to manage respectively.

Commonly, designers transform the multi-objective problem into a mono-objective one using the weighting technique [19]. The latter requires the designer to select values of weights for each objective. The so obtained mono-objective function can be written as:

$$F(\vec{x}) = \sum_{i=1}^{k} \omega_i f_i(\vec{x}) \quad (2)$$

ω_i are weighting coefficients.

However, it has been proven that this technique is not suitable and may lead to non-optimal solutions (see for instance [20]).

Due to the fact that in most cases, circuits present conflicting non-commensurable objectives, making appeal to the multi-objective techniques is mandatory. Actually, such techniques allow generating the aforementioned sets of non-dominated[1] solutions known as Pareto fronts [18].

In the literature, a large plethora of multi-objective metaheuristic were/are being used, to optimize performances of AMS/RF circuits (mainly using the equation-based approach) [2–4, 8, 21–22].

In this work we use a multi-objective metaheuristic, called *Multi-Objective Heuristic Algorithm* (MOHA) [17].

5 Application to the Optimal Design of an LC-VCO Circuit

5.1 The LC-VCO Equivalent Circuit

One of the main challenges when designing a VCO, is to obtain results between simulation and on-chip measurement as close as possible. For this propose, it is essential to use design methodologies where parasitics must be accounted for, even when simulators are used. In Fig. 3(a) a cross-coupled LC-VCO is represented. It encompasses two main blocks: the LC thank, responsible for the oscillation frequency, and the active

[1] A solution $\vec{x}$ of a MO problem is said non-dominated solution if and only if there does not exist another solution $\vec{y}$ such that $\vec{f}(\vec{y})$ dominates $\vec{f}(\vec{x})$, i.e. no component of $\vec{f}(\vec{x})$ is smaller than the corresponding component of $\vec{f}(\vec{y})$ and at least one component is greater.

circuit, which accounts for reducing the circuit losses by introducing a negative resistance. The transistor *Mb* is responsible for enforcing the current in the circuit. Most LC-VCO designs aim at achieving both minimum phase noise and power consumption for a given oscillation frequency. For instance, if low power consumption is desired, a low bias current must be delivered to the circuit. Yet, parasitic effects will have a major role in circuit behaviour, yielding to the degradation of phase noise. On the other hand, if low phase noise is required, high current is desired. The phase noise *vs* power consumption trade-off, among others, makes the VCO design a suitable applicant for simulation-based optimization based design approach.

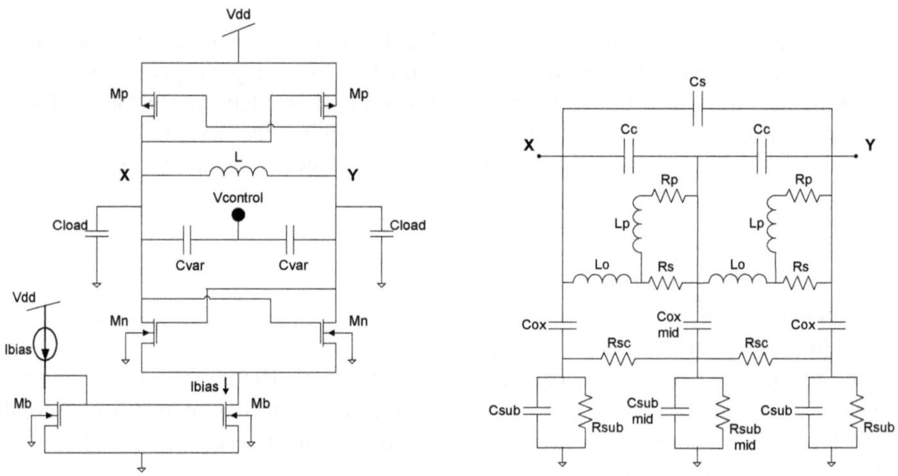

(a) The cross coupled topology (b) The inductor equivalent circuit

Fig. 3. LC – voltage controlled oscillator (LC-VCO)

As mentioned before, the LC tank is the block responsible for the oscillation frequency, and plays a major role in the LC-VCO performance. Since inductor parasitics appearing between the inductor metal tracks and the lower layers of oxide and substrate are very relevant for GHz frequencies, an inductor equivalent circuit model, known as 2-π inductor model [15], represented in Fig. 3(b), is considered. The inductor 2-π equivalent circuit. This inductor equivalent model accounts for:

- DC parameters (L_o, L_p, R_s, R_p);
- Crossover capacitance, *Cs* - that capacitance appears between the spiral and the underpass necessary to connect the inner turn to the outside of the spiral inductor;
- Metal-to-metal capacitance, *Cc* – that appears due to the proximity of inductor tracks;
- Metal-to-substrate capacitance, C_{ox};
- R_{sub} and C_{sub} that models the ohmic losses in the conductive silicon substrate;
- R_{sc} which represents the electric coupling between lines through the conductive substrate.

5.2 Simulations Results

In this section the design of several LC-VCOs for operating frequencies between 1 and 2 GHz, are addressed. The design objective functions are the minimization of both the phase noise - @ an offset frequency of 1MHz - and the power consumption. The LC-VCO characteristics and 2-π inductor circuit parameters range are given in Table 1, where the transistor channel length/width range is valid for *Mp*, *Mn* and *Mb* elements. Moreover, the capacitor *Cvar* is in fact a transistor that behaves as a capacitor (referred as MCvar in Table 2), and its channel length/width range is considered to be equal to *Mp*.

Table 1. LC-VCO and 2-π inductor circuit parameters limit

LC-VCO circuit	Oscillation frequency - f_0	1 GHz – 2 GHz
	Transistor channel length	0.45 μm – 1.6 μm
	Transistor channel width	1 μm – 500 μm
	Tank inductance - L	1 nH – 15 nH
	Tank capacitance - C_{var}	1 pF – 20 pF
	Load capacitor C_{Load}	2 pF
2-π inductor circuit	Cs, Cc, Csub, Csub_mid	0.1 pF – 20 pF
	Cox, Cox_mid	0.1 fF – 1 pF
	Lo, Lp	0.01 nH – 10 nH
	Rs	0.01 Ω – 30 Ω
	Rp	1 Ω – 1 kΩ
	Rsc, Rsub, Rsub_mid	10 kΩ – 10 MΩ

In Fig. 4 we present the two Pareto fronts (*Phase Noise vs. Power*) obtained using the proposed tool for two different scenarios: considering ideal and real inductors. The supply voltage is V_{DD} = 1.2 V and the voltage at the gate of the transistor *Mb* is equal to 0.4V. Simulations are performed using Level 49 standard CMOS Technology of 0.13μm. As expected optimization results considering ideal inductors are quite better than those with real inductors. By simple observation, it is possible to conclude that for a specific phase noise, the power consumption for the LC-VCO considering a real inductor is higher by a factor of 8. That means that results obtained for ideal inductors are too optimistic.

Reached performances and optimal parameter values corresponding to the Pareto front edge points of MOHA algorithm are in Table 2 and 3: solutions giving the maximum phase noise and the minimum power, solutions (1) and (2) in figure 4 respectively. Considering the design solution (1) in Table 2, the inductance of the inductor equivalent circuit is 3.8 nH. Now, if a simulation is done for the same transistors sizes, but considering an ideal inductor of 3.8 nH, the LC-VCO characteristics are: 1.18 mW, 2.04 GHz and -122.87 dB/Hz. These results show an error around 15% for power, 6% for the oscillation frequency and 4% for the phase noise when compared with those obtained with the inductor equivalent circuit.

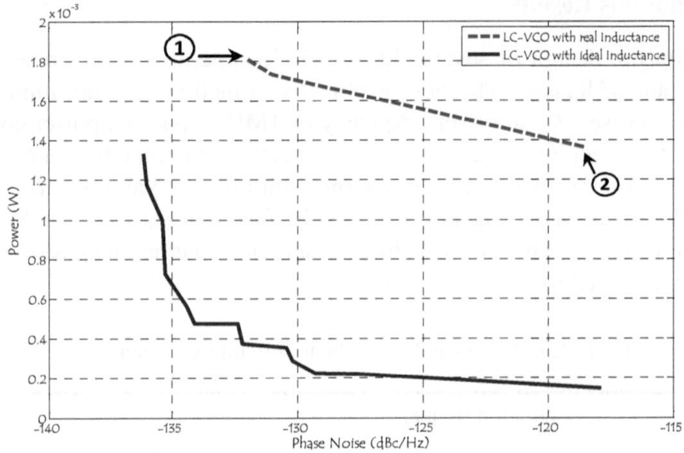

Fig. 4. LC-VCO optimization results Pareto fronts (Phase Noise vs. Power consumption)

Table 2. Optimal parameters values (LC-VCO with real inductor)

Parameters	*Lower edge of the Pareto front (solution(2))*	*Higher edge of the Pareto front (solution(1))*
Mp – width / length (μm)	176.75 / 1.57	273.75 / 0.63
Mn – width / length (μm)	256.75 / 1.57	190.25 / 0.63
Mb – width / length (μm)	240.75 / 1.57	489.25 / 0.63
MCvar – width / length (μm)	472.75 / 1.57	80.25 / 1.15
Inductance (nH)	0.62	3.83

Table 3. Reached performances (points located at the Pareto front edges)

		Lower edge of the Pareto front (solution(2))	***Higher edge of the Pareto front (solution(1))***
Real Inductor	***Phase Noise (dBc/Hz)***	-118.49	-131.99
	Power Consumption (mW)	1.36	1.80
	Oscillation Frequency (GHz)	1.94	1.45
Ideal Inductor	***Phase Noise (dBc/Hz)***	-117.87	-136.15
	Power Consumption (mW)	0.14	1.32
	Oscillation Frequency (GHz)	1.17	1.02

6 Conclusions

A multi-objective simulation based tool that allows generation the Pareto front of conflicting objectives of analog and RF circuits, is proposed. As a proof of concept, the work addressed the design of LC-VCOs as a simulation-based sizing problem.

The methodology presented deals with the complexity of the design by formulating it as a multi-objective metaheuristic optimization problem. The adoption of the 2-π model of the inductor and its use in the electrical simulation-based optimization loop, aims to obtain design solutions in which simulations are expected to be close to those in fabricated circuits.

The use of multi-objective optimization strategy, allows dealing with design trade-offs, such as power consumption - phase noise, obtaining the best design solution, supported on designers inputs.

A set of design examples showing the design of LC-VCOs for oscillation frequencies between 1 and 2 GHz is shown. The presented results highlight the differences in circuit performance when using circuits that characterize the real device behaviour rather than ideal ones, as shown for the inductor.

References

1. Gielen, G., Wambacq, P., Sansen, W.M.: Symbolic Analysis Methods and Applications for Analog Circuits: a Tutorial Overview. Proceedings of the IEEE 82(2), 287–304 (1994)
2. Fakhfakh, M., Cooren, Y., Sallem, A., Loulou, M., Siarry, P.: Analog Circuit Design Optimization through the Particle Swarm Optimization Technique. Journal of Analog Integrated Circuits and Signal Processing 63(1), 71–82 (2010)
3. Tlelo-Cuautle, E., Guerra-Gómez, I., Luis de la Fraga, G., Flores-Becerra, G., Polanco-Martagón, S., Fakhfakh, M., Reyes-García, C.A., Rodríguez-Gómez, G., Reyes-Salgado, G.: Evolutionary Algorithms in the Optimal Sizing of Analog Circuits. In: Köppen, M., Schaefer, G., Abraham, A. (eds.) Intelligent Computational Optimization in Engineering. SCI, vol. 366, pp. 109–138. Springer, Heidelberg (2011)
4. Pereira, P., Sallem, A., Fakhfakh, M., Fino, M.H., Coito, F.: A Technology-Aware Optimization of RF Integrated Inductors. In: Tlelo-Cuautle, E. (ed.) Analog Circuits: Applications, Design and Performance, pp. 213–234. Nova Science Publishers, Inc. (2012)
5. Fernández, F.V., Rodríguez-Vázquez, A., Huertas, J.L., Gielen, G.: Symbolic Analysis Techniques: Applications to Analog Design Automation. Wiley-IEEE Press (1998)
6. Gielen, G., Eeckelaert, T., Martens, E., McConaghy, T.: Automated Synthesis of Complex Analog Circuits. In: The European Conference on Circuit Theory and Design, pp. 20–23 (2007)
7. Schoofs, R., Eeckelaert, T., Steyaert, M., Gielen, G., Sansen, W.: A Continuous-Time Delta-Sigma Modulator for 802.11a/b/g WLAN Implemented with a Hierarchical Bottom-up Optimization Methodology. In: IEEE International Conference on Electronics, Circuits and Systems, pp. 950–953 (2006)
8. Tlelo-Cuautle, E., Guerra-Gómez, I., Reyes García, C.A., Duarte-Villaseñor, M.A.: Synthesis of Analog Circuits by Genetic Algorithms and their Optimization by Particle Swarm Optimization. In: Chiong, R. (ed.) Intelligent Systems for Automated Learning and Adaptation: Emerging Trends and Applications, pp. 173–192. IGI Global (2010)

9. Degrauwe, M., Nys, O., Dijkstra, E., Rijmenants, J., Bitz, S., Goffart, B., Vittoz, E., Cserveny, S., Meixenberger, C., Van Der Stappen, G., Oguey, H.: IDAC: An Interactive Design Tool for Analog CMOS Circuits. IEEE Journal of Solid-State Circuits 22(6), 1106–1116 (1987)
10. El-Turky, F., Perry, E.: BLADES: An Artificial Intelligence Approach to Analog Circuit Design. IEEE Transactions on Computer-Aided Design of Integrated Circuits and Systems 8(6), 680–692 (1989)
11. Hägglund, R., Hjalmarson, E., Wanhammar, L.: A Design Path for Optimization-Based Analog Circuit Design. In: The Midwest Symposium on Circuits and Systems, pp. 287–290 (2002)
12. Taherzadeh-Sani, M., Lotfi, R., Zare-Hoseini, H., Shoaei, O.: Design Optimization of Analog Integrated Circuits Using Simulation-Based Genetic Algorithm. In: International Symposium on Signals, Circuits and Systems, pp. 73–76 (2003)
13. Guerra-Gomez, I., Tlelo-Cuautle, E., Peng, L., Gielen, G.: Simulation-Based Optimization of UGCs Performances. In: International Caribbean Conference on Devices, Circuits and Systems (2008)
14. Sallem, A., Guerra-Gomez, I., Fakhfakh, M., Loulou, M., Tlelo-Cuautle, E.: Simulation-Based Optimization of CCIIs' Performances in Weak Inversion. In: IEEE International Conference on Electronics, Circuits, and Systems, pp. 655–658 (2010)
15. Cao, Y., Groves, R.A., Huang, X., Zamdmer, N.D., Plouchart, J.-O., Wachnik, R.A., King, T.-J., Hu, C.: Frequency-Independent Equivalent-Circuit Model for On-Chip Spiral Inductors. IEEE Journal of Solid-State Circuits 38(3), 419–426 (2003)
16. Pereira, P., Fino, M.H., Coito, F., Ventim-Neves, M.: RF Integrated Inductor Modeling and its Application to Optimization-Based Design. Analog Integrated Circuits and Signal Processing 73(1), 4–55 (2011)
17. Sallem, A., Benhala, B., Kotti, M., Fakhfakh, M., Ahaitouf, A., Loulou, M.: Simulation-Based Multi-Objective Optimization of Current Conveyors: Performance Evaluations. In: International Conference on Design & Technology of Integrated Systems in Nanoscale Era (2012)
18. Deb, K., Pratap, A., Agarwal, S., Meyarivan, T.: A Fast and Elitist Multiobjective Genetic Algorithm: NSGA-II. IEEE Transactions on Evolutionary Computation 6(2), 182–197 (2002)
19. Reyes-Sierra, M., Coello-Coello, C.A.: Multi-objective Particle Swarm Optimizers: A Survey of the State-of-the-art. International Journal of Computational Intelligence Research 2(3), 287–308 (2006)
20. Fakhfakh, M., Loulou, M., Masmoudi, N.: A Novel Heuristic for Multi-Objective Optimization of Analog Circuit Performances. Journal of Analog Integrated Circuits and Signal Processing 61(1), 47–64 (2009)
21. Kotti, M., Benhala, B., Fakhfakh, M., Ahaitouf, A., Benlahbib, B., Loulou, M., Mecheqrane, A.: Comparison Between PSO and ACO Techniques for Analog Circuit Performance Optimization. In: International Conference on Microelectronics (2011)
22. Benhala, B., Ahaitouf, A., Fakhfakh, M., Mechaqrane, A., Benlahbib, B.: Optimal Analog Circuit Sizing via Ant Colony Optimization Technique. International Journal of Computer Science and Network Security 11(6), 223–231 (2011)

Distributed Model Predictive Control for Housing with Hourly Auction of Available Energy

F.A. Barata[1] and R.N. Silva[2]

[1] Instituto Superior de Engenharia de Lisboa (ISEL), R. Conselheiro Emídio Navarro 1, 1959-007 Lisboa, Portugal
[2] Universidade Nova de Lisboa, Monte da Caparica, 2829-516 Caparica, Portugal
fbarata@deea.isel.ipl.pt, rns@fct.unl.pt

Abstract. This paper presents a distributed model predictive control (DMPC) for indoor thermal comfort that simultaneously optimizes the consumption of a limited shared energy resource. The control objective of each subsystem is to minimize the heating/cooling energy cost while maintaining the indoor temperature and used power inside bounds. In a distributed coordinated environment, the control uses multiple dynamically decoupled agents (one for each subsystem/house) aiming to achieve satisfaction of coupling constraints. According to the hourly power demand profile, each house assigns a priority level that indicates how much is willing to bid in auction for consume the limited clean resource. This procedure allows the bidding value vary hourly and consequently, the agents order to access to the clean energy also varies. Despite of power constraints, all houses have also thermal comfort constraints that must be fulfilled. The system is simulated with several houses in a distributed environment.

Keywords: DMPC, DSM, Limited resources, Energy auction.

1 Introduction

Heating accounts for a significant proportion of the world's total energy demand. The building sector alone consumes 35.3%, of which 75% is for space heating and domestic water heating. In Europe, the final energy demand for heating and cooling (49%) is higher than for electricity (20%) or transport (31%) [1]. Therefore, it is important economically, socially, and environmentally to reduce the energy consumption of buildings. New models and control techniques must be developed to move beyond standard heuristic approaches and seek to incorporate predictions of weather, occupancy, renewable energy availability, and energy price signals that can support real time energy auction markets [2], [3]. The desire approach here presented intends to take advantage from the innovative technology characteristics provided by future Smart Grids (SGs) [4]. In the smart world, simple household appliances, like dishwashers, clothes dryers, heaters, air conditioners will be fully controllable in order to achieve the network maximum efficiency. Renewable energies will be a

L.M. Camarinha-Matos, S. Tomic, and P. Graça (Eds.): DoCEIS 2013, IFIP AICT 394, pp. 469–476, 2013.

common presence and any kW provided by these technologies should not be wasted. Active demand side management (DSM) will control the loads in order to adapt them to the available renewable energy source. Therefore, how can the demand be adjusted to an intermittent energy source in a distributed network, in order to maximize the energy efficiency?

Model Predictive Control (MPC) during the last years has been granted to reduce and optimize the energy consumption in the residential sector namely to deal with temperature set points regulations [5], [6]. Model predictive control can also naturally deal with the aforementioned predictions to improve building thermal comfort, decrease peak demand and reduce total energy costs. The optimal profile of delivered energy depends on various factors which include time varying utility prices, availability of renewable energy and ambient temperature variation. The MPC have also advantage in distributed systems [7], [8]. Distributed Model Predictive Control (DMPC) allows the distribution of decision-making while handling constraints in a systematic way. DMPC strategies can be characterized by the type of couplings or interactions assumed between constituent subsystems [9]. The DMPC strategy here presented uses the method of subsystems sharing coupled constraints [9], [10].

In this context, in a scenario with distributed infrastructures that are interconnected or related with each other, makes them suitable for Multi Agent System (MAS) technology, and consequently, for the autonomous management of houses and buildings.

The paper is organized as follows. Section 2 presents the technological contribution of this paper, Section 3 presents the implemented system, with the house dynamical thermal model, hourly auction scenario and DMPC formulation. Section 4 illustrates the used methodology with simulation results and in Section 5 some conclusions are draw.

2 Relationship to Internet of Thinks

From the Internet of Things (IoT) perspective, SGs also predicts a future in which devices can communicate with one another across infrastructures much the way people communicate with one another via the web. As mentioned, simple household appliances will be linked in the grid and will be fully controllable, monitored and regulated in real time. Information will be exchanged between devices in order to manage energy demands more efficiently and incorporate the increasing amounts of renewable power from sources like the sun and wind.

Assuming that future communication infrastructures will support real time energy auction markets, the hourly auction here presented is, as far as we know, a novelty contribution for what is expected to be a nearby reality.

The work contributes with a new methodology to manage energy networks from the demand side with strong presence of intermittent energy sources. In a distributed network, the implementation of a constraint in the shared available resource consumption presented here, introduces a novelty that intends to give response to problem mentioned above. With this approach the system will try to adjust consumption to the value provided by the renewable resource maximizing the efficiency and minimizing the consumer energy costs. The profile of delivered energy depends on several factors, such as price of conventional energy and availability of renewable energy. The MAS technology can solve the efficient management of *clean* ("*green*") and *dirty* ("*red*") resources, giving the priority to "*green*".

3 Implemented System

3.1 Thermal Model of the House

The house for which MPC is designed is present in (1-3) and describes only the dominant dynamics of the house.

$$\frac{dT_{house}}{dt} = \frac{1}{C}\left(Q_{heat} - Q_{losses}\right), \tag{1}$$

$$Q_{losses} = \frac{T_{house} - T_{oa}}{R_{eq}}, \tag{2}$$

$$R_{eq} = \frac{R_{wall} R_{window}}{R_{wall} + R_{window}}, \tag{3}$$

where in (1), Q_{losses} is heat and cooling losses (kW), T_{house} the inside temperature (K), C the thermal capacitance (kJ/K), and Q_{heat} the heat and cooling power (kW). In (3) T_{oa} is the outdoor temperature (K) and the parameter R_{eq} describes the equivalent thermal resistance of all walls (including roof and ceiling) and windows that isolate the house from outside, and can be describe as a electrical parallel resistance circuit [5]. The plant model representation (1) can be rewritten and changed into a discrete model using Euler discretization with a sampling time of Δt.

$$T(k+1) = AT(k) + Bu(k) + v(k), \tag{4}$$

where $A = 1 - \frac{\Delta t}{R_{eq}C}$, $B = \frac{\Delta t}{C}$, $v = \frac{P_d}{C} + \frac{T_{oa}\Delta t}{R_{eq}C}$, u(k) is the necessary heat/cooling power, T(k) is the indoor temperature, v(k) is a disturbance signal resulting from P_d the external disturbances (kW) (*e.g.* load generated by occupants, direct sunlight, electrical devices or doors and windows aperture to recycle the indoor air), and T_{oa}, the temperature of outside air (K).

3.2 Hourly Auction Scenario

The scenario considers two types of available energy resources, the *green* and the *red*. The *green* or *clean* resource must be always consumed (is non dispatchable) and it is limited to a maximum available value and it's considered a time variable resource. In opposition, the *red* is always available and it is considered a *dirty* resource, more expensive than the *green*. Therefore, if the *green* resource is insufficient to satisfy the demand required by the houses, the *red* must be consumed with an increase in the total energy cost. To incentive the *clean* resource consumption, it is considered that the *green* energy price per kWh has a maximum auction value (0.09€/kWh) always cheaper than the *red* energy price (0.14€/kWh). The agents (one by each house) must bid in an auction (provided by the market operator), the price that they are willing to pay to consume the *green* resource. The agents make their bid in the auction with one day ahead to show how much intend to pay per kWh to consume the *green* resource in each one of the next 24 hours. The bid can be made according to the consumption needs. Each house has a known fixed 24 hours consumption profile (Fig. 1) and it is established a priority level from 1 (low) to 3 (high), that indicates how important that hour is in terms of consumption (Fig. 3). Thus, the hours with high priority levels indicates high consumption and consequently a higher bid value (Table 1). Figure 2 show the outdoor temperature forecasts for all houses.

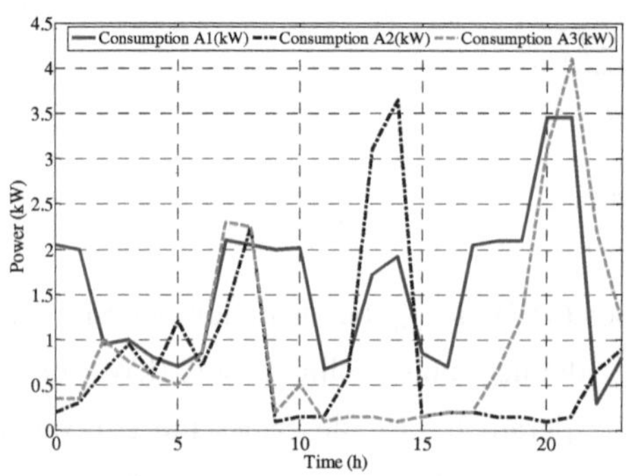

Fig. 1. Fixed consumption profile of each house (C_1, C_2 and C_3)

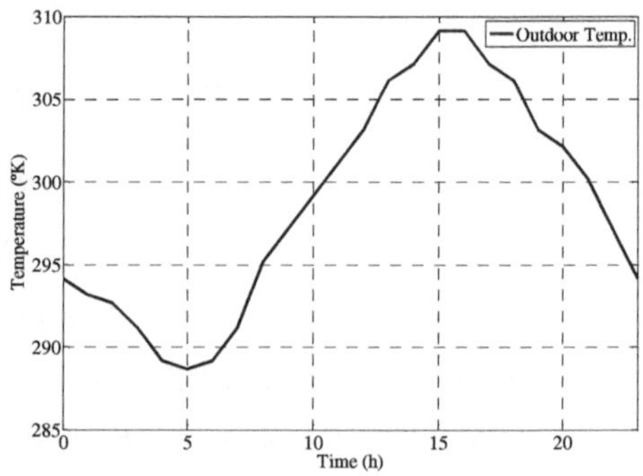

Fig. 2. Outdoor temperature forecasting (T_{oa})

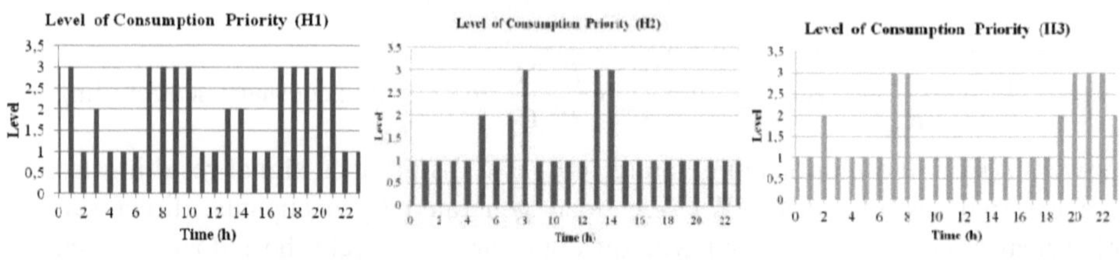

Fig. 3. Consumption priority level of each house

Table 1. Bid value for each consumption level by agent

Consumption	Priority Level	House 1	House 2	House 3
0-1 kW	1	2/5*0.09	3/5 *0.09	1/2*0.09
1-2 kW	2	7/10*0.09	4/5*0.09	2/3*0.09
>2 kW	3	8.5/10*0.09	9/10*0.09	3/4*0.09

The bid value establishes an order to access to the resource, being the green resource consumption made by the agents sequentially by that order. The first agent consumes and the remainder *green* resource is passed to the next agent as the maximum green available resource. As mentioned, when the *green* resource becomes insufficient to satisfy all the demand, the *red* is available. The *red* resource consumption implies a penalty in the final cost function (3) due to the soft constraint violation imposed by the maximum available *green* resource is exceeded.

3.3 Model Predictive Control

MPC principle of controlling house heating and cooling is to react on the heating/cooling actuators based on current measurements/estimates of temperatures in $T(k)$ and predictions of future disturbances in $v(k)$ (obtained from the weather forecast service). The MPC will explicitly take into account the constraints of heating/cooling actuators and the temperature comfort limits while minimizing the energy inserted from the actuators in the one-day-ahead period. The objectives are: minimize the energy consumption to heating and cooling; minimize the peak power consumption; maintain the zones within a desired temperature range and maintain the used power within the green available bounds. At each time step, each one of the agents must solve is MPC problem.

$$\min_{U,\overline{\varepsilon},\underline{\varepsilon},\overline{\gamma},\underline{\gamma}} \sum_{k=0}^{N-1} u_{t+k|t}^2 \Delta t + \phi \max\left\{u_{t|t}^2,\ldots,u_{t+N-1|t}^2\right\} + \rho \sum_{k=1}^{N}\left(\overline{\varepsilon}_{t+k|t}^2 + \underline{\varepsilon}_{t+k|t}^2\right) + \psi \sum_{k=1}^{N}\left(\overline{\gamma}_{t+k|t}^2 + \underline{\gamma}_{t+k|t}^2\right) \tag{5}$$

Subject to the following constraints,

$$T_{t+k+1|t} = AT_{t+k|t} + Bu_{t+k|t} + v_{t+k|t}, \tag{6}$$

$$\underline{T} - \underline{\varepsilon}_{t+k|t} \le T \le \overline{T} + \overline{\varepsilon}_{t+k|t}, \tag{7}$$

$$\underline{U}_{A_i} - \underline{\gamma}_{t+k|t} \le U \le \overline{U}_{A_i} + \overline{\gamma}_{t+k|t}, \tag{8}$$

$$\underline{\gamma}_{t+k|t}, \overline{\gamma}_{t+k|t}, \underline{\varepsilon}_{t+k|t}, \overline{\varepsilon}_{t+k|t} \ge 0. \tag{9}$$

In (5), u represents the power control inputs, ϕ is the penalty on peak power consumption, ρ is the penalty on the comfort constraint violation, ψ the penalty on the power constraint violation and N is the length of the prediction horizon. In (7), $\overline{\varepsilon}$ and $\underline{\varepsilon}$ are the vectors of temperature violations that are above and below the desired comfort zone defined by $\overline{T}$ and $\underline{T}$. In (8), coupled constraint, $\overline{\gamma}$ and $\underline{\gamma}$ are the power violations that are above or lower the maximum, $\overline{U}_{A_i}$, and minimum, $\underline{U}_{A_i}$, available *green* power for heating/cooling the space, with $\underline{U}_{A_i} = -\overline{U}_{A_i}$ .A scheme of the system implemented is shown in the next picture.

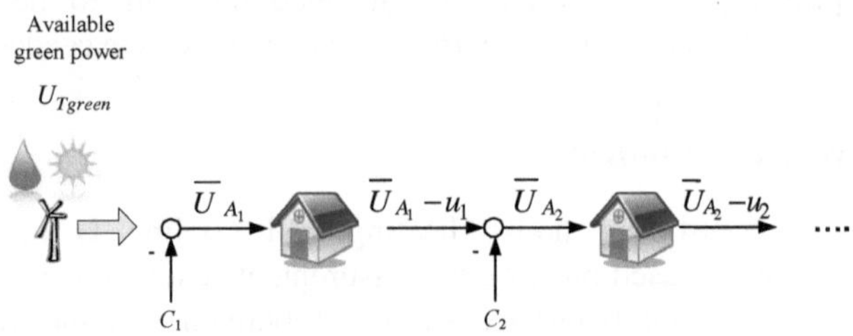

Fig. 4. Implemented system

$$C_i = [c_i(k),...,c_i(k+N)]^T , \tag{10}$$

$$U_{Tgreen} = [|u_{Tgreen}(k)|,...,|u_{Tgreen}(k+N)|]^T , \tag{11}$$

$$u_i = [|u_i(k)|,...,|u_i(k+N)|]^T , \tag{12}$$

where, for a generic Agent *i* at the control horizon, $\overline{U}_{A_i}$ represents the *green* available resource for indoor comfort , U_{Tgreen} represents the *green* available total resource, C_i the fixed consumption profile and u_i the used power to heating/cooling the space that results from the optimization program.

4 Results

The presented results were obtained with an optimization *Matlab* routine that finds a constrained minimum of a quadratic cost function that penalizes the sum of several objectives. It is considered that all houses have the same outdoor temperature presented in Fig. 1. The thermal characteristics, load disturbances profile (Fig. 4) and comfort temperature bounds are different for all houses (Table 2). Agents can also have distinct penalties on power and temperature constraints violations, they can hourly privilege comfort or cost according to consumer choice. Here, is assumed that the penalty values of each agent are always the same. Table 2 shows the used parameters.

In the figures here presented, the subtitles "*Power constraint*" represents $\overline{U}_{A_i} = \overline{U}_{A_{i-1}} - u_{i-1} - C_i$, "*Green resource*" represents $\overline{U}_{A_{i-1}} - u_{i-1}$ and "*Heating/Cooling used power*" represents u_i.

Table 2. Scenario parameters

Parameter	A_1	A_2	A_3	***Units***
R_{eq}	50	25	75	K/kW
C	9.2×10^3	4.6×10^3	11×10^3	kJ/K
ρ	100	100	300	-
ψ	500	200	300	-
Φ	2	2	2	-
Δt	1	1	1	H
N	24	24	24	-
$T(0)$	297.15	296.15	297.15	°K

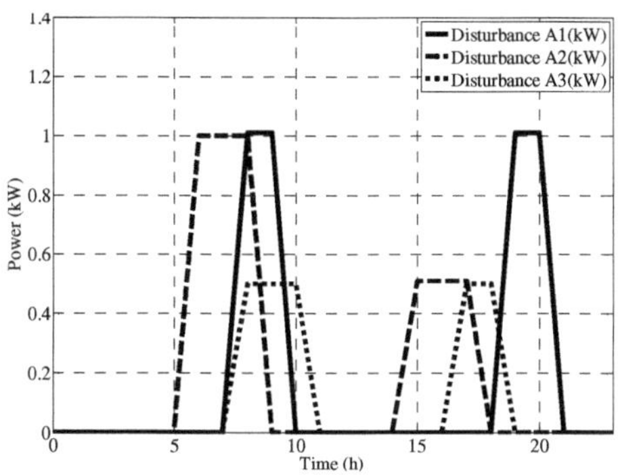

Fig. 5. Disturbance forecasting (P_d)

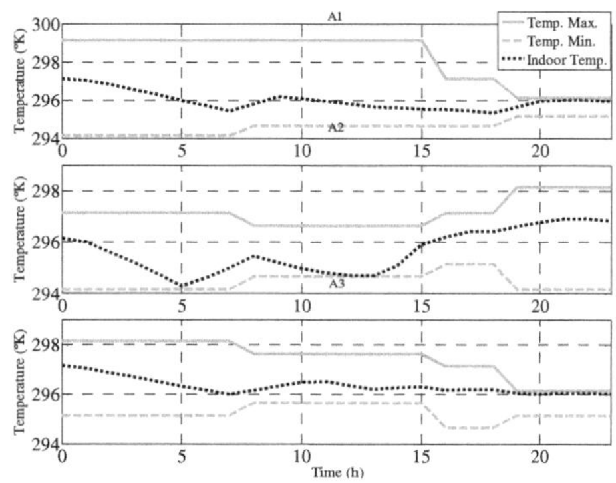

Fig. 6. A1, A_2 and A_3 indoor temperature and their constraints

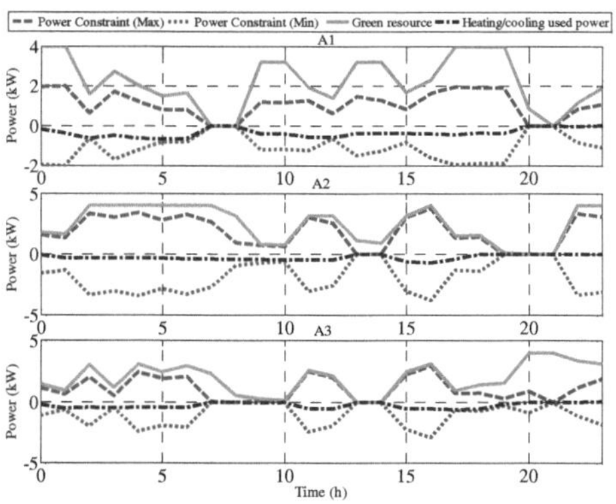

Fig. 7. Used power to heat/cool and their constraints

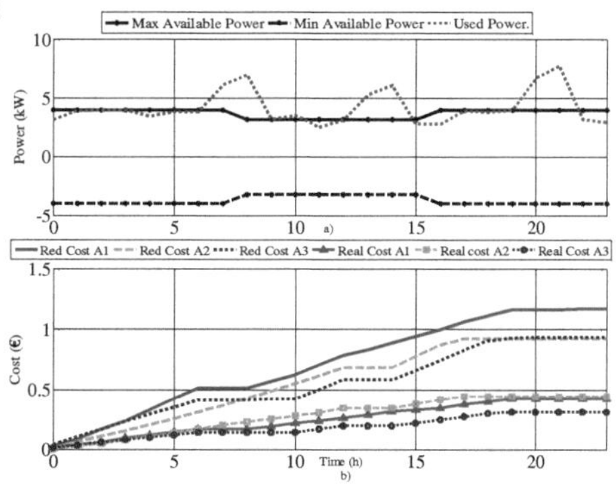

Fig. 8. a) Total available *green* resource and used power. b) Heating/cooling total cost.

In Figure 6, it can be seen that the comfort constraint is respected, the indoor temperature is always inside the comfort zone for all agents. Taking advantage of the predictive knowledge of the disturbance (P_d) and making use of the space thermal storage, it can also be seen that in both scenarios the MPC treats the indoor temperature before the disturbance beginning. The used power to heat/cool the space is maintained inside the constrained bounds. Note that when the "*Power Constraint*" is null the used power is also null, Fig. 7. The "Used Power", $U_{used}(k)=\sum_{i=1}^{3}\left[\left|u_i(k)\right|+C_i(k)\right]$, is sometimes above the daily maximum *green* available resource, meaning that the *red* resource was consumed, Fig 8(a). Figure 8(b) illustrates the effectiveness of the approach and demonstrates the advantage of the auction. For each one of the agents it can be seen that the "*Real Cost*" is much lower than the cost of not to bid in auction and only consume the *red* resource "*Red Cost*".

5 Conclusions

In this paper, a distributed MPC control technique was presented in order to provide thermal house comfort. The obtained solution solves the problem of control of multiple subsystems subject to coupled constraint that changes hourly. Each subsystem solves is own problem involving its own state predictions and the shared constraints. It could be observed through the simulations and results analysis that were obtained suitable dynamic performances. Despite access orders being changing hourly, the predictive characteristics of the implemented system were not lost, being the soft constraints, temperature and power satisfied. By changing the penalties values during the day, the implemented system also allows the consumer to shift hourly between indoor comfort and lower costs.

References

1. ESTTP. Solar Heating and Cooling for a Sustainable Energy Future in Europe
2. Rosen, C., Madlener, R.: An auction mechanism for local energy markets: Results from theory and simulation. In: Complexity in Engineering (COMPENG), pp. 1–4 (2012)
3. Ramachandran, B., Srivastava, S., Cartes, D., Edrington, C.: Distributed energy resource management in a smart grid by risk based auction strategy for profit maximization. In: 2010 IEEE Power and Energy Society General Meeting, pp. 1–7 (2010)
4. Perrod, P., Critchley, R., Catz, E., Bazargan, M.: New participants in SmartGrids and associated challenges in the transition towards the grid of the future. In: IEEE Bucharest Power Tech Conference, Bucharest, pp. 1–5 (2009)
5. Moroşan, Bourdais, R., Dumur, D., Buisson, J.: Building temperature regulation using a distributed model predictive control. In: American Control Conference (ACC), pp. 3174–3179 (2010)
6. Ma, Y., Kelman, A., Daly, A., Borrelli, F.: Predictive Control for Energy Efficient Buildings with Thermal Storage. IEEE Control System Magazine 32(1), 44–64 (2012)
7. Negenborn, R.: Multi-Agent Model Predictive Control with Applications to Power Networks. PhD Thesis, Technische Universiteit Delft, Nederland (2007)
8. Scattolini, R.: Architectures for distributed and hierarchical Model Predictive Control – A review. Journal of Process Control 19, 723–731 (2009)
9. Trodden, P., Richards, A.: Distributed model predictive control of linear systems with persistent disturbances. International Journal of Control 83(8) (2010)
10. Keviczky, T., Borrelli, F., Balas, G.: Decentralized Receding Horizon Control for Large Scale Dynamically Decoupled Systems. Automatica 42, 2105–2115 (2006)

Sensing Cloud Optimization to Solve ED of Units with Valve-Point Effects and Multi-fuels

Pedro Fonte[1,2], Claudio Monteiro[2], and Fernando Maciel Barbosa[2,3]

[1] ISEL – Lisbon Superior Engineering Institute, Lisbon, Portugal
pfonte@deea.isel.pt
[2] University of Porto
[3] INESC TEC PORTO, Porto, Portugal
{cdm,fmb}@fe.up.pt

Abstract. In this paper a solution to an highly constrained and non-convex economical dispatch (ED) problem with a meta-heuristic technique named Sensing Cloud Optimization (SCO) is presented. The proposed meta-heuristic is based on a cloud of particles whose central point represents the objective function value and the remaining particles act as sensors "to fill" the search space and "guide" the central particle so it moves into the best direction. To demonstrate its performance, a case study with multi-fuel units and valve- point effects is presented.

Keywords: Economic dispatch, Optimization, Heuristics, Cloud of particles.

1 Introduction

The Economical Dispatch (ED) problem is an important issue in the power system operation. Fundamentally it is intended to evaluate the value that each on-line unit should generate with the lowest cost, as (1), respecting the technical and load demand constrains. The ED uses as a basis the Unit Commitment (UC) solution, excluding from the optimization the generation units that are off. It is common to adopt approaches that merge the problem of UC with the problem of ED using Integer Mixed Programming optimization algorithms. This can have advantages because the non-linear detail in the ED can justify a change in the solution of the UC. On the other hand, the ED can integrate UC costs considering the fix costs of the generation [1].

$$\min \sum_{i=1}^{N_G} F_i\left(P_i\right) \tag{1}$$

Where F_i is the cost function of each unit P_i, and N_G represents the number of on-line units [2]. Over the past decade, many methods have been developed to solve the economical dispatch problem. There are traditional methods such as Gradient, Lagrangean function, Lambda-iteration method, Dynamic Programming, Newton's method, Linear Programming and Interior Point method, among others [3]. However, the generation cost functions of recent thermal units are not continues, not convex, neither differentiable due to valve-point loading effect, multi-fuel burn systems and

L.M. Camarinha-Matos, S. Tomic, and P. Graça (Eds.): DoCEIS 2013, IFIP AICT 394, pp. 477–484, 2013.

operational prohibited zones. Thus, the ED problem becomes a non-convex optimization problem with constrains, which cannot be solved directly by some of the traditional mathematical methods. Generally, dynamic programming can solve this kind of problems, but can suffer with the dimension and time needed to solve it [4-6]. Due to that several heuristic methods were proposed to solve this kind of problems such as Genetic Algorithms (GA), Simulated Annealing (SA), Taboo Search (TS), Evolutionary Programming (EP), Evolutionary Strategies (ES), Particle Swarm Optimization (PSO), Artificial Neural Networks approach with Hopfield Networks and hybrid artificial intelligence methods [7]. From the base algorithms several improved approaches and hybrid were proposed, as Improved Taboo Search [4], Fast Evolutionary Programming (FEP) and Improved Fast Evolutionary Programming (IFEP)[8], Improved Particle Swarm Optimization (IPSO)[7], PSO embedded Evolutionary Programming (PSO-EP) [9], PSO with crossover operations [7] and Fast Evolutionary programming with Swarm Direction [2]. Other techniques as Bacteria Foraging Optimization (BFO), Ant Colony Optimization (ACO) [3][11], Combination of Chaotic Differential Evolution and Quadratic Programming [12], a special class of ant colony (API) and Real Coded Genetic Algorithm (GA-API) can be considered as well [13] or Improved Genetic Algorithm with Multiplier Updating (IGA_MU) and Conventional Genetic Algorithm with Multiplier Updating (CGA_MU)[10], among many others. In this paper is proposed to solve an ED problem of units with multi-fuels and valve-point effects using the algorithm called Sensing Cloud Optimization (SCO)[14]. It is a stochastic technique based in a cloud of particles with parallel search without presenting evolutionary behaviors. There is no competition between the particles or self-adaptation of their characteristics being a purely cooperative system, since all contribute to reach the optimum value.

The heuristics techniques due to its stochastic behavior can be trapped in local minima; therefore, the added value of SCO is the dynamic mechanism to avoid, as better as possible, to be trapped in local minima [14]. From the work developed up to now, it has demonstrated precise results as well as the capacity to deal with large quantity of variables [16]. In this paper is intended to continue the research solving an ED of units with valve-point effects and multi-fuels. In future works we intend to develop an integrated set of mathematic techniques to optimise the operation of electric energy systems with significant penetration of high variability resources. The final result should be a complete methodological suit idealised to answer integrally to the dispatch of energy systems with high integration of variable power resources.

2 Contribution to Internet of the Things

The Economic Dispatch, Security Constrained Economic Dispatch (SCED) and Reliability Constraint Economic Dispatch (RCED) are some of the most important optimization problems in a power system scheduling. The result is the mix of thermal power plants production under technical and security constraints at the lowest price. The optimization of the electricity cost makes the society and economy more sustainable and competitive. In an Internet of the things perspective, the power units connected to the internet gain the capacity to communicate and share information among them. As well as, the capacity to configure and decide autonomously, when and how much produce, under an optimal perspective, optimizing the production cost and environmental impacts. This share of information allows technical decisions, maintaining the security of electrical energy supply. Beyond that, in a global perspective allows the scheduling of the decision chain since the extraction of fossil fuels, its transportation, distribution and utilization. The introduction of SCO algorithm aims to provide a tool with good precision to solve high dimensional and constrained ED problems.

3 Problem Description

Traditional ED problems characterise the cost function as quadratic, but in more recent thermal units, an absolute term is added to the quadratic cost function due to valve point effects. Beyond this, some thermal units, as combined cycle, operate with different kind of fuels and multiple cost curves, resulting an "hybrid cost function" as (2), represented with several piecewise functions reflecting the effects of fuel type k [10]. Where a_i, b_i, c_i, e_i and f_i are fuel cost coefficients of the ith unit, and P_i and P_i^{min} represent the production and minimum limit of ith unit, respectively [12].

$$F_i(P_i)=\begin{cases} a_{i1}+b_{i1}P_i+c_{i1}P_i^2+\left|e_{i1}\times\sin\left(f_{i1}\times\left(P_{i1}^{min}-P_{i1}\right)\right)\right|, & \text{for fuel 1, } P_{i1}^{min}\le P_i\le P_{i1} \\ a_{i2}+b_{i2}P_i+c_{i2}P_i^2+\left|e_{i2}\times\sin\left(f_{i2}\times\left(P_{i2}^{min}-P_{i2}\right)\right)\right|, & \text{for fuel 2, } P_{i1}^{min}< P_i\le P_{i2} \\ \vdots & \\ a_{ik}+b_{ik}P_i+c_{ik}P_i^2+\left|e_{ik}\times\sin\left(f_{ik}\times\left(P_{ik}^{min}-P_{ik}\right)\right)\right|, & \text{for fuel } k, P_{ik-1}< P_i\le P_i^{max} \end{cases} \quad (2)$$

In an Electric Power System, the total power production must equal the load demand D and power transmission losses P_L, as (3). The power losses are function of unit's power outputs and can be calculated by the losses coefficients matrix **B** [13].

$$\sum_{i=1}^{N_G} P_i = P_L + D \quad (3)$$

Simultaneously several technical operation constrains must be satisfied such as minimum and maximum power generation and ramp limits. The output of each unit must be within its maximum P_i^{max} and minimum limits P_i^{min}. The increasing ramp limit is expressed by UR_i and DR_i represents the decreasing ramp limit. Together with minimum and maximum limits, results (4), where P_i represents the power output of ith unit at a given time interval, and P_i^o the power at previous time interval,

$$\max\left(P_i^{min}, P_i^0 - DR_i\right) \le P_i \le \min\left(P_i^{max}, P_i^0 + UR_i\right) . \quad (4)$$

In some cases, the total operation range of a generating unit is not always available due to physical operation limitations, as steam valve operation, vibrations in the shaft bearings, among others, resulting in prohibited operating zones. In practical operation, the adjustment of power output P_i must avoid operating in the prohibited zones.

4 Algorithm

In [14] and [16], SCO showed appropriate heuristic characteristic to solve not convex, not differentiable and highly constrained optimisation problems. It is characterized by two distinct steps, the cloud particles fitness evaluation and a statistical analysis to determine the cloud's direction and dimension. The cloud's dimension presents a dynamic adjustment in search space in order to accelerate the convergence and to avoid to get trapped in local minima. It was introduced the concept of central particle, which tries to find the optimal value, being the remaining particles of the cloud spread around, according with a Gaussian distribution. By this way, the search space covered by the cloud can be dynamically controlled only by the variance of the distribution. These particles will act as sensors "to fill" the search space and give "signals" to the central particle moves into the best direction in the search space. The adaptive adjustment of the cloud's dimension is performed by two inverted sigmoid. The algorithm when applied to the ED problem can be described by the following 10 steps.

Step 1. Create randomly a central particle $P_{q(i)}$, with i=1..N_G dimensions, representing each generation unit according to its limits, as in (5) (If there are starting values, $P_{q(i)} = P_{(i)}^{o}$),

$$P_{q(i)} = rand\left(0,1\right).\left(P_i^{\max} - P_i^{\min}\right) + P_i^{\min} \ . \tag{5}$$

Create the remaining cloud [$N_G \times N_P$] with j = 1..N_P particles and i = 1..N_G dimensions normally distributed centered in the central particle and standard deviation $\sigma_{(i)}^{(k=1)}$.

Step 2. To each cloud particle $P_{(j)}^{(k)}$, calculate the transmission loss and evaluate each particle with (6) and retain the best fitness value $P_{(j)BEST}^{(k)}$ and its position.

$$f_{(j)}^{(k)} = \sum_{i=1}^{N_G} F_i\left(P_{(i,j)}^{(k)}\right) + \left(\sum_{i=1}^{N_G} P_{(i,j)}^{(k)} - D - P_L^{(k)}\right)^2 \tag{6}$$

In this work, to the traditional fitness function (1), a quadratic penalty has added to decrease the deviation between the power production and the sum of power demand and active losses.

Step 3. Calculate the second order regression coefficients to each dimension i, $[\beta_{0(i)}\ \beta_{1(i)}\ \beta_{2(i)}]$ and the determination coefficient $R_{q(i)}^{2(k)}$.

Step 4. Verify the convexity of polynomials. Depending of the function to be optimized and the search space covered by the cloud, the second-order fitness function can be concave indicating a maximum instead a minimum. In this case is necessary to calculate the trend point in order to ensure that the central point continues to move toward a minimum. This is done calculating the roots of the concave function subtracted by the best fitness, as indicated in (7) and choose the root closest of $P^{(k)}_{(j)BEST}$ by (8);

$$\beta_{0(i)} + \beta_{1(i)} P^{(k)}_{(i)} + \beta_{2(i)} P^{2(k)}_{(i)} - f^{(k)}_{(j)}\left(P^{(k)}_{(i)BEST}\right) = 0 \tag{7}$$

$$t^{(k)}_{p(i)} = \min\left\{\left|P^{(k)}_{(i)1} - P^{(k)}_{(i)BEST}\right|, \left|P^{(k)}_{(i)2} - P^{(k)}_{(i)BEST}\right|\right\} . \tag{8}$$

Otherwise, if it is convex, $t^{(k)}_{p(i)} = -{}^{b_i}\!/_{2a_i}$.

Step 5. Generate new central particle by (9) and verify if satisfies all constrains.

$$P^{(k+1)}_{q(i)} = t^{(k)}_{p(i)}.R^2_{q(i)} + \left(1 - R^2_{q(i)}\right).P^{(k)}_{(i)BEST} \tag{9}$$

If the central particle satisfies all constraints, then it is a feasible solution. (As remain particles acts as sensors and are not candidates to optimal solution they don't need to satisfy all constrains).

Step 6. Calculate the Euclidean distance for each particle j to the central particle, by (10)

$$d^{(k)}_{(j)} = \left\|P^{(k)}_{q} - P^{(k)}_{(j)}\right\| \tag{10}$$

Step 7. Calculate the linear regression coefficients $\left[\alpha_{o(j)}, \alpha_{1(j)}\right]$ of $P^{(k)}_{(j)} = f\left(d^{(k)}_{(j)}\right)$ and determinate the coefficient $R^{2(k)}_{\varphi(j)}$.

Step 8. Calculate the new standard deviation for each dimension i of new cloud by (11)

$$\sigma^{2(k+1)}_{(i)} = \sigma^{2(k)}_{(i)}.F_{s1(j)}.F_{s2(i,j)} . \tag{11}$$

The changing in cloud variance and, consequently, in the cloud dimension is done by two inverted sigmoid (12) and (13). In (12), $\Delta\varphi$ is calculated by (14) and (15).

$$F_{s1(k)} = 1 + \frac{\Delta\varphi - 1}{e^{\left(\frac{-8(t-tc)}{ts}\right)}} \tag{12}$$

$$F_{s2(i,k)} = 1 - \frac{(K-1)}{1 + e^{\left(-mR^2_{s(i)} + n\right)}} \tag{13}$$

$$\nabla\varphi = \frac{df\left(\varphi_{(j)}\right)}{d\left(\varphi_{(j)}\right)}\left({}^{\sigma_{\varphi(j)}}\!/_{\sigma_{fitness(j)}}\right) \tag{14}$$

$$\Delta\varphi = \frac{1}{1 + h.\nabla\varphi} \tag{15}$$

Step 9. If $k = it_{max}$ go to step 10, otherwise, $k=k+1$ and go to step 1.
Step 10. The central particle that generates the minimum value is the optimal generation power of each unit.

5 Numerical Examples and Results

One case is studied to verify the capacity to solve non-convex and constrained problems and reach the optimal values. As heuristic methods may not converge exactly to the same solution each run owing to their stochastic behaviour, their performances could not be judged by the results of a single trial. Due to that, the case study will be performed 50 times. In power systems literature, generally, the convergence is mainly related with number of iterations or generations [2], [6], [8-10] or CPU time per iteration /generation [4]. However, this way doesn't give adequate information about the computational effort to perform a task in order to have the same base of comparison [13]. Thus, in this paper to compare the computational efforts independently of the CPU or number of iterations, the numbers of objective and constraints functions evaluations are used.

5.1 Case Study

The test system consists of 10 thermal units with valve-point effects and multi-fuels without forbidden operation zones or ramp limits, neither power losses, feeding a load demand of 2700 MW. All required data is available in [10]. To solve this case, the central particle as well as, the remaining particles of the cloud will have 10 dimensions ($P_1...P_{10}$), one for each generation unit. Depending on the number of particles N_P, the dimension of cloud is [10 x N_P]. The number of iterations was limited to 200. As referred in [14] the number of particles has not excessive importance because there isn't a direct relation between the increase of its number and the increase of the performance. On the other hand, as they act as "sensors" in the search space and are used to calculate the first and second order polynomials, there is a minimum number necessary to describe the curve fitting. Due to this, after some experiences the particles number was set to 10. In (15) h was set to 1, in (12) t_c was set to 0,5 and t_s to 1 and in (13) the values of K, m and n were set, respectively to 1.02, 50 and 5.

As indicated in [10], each unit has different cost functions and operation limits depending on the fuel. The results will be the cost value obtained by the minimization of (7) subject to (4). The problem includes one objective function with 10 variables and 20 inequality constraints. In figure 1 is shown the convergence behavior being the best value reached after 1800 cost function evaluations.

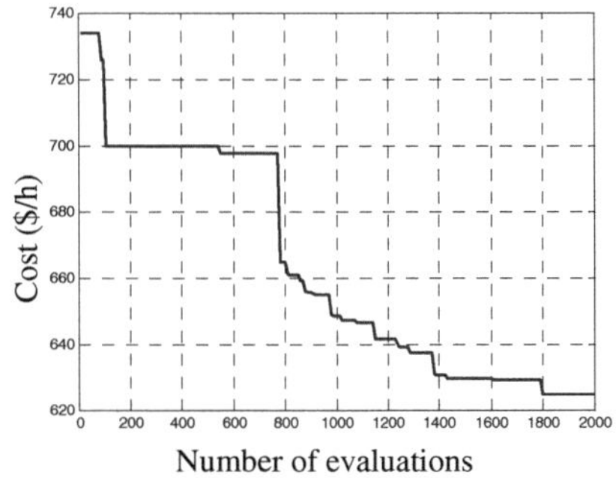

Fig. 1. Convergence behaviours of SCO for a 10-units problem

The results reached by SCO were compared with CGA_MU and IGA_MU[10] as well as with PSO-LRS, NPSO and NPSO_LRS[6], the results are shown in table 1. It can be concluded that all the proposed methods reach the same combination of fuels and SCO reached comparable values with other proposed metaheuristics.

Table 1. Results obtained (10-unis 2700 MW) (Best individual)

Power (MW)	*Method*						
	SCO	*CGA_MU*	*IGA_MU*	*PSO-LRS*	*NPSO*	*NPSO-LRS*	*fuel*
P_{G1}	225,612	222,0108	219,1261	219,1261	220,6570	223,3352	2
P_{G2}	207,103	211,6352	211,1645	211,1645	211,7859	212,1957	1
P_{G3}	281,576	283,9455	280,6572	280,6572	280,4026	276,2167	1
P_{G4}	237,912	237,8052	238,4770	238,4770	238,6013	239,4187	3
P_{G5}	271,667	280,4480	276,4179	276,4179	277,5621	274,6470	1
P_{G6}	242,476	236,0330	240,4672	240,4672	239,1204	239,7974	3
P_{G7}	284,633	292,0499	287,7399	287,7399	292,1397	285,5388	1
P_{G8}	243,321	241,9708	240,7614	240,76,14	239,1530	240,6323	3
P_{G9}	432,693	424,2011	429,3370	429,3370	426,1142	429,2637	3
P_{G10}	273,006	269,9005	275,8518	275,8518	274,4637	278,9541	1
P_T (MW)	2700	2700	2700	2700	2700	2700	
Cost ($/h)	624,65	624,72	624,52	624,23	624,16	624,13	
Evaluat.	1800	N/A	N/A	3440	3240	2120	

6 Conclusions and Future Works

This paper investigated the capability of SCO to solve highly non-convex and constrained ED problems as the case of units with valve-point effects and multi-fuels. The values reached are comparable with evolutionary methods as PSO and GA and other hybrid solutions. The compared algorithms are hybrid models or evolution from base algorithms. SCO is the first approach which in future could be improved and achieve better performances.

References

1. Catalão, J.P.S.: Electric Power Systems: Advanced Forecasting Techniques and Optimal Generation Scheduling. CRC Press (2012)
2. Gaing, Z.-L., Ting-Chia, O.: Dynamic Economic Dispatch Solution Using Fast Evolutionary Programming with Swarm Direction. In: ICIEA 2009 – 4th IEEE Conference on Industrial Electronics and Applications, vol. (1), pp. 1538–1544 (2009)
3. Hazra, J., Sinha, A.: Application of soft computing methods for Economic Dispatch in Power Systems. International Journal of Electrical and Electronics Engineering (2), 538–543 (2009)
4. Gaing, Z.-L.: Particle swarm optimization to solving the economic dispatch considering the generator constraints. IEEE Transactions on Power Systems 18(3), 1187–1195 (2003)
5. Lin, W.-M., Cheng, F.-S., Tsay, M.-T.: An improved tabu search for economic dispatch with multiple minima. IEEE Transactions on Power Systems 17(1), 108–112 (2002)
6. Selvakumar, A.I., Thanushkodi, K.: A New Particle Swarm Optimization Solution to Nonconvex Economic Dispatch Problems. IEEE Transactions on Power Systems 22(1), 42–51 (2007)
7. Lee, K.Y.: An Improved Particle Swarm Optimization for Nonconvex Economic Dispatch Problems. IEEE Transactions on Power Systems 25(1), 156–166 (2010)
8. Sinha, N., Chakrabarti, R., Chattopadhyay, P.K.: Evolutionary programming techniques for economic load dispatch. IEEE Transactions on Evolutionary Computation 7(1), 83–89 (2003)
9. Sinha, N., Purkayastha, B.: PSO embedded evolutionary programming technique for nonconvex economic load dispatch. In: IEEE PES Power Systems Conference and Exposition, vol. 1, pp. 66–71 (2004)
10. Chiang, C.-L.: Improved Genetic Algorithm for Power Economic Dispatch of Units With Valve-Point Effects and Multiple Fuels. IEEE Transactions on Power Systems 20(4), 1690–1699 (2005)
11. Serapião, A.: Fundamental of Optimization by Swarm Inteligence: an Overview, Fundamentos de Otimização por Inteligência de Enxames: uma Visão Geral. Revista Contole & Automação 20(3) (July-September 2009) (in Portuguese)
12. Coelho, S., Mariani, V.C.: Combining of Chaotic Differential Evolution and Quadratic Programming for Economic Dispatch Optimization With Valve-Point Effect. IEEE Transactions on Power Systems 21(2), 989–996 (2006)
13. Ciornei, I., Kyriakides, E.: A GA-API Solution for the Economic Dispatch of Generation in Power System Operation. IEEE Transactions on Power Systems 27(1), 233–242 (2012)
14. Fonte, P.M., Monteiro, C., Barbosa, F.M.: SCO – Sensing Cloud Optimization – A Novel Optimization Technique (in reviewing)
15. Park, J.-B., Lee, K.-S., Shin, J.-R., Lee, K.Y.: A Particle Swarm Optimization for Economic Dispatch With Nonsmooth Cost Functions. IEEE Transactions on Power Systems 20(1), 34–42 (2005)
16. Fonte, P., Monteiro, C., Barbosa, F.M.: SCO – Sensing Cloud Optimization Applied to a Non-convex and Constrained Dynamic Economical Dispatch. IEEE Transactions on Power Systems (in reviewing)

Impact of Component Losses on the Efficiency of a New Quasi-Z-Source-Based Dual Active Bridge

Viktor Beldjajev, Indrek Roasto, and Janis Zakis

Tallinn University of Technology, Tallinn, Estonia
vbeldjajev@gmail.com, {Indrek.Roasto,Janis.Zakis}@ieee.org

Abstract. The paper analyzes the impact of the component losses on the efficiency of the novel DC/DC converter. The converter is a combination of the quazi-Z-source (qZS) network and dual active bridge (DAB). In the analysis the mathematical loss models of the proposed DC/DC converter are derived and efficiency is estimated. Eventually the efficiency is verified experimentally.

Keywords: quazi-Z-source inverter, dual active bridge, power electronic transformer.

1 Introduction

During the last decades the bi-directional DC/DC converters have become key components in alternative energy systems. The aim of such converters is to control the electric power flow between the sources and loads, keeping the output voltage on the required level. This paper proposes a novel bi-directional DC/DC converter topology that consists of the quazi-Z-source network, dual active bridge and a high frequency transformer. A possible application of such converter is the isolation stage of the new power electronic transformer (PET) topology where the power flow control capability between two different voltage buses and high efficiency are the major requirements [1-3]. In general, the power flow through the transformer can be controlled by means of voltage elevation on one side or by applying a phase shift control of the inverter bridges on both sides of the transformer. The phase shift control of DAB allows the zero voltage switching to be achieved, that results in reduced switching losses and higher efficiency. In addition, the voltage boost properties of the qZS network allows the voltage on the DC bus to be kept on the constant level that results in lower current stress on the switches. This paper analyzes the converter operation under such conditions when the power flow is controlled only by the phase shift technique. First the loss models are presented according to the waveforms and steady state analysis. Also the impact of the conduction losses of the components on the efficiency characteristics is presented. The obtained efficiency is verified experimentally.

2 Relationship to Internet of Things

Nowadays, the global power system can be described by steadily growing number of renewable energy resources and emerging DC loads on the residential level.

L.M. Camarinha-Matos, S. Tomic, and P. Graça (Eds.): DoCEIS 2013, IFIP AICT 394, pp. 485–492, 2013.

Continuously developing semi-conductor technology and information internet has inspired the idea of the “Energy Internet” concept that may change energy industry from the centralized system to the client-based distributed infrastructure, improving this way the efficiency of the power grid through optimal management of the energy routers. Energy router, a PET-based device that exchanges the energy between the sources, storage devices, DC loads and end-users, is one of the most critical elements of the Energy Internet. The PET is expected to have a bi-directional power flow, high power conversion and power quality enhancement capabilities, plug and play interface and optimal energy management [4]. The paper contributes to the development of the energy router, by analyzing the suitable DC/DC converter for the power conversion.

3 Operating Principle of the Converter

The circuit of the proposed converter, including the conduction losses in the components is shown in Fig.1. Following conduction losses are considered: the collector-to-emitter voltage drop of the IGBT (U_{CE}), diode voltage drop (U_D) and the winding resistance of the inductors L_1 and L_2 (r_L). The converter can operate in both directions, in the forward and in the reverse operating mode. In the forward operating mode the energy is transferred from the HV side towards the LV side. The transistors $T_5 \dots T_8$ on the HV side and $T_1 \dots T_4$ on the LV side are switched in pairs using the positive phase shift angle φ. In the forward operating mode the diode *D*in the qZS network must be shunted with a switch *S* in order to allow the power flow to the LV side. In the reverse operating mode the energy is transferred from the LV side towards the HV side. The transistors $T_1 \dots T_4$ on the LV side and transistors $T_5 \dots T_8$ on the HV side are switched in pairs using the negative phase shift angle φ as shown in Fig.2*a*. The switch *S* must be in the opened state. The voltage on the LV side DC link can be elevated by the qZS network. Therefore a shoot-through switching state of the inverter switches is introduced, when both switches on one leg (or all 4 switches) conduct simultaneously [5]. During the shoot-through state the energy is stored in the inductors L_1 and L_2 and transferred to the DC link during the active state. The research in [6] stated that adding additional phase shift angle φ to the HV side bridge gate signals during the shoot-through state on LV side allows the power of the converter to be adjusted. Proposed technique is shown inFig.2*b*. Moreover, the additional phase shift can reduce the conduction losses in the transformer that are caused by the circulating current during the shoot through state of the low voltage side inverter (transformer is short circuited).Considering N_{TR} as the transformer’s turns ratio, the best performance of the phase shift control and the lowest current stress on the switches is achieved if the converter is designed according to the requirement

$$U_{DC} = N_{TR} U_{HV} \tag{1}$$

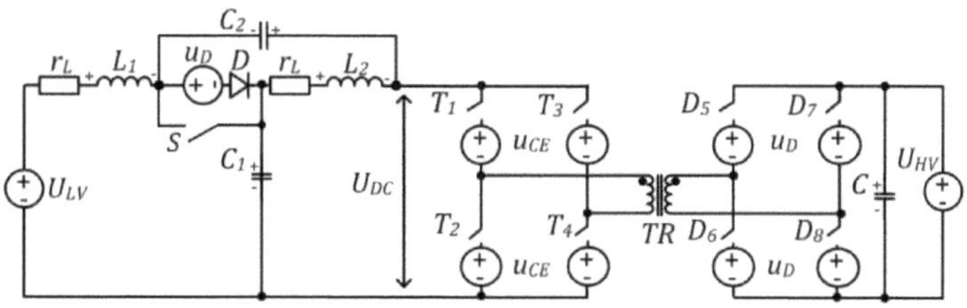

Fig. 1. Circuit of qZS-DAB converter with the losses in the components

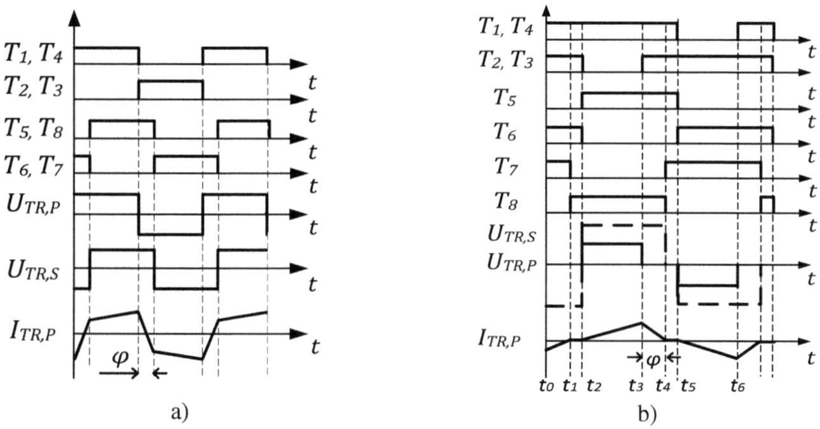

Fig. 2. Waveforms of the qZS-DAB converter in the reverse operating modes: a) operation under normal conditions, b) voltage boost mode

4 Impact of the Component Losses on the Output Power

In order to facilitate the impact of the component losses on the output power of the qZS based DAB, the mathematical models were derived based on the steady state analysis of the qZS inverter [7] and DAB [8]. Selected circuit parameters are shown in Table 1.

Table 1. The circuit parameters of the qZS-DAB converter

Parameter	Value
HV side voltage U_{HV}	90 V
LV side voltage U_{LV}	30 V
Switching frequency f_s	20 kHz
MF transformer turns ratio N_{TR}	1/3
Leakage inductance of MF transformer L_{TR}	10μH
Resistance of the inductors L_1 and $L_2 r_L$	150mΩ
IGBT saturation voltage U_{CE}	1.8 V
Diode voltage drop U_D	1.1 V

4.1 Normal Mode

In the normal mode the voltage on the DC link is at the desired level and conventional phase-shift control method is used to transfer the power. The operating period T consist of the active state of the transformer and power transferred from the low voltage side to the high voltage side can be evaluated as follows

$$P_{loss} = \frac{N_{TR}(U_{HV} - 2U_D)\cdot(U_{LV} - 2U_{CE})D_\varphi(1-|D_\varphi|)}{2f_S L_{TR}}, -1 \le D_\varphi \le 1. \quad (2)$$

where D_φ is the phase shift duration, f_S is the switching frequency and L_{TR} is the leakage inductance of the medium frequency (MF) transformer.

4.2 Voltage Boost Mode

In the voltage boost mode the voltage on the DC link can be elevated to match the desired level. The operating period T of the MF transformer consists of an active state t_A and a shoot through state t_S that can also be represented with the duty cycles

$$\frac{t_A}{T} + \frac{t_S}{T} = D_A + D_S = 1. \quad (3)$$

A phase shift D_φ is added to gate signals to adjust the power level of the converter as shown in Fig 2*b*.The power transferred from the LV side to the HV side can be evaluated as follows

$$P_{loss} = \frac{(U_{DC} - 2U_{CE})(2N_{TR}(U_{HV} - 2U_D) - (U_{DC} - 2U_{CE}))D_\varphi(1 - D_S)}{4f_S L_{TR}}. \quad (4)$$

In order to estimate the DC link voltage and thus the output and input power of the converter the equation system (5) can be composed based on the shoot-through (see Fig. 3*a*) and active (see Fig. 3*b*) switching states.

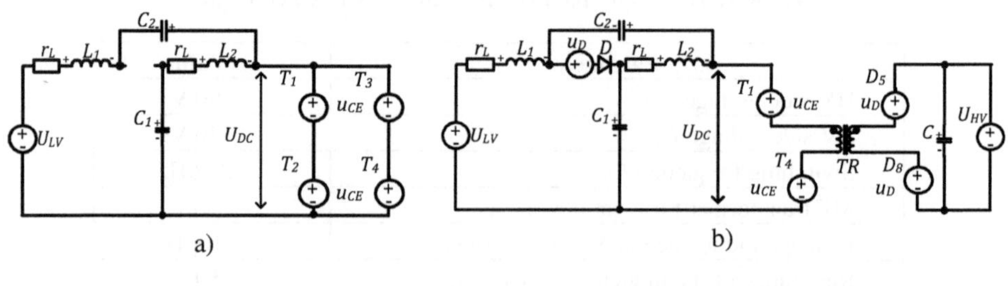

Fig. 3. Switching states of qZS in voltage boost mode: a) shoot through state, b) active state

$$\begin{cases} u_{L1} = (U_{C2} + U_{LV} - U_{CE} - I_L r_L)D_S + (U_{LV} - U_{C1} - U_D - I_L r_L)(1 - D_S) = 0 \\ u_{L2} = (U_{C1} - U_{CE} - I_L r_L)D_S + (-U_{C2} - U_D + I_L r_L)(1 - D_S) = 0 \\ U_{C1} + U_{C2} + U_D - U_{DC} = 0 \\ i_{C2} = I_L D_S + (I_{TR} - I_L)(1 - D_S) = 0 \end{cases} \tag{5}$$

Solving the equation system the DC link voltage can be obtained as follows

$$U_{DC} = \frac{U_{LV} - 2I_{TR} r_L D_S (1 - D_S) - U_D - 2U_{CE} D_S}{1 - 2D_S}, \tag{6}$$

where the I_{TR} is the average transformer current during half-period and it can be obtained from (4) and (5) in a following way

$$I_{TR} = \frac{P_{loss}}{U_{DC} - 2U_{CE}}. \tag{7}$$

As can be seen from (6) the impact of component losses on the U_{DC} depends on the shoot-through duty cycle duration. According to (7) the transformer average current depends on the transferred power and is inversely proportional to the voltage value on the DC link and voltage drop of the IGBT.

Typicallya drawback of the DAB is the increase of the current stress due to the voltage variation on the DC buses, however the DAB extended with the additional qZS network allows the current stress on the switches to be reduced. The current stress on the switchesT_1 and T_5, obtained with the simulation results for determined circuit parameters, is presented in Table 2. According to the simulation results, the current stress on the LV side switches can be reduced from 29 A in case of the DAB to 22 A when the qZS-DAB operates in the voltage boost mode. Moreover, in the voltage boost mode the current on the HV side switches can be reduced from 3A to 0.5A that in turn results in smaller conduction and switching losses.

Table 2. Current stress in the switches

	I(T$_1$)	**I(T$_5$)**
Conventional DAB	29.3 A	3.07 A
qZS-DAB (normal mode)	27.6 A	3.4 A
qZS-DAB (boost mode)	22 A	0.5 A

5 Impact of the Component Losses on the Efficiency

In order to facilitate the influence of the component losses on the efficiency of the qZS based DAB the derived mathematical models were analyzed for different component values at varying phase shift duty cycles. The output and input power was estimated and the efficiency was obtained according to the following equation

$$\eta = \frac{P_{OUT}}{P_{IN}} \cdot 100\% . \tag{11}$$

The Fig.4 depicts the influence of the U_{CE} on the efficiency of the converter for different phase shift values. It can be seen that the increase of the U_{CE} from 0…3 V decreases the efficiency from 0.95…0.73. Increasing the input voltage would result in a higher efficiency and relatively smaller impact of the U_{CE}. The influence of the active resistance of the inductors is shown in the Fig.5. For the given circuit parameters the increase of the r_L from 0…1 reduces the converter efficiency from 0.85…0.61 for normal mode and from 0.82…0.77 for the voltage boost mode. The influence of the diode voltage drop on the efficiency is shown in the Fig.6. It can be seen that the increasing the diode voltage drop U_D the efficiency decreases from 0.86…0.73.

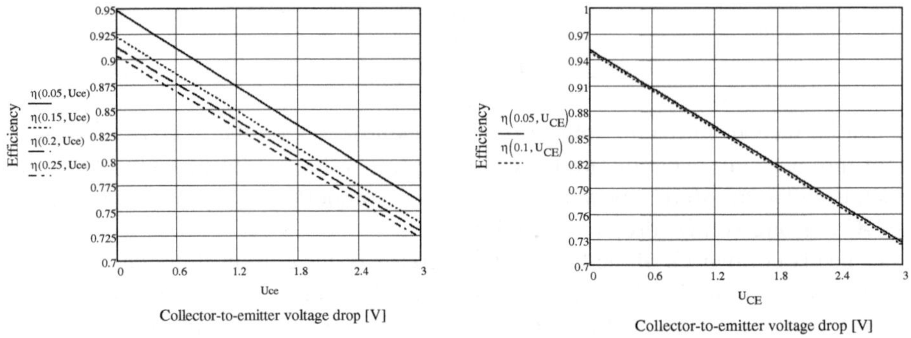

a) b)

Fig. 4. Impact of the collector-to-emitter voltage drop on the efficiency: a) normal mode; b) voltage boost mode

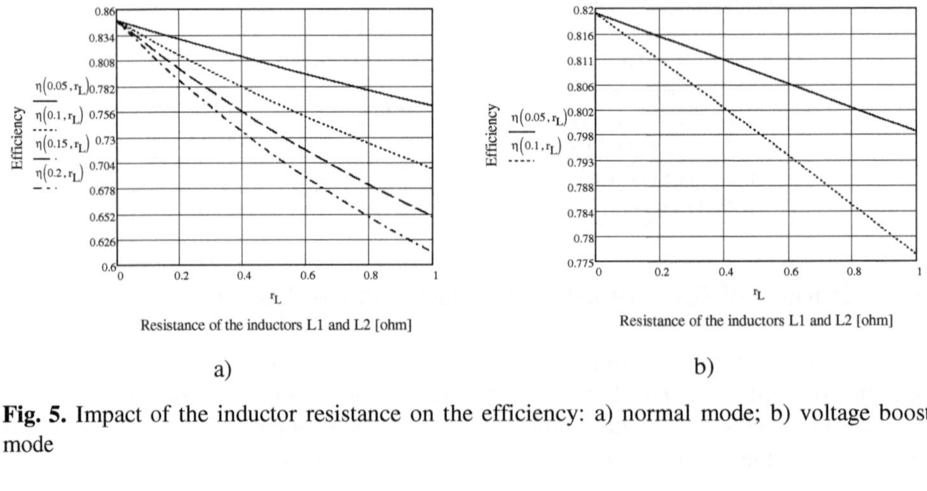

a) b)

Fig. 5. Impact of the inductor resistance on the efficiency: a) normal mode; b) voltage boost mode

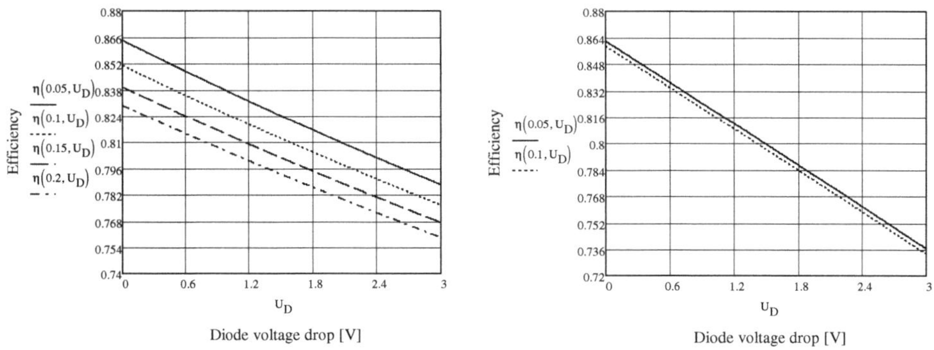

a) b)

Fig. 6. Impact of the diode voltage drop on the efficiency: a) normal mode; b) voltage boost mode.

6 Experimental Results

The experimental results were carried out on the experimental prototype for the normal and voltage boost modes. The IGBT switches IRG7Ph42ud1pbf were used in the DAB. For the normal mode, the phase shift angle of $\varphi = 10°$ was used. For the voltage boost mode the gate signals with duty cycle $D = 0.6$ were applied on the switches, that resulted in the shoot-through duty cycle $D_S = 0.1$. The phase shift in this case was maximal $D_\varphi = D_S$. The voltage and current measurement results for $P = 250$ W are shown in Fig. 7. According to the parameters listed in Table 1. and the dependency shown in chapter 5, the estimated efficiency of the converter in both modes is roughly 83 %. Measurements have shown 78 % efficiency of the converter with given parameters. Obviously the measured efficiency is lower since the switching losses and core losses were not taken into account in the analysis. Also on the background of the low input voltage the U_{CE} has relatively bigger impact than for the higher voltages.

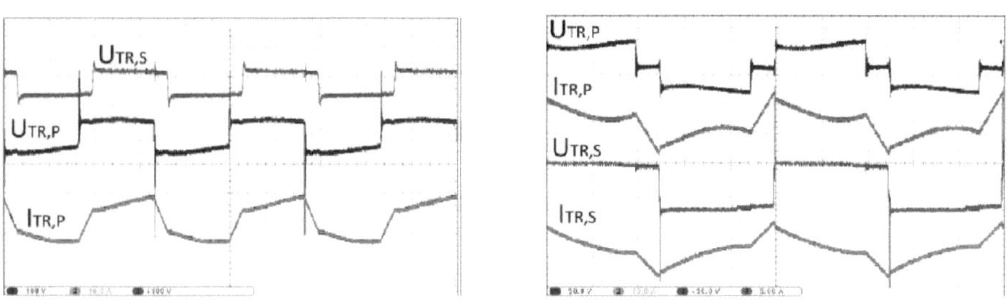

a) b)

Fig. 7. Waveforms showing input voltage (U_{LV}) and current (I_{LV}) as well as output voltage (U_{HV}) and current (I_{HV}) in normal mode (a) and voltage boost mode (b)

7 Conclusion

The paper presented a loss analysis of the new qZS based DAB. The mathematical loss models of the converter were derived for normal operating mode, using conventional DAB control, and voltage boost mode, when the shoot-through switching state was used to step up the voltage on the DC link. The mathematical models were analyzed for different circuit parameters. The results have shown that U_{CE} and U_D have significant impact on the efficiency of the converter. For this reason the components with as low values should be selected, for example MOSFET switches if the converter operates with low voltage. Moreover, the simulation results verified the decrease of the current stress on the switches. The estimated efficiency for the selected circuit parameters was roughly 83%. The experimental results in the nominal operating point showed the 78% efficiency, which is smaller in comparison with obtained efficiency, since the analysis did not consider the switching and core losses.

Acknowledgment. This research and work has been supported by Estonian Ministry of Education and Research (Project SF0140016s11) and Estonian Science Foundation (Grant ETF8687).

References

1. Beldjajev, V., Roasto, I., Lehtla, T.: Intelligent Transformer: Possibilities and Challenges, RTU, p. 14 (October 2011)
2. Heinemann, L., Mauthe, G.: The universal power electronics based distribution transformer, an unified approach. In: 2001 IEEE 32nd Annual Power Electronics Specialists Conference, PESC, vol. 2, pp. 504–509 (2001)
3. Roasto, I., Romero-Cadaval, E., Martins, J., Smolenski, R.: State of the Art of Active Power Electronic Transformers for Smart Grids. In: 38th Annual Conference of IEEE Industrial Electronics, IECON 2012, October 25-28 (2012)
4. Zhang, J., Wang, W., Bhattacharya, S.: Architecture of solid state transformer-based energy router and models of energy traffic. In: 2012 IEEE PES Innovative Smart Grid Technologies (ISGT), January 16-20, pp. 1–8 (2012)
5. Vinnikov, D., Roasto, I., Strzelecki, R., Adamowicz, M.: Step-Up DC/DC Converters With Cascaded Quasi-Z-Source Network. IEEE Transactions on Industrial Electronics 59(10), 3727–3736 (2012)
6. Krismer, F.: Modeling and optimization of bidirectional dual active bridge DC-DC converter topologies. ETH (2010)
7. Vinnikov, D., Roasto, I.: Impact of component losses on the voltage boost properties and efficiency of the qZS-converter family. In: 2011 7th International Conference-Workshop Compatibility and Power Electronics (CPE), June 1-3, pp. 303–308 (2011)
8. Mi, C., Bai, H., Wang, C., Gargies, S.: Operation, design and control of dual H-bridge-based isolated bidirectional DC-DC converter. IET Power Electronics 1(4), 507–517 (2008)

Part XVII
Telecommunications

Channel Availability Assessment for Cognitive Radios

António Furtado[2,3], Miguel Luís[1,2], Rodolfo Oliveira[1], Rui Dinis[1,2], and Luis Bernardo[1]

[1] CTS, Uninova, Dep.º de Eng.ª Electrotécnica, Faculdade de Ciências e Tecnologia, FCT, Universidade Nova de Lisboa, 2829-516, Caparica, Portugal
[2] IT, Instituto de Telecomunicações, Portugal
[3] ISR, Instituto de Sistemas e Robótica, Portugal
{a.furtado,nmal}@campus.fct.unl.pt,
{rado,rdinis,lflb}@fct.unl.pt

Abstract. This work addresses two problems related with the assessment of channel availability for cognitive radio systems. We start to characterize the performance of an energy detector for the case when PUs can change their state during the sensing period. The theoretical performance is validated through simulations and compared with a theoretical model where the PUs' state remains constant during the sensing period. The second point addressed in this work is the characterization of the channel availability, which is based on the output of the energy detector weighted by the probabilities of detection or false alarm computed in real-time. Several scenarios were evaluated and for each scenario the channel availability was correctly assessed by the SUs.

Keywords: Cognitive radio networks, Energy Sensing, System Analysis.

1 Introduction

Cognitive Radio (CR) has been proposed as a solution to alleviate the increasing demand for radio spectrum [1]. The nodes equipped with CRs, usually denominated Secondary Users (SUs), must be aware of the activity of the licensed users, denominated Primary Users (PUs), in order to dynamically access the spectrum without causing them harmful interference.

Spectrum Sensing (SS) aims at detecting the availability of vacant portions (holes) of spectrum and has been a topic of considerable research over the last years [1]. It plays a central role in CR systems. The traditional SS techniques include Waveform-based sensing (WBS) [2], a coherent technique that consists on correlating the received signal with *a priori* known set of different waveform patterns; Matched Filter-based sensing (MFBS) [3], an optimal sensing scheme where the received signal is also correlated with a copy of the transmitted one; and Cyclostationarity-based sensing (CBS) [4], a technique that exploits the periodic characteristics of the received signals, *i.e.*, carrier tones, pilot sequences, etc. MFBS assumes prior knowledge of the primary's signal, while WBS assumes that the received signal matches with one of the patterns previously known. This means that these sensing

L.M. Camarinha-Matos, S. Tomic, and P. Graça (Eds.): DoCEIS 2013, IFIP AICT 394, pp. 495–504, 2013.

techniques are not feasible in some bands, where several communication technologies may operate without *a priori* knowledge. On the other hand, CBS is impracticable for signals that do not exhibit cyclostationarity properties.

Energy-based sensing (EBS) [5], [6] is the simplest spectrum sensing technique and its main advantage is related with the fact that it does not need any *a priori* knowledge of PU's signal. At the same time, it is well known that EBS can exhibit low performance in specific comparative scenarios [7], or when noise's variance is unknown or very large. EBS has been studied in several CR scenarios, namely on local and cooperative sensing schemes [1]. More recently, several EBS schemes adopting sub-Nyquist sampling have been proposed, which are advantageous in terms of the sensing duration [8]. However, they consider that PUs only change their behavior in the beginning of the sensing period, which is a quite unrealistic assumption because they consider PU's synchronization with SUs.

In this work we characterize the performance of the EBS, not only for the scenario where PUs presents a constant behavior during the sensing period, but also for the scenarios where PUs can change their ON/OFF state, during the sensing period, which impacts on the spectrum sensing decision. In addition to the characterization of the EBS performance, we also characterize the channel availability, based on the output of the Energy Detector (ED) and for a scenario where the PUs exhibit a constant behavior (ON/OFF) during the sensing period.

The rest of this paper is organized as follows. The next section highlights the main contributions of this work. Section 3 presents the proposed system model and Section 4 validates the described probabilities of detection and false alarm and presents the simulation results of the probability of the channel availability. Finally some concluding remarks are given in Section 5.

2 Contribution to the Development of the Internet of Things

As it is well known, Internet of Things will be supported most of the times by wireless access technologies. Wireless communication technologies allow high level of device's mobility, and are indicated for scenarios where mobility is required. At the same time, wireless technologies avoid the use of physical (wired) connections, being an effective solution that will accelerate the implementation of such a network. As mentioned in the introduction, Cognitive Radio was proposed as a solution to alleviate the increasing demand for radio spectrum and, consequently, as more spectrum becomes available more wireless devices can access the network. Consequently, CR will help the development of the Internet of Things by increasing the number of wireless devices that may access the network.

3 System Overview

This work considers a cognitive radio network with a pair of PUs accessing the channel and a pair of SUs that access the channel in an opportunistic way. SUs are equipped with a single radio transceiver. However, because SUs are unable to distinguish SUs and PUs' transmissions, SU's operation cycle includes the sensing and transmission periods, which facilitates the synchronization of the sensing task. Sensing and transmission period durations are represented by T_S^{SU} and T_D^{SU} respectively, as illustrated in Figure 1.

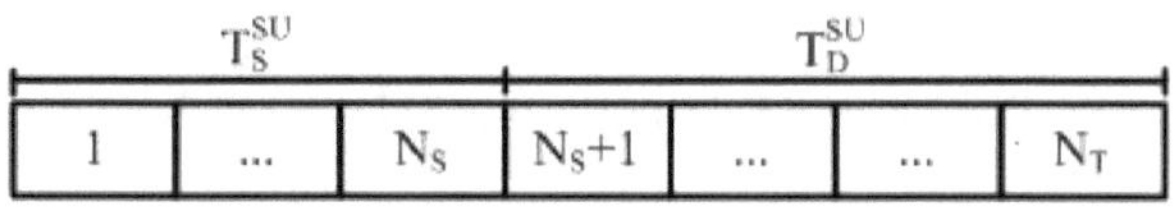

Fig. 1. SU's frame structure representing SU's operation cycle

The SU's frame, $T_F^{SU} = T_S^{SU} + T_D^{SU}$, contains N_T slots, where each slot duration is given by the channel sampling period adopted in the spectrum sensing task. The first N_S slots define the sensing period duration, and the remaining ones ($N_S + 1$ to N_T) represent the transmission period duration. It is assumed that SUs always have data to transmit and all SUs are synchronized.

3.1 Constant PU's Behavior

The assumption of constant PU's behavior adopted in several works [6], [9]-[11], indicates the case when PUs maintain their behavior during the SU's frame duration (equivalent to assume that PUs are synchronized with SU's operation cycle). Under this assumption, PUs will not change their activity state during the SU's frame, *i.e.*, the beginning of PU's transmission always matches with the beginning SU's sensing period. To distinguish between occupied and vacant spectrum bands, SUs sample the channel during the sensing period T_S^{SU}, and for each sample k two hypotheses can be distinguished

$$\begin{array}{ll} \mathcal{H}_{00}: x(k) = w(k) & k = 1,2,\dots,N_S \\ \mathcal{H}_{11}: x(k) = w(k) + s(k) & k = 1,2,\dots,N_S, \end{array} \tag{1}$$

where $s(k)$ denotes the signal transmitted by the PUs, with distribution $\mathcal{N}(\mu_s, \sigma_s^2)$. $w(k)$ is assumed to be a zero-mean additive white Gaussian noise (AWGN), *i.e.*, $w(k) = \mathcal{N}(0,1)$. The condition $\mathcal{H}_{00}$ represents the case when PUs are absent, while $\mathcal{H}_{11}$ indicates the opposite behavior. During the detection stage, each SU calculates the amount of energy received in the N_S samples $Y = \sum_{k=1}^{N_S} |x(k)|^2$, being then compared with the energy threshold γ to decide whether a PU is present or absent. Since Y follows a Chi-square distribution [9], and assuming that the number of samples N_S is large enough, the Central Limit Theorem (CLT) can be used to approximate the Chi-square distribution to a Gaussian distribution [12]:

$$Y \sim \begin{cases} \mathcal{N}(N_S, 2N_S), & \mathcal{H}_{00} \\ \mathcal{N}\big(N_S + \lambda, 2(N_S + 2\lambda)\big), & \mathcal{H}_{11} \end{cases}, \tag{2}$$

where $\lambda = \sum_{k=1}^{N_S} \big(\mu_s/(1+\sigma_s)\big)^2 = \sum_{k=1}^{N_S} \lambda'$ is the noncentrality parameter and represents the sum of N_S samples of SNR collected from the channel during the sensing period. Therefore, for a single SU, the probability of detection $P_D^{\mathcal{H}_{11}}$ and the probability of false alarm $P_{FA}^{\mathcal{H}_{00}}$ are represented by

$$P_D^{\mathcal{H}_{11}} = Pr(Y > \gamma|\mathcal{H}_{11}) = \mathcal{Q}\left(\frac{\gamma - (N_S + \lambda)}{\sqrt{2(N_S + 2\lambda)}}\right) \tag{3}$$

$$P_{FA}^{\mathcal{H}_{00}} = Pr(Y > \gamma|\mathcal{H}_{00}) = \mathcal{Q}\left(\frac{\gamma - N_S}{\sqrt{2N_S}}\right), \tag{4}$$

where $\mathcal{Q}(.)$ is the complementary distribution function of the standard Gaussian.

3.2 Non-constant PU's Behavior

Here we consider the case when PUs can randomly arrive or depart during the entire SUs' frame, which may occur during T_S^{SU} or T_D^{SU}. When the change occurs during the spectrum sensing period T_S^{SU}, two scenarios must be considered. As illustrated in Figure 2(a), during the spectrum sensing task the PU's behavior may change from active to inactive, where the first N_G slots denote the presence of PUs, and the remaining ones ($N_G + 1$ to N_S) represent the absence of PUs.

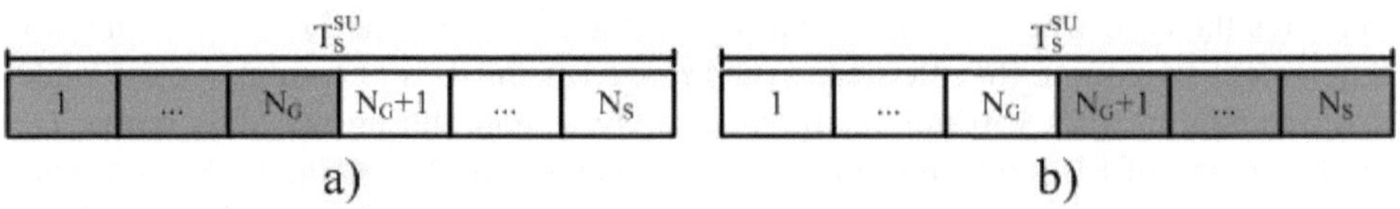

Fig. 2. PU's activity change: **(a)** from active to inactive; **(b)** from inactive to active. Gray slots denote PU's activity.

On the other hand, Figure 2(b) illustrates the opposite scenario when PUs may change from inactive to active during the sensing period. These scenarios are respectively represented by hypotheses $\mathcal{H}_{10}$ and $\mathcal{H}_{01}$. The output of the energy detector admits four hypotheses, rather than two. The output of the energy detector under the hypothesis $\mathcal{H}_{01}$ is given by

$$Y_{\mathcal{H}_{01}} = \sum_{k=1}^{N_G} |w(k)|^2 + \sum_{k=N_G+1}^{N_S} |w(k) + s(k)|^2, \tag{5}$$

where the left-hand sum, $Y_{\mathcal{H}_{01}}^a$, follows a central Chi-square distribution with N_G degrees of freedom, while the right-hand sum, $Y_{\mathcal{H}_{01}}^b$, follows a non-central Chi-square distribution with $N_S - N_G$ degrees of freedom, and a non-centrality parameter λ. If N_G and $N_S - N_G$ are large enough the CLT holds and, consequently, $Y_{\mathcal{H}_{01}}^a \sim \mathcal{N}(N_G, 2N_G)$ and $Y_{\mathcal{H}_{01}}^b \sim \mathcal{N}(N_S - N_G, 2(N_S - N_G + 2\lambda))$. Assuming that $Y_{\mathcal{H}_{01}}^a$ and $Y_{\mathcal{H}_{01}}^b$ are independent and identically distributed (i.i.d.), $Y_{\mathcal{H}_{01}}$ can be stated as the sum of two

Gaussian random variables (r.v.) $Y_{\mathcal{H}_{01}} \sim \mathcal{N}(N_S + \lambda, 2N_S + 4\lambda)$. Since PUs are only active in $N_S - N_G$ samples, λ is now given by $\lambda = (N_S - N_G)\lambda'$. Following the same rationale for the remaining hypotheses, the output of the energy detector considering the four hypotheses is written as:

$$Y \sim \begin{cases} \mathcal{N}(N_S, 2N_S), & \mathcal{H}_{00} \\ \mathcal{N}(N_S + \lambda' N_G, 2(N_S + 2\lambda' N_G)), & \mathcal{H}_{10} \\ \mathcal{N}(N_S + \lambda'(N_S - N_G), 2N_S + 4\lambda'(N_S - N_G)), & \mathcal{H}_{01} \\ \mathcal{N}(N_S + \lambda, 2(N_S + 2\lambda)), & \mathcal{H}_{11} \end{cases}. \quad (6)$$

Under the hypothesis $\mathcal{H}_{10}$ a PU is not active at the end of the sensing period and the probability of false alarm is given by

$$P_{FA}^{\mathcal{H}_{10}} = Pr(Y > \gamma | \mathcal{H}_{10}) = \mathcal{Q}\left(\frac{\gamma - N_S - \lambda' N_G}{\sqrt{2N_S + 4\lambda' N_G}}\right). \quad (7)$$

On the other hand, under the hypothesis $\mathcal{H}_{01}$ a PU is always active at the end of the sensing period. In this case the probability of detection is given by

$$P_D^{\mathcal{H}_{01}} = Pr(Y > \gamma | \mathcal{H}_{01}) = \mathcal{Q}\left(\frac{\gamma - N_S - \lambda'(N_S - N_G)}{\sqrt{2N_S + 4\lambda'(N_S - N_G)}}\right). \quad (8)$$

3.3 Channel Availability

The channel availability observed by SUs directly depends on the ED performance. Because the ED determines the spectrum occupancy depending on the N_S samples collected from the channel, the ED can also determine the probability of detection and false alarm, which can be used as a figure of merit that evaluates the quality of the ED's decision. The SUs can access the shared channel when:

- There are no PUs transmitting and the SU observes $Y < \gamma$;
- A PU is transmitting and the SU observes $Y < \gamma$.

The first situation corresponds to the case where there is no false alarm and the second situation represents a case of missed detection when a SU may interfere with the PUs.

After performing a decision, the energy detector can compute the probability of the channel being found available, which is given by the following expression:

$$P_\alpha = (1 - P_D)\mathcal{H}_1 + (1 - P_{FA})\mathcal{H}_0, \quad (9)$$

where $\mathcal{H}_1$ and $\mathcal{H}_0$ correspond to the events when the ED observes channel occupied or idle, respectively. Note that P_D and P_{FA} are the probabilities computed by the ED using the N_S samples collected from the channel, being computed using (3) and (4), respectively.

4 Performance Evaluation

This section describes a set of simulations and numerical results to validate the probabilities of detection and false alarm and assess the validity of the channel availability probability for different SNR values.

4.1 Validation of Probability of Detection and False Alarm

We have considered a scenario formed by one PU transmitter-receiver pair and one SU that observes the channel availability. The operation mode of PUs and SUs is as described in Section 3. In this simulation we consider the case when PUs can randomly arrive or depart during the entire SUs frame. The PUs can change their state ON/OFF accordingly to a uniform distribution. For each value of SNR and N_S the decision threshold was parameterized to achieve a target in terms of probability of false alarm: $P_{FA}^{\mathcal{H}_{00}} = 0.005$ was adopted for the parameterization. For the hypotheses $\mathcal{H}_{10}$ and $\mathcal{H}_{01}$ the slot N_G (which corresponds to the slot where the PU change its state) is given by a uniform distribution.

Figure 3 illustrates the theoretical probability of false alarm for the hypotheses $\mathcal{H}_{00}$ and $\mathcal{H}_{10}$ along with the simulation results for different SNR values.

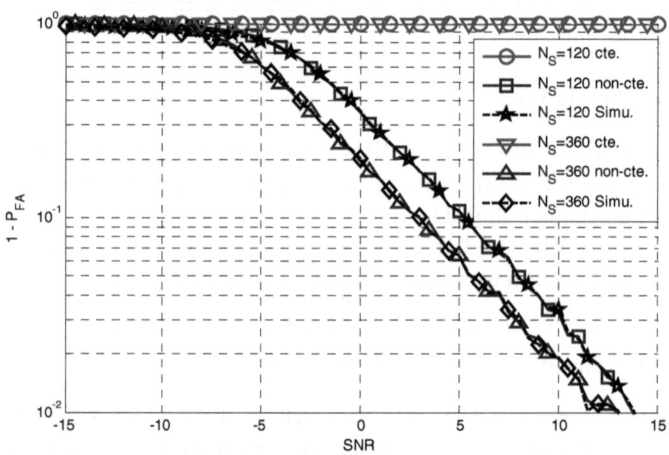

Fig. 3. Comparison of simulated and theoretical probability of false alarm

As a first remark, the results plotted in Figure 3 indicate that the simulated results are better validated by the theoretical model where a non-constant ("non-cte") PU's behavior is considered (P_{FA} is given by (7)). The curve titled "cte." represents the theoretical values given by $(1 - P_{FA})$, which represents the case when a constant PU behavior is observed (P_{FA} is given by (4)). Comparing the simulation and the theoretical results, we conclude that, if the constant model is considered in a scenario where the PUs presents a non-constant behavior, the error between the model and the real scenario is huge (for higher SNR values).

Figure 4 compares the theoretical probability of miss detection and the simulation results, where the $(1 - P_D)$ curve computed with (3) is titled "cte." and the $(1 - P_D)$ curve computed with (8) is titled "non-cte.".

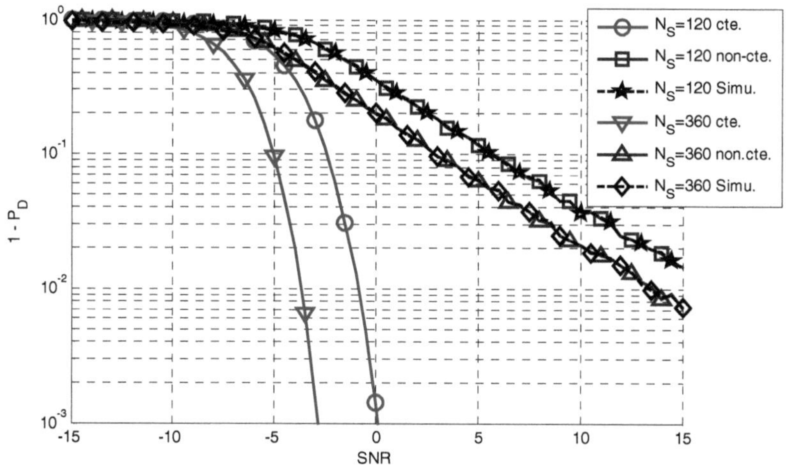

Fig. 4. Comparison of simulated and theoretical probability of detection

The plot shows that the theoretical probability expressed in (8) is successfully validated by the simulation results. On the order hand, the theoretical curve computed with (3) exhibits a high deviation from the simulated results, namely for higher values of SNR.

4.2 Probability of Channel Availability

This subsection presents the simulation results that evaluate the probability of channel availability computed in real-time by the energy detector. In the simulations, for each value of N_S and SNR, we first defined the detection probability and then the decision threshold was computed to meet the required detection probability. We consider N consecutive SU's operation cycles, and for each operation cycle the SUs calculate the probability of detection or the probability of false alarm depending on the output of the ED. If the ED decides that the channel is occupied then the SU calculates the probability of detection by using the expression (3). On the other hand if the ED decides that channel is idle, then the SU calculates the probability of false alarm by using the expression (4). The probability of a PU being active in each one of the N operation cycles is given by a uniform distribution. In the simulations we only consider a constant PU's activity (the PU remains active or inactive during all SU's operation cycle).

Figure 5 represents the average probability of channel availability (mean of the probabilities computed in each detection decision) for different values of SNR. In the simulations the PU is using the channel with a probability equal to 0.5 and the ED is parameterized for a probability of detection equal to 0.99.

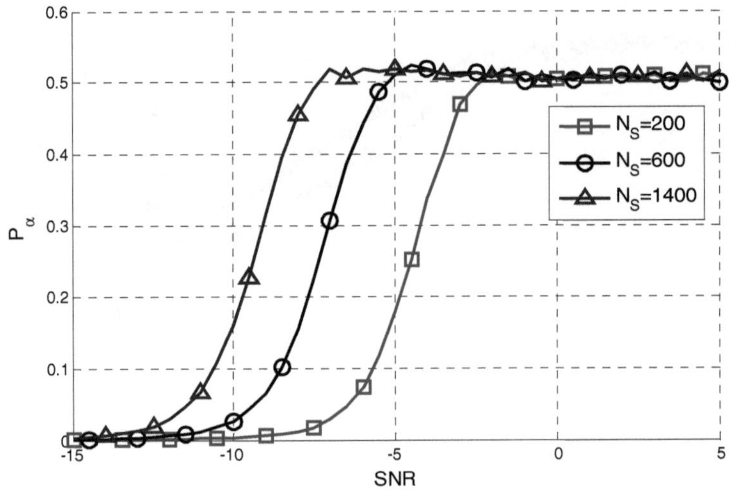

Fig. 5. Probability of channel availability

As it is possible to observe in Figure 5, for higher values of SNR the SU presents a probability of channel availability equal to 0.5, which means that the SU can accurately approximate the probability of channel availability. For lower SNR values the channel availability probability is below the real availability of the channel (0.5), which means that the SU observes the channel more occupied than it actually is. Observing the expression (9), we conclude that this error is due to the presence of false alarm events, since the probability of false alarm approaches one as SNR decrease. This observation is mainly due to parameterization method adopted to define the decision threshold (γ), which aims to achieve a detection probability greater than 0.99 and leads to a higher probability of false alarm, especially for lower values of SNR.

The probability of false alarm does not depend only on the value of SNR but also depends on the N_S samples observed by SUs. As it is possible to observe in expression (9), with the increase of N_S the probability of false alarm decreases, so the probability of channel availability is more close to the real availability of the channel.

5 Conclusions

In this work we have addressed two problems related with the assessment of channel availability for cognitive radio systems. We started to characterize the performance of an energy detector for the case when PUs can change their state during the sensing period. The theoretical performance was validated through simulations and compared with a theoretical model where the PUs' state remains constant during the sensing period. For higher SNR values, the theoretical model where the PUs' state is constant during the sensing period shows a huge deviation when compared with the scenario where PUs can randomly arrive or depart during the entire SUs frame.

The second point addressed in this work was the characterization of the channel availability, which was based on the output of the energy detector weighted by the probabilities of detection or false alarm computed in real-time. The validation results show that the probability of channel availability proposed in this work presents a good approximation as the SNR increases. For lower SNR values, the probability of channel availability exhibits a considerable error, but as demonstrated in the simulated results, the probability is almost null, meaning that it does not impacts in terms of PU's interference although it penalizes the SUs in terms of throughput.

Acknowledgments. This work was partially supported by EU COST IC0902 and the Portuguese Science and Technology Foundation under the projects PTDC/EEATEL/115981/2009, PTDC/EEA-TEL/099074/2008, PTDC/EEA-TEL/099973/2008, PTDC/EEATEL/120666/2010, PEst-OE/EEI/UI0066/2011 and SFRH/BD/68367/2010.

References

1. Yucek, T., Arslan, H.: A survey of spectrum sensing algorithms for cognitive radio applications. IEEE Communications Surveys Tutorials 11, 116–130 (2009)
2. Zahedi-Ghasabeh, Tarighat, A., Daneshrad, B.: Spectrum Sensing of OFDM Waveforms Using Embedded Pilots in the Presence of Impairments. IEEE Transactions on Vehicular Technology 61, 1208–1221 (2012)
3. Bouzegzi, A., Ciblat, P., Jallon, P.: Matched Filter Based Algorithm for Blind Recognition of OFDM Systems. In: Proc. IEEE VTC 2008-Fall, pp. 1–5 (September 2008)
4. Al-Habashna, A., Dobre, O., Venkatesan, R., Popescu, D.: Cyclostationarity-Based Detection of LTE OFDM Signals for Cognitive Radio Systems. In: Proc. IEEE GLOBECOM 2010, pp. 1–6 (December 2010)
5. Urkowitz, H.: Energy Detection of Unknown Deterministic Signals. Proceedings of the IEEE 55, 523–531 (1967)
6. Ghasemi, A., Sousa, E.S.: Optimization of Spectrum Sensing for Opportunistic Spectrum Access in Cognitive Radio Networks. In: Proc. IEEE CCNC 2007, pp. 1022–1026 (January 2007)
7. Bhargavi, D., Murthy, C.: Performance comparison of energy, matched-filter and cyclostationarity-based spectrum sensing. In: Proc. IEEE SPAWC 2010, pp. 1–5 (2010)
8. Tian, Z., Giannakis, G.: Compressed Sensing for Wideband Cognitive Radios. In: IEEE ICASSP 2007, vol. 4, pp. IV-1357–IV-1360 (April 2007)
9. Lee, W.-Y., Akyildiz, I.F.: Optimal spectrum sensing framework for cognitive radio networks. IEEE Transactions on Wireless Communications 7(10), 3845–3857 (2008)

10. Liang, Y.-C., Zeng, Y., Peh, E.C.Y., Hoang, A.T.: Sensing-Throughput Tradeoff or Cognitive Radio Networks. IEEE Transactions on Wireless Communications 7(4), 1326–1337 (2008)
11. Wang, S., Zhang, J., Tong, L.: A characterization of delay performance of cognitive medium access. IEEE Transactions on Wireless Communications 11(2), 800–809 (2012)
12. Tang, H.: Some physical layer issues of wide-band cognitive radio systems. In: Proc. IEEE DySPAN 2005, pp. 151–159 (November 2005)

Optical Strategies for Generating and Transmitting Signals in OFDM-RoF Systems

Vitor Lopes[2], Diogo Bento[2], Maria C.R. Medeiros[1], and Paula Laurêncio[1,2]

[1] IT-Institute of Telecommunications, Av. Rovisco Pais 1, 1049-001, Lisboa, Portugal
[2] Instituto Superior de Engenharia, Universidade do Algarve, 8005 Faro, Portugal
vlopes@ualg.pt, db_eee@yahoo.com, {cmedeiro,plaurenc}@ualg.pt

Abstract. Radio-over-Fiber (RoF) systems which act as backbone networks are very promising means of reducing overall costs of 60 GHz WPANs, as well leading to networks with limited intercellular interference. The main objective of this paper is to compare the performance of two systems that generate Optical Single Side Band (OSSB) OFDM signals at the mm-wave band with and without optical up-conversion. Both systems use a DD-MZM: one is a standard model, whereas the other is a DD-MZM cascaded with a phase modulator (PM). The comparative study presented here also includes practical impairments such as the finite extinction ratio of DD-MZMs.

Keywords: Radio over fiber, Orthogonal Frequency Division Multiplexing, 60 GHz, fiber chromatic dispersion, Optical Single Side Band,optical up-conversion.

1 Introduction

The 60 GHz (mm-wave) band has been allocated worldwide for short range wireless communications due to its inherent high propagation losses. The abundant unlicensed spectrum around the 60 GHz has the potential to support consumer demands for indoor wireless applications that require high bandwidth such as real-time streaming content download for high-definition TV, wireless gigabit Ethernet, etc. Due to the small coverage area, 60 GHz networks require a large number of base stations (BSs) to cover a service area. This requirement has led to the development of system architectures where functions such as routing, signal processing, handover and frequency allocation are performed at the central office (CO). The best solution for connecting the CO with BSs in such radio network is via an optical fiber network, now known as radio over fiber (RoF) [1].

Due to the high data rates of broadband wireless systems, we can have severe time-dispersion effects associated to the multipath propagation. Multicarrier (MC) modulation schemes combined with frequency-domain receiver implementations, especially the ones belonging to the orthogonal frequency division multiplexing (OFDM) class are widely used in several broadband wireless communication systems which have to deal with strong frequency-selective fading channels; as such, they are

L.M. Camarinha-Matos, S. Tomic, and P. Graça (Eds.): DoCEIS 2013, IFIP AICT 394, pp. 505–512, 2013.

considered a good contender for this application. However, the optical distribution of high bit rate mm-wave signals is susceptible to fiber chromatic dispersion, which severely limits the transmission distance. This shortcoming can be mitigated using optical single-sideband (OSSB) modulation [2]. Several techniques have been proposed to generate mm-wave optical signals. Among them, optical frequency up-conversion has the potential to provide a simple and cost efficient solution [3]. The principle of optical frequency up-conversion is to generate higher-order optical harmonics using an external nonlinear modulator, typically a dual-drive Mach-Zehnder modulator (DD-MZM) driven by a low frequency electrical signal. The efficiency of this technique relies not only on its capacity to maximize the required high-order harmonic, but also on its robustness against fiber dispersion. Therefore, the objective of this paper is to compare two systems that generate OSSB signals at the mm-wave band with and without optical up-conversion. Both systems use a DD-MZM: one is a standard model and other is a DD-MZM cascaded with a phase modulator (PM) [4].

The remainder of this paper is organized as follows. In section 2 we discuss the relationship to the Internet of things. In section 3, we present the network architecture. In section 4, we present the simulation results and finally, in section 5, we conclude the paper.

2 Relationship to Internet of Things

Nowadays, high bandwidth services are supported by wired networks based on fiber optic technologies. However, consumer habits are increasingly changing; as such, users want to be able to use their mobile terminals and enjoy the same user experience as they do when connected to their fixed networks. The abundant unlicensed frequency spectrum around 60 GHz offers the potential for multi-gigabit indoor Wireless Personal Area Networks, removing the bandwidth limitation of the present wireless networks.

The current generation of Wi-Fi systems (IEEE 802.11n) can achieve theoretically 600 Mb/s, but new radio standards allowing higher data rates have emerged or are under development. All these new standards address the unlicensed mm-wave band from 57 to 66 GHz, in this band, divided into four 2.16 GHz-bandwidth channels, it is possible to achieve bit rates up to 7 Gb/s per channel [5]. IEEE 802.15.3c[6] and WirelessHD standards were finalized in 2009 and are devoted to wireless personal area network communications.More recently, the IEEE 802.11ad group has been created by WiGig consortium: it addresses networking with wireless local area network systems and targets Wi-Fi Alliance certification. 60 GHz technologies are now mature: WirelessHD devices are already available in the market for wireless high-definition multimedia interface (HDMI) applications and IEEE 802.11ad products.

Due to the high-propagation attenuation and the fact that signals in the 60 GHz band cannot cross walls,the premises should be connected employing optical fiber technology. Consequently, 60 GHz networks require an extensive high capacity feeder network; thisdemands an optical fiber network thatprovides the required performance, while being cost-effective.

Currently, wired and wireless services are separately provided by two independent physical networks. However, with the fast evolution of wireless technology, ubiquitous and always on wireless systems in buildings are expected to emerge in the near future. Hybrid broadband access networks combining wireless and wireline Wavelength Division Multiplexed Passive Optical Networks (WDM-PONs) networks can support the requirement of these future broadband ubiquitous networks. The target of a converged hybrid architecture is to connect, via a common optical WDM-PON infrastructure, both wireline and wireless users. Hybrid wireless WDM-PON architecture integrating mm-wave wireless services within WDM-PONs will be the future access networks.

3 System Description

The architecture of the 60 GHz OFDM-RoF numerical simulation platform using the VPI Transmission Maker Simulator® is shown in figure 1 and 2. The difference between the two figures is the optical external modulator; the first one is a DD-MZM biased in quadrature, whilst the second one is a DD-MZM biased in quadrature cascaded with a phase modulator (PM), both of them generating an OSSB signal. The objective of the second configuration is to demonstrate that the effect of the finite extinction ratio of the DD-MZMs on OSSB modulation can be greatly reduced. The light source considered in the simulation is a laser with 1 MHz of line width and 5 mW of emission power with -150dB/Hz relative intensity noise (RIN). A 2.5 Gbit/s NRZ data is divided in 512 blocks. Each blockis presented as 512parallel data paths to the OFDM transmitter. These 512paths aremodulated onto 512equally-spaced subcarriers using 4 Quadrature-Amplitude Modulation(4QAM). The cyclic prefix (CP) is 12.5%. Each QAM data channel is presented to an input of an IFFT that produces a complex-valued time domainwaveform containing a superposition of all of the sub-carriers. This waveform therefore is modulated onto an electrical carrier, f_c=60GHz, using an I-Qmodulator, producing a real-valued signal. This carrier gives a band of optical OFDM sub-carriers with a bandwidth of 750 MHz centered in the optical carrier at 193.1 THz plus 60 GHz after external modulation by a DD-MZM. Both DD-MZMs were assumed to have a 6 dB insertion loss and a 30dB extinction ratio (R_{ext}). The fiber attenuation was set to 0.2 dB/km and the dispersion coefficient to 16 ps/(nm.km). At the receiver side, the signal is directly detected by an ideal high-speed PIN photodetector, amplified and each QAM channel is demodulated to produce 512 parallel data channels. These can be converted into a single data channel by parallel-to-serial conversion and down-converted to recover the 2.5 Gbit/s data. The optical field $E_{MZ}(t)$ at the output of the DD-MZM can be expressed as:

$$E_{MZ}(t) = E_i(t)\left\{\eta\left[e^{j\frac{\pi V_{A1}(t)}{VA_\pi}} + e^{j\frac{\pi V_{A2}(t)}{VA_\pi}}\right] + \xi e^{j\frac{\pi V_{A2}(t)}{VA_\pi}}\right\} \tag{1}$$

Where E_i is the optical field at the input of the DD-MZM and average power Po, VA_π is the switching voltage of the DD-MZM; η and ξ are given respectively by $\sqrt{r_1 r_2}$ and

$\sqrt{(1-r_1)(1-r_2)}-\sqrt{r_1 r_2}$, where and r_1 and r_2 are the power splitting ratios of the input and output of the Y-junctions of the DD-MZM [7] respectively.

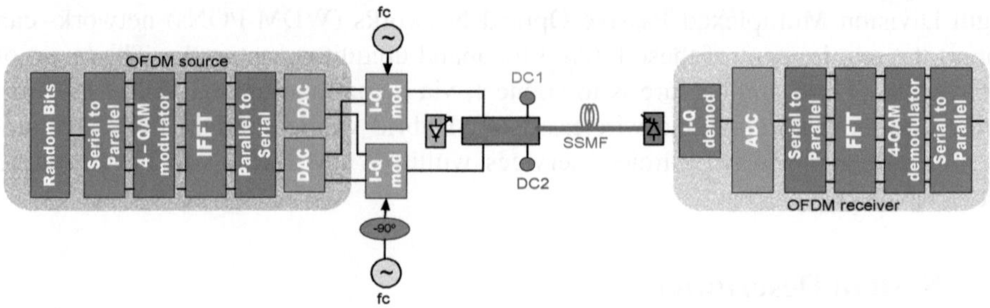

Fig. 1. System Architecture of a 60 GHz OFDM-RoF with DD-MZM

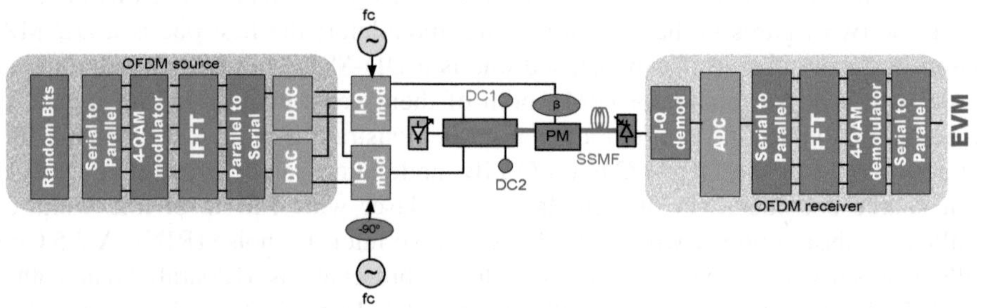

Fig. 2. System Architecture of a 60 GHz OFDM-RoF with DD-MZM cascaded with a phase modulator

The electrical voltages applied on both electrodes of the DD-MZM modulator to obtain an OSSB signal consist of the sum of the OFDM signals, $s_{OFDM}(t)$ and a dc term, given by:

$$V_{A1}(t)=s_{OFDM}(t)-VA_{\pi}/4$$
$$V_{A2}(t)=\hat{s}_{OFDM}(t)+VA_{\pi}/4 \tag{2}$$

Where $\hat{s}_{OFDM}(t)$ is the Hilbert transform of the OFDM signal.

In the case of DD-MZM cascaded with an optical phase modulator (PM) as illustrated in Fig. 3 the optical field E_o is given by

$$E_o(t)=E_i(t)\left\{\eta\left[e^{j\frac{\pi V_{A1}(t)}{VA_\pi}}+e^{j\frac{\pi V_{A2}(t)}{VA_\pi}}\right]+\xi e^{j\frac{\pi V_{A2}(t)}{VA_\pi}}\right\}e^{j\frac{\pi V_B(t)}{VB_\pi}} \tag{3}$$

While the voltage applied to the PM modulator is assumed to be $V_B(t)$; that is the modulated OFDM signal after being applied a adjustable phase shift β+π/2. The objective of this phase shift is to eliminate the residual sideband due to DD-MZM finite extinction rate (ER).

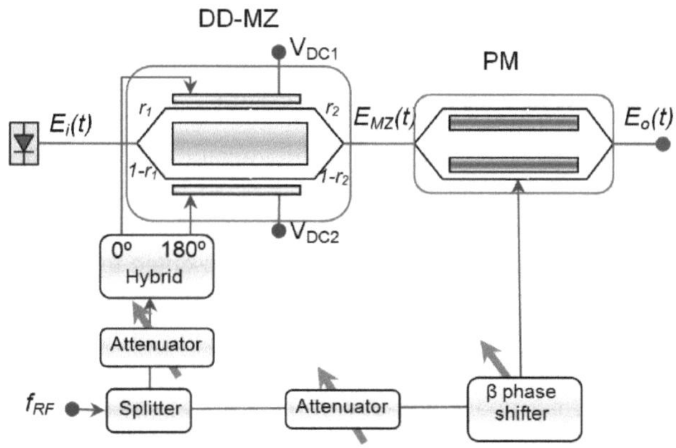

Fig. 3. Schematic of the ER compensation scheme

The principle of optical frequency up-conversion is to generate higher-order optical harmonics using an external nonlinear modulator, typically a Mach-Zehnder modulator (MZM) driven by a low frequency electrical signal, this reduces the bandwidth requirements for the optical modulator, the electrical drive signal source and the drive circuit itself. For example, for OSSB, frequency tripling is obtained by beating, upon square law detection by the photodetector, the ω_c harmonic with $\omega_c + 3\omega_m$, and all harmonics separated by $3\omega_m$, where $\omega_c$is the optical carrier frequency, and ω_m is the subcarrier frequency.

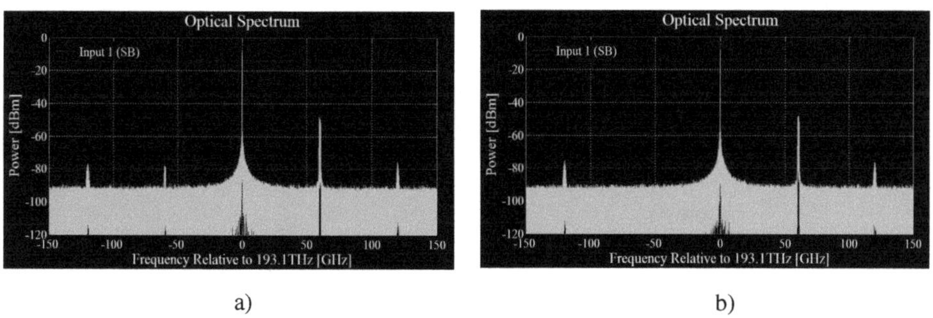

Fig. 4. Optical power spectrum at a) Dual-Arm MZM (Rext=30dB)b) Dual-Arm with PM (Rext=30dB)

For dual-arm MZM, the effect of the imbalance in the power splitting ratio generates odd low sideband harmonics as it is shown in Fig. 4a. For this situation, the imbalance leads to an optical spectrum that approaches an Optical Double Side Band (ODSB) signal. In the case of the DD-MZM cascaded with an optical phase modulator the optical spectrum approaches the ideal modulator spectrum (Fig 4b).

4 Simulation Results

4.1 Optical Modulation without Up-Conversion

In order to compare the performance of both modulator schemes when digital data is transmitted, the DD-MZM and the DD-MZM cascaded with a phase modulator OSSB systems illustrated in Fig. 1 and 2 were simulated using the VPI TransmissionMak-er™ simulation platform. The simulation parameters were presented in section 3. This includes the noise system, like RIN, shot noise and circuit noise. We named it the real circuit. We also performed an additional test to calculate the beta parameter in the phase modulator, in order to determine the angle at which the optical carrier in the lower sideband is completely suppressed. Intermodulation distortion that generates inter-subcarrier interference at fiber output arises mainly from the nonlinear characteristics of the optical modulator, the linear fiber dispersion and from the square law detection process.

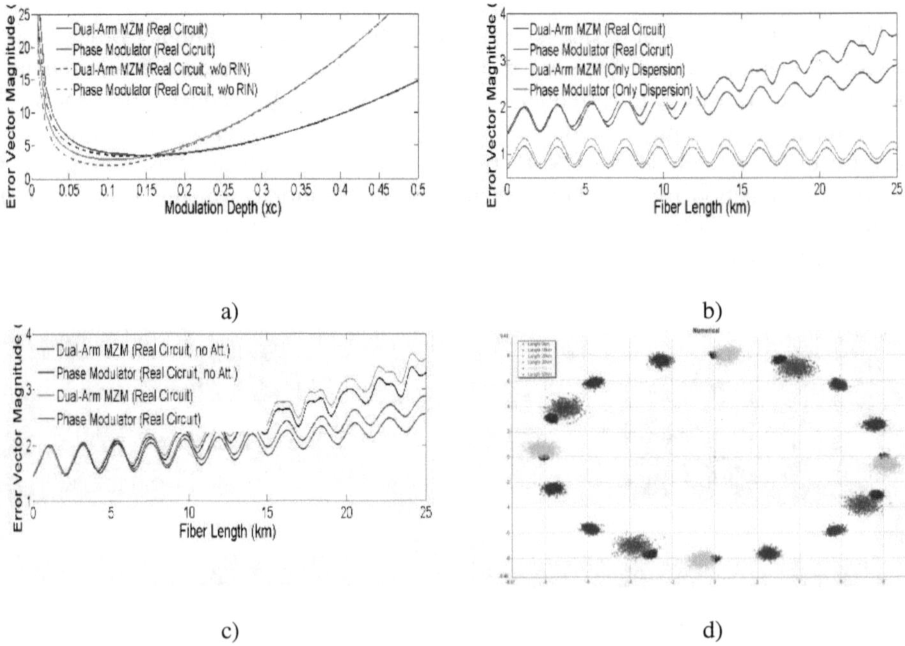

Fig. 5. EVM for DD-MZM and DD-MZM cascaded with PM versus a) modulation depth for 25km of fiber length b) fiber length real conditions versus only dispersion c) fiber length real conditions versus without attenuation d) QAM constellation diagram

Figure 5.a illustrates the Error Vector Magnitude (EVM) versus modulation depth (x_c) for 25km of fiber length. We can conclude the optimal modulation depth is similar for all tested configurations. For lower modulation depths, the system noise is dominant, whereas for higher modulation depths the intermodulation distortion between subcarriers is dominant; this leads to an optimum modulation depth x_c=0.155, in the circuit with a DD-MZM, and x_c=0.111, in the circuit with a DD-MZM, plus a

PM. From Fig 5.b, we can conclude the EVM does not change linearly with the fiber length. There are also periodic oscillations that result from the dispersion, making the EVM increase with the fiber length. This occurs due to the interplay between the laser RIN and the dispersion. When we don't have noise (only the dispersion) the EVM is periodic. Finally, in Fig.5.c, when we eliminate the attenuation we can see the EVM decreases slightly because the interplay between the RIN and the dispersion is dominant. This result can be confirmed in Fig. 5.d where the received constellation is presented (without attenuation) for different fiber lengths. From these results we concluded that OSSB based systems are robust against imperfect power splitting ratios and the use of an extra PM to eliminate the first low sideband harmonic is not necessary.

4.2 Optical Modulation with Up-Conversion

In this section we present the results obtained for the two modulation schemes studied, but using optical up-conversion with frequency tripling. The conditions are the same from last section and we use an electrical 20GHz to obtain 60GHz in frequency tripling. Comparing the performance of the systems using up-conversion with the ones from previous section, we can conclude that the results are significantly worse.

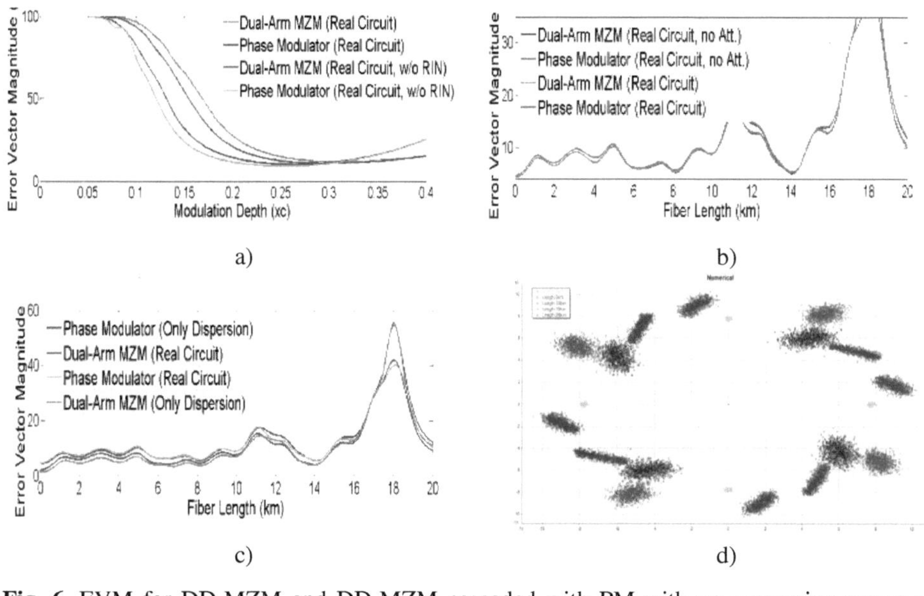

Fig. 6. EVM for DD-MZM and DD-MZM cascaded with PM with up-conversion versus a) modulation depth for 25km of fiber length b) fiber length real conditions versus only dispersion c) fiber length real conditions versus without attenuation d) QAM constellation diagram

This is explained on Fig. 6.b. and 6.c. In these figures we can see the dispersion is the main responsible for the performance degradation obtained. This occurs because up-conversion systems have different harmonics that suffer different dispersion effects when the fiber length increases.Furthermore, the beating during the photodetection process results in additional system degradation. This effect combined with the laser RIN can be seen in Fig. 6.d, where the receiver constellation is shown for different fiber lengths. We can conclude that this system needs equalization for distances beyond 15km. The use of the cascaded PM does not solve the problem.

5 Conclusions and Further Work

We concluded that OSSB based systems are robust against imperfect power splitting ratios and the use of an extra PM to eliminate the first low sideband harmonic is not necessary when 60GHz native frequency is used. When frequency tripling is used, this system needs equalization for distances beyond 15km. We also concluded that the dispersion is the mainly responsible for the performance degradation of the system when compared with the 60GHz natives. The use of the cascaded PM does not solve the problem.

In further work analytical methods will be developed to evaluate the degradation of performance of those systems caused by laser intensity noise at fiber output, intermodulation distortion and fiber dispersion. In case of optical up-conversion equalization techniques will be applied.

References

1. O'Reilly, J.J., Lane, P.M., Attard, J., Griffin, R.: Broadband wireless systems and networks: an enabling role for radio-over-fibre. Philosophical Transactions of the Royal Society a-Mathematical Physical and Engineering Sciences 358, 2297–2308 (2000)
2. Laurêncio, P., Medeiros, M.C.R.: Dynamic range of optical links employing optical single-side-band modulation. IEEE Photonics Technology Letters 15(5), 748–750 (2003)
3. Lin, C.T., Chen, J., Shih, P.T., Jiang, W.J., Chi, S.: Ultra-high data-rate 60 GHz radio-over-fiber systems employing optical frequency multiplication and OFDM formats. Journal of Lightwave Technology 28(16), 2296–2306 (2010)
4. Thakur, M.P., Medeiros, M.C.R., Laurêncio, P., Mitchell, J.E.: Optical frequency tripling with improved suppression and sideband selection. Optics Express 19, 457–468 (2011)
5. Hansen, C.J.: Wigig: Multi-Gigabit Wireless Communications in the 60GHz Band. In: IEEE Wireless Communications. Broadcom Corporation (December 2011)
6. IEEE802.15.3c WPAN Task Group 3c, `http://www.ieee802.org/15/pub/TG3c.html`
7. Lin, C.T., Chen, J., Dai, S.P., Peng, P.C., Chi, S.: Impact of Nonlinear Transfer Function and Imperfect Splitting Ratio of MZM on Optical Up-Conversion Employing Double Sideband With Carrier Suppression Modulation. Journal of Lightwave Technology 26(15), 2449–2459 (2008)

Local Oscillator Phase Noise Influence on Single Carrier and OFDM Modulations

Vitor Fialho[1,2], Fernando Fortes[2,3], and Manuela Vieira[1,2]

[1] Universidade Nova de Lisboa – Faculdade de Ciências e Tecnologia– DEE
Monte da Caparica 2829-516, Lisbon, Portugal
[2] Instituto Superior de Engenharia de Lisboa – ISEL
Rua Conselheiro Emídio Navarro,1 1959-007-Lisbon, Portugal
[3] Instituto das Telecomunicações
Av. Rovisco Pais, 1 1049 - 001 Lisbon - Portugal
{vfialho,ffortes}@deetc.isel.ipl.pt, mv@isel.pt

Abstract. This paper describes the influence of local oscillator phase noise on single and multi-carrier modulations schemes. The oscillator has the main role on modulation and demodulation process of the radio frequency channels. The synchronization of the entire system depends on the accuracy of this circuit. This work is based on the simulation of these two scenarios under influence of a non-ideal oscillator. The output of the simulation scenario results of error vector magnitude, bit error rate and symbol error rate. The relation of these metrics with the local oscillator phase noise allows the performance estimation the global system. A simple phase noise model is presented with configurable power spectral density and offset frequency.

Keywords: phase noise, radio frequency transceiver, error vector magnitude.

1 Introduction

The local oscillator (LO) is one of the building blocks of every radio frequency (RF) front-end. This device is responsible, with the mixer, for the up and down conversion of a specific channel. However the modulation and demodulation processes suffer additional effects caused by the electronic circuits and the transmission channel itself. Common standards such as IEEE 802.11, [1] IEEE 802.16[2] do not present electrical specifications of the RF building blocks but system level specifications. The RF transmitter in several standards is specified by the channel centre frequency, bandwidth, spectral mask and error vector magnitude (EVM). The RF receiver is specified with bit error rate (BER) dependency on the received RF channel power (sensitivity), which depends on the received signal to noise ratio (SNR) and noise degradation due to noise figure and LO phase noise [3].

The digital modulation scheme used on these several standards depends on the relation of BER and SNR. Therefore, according to the channel noise conditions, the chosen modulation may change from quadrature phase shift keying (QPSK) to quadrature amplitude modulation (QAM).

L.M. Camarinha-Matos, S. Tomic, and P. Graça (Eds.): DoCEIS 2013, IFIP AICT 394, pp. 513–520, 2013.

This scatter representation of the in-phase (I) and quadrature (Q) channels enables the evaluation of the global system performance, since RF impairments have a typical behavior on the received symbols [4]. Since the wireless transmission is a frequency selective fading channel, the most common standards adopted orthogonal frequency division multiplexing (OFDM) instead of a single carrier transmission. The main difference of both techniques is that OFDM use of a multicarrier technique which divides the entire bandwidth of the channel into smaller sub bands, around each subcarrier. This allows the study of LO influence on both systems and estimate the impact on the typical base band figures of merit.

With the upcoming of the 3GPP Long Term Evolution (LTE) single carrier and OFDM techniques are used on the same equipment. For the downlink transmission it is used OFDM in order to minimize multipath fading and inter-symbolic interference (ISI). In the uplink it is used a single carrier modulation technique since PAPR values are lower compared with OFDM.

Since the main study of this work is the influence of LO phase noise on the radio channel, only a generic OFDM signal is discussed. Cyclic prefix, interleaving and windowing [5] are not discussed in this work.

2 Internet of Things

The identification of several equipments, also know as things, via Internet, need to have a build in RF device. This device allows communication between the equipment and a given gateway for further identification. Therefore the study of LO performance on the global RF front-end has an important role on this area.

3 Global System Description

As depicted in figure 1 the RF front end, (transmitter and receiver), the signal delivered to, and from the mixer, are mapped in IQ channels. The channels content itself, and the modulation method used, single carrier or OFDM, does not have a direct influence on RF front end behavior. However, the RF impairments from the building blocks will influence the global system performance, namely the LO phase noise [4]. The base band signal is formatted by the base band processor that converts a digital to an analog signal for the up conversion.

3.1 Base Band Signal

In this sub-section it is presented the generic block diagram of a single carrier and OFDM signal generation.

Figure 2 presents the transmitter and receiver of a single carrier system based on IQ channels, where the digitized binary data stream is coded in QPSK or M–QAM. This is a straightforward way of mapping the symbols on an IQ channel. The square root raised cosine is used on the transmitter and receiver and allows, simultaneously, the control of the signal bandwidth and eliminate the inter-symbolic interference (ISI) [2][3][4].

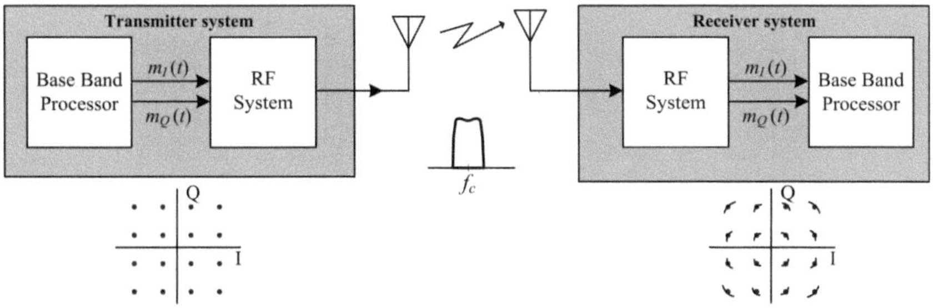

Fig. 1. Communication system based on IQ signals under phase noise influence

After the filtering process, the resulting base band analog signals, composed by $m_I(t)$ and $m_Q(t)$ are up converted and centered on the desired frequency channel, defined by the local oscillator of the transmitter (LO_{TX}), included on the RF transmitter front end block. The maximum occupied bandwidth corresponds to the symbol frequency. After the transmitter and receiver synchronization process is complete, the data channel content is down converted for filtering and decoding.

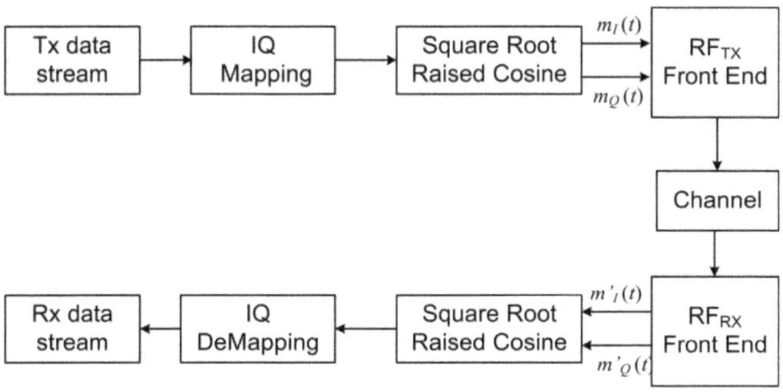

Fig. 2. Single carrier base band IQ signal generation

Figure 3 presents the transmitter and receiver OFDM block diagram. This is based on a multi-carrier modulation technique which uses orthogonal sub-carriers to transmit the base band data. In the frequency domain, since the bandwidth of a sub-carrier is designed to be smaller than the coherence bandwidth of the channel, each sub-channel is seen as a flat fading channel which simplifies the equalization process [5]. In the time domain, by splitting a high-rate data stream into a number of lower-rate data streams that are transmitted in parallel, OFDM resolves the problem of ISI in wide band communications.

The output of the digital modulation is converted from serial to parallel streaming, composed by X_0 to where X_n where n represents the IFFT dimension. The detailed explanation of the global system and building block optimizations is described in several works [5][6][7]. Since OFDM is based on orthogonal sub-carriers, where the sub-channel information is centered on, it is important to have a correct knowledge of the LO phase noise influence on the entire bandwidth, and subsequent relation with base band metrics, such as EVM, BER and symbol error rate (SER) [4].

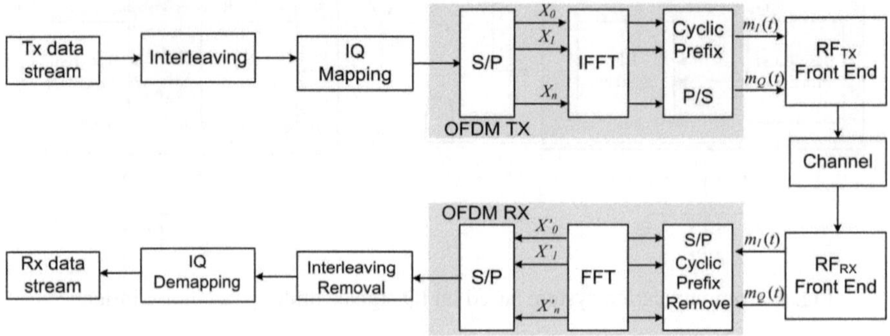

Fig. 3. Generic OFDM block diagram topology

3.2 Radio Frequency Front End

Figure 4 presents a typical RF front end based on IQ channels. The transmitter is composed by the mixer and LO_{TX} generating the $s_{RF}(t)$ centered on the desired transmitting frequency.

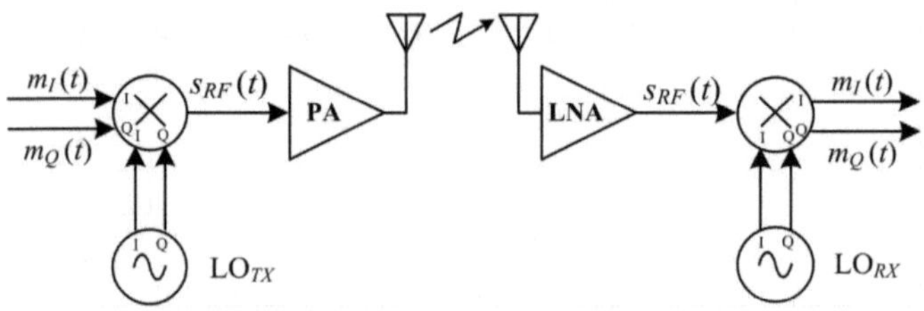

Fig. 4. Radio frequency front end based on IQ channels

The IQ base band signals are up-converted by the mixer and LO_{TX} to the desired RF channel. The power amplifier (PA) amplifies the signal to the desired transmission power. The receiver is composed by the low noise amplifier (LNA) which amplifies the RF band in which the desired RF channel is located. The filtered channel is down-converted by the mixer and LO_{RX} to an IQ base band signal for prior processing.

The RX blocks presented in figure 2 and figure 3 can be described as the RF front end presented in figure 4. As depicted this block is independent of how the base band signals are composed, as discussed in 3.1.

The RF impairments will be reflected on the received (demodulated) channel. The PA non-linearities and consequent peak peak-to-average power ratio (PAPR) is discussed by several authors [5][6][7]. As depicted, the common RF building block presented on the transmitter and receiver is the LO. Therefore it is important to study the influence of phase noise on the global system performance. Since the base band information is mapped on IQ channel it is possible to infer the global system behavior based on EVM and consequently on SER and BER.

Leeson [8], and the subsequent authors [9] established well known models that describe the phase noise variation at a specific frequency carrier offset. RF designers, specifically the LO designers, focus their optimization on specific phase noise values, typically at 1MHz and 10MHz, depending on the system application[10]. This is useful for the oscillator characterization itself. However, when the oscillator up or down converts a channel, this characterization is not complete, since the phase noise variation will not be constant in all the channel bandwidth. Therefore it is important the have the knowledge of phase noise integrated power on the channel bandwidth and not at a specific offset frequency.

4 Simulation Scenario

In this section the simulation scenarios for both systems are presented, as well as the oscillator phase noise model developed for this work.

The single carrier and OFDM systems are implemented in MatLab-Simulink with a channel bandwidth of 1 MHz.

The simulation scenario for the single carrier modulation is based on the figure 2 topology. The IQ mapping is 16-QAM with a symbol frequency of 1MHz. The square-root raised cosine has a roll-off factor of 0.2.

OFDM simulation scenario is based on the topology presented in figure 3 with a FFT and IFFT dimension of 64 points. Therefore the final symbol rate is given by $f_{samp}/64$ with a configurable number of sub-channels [1].

The RF front end is based on an oscillator with configurable phase noise power, as presented in section 4.1.

4.1 Phase Noise Model

A generic oscillator with phase noise is expressed by (1), where $\theta_{err}(t)$ corresponds to a Gaussian distributed values with variance v_{noise}^2. For low $\theta_{err}(t)$ values, (1) can be expressed by (2).

$$c(t)=\cos(2\pi f_c\, t+\theta_{err}(t))+j\sin(2\pi f_c\, t+\theta_{err}(t)). \tag{1}$$

$$c(t)=\cos(2\pi f_c\, t)+j\sin(2\pi f_c\, t+\theta_{err}(t)). \tag{2}$$

The developed phase noise model is presented in figure 5. The output values of the Gaussian noise generator are parameterized by v_{noise}^2 and applied to a low pass filter with configurable cut-off frequency, f_{cut}. This allows the evaluation of the integrated phase noise power with limited bandwidth [8]. This configuration allows further studies of filtering technique in order to perform noise shaping.

The first order low pass filter is defined is s domain as (3), where a corresponds to the pole frequency. Since the implemented simulation scenario is based on a discrete environment, it is necessary to obtain an equivalent transfer function that represents the same filter but in a discrete domain. The approximate equivalence from Laplace domain and z domain is given by (4), where T_s represents the simulation sampling time.

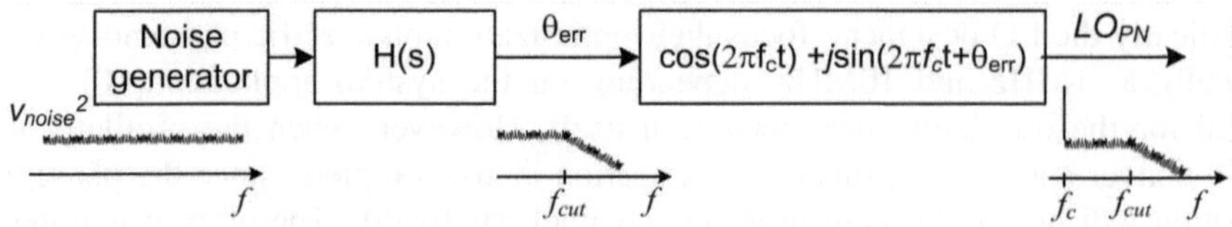

Fig. 5. Local oscillator phase noise model with configurable power and cutoff frequency

$$H(s) = \frac{1}{s + a} \tag{3}$$

$$H(z) = H(s)\bigg|_{s=\frac{2}{Ts}\cdot\frac{1-z^{-1}}{1+z^{-1}}} = \frac{Ts}{2 + aTs} \cdot \left[\frac{z+1}{z + \frac{aTs - 2}{aTs + 2}} \right] \tag{4}$$

5 Simulation Results

In this section the simulations results of single carrier and OFDM are presented under LO phase noise influence.

In order to quantify the error on both modulations techniques EVM is used since it can, without decoding the channel content, establish a relation between the transmitted and received signals, $s_{TX}[n]$ and $s_{RX}[n]$, respectively, as described by (5).

$$EVM = \sqrt{\frac{\frac{1}{N}\sum_{n=1}^{N}\left(s_{RX}[n] - s_{TX}[n]\right)^2}{\frac{1}{N}\sum_{n=1}^{N}\left(s_{TX}[n]\right)^2}} \tag{5}$$

In order to simplify the EVM extraction, both signals are normalized to their square-root mean power. Therefore, the maximum EVM obtained is coincident with the v_{noise} with a power spectral density (PSD) is given by v_{noise}^2/f_{samp}.

Since the EVM express the error, this can be estimated based on noise PSD. However, this depends on the number sub-carriers (NC), as expressed by(6).

$$EVM = \sqrt{\frac{v_{noise}^2}{f_{samp}} \cdot NC \cdot \Delta_f} \tag{6}$$

Figure 6 (a) presents the EVM variation for the OFDM system with f_{cut} frequency greater than $f_{samp}/2$ (equivalent to a constant phase noise) for different v_{noise} values corresponding to the EVM values obtained by (6). Since the variation of EVM for the presented values are very small and similar, it is presented the $\log_{10}$(EVM) instead of EVM itself.

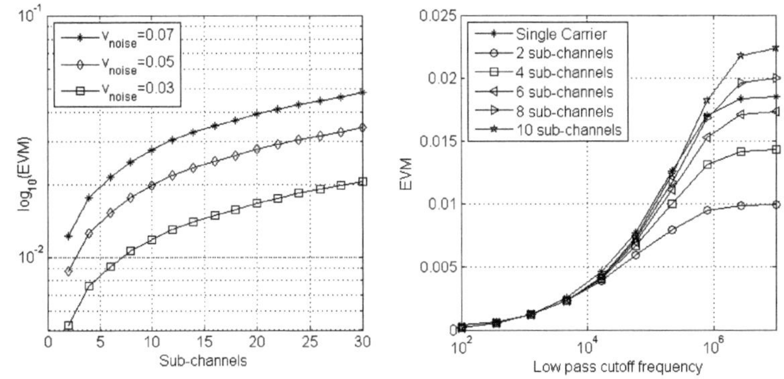

(a) (b)

Fig. 6. (a) Variation of $\log_{10}$(EVM) variation with OFDM sub-channels for different v_{noise} values; **(b)** Variation of EVM in function first order low pass filter cutoff frequency for multiple sub-channels

Figure 6 (b) presents the EVM variation for both modulations techniques with v_{noise}=50mV. For each sub-channel EVM is obtained in function of f_{cut} sweep of the low pass filter. As depicted for f_{cut} values bellow 100 kHz, the EVM values for both modulation schemes are identical.

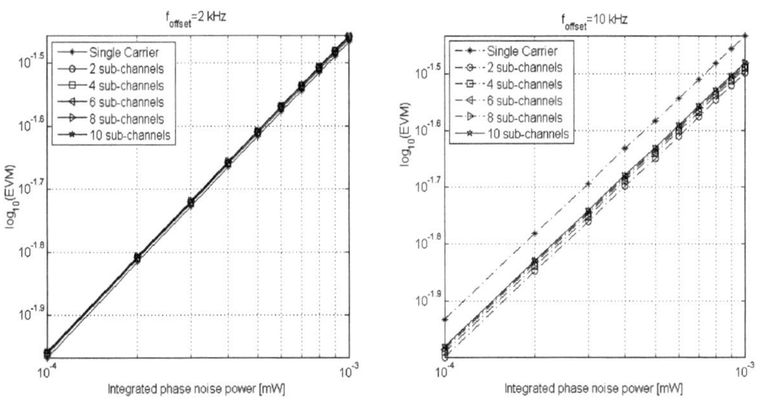

(a) (b)

Fig. 7. EVM evolution for different carrier offset frequency: **(a)** f_{offset}=2 kHz; **(b)** f_{offset}=10 kHz

Figure 7 (a) and (b) presents EVM variation in function of integrated phase noise power for two different carrier offset frequency. For lower carrier offset frequency, the EVM difference between single carrier and OFDM is identical.

For offset frequency above 10 kHz, as presented in figure 7 (b) the number OFDM sub-channels has influence on EVM, and the single carrier system has a greater degradation then OFDM

6 Conclusions

In this work a comparison between single carrier and OFDM modulation under the influence of local oscillator phase noise is presented.

A local oscillator phase noise model was presented. This model allows the configuration of the integrated phase noise power for a specific bandwidth. The final system performance is obtained by EVM. By the simulation results, it is possible to verify that the increase of sub-carriers on OFDM, for a specific phase noise may cause a greater degradation on EVM, and consequently on BER. The presented simulations allow concluding that, for oscillators with high phase noise, single carrier and OFDM has the same behavior. Therefore it is relevant, in the project of a RF system, the knowledge of the integrated phase noise power for several carrier offset frequency in order to optimize the global system performance.

This technique allows the study of different filter topologies with the purpose of perform noise shaping.

References

1. IEEE Standard for Wireless LAN Medium Access Control (MAC) and Physical Layer (PHY) Specifications: High Speed Physical Layer Extension in the 2.4GHz, IEEE Standard 802.11b-1999
2. IEEE 802.16. IEEE Recommended Practice for Local and metropolitan area networks Coexistence of Fixed Broadband Wireless Access Systems
3. Fortes, F., et al.: RF Receiver Front End for 28.5 GHz applications on a 70 GHz FT SiGe BiCMOS process. Wiley Microwave and Optical Technology Letters 52(3), 736–740 (2010)
4. Fialho, V., et al.: Test Setup for Error Vector Magnitude Measurement on WLAN Transceivers. In: 19th IEEE International Conference on Electronics, Circuits, and Systems-ICECS (December 2012)
5. Wang, et al.: OFDM or Single-Carrier Block Transmissions? IEEE Transactions on Communications 52(3) (March 2004)
6. Armstrong, J., et al.: OFDM for Optical Communications. Journal of Lightwave Technology 27(3) (Febraury 2009)
7. Jamal, M.: Study of Multiple Access Schemes in 3GPP LTE OFDMA vs. SC-FDMA. In: IEEE International Conference on Applied Electronics, AE (2011)
8. Leeson, D.B.: A Simple Model of Feedback Oscillator Noise Spectrum. IEEE Proceedings on IEEE 54(2), 329–330 (1966)
9. Lee, T.H., Hajimiri, A.: Oscillator Phase Noise: A Tutorial. IEEE Journal of Solid-State Circuits 35(3), 326–336 (2000)
10. Muh-Dey, W., Sheng-Fuh, C., Xu-Wei, L., Gyn-Wei, K.: Phase Noise Optimization of CMOS VCO with Double-Harmonic-Tuned LC Tank. In: IEEE European Microwave Conference, pp. 1054–1057 (October 2010)

Indoor Sound Based Localization: Research Questions and First Results

João Moutinho[1], Diamantino Freitas[2], and Rui Esteves Araújo[1]

[1] INESC TEC (formerly INESC Porto), Faculty of Engineering, University of Porto
[2] Faculty of Engineering, University of Porto
Rua Dr. Roberto Frias4200-465 Porto, Portugal
{jnm,dfreitas,raraujo}@fe.up.pt

Abstract. This PhD work has the goal to develop an inexpensive, easily deployable and widely compatible localization system for indoor use, suitable for pre-installed public address sound systems, avoiding costly installations or significant architectural changes in spaces. Using the audible sound range will allow the use of low cost off-the-shelf equipment suitable for keeping a low deployment cost. The state-of-the-art presented in this paper evidences a technological void in low-cost, reliable and precise localization systems and technologies. This necessity was also confirmed by the authors in a previous project (NAVMETRO®) where no suitable technological solution was found to exist to overcome the need to automatically localize people in a public space in a reliable and precise way.

Although research work is in its first steps, it already provides a thorough view on the problem while discussing some possible approaches and predicting strategies to overcome the key difficulties. Some experiments were already conducted validating some initial premises and demonstrating how to measure the signal's time-of-flight necessary to infer on distance calculations.

Keywords: context-aware systems, sound based localization, TOA, TDOA, multilateration, trilateration, multipath, IPS, PL.

1 Introduction

One of the most popular research areas in ubiquitous or pervasive computing is the development of location-aware systems. These are systems in which electronic devices provide the users some kind of information or service depending on their location. The basilar component of a location-aware system is the location-sensing mechanism. This PhD work has the goal to develop an inexpensive, easily deployable and widely compatible localization system appropriate for indoor use, suitable for pre-installed public address sound systems, avoiding costly installations or significant architectural changes in spaces. Robustness, reliability and a within one meter precision are relevant requirements for this work and its consequent application.

The most widely used location system in the world is the well-known GPS satellite-based navigation system. However, GPS doesn't work well without direct line-of-sight to satellites. Never the less, there are alternatives that use radio signals,

L.M. Camarinha-Matos, S. Tomic, and P. Graça (Eds.): DoCEIS 2013, IFIP AICT 394, pp. 521–528, 2013.

ultrasound, infrared, optical technologies, or a combination of few. They all have their own strengths and weaknesses and are very useful in their respective application domains. However these systems are not always suitable for indoor positioning since they usually imply the need of specialized hardware and infrastructure that is not typically available "in hands", turning them into prohibitively expensive solutions for wide deployment: "Good applications are those that achieve an adequate equilibrium between system requirements, technological advantages, and associated costs" [1]. Audible sound, as the signal to measure distance with its time-of-flight and consequently distance, is a possible solution for this necessary equilibrium. It's not uncommon that public indoor spaces have some kind of public address sound system with speakers uniformly distributed providing good sound coverage: train or subway stations, airports, shopping plazas, museums, etc. A subway station example is actually the starting point of this research work, as previous work has been developed by the authors in the NAVMETRO project [2] concerning acoustical localization.

2 Relationship to Internet of Things

Context-aware computing is becoming more and more necessary as technology evolves and the interaction between the user and computer is made simpler. This context-aware information can provide a specialized interface, automating tasks or may allow adjusting program settings to provide a personalized experience.

Location-aware systems address the acquisition of coordinates in a grid or at least distances to reference points. Outdoors, RADAR served for local demand and GPS satellite systems for global demand. These technologies have become ubiquitous in navigation and demonstrate the usefulness of having the user's position available. Indoor, where no satellites are visible and RADAR is not suitable due to walls and a smaller scale, no unanimous solution was yet achieved. The increasing use of the radio-frequency spectrum (Wi-Fi, Bluetooth, GSM, etc.), illumination problems, visual obstructions for cameras and the sound noise that exists in every indoor environment, complicate the problem enough so that no solution is perfect, cheap, precise or robust. Most of the existing solutions fail to be inexpensive to implement, precise, robust and costless to the user. To fill this gap, a proposed localization solution is presented based in audible sound.

The initial premise is that most indoor spaces already have public address sound systems and therefore using them would allow sparing relevant installation costs. The second premise is that everyone has a cell phone. These two assumptions together, will allow filling the low cost imposed requirement necessary for a successful wide deployment. Robustness and precision are next to follow and research and development is being conducted to overcome the several difficulties that using audible sound presents. Not many authors have approached the use of audible sound due to its intrusive characteristics in the acoustic environment, especially indoors. However, previous experience in other sound based applications [2] provided ideas and knowledge to expect overcoming the several difficulties that will present.

3 Research Questions

How to use sound to automatically localize a device indoors using a common sound installation without disturbing the acoustic environment? To successively accomplish answering this apparently simple question with complex issues, one may divide the problem into four possible sub questions.

- **How to localize a device?**
  This is probably the most common subject on the research work in this area. It relies on the established framework of localization no matter the technology or the application. From the simplest example on getting the x,y coordinates (in the 2D problem) of a certain position, until the use of techniques to infer on the distance that goes from the loudspeaker to the receiver at the device's position as shown in Fig.1.

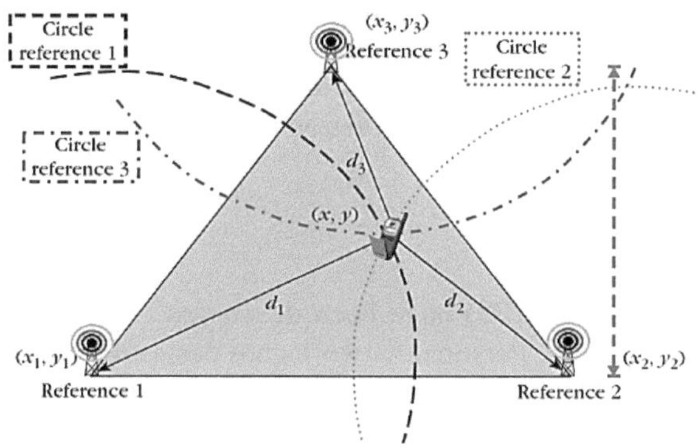

Fig. 1. Position as function of the distance from the sound sources

Localization algorithms considering this audible sound approach require measurements of range with reference fixed landmarks. Ranging can be achieved using different signal measurements such as time of arrival (TOA) or time difference of arrival (TDOA). Using sound instead of radio frequency signals can provide less numeric instability, since the speed of sound is very much smaller than the speed of light.

- **What kind of sound?**
  The choice of sound to be used as the signal to obtain position of a certain device is one of the most important decisions. One of the most obvious problem constraints is the need to disguise the necessary stimuli in the "natural" environmental sound. This is the key issue in using audible sound. Increasing environmental noise may turn this position determination not desirable for the users. An approach exploring possibilities like using high frequencies (enough to pass the loudspeaker filter and still be captured by the device microphone), sound watermarking [3] or frequency or time masking, will be performed to choose the most adequate way of successfully locating the device while not disturbing the acoustic environment.

- **Which device to use?**

To explore the most convenient scenario regarding possible applications with persons and the stated low cost requirement, the possibility of using GSM cell-phones will be investigated. This would be convenient since almost all persons use one. However many problems may arise specifically due to technology and equipment limitations, starting on its microphone (the receiver transducer), its codec (to compress the audio's digital signal), network's packet delay jitter or simply the GSM communication delay from the distance to the antennas and operator.

- **Acoustic space?**

The propagation environment may cause signal shadowing and reflections that may introduce a multipath scenario in which reflected copies of a transient signal travel longer distances than the direct line of sight. Additive combinations of reflected signals cause field intensity variations or fading. These phenomena occur in acoustic environments, and their nature and characteristics are highly dependent on the operation frequency band, and even on temperature. These errors may be reduced through a wise choice on the excitation sound and with signal processing techniques like filtering, averaging practices, and multi-observation or redundancy.

4 State-of-the-Art

Indoor positioning systems (IPS) have been developed to provide location information of persons and devices. Personal networks are designed to meet the users' needs and interconnect users' devices equipped with different communications technologies in various places to form one network. Location-aware services need to be developed in personal networks to offer flexible and adaptive personal services.

4.1 Infrared Based Systems

One example is the "Active Badge Location System" where a device produces periodic signals that will be received by a spatially wide network of sensors that localize the device's signal [4]. The Active Badge system uses diffuse IR technology to realize location sensing by estimating the location of the active badges taken along by the persons. In each located place such as a room, one or more sensors are fixed and detect the IR signal sent by an active badge. The position of an active badge can be specified by the information from these sensors, which are connected by wires and forwards the location information of the tracked active badges to a central server. Although the prices of active badges and networked sensors are cheap, the cables connecting sensors raise the cost of the Active Badge system. Interference from florescent Light and sunlight raise problems.

4.2 Ultrasound Based Systems

Ultrasound technologies like the "Active Bats" [5] or the "Crickets" [6] are also used to localize a device. Active bats, has receivers placed in a square grid under the ceiling,

1.2m apart, that are interconnected by a serial link. Mobile units subject to localization are small devices called “Bats”. They are equipped with an ultrasonic transmitter and a radio transceiver. A “Bat” that is to be located is addressed via the radio link. It then transmits a pulse of ultrasound at a known time. The Cricket location system and its extension, Cricket Compass, developed at the MIT, is a similar implementation [7]. But the general drawback of these approaches is that using standard transducers, ultrasound can usually propagate less than six meters. As a consequence a large number of precisely located reference units must be mounted.

4.3 Radio Based Systems

Due to the wide diffusion of Radio Frequency technologies, RSSI (Received Signal Strength Information) based localization techniques have been reported as a possible solution to low cost and easy to implement installations that allows tracking RF devices. It uses radio maps with a previously built signal strength model. The location of the mobile device is estimated by finding a best match from the signal strength model and the measurement with the use of deterministic and probabilistic techniques [8]. The attenuation rate is the rate α at which signal strength decreases over distance:

$$RSS \propto d^{-\propto} \quad (1)$$

As a practical rule, if $\alpha = 2$, then signal strength drops by 3dB every time distance doubles. This sub-linear attenuation rate means that the difference in signal strength between 1m and 2m is similar to the difference between 10m and 20m: exactly 3dB. Taking this into account, a constant level of noise can result in ever increasing error when signal strength is used to estimate distance; if RSSI noise is sufficient that we cannot tell the difference between 1 and 1.5m, one can't also tell the difference between 10m and 15m.

Other example approaches are the RFID base localization systems [9] that even though robust, have very low position resolution as it relies in having as many locations as RFID separate fields.

4.4 Optical Based Systems

Artificial vision based systems are also a possibility, and stereovision systems are pointing some new directions [10], but they are expensive due to the requirement of a high number of cameras and heavy processing.

Many false positives may be returned due to problems in the recognition of objects due to lighting conditions or simply due to occlusions and moving obstacles. Most techniques rely on the need to have previously trained conditions to correctly classify and track position. It is an approach that relies on controlled environments still not robust.

4.5 Inertial Based Systems

Inertial Navigation Systems (INS) together with RFID is also being explored to tackle the PL problem. The smartphone “advent” and their use of Micro-electromechanical

Systems (MEMS) allow integrating their information to get relative positioning. A possible example is an accelerometer based Positioning Scheme (APS) for tracking objects in indoor environments. The main idea of the APS is to compute an object's displacement by integrating acceleration and speed. Given the original coordinates of an object, the final position could be estimated according to the directional displacement. To overcome the "dead reckoning" problem of getting large accumulative errors, RFID techniques are used to calibrate position now and then [11].

4.6 Audible Sound Based Systems

Some audible sound-based techniques to infer on an object's or device's position are also found. Most of them use sound as a natural consequence of their operation, just like airplanes that produce noise that can be used to track them [12]. A 3-D IPS named Beep [13] was designed as a cheap positioning solution using audible sound technology. Beep uses a standard 3-D multi-lateration algorithm based on TOA measured by the Beep system sensors as a PDA or another device emits sound signals.

Other possibilities rely on having microphone arrays [14] to track some sound source positioning by AOA (angle of arrival techniques).

Another possible approach is a technique named "Acoustic Background Spectrum" where sound fingerprinting is employed to uniquely identify rooms or spaces in a passive way, just with the noise "fingerprint" of that space [15].

Even though using audible sound, these solutions aren't cheap, reliable and precise.

5 Research Contribution and Innovation

This research will investigate the use of watermarked coded audible sound in acoustic reverberant spaces with the objective of localizing a device that feedbacks the received sound to a central intelligence. Nothing was found in any previous work with these requirements and it will try to fill a void in the market of low cost, reliable and precise localization systems in public spaces. This research will have its contribution in the signal processing area where sound references will have to be processed to eliminate reverberations, noise and multipath disturbances. Coding and watermarking will also have to be explored and new techniques investigated so that information can travel along with the "natural" installation sound without being perceived by the listeners together with other concurrent information in the same acoustic space, i.e. multiuser. Exploring the use of the cell phone audio channel to close the loop will also require deep investigation so that no other hardware "tag" device is required therefore lowering the solution cost making more simple its wide dissemination.

6 Preliminary Results

Experiences were conducted with dozens of persons using only the human hearing and a cell phone to interact with a central intelligence [2]. It provided validation for the use of audible sound in a public (noisy) space. Using pleasant bird sounds (chirp like in content) as cues provided the experimental proof that one may use sound in public spaces without disturbing the comfort of people in the same space while providing good navigation references [16]. Using advanced masking techniques (frequency and time) and sound watermarking [3] one can predict an even smaller impact, validating the use of sound as the signal to measure time-of-flight to infer on distance and therefore position using multi-lateration.

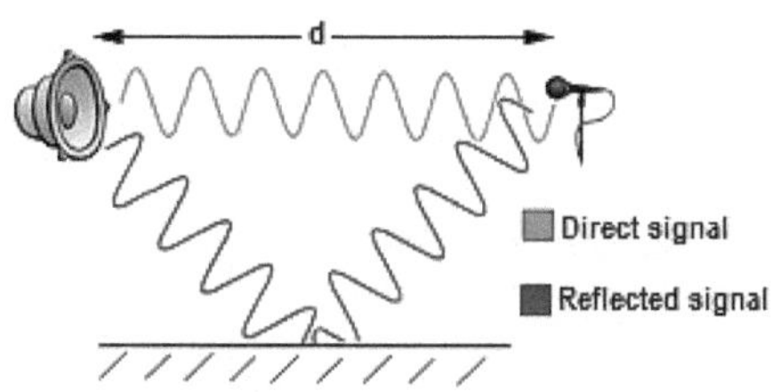

Fig. 2. Direct sound versus floor reflected sound between emitter and receiver

Results on signal processing were also already obtained in measuring time-of-flight of chirp acoustic signals in non-ideal scenarios (heavily reverberant acoustic environments). The correct wave front detection corresponding to the direct component of the emitted sound is also a compelling challenge that is necessary to overcome the precise measure time to correctly infer on distance. Using audible sound, a relatively low speed signal, is unquestionably another advantage since it requires less precision on time marking.

The front wave arrival detection instance is being performed with the use of cross correlation and Hilbert pre-envelope techniques, which has proven to be valid for measuring the time-of-flight while providing a measure for the confidence level of the measurement. In absolute distance measures, variable latency issues on the test platform are still compromising rigorous results. However, results are promising. Work-in-progress experiences are coding sound cues with embedded time code and using time difference of arrival. Results are however too preliminary.

7 Conclusions and Further Work

The presented work is still on its first developments. It is of capital importance to understand the boundaries of the problems in hands to have a good problem formulation. The research questions in hands are multidisciplinary and will require a deep and wide analysis on the subjects. A state-of-the-art was already performed and allowed to specify the requirements in hands. There was no similar approach found in the literature and therefore investigation is stepping on some new ground.

Preliminary results on a previous project showed that a similar principle with similar constraints, using a manual mechanism based in the human auditory system, provides excellent results. Turning this mechanism fully automatic do the user, while dealing with the non-ideal or controllable aspects, will be the challenge to overcome.

Acknowledgements. This work was financed through FCT with the associated PhD grant reference SFRH/BD/79048/2011.

References

1. Munoz, D., Lara, F.B., Vargas, C., Enriquez-Caldera, R.: Position Location Techniques and Applications. Academic Press (2009)
2. Moutinho, J.N.: Sistema de Apoio Acústico à Navegação Pessoal. Master's Thesis, Faculty of Engineering, University of Porto (2009)
3. Lazic, N., Aarabi, P.: Communication over an acoustic channel using data hiding techniques. IEEE Transactions Multimedia 8, 918 (2006)
4. Want, R., Hopper, A., Falcão, V., Gibbons, J.: The Active Badge Location System. ACM Transactions on Information Systems 40(1), 91–102 (1992)
5. Hazas, M., Hopper, A.: A Novel Broadband Ultrasonic Location System for Improved Indoor Positioning. IEEE Transactions on Mobile Computing 5(5) (2006)
6. Priyanthaa, N.B.: The cricket indoor location system: PhD Thesis, 199 p. MIT (2005)
7. Linde, H.: On Aspects of Indoor Localization, University of Dortmund (2006)
8. Whitehouse, K., Karlof, C., Culler, D.: A practical evaluation of radio signal strength for ranging-based localization. ACM SIGMOBILE Mobile Computing and Communications Review 11(1) (2007)
9. Ni, L.M., Liu, Y., Lau, Y.C., Patil. A.P.: LANDMARC: Indoor Location Sensing Using Active RFID. In: IEEE International Conference on Pervasive Computing and Communications (2003)
10. Krumm, J., Harris, S., Meyers, B., Brumitt, B., Hale, M., Shafer, S.: Multi camera Multi person tracking easy living. In: Third IEEE International Workshop on Visual Surveillance, Dublin, Ireland, pp. 1–8 (2000)
11. Hsu, C., Yu, C.: An Accelerometer based approach for indoor localization. In: Symposia and Workshops on UIC 2009 and ATC 2009 Conferences, pp. 223–227 (2009)
12. Blumrich, R., Altmann, J.: Medium-range localization of aircraft via triangulation. App. Acoustics 61(1), 65–82 (2000)
13. Mandal, A., Lopes, C.V., Givargis, T., Haghighat, A., Jurdak, R., Baldi, P.: Beep: 3D Indoor Positioning Using Audible Sound. In: Proc. IEEE CCNC, Las Vegas (2005)
14. Atmoko, H., Tan, D.C., Tian, G.Y., Bruno, F.: Accurate sound source localization in a reverberant environment using multiple acoustic sensors. Meas. Sci. Technol. 19 024003 10pp (2008)
15. Tarzia, S.P., Dinda, P.A., Dick, R.P., Memik, G.: Indoor localization without infrastructure using the acoustic background spectrum. In: MobiSys 2011, pp. 155–168. ACM, NY (2011)
16. Moutinho, J.N., Freitas, D.S.: Sound cue selection methods for source localization in noisy environments. In: Proceedings Inter-Noise (2010)

Part XVIII
Electronics: Devices Design

Enhancing a Layout-Aware Synthesis Methodology for Analog ICs by Embedding Statistical Knowledge into the Evolutionary Optimization Kernel

Frederico Rocha, Ricardo Martins, Nuno Lourenço, and Nuno Horta

Instituto de Telecomunicações, Instituto Superior Técnico – TU Lisbon
IST-Torre Norte, AV. Rovisco Pais, 1049-001 Lisboa, Portugal
{frocha,ricmartins,nlourenco,nuno.horta}@lx.it.pt

Abstract. This paper applies to the scientific area of electronic design automation (EDA) and addresses the automatic sizing of analog integrated circuits (ICs). Particularly, this work presents an innovative approach to enhance a state-of-the-art layout-aware circuit-level optimizer (GENOM-POF), by embedding statistical knowledge from an automatically generated gradient model into the multi-objective multi-constraint optimization kernel based on the NSGA-II algorithm. The approach was validated with typical analog circuit structures, using the UMC 0.13 μm integration technology, showing that, by enhancing the circuit sizing optimization kernel with the gradient model, the optimal solutions are achieved, considerably, faster and with identical or superior accuracy. Finally, the results are Pareto Optimal Fronts (POFs), which consist of a set of fully compliant sizing solutions, allowing the designer to explore the different trade-offs of the solution space, both through the achieved device sizes, or the respective layout solutions.

Keywords: Analog Integrated Circuits Design, Automatic Sizing, Electronic Design Automation, Evolutionary Computation, Gradient Model.

1 Introduction

In the System-on-Chip (SoC) age it is common to find devices where the whole system is integrated in a single chip, this is done to reduce production costs and increase performance. These complex integrated circuit (IC) designs are established in telecommunications, medical and multimedia applications, where blocks of Analog and Mixed-Signal (AMS), digital processors and memory blocks appear together [1]. Presently most functions in mixed-signal ICs and SoC designs are implemented using digital or digital signal processing (DSP) circuitry, where analog blocks constitute only a small part of the components, being essentially the link between digital circuitry and the continuous-valued external world. However, when integrating digital and analog circuits together on the same die, it becomes notorious that the development time of analog blocks is much higher when compared to the digital counterpart [2]. This difference is due to that analog design in general is less

L.M. Camarinha-Matos, S. Tomic, and P. Graça (Eds.): DoCEIS 2013, IFIP AICT 394, pp. 531–538, 2013.

systematic, more heuristic and knowledge intensive than digital, and the lack of maturity of the EDA that are in fact used by analog designers.

Historically, the tools for automated circuit sizing are classified as knowledge-based or optimization based, and an extensive analysis of the related work can be found in [1]. The early strategies [5] tried to systematize the design by using a pre-designed plan built with equations and a design strategy provided by the designer, which will produce component sizes that meet the performance requirements. Despite the short execution time, deriving the design plan is hard and time-consuming, and it requires constant maintenance in order to keep it up to date with technological evolution, also, the results are not optimal, suitable only as a first-cut-design.

Aiming for optimality, the next generations of sizing tools apply optimization techniques. Based on the evaluation techniques employed, the optimization-based sizing tools can be further classified into two main sub-classes, respectively, equation-based and simulation-based. The equation-based approaches [6][7] use analytic design equations to evaluate the circuit's performance during the optimization loop. The main drawbacks are that not all design characteristics can be easily mapped by analytic equations and the approximations introduced in the equations yield low accuracy designs, suited only to derive first-cut designs. The simulation-based approaches [8] [9] [10] use an electrical simulator to evaluate the circuit's performance. The strong points of this approach are generality and easy-and-accurate model, however, typified by long execution time. In order to cope with this limitation alternative approaches have been explored, e.g., use equations to derive an approximate initial solution, use parallel mechanisms that shares the evaluation load among multiple computers, use macro modeling techniques to speed up the evaluation of the circuit's performance, etc.

In this paper, a methodology to enhance the state-of-the-art layout-aware circuit-level optimizer, GENOM-POF [3], by adding circuit specific knowledge that is automatically extracted using machine learning techniques is described. The Gradient Model, here introduced, is embedded in the genetic operators of the NSGA-II [4] optimization kernel and is generated by sampling the design space, extracting and ranking the contributions of each design variable to each performance measure or objective, and, finally, building the model based on a set of gradient rules.

This paper is organized as follows: next section briefly highlights the contributions to technological innovation; then, in section 3, the enhanced GENOM-POF with Gradient Model is described; afterwards, in section 4, the achieved results are discussed; and finally, in section 5, the conclusions are presented.

2 Contribution to Internet of Things

The implementation of Internet of Things requires low power circuits using challenging integration technologies. The design of such circuits includes the design of analog-to-digital and digital-to-analog interfaces which are highly specialized and time consuming, even for expert designers. The electronic design automation is a fundamental research area supporting the designer to find optimal implementation solutions in a reduced time frame.

3 GENOM-POF Enhanced with the Gradient Model

GENOM-POF is part of the AIDA [11], an analog IC design automation framework that results from the integration of two in-house tools, GENOM-POF, and, LAYGENII [12], that performs the automatic layout generation from circuit-level specifications. Before moving to the description of how the gradient model is used to enhance GENOM-POF, the tool is reviewed and contextualized.

3.1 GENOM-POF Architecture

GENOM-POF is based on the elitist multi-objective evolutionary optimization kernel NSGA-II, and uses the industrial grade simulator HSPICE® to evaluate the performance of the design. It targets the design of robust circuits, by allowing the consideration of corner cases during optimization.

The inputs are the circuit and test-benches in the form of HSPICE® netlists, and the layout template required by LAYGEN II to instantly generate the floor plan of each of the sizing solutions. The designer also defines ranges for the optimization variables, design constraints, and optimization objectives. Then, GENOM-POF models the circuit as an optimization problem, defined by the tuple *{x - optimization variables, F(x) -objective functions, G(x) - constraint functions}* and suitable to be optimized by the NSGA-II kernel. The output is a family of Pareto optimal sized circuits that fulfill all the constraints and represent the feasible tradeoffs between the different optimization objectives.

In this work, the gradient model is integrated in GENOM-POF by embedding the extracted circuit knowledge into the evolutionary kernel operators increasing their efficiency. The enhanced GENOM-POF architecture is shown in Fig. 1, and the next section describes in detail how the integration is performed.

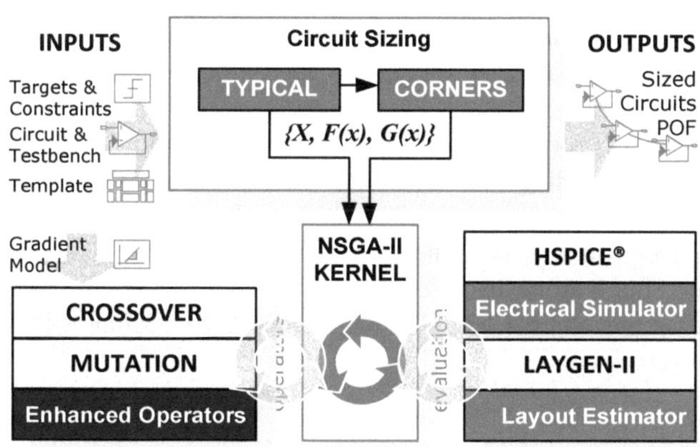

Fig. 1. GENOM-POF architecture with the integration of the Gradient Model

3.2 Gradient Model Generation

The automatic generation of the Gradient Model is based on the Design of Experiments (DOE) [13] technique to sample the circuit behavior. The gradient model is generated by sampling the circuit design space (using DOE), extracting and ranking the contributions of each design variable (input) to each design performance or objective (output), and finally, building a set of gradient rules that will be used to enhance GENOM-POF. Two approaches of DOE will be used in this work, full factorial design and fractional factorial design. The number of samples, electrical simulations, required to construct the DOE's matrix (or just matrix), of both strategies obeys to the following equation:

$$Number\ of\ simulations = B^{(n-p)} \tag{1}$$

where B is the number of points per variable or matrix base ($B > 1$), n is the number of input variables and p the number of non-elementary input variables. In the full fractional DOE the circuit is sampled in all the combinations of variables' values. For each variable (x_i), B logic levels are defined, and to each value, it is assigned a value $v_{i,b}$ derived from the variable's range according to eq. 2.

$$v_{i,b} = X_i^{Min} + \frac{(X_i^{Max} - X_i^{Min})}{2B} \times (1 + 2b), \qquad b = 0, \ldots, B-1 \tag{2}$$

The next step is to perform the statistical analysis of the experiments in order to understand which variables affect most the outputs; this is called the main effect. The main effect is the effect of one independent (input) variable on the dependent (output) variable, ignoring the effects of all other independent variables0, where $m_{i,j}$, the main effect of input variable i in the output variable j is computed according to eq. 3.

$$m_{i,j} = \sum_{k=1}^{B^{n-p}} w_{i,k} \times y_k, \qquad w_{i,k} = \begin{cases} +1\ when\ x_{i,k} \geq B/2 \\ -1\ when\ x_{i,k} < B/2 \end{cases} \tag{3}$$

where k identifies the sample and y the output measure for the sample. When the total main effect of an input variable is positive/negative, this is an indication that if the value of that input variable is increased, the value of the output will tend to increase/decrease.

Then, a refinement procedure is executed. For each output variable y_j, a new DOE matrix is constructed using the fractional factorial sampling, with the N input variables that have the larger contributions as the only elementary variables.

The refined DOE matrix is then converted to the set of gradient rules for that output variable. This is done by discarding the columns referring to non-elementary variables and transforming the levels of the elementary variables $x_i$into input gradient symbols $Si_{i,j,k}$ according to:

$$Si_{i,j,k} = \begin{cases} (+)\ when\ x_{i,k} \geq B/2 \\ (-)\ when\ x_{i,k} < B/2 \end{cases} \tag{4}$$

where k identifies the line of the matrix. The output gradient symbols So are converted from the output values as:

$$So_{j,k} = \begin{cases} (+)\ when\ y_{j,k} \geq Y_j^{Max} - \Delta_j \\ (U)\ when\ \left(Y_j^{Min} + \Delta_j\right) < y_{j,k} < (Y_j^{Max} - \Delta_j) \\ (-)\ when\ y_{j,k} \leq Y_j^{Min} + \Delta_j \end{cases} \quad (5)$$

where Y_j^{Max} and Y_j^{Min} are, respectively, the maximum and minimum values of the output y_j obtained in the DOE matrix (not the refined DOE matrix), and Δ_j is $\left|Y_j^{Max} - Y_j^{Min}\right|/3$. The meanings of the symbols are: (-) decrease; (+) increase and (U) undefined.

3.3 Gradient Model applied to the Mutation Operator

The integration of the Gradient Model into GENOM-POF is done in two fronts, first by embedding it in the evolutionary operator of mutation and second by adding the Gradient Model setup interface to the AIDA graphical user interface.

In GENOM-POF the chromosome is represented by the vector of continuous variables $= \{x_1, \dots, x_n\}$ representing the design variables. In order to speed up the convergence of the algorithm the gradient model is used to introduce design knowledge into the mutation operator.

The mutation operator in GENOM-POF uses the continuous valued operator introduced Deb and Goyal in [4]. In this operator, δ_i defined as $\delta_i = (x_i^M - x_i)/(X_i^{Max} - X_i^{Min})$, where x_i^M and x_i are the mutated and original values respectively, is the mutation perturbation applied. $\delta_i$is a random variable, with values in [-1,1] and *p.d.f.*

$$P(\delta) = 0.5\ \times (\eta + 1) \times (1 -\ |\delta|)^{\eta} \quad (6)$$

A factor of disturbance$\bar{\delta}$of δcan be obtained from an uniform random number $u\ \epsilon\ [0, 1[$ using eq. 6, which is obtained from eq. 7by solving $\int_{-1}^{\bar{\delta}} P(\delta) = u$.

$$\bar{\delta} = \begin{cases} (2u)^{\frac{1}{\eta+1}} - 1, & if u < 0.5 \\ 1 - [2(1-u)]^{\frac{1}{\eta+1}}, & if\ u \geq 0.5 \end{cases} \quad (7)$$

The mutated value, x_i^M, is given by $x_i^M = x_i + \bar{\delta}_i(X_i^{Max} - X_i^{Min})$. The gradient rules are then applied. The application of the rules follows the expression in eq. 8:

$$x_i^G = \left(1 + \mu \cdot \gamma(Si_i)\right)x_i^M \quad (8)$$

where x_i^G is the variable value after the application of the rule, $\gamma(Si_i)$ is a function of the gradient symbol defined in eq. 9, and $u\ \epsilon\ [0, c[$ is a uniformly distributed random number between 0 and c, the change rate model parameter.

$$\gamma(Si_i) = \begin{cases} +1\ when\ Si_i = (+) \\ -1\ when\ Si_i = (-) \end{cases} \quad (9)$$

The rules are selected by searching for each optimization objective if there is a rule that causes the desired effect in the corresponding response variable. Finally, fig. 2 illustrates how the automatic generated gradient model is applied to the mutation operator.

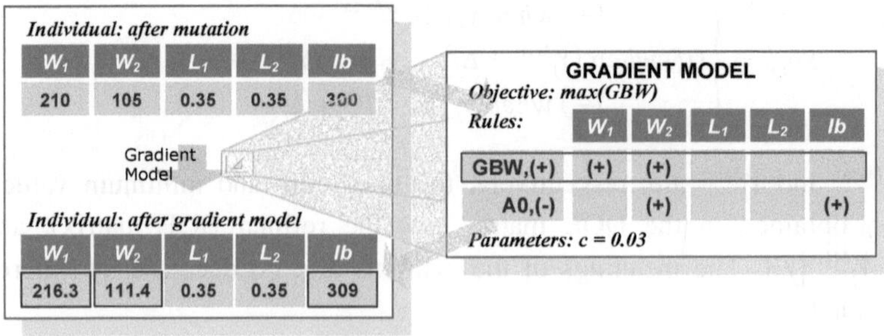

Fig. 2. Example of applying the Gradient Model to the mutation operator

4 Results

The proposed methodology was tested on an Intel® Core™ 2 Quad CPU 2.4 GHz with 6 GB of RAM and multi-threads to perform the evaluation process of each population.

The circuit used to compare the GENOM-POF with GENOM-POF integrated with Gradient Model is the single ended folded cascade amplifier, presented in Fig. 3(a)(b). For the setup of this comparison the items required were the net list and the test bench of the circuit. This case study was done considering 15 input variables, 2 objectives and 19 constraints defined in Fig. 3(c).The optimization variables are the widths and lengths of the cascade bias tensions *vbnc* and *vbpc*, and the bias current. This circuit is optimized in both GENOM-POF and GENOM-POF integrated with Gradient Model in exactly the same conditions, for a fair comparison. For this study all the 15 input variables are considered, the Gradient Model was generated with a base of two (B = 2) and considering only the design variable with larger contribution (N = 1). The extracted gradient rules for the optimization objective are shown in Fig. 3(d). The model was automatically generated in less than 5 minutes and can be reused.

Fig. 4 illustrates the improvements achieved by the proposed approach. Particularly, it shows that the Gradient Model enhanced GENOM-POF by achieving better solutions at generation 2.000 than GENOM-POF at generation 2.000 or 4.000.

The Fig. 4 also shows that even for 60.000 generations GENOM-POF does not reach the maximum DC Gain obtained by the new approach.

In order to confirm that this is not an isolated case, 20 executions with different seeds were done. The output is shown in Fig. 5, were it can be seen that the inclusion of gradient model consistently lead to better solutions. The 20 runs show an average number of points in the final POF of 51.55 for GENOM-POF and 81.70 for GENOM-POF integrated with Gradient Model. Furthermore, the normalized non-dominated area, which measures the ratio between the non-dominated and dominated area in the performance planer each an average area of 0.43 for GENOM-POF and 0.20 for GENOM-POF integrated with Gradient Model. This confirms the analysis of Fig. 5, where the GENOM-POF enhanced with the Gradient Model produces more and better solutions.

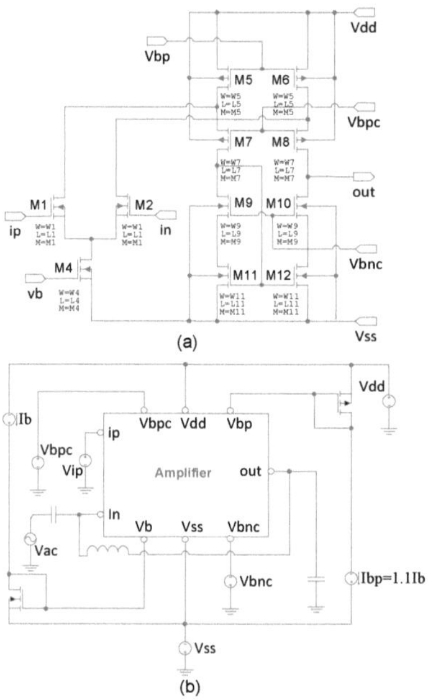

Variables	cn, cp, l1, l4, l5, l7, l9, l11, ib, w1, w4, w5, w7, w9, w11
Ranges	0.18e-6 <= l* <= 5.0e-6 0.24e-6 <= w* <= 200.0e-6 -0.4 <= cn <= 0.0 0.0 <= cp <= 0.4 30.0e-6 <= ib <= 400.0e-6
Objectives	min(area) max(a0)
Constraints	gb >= 1.2e7 a0 >= 80 55 <= pm <= 90 sr >= 1e7 ov_m(*) >= 30e-3 d_m(*) >= 1.2 osp>= 0.3 osn<= -0.3 (*) the constraint applies to: M1, M4, M5, M7, M9 and M11

(c)

Target	**Variable / Gradient**
A0, (-)	L9, (-)
A0, (+)	L9, (+)
area, (-)	W11, (-)
area, (+)	W11, (+)

(d)

Fig. 3. (a) Electrical schematic and (b) testbench of the single-ended folded cascode amplifier and(c)ranges, objectives and constraints.(d) Gradient Rules.

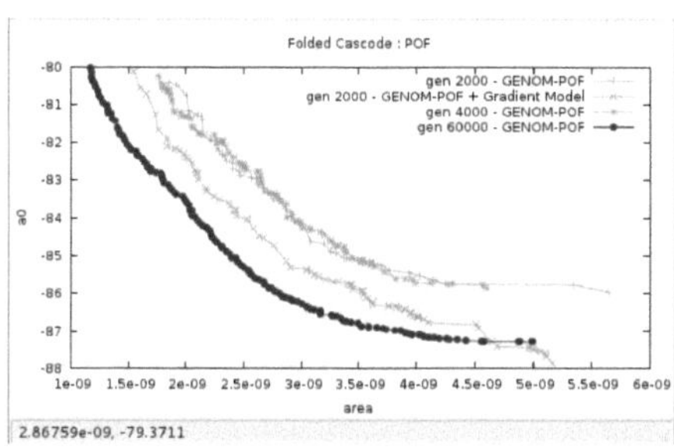

Fig. 4. Pareto Fronts: GENOM-POF (for 60.000, 4.000 and 2.000 generations) vs. Enhanced GENOM-POF (for 2.000 generations)

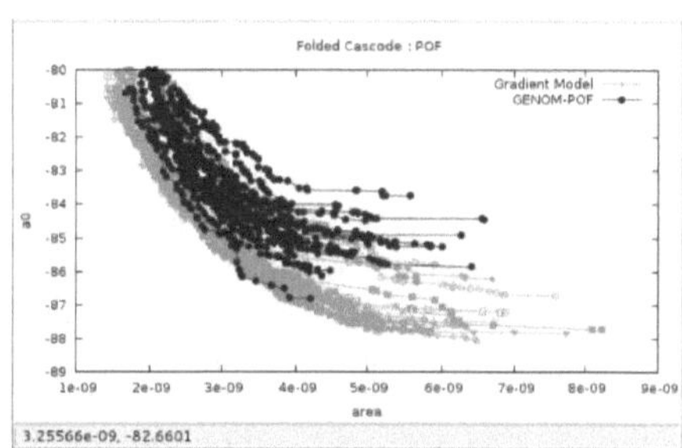

Fig. 5. GENOM-POF vs. Enhanced GENOM-POF for 20 different initial populations (for 2.000 generations)

5 Conclusions

The work presented in this paper corresponds to an innovative IC design automation approach by embedding a simple but effective design knowledge model, Gradient Model, into the evolutionary optimization kernel of a state-of-the-art sizing tool. The new technique proved to be capable to accelerate and reduce the execution time of the circuit-level optimizer GENOM-POF. This integration of the Gradient Model with GENOM-POF enhances the optimizer efficiency, forwarding the data to the desired objectives and causing a significant reduction in the number of electrical simulations. The model potential has been proved through a complex case study presented. Finally, the proposed objectives for this work were achieved and a new optimizer was created.

References

1. Gielen, G.G.E.: CAD tools for embedded analogue circuits in mixed-signal integrated systems on chip. IEE Proceeding on Computers and Digital Techniques 152(3), 317–332 (2005)
2. International Technology Roadmap for Semiconductors 2012 Edition, `http://public.itrs.net/`
3. Lourenço, N., Horta, N.: GENOM-POF: Multi-Objective Evolutionary Synthesis of Analog ICs with Corners Validation. In: GECCO 2012: Proceedings of the Fourteenth International Conference on Genetic and Evolutionary Computation Conference (July 2012)
4. Deb, K., Pratap, A., Agarwal, S., Meyarivan, T.: A fast and elitist multiobjective genetic algorithm: NSGA-II. IEEE T. Evolut. Comput. 6(2), 182–197 (2002), doi: `http://dx.doi.org/10.1109/4235.996017`
5. Degrauwe, M.G.R., Nys, O., Dijkstra, E., et al.: IDAC an interactive design tool for analog CMOS circuits. IEEE Journal of Solid-State Circuits 22(6), 1106–1116 (1987)
6. del Mar Hershenson, M., Boyd, S.P., Lee, T.H.: GPCAD: a tool for CMOS op-amp synthesis. In: International Conference on Computer-Aided Design, Digest of Technical Papers of the IEEE/ACM, November 8-12, pp. 296–303 (1998)
7. Kuo-Hsuan, M., Po-Cheng, P., Hung-Ming, C.: Integrated hierarchical synthesis of analog/RF circuits with accurate performance mapping. In: International Symposium on Quality Electronic Design, Santa Clara, California, USA, March 14-16, pp. 1–8 (2011), doi:10.1109/ISQED.2011.5770817
8. Phelps, R., Krasnicki, M., Rutenbar, R.A., et al.: Anaconda: simulation-based synthesis of analog circuits via stochastic pattern search. IEEE Transactions on Computer-Aided Design of Integrated Circuits and Systems 19(6), 703–717 (2000)
9. Barros, M.F.M., Guilherme, J.M.C., Horta, N.C.G.: Analog circuits and systems optimization based on evolutionary computation techniques. Springer, Berlin (2010)
10. Deniz, E., Dundar, G.: Hierarchical performance estimation of analog blocks using Pareto Fronts. In: Conference on Ph.D. Research in Microelectronics and Electronics Berlin, Germany, July 18-21, pp. 1–4 (2010)
11. Martins, R., Lourenço, N., Guilherme, J., Horta, N.: AIDA: Automated Analog IC Design Flow from Circuit Level to Layout. In: SMACD 2012, International Conference on Synthesis, Modeling, Analysis and Simulation Methods and Applications to Circuit Design (September 2012)
12. Martins, R., Lourenco, N., Horta, N.: LAYGEN II: Automatic Analog ICs Layout Generator based on a Template Approach. In: Proceedings of the Genetic and Evolutionary Computation Conference, Philadelphia, USA, July 7-11 (2012)
13. Montgomery, D.C.: Design and Analysis of Experiments, 5th edn. John Wiley and Sons, New York (2001)

Optoelectronic Logic Functions Based on Reconfigurable SiC Multilayer Devices

Manuel A. Vieira[1,2], Vitor Silva[1,2], Paula Louro[1,2], Manuela Vieira[1,2,3], and Manuel Barata[1,2]

[1] Electronics Telecommunications and Computer Dept, ISEL, Lisbon, Portugal
[2] CTS-UNINOVA, Quinta da Torre, 2829-516, Caparica, Portugal
[3] DEE-FCT-UNL, Quinta da Torre, 2829-516, Caparica, Portugal

Abstract. WDM multilayered SiC/Si devices based on a-Si:H and a-SiC:H filter design are approached from a reconfigurable point of view. Results show that the devices, under appropriated optical bias, act as reconfigurable active filters that allow optical switching and optoelectronic logic functions development. Under front violet irradiation the magnitude of the red and green channels are amplified and the blue and violet reduced. Violet back irradiation cuts the red channel, slightly influences the magnitude of the green and blue ones and strongly amplifies de violet channel. This nonlinearity provides the possibility for selective removal of useless wavelengths. Particular attention is given to the amplification coefficient weights, which allow taking into account the wavelength background effects when a band needs to be filtered from a wider range of mixed signals, or when optical active filter gates are used to select and filter input signals to specific output ports in WDM communication systems. A truth table of an encoder that performs 8-to-1 multiplexer (MUX) function is presented.

Keywords: Optoelectronic logic functions, Reconfigurable devices, Numerical simulation, Transfer characteristics, Coder-encoder devices, Encoder truth tables.

1 Introduction

Optical communication in the visible spectrum usually interfaces with an optoelectronic device for further signal processing. There has been much research on semiconductor devices as elements for optical communication [1, 2, 3].

Multilayered structures based on amorphous silicon technology are predictable to become reconfigurable to perform WDM optoelectronic logic functions and provide photonic functions such as signal amplification and switching [4]. They will be a solution in WDM technique for information transmission and decoding in the visible range [5]. Amplification and amplitude change are two key functionality properties outcome of a balanced interaction between wavelength of the optical signal and background wavelength and placement within a WDM link. Any change in any of these factors will result in filter readjustments. An algorithm to decode the information is presented and support new optoelectronic logic architecture.

L.M. Camarinha-Matos, S. Tomic, and P. Graça (Eds.): DoCEIS 2013, IFIP AICT 394, pp. 539–546, 2013.

2 Internet of Things

Future advances in computer speeds are contingent on augmenting intra- and inter-chip bandwidth. Increases in power efficiency per bit of data is projected to be achieved by replacing electrical interconnects with their optical counterparts in the near future. The implementation of photonic circuits for the purpose of chip-level communications requires devices such as modulators, detectors and wavelength division multiplexers (WDM).

Reconfigurable multi-rate next generation optical networks are currently investigated to handle the ever increasing growth of the Internet traffic. Reconfigurable wavelength selectors that allow for operation on a large number of wavelength channels, with dynamic response, are essential sub-systems for implementing reconfigurable WDM networks and optical signal processing.

In this paper, we demonstrate that is possible, using the amorphous Si/C technology, to produce reconfigurable SiC multilayer devices acting simultaneously as photodetectors, selectors and active filters. Therefore, the theoretical approach presented in this research is useful in examining a particular aspect of the optical communications, configuring firm activities towards innovation and combining unique resources and capabilities. Devices based on new materials for signal (de)multiplexing in the visible spectrum are an important contribute to the future development of the Internet of Things.

3 Device Configuration and Operation

The device consists of a p-i'(a-SiC:H)-n/p-i(a-Si:H)-n heterostructure with low conductivity doped layers (Fig.1). The thicknesses and optical gap of the front i'- (200 nm; 2.1 eV) and back i- (1000 nm; 1.8 eV) layers are optimized for light absorption in the blue and red ranges, respectively [6, 7]. As a result, both front and back pin structures act as optical filters confining, respectively, the blue and the red optical carriers.

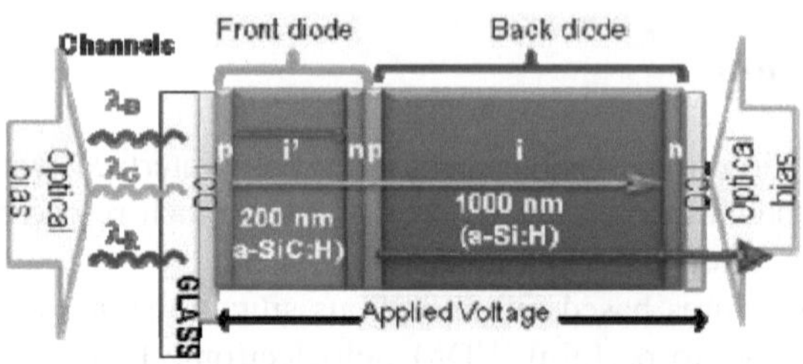

Fig. 1. Device configuration and operation

Several monochromatic pulsed lights, separately ($\lambda_{R,G,B}$ input channels) or in a polychromatic mixture (multiplexed signal) at different bit rates illuminated the device from the glass side. Steady state optical bias with different wavelength are superimposed (400nm-800 nm) from the front or from the back sides and the generated photocurrent measured at -8 V. The device operates within the visible range using as input color channels (data) the wave square modulated light (external regulation of frequency and intensity) supplied by a red (R: 626 nm; 51 μW/cm^2), a green (G: 524 nm; 73 μW/cm^2) and a blue (B: 470 nm; 115 μW/cm^2) LED. Additionally, steady state red, green, blue and violet (background) was superimposed by different LED's (R: 625 μW/cm^2, G: 515 μW/cm^2, B: 400 μW/cm^2, V: 2800 μW/cm^2).

4 Photonic Active Filters

The spectral sensitivity was tested through spectral response measurements under different frequencies, with and without steady state optical bias applied either from the front or back side.

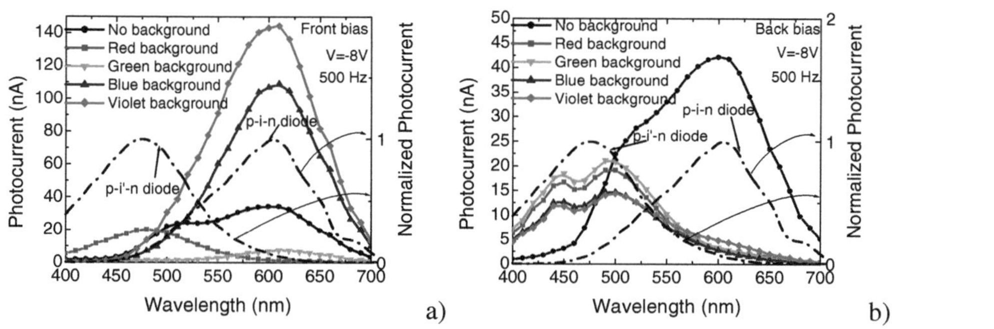

Fig. 2. Photocurrent without and with front (a) and back (b) backgrounds. The current of the individual photodiodes are superimposed (dash lines).

In Fig. 2, the spectral photocurrent (symbols) is displayed under red, green, blue and violet background and without it. In Fig. 2a the steady state optical bias was applied from the front side and in Fig. 2b from the back side. For comparison the normalized spectral photocurrent for the front, p-i'-n, and the back, p-i-n, photodiodes (dash lines) are superimposed.

Data shows that the front and back diodes, separately, presents the typical response of single p-i-n cells with intrinsic layers based on a-SiC:H or a-Si:H materials, respectively. The front diode cuts the wavelengths higher than 550 nm while the back one rejects the ones lower than 500 nm. The overall device presents an enlarged sensitivity when compared with the individual ones.

Under front irradiation the sensitivity is much higher than under back irradiation. Under front irradiation (Fig. 2a) the violet background amplifies the spectral sensitivity in the visible range while the blue optical bias only enhances the spectral sensitivity in the long wavelength range (>550 nm) and quenches it in the others. Under red bias, the photocurrent is strongly enhanced at short wavelengths and disappears for wavelengths higher than 550 nm. Under green the sensitivity is strongly reduced in all the visible spectra. In Fig. 2b, whatever the wavelength of the backgrounds, the back irradiation strongly quenches the sensitivity in the long wavelength range (>550 nm) and enhances the short wavelength range. So, back irradiation, tunes the front diode while front irradiation, depending on the wavelength used, can tune the back one.

In Fig. 3 the spectral gain, defined as the ratio between the spectral photocurrents under red (α^R), green (α^G) blue (α^B) and violet (α^V) illumination and without it is plotted at 3500 Hz. The optical bias is applied from the front side, in Fig. 3a and from the back side, in Fig. 3b.

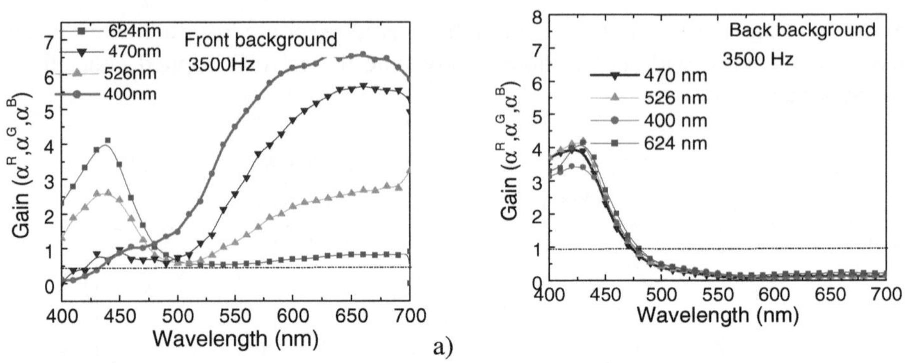

a) b)

Fig. 3. Spectral gain under red (α^R), green (α^G), and blue (α^B) optical bias, applied from the front (a) and the back (b) sides at 3500Hz

Under front bias and red irradiation, the gain is high at short wavelengths and strongly lowers for wavelengths higher than 500 nm, acting as a short-pass filter. Under green background, the device behaves as a band-stop filter that screens out the medium wavelength range (475 nm-550 nm) enhancing only the photocurrent for wavelengths outside of that range. Under blue and violet light the device works as a long-pass filter for wavelengths higher than 550 nm, blocking the shorter wavelengths. Back light, whatever the frequency, leads to a short-pass filter performance. Results show that by combining the background wavelengths and the irradiation side, the short-, medium- and long-spectral region can be sequentially tuned.

5 Encoder and Decoder Device

To analyze the device under information-modulated wave and uniform irradiation, three monochromatic pulsed lights separately (red, green and blue input channels, Fig. 4) or combined (multiplexed signal, Fig.5) illuminated the device at 6000 bps. Steady state violet optical bias was superimposed separately from the front (solid lines) and the back (dash lines) sides and the photocurrent generated measured at -8 V. The transient signals were normalized to their values without background.

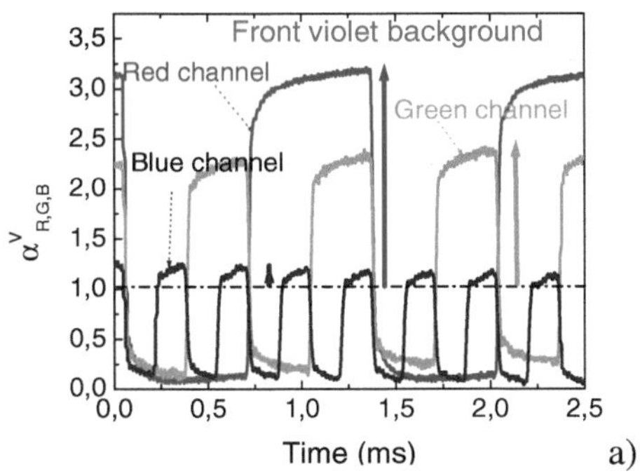

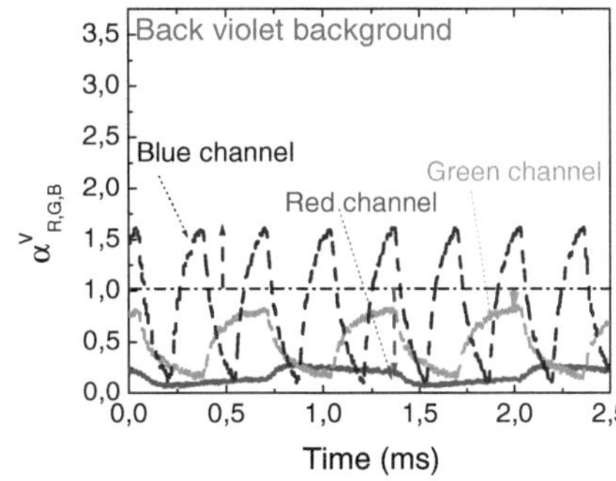

Fig. 4. Normalized red, green and blue transient signals at -8V with violet (400 nm) steady state optical bias applied from the front and the back sides

Results show that, even under transient conditions, the background side affects the signal magnitude of the color channels. Under front irradiation, it enhances mainly the light-to-dark sensitivity in the medium-long wavelength ranges. Violet radiation is absorbed at the top of the front diode, increasing the electric field at the least absorbing cell, the back diode. Taking into account Fig. 4, the red and green collections are strongly enhanced ($\alpha^V{}_G$=2.2, $\alpha^V{}_R$=3.1) while the blue one stays near its dark value ($\alpha^V{}_B$=1.1). Under back irradiation the small absorption depth of the violet photons across the back diode quenches the electric field and so, the red collection almost disappears ($\alpha^V{}_R$=0.2). Here, blue channel is absorbed across the front diode where the electric field is enhanced resulting in an increase collection of the blue channel ($\alpha^V{}_B$=1.6). Since the green channel is absorbed across front and back diodes its collection is balanced by the increased collection in the front diode and its reduction at the back one ($\alpha^V{}_G$=0.7).

For an optoelectronic digital capture system, opto-electronic conversion is the relationship between the optical inputs and the corresponding digital output levels. In Fig. 5 the normalized multiplexed signal under front (symbols) and back (lines) violet irradiation is displayed. On the top of the figure the signals used to drive the input channels are shown to guide the eyes into the ON/OFF channel states.

Under front irradiation (pin_1; $\alpha^V{}_R$>>1, $\alpha^V{}_G$>1 and $\alpha^V{}_B$~1) the 2^3 levels can be grouped into two main classes due to the high amplification of the red channel. The upper four levels are ascribed to the presence of the red channel ON, and the lower four to its absence, allowing the red channel decoder (4-to-1 multiplexer; long-pass filter function). Since under front irradiation the green channel is amplified the two highest levels, in both classes, are ascribed to the presence of the green channel and the two lower ones to its lack.

Under back irradiation (pin_2; $\alpha^V{}_R$<<1, $\alpha^V{}_G$<1 and $\alpha^V{}_B$>1, Fig. 4), the blue channel is enhanced; the green reduced the red almost suppressed. The encoded multiplexed signal has the eight sublevels grouped also into two main levels, the higher where the blue channel is ON and the lower where it is OFF (4-to-1 multiplexer; short-pass filter

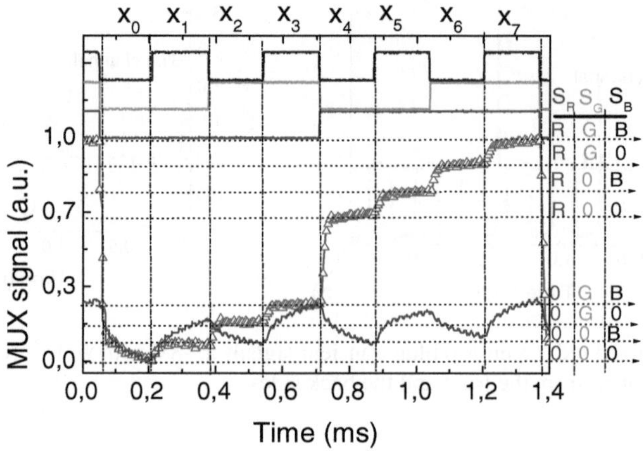

Fig. 5. Normalized multiplexed signal under front (symbols) and back (lines) violet irradiation. On the top the signals used to drive the input channels are shown to guide the eyes into the ON/OFF channel states.

function). In each main level the four existent sublevels are grouped in two classes, with and without the channel green ON. Each of those sublevels split into two near ones, attributed to the presence or absence of the red channel. If we consider this red output bit "not significant" only four separate levels (2^2) are considered and the logic MUX function is converted into a logic filter function. The blue channel is then decoded. This decoding algorithm is mapped in the right side of the multiplexed signal using a RGB code for each level (Fig. 5).

Like regular binary numbers, the binary RGB code is an arithmetic code and so, it is weighted, *i. e.* there is specific weights assigned to each bit position. Under front violet irradiation, the most significant digit, *the left most bit*, in the RGB code, is the *red* ($\alpha^V{}_R$>>1). Going from the left to right, the next is the *green* ($\alpha^V{}_G$>1) and the last is the *blue* ($\alpha^V{}_{Bpin1}$~1). Under back violet irradiation, the left most bit is the *blue,* the next is the *green* and last the *red.*

The truth tables of both encoders of Fig. 6, that perform 8-to-1 MUX function, are shown in Fig. 6. In the inputs (x_0….x_7), the index of each bit, is related to the first (highest) nonzero logic input. Here, the MUX device selects, through the violet background, one of the eight possible input logic signals and sends it to the output ($y=x_S$). The output is a three-bit [$S_2S_1S_0$] binary RGB number that may identify one of eight possible inputs.

Results show that the correspondence between the on/off state of the input channels S_R, S_G, S_B in Fig. 5 and S_2, S_1, S_0, outputs in Fig. 6, are obvious. In Fig. 6, for the input x_7 and output S_2, the first nonzero logic input is 7 ($2^2+2^1+2^0$), which corresponds an output [111]. Those OR gates are expressed, respectively, as: $S_2=x_7+x_6+x_5+x_4$ under front irradiation (Fig. 6a) and $S_2=x_7+x_3+x_5+x_1$ under back light (Fig. 6b).

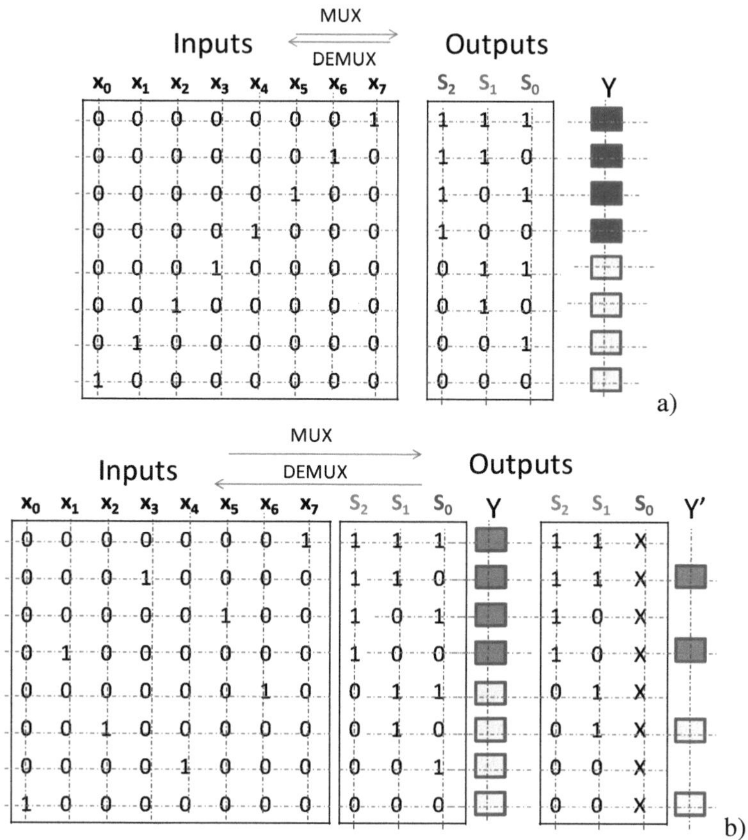

Fig. 6. Truth tables of the encoders that perform 8-to-1 multiplexer (MUX) function, under (a) front and (b) back violet irradiations (x means " not significant")

6 Conclusions

Tandem WDM a-SiC:H pi'n/pin devices are analyzed under different front and back optical bias wavelengths.

Results show that those devices, under appropriated optical bias act as reconfigurable active filters that allows optical switching and optoelectronic logic functions development. Depending on the wavelength of the external background the device acts either as a short- or a long- pass filter or as a band-stop filter.

An optoelectronic digital capture system based on this device is analyzed and the relationship between the optical inputs and the corresponding digital output levels established. A truth table of an encoder that performs 8-to-1 multiplexer (MUX) function is presented.

Acknowledgements. This work was supported by PTDC/EEA-ELC/111854/2009, PTDC/EEA-ELC/115577/2009 and PTDC/EEA-ELC/120539/2010.

References

1. Connelly, M.J.: Semiconductor Optical Amplifiers. Springer, Boston (2002) ISBN: 978-0-7923-7657-6
2. Petit, C., Blaser, M.: Photodiodes with integrated optical filters for passive optical network applications. In: Fonjallaz, P.-Y., Pearsall, T.P. (eds.) Workshop on Optical Components for Broadband Communication. Proc. of SPIE, vol. 6350, p. 63500 (2006)
3. Ibrahim, S., Luo, L.W., Djordjevic, S.S., Poitras, C.B., Zhou, I., Fontaine, N.K., Guan, B., Ding, Z., Okamoto, K., Lipson, M., Yoo, S.J.B.: Fully reconfigurable silicon photonic lattice filters with four cascaded unit cells. In: Optical Fiber Communications Conference, OSA/OFC/NFOEC, San Diego (March 21, 2010); paper OWJ5
4. Yang, H., Jung, H.-D., Zheng, Y., Huiszoon, B., van Zantvoort, J.H.C., Tangdiongga, E., Koonen, A.M.J.: OFDM Radio-over-Fibre Systems Employing Routing in Multi-Mode Fibre In-Building Networks. In: Proc. ECOC 2008, Brussels, September 21-25 (2008); paper Tu.4.F.6
5. Yang, H., Lee, S.C.J., Tangdiongga, E., Okonkwo, C.M., Boom, H.P.A., van den Breyer, F., Randel, S., Koonen, A.M.J.: 47.4 Gb/s transmission over 100m graded-index plastic optical fiber based on rate-adaptive discrete multitone modulation. Journal of Lightwave Technology 28(4), 352–359 (2010)
6. Vieira, M., Louro, P., Fernandes, M., Vieira, M.A., Fantoni, A., Costa, J.: Three Transducers Embedded into One Single SiC Photodetector: LSP Direct Image Sensor, Optical Amplifier and Demux Device. In: Betta, G.F.D. (ed.) Advances in Photodiodes, ch. 19, pp. 403–425. InTech (2011) ISBN: 978-953-307-163-3
7. Louro, P., Vieira, M., Vieira, M.A., Fernandes, M., Costa, J.: Use of a SiC:H Photodiodes in Optical Communications Applications. In: Betta, G.F.D. (ed.) Advances in Photodiodes, ch. 19, pp. 377–402. InTech (2011) ISBN: 978-953-307-163-3

Measurement of Photo Capacitance in Amorphous Silicon Photodiodes

Dora Gonçalves[1,3], L. Miguel Fernandes[1,2], Paula Louro[1,2], Manuela Vieira[1,2,3], and Alessandro Fantoni[1,2]

[1] Electronics Telecommunications and Computer Dept., ISEL, Lisbon, Portugal
[2] CTS-UNINOVA, Lisbon, Portugal
[3] DEE-FCT-UNL, Quinta da Torre, Monte da Caparica, Caparica, Portugal

Abstract. This paper discusses the photodiode capacitance dependence on imposed light and applied voltage using different devices. The first device is a double amorphous silicon pin-pin photodiode; the second one a crystalline pin diode and the last one a single pin amorphous silicon diode. Double amorphous silicon diodes can be used as (de)multiplexer devices for optical communications. For short range applications, using plastic optical fibres, the WDM (wavelength-division multiplexing) technique can be used in the visible light range to encode multiple signals. Experimental results consist on measurements of the photodiode capacitance under different conditions of imposed light and applied voltage. The relation between the capacitive effects of the double diode and the quality of the semiconductor internal junction will be analysed. The dynamics of charge accumulations will be measured when the photodiode is illuminated by a pulsed monochromatic light.

Keywords: capacitance, photodiode, amorphous silicon.

1 Introduction

1.1 Research Question and Motivation

The research question that supports the study presented in this paper is: "Is it possible to control the photo capacitance of the double pin-pin photodiode?"

The methodology used to give answer to this research question consists on the comparison of a double pin-pin photodiode based on a-Si:H/a-SiC:H with an amorphous silicon pin device and a crystalline pin diode through the measurement of capacitance-voltage and photocurrent-voltage characteristics. We are interested in the influence of the material and device structure on the photodiode capacitance in order to improve the performance of these devices for WDM applications in the visible range.

1.2 State of the Art and Related Literature. Methodological Approach

In a crystalline silicon pn junction there are two types of capacitances, related to the charge stored in the depletion layer (junction capacitance) and to the diffusion of carriers along the junction (diffusion capacitance) [1].

L.M. Camarinha-Matos, S. Tomic, and P. Graça (Eds.): DoCEIS 2013, IFIP AICT 394, pp. 547–554, 2013.

When we consider the pn junction under reverse bias conditions, there is an increase in the width and charge of the depletion layer; as the voltage across the device changes, the charge stored in the depletion layer changes as a result. We can observe a correspondence between a capacitor and the depletion layer of a pn junction; the charge stored on either side of the layer is a non linear function of the applied reverse voltage. Under a small-signal approximation the depletion capacitance, or junction capacitance, is the gradient of this curve, at the bias point. On the other hand, we can consider the depletion layer as a parallel-plate capacitor and obtain the expression for its capacitance, using the traditional procedure. When the pn junction is under forward-bias conditions, the depletion layer narrows and the depletion barrier voltage reduces; the diffusion current increases until equilibrium is achieved. In the steady state, a certain amount of excess minority carriers charge is stored in each of the bulk pn regions. If the terminal voltage changes, this charge will have to adjust, before a new steady state is achieved. This charge storage occurrence introduces another capacitive effect, which is directly proportional to the diode current: we are referring the diffusion capacitance.

Nowadays, hydrogenated amorphous silicon and hydrogenated amorphous silicon carbide have found a large application in electronic devices. In this type of pn junctions, we must have special attention to the midgap density of states and junction properties, to obtain the correct expression of the capacity. The pin configuration is frequently used in photo devices; the i-layer may be alloyed with carbon to optimize the band gap for a given application. May be the simplest and most widely applied techniques to characterize mid gap states (i.e. deep levels) in semiconductors are admittance measurements. The imaginary part of the admittance, i.e. the capacitance, is directly related to the charge trapped in or released from mid gap states, as a consequence of time-varying excitation. State occupancy is manipulated by varying the band bending associated with a Schottky barrier or pn junction. The spectral and energetic distribution of states within the space-charge region is probed by varying the frequency, temperature and bias voltage, as seen in [2].

A possible application of these type of photodiodes is in short range optical communications using the multilayered a-SiC:H heterostructure as a wavelength division demultiplexing device [3]. Nowadays optical communications demand the transmission of a huge amount of information. To increase the transmission capacity and to allow bidirectional communication over one strand fibre, wavelength division multiplexing (WDM) is used. This technique consists on combining multiple wavelength optical signals on a single optical fibre. To perform such task the WDM system uses a multiplexer at the transmitter to mix the signals and a demultiplexer at the receiver to split them apart [4, 5]. The pin-pin diode used in this analysis was designed to work as an optical filter, with selective wavelength sensitivity. Both pin structures of this device were optimized for selective collection of photo generated carriers. Band gap engineering was considered to adjust the photo generation and recombination rates of the intrinsic region of each diode, taking into account wavelength absorption and carrier collection in the visible range.

Photo capacitive effects influence the device frequency response under transient conditions. As a WDM device is intended to be used to increase the bandwidth of the optical communication system the transmission rate must be kept as high as possible. Capacitive effects result in signal degradation and induce errors at the reception end [2], which is an important limitative aspect for the device performance. In order to minimize this effect it is important to understand the mechanisms responsible for the device photo capacitance.

Capacitance values are evaluated considering the impedance of the device in study.

2 Contribution to Internet of Things

The Internet of Things is a "vision" that is being created these days. It concerns the idea of expanding communication among things. This will bring the ability of objects performance to change due to what they contact through the internet. In the particular case of optical communications in the visible range, recent research and new sensor devices, for instance WDM (wavelength-division multiplexing), will contribute to innovation and new resources and capabilities. And that will contribute to the future development of the Internet of Things.

3 Research Contribution and Innovation

Measurements of capacitance were made with different photodiodes, under different illumination conditions and changing the applied voltage. Three different devices were used in these experiments: a commercial crystalline silicon pin photodiode (Vishay Semiconductors BVP10), an amorphous silicon pin photodiode and a pin-pin structure based on amorphous silicon and amorphous silicon carbide. The a-Si:H devices were produced by PECVD (Plasma Enhanced Chemical Vapour Deposition) The pin-pin device structure consists of a p-i'(a-SiC:H)-n/p-i(a-Si:H)-n heterostructure with low conductivity doped layers. Our goal was to point out the different behaviours of the devices and to relate these differences to the structure composition and to the material used. The light incident onto the diodes was obtained by a white lamp and a colour filter covering the light source. The measurements were performed using a pulsed monochromatic light with a frequency ranging from 10 Hz up to 1 kHz. From the WDM application point of view, these frequencies are indeed very low. We have chosen this measurement condition for permitting a clear analysis about the influence of the materials and interfaces properties on the device capacitance.

4 Discussion of Results and Critical View

In Fig. 1 it is reported the CV measurement of an a-Si:H pin photodiode, under dark conditions and using illumination with different wavelengths (red: 626 nm, green: 510 nm, and blue: 430 nm). The device is biased in the range -1V up to +1V.

Results show that in dark the capacitance does not change with the applied voltage, either under reverse and forward bias. Under illumination, a different behaviour is observed, as the capacitance changes with the voltage and in the analysed bias range it takes higher values. The highest capacitances are observed under red illumination; the medium capacitance values are observed under green light; and at last the smallest ones under blue light.

Under reverse bias (-1V) the capacitance under blue and green light match the dark capacitance, while the capacitance under red illumination is slightly higher. However, as the voltage increases the capacitance under illumination increases at different rates.

This is in agreement with the classical theory of the junction capacitance.

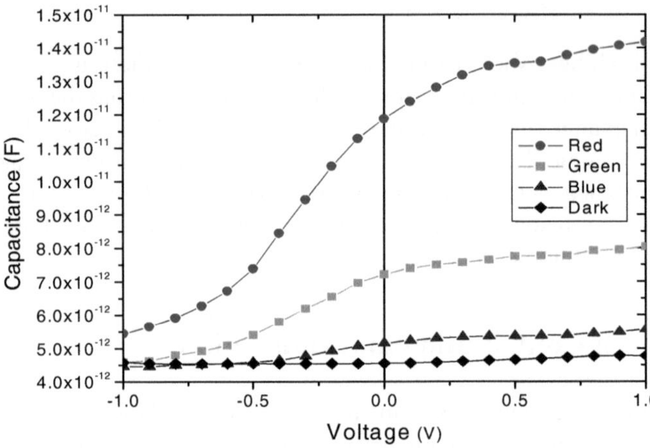

Fig. 1. Capacitance-voltage characteristics of an a-Si:H photodiode, under dark condition and under illumination with different wavelengths (red, green, and blue)

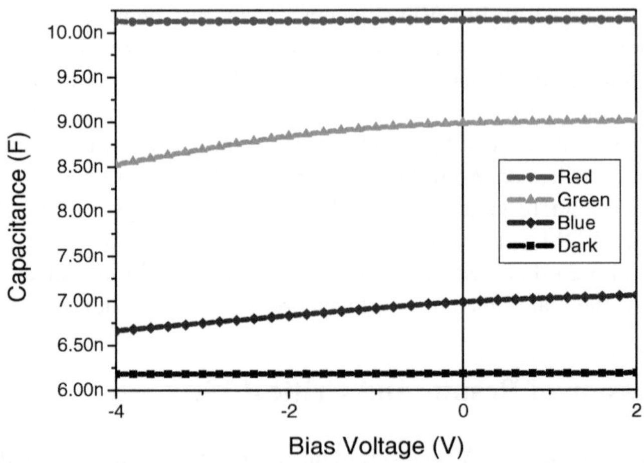

Fig. 2. Capacitance-voltage characteristics of a pin-pin photodiode, under dark condition and under illumination with different wavelengths (red, green, and blue)

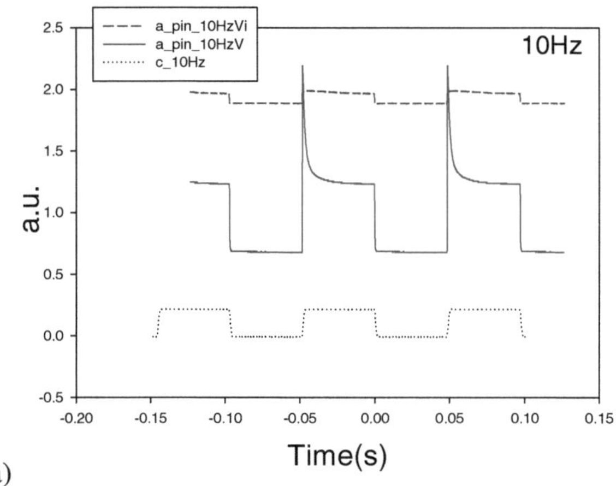

a)

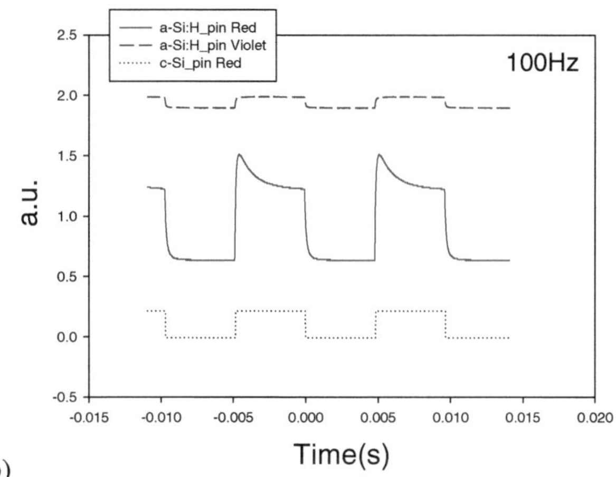

b)

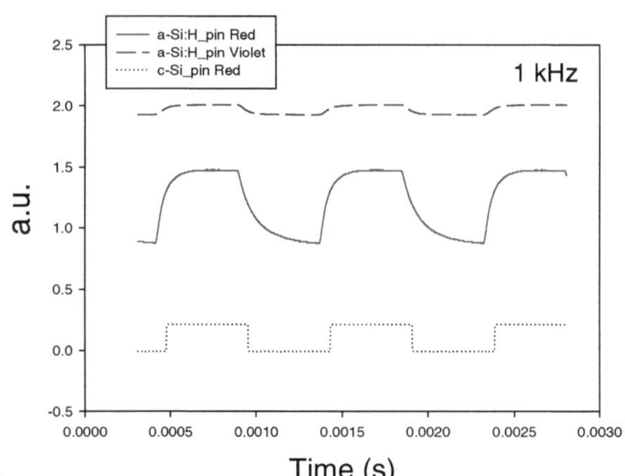

c)

Fig. 3. Comparison of the transient response of the c-Si diode under a pulsed red light and the a-SI:H pin diode under red and blue pulsed light: a) 10 Hz, b) 100Hz and c) 1 kHz

In Fig. 2 we can see the CV measurement referred to a pin-pin photodiode. The value of the light induced capacitance depends slightly on the applied bias, on the light intensity. And it varies with light wavelength greatly; the red colour corresponds to the higher values of capacitance. Comparing the pin diode with the pin-pin device we can see that both capacitances vary with applied bias and with light wavelength, being the capacitance in the pin-pin structure much higher than the pin one.

Considering imposed red light, the pin diode has a major increase in the value of capacitance, as the value of bias voltage gets higher; in the case of the pin-pin diode, the capacitance value is very high, but remains constant, regardless the bias voltage value. In the case of green and blue light, the capacitance changes are very similar.

In Fig. 3 it is reported the comparison of the transient response of the crystalline silicon diode (c-Si diode) under a pulsed red light, used as reference term, and the hydrogenated amorphous silicon diode (a-Si:H pin diode) under red and blue pulsed light, with different frequencies: 10 Hz, 100Hz and 1 kHz, namely a), b) and c).

In Fig. 3 - a), b) and c), the inferior trace of each plot shows the behaviour of the crystalline diode, under red pulsed light. We can see a similar pulsed photocurrent as a function of time, as the diode response. In the classical theory, capacitive effects of the device were not expected.

In Fig. 3 – a), b) and c), the central image in each one show the response of an amorphous pin diode under red pulsed light. A transient photocurrent, as a function of the time, is the response of this experiment. A peak value of current can be recognized; for small values of frequency, 100 Hz and specially 10 Hz, the amorphous photodiode photo current has a peak of intensity, in the beginning of each time period. It repeats the behaviour of a RC circuit in transient regime.

Using a different colour of pulsed light (violet), as we can see in the superior images of each figure 3 –a), b) and c), the amorphous silicon diode response is similar to the applied pulsed light.

In Fig. 3 – c) the frequency used is 1 kHz. The crystalline diode keeps its response stable and the amorphous photodiodes have different responses, similar to the previous plots, but on a smaller scale.

In Fig. 4 we present the time response of a double amorphous silicon device, a pin-pin diode, for different wavelengths of pulsed light. The red, green and blue imposed lights are transformed in transient photocurrents, with a maximum and a minimum values, during a time period, whose rising and decaying times are a little different. The peak values are obtained under the red colour, and the smaller ones under the blue colour, recalling the behaviour of a RC circuit in transient regime.

The charging and discharging times to and from the offset photocurrent value are different. This may lead to conclude that the two processes are made on different paths with different resistance or/and capacitance characteristics.

Observed values for the capacitance of the pin-pin are higher than for the pin device, which may be related to the presence of two double pin structures stacked

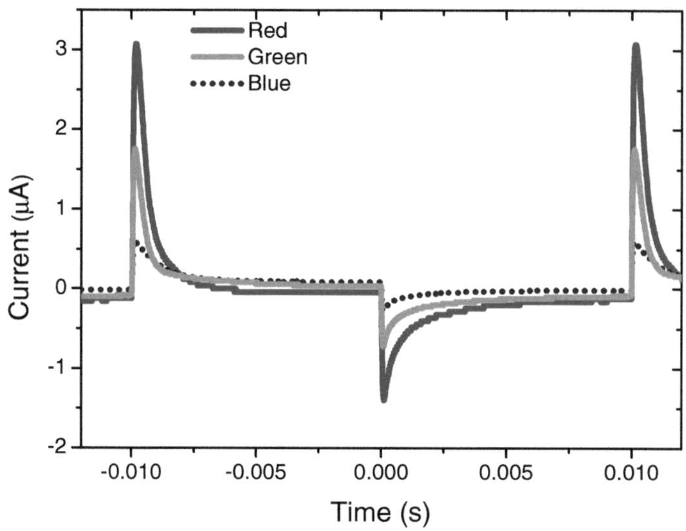

Fig. 4. Time response in a pin-pin diode, showing the photo current for different wavelengths (red, green and blue colours)

together. From this configuration results an additional internal pn junction causing a separation of the trapped charge within the two active layers of the device. Also a higher series resistance is expected in the double diode configuration.

5 Conclusions and Further Work

Diode photo capacitance varies with different parameters. Considering the crystalline diode, the photocurrent follows the imposed pulsed light showing no capacitive effect; it is independent of its frequency and colour.

In the case of an amorphous pin diode the photocurrent presents a transient behaviour that is consistent with charge and discharge of a capacitive circuit. Its highest value is present when the frequency value is low, and when the pulsed light is red.

Considering the double pin-pin diode, the time response has two peaks in a time period, one positive and the other, smaller, negative. They expose the capacitive effects in this device; capacitance values also depend on light wavelength, the highest values for the red colour. These results permitted to relate the photodiode behaviour with the device characteristics of capacitance and resistance.

In future work it will be useful to make measurements of different devices with varying thickness of the absorbing layers. Also the influence of the defect densities in the intrinsic layer should be studied. These results should be confirmed by electronic simulation based on an equivalent model of the device that is being tested. And the study will be extended to higher values of frequency, up to 1 GHz.

References

1. Sze, S.M., Ng, K.K.: Physics of Semiconductor Devices, 3rd edn. John Wiley & Sons, Inc. (2007)
2. Hegedus, S.S., Fagen, E.A.: Characterization of a-Si:H and a-SiGe:H p-i-n and Schottky Junctions by Admittance Circuit modelling. IEEE Transactions on Electron Devices 39(10) (October 1992)
3. Vieira, M., Vieira, M.A., Louro, P., Fernandes, M., Fantoni, A., Silva, V.: SiC multilayer photonic structures with self optical bias amplification. In: MRS Proceedings, vol. 1426 (2012)
4. Louro, P., Vieira, M., Vieira, M.A., Silva, T.: Mater. Res. Soc. Symp. Proc., vol. 1321 (2011)
5. Vieira, M.A., Vieira, M., Louro, P., Fernandes, M., Costa, J., Garção, A.S.: Mater. Res. Soc. Proc., vol. 1321 (2011)
6. Louro, P., Vieira, M., Vieira, M.A., Fernandes, M., Fantoni, A., Francisco, C., Barata, M.: Optical multiplexer for short range communications. Physica E: Low-dimensional Systems and Nanostructures 41, 1082–1085 (2009)

Optoelectronic Digital Capture Device Based on Si/C Multilayer Heterostructures

Vitor Silva[1,2], Manuel A. Vieira[1,2], Paula Louro[1,2], Manuela Vieira[1,2,3], and Manuel Barata[1,2]

[1] Electronics Telecommunications and Computer Dept, ISEL, Lisbon, Portugal
[2] CTS-UNINOVA, Quinta da Torre, 2829-516, Caparica, Portugal
[3] DEE-FCT-UNL, Quinta da Torre, 2829-516, Caparica, Portugal

Abstract. Combined tunable WDM converters based on SiC multilayer photonic active filters are analyzed. The operation combines the properties of active long-pass and short-pass wavelength filter sections into a capacitive active band-pass filter. The sensor element is a multilayered heterostructure produced by PE-CVD. The configuration includes two stacked SiC p-i-n structures sandwiched between two transparent contacts. Transfer function characteristics are studied both theoretically and experimentally. Results show that optical bias activated photonic device combines the demultiplexing operation with the simultaneous photodetection and self amplification of an optical signal acting the device as an integrated photonic filter in the visible range. Depending on the wavelength of the external background and irradiation side, the device acts either as a short- or a long-pass band filter or as a band-stop filter. The output waveform presents a nonlinear amplitude-dependent response to the wavelengths of the input channels. A numerical simulation and a two building-blocks active circuit is presented and gives insight into the physics of the device.

Keywords: SiC hetrostructures, Optical sensors, Optical active filters, Numerical and electrical simulations, Optoelectronic model.

1 Introduction

An optoelectronic device converts light photons to electrons that mimic the light signal in such way that data transmitted by the light beam can be received and processed further with electrical circuits. These devices have one surface over which the optical signal shines. On our device, a multilayered Si/C heterostructure [1, 2], light can shine on the two surfaces, namely back and front. This device acts as an optical filter when other fixed wavelengths superimpose the incident light data signal on the surface it shines with. By selecting a wavelength on either the red or blue part of the spectrum the device can be tuned as a filter and used as a wavelength division multiplexing-demultiplexing technique, WDM [3]. This device has been characterized with a model with its experimental verification. This paper continues this work by analyzing the digital signals which modulate the incident light beam and performs logic functions [4, 5].

L.M. Camarinha-Matos, S. Tomic, and P. Graça (Eds.): DoCEIS 2013, IFIP AICT 394, pp. 555–562, 2013.

2 Internet of Things

Internet is a well known concept that reflects the whole connection of computers forming a single international network. Behind computers are users and their communication skills looked upon this network as an excellent communication media for human interaction and thus the social network bruited. People while communicating with one another usually share ideas, emotions and objects as photographs, text, sound, video, and much more things could be handled. The Internet of Things is a growing concept since Kevin Ashton brought it forward in 1999[6]. Objects are easily identified when an RFID tag is attached to them, and its whereabouts can be known. But objects don't actually communicate. Were this possible and an Internet of physical objects can emerge. People and objects interacting; whoever leaves its home keys inside the house would not be locked outside but instead a warning signal from the keys would call to ones attention. The umbrella would also call to its attention if the weather report expects rain. An occasional party could be foreseen in the agenda and tagged clothes would shine a few leds indicating possible matches inside the cupboard and eventually for the couple, even if they were in different homes. The device we propose would certainly be integrated in objects of design which interact with people via colored visible light.

3 Device Optimization and Characterization

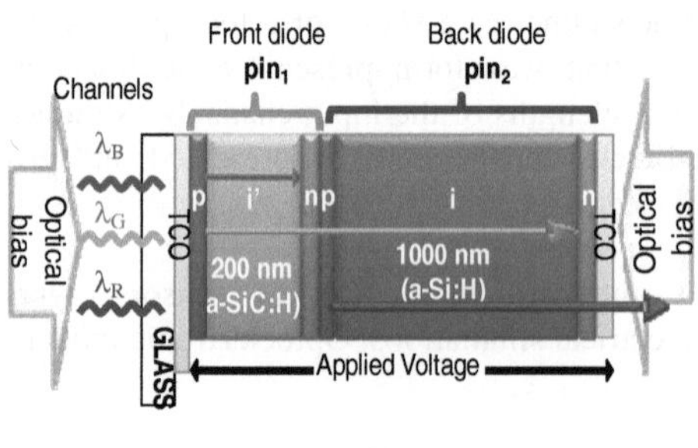

Fig. 1. Device configuration and operation

The sensor element is a multilayered heterostructure based on a-Si:H and a-SiC:H. The configuration shown in Fig. 1 includes two stacked p-i-n structures sandwiched between two transparent contacts. The thicknesses and optical gap of the front i'- (200nm; 2.1 eV) and back i- (1000nm; 1.8eV) layers are optimized for light absorption in the blue and red ranges, respectively.

Spectral response measurements without and under different optical bias and frequencies were performed in order to test the devices sensitivity.

The device operates within the visible range using as input color channels (data) the wave square modulated light (external regulation of frequency and intensity) supplied by a red (624 nm; 51 $\mu W/cm^2$), a green (526 nm; 73 $\mu W/cm^2$) and a blue (470 nm; 115 $\mu W/cm^2$) LED. Additionally, steady state violet (400 nm, 2800 $\mu W/cm^2$), red (624 nm, 652 $\mu W/cm^2$), green (526 nm, 580 $\mu W/cm^2$) and blue (470 nm, 680 $\mu W/cm^2$)

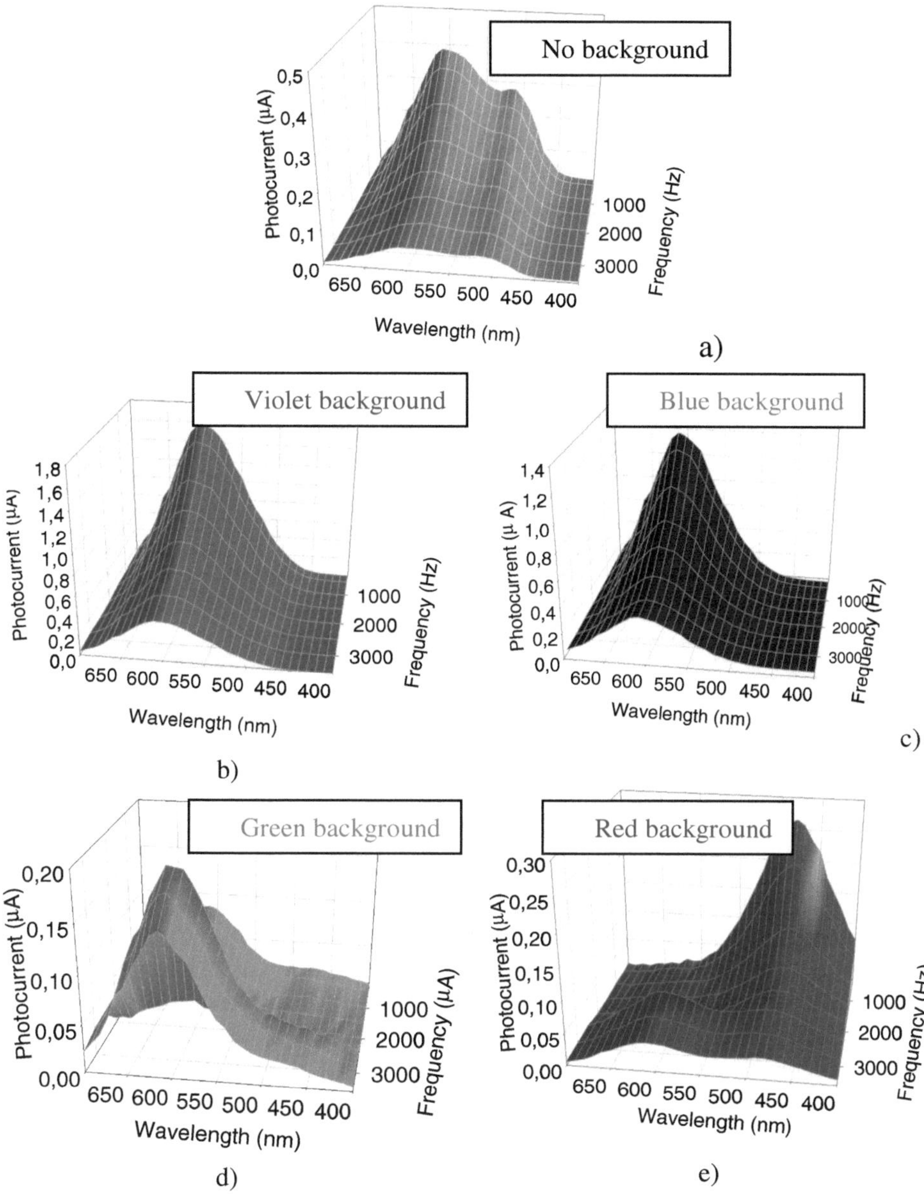

Fig. 2. Spectral photocurrent without (a) and under violet (b), blue (c), green (d) and red (e) bias control applied from the front side

illumination (optical bias) was superimposed from the front (pin_1) and back (pin_2) sides, in LEDs driven at different current values.

The spectral sensitivity was analyzed by applying different wavelengths optical bias from the front and back sides of the device (see Fig.1). Under front irradiation, in Fig. 2 it is displayed the spectral photocurrent for different frequencies without (a) and under violet (b), blue (c), green (d) and red (e) light bias control at -8V applied voltage and different frequencies (250 Hz-3500Hz).

Results show that the spectral response depends strongly on the bias control wavelength and frequency. As the bias control wavelength increases the spectral sensitivity shifts to the low wavelength spectral regions and decreases with the frequency, suggesting capacitive effect across the device.

In Fig. 3 it is displayed the spectral photocurrent at 3500 Hz, under red, green, blue and violet background irradiations (color symbols) and without it (black symbols) applied from the front (a) and back (b) diodes. For comparison the spectral photocurrent (right axis) for the front, p-i'-n and the back, p-i-n, photodiodes (dash lines) are superimposed.

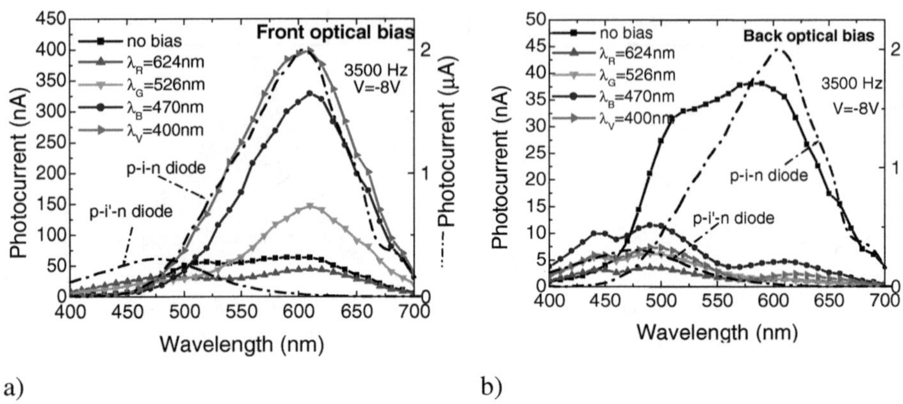

Fig. 3. Spectral photocurrent under red, green, blue and violet background irradiations (color symbols) and without it (black symbols) applied from the front (a) and back (b) diodes

Results show that the spectral sensitivity, under steady state irradiation, depends on its wavelength and on the impinging side.

Under front irradiation, the back diode photoresponse is tuned and the sensitivity strongly increases for wavelengths higher than 500 nm when compared with its value without optical bias. Here, as the background wavelength decreases the spectral response increases.

Under back irradiation the front diode photoresponse is selected. The sensitivity strongly increases in the short wavelengths range and collapse in the long wavelength region.

4 Light Filtering Effects

In Fig. 4, at 3500 Hz, it is displayed the ratio between the photocurrent under different optical bias and without it (gain) under front (symbols) and back (lines) irradiations: red (α^R), green (α^G), blue (α^B) and violet (α^V).

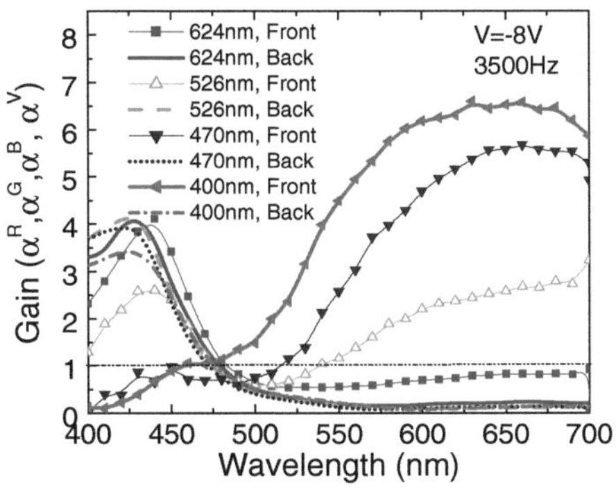

Fig. 4. Gain under front (symbols) and back (lines) irradiations: red (α^R), green(α^G), blue (α^B), violet (α^V)

Under back irradiation (lines) the gain does not depend on the wavelength of the background. The spectral sensitivity in the low wavelength range is enhanced and the device acts, always, as a short-pass filter.

Under front irradiation (symbols) the filter properties of the device depend on the background wavelength. Under red irradiation (Fig. 2e) the gain is high in the short wavelengths range acting the device as a short-pass filter. Under violet (Fig.2b) and blue (Fig.2c) irradiations the transfer function presents an enhanced gain in the long wavelength range acting as a long-pass filter. Under front green background (Fig.2d) the device, for frequencies higher than 2000 Hz, is a band-rejection active filter that works to screen out wavelengths that are within the medium range (475nm-550nm), giving easy passage at all wavelengths below and above.

Results confirm that under controlled wavelength backgrounds it is possible to fine-tune the spectral sensitivity of the device. Its sensitivity is strongly enhanced ($\alpha>1$) in a specific wavelength range and quenched ($\alpha<1$) in the others, tuning or suppressing a specific band. The sensor is a wavelength current-controlled device [7] that makes use of changes in the wavelength and impinging side of the optical bias to control the power delivered to the load. Self optical bias amplification or quenching under uniform irradiation is achieved. By using background lights with complementary light penetration depths across the pi´n/pin device and changing the irradiation side, it is possible to control the spectral response and to filter a specific wavelength band.

5 Research Question and Hypothesis

How many channels in the visible light range can be transmitted by WDM with a a-Si:H and a-SiC:H sensor selecting the desired channel with visible light?

Light shining on one of the surfaces, the selector, can be used to distinguish between two signals with different wavelengths. For n different selecting frequencies there would be 2^n different signals to be identified. Value of n is then limited by the bandwidth of each channel. By changing the wavelength of the selector the output changes due to the non linear gain of the device which will reduce the maximum number of distinguishable channels. This limitation can be overcome and the number of identifiable channels may increase if the output of the selecting wavelengths are combined with an algorithm and signal modulation. Experimental results are presented in this paper with four different visible leds which are commonly available in the market. These results were obtained with a maximum bit rate of 6000bps. Although this is a very low throughput we are not at this stage worried with this aspect, but to study the recovery of several input signals with only one sensor. In this paper we present the gains of each of the four channels. This represents the influence of each of the selectors upon each input signal wavelength.

6 Photonic Active Filters

To analyze the device under information-modulated wave and uniform irradiation, three monochromatic pulsed lights separately (red, green and blue input channels) or their combination (MUX signal) illuminated the device. Steady state red (652μWcm^{-2}), green (515μWcm^{-2}), blue (680μWcm^{-2}) and violet (2800μWcm^{-2}) optical bias was superimposed separately from the front (pin_1) and the back (pin_2) sides and the photocurrent measured at -8 V.

In Table 1 the gains (α) of the individual channels, defined as the ratio between photocurrents under irradiation (ON state) and without it (OFF state), for the red, green, and blue input channels are presented. Here, the superscripts are related to the background wavelength (R, G, B, V) and the subscripts ($Rpin_{1,2}$,$G\ pin_{1,2}$ $B\ pin_{1,2}$) to the channel color and irradiation side.

Results show that, even under transient conditions, the effect of the background wavelength and impinging side presents the same nonlinear dependence as in Fig. 4. The morphology of filter results from the interaction of the electric field under applied optical bias (red, green, blue, violet) and the transient electric field induced by the input channels (red, green, blue and violet). Under back irradiation the small absorption depth of the violet photons across the back diode quenches the electric field there and so, the red collection (Red ON) almost disappears ($\alpha^V_{R,pin2}$<<1). Blue channel is absorbed across the front diode where the electric field was enhanced resulting in an increase collection of the blue channel ($\alpha^V_{B,pin2}$>1). Since the green channel is absorbed across front and back diodes its collection is balanced by the increase collection in the front diode and its reduction at the back ($\alpha^V_{G,pin2}$~1). The front violet background is absorbed at the surface of the front diode, increasing the electric field at the back diode, where the red and part of the green channels generate optical carriers. So, the collection is strongly enhanced ($\alpha^V_{R,pin1}$>>1, $\alpha^V_{G,pin1}$>1) while the blue collection stays near its dark value ($\alpha^V_{B,pin1}$~1).

Polychromatic combinations of the same red, green, blue and violet input channels whose gain is presented in Table I was used to generate a multiplexed (MUX) signal.

In Fig. 5 the filtered signals under front (pin_1) and back (pin_2) violet light control are displayed. The signals were normalized to the maximum intensity under violet front irradiations to suppress the dependence on sensor and LEDs positioning. The bit sequences used to transmit the information are shown at the top of the figures.

Table 1. Gains ($\alpha^{R,G,B,V}_{R,G,B,pin1,2}$) at the input red, green, and blue channels wavelengths

Channels	α^R	α^G	α^B	α^V
$\alpha_{R,pin1}$	0,62	1,28	3,00	5,13
$\alpha_{R,pin2}$	0,12	0,42	0,47	0,34
$\alpha_{G,pin1}$	0,63	0,85	1,52	2,40
$\alpha_{G,pin2}$	0,59	0,62	0,67	0,67
$\alpha_{B,pin1}$	1,39	1,00	0,81	1,11
$\alpha_{B,pin2}$	2,22	1,94	1,97	1,64
$\alpha_{V,pin1}$	11,57	5,80	1,60	1.00
$\alpha_{V,pin2}$	11,57	13,90	1,20	10,42

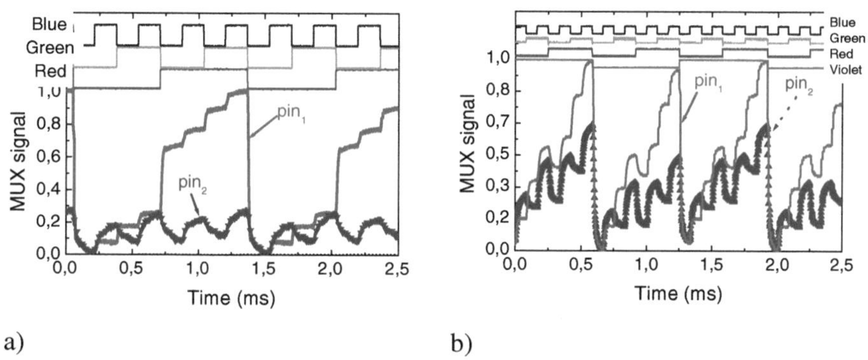

Fig. 5. Filtered output signals under front (pin_1; lines) and back (pin_2; symbols) violet irradiation. a) Red, Green and Blue channels. b) Red, Green, Blue and violet channels. On the top of the figures the optical signals used to transmit the information guide the eyes.

Different gains for the RGB channels were obtained (Table 1). Due to this wavelength non-linearity under front violet background, the encoded multiplexed signal presents as many levels as the possible RGB combinations, in a maximum of 2^3 (eight-level encode).Those levels can be grouped into two main classes due to the high amplification of the red channel ($\alpha^V_{R,pin1}>>1$). The upper levels are ascribed to the presence of the red channel and the lower to its absence allowing the red channel decoder. Since under front irradiation the green channel is amplified ($\alpha^V_{G,pin1}>1$), the highest levels, in both classes, are ascribed to the presence of the green channel and the lower ones to its lack (long-pass filter). Under back irradiation the red channel is suppressed ($\alpha^V_{R,pin2}<<1$), the blue enhanced ($\alpha^V_{B,pin2}>1$) and the green reduced($\alpha^V_{G,pin2}<1$), so the encoded multiplexed signal presents a maximum of four separate levels (2^2). The highest levels correspond to the presence of the blue channel ON with or without the green ON respectively, and the other to its absence. The blue channel is then decoded using this simple algorithm (short-pass filter).

Results show that by means of violet optical bias control, the MUX photonic function may be modified, giving reconfiguration. This new method, based on wavelength background processing techniques enables the optical routing.

7 Conclusions

Integrated photonic filters based on SiC multilayer devices are analyzed and its transfer functions presented showing that the light-activated pi'n/pin a-SiC:H devices combine the demultiplexing operation with the simultaneous photodetection and self-amplification of an optical signal. Reconfiguration is provided by background wavelength control. Results show that the background wavelength and irradiation side control the output signal. By using background lights with complementary light penetration depths and changing the irradiation side, it is possible to change the spectral response and to filter a specific wavelength band acting either as a short- or a long- pass band filter or as a band-stop filter.

Acknowledgements. This work was supported by PTDC/EEA-ELC/111854/2009, PTDC/EEA-ELC/115577/2009 and PTDC/EEA-ELC/120539/2010.

References

1. Vieira, M., Louro, P., Fernandes, M., Vieira, M.A., Fantoni, A., Costa, J.: Three Transducers Embedded into One Single SiC Photodetector: LSP Direct Image Sensor, Optical Amplifier and Demux Device. In: Betta, G.F.D. (ed.) Advances in Photodiodes, ch. 19, pp. 403–425. InTech (2011) ISBN: 978-953-307-163-3
2. Louro, P., Vieira, M., Vieira, M.A., Fernandes, M., Costa, J.: Use of a SiC:H Photodiodes in Optical Communications Applications. In: Betta, G.F.D. (ed.) Advances in Photodiodes, ch. 19, pp. 377–402. InTech (2011) ISBN: 978-953-307-163-3
3. Iguchi, Y., Yamabayashi, N.: Novel rear-illuminated 1.55 μm-photodiode with high wavelength selectivity designed for bi-directional optical transceiver. In: Proc. 2nd Int. Conf. on InP and Related Mater, vol. 317, pp. 317–320 (2000)
4. Bas, M.: Fiber Optics Handbook, Fiber, Devices and Systems for Optical Communication, ch. 13. McGraw-Hill, Inc. (2002)
5. Randel, S., Koonen, A.M.J., Lee, S.C.J., Breyer, F., Garcia Larrode, M., Yang, J., Ng'Oma, A., Rijckenberg, G.J., Boom, H.P.A.: Advanced modulation techniques for polymer optical fiber transmission. In: ECOC 2007 (Th 4.1.4), Berlin, pp. 1–4 (2007)
6. Ashton, K.: That 'Internet of Things' Thing. RFID Journal (2012), `http://www.rfidjournal.com/article/view/4986` (accessed October 27, 2012)
7. Vieira, M.A., Vieira, M., Costa, J., Louro, P., Fernandes, M., Fantoni, A.: Double Pin Photodiodes with Two Optical Gate Connections for Light Triggering: A Capacitive Two-phototransistor Mode. Sensors & Transducers Journal 10 (special issue), 96–120 (2011) ISSN: 1726-5479

Part XIX
Electronics: Amplifiers

Design of a Fully Differential Power Output Stage for a Class D Audio Amplifier Using a Single-Ended Power Supply

Pedro V. Leitão, João L.A. de Melo, and Nuno Paulino

Departamento de Engenharia Electrotécnica (DEE), UNINOVA/CTS
Faculdade de Ciências e Tecnologia (FCT)-Universidade Nova de Lisboa (UNL)
2829-516, Caparica, Portugal
pvicenteleitao@gmail.com, research@joaodemelo.com,
nunop@uninova.pt

Abstract. This paper presents a full-bridge discrete power output stage for a Class D audio amplifier using a single-ended power supply. The circuit receives a digital control signal with 5 V of amplitude and it generates a floating differential output voltage up to 20 V of amplitude from a single 20 V power supply voltage. Using as control signal a pulse density modulation (PDM) wave generated by an optimized 3rd order continuous-time (CT) sigma delta modulator (ΣΔM), the system achieves a signal-to-noise-plus-distortion ratio (SNDR) of 83.1 dB, total harmonic distortion (THD) of -89 dB and a power efficiency of 92 %, while delivering 11 W over an 8-Ω load with a signal bandwidth of 20 kHz and a sampling frequency of 1.28 MHz.

Keywords: Class D Audio Amplifier, Output Stage, Gate Driver.

1 Introduction

There has been an ongoing tendency to increase the energy efficiency of every electronic device as global sustainability starts to be a major concern. Audio amplifiers are not different as they play a big role regarding power consuming in electronic systems. In Class D amplifiers the output stage devices operate as switches, resulting that when the CMOS output transistors are conducting there is a small drain-to-source voltage drop causing the dissipated power by them to be close to zero. Comparing with the typical Class AB amplifier, which has a maximum theoretical efficiency of 78.5 %, the Class D amplifier can theoretically achieve a power efficiency of 100 % [1].

The performance of a Class D audio amplifier output stage is mainly determined by the output stage configuration and the quality of the output switching [2]. The full-bridge configuration versus the half-bridge configuration can cancel out even order harmonic distortion, DC offset, allows the use of a single power supply (which doubles the available power supply voltage) and the use of a 1.5-bit quantization scheme. This quantization scheme will provide current to the load only when needed, decreasing the switching activity thus causing the power efficiency to increase while

L.M. Camarinha-Matos, S. Tomic, and P. Graça (Eds.): DoCEIS 2013, IFIP AICT 394, pp. 565–572, 2013.

decreasing the electromagnetic interference (EMI) [3]-[4]. The output power transistors connect the output nodes either to the positive power supply or ground, therefore it is important to guaranty that they do not conduct simultaneously, in order to prevent large current spikes from the positive power supply to ground from occurring. This results in a need to introduce non-overlapping time which is called *dead time* in the control signals and is one of the most dominant sources of distortion in Class D audio amplifiers [6]. In general, in a Class D output stage, the power output transistors are implemented using two identical NMOS transistors in a totem-pole configuration. These are preferred due to their small area and capacitance. Nonetheless, in order to drive the high side NMOS power transistor it is necessary additional circuitry using more complicated techniques (e.g. bootstrap technique) and timed circuitries in order to correctly activate the necessary transistors [7]-[9].

The proposed output stage presented in this paper, depicted in Fig. 1, relies on a full-bridge configuration and in a PMOS-NMOS output totem-pole arrangement which enables the high side gate driver to have a simpler circuitry. The goal is to achieve a THD close to -90 dB (0.0032 %) and power efficiency higher than 90% while using a single 20 V power supply over an 8-Ω load.

The system behavior is as follows. The analog input signal is modulated into a digital signal through pulse-width modulation (PWM) or pulse-density modulation (PDM), then this signal is level-shifted in order to drive the output CMOS power transistors and finally the differential output voltage is filtered by a low-pass filter, which reconstructs the signal before the loudspeaker in order to eliminate the high frequency content. The non-overlapping time module provides the necessary time delay in order to prevent simultaneous conduction between the output transistors. The modulation part of the system has a feedback from the output stage which enables the system to correct the output stage errors and to increase the power supply rejection ratio.

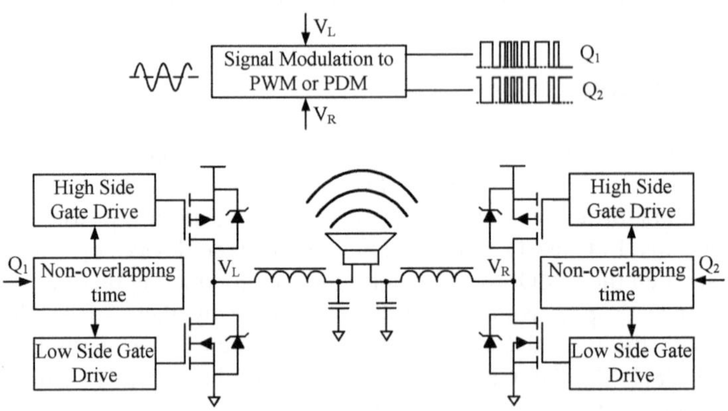

Fig. 1. Class D audio amplifier with full-bridge configuration

2 Relationship to Internet of Things

The Internet of Things refers to uniquely identify objects and their virtual representations in an Internet-like structure. If all objects and people in daily life were equipped with radio tags, they could be easily identified and inventoried by computers. In a social context, audio amplifiers, especially Class D amplifiers, being high efficiency are able to aid visual conditioned people in their day-to-day living by identifying the object/person as they come by with an extended life battery.

3 Class D Audio Amplifiers Output Stage

3.1 Low Side Gate Driver

The power output NMOS transistor gate can be simply driven by a CMOS inverter. Controlling the inverter switching with timed pulses can significantly decrease the power consumption. Fig. 2(a) shows the proposed low side driver and Fig. 2(b) shows the simulated voltages and currents in the circuit.

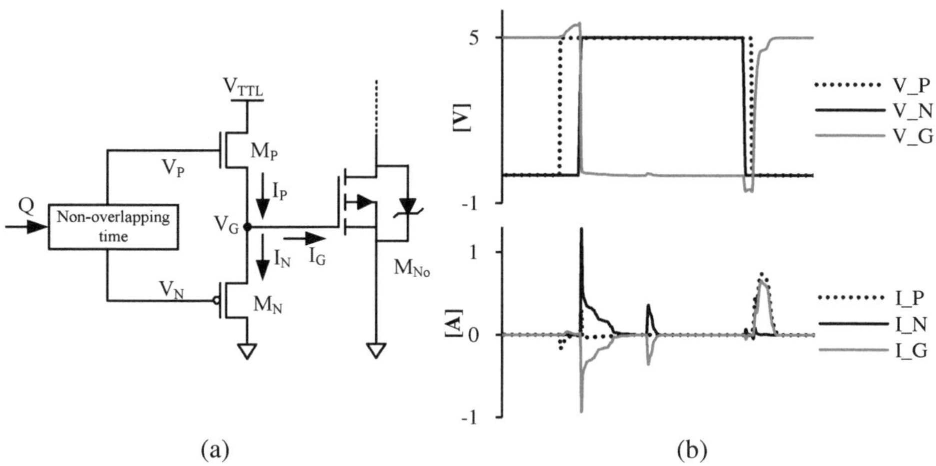

Fig. 2. (a) Low side gate driver and power NMOS transistor (b) Simulated voltage levels and current flow in the low side gate driver during the M_{NO} switching

In order to achieve high efficiency the drain current of M_P (I_P) and M_N (I_N) should only flow to the gate of M_{NO} (I_G). This can be achieved by using the correct non-overlapping value, as represented in Fig. 2(b). Otherwise, a large current can flow from the power supply through the inverter (M_P and M_N).

3.2 High Side Gate Driver

The starting point for this circuit was the PMOS gate driver described in [10] and depicted in the box of Fig. 3(a). The circuit operation is as follows: when P_1 is high the transistors M_1 and M_2 are on, forcing the base-to-emitter voltage drop of transistor Q_1 to be zero, turning it off. M_2 will also push-down the gate voltage of transistor M_{PLeft} to zero, turning it on. When P_1 is low, the transistor Q_1 will turn on due to the resistor R_1 which will pull-up the gate voltage of M_{PLeft}.

This means that the M_{PLeft} turn-off speed is dependent on R_1 which imposes a severe trade-off: when M_{PLeft} is on, a DC current $V_{CC}/(R_1+R_{DS(M1)})$ will flow through R_1 and M_1, creating an undesired power dissipation. Increasing R_1 in order to reduce this current will slow down the M_{PLeft} turn-off speed as this resistor bias Q_1. This classifies the gate drive as Class A when the M_{PLeft} is off, as it is necessary to have a constant DC current.

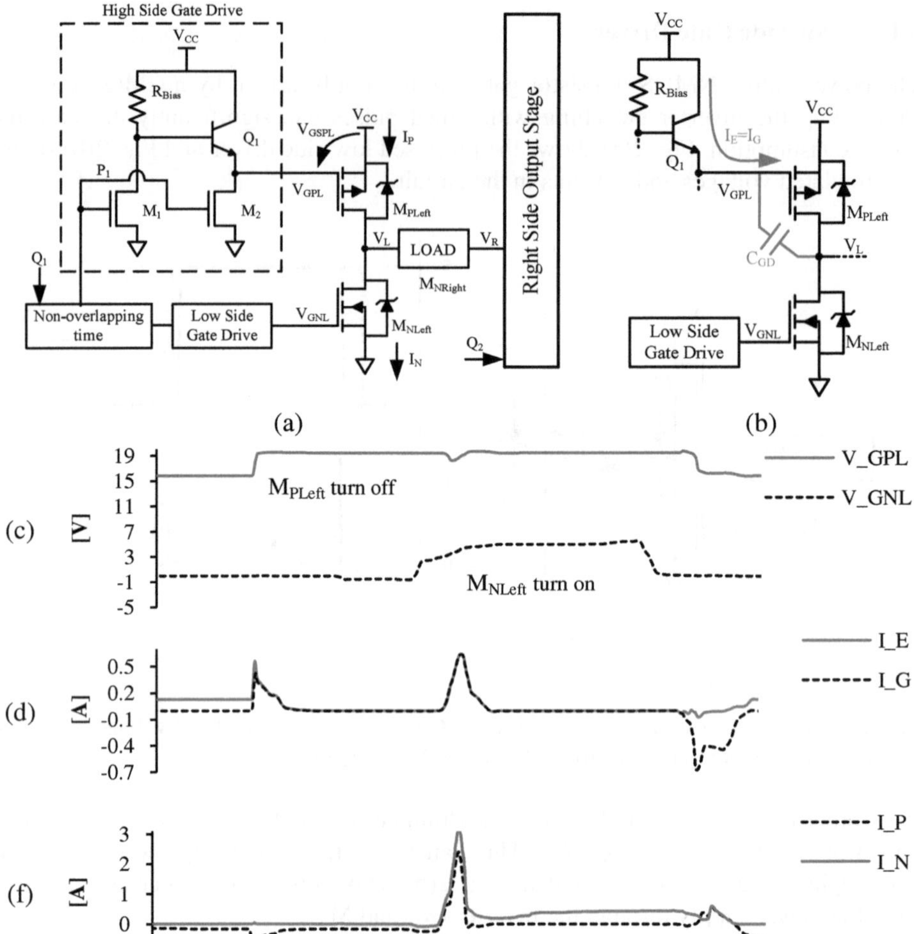

Fig. 3. (a) PMOS gate driver [10] (in box) in full-bridge configuration (b) PMOS gate driver [10] representation with parasitic capacitance while M_{NLeft} turn on and an example showing the simulated (c) V_{GPL} and V_{GNL} voltages (d) I_E and I_G currents (f) I_P and I_N currents

By inspecting the circuit it is possible to see that the M_{PLeft} gate-to-source voltage, V_{GSPL}, during M_{PLeft} turn-off, is equal to $V_{RBias}+V_{BE(Q1)}$. This means that M_{PLeft} is not fully turned off as V_{GSPL} is not exactly zero. Fig. 3(b) shows the high side gate driver with parasitic capacitances during the M_{PLeft} turn off and M_{NLeft} turn on, and on Fig. 3(c), (d) and (f) an example with the simulated transitory waves is also shown. As M_{PLeft} is turning off, V_{GPL} value increases to V_{CC} - V_{R1}+V_{BE} which in the shown example is 19.5 V, with a 20 V power supply. After the non-overlapping time the V_{GNL} voltage increases to 5 V turning M_{NLeft} on. This will pull down the node V_L, forcing the V_{GPL} node also to be pulled down, due to the parasitic capacitance C_{GD} of M_{PLeft} being large (as represented in Fig. 3(b)). This will cause the V_{GSPL} voltage to increase, falsely turning M_{PLeft} on and creating large current spikes between V_{CC} and ground through transistors M_{PLeft} and M_{NLeft} because they are simultaneously conducting. This effect is called *shoot-through current* [5] and will decrease the efficiency significantly, introduce harmonic distortion and even destroy the output transistors if Q_1 takes too long to respond to this variation.

This problem is bypassed by the proposed high side gate driver presented in this paper.

3.3 Proposed High Side Gate Driver

The new full-bridge output stage proposed in this paper is presented in Fig. 4. Two CMOS inverters were placed at the gate of the output PMOS. In order to achieve lower power dissipation, especially in R_{Bias}, one zener diode was inserted in each branch of the circuit. The zener voltage of these diodes should be V_{CC} – 5 V so that the V_{GS} voltage drop of the output PMOS transistor, the R_{Bias} voltage drop and the CMOS inverters only varies from 5 V to V_{CC}. This will cause the V_{GS} voltage drop of the PMOS to be constant thus improving the performance of the circuit while simplifying the voltage supply scaling, as only the zener diode needs to be scaled.

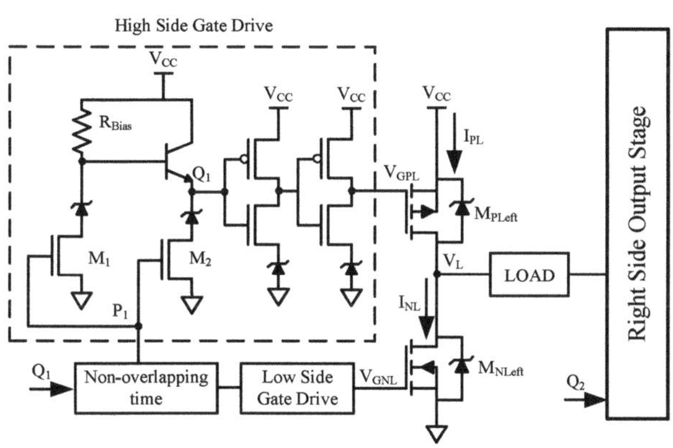

Fig. 4. Proposed PMOS gate driver (in box) in full-bridge configuration

The inverters placed at the gate of the output PMOS will correctly pull-up V_{GPL} to V_{CC} so that it will be correctly turned off. The correctly timed delay will prevent shoot-through current.

An example for the switching waves of the output stage is shown on Fig. 5. The circle in Fig. 5(b) shows a commonly found effect in Class D output stages. As V_L voltage rises to V_{CC}, a parasitic current goes from V_{CC} through the parasitic capacitance C_{GD} of the NMOS power transistor. This effect is rapidly compensated by the low side gate driver.

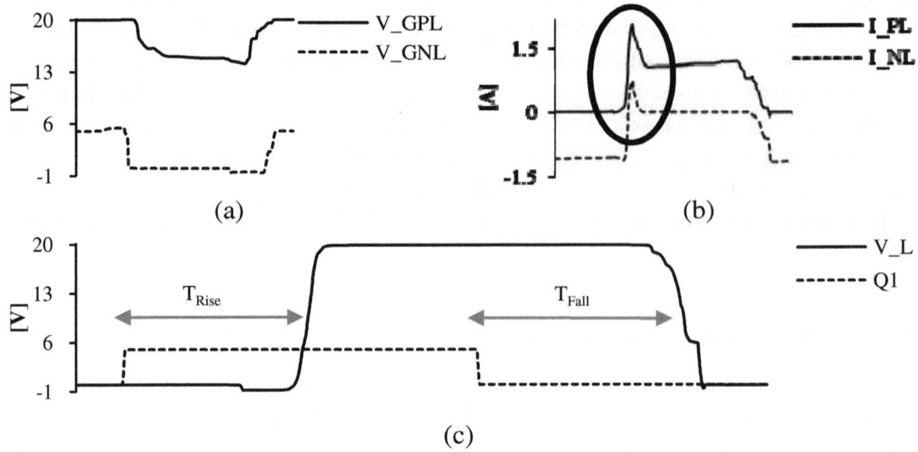

Fig. 5. Simulated example for (a) switching voltage levels at the gate of the output power transistors (b) current in the output power transistors (c) input voltage (Q_1) and output voltage (V_L) with the T_{Rise} and T_{Fall} delay represented

4 Class D Audio Amplifier with a 3rd Order 1.5-Bit Continuous-Time (CT) Sigma Delta (ΣΔ) Modulator

The presented output stage was simulated with a 3rd order 1.5 bit ΣΔM optimized for Class D audio amplifiers with distributed feedback and local resonator feedback, described in [11]. The output stage was design to handle 20 V power supply, 20 kHz low-pass filter, a sampling frequency of 1.28 MHz and to drive an 8-Ω load.

Fig. 6(a) shows the P_{RBias} dissipation plot during continuous conduction and the Fig. 6(b) shows the T_{Rise} and T_{Fall} delay of the output stage with the R_{Bias} increase. From these plots it is possible to conclude that the output stage power efficiency will increase as R_{Bias} increases, but the ΣΔM will have to handle the output stage increased delay. The used 3rd order ΣΔM was optimized using 100 ns and 150 ns delay while using the same goal function. This allowed the use of R_{Bias} values between 60 to 100 Ω and 160 to 200 Ω, respectively. Table 1 shows the system performance summary for these optimizations and Fig. 7 shows the fast Fourier transform (FFT) of the output sinusoidal wave while applying a 0 dBV and 2 kHz input signal.

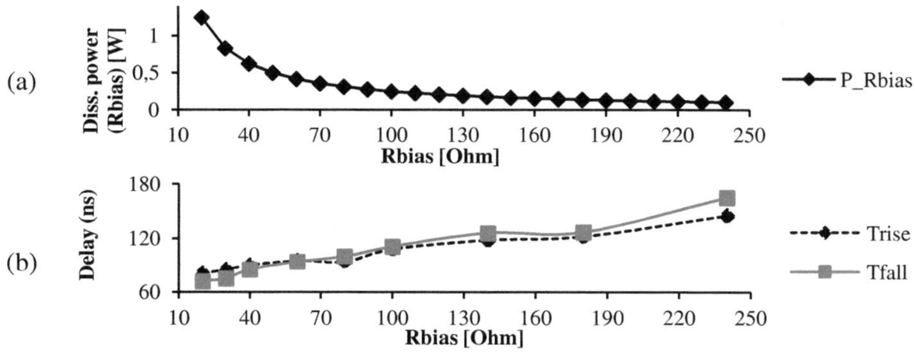

Fig. 6. (a) P_{Rbias} plot ($P_{Rbias} = 5^2/R_{Bias}$) (b) T_{Rise} and T_{Fall} delay

Table 1. System performance summary with 100 ns output stage delay optimization (#1) and 150 ns delay optimization (#2)

Optm.	R_{Bias} (Ω)	SNDR (dB)	THD (dB)	HD2 (dB)	HD3 (dB)	η (%)	P_{Load} (W)	Delay (ns)	
								Up	Down
#1	60	82.21	-84.96	-90.3	-89.3	87.90	13.36	100	95
	80	81.92	-84.76	-90.8	-88.8	89.21	13.36	105	100
	100	83.09	-86.38	-93.1	-89.2	89.76	13.36	113	110
#2	160	83.62	-89.70	-98.4	-101.5	91.73	11.29	125	135
	180	83.41	-87.08	-97.1	-100.4	91.69	11.30	135	140
	200	83.14	-88.94	-99.9	-106.3	92.20	11.30	140	150

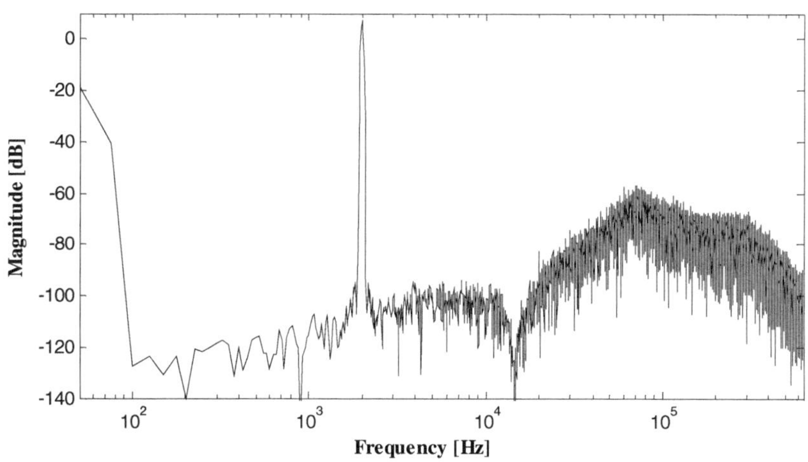

Fig. 7. Spectrum of the output signal (using a Blackman-Harris window and 2^{16} points) with R_{Bias} equal to 200 Ω and SNDR of 83 dB

These results show that using a higher R_{Bias} value the system will operate with higher power efficiency, higher SNDR and lower harmonic distortion, but will deliver less energy to the output load. This is due to the increased T_{Rise} and T_{Fall} delay which will result in a smaller active time in the output PDM wave, thus reducing the maximum output power. The best compromise between SNDR, harmonic distortion and power efficiency has been found to be with R_{Bias} equal to 200 Ω where the system achieves a SNDR of 83 dB, THD of -89 dB and power efficiency ($P_{Load}/P_{delivered}$) of 92%.

5 Conclusions

A full-bridge discrete power output stage for a Class D audio amplifier using a single 20 V power supply voltage was presented. The circuit is controlled by a digital 5 V of amplitude PDM signal, produced by an optimized 3rd order ∑Δ modulator with the output stage inside the closed loop. This system is capable of driving an 8-Ω load while achieving a SNDR of 83 dB and a THD of -89 dB (0.0035 %). The use of a PMOS-NMOS power output inverter simplified the design while still achieving power efficiency of 92% for an 11.3 W power delivered to the load. The use of 1.5-bit quantization scheme and the lossless LC low-pass filter allowed to reduce the EMI distortion (max. of -60 dB at 80 kHz).

References

1. Duncan, B.: High Performance Audio Power Amplifiers, Newnes, Great Britain (1996)
2. Berkout, M.: A Class D Output Stage with Zero Dead Time. In: Proc. IEEE Solid-State Circuits Conference (ISSCC), pp. 134–135 (2003)
3. Pan, J., Yao, L.B., Lian, Y.: A Sigma-Delta Class D Audio Power Amplifier in 0.35um CMOS Technology. In: Int. SoC Design Conference, pp. 24–25 (2008)
4. de Melo, J., Paulino, N.: Design of a 3rd order 1.5-bit continuous-time (CT) Sigma-Delta (SD) modulator optimized for Class D audio power amplifier. J. Microelectronics and Computer Science 1(2) (2011)
5. Gaalaas, E.: Class-D audio amplifiers: what, why, and how. Analog Dialogue 40 (2003); Analog Devices
6. Mosely, I., Mellor, P., Bingham, M.: Effect of Dead Time on Harmonic Distortion in Class D Audio Power Amplifiers. Electronics Letters 35(12), 950–952 (1999)
7. Morrow, P., et al.: A 20-W Stereo Class D Audio Output Stage in 0.6 mm BCDMOS Technology. IEEE J. Solid-State Circuits 39(11), 1948–1958 (2004)
8. Berkhout, M.: An Integrated 200-W Class-D Audio Amplifier. IEEE J. Solid-State Circuits 38(7), 1198–1206 (2003)
9. Berkhout, M., Dooper, L.: A 3.4W Digital-In Class D Audio Amplifier. In: Proc. ESSCIRC (ESSCIRC), pp. 87–90 (2011)
10. Yi, H., Hong, S.: Design of L-band high speed pulsed power amplifier using ldmos fet. In: Progress In Electromagnetics Research M, vol. 2, pp. 153–165 (2008)
11. de Melo, J.L.A., Nowacki, B., Paulino, N., Goes, J.: Design methodology for sigma-delta modulators based on a genetic algorithm using hybrid cost functions. In: Proc. IEEE Int. Symp. Circuits Systems (ISCAS), pp. 301–304 (May 2012)

A 1.2 V Low-Noise-Amplifier with Double Feedback for High Gain and Low Noise Figure

Ivan Bastos[2], F. Querido[2], D. Amoêdo[2], Luis B. Oliveira[2], J.P. Oliveira[2], João Goes[2], and Manuel M. Silva[1]

[1] INESC-ID Lisboa
Tech. University of Lisbon
Lisbon, Portugal
manuel.silva@inesc-id.pt
[2] CTS-UNINOVA, Dep. Eng. Electrotécnica, Faculdade de Ciências e Tecnologia, Universidade Nova de Lisboa
Caparica, Portugal
bastos356@gmail.com, l.oliveira@fct.unl.pt

Abstract. In this paper we present a balun low noise amplifier (LNA) in which the gain is boosted using a double feedback structure. The circuit is based in a conventional Balun LNA with noise and distortion cancellation. The LNA is based in two basic stages: common-gate (CG) and common-source (CS). We propose to replace the resistors by active loads, which have two inputs that will be used to provide the feedback (in the CG and CS stages). This proposed methodology will boost the gain and reduce the NF. Simulation results, with a 130 nm CMOS technology, show that the gain is 23.8 dB and the NF is less than 1.8 dB. The total power dissipation is only 5.3(since no extra blocks are required), leading to an FOM of 5.7 mW^{-1} from a nominal 1.2 supply.

Keywords: CMOS LNAs, Noise canceling, Wideband LNA.

1 Introduction

Nowadays, there is a high demand for wireless communications, which includes Industrial, Scientific, and Medical (ISM) and Wireless Medical Telemetry Service (WMTS) applications [1]. These low cost applications require low power, low voltage transceivers fully integrated in a single chip [2-4]. The LNA that is a key block in these systems will be investigated in this paper.

Wideband LNAs with high gain and low noise figure (NF), using noise and distortion cancelation have been proposed [5-7]. But, these circuits have large power dissipation for high gain and low noise figure.

In this paper our main goal is to design a very low area and low cost LNA, with very high gain and low NF using a 1.2 V supply. This is obtained by replacing the load resistors by transistors biased close to saturation. In [7]a circuit operating at 1.2 V with controllable gain was proposed. In this paper we investigate the possibility of introduce a double feedback technique to boost the gain and reduce the noise figure (NF).

L.M. Camarinha-Matos, S. Tomic, and P. Graça (Eds.): DoCEIS 2013, IFIP AICT 394, pp. 573–581, 2013.

Equations for gain and noise figure are presented, which can be used to optimize the circuit performance. A circuit prototype in a 130 nm standard CMOS technology at 1.2 V have been designed and simulated to demonstrate the proposed technique.

The circuit prototype has gain of 23.8 dB and NF below 2 dB, dissipating only 5.3 mW, leading to a FOM of 5.7 mW^{-1}, which is, to the authors' knowledge, the best FOM in the literature for LNAs with a nominal 1.2 V supply.

2 Contribution for Internet of Things

Recently, more devices are being embedded with sensors and actuators with the ability to communicate and exchange information, creating a cloud environment. The physical communication plays a critical role in portable wireless devices equipped with transceivers where power consumption, immunity to noise, and signal amplification are important parameters to ensure a reliable and efficient communication in a crowded channel environment. With this goal in mind the design of RF front-end blocks for low power applications, in CMOS technology, will contribute towards the achievement of more cheap and robust devices. In this paper we will focus on the design of low power LNAs.

3 Balun LNA with Noise Cancellation

In a receiver path, since typically, the antenna and RF filters are single-ended, it is very important to have a LNA with single-ended configuration input. A differential signal in the receiver is preferred to reduce harmonic distortion and to reject power supply and substrate noise [6]. Traditionally, we have an external balun that converts single-ended signals to differential, but it introduces losses and degrades the receiver NF. A balun LNA is a very good solution to convert a single-ended to a differential signal, which simplifies the design avoiding he external balun [6].

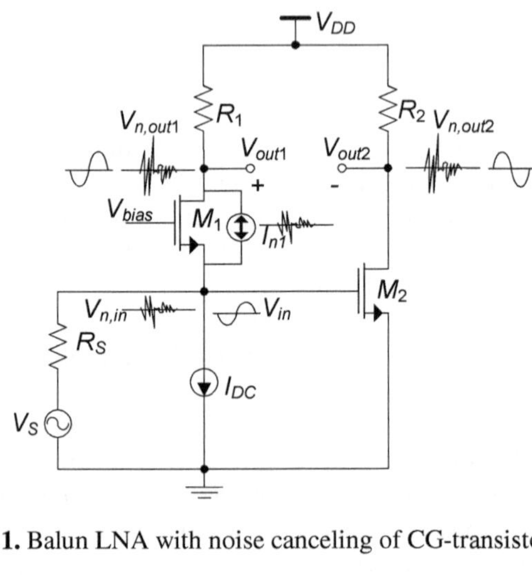

Fig. 1. Balun LNA with noise canceling of CG-transistor [6]

A balun LNA, in which the thermal noise of CG-transistor is canceled because this noise appears in phase at two outputs and their gains are in opposition, is proposed in [6]. The gain is doubled and the noise is reduced when the output signals are balanced. It can also be shown that the distortion introduced by M_1 is also cancelled.

The differential voltage gain of the LNA is obtained from the difference of the common-gate (CG) stage and the common-source (CS) stage gains:

$$A_v|_{Diff} = g_{m1}(R_1//r_{ds1}) + g_{m2}(R_2//r_{ds2}) . \tag{1}$$

where, r_{ds} is the transistors output resistance and g_m is the transconductance.

The input impedance is given, approximately:

$$Z_{in} = \frac{1}{g_{m1}} . \tag{2}$$

Note that the body and source of M_1 are connected to eliminate the body effect.

4 Proposed Circuit

Considering the traditional CG-CS LNA circuit (Fig. 1) as reference, we investigate a topology using active loads, by replacing the resistors by transistors, biased in the triode region, which behave, approximately, as linear resistors [7]. In order to enhance the gain, while maintaining a low noise figure, we investigate first the possibility to use local Feedforward and Feedback (FF), as shown in Fig. 2a).

Taking the advantage of using transistors, instead of resistors, we applyV_{in} on the gate of transistor to the gate of M_4, which is amplified and added to V_{out2}. The resulting signal is amplified through $M_3$by feedback and added to V_{out1}. With this structure we have a significant increase in the gain, mainly in the CG stage, which need to be carefully designed to ensure 50 Ohms input match. In this case the thermal noise of M_1 is only partially cancelled, which degrades the LNA noise figure. To overcome this issue we propose a new circuit approach in which we apply a Double Feedback structure (DF), as shown in Fig. 2b).

This proposed circuit boost the gain, and reduce the noise of M_1that appears with the same level on the LNA outputs (load transistors $M_3$and M_4), while the output signals remain balanced. This circuit is more simply and completely symmetrical, and therefore, is expected to achieve the best performance results.

In the feedback is used a high pass RC coupling, as shown in Fig. 2. With these connections, the parasites capacitances of M_3 and M_4 will reduce the bandwidth (the gate-source and gate-drain capacitances), but the main goal is achieved: high gain and low NF.

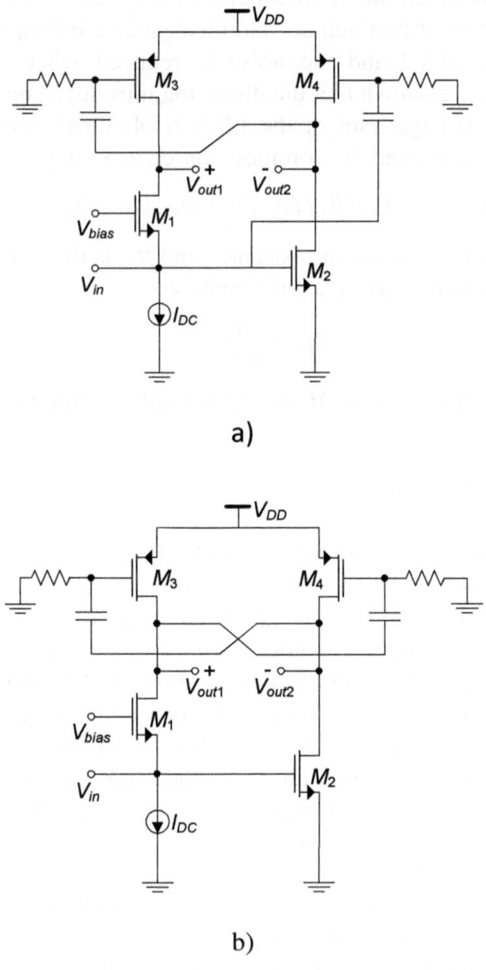

Fig. 2. a) LNA using FF. **b**) Proposed LNA using DF

In order to provide some circuit insight, we derive here the equations for gain (CG and CS stages) and LNA input impedance (for the proposed DF case):

$$\frac{V_{out1}}{V_{in}} = \frac{g_{mCG}g_2+g_{m2}g_{m3}}{g_1g_2-g_{m3}g_{m4}} . \tag{3}$$

$$\frac{V_{out2}}{V_{in}} = -\frac{g_{m2}g_1+g_{mCG}g_{m4}}{g_1g_2-g_{m3}g_{m4}} . \tag{4}$$

where,

$$g_1 = g_{ds1} + g_{ds3} .$$

$$g_2 = g_{ds2} + g_{ds4} .$$

Using (3) and (4), we obtain the LNA differential gain,

$$A_v|_{Diff} = \frac{V_{out1}-V_{out2}}{V_{in}} = \frac{g_{mCG}(g_{m4}+g_2)+g_{m2}(g_{m3}+g_1)}{g_1 g_2 - g_{m3} g_{m4}}. \quad (5)$$

The input-impedance is given by

$$Z_{in} = \frac{g_1 g_2 - g_{m3} g_{m4}}{g_{mCG}[g_2 g_{ds3} - g_{m3} g_{m4}] - g_{m2} g_{m3} g_{ds1}}. \quad (6)$$

Using equations (5) and (6), we can optimize the circuit performance in order to increase the gain, minimizing the impact in the input match.

From [6, 7], if it is assumed that $g_{m1} = g_{m2} = g_m$, the noise factor is:

$$F_{LNA} = 1 + \frac{k_f}{8kTR_S c_{ox} f^{\alpha_f}}\left(\frac{1}{W_1 L_1} + \frac{1}{W_2 L_2}\right) + \frac{\gamma}{2R_S g_m} + \frac{1}{R_S r_{ds} g_m^2}. \quad (7)$$

where k is Boltzmann's constant, c_{ox} is the oxide gate capacitance per unit area, W_i and L_i are the transistor dimensions, T is the absolute temperature, γ is the excess noise factor, k_f and α_f are intrinsic process parameters, which depend on the size of the transistors [8, 9]. With the proposed circuit there is additional noise due to the double feedback structure, however, this can be minimized by proper design.

From [6], to improve the noise figure, the g_{m2} should be greater than g_{m1}, while the g_{ds4} is increased to keep the output signals balanced.

$$g_{m2} = n \cdot g_{m1}.$$

$$g_{ds4} = n \cdot g_{ds3}.$$

The optimal value of n is obtained by simulations.

5 Simulation Results

The circuit prototype is designed using a $130\ nm$ CMOS standard technology with $1.2\ V$ supply. The circuit parameters are given on Table 1. The length of each transistor channel is the minimum to maximize speed, and V_{bias} is defined in $760\ mV$ to define the biasing current at the CS stage.

Table 1. LNA circuit parameters using DF

	I_D (mA)	W (μm)	r_{ds} (Ω)	g_{ds} (mS)	g_m (mS)
M_1	2	139.2	472	2.12	30.80
M_2	2.43	358.4	357	2.80	44.23
M_3	2	13.14	236	4.23	2.07
M_4	2.43	16.4	186	5.38	2.48

In Table 2 we compare the theoretical results with simulations. We use equation (1) to traditional LNA with resistors[6] and using MOS in triode [7], and equation (5) for the proposed DF approach.

Table 2. Gain (dB) for different topologies

	Res	MOS	DF
Theoretical	19.07	20.45	23.78
Simulation	19.03	20.50	23.81

In Table 3 we compare the simulation results for the traditional case with resistors, MOS transistors, and two approaches capable to boost the gain: 1) using feedback and feedforward (FF), and 2) the proposed case of double feedback (DF).

In order to investigate the influence of double feedforward in the LNA key parameters: gain, noise figure, linearity, and frequency band, several simulations are presented in Table 3. For a better comparison of the obtained results, the following figure of merit is used [10]:

$$\mathrm{FOM}[\mathrm{mW}^{-1}] = \frac{Gain}{(NF-1)P_{DC}[\mathrm{mW}]}. \tag{8}$$

Table 3. Circuit Simulations for different topologies

	$Gain$ (dB)	NF (dB)	$IIP3$ (dBm)	$Power$ (mW)	$Band$ (GHz)	FOM (mW^{-1})
Res	19.03	< 2.34	3.62	5.16	0.2-7.4	2.43
MOS	20.50	< 2.02	-3.47	5.30	0.2-6	3.37
FF	23.28	< 2.39	-11.03	5.51	0.1-2.5	3.61
DF	23.81	< 1.79	-9.92	5.32	0.1-2	5.72

In Table 3 is shown that the DF approach has the highest gain and the lower NF, leading to the highest FOM. The disadvantages are the increase of the circuit non-linearity and the reduction of the available bandwidth.

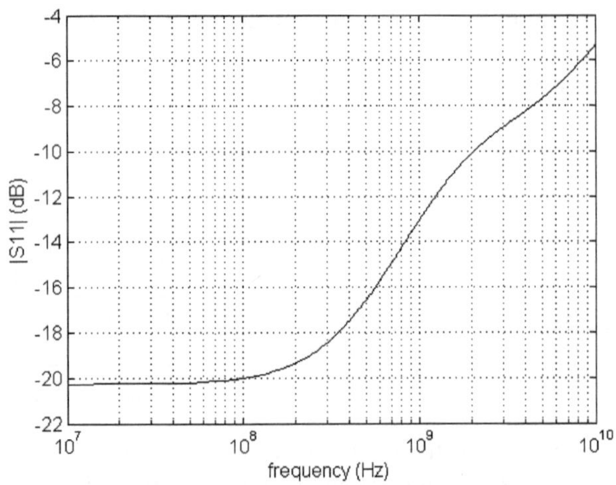

Fig. 3. Simulated LNA S11 parameter

In figs. 3 to 5, the simulation results for the input match (S11), gain, and NF, for the proposed circuit (DF) are presented.

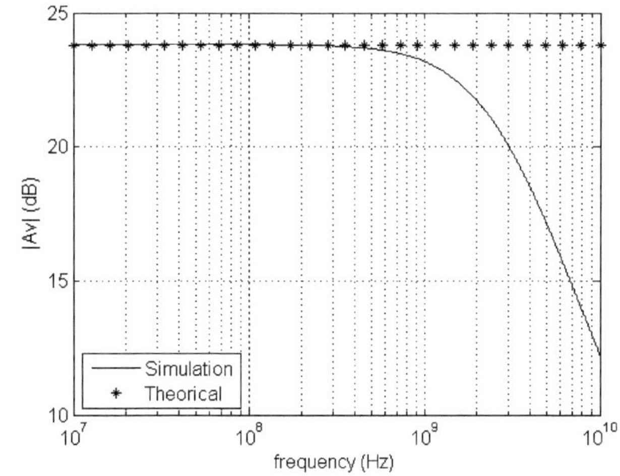

Fig. 4. Simulated LNA Gain

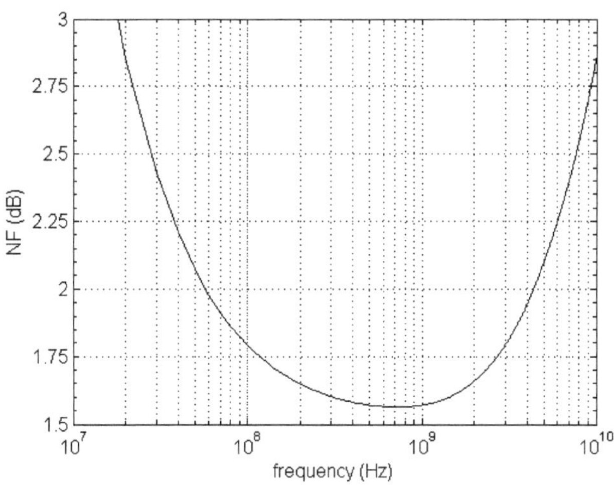

Fig. 5. Simulated LNA Noise Figure

Comparing these results with state-of the art inductorless LNA (table4),we observe that our circuit is very good in terms of gain and NF, and has very low power, which leads to the best FOM (please note that the results are obtained by simulation, some degradation is to be expected in the fabricated circuit; some of the references in table 4 are from measurements).

Table 4. Comparison with state-of-the art LNAs

Ref	Tech (nm)	Band (GHz)	Gain (dB)	NF (dB)	IIP3 (dBm)	Power (mW)	FOM (mW^{-1})
[6]^M	65	0.2-5.2	13-15.6	< 3.5	>0	14	0.4
[11]^M	90	0.5-8.2	22-25	< 2.6	-4/-16	42	0.5
[12]^M	90	0.8-6	18-20	< 3.5	>-3.5	12.5	0.6
[13]^S	90	0.1-1.9	20.6	< 2.7	10.8	9.6	1.3
[14]^S	130	0.2-3.8	11.2	< 2.8	-2.7	1.9	2.1
[15]^M	180	0.5-0.9	16	< 4.3	-	22	0.2
[16]^M	180	0.1-0.9	15	< 4.2	-	10	0.3
[7]^S	130	0.2-5	20.4	2.6	-10.9	4.8	2.7
[17]^S	130	0.2-6.6	19.8	< 1.8	1.6	4.8	3.5
This Work^S	**130**	**0.1-2**	**21.7-23.8**	**1.6-1.8**	**-9.7**	**5.3**	**5.7**

(^{S}) - Simulation results. (^{M}) - Measurement results.

The proposed circuit approach is especially interesting in low power and low voltage biomedical applications [1]. Since in these applications low power is the key aspect and some non-linearity can be tolerated. There are ISM bands in 450 MHz and 900 MHz and WMTS band in 600 MHz and 1.4 GHz, for which this circuit can be a good alternative to the conventional solutions.

6 Conclusions

In this paper we present a low voltage and low power wideband balun LNA with DF for high gain and low NF. A circuit prototype operating at 1.2 V is presented in a 130 nm CMOS technology, which validates the proposed methodology. Simulation results show that the gain of the balun LNA is 23.8 dB, and the NF is below 2 dB for a power consumption of 5.3 mW. The proposed circuit is especially useful for low power and low voltage operation in biomedical applications (ISM and WMTS bands).

The proposed circuit, with 1.2 V supply, to the best of the authors' knowledge, has the high FOM (5.7 mW^{-1}) when compared with CMOS LNAs in the literature.

References

1. Iniewski, K.: VLSI Circuits for Biomedical Applications. Artech House (2008)
2. Razavi, B.: RF Microelectronics. Prentice-Hall (1998)
3. Lee, T.H.: The Design of CMOS Radio Frequency Integrated Circuits, 2nd edn. Cambridge University Press (2004)
4. Crols, J., Steyaert, M.: CMOS Wireless Transceiver Design. Kluwer (1997)
5. Bruccoleri, F., Klumperink, E., Nauta, B.: Wide-band CMOS low-noise amplifier exploiting thermal noise canceling. IEEE J. Solid-State Circuits 39(2), 275–282 (2004)

6. Blaakmeer, S., Klumperink, E., Leenaerts, D., Nauta, B.: Wideband Balun-LNA with Simultaneous Outputs Balancing, Noise-Canceling and Distortion-Canceling. IEEE J. Solid-State Circuits 43(6), 1341–1350 (2008)
7. Bastos, I., Oliveira, L.B., Goes, J., Silva, M.: Balun LNA with continuously controllable gain and with noise and distortion cancellation. In: IEEE Int. Symposium Circuit and Systems (ISCAS 2012), pp. 2143–2146 (May 2012)
8. Chew, K.W., Yeo, K.S., Chu, S.F.: Effect of technology scaling on the 1/f noise of deep submicron MOS transistors. Solid-State Electron 48, 1101–1109 (2004)
9. Manghisoni, M., Ratti, L., Re, V., Speziali, V., Traversi, G.: Noise Characterization of 130 nm and 90 nm CMOS Technologies for Analog Front-end Electronics. In: IEEE Nuclear Science Symposium Conference, vol. 1, pp. 214–218 (2006)
10. Linten, D., et al.: Low-power 5 GHz LNA and VCO in 90 nm RF CMOS. In: 2004 Symposium on VLSI Circuits, Digest of Technical Papers, pp. 372–375 (June 2004)
11. Zhan, J.-H.C., Taylor, S.S.: A 5 GHz resistive-feedback CMOS LNA for low-cost multi-standard applications. In: IEEE ISSCC 2006 Dig. Tech. Papers, pp. 200–201 (February 2006)
12. Bagheri, R., Mirzaei, A., Chehrazi, S., Heidari, M.E., Lee, M., Mikhemar, M., Tang, W., Abidi, A.A.: An 800-MHz–6-GHz software-defined wireless receiver in 90-nm CMOS. IEEE J. Solid-State Circuits 41(12), 2860–2876 (2006)
13. Mak, P.-I., Martins, R.: Design of an ESD-Protected Ultra-Wideband LNA in Nanoscale CMOS for Full-Band Mobile TV Tuners. IEEE Trans. Circuits Systems I 56, 933–942 (2009)
14. Amer, A., Hegazi, E., Ragai, H.: A low power wideband CMOS LNA for WiMax. IEEE Trans. Circuits Systems II 54(1), 4–8 (2007)
15. Xiao, J., Mehr, I., Silva-Martinez, J.: A High Dynamic Range CMOS Variable Gain Amplifier for Mobile DTV Tuner. IEEE J. Solid-State Circuits 42(2), 292–301 (2007)
16. Han, K., Zou, L., Liao, Y., Min, H., Tang, Z.: A wideband CMOS variable gain low noise amplifier based on single-to-differential stage for TV tuner applications. In: IEEE Solid-State Circuits Conference, A-SSCC 2008, November 3-5, pp. 457–460 (2008)
17. Bastos, I., Oliveira, L.B., Goes, J., Silva, M.: MOSFET-only wideband LNA with noise cancelling and gain optimization. In: IEEE Mixed Design of Integrated Circuits and Systems (MIXDES), June 24-26, pp. 306–311 (2010)

A Switched-Capacitor Band-Pass Biquad Filter Using a Simple Quasi-unity Gain Amplifier

Hugo Serra, Nuno Paulino, and João Goes

Centre for Technologies and Systems (CTS) – UNINOVA
Dept. of Electrical Engineering (DEE), Universidade Nova de Lisboa (UNL)
2829-516 Caparica, Portugal
has14926@campus.fct.unl.pt, {nunop,jg}@uninova.pt

Abstract. This paper presents a switched-capacitor (SC) band-pass biquad using a simple quasi-unity gain amplifier. In sub-nanometer CMOS technologies the intrinsic gain of the transistors is low; this increases the difficulty of designing high gain amplifiers. The proposed SC filter is based on the Sallen-Key biquad and it requires only a simple low gain amplifier. A differential filter circuit, including a suitable amplifier based on a fully-differential voltage-combiner is presented and analyzed. The correct functionality of this circuit is validated through electrical simulations of a second-order band-pass filter. These simulations show that, for a clock frequency of 100 MHz, the frequency response of the circuit is similar to the corresponding prototype filter.

Keywords: Analog circuits, band-pass Sallen-Key, switched-capacitor circuits, voltage-combiner amplifier.

1 Introduction

Analog filters are extremely important blocks in several electronic systems, such as RF transceivers or sigma delta modulators. They allow selecting between signals with different frequency and eliminating unwanted signals.

The scaling-down of transistors in advanced deep-submicron CMOS technologies results in the reduction of the intrinsic gain (g_m/g_{ds}) [1] and in an increase in the variability, making the design of high gain amplifiers increasingly difficult, especially for larger bandwidths. This limitation has large impact in the performance of filter circuits.

This paper proposes the design of filter circuits using low gain amplifiers, in order to avoid the difficulty of designing high gain amplifiers with large bandwidth. The SC filter circuit described in this paper is based on a band-pass Sallen-Key biquad [2] which does not require high gain amplifiers. This filter topology simplifies the design of the amplifier although it also eliminates the virtual ground node from the circuit. Without this node, parasitic insensitive SC branches cannot be used. Due to modern parasitic extraction software which can reliably predict the values of parasitic capacitances, the historical disadvantage of parasitic sensitive SC branches (parallel SC) is no longer critical, thus allowing their influence to be compensated during the design phase of the filter.

L.M. Camarinha-Matos, S. Tomic, and P. Graça (Eds.): DoCEIS 2013, IFIP AICT 394, pp. 582–589, 2013.

The paper is organized as follows. As required, Section 2 shows the relationship between the work presented in this paper and the Internet of Things topic. Section 3 shows a brief state of the art about typical SC circuits. Section 4 describes and analyzes the biquadratic (biquad) section implemented in this paper. Section 5 describes and analyzes the low gain amplifier used in the biquadratic section. In Section 6 the simulation results of second-order band-pass SC filter are given. Section 7 draws the main conclusions from the work carried out in this paper.

2 Relationship to Internet of Things

To have an Internet of things it is necessary to have electronic systems associated to objects (things) that need to be connected to the Internet. In order for these systems to be smaller and to have a lower cost, it is important to use the concept of system-on-a-chip (SoC). This means that a single die is used to build the entire system thus reducing the size, cost, and power dissipation of the system.

The technology used in a SoC is selected in order to maximize the performance of the digital circuits of the system; this means that the analog circuits in the system have to be designed using advanced nanometer (nm) CMOS technologies. This can be a problem because the transistors in these technologies are not optimized for working in analog circuits; in particular they have low intrinsic gain which makes the design of high gain amplifiers particularly difficult. This paper, describes the design of a band-pass filter using low gain amplifiers, which facilitates the use of advanced nm CMOS technologies, thus addressing one of the problems associated to the Internet of things.

3 Switched Capacitor Circuits

Interests in SC networks started in the late 70s due to the possibility of implementing analog filters using monolithic integrated circuit (IC) technology and because it is possible to obtain a good accuracy in the ratio between two capacitor values, as opposed to the low accuracy of the absolute values of resistors and capacitors. Also, since a high value resistor can be generated using small on-chip capacitors and a high frequency clock, the area occupied by a SC filter in an integrated circuit is typically smaller than the area occupied by an equivalent RC filter.

SC circuits can be implemented using different types of SC branches some of which, when using high gain amplifiers, are insensitive to parasitic capacitances [3][4]. However, this type of approach becomes harder to implement with the decrease of the intrinsic gain of transistors.

4 Biquadratic Sallen-Key Based Circuit

The single-ended configuration of the band-pass SC Sallen-Key based topology is shown in Fig. 1. This circuit is obtained by replacing the resistors with parallel SC branches in the band-pass Sallen-Key circuit. An additional capacitor (C_3) was added to the circuit to facilitate the process of compensating the input parasitic capacitance of the amplifier.

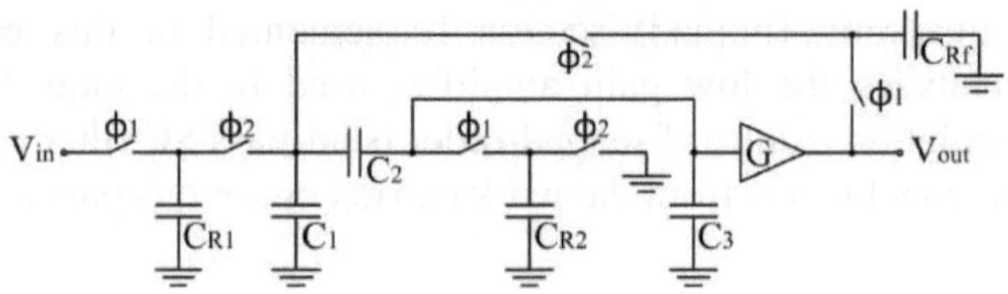

Fig. 1. Band-pass SC biquad filter in single-ended configuration

The transfer function (1) was obtained from the analysis of the previous circuit from a charge conservation perspective, considering that the circuit's output is sampled at the end of phase Φ_1.

$$H(z) = V_{out}(z)/V_{in}(z) = G\,(z-1)d/(a - b\,z + c\,z^2) \tag{1}$$

where,

$$a = \big(C_1(C_2+C_3) + C_3(C_{R1}+C_{Rf}) + C_2(C_3+C_{R1}+C_{Rf})\big)\big(C_2(C_3+C_{R2}) + C_1(C_2+C_3+C_{R2})\big)$$

$$b = C_1\big(C_3^2(C_{R1}+C_{Rf}) + C_2^2(4C_3+C_{R1}+C_{R2}+C_{Rf}+G\,C_{Rf}) + C_2C_3(2(2C_3+C_{R1}+C_{R2}+C_{Rf}) + G\,C_{Rf})\big) + C_1^2(C_2+C_3)(2(C_2+C_3)+C_{R2}) \tag{2}$$

$$c = \big(C_2(C_3+C_{R2}) + C_1(C_2+C_3+C_{R2})\big)\big(C_1(C_2+C_3) + C_3(C_{R1}+C_{Rf}) + C_2(C_3+C_{R1}+C_{Rf})\big)$$

$$d = G\big(C_2C_3 + C_1(C_2+C_3)\big)C_2C_{R1}$$

Since the filter is implemented using parasitic sensitive branches, the capacitor values must be adjusted to compensate for the parasitic capacitances present in the circuit. Fig. 2 shows the filter schematic considering these parasitic capacitances introduced by the switches and the amplifier.

Because all parasitic capacitances are in parallel with existing capacitors, their influence can be directly compensated by changing the values of these capacitors. The compensated capacitor values of filter in consideration can be obtained from (3).

$$C_{1x} = C_1 - C_{M2s} - C_{M5d} \qquad C_{3x} = C_3 - C_{Bin} - C_{M3d} \qquad C_{Rfx} = C_{Rf} - C_{M5s} - C_{M6d}$$
$$C_{R1x} = C_{R1} - C_{M1s} - C_{M2d} \qquad C_{R2x} = C_{R2} - C_{M3s} - C_{M4d} \tag{3}$$

An approximation of the parasitic capacitances of the switches and of the input capacitance of the amplifier is obtained from the DC simulation of the circuit. The drain capacitance of transistor M_1 is neglected since it doesn't alter the circuits transfer function. The amplifier's output capacitance can also be neglected since it is connected to a low impedance node.

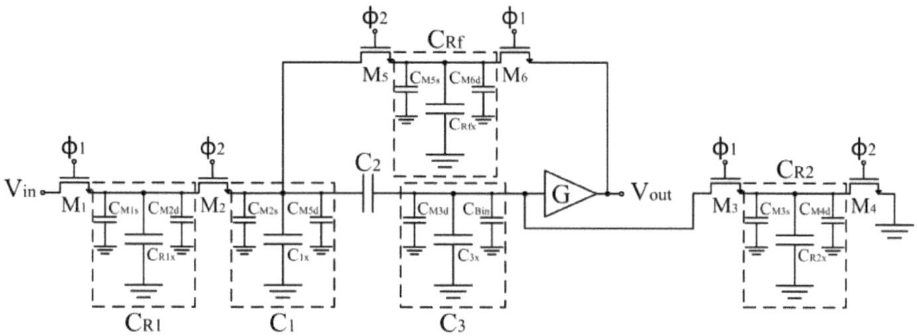

Fig. 2. Band-pass SC filter in single-ended configuration considering parasitic capacitances

In order to cancel the even harmonics and reduce the distortion due to charge injection from the switches, the differential configuration was obtained (Fig. 3). Assuming that in this configuration the voltage drop across the capacitors is two times larger than in the single-ended configuration, the capacitors must have half the capacitance of the single-ended capacitors.

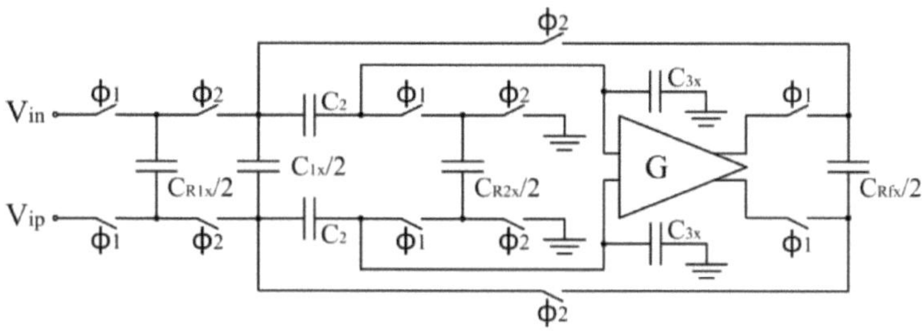

Fig. 3. Band-pass SC filter in differential configuration

Note that in order to maintain the common mode voltage relatively constant within the circuit, $C_{R2x}/2$ is connected to this common mode voltage during phase Φ_2 instead of ground.

5 Voltage-Combiner Amplifier

The single-ended configuration of the voltage-combiner amplifier is shown Fig. 4. Notice that M_1 and M_2 devices act, respectively, as common-source and common-drain devices.

In single-ended configuration this circuit acts as buffer and has a gain below unity. Doubling the circuit and connecting the sources of transistor M_1 (node V_{dif}), a differential pair is formed and the circuit becomes a low gain amplifier.

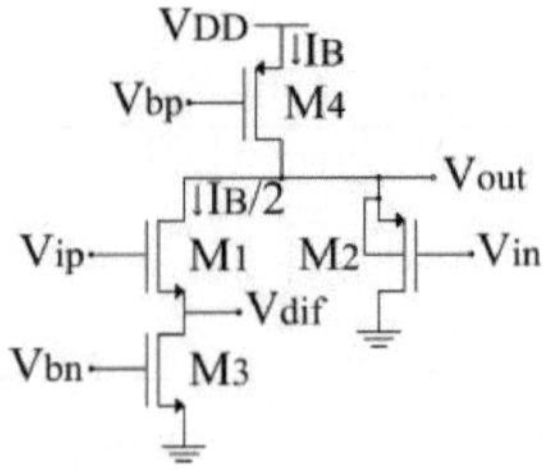

Fig. 4. Voltage-combiner in single-ended configuration

The low frequency open-loop gain of the circuit shown in Fig. 4 is given by:

$$G = \frac{g_{m1}g_{ds3} + g_{m2}(g_{m1} + g_{mb1} + g_{ds1} + g_{ds3})}{(g_{m2} + g_{ds2} + g_{ds4})(g_{m1} + g_{mb1} + g_{ds3}) + g_{ds1}(g_{m2} + g_{ds2} + g_{ds3} + g_{ds4})} \quad (4)$$

In order to improve the linearity of the amplifier, a small amount of source degeneration is used [5]-[7] in the differential pair formed by common-source transistors M_1, as shown in Fig. 5, using two MOS transistors operating in the triode region (M_5 and M_6) which exhibit higher linearity than transistors operating in the saturation region. The gain of this circuit can be easily adjusted, through design, to vary between 0.879 and 1.637, for a common mode voltage of 600 mV.

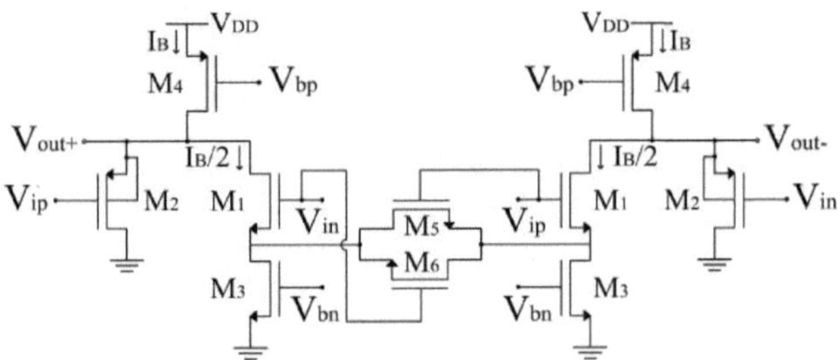

Fig. 5. Voltage-combiner with source degeneration using MOS transistors

Depending on the width used on transistors M_5 and M_6, the gain of the amplifier and its linearity will vary. To improve the amplifiers linearity, the width of both transistors should be lowered (increasing r_{ds}) until the optimum point is found. As a consequence the gain of the amplifier will decrease with the decrease of these transistors width. Since lowering the gain makes the design of the filter harder, the widths of M_5 and M_6 were chosen so that the gain is larger than 1.2. The simulated gains of the amplifiers are shown in Fig. 6 and Fig. 7.

The amplifier was sized in order to maximize both gain and GBW. The sizing used is shown in Table 1.

Table 1. Transistor sizes used in the voltage-combiner amplifier

Devices	M_1	M_2	M_3	M_4	M_5 and M_6
W [μm]	16	21.36	4.8	24	20
L [nm]	120	120	120	120	120

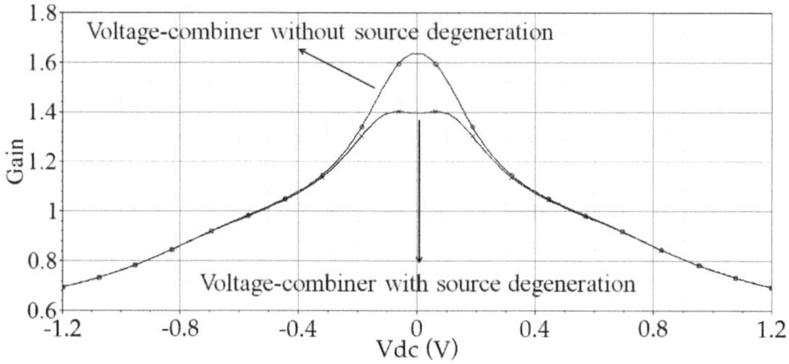

Fig. 6. Low-frequency gain of the voltage-combiner amplifier as a function of the differential input voltage

The frequency response of the amplifier is shown in Fig. 7.

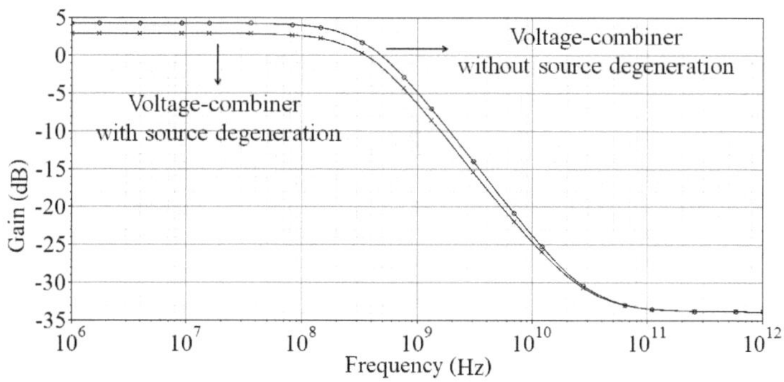

Fig. 7. Bode diagram (amplitude) of the voltage-combiner amplifier

6 Simulation Results

The biquadratic section described in Section 4 and the amplifier from Section 5 were designed in a standard 1.2 V 130 nm CMOS technology, in order to implement a biquadratic second-order band-pass Butterworth SC filter, with a central frequency of 1 MHz, a pass band of 500 kHz, and a clock frequency of 100 MHz.

The current drawn by the amplifier is determined in order for the settling error to be below 0.1%. To achieve this, it is necessary to have a certain gain-bandwidth product (GBW) that can be calculated from (5). Since the clock frequency is 100 MHz, the GBW of the amplifier has to be higher than 220 MHz to ensure a settling error below 0.1%. The transistors sizes used for the voltage-combiner amplifier that are shown in Table 1 were chosen in order to satisfy the following condition.

$$e^{-\mathrm{GBW[rad/s]}T_s/2} < 0.1\% \Leftrightarrow \mathrm{GBW[Hz]} > -\ln(0.1\%)F_s/\pi \tag{5}$$

Based on the gain of the voltage-combiner amplifier, the filter was initially designed from an ideal standpoint (assuming ideal switches, capacitors and amplifier) and then using the parasitic capacitances values obtained from an operating bias point (DC) simulation of the real circuit, the parasitic capacitances were compensated. Table 2 shows the values of the capacitors used in the single-ended configuration of the filter and the values of the parasitic capacitances. Note that the values of the parasitic capacitances are an approximation since due to the common mode voltage variation within the filter, these capacitance values will slightly vary.

Table 2. Filter gain, parasitic capacitances and single-ended capacitor values

Gain [V/V]	C_1 [pF]	C_2 [pF]	C_3 [fF]	C_{R1} [fF]	C_{R2} [fF]	C_{Rf} [fF]	C_{Bin} [fF]	C_{pd} [aF]	C_{ps} [aF]
1.3965	2	4	300	72.56	90.26	491.75	68	160.34	150.59

To determine the frequency response of the SC filter, an impulse is applied to the input, allowing charge into the circuit exclusively during one single clock phase (Φ_1), charging capacitors. Using this input causes the circuit to produce its impulse response. This signal is then sampled at intervals of $1/F_s$, at the end of clock phase Φ_1, until the impulse response is zero. These samples can be used to compute the frequency response of the filter shown in Fig. 3.

The frequency responses of the prototype filter, ideal circuit, and real circuit of the second-order filter are shown in Fig. 8. The attenuations at the filter bandwidth, as well as the amplifier's parameters and IM2 distortions are shown in Table 3.

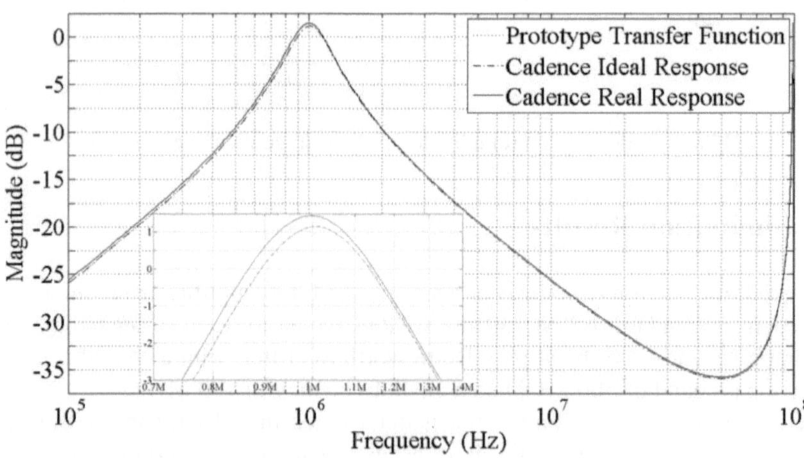

Fig. 8. Frequency response of the second-order band-pass filter

Table 3. Simulation results

Filter Attenuation @ Filter Bandwidth			Voltage-Combiner Amplifier		IM2 @ 200 mV_{pp}	
Prototype [dB]	-2.58	-2.29	DC Gain [dB]	2.90	Real Circuit [dB]	-52.11
Ideal Circuit [dB]	-2.58	-2.29	Power [µW]	244.33		
Real Circuit [dB]	-2.07	-2.17	GBW [MHz]	349.8		

7 Conclusion

This paper presented a solution to implement SC band-pass filters without the need of high gain amplifiers. The solution proposed using a low gain voltage-combiner amplifier and parasitic sensitive branches that, due to not having a virtual ground node in the circuit, require the compensation of parasitic capacitances during the design process. This technique simplified the design of the amplifier, reducing the total power consumption, and the silicon area of the overall filter. Simulation results, of the second-order band-pass SC filter, showed that for a clock frequency of 100 MHz, it's possible to obtain a frequency response similar to the one using ideal components. During the simulation phase it was seen that most of the distortion of the filter is introduced by the amplifier circuit, making it necessary to use other techniques, in addition to source degeneration, in order to reduce the distortion introduced by this circuit and also to make it less sensitive to process variation.

References

1. Perez, A.P., Maloberti, F.: Performance enhanced op-amp for 65nm CMOS technologies and below. In: IEEE Int. Symp. Circuits and Systems (ISCAS 2012), pp. 201–204 (May 2012)
2. Sallen, R.P., Key, E.L.: A practical method of designing RC active filters. IRE Trans. Circuit Theory CT 2, 74–85 (1955)
3. Hosticka, B.J., Brodersen, R.W., Gray, P.R.: MOS sampled data recursive filters using switched capacitor integrators. IEEE J. Solid State Circuits SC-12(6), 600–608 (1977)
4. Martin, K.: Improved circuits for the realization of switched-capacitor filters. IEEE Trans. Circuits and Systems SC-27(4), 237–244 (1980)
5. Acosta, L., Carvajal, R.G., Jimenez, M.: A CMOS transconductor with 90 dB SFDR and low sensitivity to mismatch. In: IEEE Int. Symp. Circuits and Systems (ISCAS 2006), pp. 69–72 (May 2006)
6. Krummenacher, F., Joehl, N.: A 4-MHz CMOS continuous-time filter with on-chip automatic tuning. IEEE J. Solid-State Circuits 23(3), 750–758 (1988)
7. Kuo, K.C., Leuciuc, A.: A linear MOS transconductor using source degeneration and adaptive biasing. IEEE Trans. Circuits and Systems II: Analog and Digital Signal Processing 48(10), 937–943 (2001)

Design of Cascode-Based Transconductance Amplifiers with Low-Gain PVT Variability and Gain Enhancement Using a Body-Biasing Technique

Nuno Pereira[2], Luis B. Oliveira[1,2], and João Goes[1,2]

[1] Centre for Technologies and Systems (CTS) – UNINOVA
[2] Dept. of Electrical Engineering (DEE), Universidade Nova de Lisboa (UNL)
Campus FCT/UNL, 2829-516, Caparica, Portugal
nrf.pereira@gmail.com, l.oliveira@fct.unl.pt, jg@uninova.pt

Abstract. A body-biasing compensation scheme based on two proportional-to-absolute-temperature (PTAT) circuits is proposed to reduce the PVT variability of the DC gain of cascode amplifiers. A brief description of a basic PTAT is given as well as its application to cascode-based operational transconductance amplifiers. Simulation results show that the proposed compensated circuit amplifier exhibit a (DC) gain variability smaller than the original (uncompensated) circuit, while reaching a gain enhancement of about 3 dB.

Keywords: amplifier, body-biasing, CMOS analog circuits, PVT compensation.

1 Introduction

In order to increase speed and reduce area, MOS devices are scaled down. However, this leads to short channel effects, thus reducing the intrinsic gain. Therefore, due to the supply reduction, high gain OpAmps are harder to design and cascode techniques are more difficult to employ [1]. In analog CMOS circuit design, the transistor is preferentially used in saturation. The intrinsic gain of a MOS transistor is given by

$$A_{vi} = g_m / g_{ds} \tag{1}$$

where g_m is the transconductance and g_{ds} is the output conductance. For a MOS device in saturation, assuming an approximate square law for the drain current, the transconductance is given by

$$g_m = \mu C'_{ox}(W/L)(V_{GS} - V_t)^2 \tag{2}$$

where μ is the carrier mobility, C'_{ox} is the oxide capacitance per unit of area, W and L are the width and length of the transistor, V_{GS} is the biasing gate source voltage and V_t is the threshold voltage. For a short channel, the increase of V_{DS} leads to the decrease of V_t, thus increasing the drain current. This gives rise to an increase of g_{ds}. So, despite the increase of transconductance with scaling, the intrinsic gain is

L.M. Camarinha-Matos, S. Tomic, and P. Graça (Eds.): DoCEIS 2013, IFIP AICT 394, pp. 590–599, 2013.

reduced [1]. This is illustrated in [2]. The temperature variation leads to a decrease in both the carrier mobility and the threshold voltage. In regards to the latter, it may decrease from 2 mV up to 4 mV for every 1° C rise [3]. These factors will have a direct effect on the transconductance of a transistor, as seen in (2). Also, process corners refer to the variation of fabrication parameters used in applying an integrated circuit design to a wafer. If a circuit running on devices fabricated at these process corners does not function as desired, the design is considered to have inadequate design margin. In this paper we present a circuit technique to reduce gain variability with temperature, supply and process variations in cascode amplifiers using a body-suitable biasing circuitry and at the same time increase the overall amplifier gain by about 3 dB.

2 Relationship to Internet of Things

Operational and transconductance amplifiers are, most probably, the most active building-block in analog and mixed-signal integrated circuits used in wireless and wire line communication systems. The work presented in this paper can contribute to the future development of the Internet of Things, since it can provide improved energy efficient circuit amplifier architectures, robust to temperature, supply and process variations. This ensures that a circuit manufactured to work, will perform as expected, regardless of the environment conditions.

3 Cascode Amplifiers

A single-stage operational transconductance amplifier (OTA) has usually a cascode configuration [4]. We consider a traditional cascode amplifier, shown in Fig. 1.

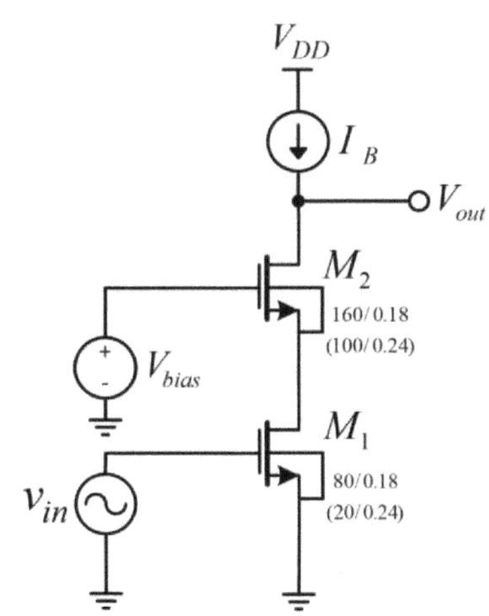

Fig. 1. Cascode Amplifier. All transistor sizes (*W/L*) are in μm. Sizes without brackets are for a 65 nm technology and those within brackets are for a 0.13 μm one.

The gain of this circuit is given by

$$A_v = g_{m1} \cdot \left[\frac{1}{g_{ds1}} \cdot \left(\frac{g_{m2}}{g_{ds2}}\right)\right] || R_B \tag{3}$$

where R_B is the resistance of the biasing current source, whose value is 100 μA and 150 μA for 65 nm and 130 nm technology, respectively. The positive power supply voltage is of 1.2 V.

Fig. 2 shows the variations of g_m, g_{ds}, and gain of transistor M_2 (g_m/g_{ds}) for 65 nm and 130 nm.

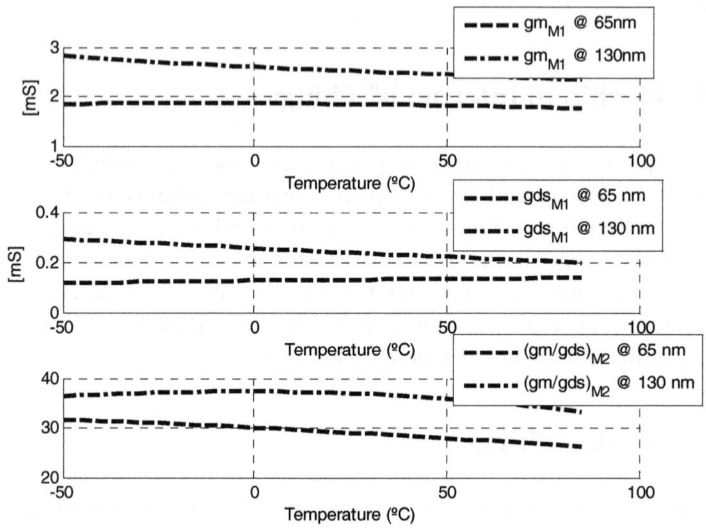

Fig. 2. Parameter variation with temperature for both node technologies

4 Proposed Solution

Fig. 3 presents the basic proportional-to-absolute-temperature (PTAT) circuit used in this paper, known as "constant transconductance" bias circuit, proposed in [4, 5].

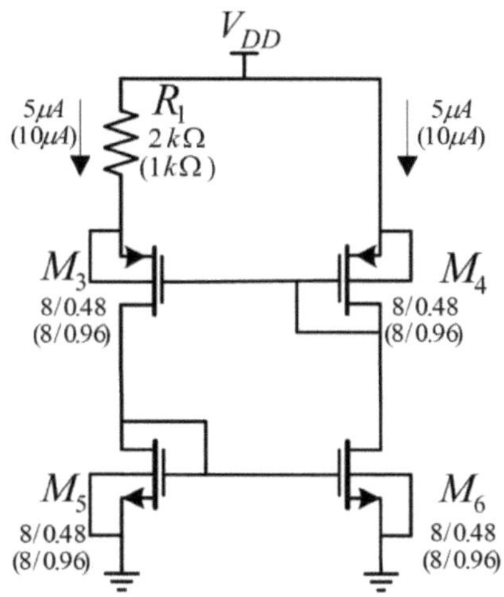

Fig. 3. Constant Transconductance Bias circuit. All transistor sizes (*W/L*) are in μm. Sizes without brackets are for a 65 nm technology and those within brackets are for a 0.13 μm one. The supply voltage is of 1.2 V.

We investigate the variation with temperature of the voltages V_A, V_B, and V_C, when all the transistors are matched.

The variation of V_A is irrelevant, since it varies less than 5 mV for the temperature range (-50 °C to +85 °C). For both technology nodes, V_B decreases with temperature while V_C increases, as shown in Fig. 4.

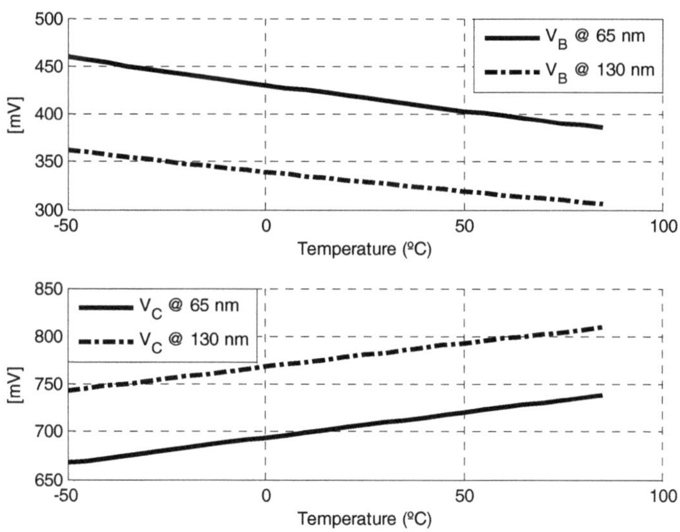

Fig. 4. Voltage variation at nodes V_B and V_C (for both node technologies)

The cascode amplifier (Fig. 1) with body-biasing by the circuit of Fig. 3 is shown in Fig. 5. The I_B current source is replaced with a current mirror, and every biasing voltage source is replaced with a MOSFET in diode configuration.

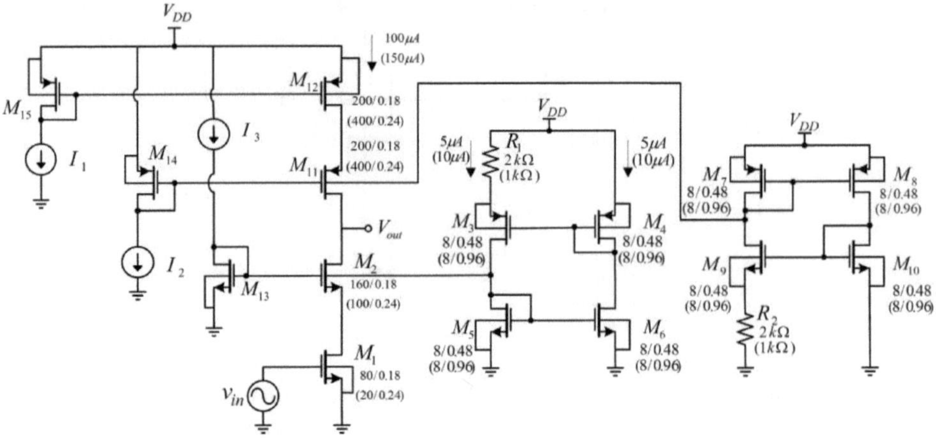

Fig. 5. Telescopic-cascode with body-biasing circuitry. All transistor sizes (*W/L*) are in μm. Sizes without brackets are for a 65 nm technology and those within brackets are for 0.13 μm.

In section 3 we have plotted the variations of g_m and g_{ds} in 65 nm and 130 nm. The main objective is to choose one of these voltage variations and apply it to the bulk of the M_2 transistor of the cascode circuit, biasing it in order to reduce the variability of the intrinsic gain, thus narrowing the variability of the overall gain, and increase the total gain of the circuit. The voltage V_B is used, since it decreases with temperature (for either technology), as seen in Fig. 4. If it is applied to the bulk of the M_2 transistor, it reduces the amplifier gain variability.

After replacing the current source with the current mirror, the MOSFETS that compose it also needed to have a temperature-independent behavior. Thus, the bulk of M_{11} should be connected to a symmetrical voltage to that applied to the bulk of M_2. In order to do so, a mirrored version of the "constant transconductance" bias circuit was designed.

Fig. 6 presents the variations of the transconductances of the $M_{1,2}$ transistors, when using body-biasing compensation. As it can be seen, the intrinsic gain of the M_2 transistor has a lower variation than that obtained when no body-biasing technique is used (Fig. 2), for both node technologies. Furthermore, the g_{ds} of the M_1 transistor also reduces its value while g_m increases, thus boosting the gain.

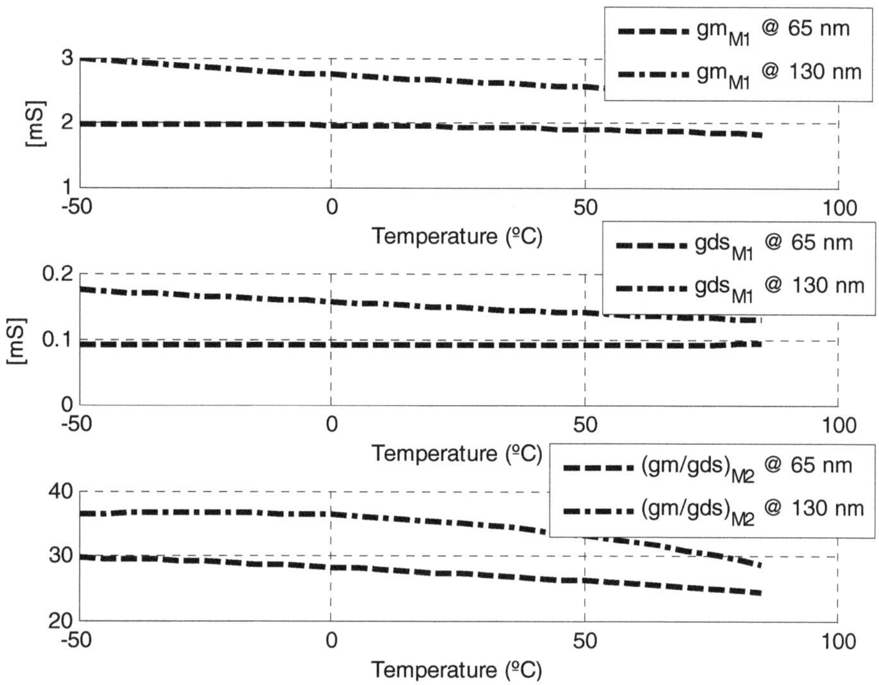

Fig. 6. Parameter variation with temperature with body-biasing compensation for both node technologies

5 Simulations and Results

All simulations are performed using BSIM3v3 models for standard logic 65 nm and 130 nm CMOS technology. A supply voltage of 1.2 V is used and the process variations considered are slow-slow (ss), typical-typical (tt) and fast-fast (ff). The temperature range is from -50 to +85 ºC.

In Fig. 7, the gain of the circuit is plotted, for both technology nodes, with and without body-biasing technique. For both technology nodes the gain variability is lower when using the body-biasing circuitry. Furthermore, there is an enhancing of the gain by about 3 dB, as expected, as it was explained in section 4.

With regard to supply variation, Fig. 8 shows that with a variation of ± 5% (1.14 V to 1.26 V) of the supply voltage, the body-biasing circuitry leads to a lower variability for the gain (below ± 0.5 dB). Some traces can be superimposed.

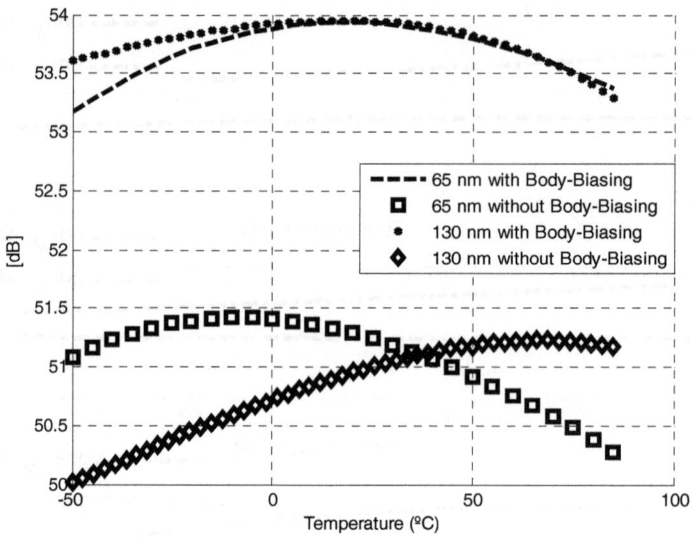

Fig. 7. Overall gain variability with temperature for both technologies node (with and without body-biasing technique)

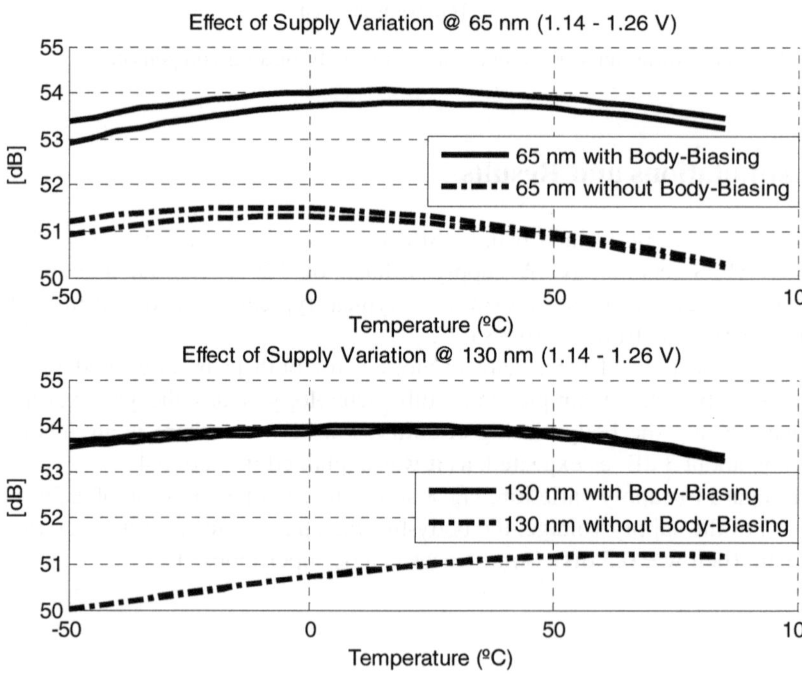

Fig. 8. Overall gain variability with temperature & supply variation for both technologies node (with and without body-biasing technique)

Concerning process variation, Fig. 9 and 10 show that for the processes considered (ss, tt and ff), there is always an enhancement of the gain by about 3 dB. For most of the cases, the (DC) gain variability is below ± 0.5 dB (except for process ff, at 130 nm node technology, where the variability is higher than that without body-biasing).

This body-biasing circuitry requires only an extra current consumption of about 5% to 7% (for both technologies) of the total current consumption of the amplifier. This results in low power consumption, for both node technologies, from a 1.2 V supply.

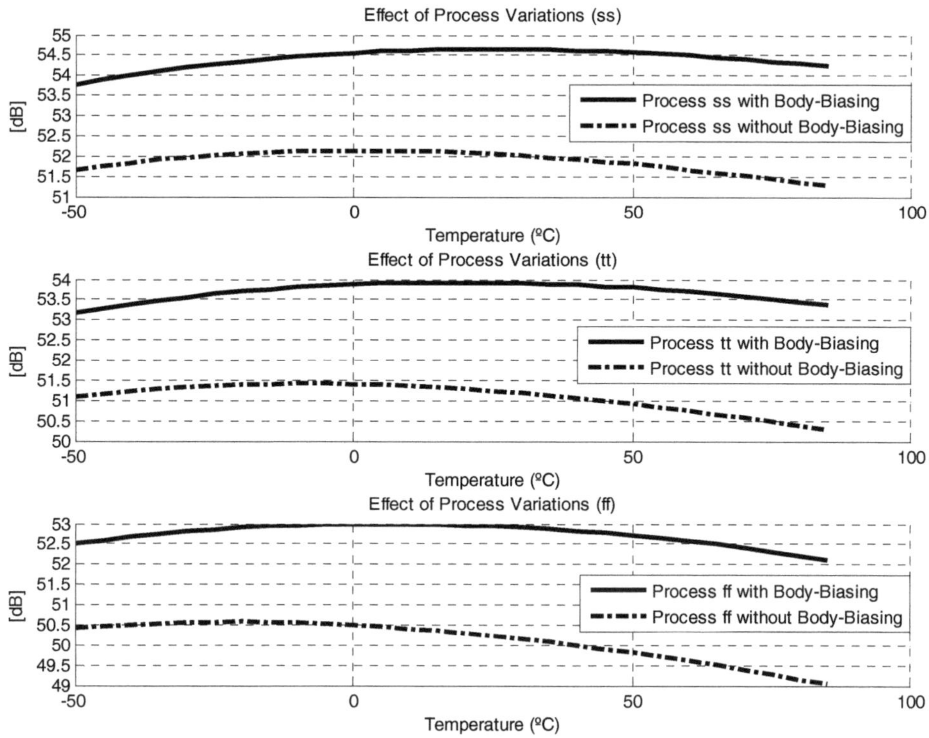

Fig. 9. Overall gain variability with temperature & process variation for 65 nm (with and without body-biasing)

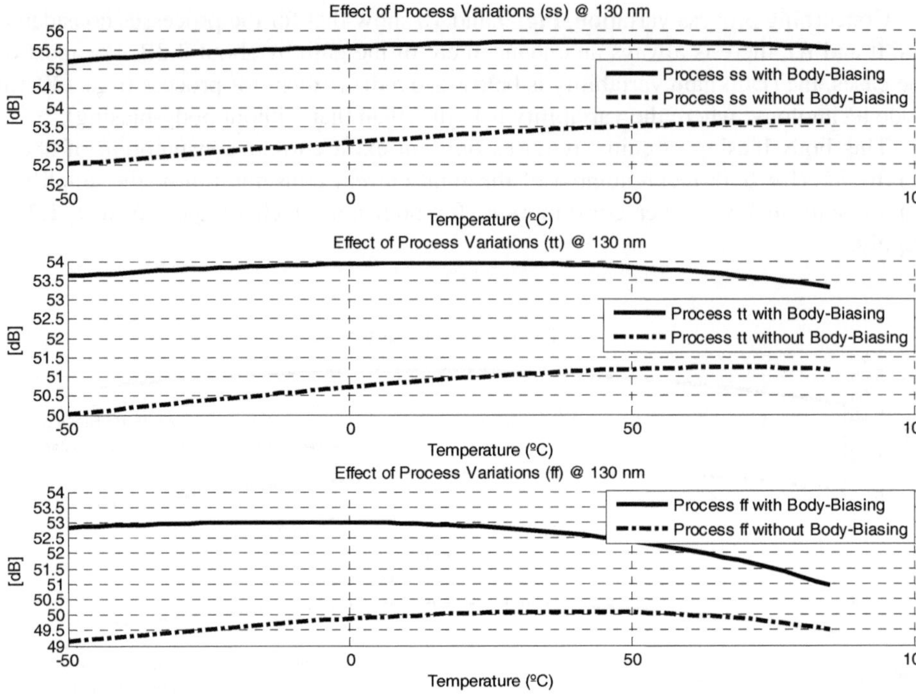

Fig. 10. Overall gain variability with temperature & process variation for 130 nm (with and without body-biasing)

6 Conclusions

This paper presented a simple circuit technique to reduce gain variability with temperature, supply, and process variations in cascode amplifiers, using a body-biasing scheme, and at the same time, enhance the overall gain of the amplifier.

Simulation results of a standard telescopic-cascode amplifier, in two different nanoscale CMOS technologies (130 nm and 65 nm) have shown that it is possible to obtain process-supply-and-temperature-compensation.

The simulated DC gain has low variability (below ± 0.5 dB for all supply-and-temperature variations) while it is enhanced, over all PVT corners by 3 dB.

The authors are unaware of the existence of similar body-biasing compensation schemes to reduce the PVT variability of the DC gain, up to this date. Therefore, it was impossible to perform a comparison with other body-biasing schemes.

References

1. Oliveira, J.P., Goes, J.: Parametric Analog Signal Amplification Applied to Nanoscale CMOS Technologies, 2012th edn. Springer (2012)
2. Pekarik, J., Greenberg, D., Jagannathan, B., Groves, R., Jones, J.R., Singh, R., Chinthakindi, A., Wang, X., Breitwisch, M., Coolbaugh, D., Cottrell, P., Florkey, J., Freeman, G., Krishnasamy, R.: RFCMOS technology from 0.25μm to 65nm: the state of the art. In: Proc. of the IEEE Custom Integrated Circuits Conf., pp. 217–224 (October 2004)
3. Weste, N.H.E., Eshraghian, K.: Principles of CMOS VLSI Design, 2nd edn. Addison Wesley (1994)
4. Carusone, T.C., Johns, D.A., Martin, K.W.: Analog Integrated Circuit Design, 2nd edn. Wiley (2011)
5. Razavi, B.: Design of Analog CMOS Integrated Circuits, 1st edn. McGraw-Hill Science/Engineering/Math (2000)

Part XX
Electronics: RF Applications

Comparison of Feedback Influence on Ring Oscillator Performance for IR-UWB Pulse Generator in 0.13 μm and 0.18 μm CMOS Technologies

Jelena Radic, Alena Djugova, Laszlo Nagy,
Mirjana Videnovic-Misic, and Ljiljana D. Zivanov

Dep. for Power, Electronics, and Communications Engineering,
Faculty of Technical Sciences, University of Novi Sad,
Trg Dositeja Obradovica 6,
21000 Novi Sad, Serbia
{jelenar_,alenad,lnadj,mirjam,lilaziv}@uns.ac.rs

Abstract. A CMOS three-stage ring oscillator is examined in UMC 0.13 μm and 0.18 μm technologies. The influence of PMOS transistor and resistor, as inverter feedbacks, on the ring oscillator frequency and the peak-to-peak amplitude is investigated in both technologies. Furthermore, as the ring oscillator usually drives a buffer in pulse generator/transmitter chain, dependence of its Figures of Merit on the buffer feedback is presented in the paper. Simulation results showed that the ring oscillator frequency is strongly dependent on the inverter feedback. The presented techniques can be used to increase (resistive feedback) and control (PMOS transistor feedback) the ring oscillator frequency. As the ring oscillator is a part of an IR-UWB (Impulse Radio Ultra Wide Band) pulse generator, its oscillating frequency determines the spectrum central frequency and has significant effect on spectrum fitting within UWB FCC mask.

Keywords: CMOS process, impulse radio ultra wideband, PMOS transistor feedback, pulse generator, resistive feedback, ring oscillator.

1 Introduction

Since the Federal Communications Commission (FCC) allocated the 3.1 – 10.6 GHz unlicensed spectrum for commercial ultra wideband (UWB) application in February 2002 [1], several technologies have been developed to satisfy the communication market requirements such as Multi-Band Orthogonal Frequency Division Multiplexing (MB-OFDM), Direct Sequence (DS) and Impulse Radio (IR) [2 – 5]. As a carrier-less approach which makes use of ultra-short duration pulses (pulse duration is less than 1 ns) that can yield ultra-wide bandwidth signals, IR-UWB technology stands out from the rest by providing very simple (without mixer and power amplifier), low-cost and low-power UWB transmitter realizations [6]. Besides wide bandwidth, it offers a great resistance to multipath fading that usually plagues for

L.M. Camarinha-Matos, S. Tomic, and P. Graça (Eds.): DoCEIS 2013, IFIP AICT 394, pp. 603–610, 2013.

narrow-band systems, and allows multiple access and robustness to interference [7]. Another advantage of the IR-UWB radios is high precision ranging [8], with a potential for centimeter accuracy in indoor environments. Consequently, the approach appears to be good candidate for very high data rate short-range communications, low data rate communications related to localization or/and positioning systems, sensor networks, biomedical and many other cutting edge Internet of Things applications.

A pulse generator plays the core role in an UWB system as it produces pulse train which spectrum has to fully comply with and efficiently use the FCC mask. Therefore, it is extremely challenging to design UWB transmitter that satisfies such demanding requirements while achieving wide bandwidth, low-power, low cost and simple architecture. Besides, it is very desirable to provide pulse generator spectrum tuning to enable compensation due to process, voltage and temperature (PVT) variations, and antenna and communication channel performance changes.

As an essential part of analogue-digital pulse generator, a ring oscillator is investigated in this work. In addition to the standard switched three-stage ring oscillator, topologies with resistors and PMOS transistors, introduced as feedbacks in the ring oscillator inverters, are examined in 0.13 μm and 0.18 μm UMC CMOS technologies. A novel approach that uses the resistive feedback in the buffer stage to increase ring oscillator frequency without degrading its peak-to-peak amplitude is proposed. Influence of negative feedback on the ring oscillator performance and comparison of results obtained using considered processes are presented.

2 Relation to Internet of Things

After the computer in the 1940s and the Internet in the 1970s, the Internet of Things (IoT) represents the third revolution in information and communication technology (ICT). With main vision of "Everything connects", its development depends on dynamic technology evolutions in a set of multidisciplinary and interdisciplinary fields, ranging from the material and devices, sensor technology, to electronic system design and wireless communications. To fulfill the IoT demands, future micro-devices must provide a variety of advanced futures such as sensing, processing, communication, positioning, and capability to be integrated into everyday objects. To accomplish this, reliable-operating and maintenance-free wireless networks with low-power and low-cost radio transceivers are essential. Many advantages of the IR-UWB systems provide them leading role in wireless communications and sensor networks, and thus precedence in IoT applications. Wideband signals offer robust communications and high-precision positioning; duty-cycled operations allow link scalability and energy-efficient solutions; and baseband-like architecture facilitates extremely simple and low-power transmitters that can be integrated in various objects.

Thought it is not required, ability of UWB transmitter spectrum tuning represents very desirable quality that allows compensation for PVT variations and communication channel performance changes, and furthermore interference avoidance. Methods proposed in this work provide control and increase in the ring oscillator frequency. Since this parameter determines the center frequency of the ring oscillator-based pulse generator output spectrum, such ideas can be used to enable the IR-UWB transmitter spectrum adjustment.

3 Standard Three-Stage Ring Oscillator Design

It is already mentioned that the pulse generator is one of the most important parts of the IR-UWB system. The oscillator-based transmitter that contains a switched ring oscillator represents the most commonly used topology, presented in Fig. 1 [9]. The glitch generator turns on/off the ring oscillator approximately defining duration of its oscillation and so the width of the transmitter output pulse. The pulse generator spectrum center frequency is determined by the ring oscillator frequency. The buffer stage isolates the switched oscillator from the output filter, and improves the UWB transmitter current driving capability. The pulse shaping, usually the band-pass filter additionally accommodates spectrum fitting within the FCC mask.

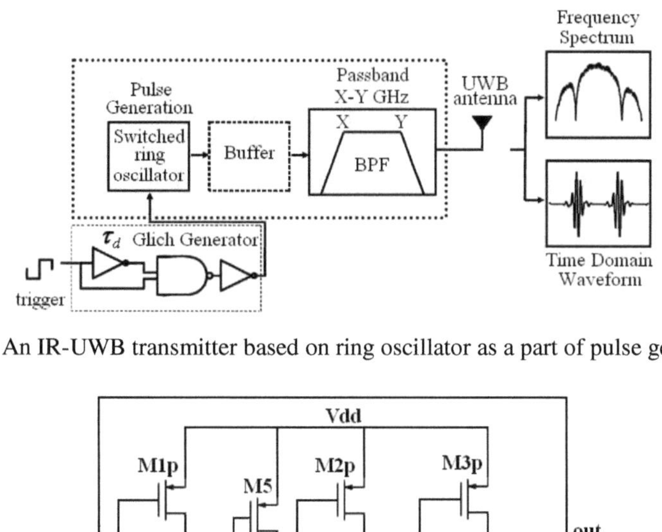

Fig. 1. An IR-UWB transmitter based on ring oscillator as a part of pulse generator

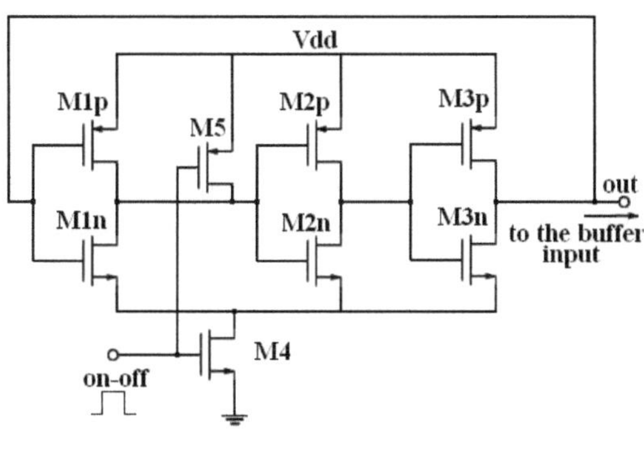

Fig. 2. The standard three-stage ring oscillator architecture

The switched ring oscillator topology is shown in Fig. 2. It consists of the standard three-stage ring oscillator (inverter stages M_1–M_3) and oscillation-enabling switches (transistors M_4 and M_5). The ring oscillator is turned on by transistor M_4 at the *on-off* signal (produced by the glitch generator) rising edge. The inverter stages output voltage values are determined by the size ratio of the corresponding PMOS and NMOS transistors. During the *on-off* signal low level, the transistor M_4 is switched off, and no signal is generated at the oscillator output. Furthermore, transistor M_5 connects the M_1 transistor output (the M_2 transistor input) to V_{dd} providing the oscillation start from the same initial state at the rising edge the *on-off* signal.

4 Ring Oscillator in 0.13 μm and 0.18 μm CMOS Technologies

The proposed design has been simulated in mixed mode/RF 0.13 μm and 0.18 μm UMC CMOS technologies using Cadence's Spectre RF Simulator. Supply voltages V_{dd} of these technologies are 1.2 V and 1.8 V, respectively.

The ring oscillator working frequency depends directly on transistors sizes. The period of the oscillation T rises proportionally with increase in transistors sizes, while the oscillating frequency decreases (f_0=1/T). For the same NMOS and PMOS ring transistors size (channel width/length W/L=3.6 μm/0.12 μm) in 0.13 μm CMOS technology, the oscillation frequency of 7.65 GHz has been obtained. However, for the smallest NMOS transistors (W/L=25 μm/0.18 μm) and approximately two times larger PMOS transistors (W/L=45 μm/0.18 μm) in 0.18 μm CMOS technology, the f_0 parameter of 3.77 GHz has been achieved. It should be noted that in inverter topology PMOS transistor is usually equal or two times larger than NMOS transistor to compensate the difference in PMOS and NMOS carrier mobility and set V_{TH} closer to V_{dd}/2. Although former technology allows considerably higher ring oscillator frequency, 0.18 μm technology is much more cost effective [10], and attractive from the IC fabrication costs point of view. Nevertheless, obtained values are not high enough to utilize the UWB higher-band (6 – 10 GHz) more effectively (the center frequency of at least 8 GHz is required).

4.1 Influence of the Inverter Resistive Feedback

To increase the oscillation frequency at a given DC current, the resistive feedback is used in each inverter stage, as proposed in [9]. Resistors were connected between nodes A–A', B–B', and C–C', as shown in Fig. 3. Resistors R influence on the performance (the frequency f_0 and peak-to-peak amplitude V_{pp}) of the ring oscillator

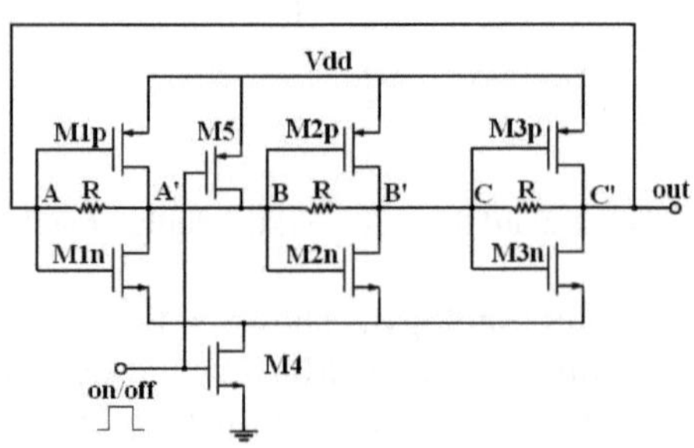

Fig. 3. Architecture of the ring oscillator with the resistive feedbacks

Table 1. Ring oscillator performance dependence on the inverter feedback resistor value

Tech.	0.13µm CMOS					0.18µm CMOS				
R (kΩ)	4	6	8	10	∞	0.5	1	2	4	∞
f_0 **(GHz)**	9.80	9.10	8.60	8.20	7.65	5.60	4.47	4.09	3.92	3.77
V_{pp} **(V)**	0.44	0.62	0.69	0.73	0.88	0.60	1.18	1.38	1.48	1.57

designs in used technologies is shown in Table 1. It can be noted that with decrease in the resistor value the oscillation frequency increases, but the peak-to-peak amplitude V_{pp} of the ring oscillator output signal is reduced, as expected. This behavior is caused by the reduction of the inverter closed loop gain with decrease in the feedback resistor value. Furthermore, the V_{pp} parameter and the time for reaching the peak signal values are reduced resulting in f_0 parameter increase. Moreover, it can be seen that the oscillation frequency was considerably increased in both technologies, while higher f_0 values are obtained in the scaled UMC technology, as expected. But, the V_{pp} parameter values achieved for the lowest resistors values (4 kΩ in case of 0.13 µm, and 0.5 kΩ in 0.18 µm CMOS process) are not large enough to drive properly the subsequent stage (usually a buffer) in the pulse generator chain. This is the main reason why higher feedback resistor values are used in 0.13 µm CMOS architecture (otherwise V_{pp} values even lower than 440 mV would be obtained). On the other hand, if higher resistor values are used in 0.18 µm technology (the same as in 0.13 µm) the ring oscillator frequency would not be increased comparing to the initial value of 3.77 GHz. Therefore, it should be emphasized that the resistor value needs to be selected considering the trade-off between the oscillation frequency and the required V_{pp} value.

Although the f_0 parameter can be significantly increased by proposed technique, this method does not allow the ring oscillator frequency to be tuned. It is already mentioned that even if it is not required in the IR-UWB technology, the ability of the frequency adjustment is very desirable advantage.

4.2 Influence of Inverter PMOS Transistor Feedback

To enable electrical tuning, the topology with PMOS transistors as inverter feedbacks has been proposed, Fig. 4. Influence of the PMOS gate control voltage V_{ctrl} on the performance of the ring oscillator architectures in 0.13 µm and 0.18 µm technologies is summarized in Table 2.

In addition to higher f_0 values, the 0.13 µm technology offers noticeably wider tuning range (3.15 GHz) compared to frequency range (1.46 GHz) available in 0.18 µm CMOS process. For V_{ctrl} parameter value equal to zero, the same situation as in the previous method can be noticed; the achieved V_{pp} values are below acceptable limits. This can be explained by the fact that low resistance of PMOS feedback transistor, operating in deep triode region, reduced significantly inverter closed loop gain. As the presented topology has the PMOS transistor feedback in each inverter

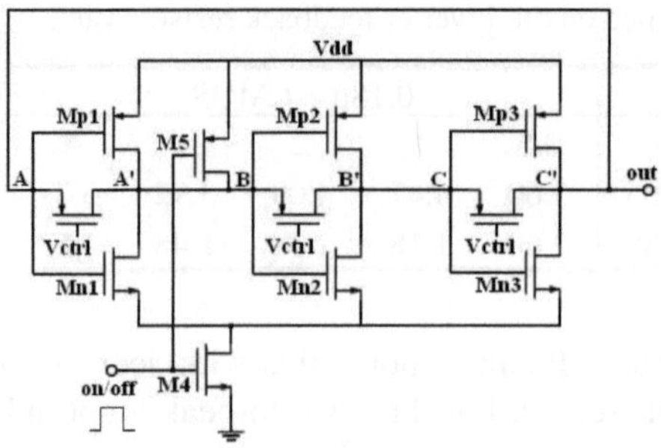

Fig. 4. Topology of the ring oscillator with PMOS transistor feedbacks

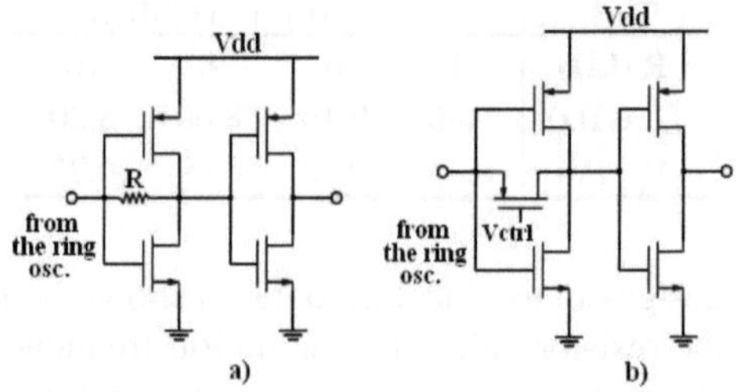

Fig. 5. Two-stage buffer with: (a) Resistive feedback (b) PMOS transistor feedback

Table 2. Ring oscillator performance dependence on the gate control voltage of the PMOS transistor in the inverter feedback

Tech.	0.13μm CMOS					0.18μm CMOS				
V_{ctrl} **(V)**	0	0.3	0.6	0.9	1.2	0	0.3	0.6	1.2	1.8
f_0 **(GHz)**	8.75	6.70	5.85	5.60	5.60	4.29	3.40	3.13	2.84	2.83
V_{pp} **(V)**	0.11	0.72	0.90	0.94	0.94	0.46	1.09	1.34	1.56	1.57

stage, the effect is much more intense resulting in the unsuitably low V_{pp} value. It should be pointed out that the maximum oscillation frequency and the available frequency range depend on the ring oscillator transistors sizes. Increase in the transistors sizes reduces the observed Figures of Merits (FOM).

4.3 Buffer Feedback Influence

The method of increasing the ring oscillator frequency by using the inverter resistive feedback is already known in literature, [9]. To the best of our knowledge, the influence of the buffer feedback on the ring performance is still not examined. The two-stage buffer topologies with resistive and PMOS transistor feedback are proposed in Fig. 5. The first stage provides control/increase in the ring oscillator frequency, while the second buffer stage prevents this effect to change the output filter shaping characteristics. Dependences of performance of the ring oscillator designs in 0.13 μm and 0.18 μm UMC technologies on the buffer resistive and PMOS transistor feedbacks are given in Tables 3 and 4, respectively.

Table 3. Ring oscillator performance dependence on the buffer feedback resistor value

Tech.	0.13μm CMOS					0.18μm CMOS				
R **(kΩ)**	2.32*	4.0	6.0	8.0	10.0	0.25	0.5	1.0	2.0	4.0
f_0 **(GHz)**	9.05	8.50	8.30	8.15	8.10	4.27	4.11	3.98	3.87	3.82
V_{pp} **(V)**	0.72	0.79	0.82	0.84	0.85	1.19	1.35	1.44	1.50	1.53

*Minimal resistance value for the RNHR resistor model.

Table 4. Influence of the buffer feedback transistor V_{ctrl} parameter on the ring oscillator performance

Tech.	0.13µm CMOS					0.18µm CMOS				
V_{ctrl} **(V)**	0	0.3	0.6	0.9	1.2	0	0.3	0.6	1.2	1.8
f_0 **(GHz)**	8.90	8.25	7.70	7.45	7.4	3.94	3.74	3.63	3.55	3.55
V_{pp} **(V)**	0.71	0.82	0.87	0.87	0.88	1.17	1.27	1.37	1.47	1.47

Varying R from 10 kΩ to 2.32 kΩ in 0.13 µm technology, the oscillation frequency is increased from 8.10 GHz to 9.05 GHz, followed with decrease in V_{pp} parameter from 0.85 V to 0.72 V. Reducing R from 4.0 kΩ to 0.25 kΩ in 0.18 µm CMOS ring oscillator design, the f_0 parameter rises from 3.82 GHz to 4.27 GHz simultaneously with reduction in V_{pp} amplitude from 1.53 V to 1.19 V. This technique uses feedback resistor values lower than in method presented in section 4.1 because obtained V_{pp} values are higher than $V_{dd}/2$. It can be noticed that comparing to the mentioned approach the proposed technique allows smaller increase in the ring oscillator frequency. However, it should take into account that this method uses only one resistor in contrast to three additional resistors utilized in the inverter feedback architecture, and at the same time provides higher V_{pp} parameter values for the same ring oscillator frequency. The three-stage ring oscillator period T is determined by the propagation time of a signal transition through the complete oscillator chain ($T=6\times t_p$, where t_p is the gate propagation delay). An inverter signal propagation time is largely determined by the strength of the driving gate, and the load presented by the output node itself, which sums the contributions of the connecting gates and the wiring parasitic. Varying the buffer feedback resistor changes its feedback impedance, which further modifies the buffer input impedance (Miller's theorem) leading to the change in the ring oscillator load, propagation time and the ring oscillator period/frequency.

Approach presented in Fig. 5b shows similar influence on the ring oscillator FOM. Reducing V_{ctrl} parameter from V_{dd} to zero the oscillation frequency can be tuned in the frequency ranges narrower than in technique presented in section 4.2 (especially in 0.18 µm CMOS process), but with higher obtainable V_{pp} parameter values. Moreover, the method requires only one resistor. Additionally, it can be noticed that the frequency range available in 0.13 µm CMOS technology is extensively wider than in the other considered process.

5 Discussion and Conclusion

Dependences of the three-stage ring oscillator performance on resistors and PMOS transistors, introduced as feedbacks in the oscillator inverters and buffer, are investigated in 0.13 µm and 0.18 µm UMC CMOS technologies. Simulation results in both technologies confirm strong dependency of the ring oscillator frequency and the peak-to-peak amplitude on the inverters and buffer feedbacks. Considering the presented methods, higher f_0 parameter values and wider available frequency range can be obtained with topologies that use feedbacks in the ring oscillator inverters.

However, these architectures require additional three resistors/PMOS transistors that occupy more area and thus increase chip cost. The topology with buffer feedback would be the best choice in case of the medium operating frequency range applications as it uses only one additional component and gives higher V_{pp} parameter values. Research presented here shows that the 0.13 μm process offers much more promising performance. Moreover, the layout designs dimensions are approximately two times smaller than in 0.18 μm technology. Nevertheless, as the main drawback of the former technology is higher IC fabrication cost, the later process is chosen as optimal solution in this work. The ring oscillator-based pulse generator designs in 0.18 μm technology that combine proposed methods are presented in [11], [12].

Acknowledgments. This work was supported in part by the Ministry of Education and Science, Republic of Serbia, on the project no. TR-32016, and by FP7 project no. 256615.

References

1. First Report and Order: Revision of Part 15 of the Commission's Rules Regarding Ultra-Wideband Transmission Systems Federal Communications Commission (FCC), ET Docket 98-153 (Adopted February 14, 2002, Released April 22, 2002)
2. Multi-Band OFDM Physical Layer Proposal for IEEE 802.15 Task Group 3a, IEEE P802.15-03/268r2 (2004)
3. DS-UWB Physical Layer Submission to 802.15 Task Group 3a, IEEE P802.15 (2004)
4. Win, M.Z., Scholtz, R.A.: Impulse radio: how it works. IEEE Communications Letters 2(2), 36–38 (1998)
5. Ghavami, M., Michael, L.B., Kohno, R.: Ultra Wideband Signals and Systems in Communications Engineering. John Wiley&Sons Ltd. (2004)
6. Siwiak, K., McKeown, D.: Ultra-Wideband Radio Technology. John Wiley&Sons Ltd. (2004)
7. Fernandes, J.R., Wentzloff, D.: Recent Advances in IR-UWB Transceivers: An Overview. In: IEEE International Conference on Circuits and Systems, pp. 3284–3287. IEEE Press, Paris (2010)
8. Lee, J.-Y., Scholtz, R.A.: Ranging in a dense multipath environment using an uwb radio link. IEEE Journal of Selected Areas in Communications 20(9), 1677–1683 (2002)
9. Sim, S., Kim, D.W., Hong, S.: A CMOS UWB pulse generator for 6–10 GHz applications. IEEE Microwave and Wireless Components Letters 19(2), 83–85 (2009)
10. Europractice (prototyping) miniasic prices, `http://www.europractice-ic.com/docs/MPW2012-miniasic-v8.pdf`
11. Radic, J., Djugova, A., Nagy, L., Videnovic–Misic, M.: A Low Power 3.1-7.5 GHz Tunable Pulse Generator for Impulse Radio UWB. In: IEEE International Symposium on Intelligent Systems and Informatics, SISY, pp. 425–428. IEEE Press, Subotica (2012)
12. Nagy, L., Radic, J., Djugova, A., Videnovic–Misic, M.: Ultra Low-Power Low-Complexity Tunable 3-10 GHz IR-UWB Pulse Generator. Journal of Microelectronics, Electronic Components and Materials, MIDEM 42(3), 185–191 (2012)

A Low-Voltage CMOS Buffer for RF Applications Based on a Fully-Differential Voltage-Combiner

S. Abdollahvand, R. Santos-Tavares, and João Goes

CTS-UNINOVA / Dept. de Eng. Electrotécnica e Computadores,
Faculdade de Ciencias e Tecnologia, Universidade Nova de Lisboa
FCT Campus, 2829-517 Caparic, Portugal
S.Abdollahvand@IEEE.org, {rmt,jg}@univova.pt

Abstract. This paper presents a new CMOS buffer circuit topology for radio-frequency (RF) applications based on a fully-differential voltage-combiner circuit, capable of operating at low-voltage. The proposed circuit uses a combination of common-source (CS) and common-drain (CD) devices. The simulation results show good levels of linearity and bandwidth. To improve total harmonic distortion (THD) a source degeneration technique is used. The proposed circuit has been designed in a 130nm logic CMOS technology and it achieves a simulated gain of 1.54 dB, a bandwidth of 1.14 GHz for a total power dissipation of 13.34 mW, when driving an RF active probe (with 0.8 pF in parallel with 200 kΩ).

Keywords: buffer, common-source, common-drain, RF, CMOS, low-voltage.

1 Introduction

In recent years, there has been an increasing trend in incorporating complete systems in longer lasting battery-powered equipment which requires low power dissipation circuits [1]. Particularly, buffer circuits used as an important functional block of high performance communication systems, e.g. RF applications like drivers of passive switched-capacitor down-converter circuits. Circuit For this purpose is essential to provide good linearity in terms of both output level and intercept-point performance. Provided that the transient response of the buffer is fast enough, which implies a bandwidth greater than 1 GHz, errors will be minimal. Low levels of total harmonic distortion (THD) are also essential for compatibility with communication standards. Moreover, traditionally, buffers present a gain near equal to unit but always with some signal attenuation. Some examples are the elastic source-follower or the enhanced voltage-follower as proposed in [2].

The objective of this paper is to propose a novel voltage buffer based on a voltage-combiner topology, i.e. a common-source/common-drain structure. It achieves a gain that can be designed in the range from 0 dB to 6 dB; the input impedance is equally high; and the parasitic capacitance of its inputs is low. For improving THD the source degeneration method has been used. The supply voltage variability was studied for 1.2 V ± 5% for three levels, i.e.,1.14 V, 1.2 V and 1.26 V. Since the voltage gain of

L.M. Camarinha-Matos, S. Tomic, and P. Graça (Eds.): DoCEIS 2013, IFIP AICT 394, pp. 611–618, 2013.

the proposed circuit is higher than the unit, it is expected to compensate for the gain loss due to linearization of the circuit.

2 Internet-of-Things

The buffer circuit presented in this paper is an essential part of RF transceiver integrated circuits (a paramount building block in modems) and it can be used either for driving a passive down-converter or for on-chip built-in self-testing purposes. This circuit can be used as an interface to environmental data signals collectors such as temperature smart sensors. The circuit reading environmental data, passes the information to an Internet Web server to display temperature information. This way it can contribute to Internet of Things.

3 Proposed Circuit Basic Concept

The basic idea of voltage-combiner (VCOM) technique is depicted in Fig. 1. It basically employs a combination of NMOS transistors in configuration of common-source, M_Y, and common-drain, M_X. High input impedance is equally accomplished.

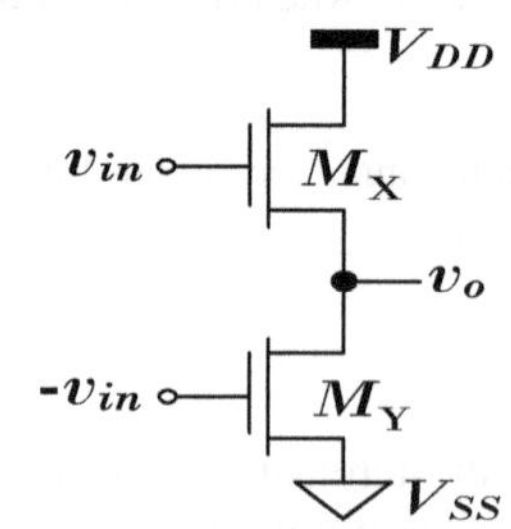

Fig. 1. VCOM circuit conceptual idea

After simplifying the small signal equivalent (differential-mode, DM) of the VCOM and substituting the components by their Y-parameter equivalents, the behavioral signal path model [3] is extracted and illustrated in Fig. 2 (for simplicity only half the circuit is shown).

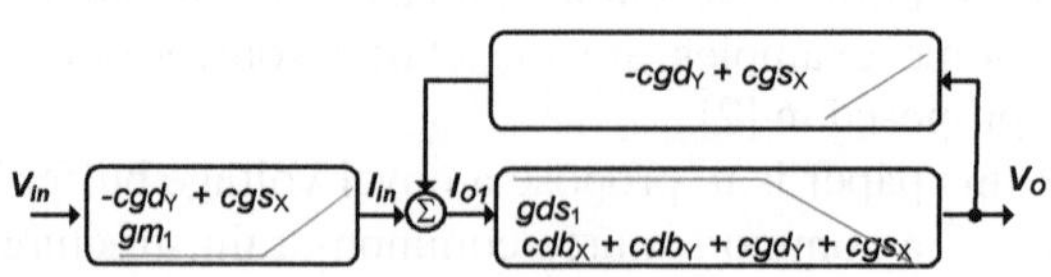

Fig. 2. Behavioral signal path model of the voltage combiner stage (for simplicity only half the circuit is shown)

This model permits large insight in the small-signal behavior of the amplifier and is a fundamental tool in the extraction of the differential gain transfer function. It is possible to verify: the Miller effect through parasitic capacitance cgd_1, the pole(s) and zero(s), and the order of the transfer function (in this case, 1st order).

Using the behavioral signal path model described in Fig. 2 and writing the equations for I_{O1} and V_O, it is possible to extract the transfer function of the VCOM. For the sake of simplicity, minor simplifications were used in the derived equations:

$$gds_1 = gds_Y + gds_X \tag{1}$$

$$cdb_1 = cdb_Y + cdb_X \tag{2}$$

Body effect of transistors M_Y and M_X has been neglected, for the sake of simplicity, But it can be easily included into the equations.

$$\begin{aligned} \mathrm{TF}_{VCOM} &= \\ &= \frac{gm_Y + gm_X + (cgs_X - cgd_Y)\cdot s}{gm_X + gds_1 + (cdb_1 + cgs_X + cgd_Y)\cdot s} \end{aligned} \tag{3}$$

From the transfer function it is possible to obtain the low-frequency open-loop gain (DC gain) of the VCOM stage, A_{VCOM},

$$A_{VCOM} = \frac{gm_Y + gm_X}{gm_X + gds_1} \tag{4}$$

considering that $gm_X >> gds_1$, a good approximation can be given by,

$$A_{VCOM} \approx \left(1 + \frac{gm_Y}{gm_X}\right), \qquad |A_{VCOM}| > 1 \tag{5}$$

Sizing the circuit to attain $gm_{13} \approx gm_{12}$, 6 dB are added in the overall DC gain of the amplifier.

Also, the dominate pole, ω_{p1VCOM}, is computed using:

$$\omega_{P1_{VCOM}} = \frac{gm_X + gds_1}{cdb_1 + cgs_X + cgd_Y} \tag{6}$$

The gain-bandwidth product performance parameter of this buffer (GBW) can be expressed by

$$\mathrm{GBW}_{VCOM} = \frac{gm_Y + gm_X}{cdb_1 + cgs_X + cgd_Y}. \tag{7}$$

4 Complete Circuit Description, Linearization and Simulation Results

4.1 Circuit Description

The proposed fully-differential voltage combiner buffer circuit (VCOM) is shown in Fig. 3.

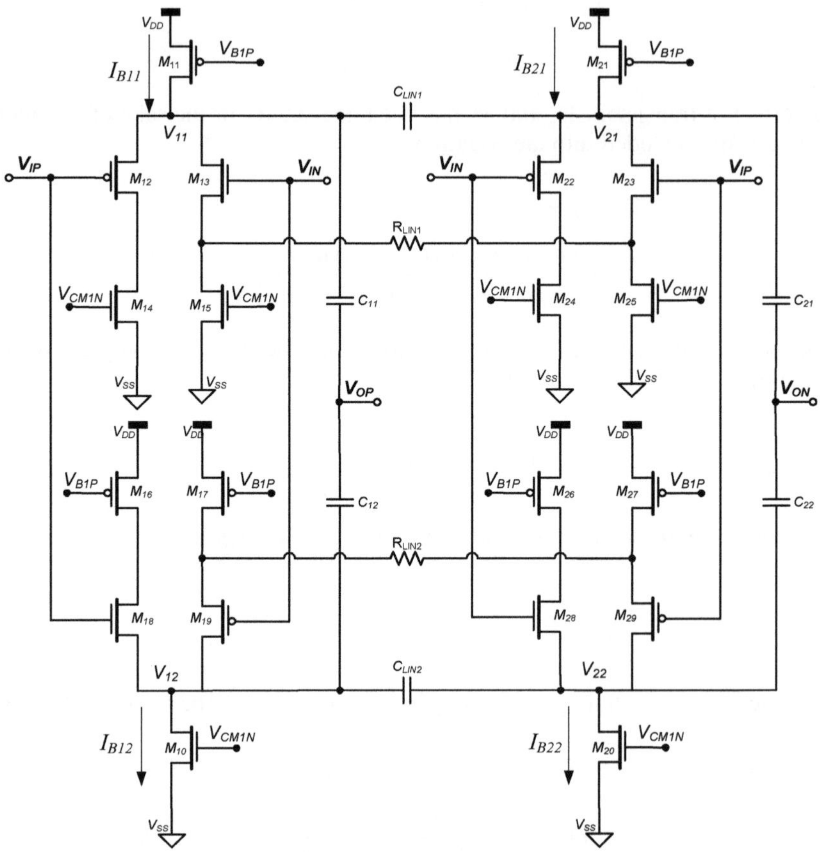

Fig. 3. Proposed input fully differential voltage combiner circuit (VCOM) (Biasing circuitry not shown for simplicity)

It consists of a cross-coupled (i.e. NMOS and PMOS) fully-differential buffer. The differential voltage-combiner is represented by M12, M13, M19 and M18 in CD and CS configurations, respectively. The remaining transistors are simple PMOS current-sources (M11, M16, M17) and NMOS current-sources (M14, M15 and M10). The circuit structure is duplicated to complete the cross-coupled fully-differential circuit. As stated before, in order to linearize the circuit, following the source degeneration resistors presented earlier, two resistors, RLIN1 and RLIN2, were added to improve the THD of both the odd and even harmonics. Furthermore, two capacitances, CLIN1 and CLIN2, were also added to filter the even harmonics. For the output signal AC coupling were used four capacitances: C11, C12, C21, and C22. To minimize the power consumption and area, transistors widths and current in the biasing circuit are scaled down by a factor of ten with respect to the main amplifier circuit.

4.2 Linearization Technique

Nonlinearity of the MOS transistors limits the circuit linearity (dynamic related with THD) between -40 dB and -60 dB [4]. In order to reach better linearity performance, resistor-based degeneration can be used [5]. The resistive source degeneration (RLIN1 and RLIN2) method is used to enhance the linear range of the transconductor (CS devices, M13, M23, M19 and M29, operating in a degenerated differential-pair fashion) circuit through transconductance reduction. The degeneration resistor increases the source terminal of the transistors, reducing the drain current [6]. Furthermore, in order to reduce odd harmonics, capacitive coupling has also been employed (through capacitors CLIN1 and CLIN2).

4.3 Simulation Results

The circuit proposed here (the circuit shown in Fig. 1) was designed in a 130 nm high-speed CMOS technology (Lmin = 120 nm). The mobility and threshold parameters (Level 2), KN, KP, VTN and VTP parameters of the devices are, respectively, 525 mAV^{-2}, 145 mAV^{-2}, 0.38 V and -0.33 V. For VCMI, the value of 550 mV was used. The linearization elements, source degeneration resistors are 75 Ω and the capacitances are 0.2 pF. Three voltage values were used as supply voltage of the circuit: 1.14 V, 1.2 V, and 1.26 V.

Table 1 shows the summary of the simulation results of the relevant performance parameters.

Table 1. Key simulated performance parameters

Technology	130 nm
Supply Voltage	1.2 V
DC Gain	1.54 dB
f_{-3dB}(@ *CL* = 0.8 pF and *RL* = 200 kΩ)	1.14 GHz
Total input capacitance, C_{pi}	89.64 fF
Total current	11.40 mA
Total power dissipation	13.34 mW

The simulation results were obtained using HSPICE simulator. In nominal conditions, using a supply voltage of 1.2 V, the simulated amplifier achieves a DC gain of about 1.54 dB, a bandwidth with a frequency cutoff, f_{-3dB}, of 1.14 GHz and a power dissipation of 13.34mW.

Figure 4 shows the bode plot indicating the DC gain and bandwidth for a voltage supply of 1.2 V, 1.54 dB and 1.14 GHz, respectively. In order to compare the key performance parameters (KPP) of the buffer over a wide positive-power supply variation [1.14 V to 1.26 V]: distortion (THD) (Fig. 5), DC Gain (Fig. 6), cutoff bandwidth (Fig. 7) and dissipated power (Fig. 8). The KPP results are evaluated for the linearized circuit, which employs resistive and capacitive source degeneration (Fig. 3).

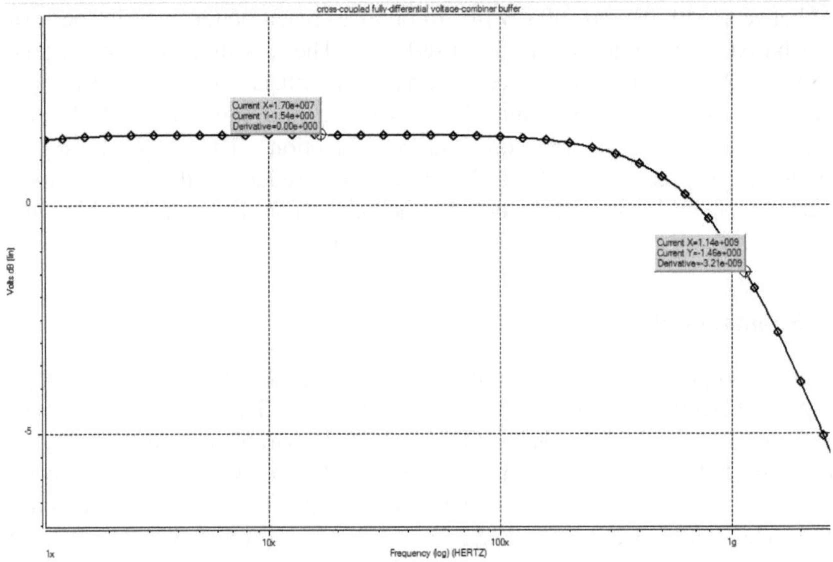

Fig. 4. Bode Plot @ Vdd = 1.2 V

Figure 5 shows the results of the THD of the linearized circuit for a fully differential input signal range from 20 mV to 400 mV with an input frequency of 250MHz.The |THD| value is above 42 dB for input amplitude voltage until 120 mV and a supply voltage of 1.26 V. For a supply voltage of 1.14 V the input amplitude voltage can go up to 200 mV for the |THD|. The circuit presents a better THD value for lower supply voltage.

The DC gain increases approximately logarithmically (i.e. linearly in dB) with the supply voltage, from 0.66 dB to 2 dB, as shown in the Fig. 6.

Figure 7 displays the cutoff bandwidth results. The better results are for the lower power supply voltage. Also, the input signal amplitude does not influence the bandwidth result. The power dissipation is only dependent on power supply voltage, as expected. Figure 8 shows that higher supply voltages originate higher power supply losses. Again, the variation of input signal amplitude does not influence the power dissipation.

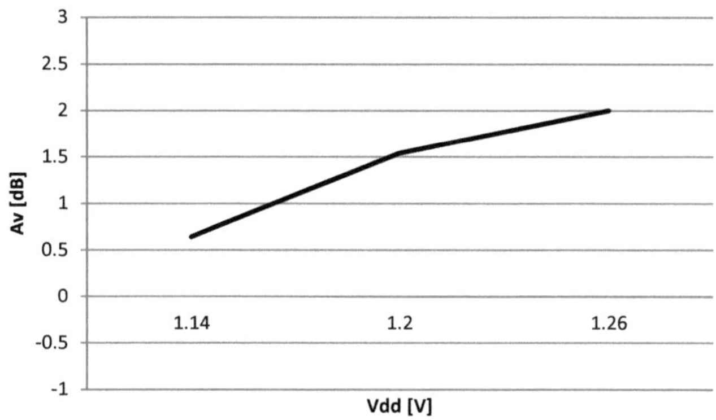

Fig. 6. Low-frequency (DC) gain vs. supply voltage

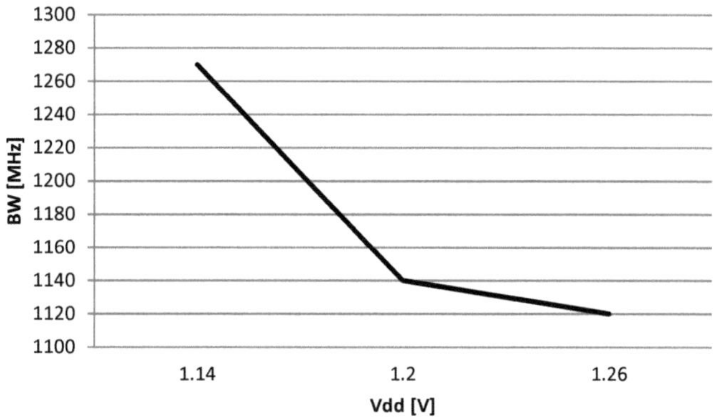

Fig. 7. Cutoff bandwidth vs. supply voltage

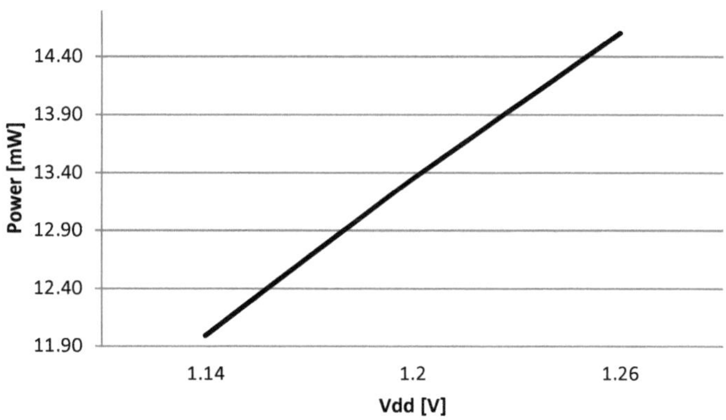

Fig. 8. Power dissipation vs. supply voltage

5 Conclusions

This paper presented a new CMOS buffer circuit topology for RF applications based on a fully-differential voltage-combiner circuit, operating at low-voltage. The proposed circuit uses a combination of CS and CD transistors. The simulation results show good levels of linearity and bandwidth. For improved THD the source degeneration method was employed. Using the proposed CS and CD fully-differential structure, the circuit achieves, in nominal conditions, a simulated gain of 1.54 dB, a bandwidth of 1.14 GHz for a total power dissipation of 13.34 mW.

Acknowledgments. This work was supported by national funds through FCT - Fundaçãopara a Ciência e Tecnologia under projects PEst-OE/EEI/UI0066/2011 and IMPACT (PTDC/EEA-ELC/101421/2008).

References

1. Tavares, R., Vaz, B., Goes, J., Paulino, N., Steiger-Garcao, A.: Design and optimization of low-voltage two-stage CMOS amplifiers with enhanced performance. In: IEEE Int. Symp. Circuits and Systems (ISCAS 2003), vol. 1, pp. I-197–I-200 (2003)
2. Carmona, J., Cortadella, J., Kishinevsky, M., Taubin, A.: Elastic Circuits. IEEE Transactions on Computer-Aided Design of Integrated Circuits and Systems (TCAD) 28(10) (2009)
3. Leyn, F., Daems, W., Gielen, G., Sansen, W.: A behavioral signal path modeling methodology for qualitative insight in and efficient sizing of CMOS opamps. In: CAD, Dig, of Tech. Papers, pp. 374–381 (1997)
4. Leuciuc, A., Zhang, Y.: A highly linear low-voltage MOS transconductor. In: IEEE Int. Symp. Circuits and Systems (ISCAS 2002), vol. 3, pp. III-735- III-738 (2002)
5. El mourabit, A., Sbaa, M.H., Alaoui-Ismaili, Z., Lahjomri, F.: A CMOS Transconductor with High Linear Range. In: ICECS 2007, pp. 1131–1134 (2007)
6. Kuo, K.-C., Leuciuc, A.: A linear MOS transconductor using source degeneration and adaptive biasing. IEEE TCAS II: Analog and Digital Signal Processing 48(10), 937–943 (2001)

Using Variable Width RF Integrated Inductors for Quality Factor Optimization

Pedro Almeida, Pedro Pereira, and Helena Fino

Departamento de Engenharia Electrotécnica, Faculdade de Ciências e Tecnologia, FCT, Universidade Nova de Lisboa, 2829-516 Caparica, Portugal
pedrodalmeida77@gmail.com, {pmrp,hfino}@ieee.org

Abstract. The advancement of CMOS technology led to the integration of more complex functions. In the particular of wireless transceivers, integrated LC tanks are becoming popular both for VCOs and integrated filters [1]. For RF applications the main challenge is still the design of integrated inductors with the maximum quality factor. For that purpose, tapered, i.e., variable width inductors have been proposed in the literature. In this paper, analytical expressions for the determination the pi-model parameters, for the characterization of variable width integrated inductors are proposed. The expressions rely exclusively on geometrical and technological parameters, thus granting the rapid adaptation of the model to different technologies. The results obtained with the model are compared against simulation with *ASITIC*, showing errors below 10%. The model is then integrated into an optimization procedure where inductors with a quality factor improvement in the order of 20-30% are obtained, when compared with fixed width inductors.

Keywords: Inductor layout optimization, variable metal width, integrated RF inductor.

1 Introduction

During the last years the worldwide market on communications has experienced an ever-growing demand for integrated systems with scaling down dimensions and increased functionality. In the particular case for RF communication circuits, integrated spiral inductors are widely used, notwithstanding their poor performance and their subsequent negative impact on the circuit efficiency at high frequencies. This poor performance is due to the large effect the technology parasitics have on the small value of inductance usually required. As a result, significant effort has been employed in investigating silicon planar inductors, their associated models and methods of improving their performance [1-3].

Regarding layout optimization, non-uniform metal width is proposed, as a way of increasing the inductor quality factor [4, 5]. The main objective of this methodology is to reduce the influence of magnetically induced losses in the inner turns of the spiral where the magnetic field reaches its maximum. By reducing the line width toward the center of the spiral a minimization of the series resistance of the inductor

L.M. Camarinha-Matos, S. Tomic, and P. Graça (Eds.): DoCEIS 2013, IFIP AICT 394, pp. 619–627, 2013.

coil, taking into account both Ohmic losses due to conduction currents, and magnetically induced losses due to Eddy currents, is obtained. Although several promising results have been reported, a fully analytical characterisation of the inductor, leading to perfect understanding of the device performance limitations, is needed so that optimized designs may be obtained.

In this paper analytical expressions for the evaluation of the inductor model parameters are proposed. These expressions rely exclusively on technological parameters and on the geometrical characterization of each inductor segment. The proposed model is used for the optimization-based design of several inductors, where the advantage of using tapered topologies is well pointed out.

The remaining of the paper is organized as follows. The novelty introduced by this paper, is highlighted in Section 2. In this section the basic inductor model is introduced and then, the adaptation of the model for variable width inductors is carefully described. Section 3 is dedicated to the description of several working examples. Finally, conclusions are offered in Section 4.

2 Relationship to Internet of Things

Internet of things relies on the interconnections of a large number of heterogeneous cooperating devices. The development of these devices has been made possible due to the rapid evolution of electronic technologies, enabling the implementation complex functions, in smaller and more rapid circuits. To cope with the necessity of minimizing the power consumption of such systems, new design methodologies must be adopted. In the particular case of communications services, RF integrated inductors are becoming popular elements. Yet, designing integrated inductors for RF applications is a challenging process where a set of correlated geometrical parameters must be obtained, leading to the need of using optimization-based design methodologies.

The main objective of the work described is the optimization of the spiral inductors quality factor, by using variable width inductors.

The novel contributions of this paper are as follows:

- It proposes a set of analytical expressions for the evaluation of the pi-model parameters, for variable width integrated inductors. The proposed expressions depend exclusively on the technological parameters and on the geometric characterization of each segment of the inductor.
- It proposes an efficient optimization-based design for nano-CMOS planar spiral inductor, based in analytical models, instead of using an electromagnetic simulation based approach;

Since the proposed equations are an extension of the model used for fixed width inductors, next sub-section gives a brief description of this model. In subsection 2.2 the adequacy of the model to variable width inductors is carefully described.

2.1 Planar Spiral Inductor Pi-Model

Several integrated inductor models have been introduced in the last years, as illustrated in Fig.1 [6]

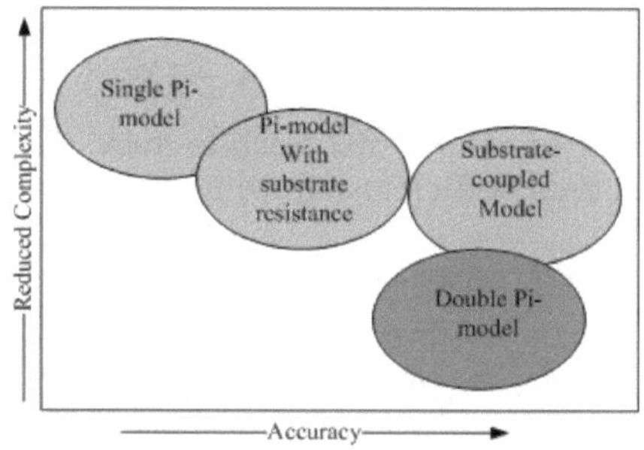

Fig. 1. Planar inductor lumped element models complexity versus accuracy trade-off

The simplest one is the pi-model, which is widely used for inductors operating in a frequency range up to a few GHz. For the sake of simplicity, the pi-model, illustrated in Fig. 2.a., is adopted, where L_s, and R_s. account for the inductance and resistance of the spiral. The overlap between the spiral and the underpass allows direct capacitive coupling between the two terminals of the inductor. The feed-through part is modelled by C_s. Capacitor C_{ox} represents the capacitance between the spiral and the substrate. Finally, C_{si} and R_{si} account for the silicon substrate capacitance and resistance, respectively.

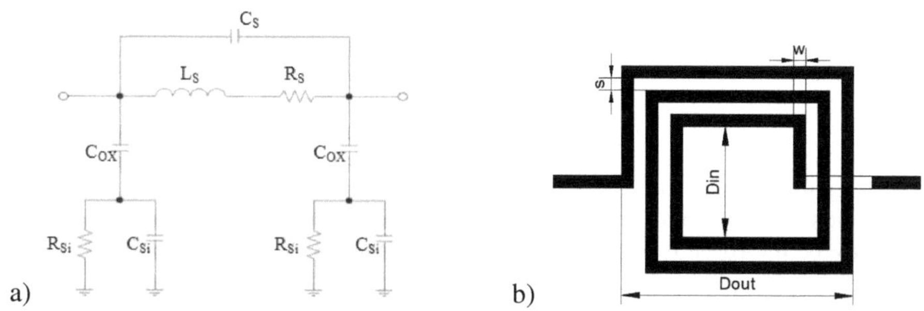

Fig. 2. a) Planar inductor pi-model; b) Geometric parameters for a square inductor

For the evaluation of the inductance, L_S, several approaches have been proposed, based either on fitting processes to experimental values [7] or through physics-based equations [8], where

$$L_s = k_1 \mu_0 n^2 d_{avg}/(1 + k_2 \rho). \quad (1)$$

Given that,

$$\rho = (d_{out} - d_{in})/(d_{out} + d_{in}). \tag{2.a}$$

$$d_{avg} = 0.5(d_{out} + d_{int}). \tag{2.a}$$

$$d_{out} = d_{in} + 2nw + 2(n-1)s. \tag{2.b}$$

Where n is the number of turns, s is the track-to-track distance, and w is the track width. Finally, k_1 and k_2, are coefficients allowing the model to be adapted to several inductor shapes.

The evaluation of the spiral resistance, R_s, is obtained by [9]

$$R_s = \sqrt{{R_{dc}}^2 + {R_{ac}}^2}. \tag{3}$$

Where,

$$R_{dc} = l/\ (\sigma wt). \tag{4.a}$$

$$R_{ac} = kl/[2\sigma\delta(w + t)]. \tag{4.b}$$

And σ and t are the metal conductivity and thickness, respectively. The metal length, l, is obtained with [10]

$$l = N_{sides} d_{avg} n \tan(\pi/N_{sides}). \tag{5}$$

And the skin depth, δ, may be determined by [11]

$$\delta = 1/\sqrt{\sigma\mu\pi f}. \tag{6}$$

For the evaluation of the capacitance, C_s, all overlap capacitances are considered and given by [12].

$$C_s = n_c w^2\, \varepsilon_{ox}/t_{oxM1-M2}. \tag{7}$$

Where ε_{ox} is the oxide permittivity, n_c is the number of overlaps and $t_{oxM1-M2}$ is the oxide thickness between the spiral upper and lower metal. The parasitic capacitance, C_{ox}, between the spiral metal and the silicon substrate, is estimated with [12]

$$C_{ox} = 0.5lw\, \varepsilon_{ox}/t_{ox}. \tag{8}$$

Where t_{ox} is the thickness of the SiO_2 between the inductor and the substrate and lw defines the area of the spiral. Finally the Substrate resistance, R_{si}, and capacitance C_{si}, are obtained with 11

$$R_{si} = 2h_{si}/(l\ w\ \sigma_{si}). \tag{9}$$

$$C_{si} = 0.5lw\, \varepsilon_o \varepsilon_r/h_{si}. \tag{10}$$

Where σ_{si} and h_{si} are the substrate conductivity and height, respectively.

This model has been extensively used in an optimization-based tool for inductor design [13] yielding solutions with very good accuracy when compared with results obtained with ASITIC .

2.2 Variable Width Inductor Model

Variable width inductor layout has been proposed as a way of maximizing the quality factor of integrated inductors. In this work, square inductors, where each segment shows a width increment of Δw, as illustrated in Fig.3. , are considered.

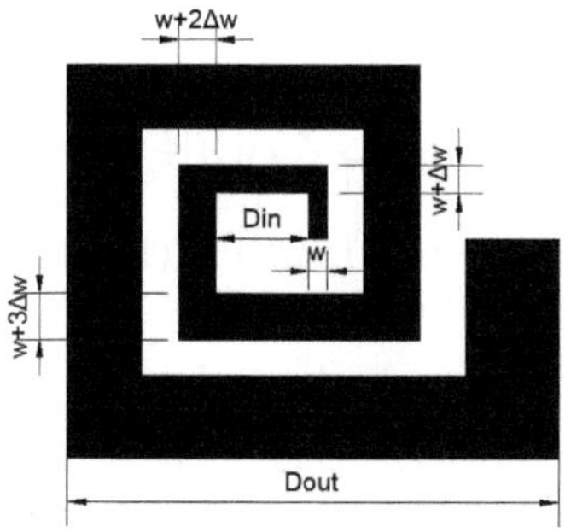

Fig. 3. Variable width square inductor

Since each segment will show an increment on both width and length new equations will be proposed, relying on the dimensions of each segment of the inductor. The width of a segment may be evaluated with

$$w_i = w + n_i \Delta w. \tag{11}$$

Where n_i is the number of the segment, and w is the initial metal width.

In this work, the basic structure for supporting all the inductor segments characterization is a matrix of five columns (one for each segment per turn) and n lines (one per inductor turn). Initially, a matrix containing every segment length, l_{ij} as

$$l_{ij} = \begin{bmatrix} l_{11} & l_{12} & l_{13} & l_{14} & l_{15} \\ l_{21} & l_{22} & l_{23} & l_{24} & l_{24} \\ \cdots & \cdots & \cdots & \cdots & \cdots \\ l_{n1} & l_{n2} & l_{n3} & l_{n4} & l_{n5} \end{bmatrix}. \tag{12}$$

is generated. Then a matrix containing every segment width, w_{ij}

$$w_{ij} = \begin{bmatrix} w_{11} & w_{12} & w_{13} & w_{14} & w_{15} \\ w_{21} & w_{22} & w_{23} & w_{24} & w_{24} \\ \cdots & \cdots & \cdots & \cdots & \cdots \\ w_{n1} & w_{n2} & w_{n3} & w_{n4} & w_{n5} \end{bmatrix}. \tag{13}$$

is generated.

Regarding the pi-model inductance, L_S, a new way for evaluating d_{out} must be adopted. Considering the basic matrixes in (12), (13), d_{out} may be obtained by

$$d_{out} = l_{n4} + w_{n5}. \tag{14}$$

For the evaluation of R_S, (4.a) and (4.b) are replaced by

$$R_{dc} = \frac{1}{\sigma t} \sum_{i=1}^{n} \sum_{j=1}^{5} \frac{l_{ij}}{w_{ij}}. \tag{15.a}$$

$$R_{ac} = \frac{k}{2\sigma\delta} \sum_{i=1}^{n} \sum_{j=1}^{5} \frac{l_{ij}}{w_{ij} + t}. \tag{15.b}$$

For this purpose the two new matrix (15.a) and (15.b) are generated containing each segment geometrical information for R_{dc} and R_{ac}

$$\frac{l_{ij}}{w_{ij}} = \begin{bmatrix} \frac{l_{11}}{w_{11}} & \frac{l_{12}}{w_{12}} & \frac{l_{13}}{w_{13}} & \frac{l_{14}}{w_{14}} & \frac{l_{15}}{w_{15}} \\ \frac{l_{21}}{w_{21}} & \frac{l_{22}}{w_{22}} & \frac{l_{23}}{w_{23}} & \frac{l_{24}}{w_{24}} & \frac{l_{25}}{w_{25}} \\ \cdots & \cdots & \cdots & \cdots & \cdots \\ \frac{l_{n1}}{w_{n1}} & \frac{l_{n2}}{w_{n2}} & \frac{l_{n3}}{w_{n3}} & \frac{l_{n4}}{w_{n4}} & \frac{l_{n5}}{w_{n5}} \end{bmatrix}. \tag{16}$$

$$\frac{l_{ij}}{w_{ij} + t} = \begin{bmatrix} \frac{l_{11}}{w_{11}+t} & \frac{l_{12}}{w_{12}+t} & \frac{l_{13}}{w_{13}+t} & \frac{l_{14}}{w_{14}+t} & \frac{l_{15}}{w_{15}+t} \\ \frac{l_{21}}{w_{21}+t} & \frac{l_{22}}{w_{22}+t} & \frac{l_{23}}{w_{23}+t} & \frac{l_{24}}{w_{24}+t} & \frac{l_{25}}{w_{25}+t} \\ \cdots & \cdots & \cdots & \cdots & \cdots \\ \frac{l_{n1}}{w_{n1}+t} & \frac{l_{n2}}{w_{n2}+t} & \frac{l_{n3}}{w_{n3}+t} & \frac{l_{n4}}{w_{n4}+t} & \frac{l_{n5}}{w_{n5}+t} \end{bmatrix}. \tag{17}$$

Finally (8) , (9) and (10) are replaced by

$$C_{ox} = \frac{1}{2} \cdot \frac{\varepsilon_{ox}}{t_{ox}} \sum_{i=1}^{n} \sum_{j=1}^{5} l_{ij} w_{ij}. \tag{18}$$

$$R_{si} = \frac{2}{G_{sub}} \sum_{i=1}^{n} \sum_{j=1}^{5} \frac{1}{l_{ij} w_{ij}}. \tag{19}$$

$$C_{si} = \frac{1}{2} \cdot C_{sub} \sum_{i=1}^{n} \sum_{j=1}^{5} l_{ij} w_{ij} \, . \tag{20}$$

For which an auxiliary matrix containing the area of each segment is considered.

3 Variable Inductor Working Example

In order to confirm the validity of the proposed layout optimization method, a set of square spiral inductors has been designed. Three examples, considering the design of 1nH, 1.5nH and 2nH inductors at a working frequency of 1GHz will be presented. In all examples the technological parameters shown in Table 1 were used

Table 1. UMC130 – Technological Parameters

Parameter	Value	Parameter	Value
ε_o	8.85e-12	t_{ox}(μm)	600
ε_r	1.0	C_{sub} (F/m^2)	4.0e-6
σ(Ωm)	1/2.65e-8	G_{sub} (S/m^2)	2.43e5

For the model validation a comparison between results obtained with variable width design, against fixed width designs, for an approximately equal area, is presented. The layout parameters were evaluated according to the constraints in Table 2

Table 2. Spiral Inductor Design Constraints

Parameter	Min	Step	Max
w (μm)	5.0	0.5	100.0
Δw (μm)	1	0.25	-
s(μm)	1.5	-	-
d_{in} (μm)	20.0	0.5	200.0
N	1.5	1.0	15.5
d_{out}	-	-	500

A. Example 1 – Inductor with 1 nH

In this example a spiral inductor of 1nH was considered. The results obtained for a fixed width layout as well as for Δw of 1.25μm, 1.5μm and 1.75μm are represented in Table 2. Also in the same table, the simulation results obtained with ASITIC for each case are represented.

Table 3. Optimization Results for 1nH Inductor with several Δw

Ind	Δw (μm)	w (μm)	d_{in} (μm)	n	d_{out} (μm)	L (nH) Model	L (nH) Asitic	ε_L	Q Model	Q Asitic	ε_Q	Q. Impr.
1	0	13.8	68.8	2.5	157	1	1.07	6.3%	**6.94**	6.28	10.5%	
2	1.25	7.5	41.3	3	154	0.998	1.06	5.4%	8.07	8.24	1.7%	16%
3	1.5	7.25	39.8	3	161	0.999	1.04	4.3%	8.57	8.53	0.5%	23%
4	1.75	5.75	40.3	3	162	1.01	1.04	2.5%	**8.52**	8.16	4.4%	**23%**

In the last column the relative improvement in the quality factor from using incremental width is given.

B. Example 2 – Inductor with 1.5nH

In this example a spiral inductor of 1.5nH was considered. The results obtained for a fixed width layout as well as for Δw of 2μm, 2.5μm and 3.0 μm are represented in Table 2. Also in the same table, the simulation results obtained with ASITIC for each case are represented.

Table 4. Optimization Results for 1.5 nH Inductor with several Δw

Ind	Δw (μm)	w (μm)	d_{in} (μm)	n	d_{out} (μm)	L (nH) Model	L (nH) Asitic	ε_L	Q Model	Q Asitic	ε_Q	Q impr
1	0	20.3	10.3	2.5	231	1.5	1.62	7.6%	**10.1**	9.27	9.0%	
2	2	11.5	61.3	3	233	1.5	1.58	5.2%	12.29	12.61	2.6%	22%
3	2.5	6.25	64.3	3	221	1.5	1.52	1.6%	11.07	10.59	4.6%	9%
4	3	8.75	58	3	253	1.51	1.54	1.8%	**13.44**	12.74	5.5%	**33%**

C. Example 3 – Inductor with 2nH

In the last example a spiral inductor of 2.0nH was considered. The results obtained for a fixed width layout as well as for Δw of 2μm, 2.5μm and 3.0μm are represented in Table 2. Also in the same table, the simulation results obtained with ASITIC for each case are represented.

Table 5. Optimization Results for 1.5 nH Inductor with several Δw

Ind	Δw (μm)	w (μm)	d_{in} (μm)	n	d_{out} (μm)	L (nH) Model	L (nH) Asitic	ε_L	Q Model	Q Asitic	ε_Q	Q impr
1	0	25.8	137	2.5	297	2	2.16	7.6%	**12.72**	11.62	9.4%	
2	2.75	12.5	85	3	296	2	2.1	4.8%	15.03	15.21	1.2%	18%
3	3	9.75	86.3	3	288	2	2.05	2.6%	14.39	14.19	1.4%	13%
4	3.75	10.3	80.5	3	317	2	2.05	2.6%	**16.14**	15.34	5.2%	**27%**

4 Conclusions

In this paper the analytical expressions for the evaluation of integrated inductor pi-model parameters was proposed. This expressions consider variable width square inductors, were each segment shows a constant increment of width. The validity of the model was shown through three working examples considering the design of 1.0nH, 1.5nH and 2.0nH inductors at a working frequency of 1GHz. From the examples presented a quality factor improvement in the order of 20% to 30% may be obtained, by using variable width.

Should higher frequency of operation be envisaged, a more accurate inductor, such as 2-pi-model should be used. The adaptation of the proposed equations to the 2-pi-model is under test.

References

1. Krout, I., Mnif, H., Fakhfakh, M., Loulou, M.: Optimizing LC VCO Performances Through a Heuristic. Wseas Transactions on Electronics 5, 274–281 (2008)
2. Lai, P.-W., Dobos, L., Long, S.: A 2.4GHz SiGe low phase noise VCO using on chip Tapped inductor. In: Solid-State Circuits Conference - ESSCIRC, pp. 505–508 (2003)
3. Leung, L.L.K., Luong, H.C.: A 1V9.7mW CMOS frequency synthesizer for IEEE 802.11a transceivers. IEEE Trans. Microw. Theory Tech. 56, 39–48 (2008)
4. Gee, W. A.: CMOS Integrated LC Q-enhanced RF Filters for wireless Receivers, PhD. Thesis, Georgia Institute of Technology (2005)
5. Patrick Yue, C., Simon Wong, S.: Physical modeling of spiral inductors on silicon. IEEE Transactions on Electron Devices 47, 560–568 (2000)
6. Lai, I., Fujishima, M.: Design and Modeling of Millimeter-Wave CMOS Circuits for Wireless Transceivers - Era of Sub-100nm Technology. Springer (2008)
7. Mohan, S.S., del Mar Hershenson, M., Boyd, S.P., Lee, T.H.: Simple Accurate Expressions for Planar Spiral Inductances. IEEE Journal of Solid-State Circuits 34(10) (October 1999)
8. Jenei, S., Nauwelaers, B.K.J.C., Decoutere, S.: Physics-Based Closed-Form Inductance Expression for Compact Modelling of Integrated Spiral Inductors. IEEE J. Solid-State Circuits 37(1), 77–80 (2002)
9. Choi, Y.S., Yoon, J.B.: Experimental Analysis of the Effect of Metal Thickness on the Quality Factor in Integrated Spiral Inductors for RF ICs. IEEE Electron. Device Lett. 25(2), 76–79 (2004)
10. Allstot, D.J., Choi, K., Park, J.: Parasitic-Aware Optimization of CMOS RF Circuits. Kluwer Academic Publishers (2003) ISBN 1-4020- 7399-2
11. Park, M., Kim, C.H., Kim, C.S., et al.: Frequency-Dependent Series Resistance of Monolithic Spiral Inductors. IEEE Microwave Guided Lett. 9(12), 514–516 (1999)
12. Murphy, O.H.: Advanced Physical Modelling of Multilayer Inductors for CMOS RF Front-End Applications, PhD Thesis, University College Cork - National University of Ireland (October 2005)
13. Pereira, P., Fino, M.H., Coito, F., Ventim-Neves, M.: ADISI- An efficient tool for the automatic design of integrated spiral inductors. In: The 16th IEEE International Conference on Electronics, Circuits, and Systems (ICECS 2009), pp. 799–802 (2009)

Part XXI
Electronics: Applications

Use of a-SiC:H Semiconductor-Based Transducer for Glucose Sensing through FRET Analysis

Paula Louro[1,2], Vitor Silva[1,2], Manuel A. Vieira[1,2], A. Karmali[4], and Manuela Vieira[1,2,3]

[1] Electronics Telecommunications and Computer Dept, ISEL, Lisbon, Portugal
[2] CTS-UNINOVA, Quinta da Torre, 2829-516, Caparica, Portugal
[3] DEE-FCT-UNL, Quinta da Torre, 2829-516, Caparica, Portugal
[4] CIEB-ISEL, R. Conselheiro Emidio Navarro, 1, Lisbon, Portugal

Abstract. Glucose sensing is an issue with great interest in medical and biological applications. One possible approach to glucose detection takes advantage of measuring changes in fluorescence resonance energy transfer (FRET) between a fluorescent donor and an acceptor within a protein which undergoes glucose-induced changes in conformation. This demands the detection of fluorescent signals in the visible spectrum. In this paper we analyzed the emission spectrum obtained from fluorescent labels attached to a protein which changes its conformation in the presence of glucose using a commercial spectrofluorometer. Different glucose nanosensors were used to measure the output spectra with fluorescent signals located at the cyan and yellow bands of the spectrum. A new device is presented based on multilayered a-SiC:H heterostructures to detect identical transient visible signals. The transducer consists of a p-i'(a-SiC:H)-n/p-i(a-Si:H)-n heterostructure optimized for the detection of the fluorescence resonance energy transfer between fluorophores with excitation in the violet (400 nm) and emissions in the cyan (470 nm) and yellow (588 nm) range of the spectrum. Results show that the device photocurrent signal measured under reverse bias and using appropriate steady state optical bias, allows the separate detection of the cyan and yellow fluorescence signals.presented.

Keywords: Glucose sensing, FRET, nanosensors, fluorescence proteins, a-SiC:H photodiodes, multilayered heterostructures, optical sensors.

1 Introduction

There is a great scientific and medical interest to develop suitable technology for continuous in vivo glucose monitoring in patients with diabetes mellitus. There are several problems with existing devices based on electrochemistry which have encouraged alternative approaches to glucose sensing in recent years. An approach based on fluorescence intensity and lifetime, has special advantages, including sensitivity and the potential for non-invasive measurement when near infrared light is used. Such techniques include measuring changes in fluorescence resonance energy

L.M. Camarinha-Matos, S. Tomic, and P. Graça (Eds.): DoCEIS 2013, IFIP AICT 394, pp. 631–638, 2013.

transfer (FRET) between a fluorescent donor and an acceptor within a protein which undergoes glucose-induced changes in conformation [1]. First fluorescence excitation is performed at the donor wavelength. As the protein changes its conformation the donor and acceptor fluorophores, attached to the protein, change their distance. When the distance between fluorophores is reduced the donor transfers energy to the acceptor through nonradiative dipole–dipole coupling. The energy transfer causes on one hand a decrease in intensity at the donor emission wavelength and on the other an increase in intensity at the acceptor emission wavelength. Such variation in the emission spectrum can be used to detect and measure the concentration of glucose in the sample.

In this paper we use a-SiC:H optical devices to detect the cyan and yellow emission signals of fluorescent proteins. The transducer consists of two heterostructures based on a-SiC:H/a-Si:H optimized for the detection of the fluorescence emissions at wavelengths 470 nm (cyan) and 588 nm (yellow) [2].

2 Contribution to Internet of Things

Internet of Things refers to the concept of expanding the communication among things. This will bring the ability of objects behaviour to change due to what they access through the Internet. In a FRET based sensor with a wireless system, a warning or emergency message could be generated and sent in a multicast network. This same message would be received by the nearest hospital and by a patient transportation system and also by the patient's doctor cell phone. Another scenario would be a feedback response based on the transmitted values which would tell the patient what to do.

Devices based on new materials for glucose sensing are an important contribution to the future development of the Internet of Things.

3 Materials and Methods

The following recombinant plasmids were purchased from Addgene (USA): 13561 pRSET FLIPglu-2uDelta13 and 13660 pRSET FLIPglu-3.2mDelta13. Luria Bertani culture medium, iminodiacetic acid, epichlrohydrin, butanodioldiglycidyl ether and ampicillin were obtained from Sigma Chemical Company (USA) and Sepharose 6B, Sepharose CL 4B, and Sepharose CL 6B were purchased from GE HealthCare (Sweden).

All cell-free extracts of recombinant E. coli strains were assayed for protein by using the Coomassie blue dye binding method.

The recombinant plasmids were used to transform E. coli DH5αstrain which was grown overnight in Luria Bertani (LB) culture medium containing 100 μg/ ml of ampicillin at 37ºC and 180 rpm as described previously [1]. The cultures were harvested by centrifugation at 10.000 rpm for 1 min. at 4ºC and the cells were ressuspended in 2 volumes of 20 mM phosphate buffer pH 7.0 containing 1 mM benzamidine and sonicated five times for 30 s at 160 W with an interval of 5 min. The cell-suspension was centrifuged at 20.000 rpm for 1 h to remove cell debris and the supernatant was frozen at -20ºC for further studies. The cell-free extract was used as the source of glucose binding protein which was purified by immobilized metal affinity chromatography (IMAC) on immobilized copper chelates as described previously [3].

Fluorescence measurements were carried out in a Jasco FP-8300 spectrofluorometer equipped with the standard 10 mm cuvette holder.

Wavelength scans were performed on all glucose binding protein variants to determine excitation and emission peaks. Spectra were collected from 450-700 nm using the following configuration: bandwidth (excitation/emission) – 5 nm; excitation wavelength - 433 nm, data pitch – 0.2 nm, scanning speed – 125 nm/min.

Unless otherwise stated, the fluorescent intensity per unit protein was determined with the samples diluted in 20 mM sodium phosphate buffer pH. 7.0 with excitation and emission set at 433 nm and 485 and 528 nm, respectively. Glucose solutions were prepared in 20 mM phosphate buffer pH 7.0 containing 1 mM benzamidine which were added to the cell-free extract containing the glucose binding protein variants to a final concentration of 100 mM as follows: pRSET FLIPglu-2uDelta13 and pRSET FLIPglu-3.2mDelta13. Representative scans of the prepared glucose nanosensors are shown in Figure 1.

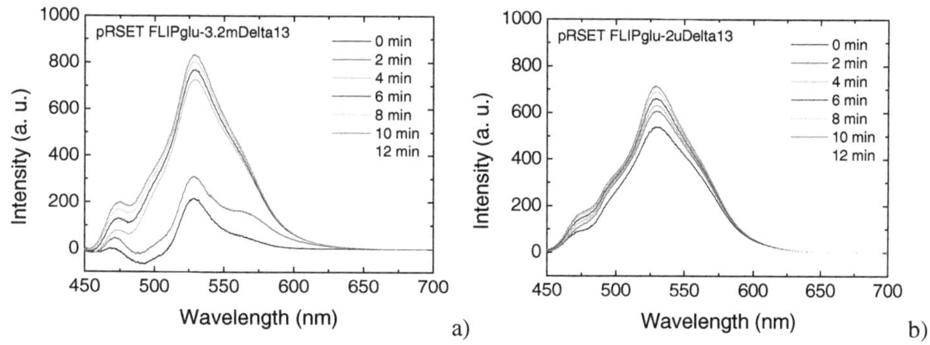

Fig. 1. Emission spectra for glucose nanosensors scanned during 12 minutes for: a) 13660 pRSET FLIPglu-3.2mDelta13 and b) 13561 pRSET FLIPglu-2uDelta13

In both spectra it is clear the presence of fluorescent emissions with peaks at the cyan and yellow regions. The observed spectrum changes along the reaction time are due to the induced conformation changes of the fluorescent proteins that stabilize after a few minutes. The correlation of these changes with time is probably related to the use of cellular extracts in our samples. The saturation curves are related to the glucose sensor concentration of the sample. Results show also that when the reaction is complete and saturation occurs the emission spectrum of both sensors exhibit similar trends. However, the cyan peak is more evident in glucose nanosensor pRSET FLIPglu-3.2mDelta13.

4 Device Configuration

The device described herein works from 400 to 700 nm which makes it suitable for FRET applications using fluorophore operating in the visible spectrum. The device is a multilayered heterostructure based on a-Si:H and a-SiC:H. The configuration of the device includes two stacked p-i-n structures between two electrical and transparent contacts (Fig. 2). Both front (pin1) and back (pin2) structures act as optical filters confining, respectively, the short and the long optical carriers, while the intermediate wavelengths are absorbed across both [4,5].

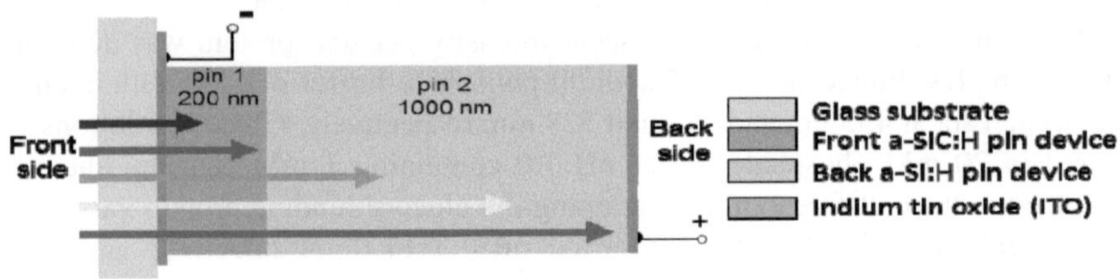

Fig. 2. Simplified schematic diagram of the structure of the device

The device was operated within the visible range using as optical signals to simulate the emitted fluorescent signals of the recombinant plasmids from E. coli strains (yellow and cyan) the modulated supplied by a cyan and a yellow LED with wavelengths of 470 nm and 588 nm, respectively. Steady state optical bias of different wavelength from the front and back sides was also used for light soaking the device.

4.1 Spectral Sensitivity

Figure 3 displays the spectral photocurrent, measured along the visible spectrum, under reverse bias without and with optical light bias focusing the device from back and front sides.

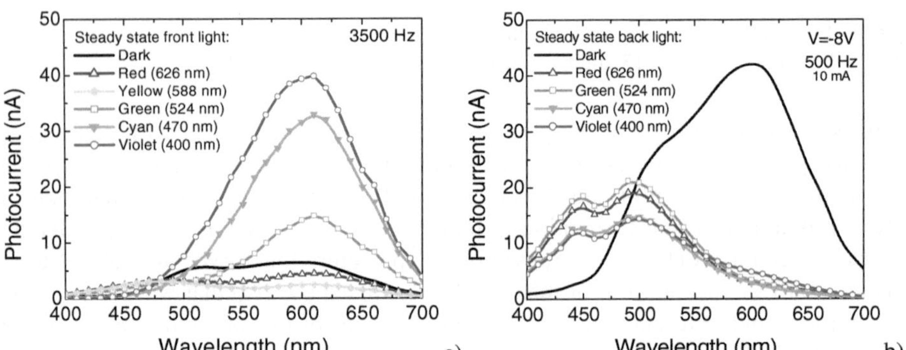

Fig. 3. Spectral photocurrent under dark conditions and using steady state light of different wavelengths by: a) front and b) back sides of the device

Results of Fig. 3a) show that the use of steady state light bias from the front side, i.e., from the a-SiC:H structure, enhances the photocurrent in the range of wavelengths longer than 500 nm when using background light of short wavelength (400 nm, 470 nm and 524 nm). For optical bias of longer wavelengths (588 nm and 626 nm) the photocurrent signal is decreased. On the remaining part of the spectrum an opposite behavior is observed as the amplification effect occurs for longer light bias wavelengths and the reduction of the signal for shorter wavelengths. Thus, the use of a 400 nm light source as optical bias amplifies both cyan and yellow fluorescent. On the other hand, the optical biasing from the back side (a-Si:H) is not so much dependent on the background light wavelength (Fig. 3b). Generally, in the range of wavelengths shorter than 480 nm it is observed an increase of the signal for every background, and a decrease in the complementary range. The enhancement of the photocurrent signal using different optical bias for the background light result is an effective method for tuning the device sensitivity and detecting the emission fluorescent wavelengths.

4.2 Transfer Function of Each Fluorescent Optical Signal

The photocurrent signal obtained at reverse electric bias (-8 V) with (dark lines) and without background light (violet at 400 nm from the front side and cyan at 470 nm from the back side) is displayed in Fig. 4 for simulated fluorescent optical signals (cyan and yellow emissions, respectively, at 470 nm and 588 nm).

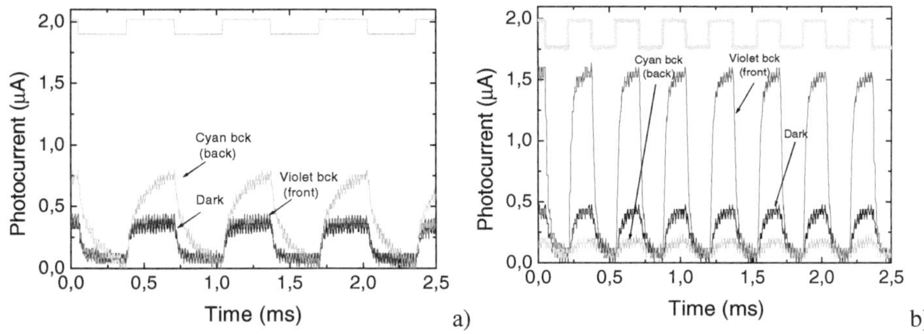

Fig. 4. Photocurrent signal obtained at reverse bias without and under different background lights with the: a) cyan (470 nm) and b) yellow (588 nm) signals

As already indicated by the spectral photocurrent measurements of Fig. 4 the same wavelength is focusing on the device back side. The use of violet light from the front side has a negligible effect on this signal. For the yellow emission at 588 nm the effect of the background light is opposite. The violet steady state light from the front side causes a large amplification of the photocurrent (amplification factor about 4 times) and at the same time the cyan light from the back side reduces the signal by half. This amplification mechanism is useful for the determination of each fluorescent signal contribution to the whole output signal. With cyan background light from the back side the cyan fluorescent signal is enhanced and the yellow one reduced. On the other hand violet background from the front side will increase the photocurrent component due to the yellow signal.

4.3 Transfer Function of a FRET Optical Signal

Several time dependent wavelength combinations of both fluorescent pulsed input optical signals (at 470 nm and 588 nm) were used to generate a multiplexed signal in the device. The output photocurrents, with and without optical background light are displayed in Figure 5. The reference level was assumed to be the signal when all the input optical signal channels are OFF. At the top of the figure, the individual optical signals are displayed to guide the eyes in relation to the different ON-OFF states. In both chromatic sequences it is clearly observed that the shape of the photocurrent signal measured under dark (black line) exhibits four threshold levels, each assigned to the correspondent optical conditions of the input signals. The ON-ON state corresponds to the maximum intensity of light bias, while the ON-OFF and OFF-ON to a lower intensity and the OFF-OFF to the dark conditions.

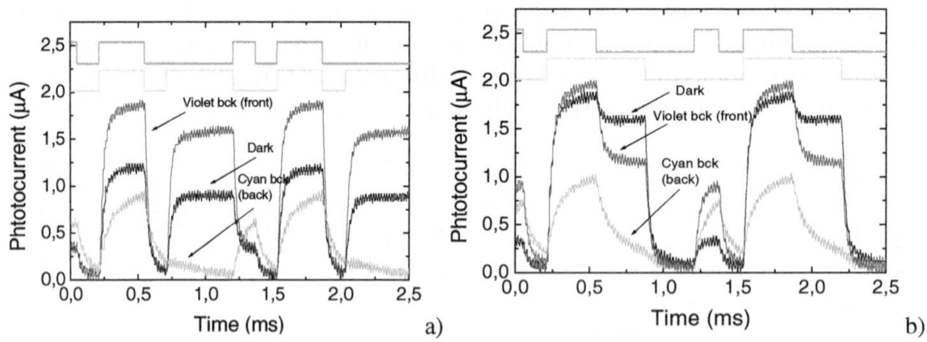

Fig. 5. Output photocurrent signals without (Φ_L=0) and with optical bias (violet light at 400 nm from the front side, and cyan light at 400 nm from the back side) of two different optical signals waveforms (shown at the top of the figure)

However, when the device is optically biased the output photocurrent changes enhancing the presence of each input fluorescent optical signal. In both sequences of Fig. 5 the shape of the photocurrent, measured by soaking the device with violet light by the front side, follows the shape of the yellow fluorescent signal). Whenever the yellow optical signal is ON, the measured photocurrent is higher than its correspondent without optical bias. On the other hand, the use of cyan optical bias by the device back side results in an output signal that follows the cyan fluorescent emission, which allows the recognition of the presence of this fluorescent signal. Thus, the proper optical bias of the device through the choice of light wavelength and the device soaking light side results in the correct tuning of each fluorescence signal. These features allow immediate decoding of both yellow and cyan emitted signals.

5 Electrical Model

Based on the experimental results and device configuration an optoelectronic model was developed [6]. The device was modeled by a two single-tuned stages circuit with two variable capacitors and interconnected phototransistors through a resistor (Fig. 6a). Two optical gate connections ascribed to the different light penetration depths across the front and back phototransistors were considered to allow independent yellow and cyan optical signals transmission.

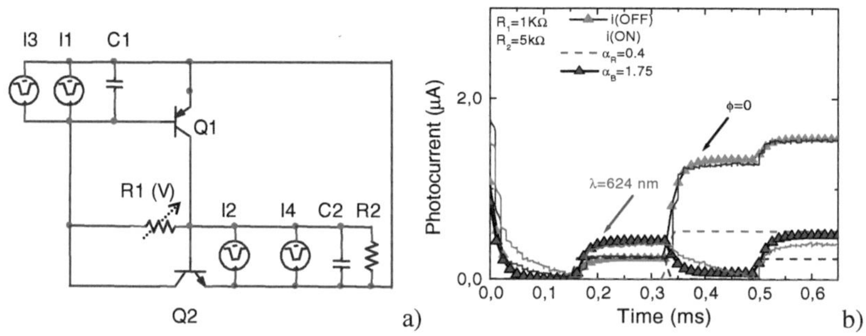

Fig. 6. a) Simplified ac equivalent electrical circuit of the device. b) Simulated photocurrent signal (symbols), input channels (dash lines) and experimental signal (solid lines): under reverse bias (-8 V) and red optical bias (624 nm).

The operation is based upon the following principle: the flow of current through the resistor connecting the two front and back transistor bases is proportional to the difference in the voltages across both capacitors (charge storage buckets). So, it uses a changing capacitance to control the power delivered to the load acting as a state variable filter circuit. In Figure 6b the simulated current without and under red backgrounds is displayed (symbols) using the same test signal of Fig. 5a. The input channels are also displayed (lines). To simulate the red background, current sources intensities (input channels) were multiplied by the on/off ratio between the input channels with and without red optical bias. Good agreement between experimental and simulated data was observed. The four expected levels, under reversed bias, and their reduction under red irradiation, are clearly seen. Under red background the expected optical amplification of the cyan channel and the quenching of the yellow one were observed due to the effect of the active multiple-feedback filter when the back diode is light triggered.

6 Conclusions

We present a novel device semiconductor based on an a-SiC:H p-i-n/p-i-n heterostructures for the detection of optical signals near the cyan and yellow regions of the visible spectrum, which can be used for the detection of the emission signals used in the FRET technique. We used a commercial spectrofluorometer to measure the fluorescent signal emitted by recombinant plasmids from E. coli strain, which showed emission peaks at the cyan and yellow bands of the visible spectrum accordingly to the conformation changes suffered by the fluorescent proteins. Simultaneously different modulated optical signals were used to simulate the emission signals of the glucose nanosensors. The output emission spectrum was analyzed by reading out the photocurrent generated by the device under reverse bias and using different wavelengths of background light to soak the device from the front and back side. The influence of the fluorescent emission optical signals on the output photocurrent was also analyzed. Future work comprises the use of lower power intensities for the simulated emission signals in order to reach the same range of the ones produced during the FRET phenomenon. In a further stage tests with the emission fluorescence signals of the prepared recombinant plasmids from *E. coli* strain with the semiconductor device and the quantification of the glucose concentration from the changes in the measured spectra.

Acknowledgements. This work was supported by PTDC/EEA-ELC/111854/2009, PTDC/EEA-ELC/115577/2009 and PTDC/EEA-ELC/120539/2010.

References

1. Deuschle, K., Okumoto, S., Fehr, M., Looger, L.L., Kozhukh, L., Frommer, W.B.: Protein Sci. 14(9), 2304–2314 (2005)
2. Vieira, M.A., Vieira, M., Costa, J., Louro, P., Fernandes, M., Fantoni, A.: Sens. Transd. 9, 96–120 (2010)
3. Martins, S., Andrade, J., Karmali, A., Serralheiro, L.J.: Molecular Recognition 19, 340–347 (2006)
4. Louro, P., Vieira, M., Vieira, M.A., Fernandes, M., Fantoni, A., Francisco, C., Barata, M.: Physica E: Low-dimensional Systems and Nanostructures 41, 1082–1085 (2009)
5. Louro, P., Vieira, M., Fernandes, M., Costa, J., Vieira, M.A., Caeiro, J., Neves, N., Barata, M.: Phys. Status Solidi. C 7(3-4), 1188–1191 (2010)
6. Vieira, M., Fernandes, M., Louro, P., Fantoni, A., Vieira, M.A., Costa, J., Barata, M.: Phys. Status Solidi C 8(3), 1079–1082 (2011)

Design of a 3rd Order 1.5-Bit Continuous-Time Fully Differential Sigma-Delta (ΣΔ) Modulator Optimized for a Class D Audio Amplifier Using Differential Pairs

Nuno Pereira[2], João L.A. de Melo[1,2], and Nuno Paulino[1,2]

[1] Centre for Technologies and Systems (CTS) – UNINOVA
[2] Dept. of Electrical Engineering (DEE), Universidade Nova de Lisboa (UNL)
Campus FCT/UNL, 2829-516, Caparica, Portugal
nrf.pereira@gmail.com, research@joaodemelo.com, nunop@uninova.pt

Abstract. This paper presents a 3rd order 1.5-bit Continuous-Time Fully Differential ΣΔ modulator with distributed feedback for a class D audio amplifier, using BJT differential pairs to implement the integrator stages. By relying on simple gain blocks instead of operational amplifiers to build the loop filter, a simpler overall circuit is obtained, where the non-ideal effects are embedded in the loop filter transfer function. This leads to a more difficult design process for the loop filter circuit, solved through the use of an optimization procedure based on genetic algorithms. Simulations of the electrical circuit show that it is capable of achieving a SNDR value of 73.4 dB and THD+N of about -80 dB for a signal bandwidth of 20 kHz and a sampling frequency of 1.28 MHz.

Keywords: Audio, Continuous-Time Sigma-Delta (ΣΔ), Class D Amplifier, Differential Pair.

1 Introduction

Over the years, there is a growing concern with the energy efficiency of electronic appliances, due to the global sustainability issue. Audio amplifiers are one example where the efficiency can be improved. They amplify input audio signals in order to drive output elements with suitable volume and power levels, with low distortion.

Class AB audio amplifiers have a maximum theoretical efficiency of 78.5% [1]. Class D amplifiers, due to their output power devices operating as switches, can reach an efficiency of 100% in theory. Given this fact, Class D amplifiers are the best solution in terms of efficiency for audio power amplifiers.

In order to generate the digital control signal for the power output devices of a Class D amplifier, it is necessary to convert the input analog signal into a digital signal. To do this, an Analog-to-Digital Converter (ADC) is employed. Sigma-Delta modulators (ΣΔMs) poise themselves as the best option for low frequency, high-resolution applications, given their native linearity, robust analog implementation and reduced anti-aliasing filtering requirements [2], [3].

L.M. Camarinha-Matos, S. Tomic, and P. Graça (Eds.): DoCEIS 2013, IFIP AICT 394, pp. 639–646, 2013.

ΣΔMs work by using negative feedback to reduce the quantization error, where a filter circuit is placed before the quantizer in order to define the frequency band where the quantization error is attenuated. This filter is traditionally built using ideal integrator stages, which are implemented with operational amplifiers (Op-Amps) in an integrator configuration [4]-[6]. These Op-Amps require a large DC gain and bandwidth in order for the behavior of the integrator circuits to be close to the ideal integrator behavior. This can result in a complex Op-Amp circuit that is difficult to design and can dissipate a lot of power. If the ΣΔM is built using discrete components, it is also difficult to find fully differential Op-Amps as discrete components, resulting in a circuit that uses a single ended topology with all the disadvantages associated. By replacing the Op-Amps with differential pairs, it is possible to build an equivalent filter circuit for the ΣΔM using lossy integrators. The finite gain and bandwidth of the differential pairs can be accommodated during the filter design process.

This paper presents a 3rd order 1.5-bit fully differential continuous-time (CT) ΣΔM with distributed feedback for use in a Class D full-bridge audio power amplifier, where the CT integrators are based on bipolar-junction-transistor (BJT) differential pairs. Since this is a CT-ΣΔM, discrete components are used, favoring the use of BJTs over CMOS technology.

2 Relationship to Internet of Things

The work presented in this paper can contribute to the future development of the Internet of Things, since it can provide energy efficient communication, through the reliable transmission of information, when this information is an audio signal.

As stated in chapter I, Class D amplifiers that use ΣΔMs to transform the analog input signal into a digital signal, tend to achieve nearly 100% efficiency, while preserving linearity and providing a robust analog implementation.

A high performance can be attained through the use of ΣΔMs, since they trade speed for accuracy. This high performance is achieved with low sensitivity to analog component imperfections and without requiring component trimming [7].

3 Class D Amplifiers Using Sigma-Delta Modulation

The basic block diagram of a Class D amplifier is shown in Fig. 1. The input audio signal is modulated into a digital control signal which drives the power devices in the output stage. This signal can be modulated, normally, using pulse-width modulation (PWM) or pulse-density modulation (PDM). The output stage can be implemented using a half-bridge or a full-bridge topology.

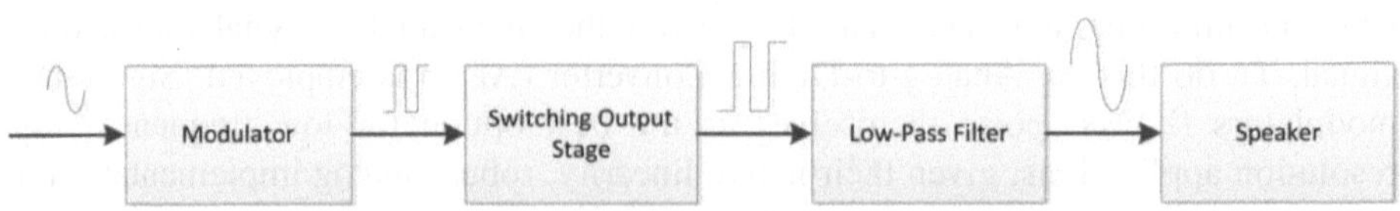

Fig. 1. Class D open-loop amplifier block diagram

When using a ΣΔM, the digital control signal is modulated using a PDM signal. The number of pulses in a given time window is proportional to the average value of the input audio signal during that time interval. The quantization error generated in this process is averaged out, since it is fed back negatively in the ΣΔ process loop.

Class D amplifiers dissipate less power than traditional Class A/B/AB amplifiers, since the output stage devices operate as switches that alternate between the positive and negative power supplies (thus generating a train of voltage pulses). Therefore, they have zero current when in the "off" state and low voltage when conducting.

The output lowpass filter is used to remove high frequency components (that would increase the electromagnetic energy radiated by the amplifier) of the output signal, that occur due to the binary switching of the output devices.

ΣΔMs are A/D and D/A converters that operate with sampling frequencies much larger than the Nyquist frequency, trading conversion time for resolution. They are capable of increasing the signal-to-noise ratio (SNR) by filtering out quantization noise outside of the signal bandwidth. ΣΔMs are a feedback-type system, and as such, could become unstable. ΣΔMs with a maximum order of 2 are inherently stable. The stability of higher-order 1-bit modulators is of critical concern, since they include a 1-bit quantizer which only has 2 feedback levels.

4 3rd Order Continuous-Time ΣΔ Modulator Based on Differential Pairs

When designing a ΣΔM, choosing the order and the sampling frequency are the first steps. As stated in section 3, ΣΔMs with orders higher than 2, if not designed properly, could result in an unstable system. In [8], a design procedure is presented that determines the optimal coefficients for a stable high-order ΣΔM built using ideal integrator blocks.

The goal of this paper is to design a ΣΔM optimized for an audio amplifier. A signal bandwidth of 20 kHz is chosen, since this is a standard value used for audio signals. To reduce the influence of the non-ideal effects in the output devices during the switching, a low sampling frequency must be used.

An ideal 3rd order ΣΔM (assuming that will be stable) with an oversampling ratio (OSR) value of 32 could theoretically produce a signal-to-noise-plus-distortion ratio (SNDR) value of around 95 dB. For the considered signal bandwidth, this results in a sampling frequency f_s of 1.28 MHz. However, due to the inherent instability of the modulator, the SNDR value could drop to 64 dB [8].

4.1 1.5-Bit ΣΔ Modulator with Distributed Feedback and Ideal Integrators

The block diagram of a 3rd order 1.5-bit ΣΔM with distributed feedback is shown in Fig. 2. The signal transfer function (STF) and the noise transfer function (NTF) of this modulator are given by (1). As expected they are a 3rd order transfer function.

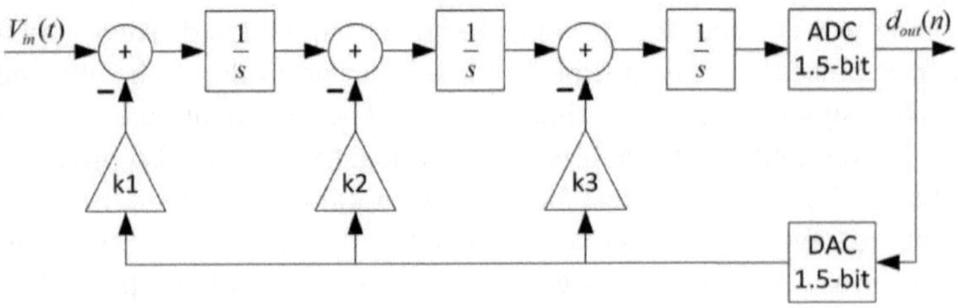

Fig. 2. Block diagram of a 3rd order 1.5-bit ΣΔM with distributed feedback

$$STF = \frac{1}{s^3+k_3 \cdot s^2+k_2 \cdot s+k_1} \qquad NTF = \frac{s^3}{s^3+k_3 \cdot s^2+k_2 \cdot s+k_1} \tag{1}$$

The values of the coefficients k_1, k_2 and k_3 are determined, through numeric computing software (e. g. *MATLAB*®), in order to implement the desired filtering function e. g. a Butterworth filter. The cut-off frequency of the filter is selected as a trade-off between stability of the modulator (lower cut-off frequency value) and increased SNDR value of the modulator (larger cut-off frequency value) [8].

By using a 1.5-bit quantizer (3 levels) instead of a 1-bit quantizer (2 levels), the linearity of the feedback path in the modulator is improved. This results in a more stable loop and in a decrease of unnecessary switching of the output stage. This will lead to a higher SNDR value than what would be obtained if a 3rd order 1-bit ΣΔM was used.

4.2 Integrating Differential Pairs

An inherent advantage of differential pairs is that they are a fully-differential circuit. Due to its symmetry, the differential output voltage of this circuit does not depend on the input common-mode voltage, leading to a high common-mode rejection ratio (CMRR). However, although the output is independent from the input common-mode voltage, the differential pair transistors must be biased to operate in the active region. This imposes limits to the input common-mode range [9]. If exceeded, the circuit will present nonlinearities, leading to distortion.

Differential pairs have an advantage that results from their symmetry: even order harmonics tend to be cancelled, since even order terms are canceled. Thus, the quality of the circuit will be determined by the third order harmonic (and subsequent odd harmonics). The basic building block for the ΣΔM will be the integrating differential pair circuit presented in Fig. 3. The supply voltages used were of ± 5V.

Both capacitors $C_{1,2}$ perform the integration operation, while resistors $R_{C1,2}$ define both the gain and the output common mode DC voltage, set in this circuit to 2.5 V (half of the positive supply voltage), to ensure that following integrator stages and their BJT are properly biased. Feedback resistors $R_{fb1,2}$ add the output signal to the input signal (V_{in}). Both input resistors ($R_{b1,2}$) limit the voltage applied to the base of the BJT, ensuring that it is low enough to prevent saturation. The biasing current source, I_{EE}, is implemented by a basic current mirror.

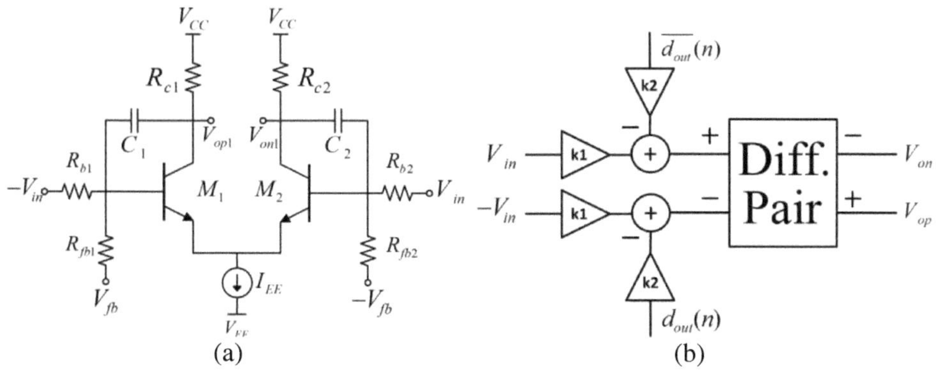

Fig. 3. Schematic of the integrator Differential Pair (NPN) (a) and Symbolic View (b)

Using the Kirchhoff's current law (KCL) applied to the small signal model; it is possible to obtain the output differential voltage equation of this circuit (2).

$$V_{out} = \frac{2R_c(V_{fb}R_b - V_{in}R_{fb})(sC_{int}r_\pi - \beta)}{R_{fb}r_\pi(1+sC_{int}R_c) + R_b\left(r_\pi + sC_{int}R_c r_\pi + R_{fb}\left(1+sC_{int}(\beta R_C + R_c + r_\pi)\right)\right)} \tag{2}$$

Capacitors C_1 and C_2 behave like Miller capacitors, introducing an additional zero to the circuit. Since the sampling frequency value is low, this zero does not cause a problem because its value is much larger than f_s.

The common-mode output voltage of the circuit is larger (2.5 V) than the common mode input voltage (0 V). Therefore, it is necessary to use a complementary version of the circuit in Fig. 3 based on PNP transistors. This complementary circuit has a common mode input voltage of 2.5 V and a common mode output voltage of 0 V. Thus, a PNP integrator stage should be preceded by a NPN integrator stage and followed by another NPN integrator stage and so forth.

In order to reduce distortion in the circuit, the signal amplitude at the base of each BJT should be small (below 50 mV [9]) to ensure that no transistor saturates (thus reducing distortion). This condition must be met during the filter design process, using available degrees of freedom from the design variables.

4.3 1.5-Bit ADC

To achieve 1.5-bit quantization (three levels), the integrator output voltages must be compared with a certain threshold voltage ($\Delta V_o - V_t > 0$). Since the design in question is fully-differential, the circuit in Fig. 4 was used in order to avoid using many comparators. The threshold voltage is generated through a voltage divider between the V_o voltages and two reference voltages (here denoted as $V_R^{+,-}$). The logic codification of the 1.5-bit quantizer is shown next to the 1.5-bit ADC (Fig. 4). The equation that represents this ADC is given by (3).

$$\Delta V_t = \Delta V_o \frac{R_2}{R_1+R_2} - \Delta V_R \frac{R_1}{R_1+R_2} \tag{3}$$

Rearranging the right side of (3), it follows that,

$$\frac{R_2}{R_1+R_2}\left(\Delta V_o - \Delta V_R \frac{R_1}{R_2}\right) \tag{4}$$

which is similar to $\Delta V_o - V_t$. Thus,

$$\Delta V_R \frac{R_1}{R_2} = V_t \tag{5}$$

From (5), it is possible to obtain the relation between R_1 and R_2, for a given V_t, V_R^+ and V_R^-. This V_t voltage is determined through simulations of the proposed architecture in order to obtain the optimum SNDR value.

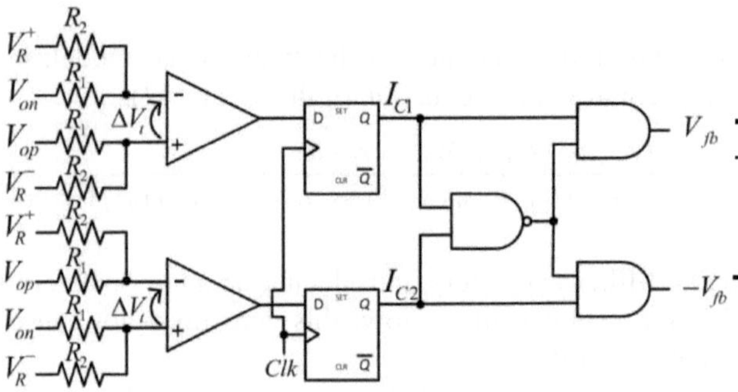

I_{C1}	I_{C2}	State	V_{fb}	$-V_{fb}$
0	0	x	0	0
0	1	+1	0	1
1	0	-1	1	0
1	1	0	0	0

Fig. 4. Fully Differential 1.5 bit ADC

4.4 Filter Design

For a 3rd order ΣΔM, three integrator stages are used, resulting in the circuit of Fig. 5. The denominator of the transfer function of this architecture will be similar to the one in Eq. (1). The transfer function of the circuit is very complex and impossible to be fully presented in this paper. By equating this denominator to the coefficients obtained when designing the 3rd order Butterworth filter with a certain cut-off frequency, it is possible to obtain the optimal values for the feedback resistors ($R_{fb1,2}$). In order to do so, some component values have to be assumed, like the capacitors and the R_b resistors.

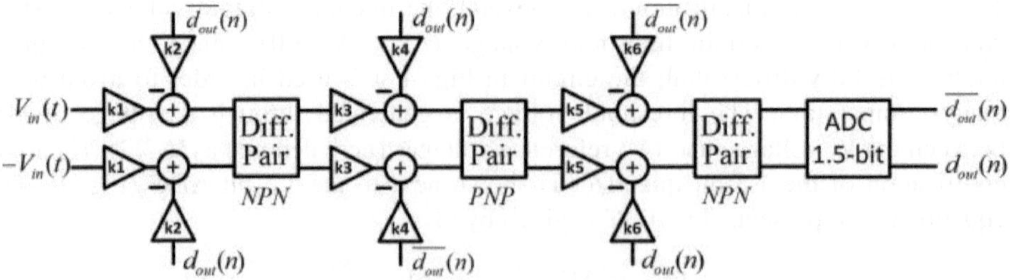

Fig. 5. 3rd Order 1.5-bit CT Fully Differential ΣΔM implementation

Although fairly accurate, this is a long and time-consuming method. So, the sizing of the 3rd order ΣΔM was performed through the use of a genetic algorithm, proposed in [10], where quantization noise, voltage swing variations and stability of the modulator are taken into account. This sizing also considers sensitivity to component variations. The values for the components of the circuit obtained after the sizing process are shown next in Table I, the biasing current is 5 mA. The reference voltages of the 1.5-bit ADC (V_R^+ and V_R^-) are +5 V and -5 V respectively.

Table 1. Component Values

Component	Value	Units
$C_{1,2} = C_{3,4} = C_{5,6}$	0.47	nF
$R_{b1,2}$	42.136	kΩ
$R_{b3,4}$	1.227	kΩ
$R_{b5,6}$	3.741	kΩ
$R_{fb1,2}$	67.634	kΩ
$R_{fb3,4}$	24.673	kΩ
$R_{fb5,6}$	27.694	kΩ
$R_{c1,2} = R_{c5,6}$	1	kΩ
$R_{c3,4}$	3	kΩ
R_1	54	Ω
R_2	5	kΩ

5 Simulations and Results

Fig. 5 shows the circuit implementation of the proposed architecture and its sizing is shown in Table I. Electrical simulations of the complete circuit were performed for different input amplitudes. The resulting output spectrum for an input signal of 1 V (0.707 V_{rms}) is shown in Fig. 6 (a) and the SNDR as a function of the input voltage is plotted in Fig. 6 (b).

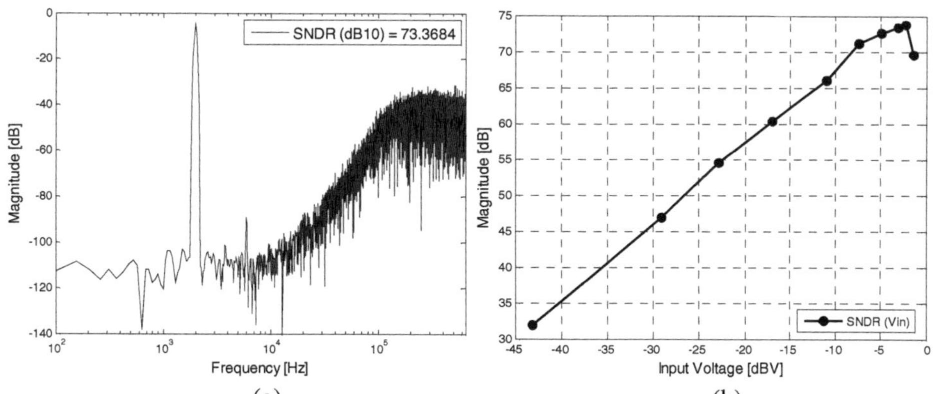

(a) (b)

Fig. 6. Output spectrum of the architecture (Blackman-Harris window with 2^{16} points) (a) and obtained SNDR as a function of the input voltage (dBV) (b)

The electrical simulation results show that for a signal bandwidth of 20 kHz and a sampling frequency of 1.28 MHz, a maximum SNDR value of 73.4 dB was obtained. The Total Harmonic Distortion plus noise (THD+N) obtained is about -80 dB.

6 Conclusions

This paper presented a 3rd order 1.5-bit CT Fully Differential ΣΔM, where the integrators that compose the circuit were realized through basic BJT differential pairs, against traditional Op-Amp methodology that is more costly.

When compared to ΣΔMs implemented with Op-Amps, this architecture can reach a similar performance and this poses as an improvement, since cheaper circuits can be used while designing a more efficient circuit.

Simulation results of the electrical circuit show that a SNDR of 73.4 dB is obtained, for a signal bandwidth of 20 kHz and a sampling frequency of 1.28 MHz.

References

1. Sedra, A.S., Smith, K.C.: Microelectronic Circuits. Oxford University Press, USA (2003)
2. Carusone, T.C.: Analog Integrated Circuit Design. Wiley (2011)
3. Schreier, R., Temes, G.C.: Understanding Delta-Sigma Data Converters. John Wiley & Sons (2004)
4. Yan, S., Sanchez-Sinencio, E.: A continuous-time sigma-delta modulator with 88-dB dynamic range and 1.1-MHz signal bandwidth. IEEE Journal of Solid-State Circuits 39(1), 75–86 (2004)
5. Wu, R., Long, J.R., van de Gevel, M., Glassche, G.: A fifth-order continuous-time sigma-delta modulator with 62-dB dynamic range and 2MHz bandwidth. In: Research in Microelectronics and Electronics Conference, PRIME 2007, Ph.D, pp. 17–20 (2007)
6. de Melo, J., Paulino, N.: Design of a 3rd order 1.5-bit continuous-time (CT) Sigma-Delta (SD) modulator optimized for Class D audio power amplifier. J. Microelectronics and Computer Science 1(2) (2011)
7. Brigati, S., Francesconi, F., Malcovati, P., Tonietto, D., Baschirotto, A., Maloberti, F.: Modeling sigma-delta modulator non-idealities in SIMULINK(R). In: Proceedings of the 1999 IEEE International Symposium on Circuits and Systems, ISCAS 1999, vol. 2, pp. 384–387 (1999)
8. Norsworthy, S.R., Schreier, R., Temes, G.C.: Delta-Sigma Data Converters: Theory, Design, and Simulation, 1st edn. Wiley-IEEE Press (1996)
9. de, M., Silva, M.: Circuitos com transistores bipolares e MOS. Fundação Calouste Gulbenkian (2003)
10. de Melo, J.L.A., Nowacki, B., Paulino, N., Goes, J.: Design methodology for sigma-delta modulators based on a genetic algorithm using hybrid cost functions. In: Proc. IEEE Int. Symp. Circuits Systems (ISCAS 2012), pp. 301–304 (2012)

System-Level Optimization of a DAC for Hearing-Aid Audio Class D Output Stage

Peter Pracný, Ivan H.H. Jørgensen, and Erik Bruun

Department of Electrical Engineering, Technical University of Denmark,
2800 Kgs. Lyngby, Denmark
pp@elektro.dtu.dk

Abstract. This paper deals with system-level optimization of a digital-to-analog converter (DAC) for hearing-aid audio Class D output stage. We discuss the ΣΔ modulator system-level design parameters – the order, the oversampling ratio (OSR) and the number of bits in the quantizer. We show that combining a reduction of the OSR with an increase of the order results in considerable power savings while the audio quality is kept. For further savings in the ΣΔ modulator, overdesign and subsequent coarse coefficient quantization are used. A figure of merit (FOM) is introduced to confirm this optimization approach by comparing two ΣΔ modulator designs. The proposed optimization has impact on the whole hearing-aid audio back-end system including less hardware in the interpolation filter and half the switching rate in the digital-pulse-width-modulation (DPWM) block and Class D output stage.

Keywords: Sigma-Delta Modulator, Digital-to-Analog Converter, Interpolation Filter, Class D, Hearing Aid, Low Voltage, Low Power.

1 1 Introduction

The hearing-aids of today are devices where strict specifications are applied. High audio quality and the need for longer operation time combined with the desire to shrink the size of the hearing-aid devices to make it virtually invisible leaves less space for the battery and integrated circuits. These demands contradict each another, making the current consumption of the electronics inside the hearing-aid one of the crucial parameters for the design. To find the optimum balance between the design parameters in every part of a hearing-aid device is therefore of vital importance. This includes the back-end of the audio signal processing path (see Fig.1). As part of the digital-to-analog conversion a digital ΣΔ modulator is usually used in audio applications. Due to the oversampling nature of the ΣΔ modulator an interpolation filter is needed prior to the modulator. In the case of a multi-bit ΣΔ modulator, to be able to connect the output of the ΣΔ modulator to the input of the Class D output-stage a DPWM block that turns the ΣΔ signal into pulse width modulation, is needed. The Class D output stage is usually implemented as an H-bridge (schematic in Fig.1 is simplified). This paper deals with optimization of such a back-end system resulting in considerable power savings compared to the design of [1]. In Section 2, design specifications for the ΣΔ modulator

L.M. Camarinha-Matos, S. Tomic, and P. Graça (Eds.): DoCEIS 2013, IFIP AICT 394, pp. 647–654, 2013.

intended for hearing-aid application are discussed. A figure of merit (FOM) that allows relative comparison of ΣΔ modulators and estimation of power savings is introduced here too. In Section 3, optimization approach is proposed. In Section 4 two ΣΔ modulator designs are compared as an example of the optimization approach. Finally, the conclusions can be found in Section 5.

2 Contribution to Internet of Things

In the future the Internet will become even more important part of our daily life. Multimedia information will be delivered to our portable electronic devices. This will require audio readout "on the fly" of the internet content, low power digital signal processing and amplification inside an ear-plug. In this work we propose how such signal processing can be done in a more power efficient way.

3 Design Specifications and Figure of Merit

A thorough discussion on hearing-aid audio back-end system specification and the ΣΔ modulator is provided in [1]. The band-width (BW) of high-end hearing aids is a trade-off between ensuring sufficient sound quality and the limited power available and is normally around 10 kHz. In order to fulfill the Nyquist criterion the sampling frequency at the input of the back-end system is $f_{s\_in} > (2 * BW) = 20$ kHz. In the case of this work we use 44.1 kHz / 2 = 22.05 kHz [1]. Also in this work we assume ideal 16 bit quantization of the back-end system input signal [2]. This results in signal-to-quantization-noise ratio (SQNR) = 98 dB. The input signal of the back-end is then up-sampled using an interpolation filter [2] and passed to the ΣΔ modulator. The interpolation filter in [2] consists of 4 stages (FIR filter, half-band filter, 1st order Sinc filter, 3rd order Sinc filter) and is used to up-sample the input signal 64 times. Another requirement in this work is the signal-to-noise-and-distortion ratio (SNDR) at the total output of the back-end of 90 dB. The interpolation filter and the ΣΔ modulator are designed to keep the quality of the audio signal at SNDR = 98 dB so that a margin of 8 dB is left for the performance reduction introduced by the output stage. Note that we are dealing with a digital ΣΔ modulator in this work and we treat it as a digital filter. This allows us to adopt the idea for a FOM from [3, 4] by counting the number of adders in the design. Unlike in the case of the interpolation filter the number of bits does not have to be the same for all the adders in a ΣΔ modulator. We have to take this fact into account and propose FOM so that the number of bits of individual adders is included. This leads us to

$$FOM = \sum_i b_i \, . \, OSR_i \, . \tag{1}$$

Where i is the number of adders in the ΣΔ modulator block, b_i is the number of bits used in individual adders and OSR_i is the oversampling used for the individual adders. In the case of the ΣΔ modulator block OSR_i is the same for all the adders. Since this FOM accounts for the majority of the cells needed to implement the ΣΔ modulator it is roughly proportional to the power of the ΣΔ modulator and is a valuable tool when choosing between designs in early design phase. The lower the FOM the better the design is. The above mentioned specifications and FOM will be used in the next sections when optimizing the back-end system including the ΣΔ modulator and comparing it to previous design of [1].

4 Design Optimization Approach

The idea behind the optimization of the ΣΔ modulator and the entire back-end design compared to [1] is to decrease the OSR of the modulator from 64 to 32 and increase its order from 3 to 6. By performing these changes in the ΣΔ modulator we aim to reduce the switching frequency of the Class D output stage and the DPWM block by 50% as this frequency is the same as the operating frequency of the ΣΔ modulator. With the Class D output stage being the main power consumer in the back-end system due to the large output transistors and low output impedance, this will result in considerable power savings. Moreover these changes will have positive impact on the interpolation filter too as oversampling by 32 only is needed compared to oversampling by 64 in [1]. This saves one stage performing oversampling by a factor of 2 in the interpolation filter of [2]. Using the same idea as in Eq.(1) for the FOM of the interpolation filter we calculate FOM = 113 for the whole interpolation filter out of which FOM = 24 goes for the stage that will be saved by our optimization. This is improvement of hardware/power saving by 21% in the interpolation filter.

With a signal with SQNR = 98 dB at the ΣΔ modulator input any oversampling (OSR) and noise-shaping order in the ΣΔ modulator providing better SQNR than 98 dB is denoted in this work as overdesign. However, just like in the case of FIR filters in [5], overdesign can allow very coarse quantization of the ΣΔ modulator coefficients leading to lower amount of adders used and thus reducing the power consumption. Keeping this in mind we compare two cascade-of-resonator-with-feedback (CRFB) ΣΔ modulator designs with the same performance. We chose the designs so that the same peak-SQNR is achieved in both cases. To ensure a simple interpolation filter, only factors of integer power of two are considered in this work. In Fig.2 peak-SQNR is plotted as a function of OSR for orders N = 1 to 8 when 3 bit quantizer is used. This figure shows that the following parameter combinations achieve peak-SQNR of approx. 106 dB: OSR = 64, order = 3, 3 bit quantizer (see [1]), OSR = 32, order = 6, 3 bit quantizer (optimized). With SQNR = 98 dB needed at the ΣΔ modulator output we leave margin for coarse coefficient quantization (as proposed in [5] for digital FIR filters) and we overdesign the ΣΔ modulator to reach 106 dB peak-SQNR. Again, for the sake of comparison both of these designs use 3 bit quantizer. As explained in [1], the number of bits used in the quantizer is one of the factors that decide the clock frequency of the DPWM block. Increasing the number of bits in the quantizer can result in clock frequency that is not available in hearing aids. For this reason we keep the number of bits in the quantizer the same as in [1] and limit the design freedom in this case to OSR and the order of the ΣΔ modulator. Moreover the maximum stable amplitude at the modulator input is the same in both cases, -1 dBFS. The noise-transfer-functions (NTF) of both ΣΔ modulators can be seen in Fig. 3. Having the same performance in both designs allows us to compare these designs using the FOM of Eq. (1). The only block of the back-end system that remains to be investigated to see whether or not this optimization approach is reasonable is the ΣΔ modulator. We discuss this in the next section.

5 ΣΔ Modulator Optimization

The two ΣΔ modulator structures we used in this work can be seen in Fig. 4. A 3 bit quantizer is used in both cases, the order of the modulator was increased from 3 (Fig. 4a) to 6 (Fig. 4b) and the OSR was decreased from 64 to 32. The ΣΔ modulator performance is the same in both cases. For both designs a model using floating-point arithmetic and a model using fixed-point arithmetic was built and simulated in Matlab. The fixed-point arithmetic model performs digital operations exactly as a VHLD design does. Thus the fixed-point arithmetic model can be directly used to judge the complexity of the filter. The FOM used in this work depends on the number of the adders and the number of bits used in the individual adders. This means that one way to improve the FOM in the 6th order modulator is to keep the number of adders as low as possible. For this reason in this work, we adopt the FIR filter over-design approach from [5] and use it for the ΣΔ modulator designs. Peak-SQNR of 98 dB needed at the modulator output allows us to use the 8 dB margin achieved by the modulator being overdesigned to reach approx. peak-SQNR of 106 dB to coarsely quantize the coefficients. Using coarse quantization of the coefficients reduces the peak-SQNR from approx. 106 dB to 98 dB – still within specification with lower number of adders used than in direct design. To confirm our optimization approach, we design both ΣΔ modulators in two versions: Version 1: with high-precision coefficients and adders to achieve peak-SQNR = approx. 106 dB (see Fig. 4). Version 2: with coarsely quantized coefficients and adders to allow peak-SQNR = 98 dB (see Fig. 4). The list of coefficients used for the 3rd order modulator can be seen in Tab. 1 and the list of coefficients for the 6th order modulator in Tab. 2. The number of bits used for the internal integrators can be seen in Fig. 4 for both Version 1 and Version 2. Taking the Matlab fixed-point models and calculating the FOM according to Eq. 1 gives data and FOM in Tab. 3, clearly showing that the FOM of the 6th the order modulator with OSR = 32 compared to 3rd order modulator with OSR = 64 of [1] remains approximately the same after the back-end system optimization in both high-precision and coarsely quantized case. This can be predicted by looking at Fig. 4. The OSR of the 6th order modulator in Fig. 4a is half compared to the 3rd order modulator in Fig. 4b but the area is doubled. To have lower power consumption in the Class-D output stage and have larger area of the ΣΔ modulator is reasonable tradeoff since the ΣΔ modulator is completely digital and thus easily scales with technology. The same cannot be said about the Class-D output stage. Expressing the current consumption of the back-end as sum of the currents needed in individual blocks we write

$$I_{total} = I_{int} + I_{SDM} + I_{DPWM} + I_{dr} \tag{2}$$

Table 1. 3rd order ΣΔ modulator coefficient list

Quantization		*Version 1*		*Version 2*	
Coeff.	*Value*	*Shift/Add*	*Adders*	*Shift/Add*	*Adders*
a_1	1/8	2^{-3}	0	2^{-3}	0
a_2	0.3446	$2^{-2}+2^{-4}+2^{-5}$	2	2^{-2}	0
a_3	0.3941	$2^{-2}+2^{-3}+2^{-6}$	2	$2^{-2}+2^{-3}$	1
b_1	1/8	2^{-3}	0	2^{-3}	0
c_1	1/2	2^{-1}	0	2^{-1}	0
c_2	1/2	2^{-1}	0	2^{-1}	0
c_3	1.4063	$2^{0}+2^{-2}+2^{-3}+2^{-5}$	3	$2^{0}+2^{-2}$	1
g_1	0.0029	$2^{-9}+2^{-10}$	1	2^{-9}	0

Table 2. 6th order ΣΔ modulator coefficient list

Quantization		*Version 1*		*Version 2*	
Coeff.	*Value*	*Shift/Add*	*Adders*	*Shift/Add*	*Adders*
a_1	1/16	2^{-4}	0	2^{-4}	0
a_2	0.1542	$2^{-3}+2^{-6}+2^{-7}$	2	2^{-3}	0
a_3	0.1705	$2^{-3}+2^{-5}+2^{-7}$	2	$2^{-3}+2^{-5}$	1
a_4	0.2532	2^{-2}	0	2^{-2}	0
a_5	0.5544	$2^{-1}+2^{-5}+2^{-7}$	2	$2^{-1}+2^{-5}$	1
a_6	0.6353	$2^{-1}+2^{-3}$	1	$2^{-1}+2^{-3}$	1
b_1	1/16	2^{-4}	0	2^{-4}	0
c_1	1/8	2^{-3}	0	2^{-3}	0
c_2	1/8	2^{-3}	0	2^{-3}	0
c_3	1/4	2^{-2}	0	2^{-2}	0
c_4	1/2	2^{-1}	0	2^{-1}	0
c_5	1/2	2^{-1}	0	2^{-1}	0
c_6	0.8791	$2^{0}-2^{-3}$	1	$2^{0}-2^{-3}$	1
g_1	0.0044	$2^{-8}+2^{-12}$	1	2^{-8}	0
g_2	0.0168	$2^{-6}+2^{-10}$	1	2^{-6}	0
g_3	0.0167	$2^{-6}+2^{-10}$	1	2^{-6}	0

Table 3. Modulator comparison

Modulator Order	*Quant. bits*	*OSR*	*Adders*	*Peak-SQNR [dB]*		*FOM*
				ideal	*quantized*	
3	3	64	18	106	106 (Version 1)	296
6	3	32	29	105	105 (Version 1)	303
3	3	64	12	106	98 (Version 2)	193
6	3	32	22	105	98 (Version 2)	192

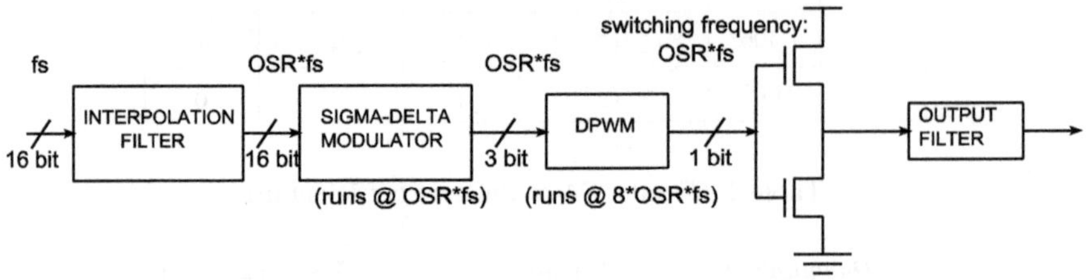

Fig. 1. Simplified schematic of the back-end of audio signal processing chain: interpolation filter, ΣΔ modulator, Class-D output-stage and output filter

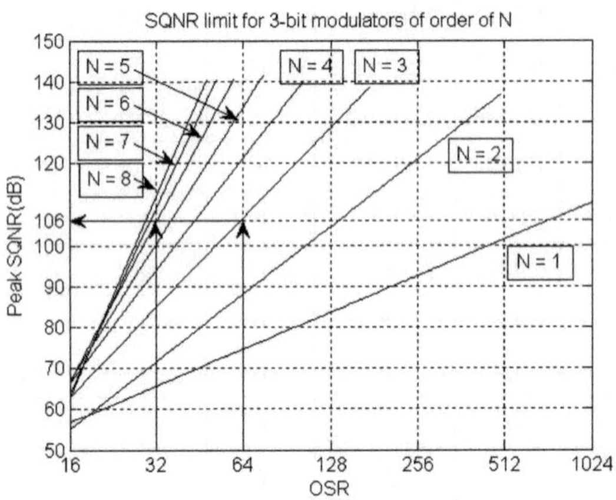

Fig. 2. Peak SQNR versus OSR for ΣΔ modulator orders N = 1 to 8 with 3 bit quantizer

Where I_{int} is the current needed in the interpolation filter (see Fig. 1), I_{SDM} is the current of the ΣΔ modulator, I_{DPWM} is the current of the DPWM block and I_{dr} is the current of the Class-D driver (output-stage). In Section 3 We explained that I_{dr} and I_{DPWM} will be lowered by 50% and I_{int} by 21% by the optimization. Table 3 shows that

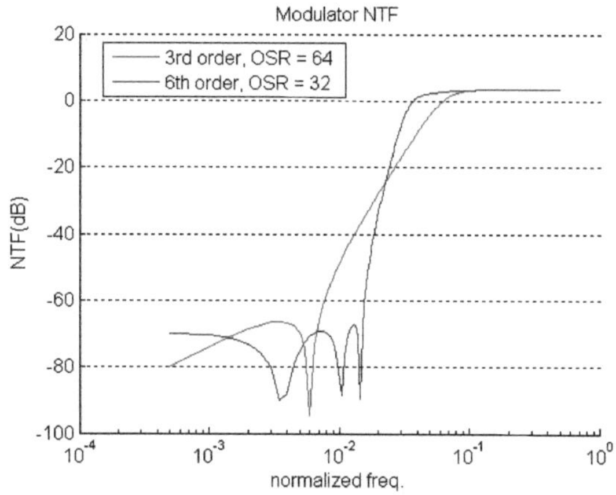

Fig. 3. ΣΔ modulator NTF in the case of (red) 3rd order modulator (frequency is normalized to 64xfsin) and (blue) 6th order modulator (frequency is normalized to 32xfsin)

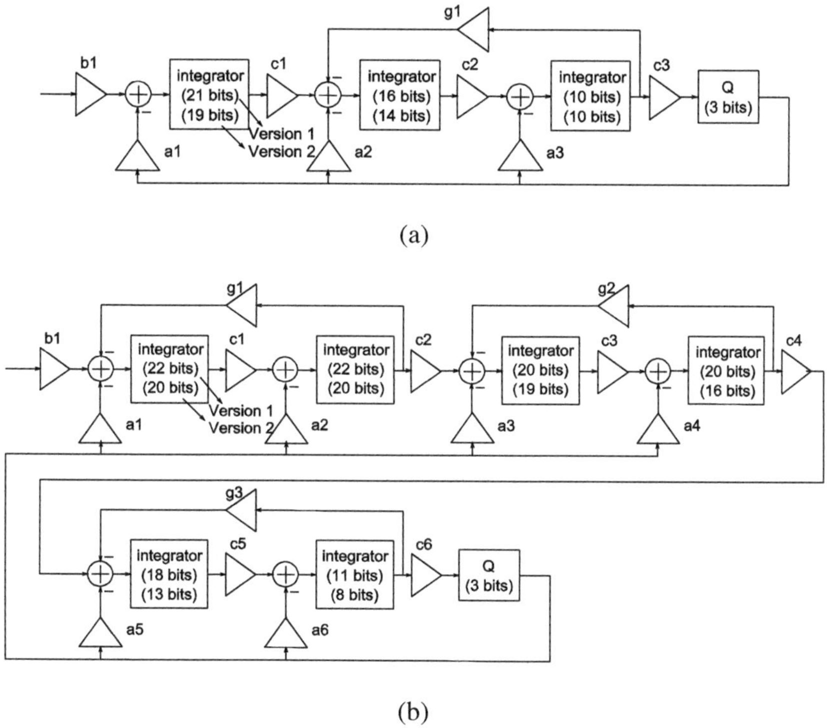

Fig. 4. Simplified ΣΔ modulator CRFB schematic (a) 3rd order modulator, OSR = 64 and (b) 6th order modulator, OSR = 32

I_{SDM} will remain approximately the same. Thus in total there are considerable power savings achieved by the proposed optimization approach. In future work, the OSR in the 6th order modulator being halved compared to [1] will allow us to increase the number of bits in the quantizer from 3 to 4 while keeping the maximum sampling frequency in the system. This will help to increase the maximum stable amplitude of the ΣΔ modulator – a crucial parameter in hearing-aid application.

6 Conclusion

In this paper we optimized the back-end of the audio signal processing path in hearing-aid application on system-level compared to the design of [1]. A figure of merit was introduced to decide early in the design process whether or not the optimization approach is reasonable. The optimization approach uses combination of the increase of the ΣΔ modulator order and the decrease of its OSR. Further savings are achieved by peak-SQNR overdesign and subsequent coarse quantization of the ΣΔ modulator coefficients. This approach leads to simplified interpolation filter, reduces the frequency of the DPWM block and the switching-rate of the Class-D output stage by 50%. The power consumption of the ΣΔ modulator is kept as in [1]. Overall the power of the entire back-end system is optimized showing that trading higher order for lower OSR in the ΣΔ modulator DAC is an approach to be considered in low-voltage, low-power, portable audio applications.

References

1. Pracný, P., Bruun, E.: ∑ Δ Modulator System-Level Considerations for Hearing-Aid Audio Class-D Output Stage Application. In: Proc. 2012 8th Conf. on Ph.D. Research in Microelectronics and Electronics (PRIME), pp. 103-106, Aachen (June 2012)
2. Pracný, P., Llimós, M.P., Bruun, E.: Interpolation filter design for hearing-aid audio class-D output stage application. In: Proc. 19th IEEE Int. Conf. on Electronics, Circuits, and Systems, Seville, Spain (December 2012)
3. Anzova, V.I., Yli-Kaakinen, J., Saramäki, T.: An algorithm for the design of multiplierless IIR filters as a parallel connection of two all-pass filters. In: Proc. IEEE Asia Pacific Conf. on Circuits and Systems, pp. 744–747. Singapore (December 2006)
4. Stoyanov, G., Nikolova, Z., Ivanova, K., Anzova, V.: Design and realization of efficient IIR digital filter structures based on sensitivity minimizations. In: Proc. 8th Int. Conf. on Telecommunications in Modern Satellite, Cable and Broadcasting Services, pp. 299–308. Niš, Serbia (2007)
5. Mehboob, R., Khan, S.A., Qamar, R.: FIR filter design methodology for hardware optimized implementation. IEEE Trans. Consumer Electronics 55(3), 1669–1673 (2009)

Author Index

MIX
Papier aus verantwortungsvollen Quellen
Paper from responsible sources
FSC® C105338

If you have any concerns about our products,
you can contact us on
ProductSafety@springernature.com

In case Publisher is established outside the EU,
the EU authorized representative is:
Springer Nature Customer Service Center GmbH
Europaplatz 3, 69115 Heidelberg, Germany

Printed by Libri Plureos GmbH
in Hamburg, Germany